Mechanisms of Mineralization of Vertebrate Skeletal and Dental Tissues

Irving M. Shapiro • William J. Landis

Mechanisms of Mineralization of Vertebrate Skeletal and Dental Tissues

Springer

Irving M. Shapiro
Department of Orthopaedic Surgery
Sidney Kimmel Medical College
Thomas Jefferson University
Philadelphia, PA, USA

William J. Landis
Department of Preventive and Restorative
Dental Sciences, School of Dentistry
The University of California
at San Francisco
San Francisco, CA, USA

ISBN 978-3-031-34302-5 ISBN 978-3-031-34304-9 (eBook)
https://doi.org/10.1007/978-3-031-34304-9

This Springer imprint is published by the registered company Springer Nature Switzerland AG
The registered company address is: Gewerbestrasse 11, 6330 Cham, Switzerland

Paper in this product is recyclable

Prologue

The mineralization of tissues in humans and other vertebrates is a process critical to life itself. The deposition principally of two inorganic chemical species, calcium and phosphorus, in the form of the mineral, apatite, associated with the major soft, organic tissue, collagen, creates one of the most interesting and important organic-inorganic reactions and several composite structures in the vertebrate body—bone, dentin, cementum, calcifying cartilage, the enthesis, and, in some avian species, calcifying tendon. In addition, the same two inorganic species interact with an organic matrix absent of collagen in tooth enamel to produce apatite bearing a hardness level which ranks among the highest known in Nature. Thus, the mineralized status of the skeleton and teeth of vertebrates provides a wide spectrum of evolutionary, molecular, structural, developmental, and functional advantages which enhance adaptation to a changing environment. From this perspective, mineralization is widely appreciated as responsible and necessary for the mechanical support of the body, for example, as well as the ability to tear, grind, and crush food in mastication and feeding. However, mineralization is also required in several other roles equally indispensable and essential to life. These include the basis for movement, locomotion, and flight; energy storage; accommodation of fluid transport and flow; defense against predators; soft organ and tissue protection; maintenance of balance; ion, trace element, and small molecule storage; toxic ion species accumulation and disposal; and other vital and fundamental functions.

How does mineralization in vertebrates occur? What are its means of formation which lead to its extraordinary breadth of functions underlying the core of life? This book, *Mechanisms of Mineralization of Vertebrate Skeletal and Dental Tissues*, addresses these two principal questions through its ten chapters of basic and clinical science and defines in some detail molecular biological, biochemical, and structural processes by which mineral is formed in vertebrate skeletal and dental tissues. Each chapter begins with a quotation representing the context of the major chapter topic, proceeds with a comprehensive explanation of the scientific basis of the topic, and concludes with a perspective which presents a chapter summary and an assessment of current research work and direction.

The opening chapter of the book, "In the Beginning," considers events in the inhospitable environment of early Earth formation and the chemical reactions which led to the genesis of life, particularly those which are catalyzed by the presence of mineral species, including apatite, and their possible roles in this remarkable occurrence and origin of being. The chapter lays the groundwork for subsequent discussion in the book of the development of multicellular organisms from unicellular ancestors and their evolutionary pathways for energy storage, cell-cell interaction, and intra-tissue communication, for example.

The avenues and processes just mentioned required the involvement of two elemental chemical species, calcium and phosphorus, for critical protection against predation and as determinants to enhance metabolism, survival, and genetic variation. Calcium and phosphate ions and their solution and saturation character in an aqueous environment as well as the fundamental physical-chemical principles governing mineralization are examined in Chapter 2, "The Enigma of Edith: Free Energy, Nucleation, and the Formation of Mineral." Here, these foundation stones derived from investigations conducted in vitro in cell-free aqueous systems are described to address homogeneous and heterogeneous nucleation theory and to demonstrate that free energy changes induce calcium and phosphate ions to form clusters which transition through an amorphous and other possible intermediate phases to crystalline apatite during vertebrate mineralization. The chapter also highlights the nature of the putative precursor mineral phases as well as apatite and structural aspects of bone and tooth mineral.

Chapter 3 provides an understanding of the "Form and Function of Tissues which Undergo Mineralization," and it details the origin, development, structure, and role of bone, calcifying cartilage, dentin, cementum, and enamel in comprising the vertebrate skeleton and dentition. In this regard, there is discussion of the origin of bone and the interplay of osteogenic cells in the lateral and paraxial mesoderm in association with cells derived from the neural crest to form the vertebrate skeleton. With respect to the dental tissues, the interdependence between dentinogenesis and amelogenesis during tooth development is considered. Another theme of this chapter is how lineage tracing studies can be used to identify the origin and fate of osteogenic and chondrogenic cells in complex tissues like those residing in the epiphyseal growth plate. Several principal genes, proteins, and other molecules involved in the formation and mineralization of the respective tissues are introduced to establish the basis for presentation of the topics of subsequent book chapters.

A powerful approach to determining how cells regulate mineralization is to evaluate the expression of genes which characterize mineralizing tissues. The literature indicates that gene regulatory networks define critical stages of skeletogenesis and tooth morphogenesis. For example, in bone-forming cells of the mineralizing deer antler, the deletion of a single transcription factor (SOX9) causes changes in over a thousand genes, transcription factors, enhancers, microRNAs, and long, noncoding RNAs. This and other studies emphasize the importance in hard tissue formation of a multitude of transcription factors which include RUNX2, SOX8/9, SMAD3, and several others which target parathyroid hormone-related protein, WNT/β-catenin, hedgehog proteins, members of the bone morphogenetic protein family, fibroblast growth factor, and solute carrier proteins, all important to the critical control and regulation of vertebrate mineralization. These and additional genes, transcription factors, and targeted proteins are elaborated in Chapter 4, "Genes and Gene Networks Regulating Mineralization," to provide a background to the complexity of the temporal, developmental, and tissue-specific events which comprise gene regulatory networks, systems promoting vertebrate skeletal and dental tissue formation and mineralization.

The results of genomic and proteomic investigations into vertebrate mineralization are fundamental, informative, and most important but, by themselves, do not provide details concerning the mechanism(s) by which skeletal and dental cells manage the massive amounts of calcium and phosphate ions required to build the mineralized phases of vertebrate hard tissues. Chapter 5, "Calcium and Phosphate Ion Uptake, Distribution, and Homeostasis in Cells of Vertebrate Mineralized Tissues," concerns mechanisms by which calcium and phosphate ions are controlled and regulated during their entry into osteoblasts, odontoblasts, ameloblasts, and other cells of hard tissues. Control and regulation of the ions are discussed in the context of the formation of palisades of the cells (except for the cells of cartilage and mineralizing entheses) linked by tight junctions which provide selective passage to the free flow of ions between the cells. Additionally, ion transport is described through modulation by the membranes of the cells or, more precisely, by a host of transmembrane channels, gates, and ion transporters and exchangers, which protect the cells from large ionic shifts that would alter the activity of critical metabolic pathways. Within the cells, the homeostasis of ions, particularly calcium, is detailed in terms of their trafficking in association with specific cytosolic binding proteins and their accumulation and/or storage in the endoplasmic reticulum, mitochondria, endosomes, and other cellular membrane-invested constituents.

The secretion of calcium and phosphate ions from the cells of vertebrate mineralized tissues is the subject of Chapter 6, "Calcium and Phosphate Ion Efflux from Cells: The Roles of Matrix Vesicles, Extracellular Vesicles, and Other Membrane-invested Transporters in Vertebrate Hard Tissue Mineralization." In part, this chapter addresses how mineral ions are packaged in the cells and transported to the extracellular matrices of mineralized tissues. The details of these processes are discussed in the context of calcium ion release from mitochondrial and endoplasmic reticulum stores for association with vesicles derived from the endosomal and autophagic pathways. These latter two important intracellular systems are further considered with respect to their roles in forming a variety of organelles carrying and transiting proteins, lipids, and other molecules or ions or ion clusters throughout the cells or toward the cellular plasma membrane. Finally, the numerous organelles, including lysosomes, autophagosomes, acidocalcisomes, exosomes, microvesicles, matrix vesicles, and several other membrane-enclosed transporters, are addressed functionally to ensure cell survival through autophagy and to contribute to extracellular mineralization through downstream endosomal pathway transitions.

Expression of genes leads to the synthesis and secretion of their counterpart proteins and other molecules. Of these, the most important and highly abundant is the family of collagens. Several collagen types associate with mineral ions to form vertebrate mineralized tissues. Chapter 7, "Collagen-based Mineralization of Bones, Teeth, and Other Vertebrate Skeletal Tissues," describes the character of collagen, its structure, composition, assembly, and interaction with mineral ions based on the predominant type I collagen species in bone, dentin, cementum, and avian tendon.

In a related manner, the nature of type X collagen is also detailed in its role in the mineralization of epiphyseal cartilage. Completing this chapter is current understanding of the means by which non-collagenous proteins such as bone sialoprotein, osteopontin, and osteocalcin may mediate extracellular mineralization as well as recent results demonstrating an extensive, complex nanochannel network proposed to provide transport and delivery of ions and small molecules throughout the extracellular matrices of calcified cartilage and bone.

Bone sialoprotein, osteopontin, and osteocalcin are but three of an extensive number of non-collagenous proteins which participate in the mineralization of vertebrate hard tissues. Chapter 8, "The Role of Non-collagenous Proteins and Other Matrix Molecules in Vertebrate Mineralization," expands details and functional descriptions of such constituents, including the family of SIBLINGS (Small Integrin-Binding Ligand, N-Linked Glycoproteins), SLRPS (Small Leucine-Rich Repeat Proteoglycans), matricellular proteins, and several additional proteoglycans and proteins comprising the skeletal and dental tissues, including enamel. In this instance, these matrix molecules participate in multiple, specific interactions which lead to comprehensive oversight and careful and strict regulation, control and coordination of extracellular matrix formation and deposition of mineral in vertebrate tissues.

The two closing chapters of this book consider the importance of systemic as well as local agents in the control of apatite development in the extracellular matrix in vertebrate tissues while preventing aberrant mineral precipitation. Such a need is met by the activities of inhibitory molecules which bind mineral ions or their clusters or block mineral deposition in the tissue fluid which is normally supersaturated in both calcium and phosphate ions. Chapter 9, "Local and Systemic Regulation of Mineralization: Role of Coupling Factors, Pyrophosphate, Polyphosphates, Vitamin D, Fetuin, Matrix Gla Protein, and Osteopontin," details the structure, secretion, and role of certain of these potent inhibitors and the manner by which they may regulate cellular function and interaction and ultimately mineral deposition in vertebrate hard tissues. Chapter 10, "Aberrant Mineral Deposition in Soft and Hard Tissues," describes several disorders of mineralization which occur in soft as well as hard tissues, including dystrophic mineralization, heterotopic ossification, microcalcifications associated with breast cancer, and defective mineralization resulting from mutations in extracellular matrix proteins. These pathologies are detailed in terms of their putative or known etiologies; their systemic and/or local effects on critical genes, transcription factors, hormones and other proteins which control the availability and access of calcium and phosphate ions to cells and extracellular matrices; and their implications that mineralization is fundamentally driven by cellular, rather than extracellular, processes.

The ten book chapters described briefly above constitute a continuous thread of information which binds together the science of vertebrate mineralization from its evolutionary origins and physical-chemical principles to its presence in both healthy and diseased human tissues. Included in the chapters are insights into gene expression and protein synthesis and secretion; ion and protein transit into, through and out of cells; major constituent molecules and their extracellular interactions; and regulatory and control processes which define the mechanism of mineralization of the vertebrate skeleton and dentition. Clearly, while many of these findings are incomplete and innumerable questions are yet to be addressed, current research suggests that mineralization is a cell-mediated process, albeit one which may well differ in specifics from tissue to tissue. In this respect, a governing feature of mineralization is its initiation by the release of calcium-loaded exosomes and calcium phosphate ion complexes from constituent cells into an extracellular matrix constructed with a particular molecular architecture. Molecules resident within the extracellular matrix and those locally secreted by hard tissue cells, aided by circulating proteins, regulate the subsequent development of mineral. All of this wide spectrum of activity reflects the genetic programming of the cells which populate these tissues and the physical-chemical laws which guide hydrated ion interactions in confined spaces.

Introduction

If we two authors had realized at the outset that writing a book about the deposition of mineral in vertebrate tissues would be so daunting, this task might not have been undertaken. The research literature on the subject of skeletal and dental tissue mineralization is enormous and encompasses topics as varied as the physical chemistry of dilute solutions, the function of the endomembranous system of the cells of bone and cartilage, and the enigma of the mineralization activity of the cells which form dental enamel. Discovery and publication in these and other related areas continue daily.

The idea for a book devoted solely to understanding the mechanisms of normal and aberrant mineralization in vertebrates originated from extended discussions between one of the authors (Dr. Shapiro) and his colleague and friend at the University of Pennsylvania, Dr. Ellis Golub. These two individuals spent hours upon hours considering topics which should be included in a book of this type. With the unfortunate death of Dr. Golub, Dr. Shapiro was able to persuade Dr. William Landis, who has made important contributions to molecular biological and structural aspects of the mineralization process, to assume an authorship role. The two of us have compiled what we believe is a detailed, comprehensive, and informative work on the process of vertebrate mineralization, to be used by anyone interested in this field fundamentally important to human well-being and life survival.

A major source of inspiration for the book has been the research and writings of Drs. William and Margaret Neuman. Over 60 years ago, these two scientists were the first to describe the mineralization event on the basis of physical chemistry and metabolism while at the same time suggesting that hydroxyapatite may have been a mineral associated with the origin of life itself. For these reasons and for their profound insight into the role and overarching importance of vertebrate mineralization, the authors dedicate this book to Drs. William and Margaret Neuman.

The authors have several additional personal dedications to make as the proper acknowledgment of forbearance, time, faith, infinite tolerance, good humor, and other qualities which were essential for completion of this work. Dr. Shapiro dedicates this book to Dr. Joan Poliner Shapiro, whose laughter fills his days with joy, and to Dr. Susan Hope Shapiro and Mr. Mark Edlitz, who, along with his grandchild, Jessie, provide love and caring which permeates his very soul. Dr. Landis dedicates this book to Ms. Jane Hogan Landis for being the ultimate support and guiding light to her husband over the duration of far more years of writing than she and he ever imagined. Both authors thank the late Dr. Ellis Golub for his friendship and contributions to the science of mineralization and Dr. Russell Neuman for his kindness in providing the images of his parents appearing in Chapter 1.

A book of this detailed and comprehensive nature could not have been written without the bedrock foundation of nearly uncountable journal articles and chapters of other books and publications which constitute the fields of both vertebrate and invertebrate mineralization. The two authors here have relied heavily on these innumerable literature entries. To begin an attempt of gratitude, the authors, first and foremost, would quote Mr. Leonard Cohen, the noted poet, musician and philosopher, who said, "There is a crack in everything, that's how the light gets in." In this regard, an army of scientists and scholars has helped enlarge the crack and enlightened our knowledge of tissue mineralization. Their published studies are cited in the reference lists at the end of each of the chapters. In the same vein, the authors must also apologize to the host of other investigators whose work has significantly expanded the knowledge of mineralization, but whose findings could not be included in the bibliography because of limitations in the space for each book chapter.

Additionally, this book would not have been completed without the enormously generous and knowledgeable contributions of a great many other colleagues, who took the burden and trouble to read and question or comment upon sections of the chapters and to discuss points of concern and provide expertise and their final resolution. The personal insights, time, and efforts of these friends and collaborators are also deeply appreciated and gratefully acknowledged by the authors of this book. For their tremendously careful scientific critique and advice, the authors wish to thank the following individuals: Tim Arnett, Ph.D. (University College London, UK), Alan Boyde, B.D.S., Ph.D. (Queen Mary University of London, UK), Colin Farquharson, Ph.D. (University of Edinburgh, UK), George Feldman, D.M.D., Ph.D. (Thomas Jefferson University, USA), Andrzej Fertala, Ph.D. (Thomas Jefferson University, USA), Peter Fratzl, Ph.D. (Max Planck Institute of Colloids and Interfaces, Golm, Germany), Theresa Freeman, Ph.D. (Thomas Jefferson University, USA), Marc Grynpas, Ph.D. (Samuel Lunenfeld Research Institute, Toronto, Canada), Noreen Hickok, Ph.D. (Thomas Jefferson University, USA), Masahiro Iwamoto, D.D.S., Ph.D., and Motomi Iwamoto, D.D.S., Ph.D. (University of Maryland, USA), Rodrigo Lacruz, Ph.D. (New York University, USA), Henry Margolis, Ph.D. (University of Pittsburgh, USA), Thomas Neil, Ph.D. (Thomas Jefferson University, USA), Nita Sahai, Ph.D. (University of Akron, USA), James San Antonio, Ph.D. (Biocordia, Media, PA, USA), Jason Shames, M.D., M.B.S. (Thomas Jefferson University, USA), Malcolm Snead, D.D.S., Ph.D. (University of Southern California, USA), Ryan Tomlinson, Ph.D. (Thomas Jefferson University, USA), and Koen van de Wetering, Ph.D. (Thomas Jefferson University, USA).

Besides those scholars noted above, the authors wish to acknowledge Bradley Snyder, M.S. (Thomas Jefferson University, USA), and Tengteng Tang, Ph.D. (McMaster University, Hamilton, Ontario, Canada), for help with the book figures. Dr. Tang was also responsible for formatting the entirety of the references and the book chapters, a thankless set of tasks whose completion is deeply appreciated. The authors would also like to thank Makarand Risbud, Ph.D. (Thomas Jefferson University, Philadelphia, PA, USA), Maurizio Pacifici, Ph.D. (The Children's Hospital, Philadelphia, PA, USA), Alan Boyde, B.D.S., Ph.D. (Queen Mary University of London, UK) and Elmar Gabriel, Ph.D. (University of Würzburg, Germany), for their friendship, support, encouragement, inspiration, and constant source of new and exciting scientific information. The authors thank as well Mr. Benjamin Edlitz (New Paltz, NY, USA) for valuable discussion and debate concerning the evolution of Paleozoic fish and in particular for his detailed drawing of an ostracoderm (Figure 1.9). With deepest gratitude, Dr. Landis acknowledges the kindness, collegiality, collaboration, friendship, and extensive training and education in orthopedic research and scientific study in general he received from the late Melvin J. Glimcher, M.D., Ph.D., in the Department of Orthopedic Surgery at the Children's Hospital and Harvard Medical School, Boston, MA, USA. Additionally, Dr. Landis is grateful to the late Dennis S. Weiner, M.D., from the Department of Orthopedic Surgery at the Akron Children's Hospital, Akron, OH, USA, and to Noritaka Isogai, M.D., Ph.D., in the Department of Plastic and Reconstructive Surgery, Kindai University, Osaka, Japan, each of whom contributed to extensive clinical orthopedic insight into mineralization and to long-standing research collaboration and publication. Lastly, the authors thank their publisher, Springer, for immense patience in repeatedly extending the deadline for publication of this book.

In writing these book chapters, the authors have brought together findings from an extraordinary number of investigations focusing on processes by which skeletal and dental cells manage the transport of ions forming apatite mineral in the vertebrate hard tissues of bone, calcifying cartilage and tendon, dentin, cementum, and enamel. The book has two principal themes, the first, to record in each of its chapters salient observations concerning key aspects of the structure, composition, molecular and cell biology, and genomic, proteomic, endocrinal, metabolic, and environmental nature of the vertebrate mineralized tissues; and the second, to use information obtained from these and other studies to relate cellular activities to the physical-chemical events which result in nucleation, growth, and development of apatite. In other words, the book attempts to explain how biological mineral is formed, intertwined with the specific organic extracellular matrix of each of the hard tissues.

A major objective of the book is its search for a unifying mechanism underlying the means by which mineralization takes place across the diverse spectrum of the vertebrate hard tissues. Like many other scientific quests, complexities breed complexities and simplistic answers to questions and problems are seldom acceptable. Nevertheless, based on new studies utilizing state-of-the-art techniques, it is now plausible to consider certain generalized features of vertebrate (as well as invertebrate) mineralization: (1) Cells may form tissue palisades (osteoblasts in bone, odontoblasts in dentin, and ameloblasts in enamel, for instance) linked by tight junctions to create barriers which regulate the free flow of ions from the systemic circulation to the mineralization front; (2) there is strict regulation of ions throughout their trafficking intracellularly as they are secreted into the extracellular matrix and localized to form the mineralizing front; and (3) the formation of apatite, that is its nucleation, growth, and development, is governed by interactions with specific proteins and other molecules comprising the extracellular matrices together with the physical-chemical laws which dictate hydrated ion interactions in confined spaces. A summary conclusion from individual findings of a host of hard tissue studies suggests that mineralization is not a simple or single default event. Rather, mineralization is an active, multistep process mediated by the cell and dependent on the nature of the proteins and other components of the extracellular matrix and the physical-chemical laws defining mineral ion nucleation, growth, and development. The book explores and elaborates these several characteristics describing vertebrate mineral formation.

The authors anticipate that the presentation and interpretation of data in this book generate ideas which will act as springboards for further scientific investigation. As such, the book will serve as a useful text for basic scientists and scholars working in the many areas which comprise hard tissue research, as well as undergraduate and graduate students, postdoctoral fellows, and those who are contemplating working in the field of biomineralization or who need a review of a specific mineralization topic. It should be noted, however, that the target audience of the book is not confined to the basic scientist alone, as the import of much of the information contained in the book encompasses the interests of clinicians and clinical scientists in the fields of orthopedic surgery, dentistry, endocrinology, and genetics, among many other areas. The individuals in such clinical disciplines work tirelessly to understand and expand the repertoire of treatments available to relieve the problems associated with abnormal mineralization of bone, cartilage, tendon, tooth, and numerous soft tissues, whose widespread conditions afflict a huge percentage of the global population. The authors are hopeful that the content of material in this book will lead to advances and key breakthroughs in scientific understanding of mineralization processes as well as in approaches and treatment of yet unsolved mineralization pathologies.

Contents

About the Authors

Irving M. Shapiro, B.D.S., M.Sc., Ph.D. is the Gertrude and Anthony DePalma Professor of Orthopaedic Surgery and Vice-Chair (Basic Science) in the Department of Orthopaedic Surgery in the Sidney Kimmel Medical College of Thomas Jefferson University, Philadelphia. Dr. Shapiro earned his dental qualifications from the (Royal) London Hospital and received the degree of Ph.D. from the University of London. He was awarded a Nuffield Fellowship for Research at the Forsyth Dental Center, Harvard School of Dental Medicine. Subsequently, he was selected as Assistant Professor in the Department of Biochemistry, School of Dental Medicine, at the University of Pennsylvania. Later, he was appointed Full Professor and the Chairman of the Department and then Director of the newly formed Institute for Environmental Studies. In 2001, Dr. Shapiro was recruited to Thomas Jefferson University as the Director of the Division of Orthopaedic Research in the Department of Orthopaedic Surgery. During his career, Dr. Shapiro has authored over 350 publications, co-edited four books, including *The Intervertebral Disc*, chaired two Gordon Conferences, and co-chaired meetings on the Growth Plate and the Spine. He has been a long-time member and Chair of NIH Study Sections and has been continually funded through his career. His scientific standing has been acknowledged through a Fogarty Senior International Fellowship, Lifetime Research Achievement Award of the Philadelphia Spine Research Society, Gold Medal of the World Dental Congress, elevation to a Fellow of the American Association for the Advancement of Science, recipient of the Distinguished Alumni Award of the University of Pennsylvania, and the Jefferson Career Research Achievement Award. Studies of bone and spine development by Dr. Shapiro have included publications in Science, Nature, Lancet, and Nature Reviews. His investigations of the growth plate have lent new understanding to the basic mechanisms responsible for bone growth, as well as the role of ionic constituents, local nutrition, and oxygen tension during the maturation of cells of growth plate cartilage. His scientific work has elucidated the role of autophagy and apoptosis in the survival and maturation of chondrocytes during bone formation and the role of hypoxia and the genes regulated by hypoxia in formation and maintenance of maturing bone. More recently, his studies have focused on the intervertebral disc and events which lead to its degeneration. He established the importance of hypoxia-inducible factor in maintenance of disc cell function and signaling pathways involved in the conservation of osmotic pressure in the disc. Continuing work initiated at the University of Pennsylvania, he created a class of new antimicrobial surfaces for use in joint replacement therapy. These latter studies resulted in the Frank Stinchfield Award from the Hip Society and the Jeannette Wilkins Award from the Musculoskeletal Infection Society, as well as financial support from the Jefferson/IBX partnership to bring his technology to the clinic .

William J. Landis, Ph.D. has taught and conducted basic and clinical science research to understand the growth and development of the skeleton and teeth of humans and other vertebrates. He holds a B.S. degree in Physics from the University of Massachusetts, Amherst, MA, and S.M. and Ph.D. degrees in Biology and Biophysics, respectively, from the Massachusetts Institute of Technology, Cambridge, MA. After taking his doctorate, Dr. Landis began a postdoctoral fellowship at the Children's Hospital and the Harvard Medical School, Boston, and there was promoted ultimately to the rank of Associate Professor of Anatomy and Cellular Biology. In 1998, he moved to the Northeast Ohio Medical University (formerly the Northeastern Ohio Universities Colleges of Medicine and Pharmacy) in Rootstown, where he was Professor and Chair of the Department of Biochemistry and Molecular Pathology. In May, 2010, he relocated his laboratory to the University of Akron where he held an endowed professorship as the G. Stafford Whitby Chair in Polymer Science. In January, 2017, he received a 2016–2017 Presidential Chair Award from the University of California, San Francisco (UCSF), and accepted an appointment as Visiting Professor there in the Department of Preventive and Restorative Dental Sciences in the School of Dentistry. In July, 2017, he was appointed professor and interim chairman of the Division of Biomaterials and Bioengineering in the Department of Preventive and Restorative Dental Sciences at UCSF, a position he held until the end of June 2018, when he retired and moved to Massachusetts. He currently holds appointments as adjunct full professor in both the UCSF School of Dentistry and the Division of Polymer Science and Polymer Engineering at the University of Akron. He continues to lecture, conduct research, and publish collaboratively, most recently at the Max Planck Institute of Colloids and Interfaces, Potsdam, Germany, and McMaster University, Hamilton, Ontario, Canada. He is a member of Phi Beta Kappa, Sigma Xi, and Alpha Omega Alpha (national medical honor society); was a Fulbright Scholar for study at the Weizmann Institute, Rehovot, Israel; has won a Kappa Delta Award for outstanding research from the American Academy of Orthopaedic Surgery and a number of additional national and international prizes; was elected a Fellow of the Microscopy Society of America; was recognized by the International Association for Dental Research with the 2017

Distinguished Scientist Award for Basic Research in Biomineralization; and was an editorial board member for five scientific journals. His research into aspects of bone and cartilage molecular biology, structure and function; tissue engineering of bone and cartilage; and the effects of mechanical forces on mineralized tissues was supported over many years by funding from the National Institutes of Health, the National Aeronautics and Space Administration, and other federal, state, and local agencies. His studies have led to the development and application of several analytical techniques, including high voltage electron microscopic tomography and laser capture microanalysis, to mineralized tissues; advancement of the avian leg tendon as a model for vertebrate mineralization; and novel insight into the mechanism of collagen-mineral interaction at atomic and molecular scales. Dr. Landis has published over 175 peer-reviewed papers and book chapters and has edited five books in the broad field of biomineralization .

In the Beginning

1

It is unimportant whether the reader is pleased, pained, or rendered righteously indignant. The book is aimed at the inherently curious: the clinician who wonders why, the anatomist who wonders where, the chemist who wonders how, and just any investigator who wonders whether.

William and Margaret Neuman, The Chemical Dynamics of Bone Mineral, Chicago: University of Chicago Press (1958)

I. M. Shapiro, W. J. Landis, *Mechanisms of Mineralization of Vertebrate Skeletal and Dental Tissues*, https://doi.org/10.1007/978-3-031-34304-9_1

An mRNA strand with strings of codons is a relatively small and simple molecule possessing limited storage capacity. It can store small amounts of genetic information; the capability of a ribozyme as an enzyme is severely limited. However, this catalytic deficiency is compensated by a variety of enzymes that are available in the hydrothermal vent environment. The short life of mRNA makes the protocell very responsive to changing conditions in the environment. Later, more durable DNA would emerge to become the molecule of choice for the large storage of genetic information regarding protein synthesis, replacing mRNA from its main function. The new generation of mRNA is created by the transcription of DNA. mRNA becomes a daughter of DNA to carry out its specific instruction of translation and protein synthesis. [1]

Glossary

Abiogenic/prebiotic Before life.

Amphipathic A molecule which has both hydrophilic and hydrophobic groups.

Carbonaceous chondrites Meteorites formed in the early solar system which contain metal silicates, oxides, or sulfides as well as organic compounds.

Chirality The property of molecules having the same structure but containing an asymmetric carbon atom connected to four different groups. As a result, such molecules form mirror images of each other which are not superimposable.

Hadean A period of geologic time which commenced about 4–4.6 billion years ago and was succeeded by the Archean period, which lasted another 1.5 billion years. During the Hadean period, there was cooling of the surface temperature of the Earth together with the formation of solid rock and possibly the beginning of life.

Nucleotide A heterocyclic purine or pyrimidine base linked by an ester bond to a phosphoryl group through a sugar (five-carbon ribose or deoxyribose). Nucleosides are purines or pyrimidines linked to a five-carbon sugar (no phosphate group).

Phosphorite Phosphate-containing rocks usually in the form of hydroxyapatite or fluorapatite.

Protocell A primitive self-replicating cell-like structure enveloped in an external amphipathic membrane.

Purine A heterocyclic aromatic organic compound which consists of a pyrimidine ring fused to an imidazole ring. Common purines include adenine and guanine.

Pyrimidine A heterocyclic pyrimidine-like compound which includes thymine, uracil, and cytosine. Thymine is found in DNA (not RNA).

Ribozymes Complex RNA molecules, comprised of a number of bases, which can catalyze the cleavage or ligation of RNA and facilitate peptide bond formation.

RNA World hypothesis A hypothesized stage in the development of life on Earth (about four million years ago) in which RNA/ribozymes became the first self-replicating molecules with the ability to store genetic information and synthesize proteins.

Vesicle A microscopic membrane-bound structure usually bounded by one or more leaflets of proteins, lipids, surfactants, or amphipathic hydrocarbons.

Gene/Protein Nomenclature
Symbols for genes are written in uppercase and italicized (for example, *COLIV*), whereas symbols for proteins are given in uppercase but not italicized (for example, COLIV). For simplicity, the chapter does not distinguish between the same genes/proteins found in animals and humans.

Abbreviations and Proper Names for Genes/Proteins and Other Molecules

ATP	Adenosine triphosphate
COLIV	Type IV collagen
DNA	Deoxyribonucleic acid
mRNA	Messenger RNA
RNA	Ribonucleic acid
VAV2	Member of the Dbl (diffuse B-cell lymphoma) family of genes/proteins, so named for the sixth letter of the Hebrew alphabet
WASLb	Wiskott-Aldrich syndrome b

1.1 Prebiotic Chemistry and Minerals as Bioactive Surfaces

Many prominent scholars believe that life emerged on Earth approximately 4.5 billion years ago in an atmosphere thought to be rich in carbon dioxide and nitrogen, possibly containing low levels of methane and other gases, and one in which oxygen had yet to make an appearance [2]. At this very early, prebiotic period in Earth history, the slightly acidic Hadean-Archean Ocean covered much of the world, with limited land mass above sea level. Under these conditions, seemingly random chemical processes culminated to generate the "protocell," a self-replicating and energy-generating structure, the forerunner of a living cell.

Although events which preceded the appearance of life on Earth are of considerable chemical and biological interest, on first reflection they seem far removed from the mechanism of mineralization of bones and teeth. However, a role for minerals like apatite and calcite in prebiotic chemistry was enunciated clearly more than 50 years ago by the founders of skeletal chemistry, William and Margaret Neuman, when they commented, "In the beginning, there was apatite" (see Box 1.1 for additional information) [3]. In retrospect, this intriguing statement was somewhat naive as apatite is only a minor component of some rocks, and the prebiotic Earth contained an abundance of other minerals such as olivine, pyroxene, calcite, silica, clays, and pyrite, many of which were more reactive than apatite. Moreover, from a biological viewpoint, apatite is just one among many minerals associated with the mineralization of both pro- and eukaryotic organisms. In this introductory context, the goal of this first book chapter is to bring together some ideas about the origins of life and the catalyzing role of mineral in the formation of critical molecules required for the initiation and subsequent evolution of the skeletal tissues.

Over 60 years ago, Stanley Miller, a graduate student in the laboratory of Harold Urey at the University of Chicago, devised an experimental approach to understanding how life originated on Earth. Miller hypothesized that the atmosphere of the

Box 1.1

Drs. William F. Neuman and Margaret Neuman. Dr. William (Bill) Neuman was chair of the Department of Radiation Biology and Biophysics at the University of Rochester, Rochester, NY. He worked with the Atomic Energy Commission and was particularly concerned with radiation protection. In 1965, he reported on the effects of space flight on the musculoskeletal system of the Gemini 7 astronauts. This work catalyzed his interest in the biochemistry of bone, especially mechanisms of action of osteoblasts in mediating the mineralization process of vertebrates. He was the recipient of many honors, including the Eli Lilly award, the Claude Bernard Medal, and the Kappa Delta research award from the American Academy of Orthopedic Surgeons. In 1958, together with his wife, Dr. Margaret Neuman, he co-authored the book, *The Chemical Dynamics of Bone Mineral*. While working as a biochemist in her own right, Margaret Neuman performed important work concerning the cell biology of bone, especially in relation to bone-seeking radioactive elements. Many of her papers were directed toward understanding mechanisms controlling bone cell function, factors regulating bone-blood homeostasis, and the origin of life. Images are a gift from Russell Neuman, Ph.D.

primitive Earth could, under appropriate conditions, support the abiogenic (chemical) synthesis of organic molecules required for life [4]. In other words, the formation de novo of compounds like sugars and amino acids would provide the building blocks for the eventual creation of primitive living organisms. To test his hypothesis, Miller set up a circulating water system in a gaseous environment which was thought to mimic that of the primitive Earth: methane, ammonia, and hydrogen. The system was then perturbed by periodic electrical discharges into the gaseous milieu. After a week, he reported that the water became turbid as a consequence of trace amounts of the amino acids, glycine, α-alanine, β-alanine, and aspartic acid among many other organic molecules. The results of this experiment directly supported the concept that organic syntheses were possible from very simple agents. Additionally, Miller found that the amino acids were bound to silica particles released from the glass used for the experiment [4]. This latter observation suggested for the first time that the surface chemistry of a mineral, silica in this case, might be involved in reactions leading to the synthesis of biogenic molecules, and the finding confirmed a number of related ideas which had been propounded earlier. Such mineral–organic interfacial events in the origins of life have been reviewed [5].

Expanding this catalysis concept further, this chapter presents views detailing how rock surfaces and chemicals percolating from rocks, the gases in the changing atmosphere, molecular constituents of tidal pools and hydrothermal vents, as well as sources of external energy conspired to initiate life. Such ideas rest on accumulating evidence that complex organic molecules, especially self-replicating ribonucleic acid (RNA) with enzymatic properties, could originate in the primitive gaseous Earth environment. It is important to note that this hypothesis is only one consideration for the origin of life, and alternative proposals, such as "Metabolism-first" and the "RNA-peptide coevolution world," also exist.

The present chapter focuses on the RNA World hypothesis, in which the functions of RNA molecules relate directly to the most basic needs of life, protein synthesis and reproduction. In addition, the chapter considers the importance of molecular barriers to separate the contents of the "protocell" from the external aqueous environment. A barrier function could be achieved by the self-assembly of hydrophobic compounds to form membranes, an action which also provides a mechanism by which metabolic energy could be generated by a chemiosmotic gradient. Finally, although little is known of events which convert the non-vital protocell into a living tissue, the chapter will examine the most primitive organisms in the fossil record and speculate on means which may have contributed to the evolution of bones and teeth.

1.2 Chemical Syntheses on Bioactive Surfaces

In the natural world, the energy required for the type of abiotic chemical syntheses mentioned above as occurring on mineral surfaces could come from a number of sources. These would include thermal changes, ultraviolet (UV) light, electrical energy, and ocean waves. Volcanic activity would also provide heat for chemical reactions, and the generation of rocks like basalt, most prominently, could enhance polymerization reactions. The reactivity of the mineral itself is of principal importance and dependent on a two-dimensional surface constellation of charged atoms, which incidentally can differ substantially from the arrangement of the underlying atoms. In an aqueous environment, protons, hydroxyl ions, and surface-bound cations and anions form a charged interface, an electrical double layer. The adherent ions associate with dissolved ions and molecules of the bulk water phase (seawater or freshwater) through hydrogen bonds. Charged organic molecules bind to the electrical double layer through a pH dependency. Thus, when protonated, the mineral surface generates a net positive charge, while an aqueous phase with a high pH forms a negatively charged surface (Fig. 1.1). In this regard, investigators knowledgeable of these material characteristics and experienced in applications of column chromatography have utilized the mineral, hydroxyapatite, for example, to bind and separate charged molecules like nucleotides (usually not in seawater) from proteins and other contaminants [6].

Accordingly, in the pristine setting of the desolate beaches of the prebiotic world, one can imagine that the incoming tide would bathe rocks in a very dilute solution of charged ions and the occasional organic molecule. Some of these molecules would loosely bind to the crystalline mineral of the rock faces. As the tide receded and drying occurred, simple, adherent water-soluble molecules such as citrate, formaldehyde, and hydrogen cyanide as well as ions such as magnesium and pyrophosphate could undergo a self-assembly process, with such species binding and accumulating on charged arrays of the regularly spaced atoms comprising the various mineral crystal faces. Intramolecular bonds with adjacent molecules would create a stabilized cross-linked molecular film, leading to self-organization of groups of molecules which could be utilized to form membrane-like structures on a mineral face (Fig. 1.1) [7]. Moreover, while the bonds would tend to be weak, they would be nonetheless resistant to removal or elution by the returning tide [8]. William and Margaret Neuman [3] noted the importance of hydrated mineral and predicted that "in the beginning" minerals served as "beach or crystal tide catalysts." They also commented that a mineral like apatite is "ubiquitous in occurrence," forms "microcrystalline structures with extremely high surface area to mass ratios," and is "stable thermodynamically with high adsorptive capacities" [3]. A combination of these characteristics would greatly enhance both molecular adsorption and physical reactivity on mineral surfaces.

Work by Greenwell and Coveney [9] drew attention to the importance of surface events in supporting abiotic chemical synthesis. They reported that layers of surface hydroxyls ("layered double hydroxides") bound charged ions and organic molecules and promoted chemical reactions [9]. Other studies showed that exposure of mineral to a "soup" of amino acids led to surface adsorption and concentration of the molecules, including aspartic acid, glutamic acid, and glycine [10, 11]. In addition, while polymerization reactions proceed very slowly in solution, such reactions of both amino acids and nucleotides can be "catalyzed" by mineral [12]. In this regard, the surface of apatite, for example, may present a template for the selective adsorption and polymerization of purines, pyrimidines, and nucleotides [13].

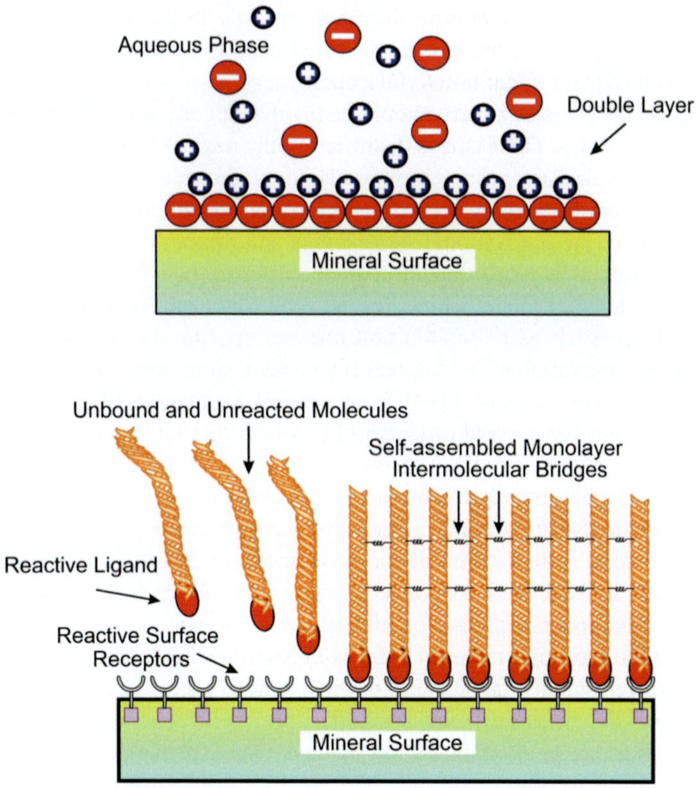

Fig. 1.1 Schematic of mechanisms by which charged ions and molecules may be bound to a mineral surface. Top panel: Interactions between charged ions and an apatitic surface. Depending on the plane of section, apatite may present a net positive or net negative charged surface. Shown here is a surface of net negative charge (attributable to predominantly phosphate or hydroxyl ions) which binds ions of opposite (positive) charge. In this manner, a double layer of charged ions is formed on the mineral surface. Bottom panel: Formation of a self-assembled monolayer on a mineral surface. The mineral surface can be naive, coated, or functionalized with protein. In the case of a surface functionalized with receptor molecules, ligand molecules may bind to receptors. The process of ligand–receptor binding leads to formation of domains of ligand molecules which may then interact with each other to develop cross-links and intermolecular bridges. Such structures stabilize the ligand molecules and promote their alignment and orientation, for example, in this self-assembled monolayer on the mineral surface

While the significance of the prebiotic steps discussed above is still energetically debated, the beach and tidal mechanisms described here emphasize a key concept: mineral serves to direct and even participate in critical molecular events. The most obvious development of this scenario is that simple amino acids would be present among the chemical species synthesized and these could then develop bonds to form peptides and even proteins. Conventionally, these biological catalysts promote biochemical reactions which often lead to the synthesis of complex metabolic intermediates and molecules like deoxyribonucleic acid (DNA) and RNA. While this finding is valid for living cells, unanswered questions remain as to how these protein catalysts may be formed and how they can replicate in the abiotic world. Stated otherwise, these proteins, once formed, would need to synthesize copies of themselves for subsequent generations. Of course, such a situation also raises a more critical concern of whether apatite could be involved in any way with the genesis of replicative molecules. This issue is discussed in the next section of this chapter.

Besides the volcanic rocks and minerals noted above, additional organic compounds may have been delivered to the Earth by comets and asteroids. Infrared spectra generated by the National Aeronautics and Space Administration (NASA) Dawn spacecraft as it journeyed through the asteroid belt for over 11 years and orbited Ceres, a dwarf planet which lies between Mars and Jupiter, lend support to this possibility. Analysis of visible and infrared spectra showed the presence of long-chain organic (aliphatic) compounds as well as ammonia-containing hydrated minerals and organic molecules [14]. Thus, organic compounds liberated from planetary detritus, over the eons of time, could generate the molecules and macromolecules associated with self-replication and life.

As an aside, meteorites have been shown to contain a rich mix of organic compounds, in particular mono- and di-carboxylic acids, ribose, and even amino acids [15]. Furthermore, carbonaceous chondrites contain carboxylic acids, amino acids, hydroxy acids, sulfonic acids, phosphonic acids, and poly-hydroxy compounds [16, 17]. All these compounds are basic building blocks of more complex tissue macromolecules. One particular meteorite, the Murchison meteorite, contained a mixture of chirally right- and left-handed amino acids, with an excess of left over right [18]. This finding is of considerable interest as the amino acid constituents of proteins of living organisms are predominantly left-handed. One explanation for this homochirality is that minerals like calcite and quartz are chiral-selective [19, 20]. Hazen and Sholl commented that, "Minerals provide Earth's most chirally biased environments, and thus may represent the most likely loci for prebiotic chiral selection and amplification" [19].

Minerals, then, whether terrestrial or extraterrestrial, have a number of functions related to the origin of life. They provide a surface for adsorption and concentration of organic compounds and inorganic ions; they afford a surface for chemical reactions; they are a source of chemical reagents; and, for some minerals, they possess a chirality which dictates structural features of the adsorbed reagent. This multitude of roles raises the question as to whether minerals enhance reactions leading to the synthesis of more complex molecules, especially those associated with membrane function, replication, and information storage.

1.3 Synthesis of Nucleosides and Nucleotides from Simple Molecules: Influence of Clays and Other Minerals

One of the tenets of the origin of life is the need to generate and concentrate simple organic molecules. Subsequently, these molecules could be used to build a more complex information-carrying molecule which could also serve as a chemical catalyst. While time was of little or no consequence in the prebiotic world, there is no doubt that these syntheses would be enhanced in this primitive milieu if there were both a source of energy (heat, UV light, osmotic pressure, or electrical or chemical potential) and a material surface for the development of self-assembled organic molecules.

For organic syntheses, the starting reagents are the gases present in the atmosphere. Although the composition of the atmosphere during an abiotic-protocell period can never be determined, the gas composition used by Miller in his experiment was unlikely [4]. There are valid reasons for assuming that the Hadean-Archean atmosphere contained substantial amounts of carbon dioxide, carbon monoxide, and nitrogen with just trace amounts of ammonia [21]. Also, hydrothermal systems, volcanic activity, and material from comets which rained down on the Earth in this period of its abiogenic history released point sources of methane, sulfur dioxide, and hydrogen sulfide.

Under specific conditions, gases can form substrates for the synthesis of metabolic intermediates. For example, over 50 years ago, Oró and Guidry showed that hydrogen cyanide with ammonia condenses to form adenine, a major component of DNA, RNA, and adenosine triphosphate (ATP) [22]. A modification of the ideas concerning the initial biogenic molecules was the report that formamide, a product first identified by Miller and Urey [23], was a critical agent in the formation of nucleotides. Formamide occupies an important position in studies of the origin of life as it has been shown to be present universally, even in forming stars and comets [24]. Formamide, synthesized from hydrogen cyanide, can also generate small amounts of adenine, hypoxanthine, uracil, and cytosine (Fig. 1.2) [25, 26].

Aside from formamide, other simple molecules such as cyanamide, cyanoacetylene, glycolaldehyde, and glyceraldehyde form pyrimidine ribonucleotides [27], and chemically masked sugars with purine precursor molecules, aminooxazole, and 5-aminoimidazoles promote the synthesis of pyrimidines [28]. Genesis of some of RNA ribonucleotides is achieved using hydrogen cyanide and hydrogen sulfide, together with formaldehyde and acetylene [29, 30].

Another obvious requirement for the formation of a polynucleotide (sugar phosphate with a purine or pyrimidine base) is the expenditure of energy. Biologically, this need is commonly provided by ATP and catalyzed by enzymes which lower the free energy (ΔG) of formation of a particular product, ribose phosphate or glucose phosphate, for instance. It is possible that microdroplets (aerosols) provide the energy for these and other reactions. The confined environment of the droplets and evaporation at the air–solvent interface cause an extraordinary increase in efficiency of product formation with rates of reaction which may vary from a thousand to a million times faster than reaction rates in bulk solutions [31, 32]. For example, since the phosphorylation of the sugar ribose in water is energetically unfavorable, an input of energy is required to form ribose phosphate. In this case, the Gibbs free energy change is +5.4 kcal/mol (for a more detailed discussion of Gibbs free energy,

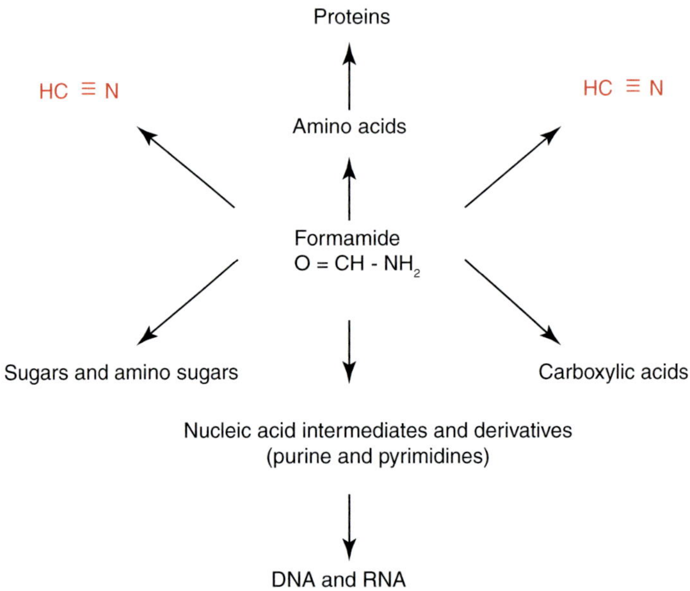

Fig. 1.2 Central role of formamide in the synthesis of proteins and nucleic acids (RNA and DNA). Formamide can be synthesized from gaseous hydrogen cyanide (HCN) (red) and is a key intermediate in the synthesis of purines and pyrimidines, the building blocks of RNA and DNA. Formamide can also be converted into products critical for life processes—carboxylic acids, sugars, amino sugars, and amino acids. Amino acids are utilized by cells to synthesize proteins. (Diagram adapted from Pino et al. [25] with permission)

see Chap. 2). The microdroplet technique converts an energetically unfavorable reaction to a more favorable one. Thus, Nam and colleagues reported that ribose could be phosphorylated in a droplet with a calculated ΔG value of −1.1 kcal/mol [33]. Moreover, these workers found micron-sized droplets of ribose, phosphoric acid, and uracil could form the ribonucleoside, uridine, at room temperature [33]. Translating these experiments to events which would occur in the prebiotic world, microdroplets could be generated by waves striking rocks and, as these droplets evaporated in the heated environment of the Earth, individual nucleosides could form and accumulate on mineral surfaces (Fig. 1.3).

Aside from microdroplets, it has been thought for some time that mineral deposits at underwater hydrothermal sites could provide heat and chemical energy necessary for prebiotic chemical syntheses. This idea was strengthened by the discovery of the "Lost City" hydrothermal field, located in the Atlantic Ocean close to the Mid-Atlantic Ridge [34]. Unlike conventional high-temperature hydrothermal vents, which receive energy from undersea volcanoes, the heat generated at the Lost City hydrothermal field is dependent on reactions between rocks of the ocean floor and seawater; hence, the heat produced at this site is much lower than that reported

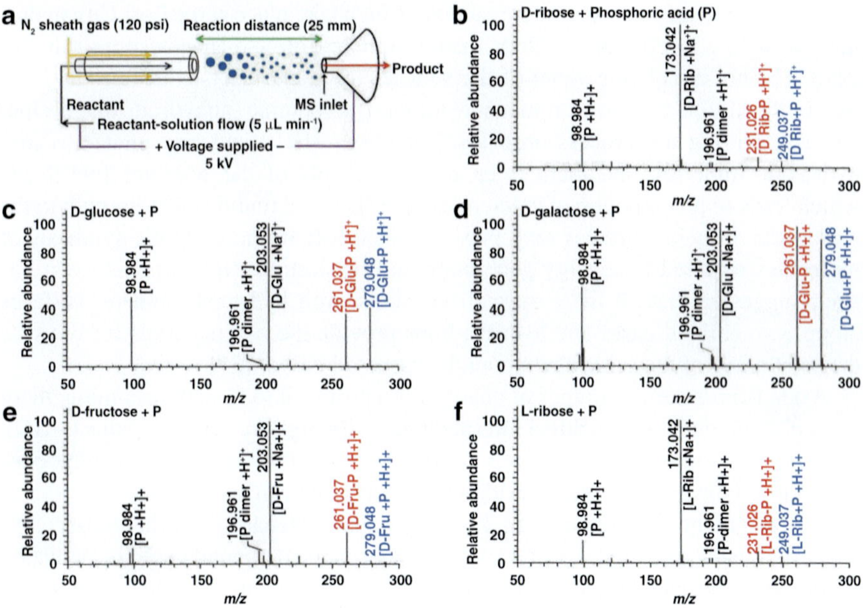

Fig. 1.3 Mass spectra of sugar phosphates produced from microdroplet reactions. The possibility exists that microdroplets (spray) generated by physical or chemical means could facilitate the formation of simple organic molecules. (**a**) Schematic of the system used to generate microdroplets. The phosphorylation reactions were performed by mixing sugars and phosphoric acid and microdroplets were prepared with a nitrogen gas nebulizer. Following electrospray ionization, peak fragments were detected by high-resolution mass spectrometry (MS). (**b–f**) Mass spectra of each of the sugar phosphorylation reactions: In each spectrum, numbers in red and lettering define peaks (*m/z*, mass/charge number) of (**b**) D-ribose-phosphate (D-Rib-P), (**c**) D-glucose-phosphate (D-Glu-P), (**d**) D-galactose-phosphate (D-Gal-P), (**e**) D-fructose-phosphate (D-Fru-P), and (**f**) L-ribose-phosphate (L-Rib-P). Blue numbers and lettering denote the aggregated species between each sugar and phosphate. Black numbers and lettering denote the species present within the reactants. (Image from Nam et al. [33] with permission)

for high-temperature hydrothermal vents, and the seawater percolating through this rift is alkaline (pH 10) [34].

A prominent characteristic of the Lost City type of hydrothermal field is the presence of long spires of precipitated calcium carbonate [34, 35]. Hot seawater permeating through these encrustations of porous chalk causes the release of a variety of elements, including transition metals. The huge surface area of the microporous spires provides numerous sites for the binding and concentration of organic molecules, while the proton and pH gradient between the alkaline vent fluid and the surrounding sulfide-rich seawater is sufficient to catalyze chemical reactions [35]. In theory, these conditions could provide the compartmentalization and the chemical energy required for the synthesis of critical metabolites and, at least conceptually, for the generation of prebiotic reactions noted above [36].

One example of rock–seawater reactions which take place at the Lost City hydrothermal field concerns the seafloor mantle mineral, olivine (magnesium-iron silicate). The hot circulating water of the vent environment reacts with ferrous (Fe^{2+}) iron in the olivine to convert it to a new mineral, serpentine (anhydrous magnesium silicate). During the process, iron loses electrons (Fe^{3+}) and precipitates as iron hydroxide minerals, and there is an increase in pH of the seawater (pH 9–11) which leads to precipitation of calcium salts [37]. These findings add strength to the notion that mineral surfaces may play a critical role in the prebiotic synthesis of molecules required for energy generation and replication. Not surprisingly, it has been suggested that, at these specialized vents, iron hydroxide mineral surfaces along with emitted gases and hydrocarbons provide the raw materials for the synthesis of many of the critical compounds discussed earlier [35].

Aside from external sources of energy for a protocell to be self-sustaining, there is the need to generate chemical intermediates with high free energy status ($-\Delta G$). Common substrates, including carbon dioxide and carbon monoxide, can be converted to glyoxylate; the photo-oxidation of acetylene can also generate this three-carbon fragment requirement. In his "glyoxylate scenario," Eschenmoser [38] suggested that two molecules of such a small organic compound could be activated to form oxaloacetate. Decarboxylation (removal of COO^-) from oxaloacetate would then lead to pyruvate, a key intermediate in the synthesis of citrate and the formation of acetyl CoA. This organic compound is a required substrate for lipid synthesis, an intermediate in the production of a number of amino acids, and a component of energy-generating pathways in a cell. Accordingly, glyoxylate could provide the starting material for the development of biosynthetic pathways leading to the generation of critical molecular intermediates of the energy-generating tricarboxylic acid cycle as well as a substrate for the eventual synthesis of other small molecules like amino acids and lipids (Fig. 1.4).

Even if mineral spires of hydrothermal fields enhanced chemical reactions, the products would be lost in the bulk aqueous medium (seawater or freshwater) without a mechanism for packaging them. To overcome such a packaging need, the protocell would require an enveloping membrane containing both the substrates and products of reactions. How the membrane of a protocell may form has prompted considerable thought and speculation [36, 39]. If life originated on mineral spires at hydrothermal vents like those in the Lost City, the surface of the mineral could act like a membrane. This membrane would separate the high pH of the internal structure of the spire from the much lower pH of the surrounding seawater. In this regard, Branscomb and Russell noted that the alkaline vents could serve as "mineral mitochondria-in-waiting," with "proton motive force and redox-driving disequilibria … correct in strength, chemical character, and vectorial orientation" [40].

A more conventional membrane, formed to accumulate both substrates and products of reactions, would most likely consist of adsorbed amphipathic hydrocarbon chains, comprised of methylene (CH_2) groups (conceivably ten or greater carbons atoms in length) with head groups which are very weak acids or hydroxyls [39]. There is also the possibility that some prebiotic membranes contained intermediates in the synthesis of cholesterol, simple phospholipids or perhaps glycerol phosphate [41, 42].

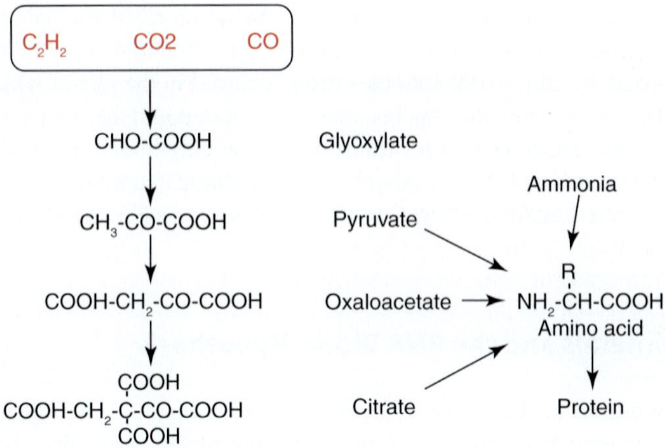

Fig. 1.4 Metabolism of glyoxylate, a key intermediate in the synthesis of proteins and critical metabolites. According to Eschenmoser [38], glyoxylate can be synthesized from acetylene (C_2H_2), carbon dioxide (CO_2), and carbon monoxide (CO) (red). These gases are thought to be present in the atmosphere of the early Earth. Gaseous ammonia is the presumptive source of nitrogen comprising the amino acids

In meteorites, short alkyl chains with a functional phosphonate group along with inorganic phosphate have been reported [43]. Of course, these molecules are substantially different from those present in the known plasma membranes of cells where triglycerides and phospholipids predominate and the most common carbon chains are 16 or 18 carbon atoms in length. Nevertheless, if dried on a mineral surface and at high pH, these amphiphiles form single-layer micelles, in which there is no interior cavity. However, some minerals would autocatalyze the micelles to form into larger bimolecular vesicular protocells which can enclose other hydrophilic molecules and, possibly, the solubilized mineral ions in its interior aqueous environment [44].

From these observations, the substrates for the synthesis of metabolic intermediates were likely all present in the evolving primitive atmosphere and/or dissolved in the Hadean Sea or surface freshwater. There was also the possibility that even more complex molecules would have been delivered to the Earth through meteor bombardment [45]. Laboratory studies show that small molecules which existed in the prebiotic atmosphere can be utilized to generate nucleosides, nucleotides, and other biologically important molecules [46, 47]. However, unless they were concentrated, the Hadean Sea would disperse these molecules, making them unavailable for further chemical reactions. One concept to overcoming such a dilution effect is to assume that molecular products were bound and concentrated on solid mineral surfaces. Further synthetic reactions to generate more complex molecular structures could be initiated by the energy supplied by the heat of volcanic eruptions, hydrothermal vents [30, 48], and perhaps the generation of sprays of microdroplets onto rocks and other surfaces. Moreover, some minerals would be expected to enhance the self-assembly of hydrocarbons and thus facilitate formation of protocell membranes. Such membranes could subsequently produce vesicles to envelope and accumulate hydrophilic molecules, especially those required for metabolism and nucleotide synthesis. Indeed, Hanczyc and colleagues found that, in the presence of a small quantity of clay, a 100-fold elevation occurred in the rate at which vesicles formed [44]. Additionally, the clay became incorporated into the vesicle interior and thereby provided a mechanism for transport of any clay-bound molecules into the enclosed vesicle space [44]. Thus, there is the likelihood that mineral surfaces could potentially play a number of different roles in early events leading to the generation of the protocell.

1.4 Minerals and the RNA World Hypothesis

Further discussion of the possible involvement of minerals in the genesis of life rests on a deeper consideration and understanding of the molecular occurrence(s) fundamentally initiating life history on Earth. Assuming that this event(s) was not begun by an extragalactic germ of DNA, many Origin of Life scholars believe that RNA was the "prebiotic bullet" which spawned life although others support alternative ideas such as "Metabolism-first" [49] or "RNA-peptide coevolution" hypotheses [1], as mentioned earlier in this chapter. The concept that RNA was the critical determinant to the evolution of life on Earth is known as the RNA World hypothesis, and it has revolutionized much of current thinking concerning the generation of information-rich molecular biological structures.

The RNA World hypothesis derives from the notion that a protein (DNA polymerase) promotes the formation of a complementary strand of DNA, a reaction which requires the presence of low molecular weight deoxyribose nucleotide bases and priming by an existing strand of messenger RNA (mRNA). Thus, from the point of view of both transcription and translation, RNA occupies a central, albeit separate, role in the information storage function of DNA in living cells. The RNA World hypothesis holds RNA in a different context from DNA in that RNA is considered as a self-replicating molecule with properties of an enzyme. In a biblical analogy, instead of protein begetting protein or DNA encoding RNA encoding protein, the RNA World hypothesis maintains that RNA both generates protein and, by inference, serves as a replicator molecule [50–52].

The conception that RNA occupied such a key position was largely based on the chance finding that an RNA molecule, a ribozyme, could act as an enzyme (a function conventionally filled by proteins) which catalyzed chemical reactions and served as an information storage molecule [53, 54]. Its discovery was not unexpected since the ribosome of living cells can be viewed as a ribozyme complex and, during protein synthesis (translation), mRNAs instruct the ribosomes to synthesize protein molecules [55]. The search for this type of molecule was initiated when Ekland and colleagues [56] isolated an RNA ligase (an enzyme which joins nucleotides together) from randomly generated RNA sequences (Fig. 1.5). In the presence of the appropriate template, RNA and nucleoside triphosphates, the ribozyme extends an RNA primer by up to six mononucleotides through 3′,5′-phosphodiester linkages (Fig. 1.6) [57, 58]. These and other considerations provided the scientific

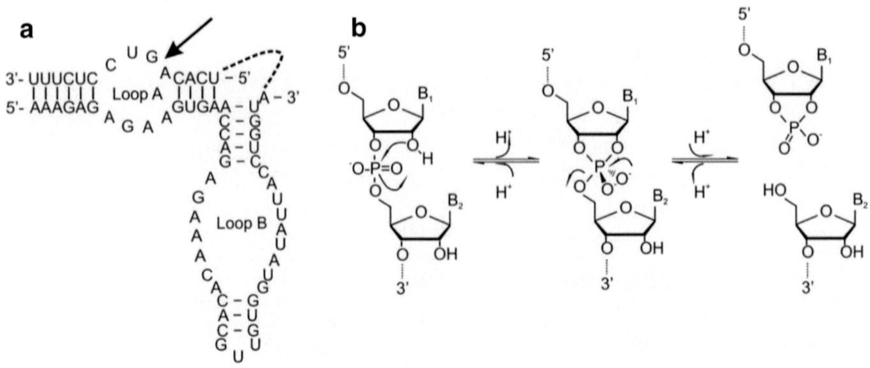

Fig. 1.5 Structure and activity of a hairpin ribozyme. (**a**) The nucleotide sequence of the ribozyme. The structure of this molecule is maintained by hydrogen bonds (Watson-Crick pairing) which link uridine (U) with adenine (A) and guanine (G) with cytosine (C). The ribozyme contains two loop regions (A and B). Loop A is part of a helix-loop-helix domain which includes an RNA substrate and the ribozyme region recognizing it. Loop B consists of two nucleotide strands of the ribozyme principally responsible for catalytic action. Between the loops and just outside loop A, there is a possible nucleotide-binding region (dotted line) which spans the 5′-to-3′ region of substrate strand and loop B, respectively. The reaction site (site of cleavage or ligation) of the substrate (arrow) is located close to loop A. (**b**) Events which lead to cleavage or formation of a phosphodiester bond between substrate nucleotides. The far right aspect of (**b**) shows two nucleotides of the substrate. Nucleotides are added to the substrate by hydrogen abstraction and the formation of an unstable intermediate pentavalent phosphate bond. Rearrangement of the pentavalent bond and the formation of a phosphodiester linkage promote the extension of the nucleotide (ligation; far left aspect). The reverse cleavage reaction follows the same pathway in the opposite direction. (Schematic from Muller [57] with permission)

basis for the formulation of the RNA World hypothesis in which RNA would satisfy all the principal requirements for life, functioning as both an enzyme for protein synthesis and a multiplicative strand for RNA generation [56]. Orgel [59] nicely summarized these and other concepts when he wrote that there is the likelihood that "the whole biosphere descended from one or a few replicating polynucleotides that formed on the primitive Earth about 4 billion years ago."

Since RNA itself is at the center of the RNA World hypothesis, it is important to consider how this molecule could be formed in the prebiotic landscape and whether there is a mineral requirement for its synthesis. Cairns-Smith provided some support for a role for minerals by suggesting that the natural alignment of atoms in a dense substrate not only bound and concentrated organic compounds but also served as a template for the synthesis of purines and pyrimidines [60]. In this regard, numerous studies have evaluated the role of possible mineral templates: clay-like montmorillonite, phyllosilicates, titanium dioxide, and a range of microcrystalline minerals [61–63]. If heated with formamide (CH_3NO) at 160 °C for 2 days, the agents mentioned above, as well as calcium carbonate and "Cosmic Dust" analogs, all supported the synthesis of purines and even pyrimidines, primary constituents of RNA [63]. Likewise, when formamide was irradiated with a proton beam in the presence of powdered meteorites, all the nucleotide bases (cytosine, uracil, adenine,

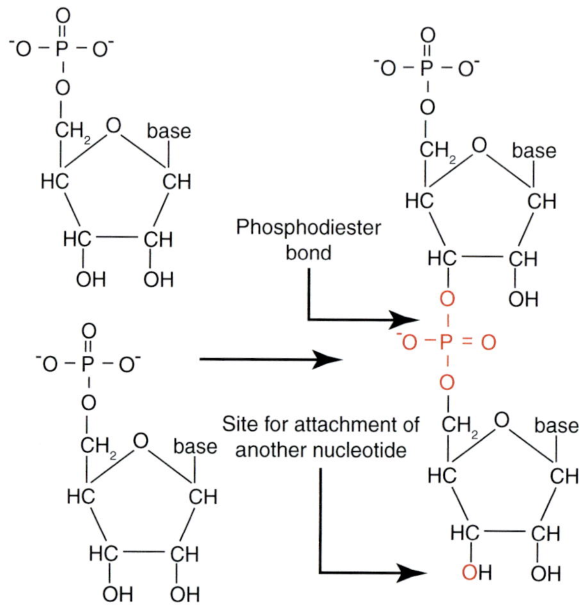

Fig. 1.6 Formation of polynucleotides. Complex (poly)nucleotides may be synthesized from individual molecules through condensation reactions which form phosphodiester bonds. The structure of two nucleotide molecules is shown in the left half of the diagram, in which each nucleotide is comprised of a phosphate group linked to a sugar ribose at the 5′ carbon position. In the right half of the diagram, the two nucleotides condense to form a 3′,5′-phosphodiester bond. The attachment site for each additional nucleotide molecules is through an oxygen bond at the 3′ carbon position of the sugar ribose. This type of reaction could be used to build a nucleotide oligomer, the basis of a ribozyme

guanine, and thymine) together with some intermediates were formed [63] (Table 1.1).

Acevedo and Orgel presented a different concept, in that they viewed synthesis as solid-phase chemistry in which a clay or apatite was used as a catalytic surface for ribonucleotide polymerization [64]. In experiments using clay (montmorillonite) or mineral (apatite) substrates, they reported that chains of polyadenylates containing more than 20 monomers could be developed and subsequently eluted from the support [64]. In another study, Joshi and colleagues [61] described how montmorillonite promoted the formation of long chains (50 nucleotides in length) of RNA molecules at near room temperature over a two-week period. With the generation of a single-stranded polynucleotide, a complimentary assortment of RNA molecules can form, a result adding strength to the notion that some of these molecules could become informational [61]. Whether the clay influenced the activity of the RNA has not been determined [65].

Table 1.1 Synthesis of nucleosides, nucleoside intermediates, and sugars from irradiated formamide and meteorite powder

Product[a]	Canyon diablo	Campo del cielo	NWA 4485	NWA 2828	Gold Basin	Dhofar 59
Uracil (**1**)	–	–	0.1	0.6	0.4	–
Cytosine (**2**)	0.3	0.1	0.23	0.32	0.28	–
Thymine (**3**)	–	–	0.05	0.05	0.08	–
Adenine (**4**)	–	–	0.2	–	0.04	0.13
Guanine (**5**)	–	–	–	0.2	0.05	1.1
Purine (**6**)	–	–	–	0.4	0.6	–
Isocytosine (**7**)	0.7	1.6	0.75	1.2	2.2	–
Hypoxanthine (**8**)	3.0	6.58	–	–	3.8	4.48
2,6-Diaminopurine (**9**)	–	–	–	0.4	0.05	–
Orotic acid (**10**)	–	–	–	–	–	–
4,6-DHP (**11**)	0.5	0.1	–	–	–	–
4(3H)pyrimidone (**12**)	–	–	–	0.1	–	–
3(OH)pyridine (**13**)	–	–	0.25	–	–	–
AHMN (**14**)	–	–	–	–	0.06	–
DAMN (**15**)	–	–	0.05	–	0.12	0.35
4-AMI (**16**)	–	–	0.19	0.15	0.2	0.05
Ribose (**17**)	–	–	1.01	0.16	–	–
2′-Deoxyribose (**18**)	–	–	2.3	0.98	0.1	1.28
Glucose (**19**)	–	–	–	0.12	–	–
2′-Deoxyglucose (**20**)	–	–	0.05	0.66	0.08	0.27
Galactose (**21**)	–	–		0.7	–	–
Mannose (**22**)	–	–	–	–	0.03	–
Inositol (**23**)	–	–	2.88	–	–	–
Arabitol (**24**)	–	–	–	0.67	–	0.39
Uridine (**25**)	–	–	0.54	–	0.03	–
Cytidine (**26**)	–	–	–	–	–	–
Adenosine (**27**)	–	–	–	–	–	0.58
Thymidine (**28**)	–	–	–	–	–	–

Table 1.1 (continued)

Product[a]	NWA 1465	Chelyabinsk	Orgueil	NWA 5357	Al Haggounia
Uracil (**1**)	0.4	0.17	0.6	0.2	1.2
Cytosine (**2**)	0.35	–	0.47	0.5	0.7
Thymine (**3**)	0.1	–	0.4	–	–
Adenine (**4**)	1.3	0.05	1.2	0.6	0.54
Guanine (**5**)	1.4	–	0.8	0.08	0.57
Purine (**6**)	–	–	–	–	–
Isocytosine (**7**)	3	–	6.7	0.5	0.2
Hypoxanthine (**8**)	3.2	1.47	0.2	–	0.53
2,6-Diaminopurine (**9**)	0.5	–	3.2	–	0.48
Orotic acid (**10**)	0.5	–	–	–	0.54
4,6-DHP (**11**)	–	–	0.1	0.03	0.75
4(3H)pyrimidone (**12**)	0.004	–	0.05	0.39	–
3(OH)pyridine (**13**)	–	–	0.26	–	–
AHMN (**14**)	–	–	0.22		0.13
DAMN (**15**)	–	–	0.93	0.47	0.03
4-AMI (**16**)	–	–	–	–	0.52
Ribose (**17**)	0.3	–	–	–	0.1
2′-Deoxyribose (**18**)	0.44	0.05	0.43	–	0.8
Glucose (**19**)	–	–	0.31	–	–
2′-Deoxyglucose (**20**)	1.06	–	1.68	–	–
Galactose (**21**)	0.8	–	0.42		–
Mannose (**22**)	0.44	–	–	–	–
Inositol (**23**)	–	–	–	–	–
Arabitol (**24**)	–	–	–	–	–
Uridine (**25**)	0.96	0.01	1	–	0.45
Cytidine (**26**)	2.2	–	0.03	–	–
Adenosine (**27**)	–	0.76	–	–	0.49
Thymidine (**28**)	–	–	–	–	0.77

A listing of products formed from reacting formamide with powders obtained from a number of different meteorites collected on striking the Earth. Experimentally, formamide (2.5 mL) was irradiated with a proton beam (170 MeV, 3 min, 6 Gy) to mimic solar radiation. Irradiation was performed in the presence of the powdered meteorites shown in the top row. Resulting samples were chemically analyzed for nucleosides, sugars, and metabolic intermediates. Data suggest that formamide could serve as a precursor to molecules critical for cell function. Quantities of products ([a] nucleosides, sugars, and metabolic intermediates) are given in micrograms per mL formamide. Abbreviations: *4,6-DHP* 4,6-dihydroxypyridine; *AHMN* 2-amino-2-hydroxymalonitrile; *DAMN* diaminonitrile; *4AMI* 4-aminoimidazole. Table from Saladino et al. [63] with permission

To conclude this short summary of the RNA World hypothesis and in line with the comments of the Neumans [3], minerals appear to play a pivotal role in the emergence of life on Earth. It is not known whether nucleotide synthesis and ribozyme formation are facilitated through solid state chemistry or surface reactions or interactions with charged sites on minerals. In all cases, however, energy must be supplied to the individual nucleotides to form the characteristic phosphodiester bonds of a ribozyme or, in the case of adsorbed amino acids, peptide bonds. In the next section of this chapter, sources of energy are considered with respect to the availability of phosphate and pyrophosphate anions and events which underscore the origin of life. In Chap. 2, this discussion is continued by reflecting on the chemistry of various phosphate species.

1.5 Phosphorus-containing Minerals, the Driving Force for Prebiotic Chemical Reactions

The ionic composition of the Hadean Ocean is of, course, unknown. However, the notion that it contained appreciable quantities of calcium ions, presumably released from rocks and sediments, is not unreasonable since the sea supported the life of an extraordinary number of calcium carbonate-containing organisms [66, 67]. In contrast, the level of soluble phosphorus-containing compounds in seas, estuaries, and pools was thought to be vanishingly small, a consequence suggested as attributable to limited lithospheric plate tectonics of the Earth with reduced rock recycling [68]. As crustal movements increased and the level of atmospheric oxygen rose, there was the appearance of rocks which contained minerals in which the cations "would otherwise be uncommon under the reducing conditions of the early earth" [68]. The most frequently occurring phosphorus-containing minerals (phosphorites) would include apatite [$Ca_{10}(PO_4)_6(OH)_2$], whitlockite [$Ca_{18}Mg_2(HPO_4)_2(PO_4)_{12}$], and possibly struvite [$MgNH_4PO_4 \cdot 6H_2O$] [69, 70]. However, the solubility of these minerals in seawater is very low, and, hence, the rate of phosphate ion release is very slow. The rate would increase as the seas became acidified and further enhanced by high temperatures such as those reached during volcanic eruptions [71]. These extraction processes were important in providing phosphate and pyrophosphate for chemical syntheses and would support ideas that life began in conjunction with volcanic eruptions together with the formation of porous apatitic surfaces.

It should also be mentioned that an alternative source of phosphorus is the iron-nickel-phosphorus rock, schreibersite [$(Fe,Ni)_3P$], much of which was delivered to the Earth through meteor bombardment [72, 73]. Pasek and Lauretta [72] commented that this mineral provided as much as 5% of the total surface phosphorus of the Earth. Corrosion of schreibersite in water provides orthophosphate (PO_4^{3-}), phosphite (HPO_3^{2-}), hypophosphite ($H_2PO_2^{-}$), and pyrophosphate ($P_2O_7^{4-}$) anions [74].

Whether release of ions was linked to volcanic eruptions or meteoric bombardment, the electronic structure of phosphorylated compounds provides the chemical energy for the formation of the building blocks of RNA-based replicator molecules and the polymerization of nucleoside precursors into RNA molecules. How these reactions take place is unknown. One possibility is that the starting nucleotides are in a cyclic configuration (see cyclic guanosine monophosphate or cyclic guanine monophosphate, Fig. 1.7), having an ester bond which links together (or ligates) the 2′ or 3′ carbon atom of a first ribose sugar with the 5′ carbon atom of a second ribose. A study by Bowler and colleagues indicated that the 3′,5′-phosphodiester would be preferentially formed from nucleoside 2′- and 3′-phosphates [75]. In the cyclic conformation, the oxygen-phosphate (that is, the phosphodiester) bond in the 3′,5′ configuration has sufficient chemical energy to drive the synthesis first of an RNA dimer and then more complex oligomeric structures [76, 77]. In either case, the strong negative charge of the phosphate anion and a highly reactive mineral surface would enhance interactions with oppositely charged cations and their molecules. As a result and depending on the solution pH, there would be binding and accumulation of charged products of numerous reactions [78].

If life first began at the Lost City, at sites of volcanic eruptions or in tidal estuaries, or if it were brought to Earth by meteorites, its subsequent evolution, once initiated, was closely associated with mineral structures. However, for the next two billion years, aside from bacteria and probably certain species of plankton and algae, the Earth remained essentially metazoan-free. Over the millennia and with the appearance of metazoans, the planet was transformed, leaving subtle clues to mechanisms which ultimately governed the formation of bones and teeth.

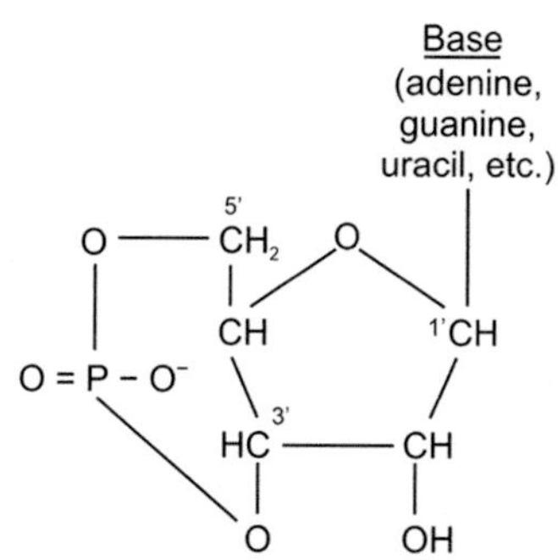

Fig. 1.7 Chemical structure of a cyclic nucleotide. While much emphasis has been placed on conventional forms of nucleotides, cyclic nucleotides are a group of important ligands for cell receptors. Shown above is a cyclic nucleotide in which a phosphodiester bond spans the 3′ to 5′ carbons on the ribose sugar. Since the phosphodiester bond is energy-rich, it can be used to generate polynucleotide structures such as those present in RNA or DNA

1.6 Skeletonization: A Critical Evolutionary Step

If, as the Neumans [3] and others proposed, minerals play a significant role in abiotic chemistry, the presence of an insoluble or partially soluble inorganic phase was undoubtedly important in the succeeding evolution of living organisms. The fossil record indicates that, during the Cambrian period, there was a rapid increase in the diversity of multicellular organisms, an event termed the "Cambrian explosion" [79]. Possibly the paucity of phosphorus and the abundance of calcium in the Cambrian oceans first promoted the evolution of calcium carbonate-based shells, spines, and other assemblies. The tissue deposition of mineral was so widespread that it was evident in not only eukaryotes, but also prokaryotes, including bacteria and archaea [80]. Based on phylogenetic analysis of fossils and extant animals, coupled with structural studies, it is generally recognized that species diversification with major skeletal and dental changes occurred in less than a 30-million-year period since the beginning of the Cambrian era [81, 82]. During this time, there is also abundant evidence of the appearance of both bone- and cartilage-like tissues as well as primitive dental tissues [83–85]. Moreover, the record suggests that the mineralization step was so critical that it was expressed a number of different times through the millennia (described as an example of convergent evolution). As a result, invertebrates generated exoskeletons and intracellular inclusions of many types of minerals, commonly calcium carbonate as noted above. In contrast, the vertebrate endoskeleton, exoskeleton, and dentition were based on the calcium phosphate class of minerals, apatite. Thus, while invertebrates like mollusks expressed neither bones nor true teeth, mechanisms of vertebrate mineral deposition developed very early and included the formation of extracellular matrices which were rich in macromolecular proteoglycan aggregates. These matrix constituents together with certain fibrillar proteins were similar chemically to those of present-day bone and cartilage [84, 86].

The closing pages of this chapter briefly review certain events closely linked to the appearance of mineralized tissues through the lens of over 500 million years of evolution. More detailed discussion of the evolution of mineralized tissues has been presented [87]. Little is known concerning the change from the protocell to the living cell, but currently certain molecular changes are being revealed that accompany the development from a basic protozoan to a more advanced metazoan. Studying the amoeboid, *Capsaspora owczarzaki*, an evolutionarily simple organism which forms multicellular aggregates embedded in an extracellular matrix, Sebé-Pedrós et al. [88] showed that the protozoan–metazoan transition involved changes in transcription factor networks, expression of specific enhancer and promoter elements, and non-coding RNA. In other words, extensive genomic expansion and organization occurred. Further, the multicellular form of *C. owczarzaki* elaborates an extracellular matrix comprised of both fibrillar collagens and type IV collagen (COLIV) [88]. The *COLIVA* gene encodes Goodpasture antigen-binding protein-1, which phosphorylates COLIV to generate a basement membrane [89, 90]. This basement membrane collagen along with binding proteins such as laminin, integrins and the proteoglycan, perlecan, promotes a wide array of functions which include both cell adhesion and migration. As yet, while genes which may foster mineral development have not been identified in this transitional organism, *C. owczarzaki* nevertheless provides insight into some evolutionary changes leading to early collagen formation and the more complex functionality of basic cellular structure.

Kumar and Hedges [91] and Kutschera and Niklas [92] regarded skeletonization and the mineralization process as a critical, driving evolutionary force. When evolutionary pressures prompted the development of an endoskeleton, this profound change stimulated muscular development and augmented mobility. Such a modification also enhanced protection and promoted predation, while the development of teeth advanced digestion and furthered predation. Together, promotion of predation and protection from predation served to catalyze speciation and organism radiation.

Since understanding the mechanism of mineralization relates directly to tissues like bone, calcified cartilage, dentin, enamel, and others, it is worthwhile to comment briefly on speciation and the evolution of a skeleton (Fig. 1.8). Speciation was not an entirely random process since, once an organism acquired new features which differed from a common ancestor, subsets of those organisms continued to evolve to form a family with a number of shared basic features which characterized all organisms in the subgroup. Among the most primitive vertebrate fish, Agnatha formed a grouping of creatures which were defined by the absence of a jaw [96]. A major evolutionary advance was the acquisition of hinged jaws. Animals in this grouping, gnathostomes, were able to feed on a wide variety of foods, thereby enhancing further speciation and the evolution of several major subgroups. One subgroup was built around cartilage, the chondrichthyans. These fish were present in the Devonian period (416–350 million years ago) [97] and co-existed along with placoderms and ostracoderms (Fig. 1.8). Extant examples of these fish are in the grouping, elasmobranchs (sharks and rays). While these presently living chondrichthyans do not ossify their cartilage, a thin, discontinuous mosaic of mineralized tiles or tesserae covers their cartilaginous endoskeleton [98] and, further, a bone-like tissue is present in the neural arches of their vertebrae [93]. Another major taxonomic grouping was characterized by a bony skeleton. These osteichthyans (bony fish) exhibited cortical as well as trabecular bone. Examples of fish in this class include salmon and herring [93–95]. Thus, jaws, bone, and cartilage formed separate but common biological features of these two groups, and the evolution of the jaws was paralleled by the appearance of teeth.

From the past, the earliest examples of mineralized structures appeared among vertebrates in conodonts (Greek, "konos" or "cone" and "odont" or "tooth"), a group of now long-extinct jawless animals morphologically resembling

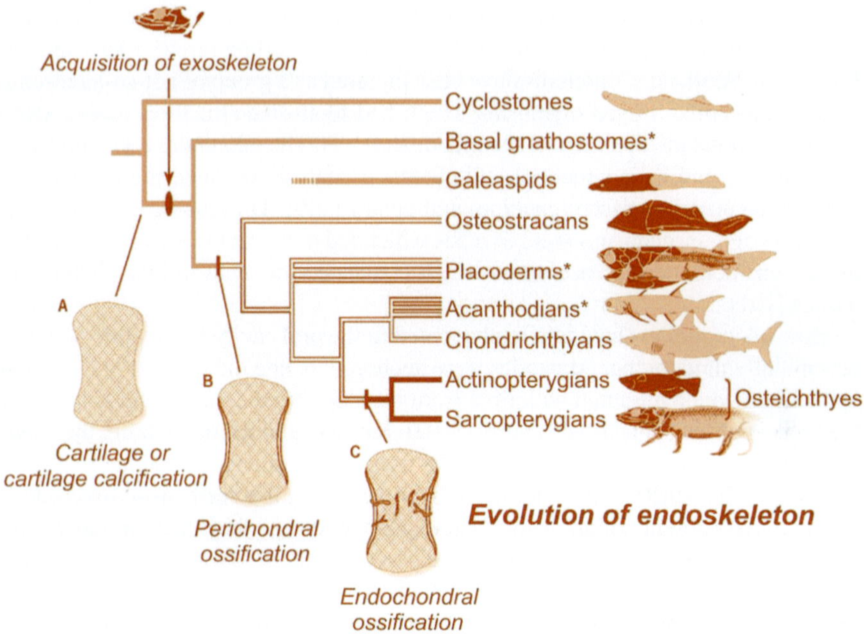

Fig. 1.8 A diagram showing the evolution of the vertebrate skeleton based on phylogenetic considerations. At least three types of skeletal tissues, cartilaginous, perichondral/intramembranous, and endochondral, have been documented and are shown here as cross-hatched barbell-shaped figures (**a**–**c**, respectively). Dermal skeletal tissues are also known and include tooth dentin and dermal bone, which is formed from intramembranous ossification [93]. (**a**) In the superclass, Agnatha, of now extinct jawless fish, cartilaginous bones were present in basal gnathostomes, galeaspids, and some cyclostomes. These tissues could exist in both mineralized and unmineralized forms. Some of the fish within heterostracans, a class of Agnatha, developed a mineralized exoskeleton of dermal bone-like tissue. In the mineralized head and thorax regions of these fish, dentin tubercles appeared, covered with an enamel-like tissue [94]. Notably, conodonts, a phylogenetically early group of Agnathan fish, possessed primitive mineralized teeth but no external or internal skeletal elements [94]. (**b**) Perichondral-derived bone (intramembranous bone) is found in certain osteostracans, placoderms, non-osteichthyes-jawed vertebrates, and examples of modern-day fish. Perichondral/intramembranous bone appears in some of these jawed fish as skull and dermal bones. In specific instances, exoskeletal bones of fish in these classes were decorated with denticles of enameloid and dentin-like tissues. The primitive group of placoderms, containing both perichondral/intramembranous and dermal bones, was characterized by a dense protective coating armor of bone on their head and thorax, together with a jaw which allowed predation. Placoderms and osteostracans developed both endoskeletons and dermal skeletons through intramembranous ossification, and certain of the osteostracan-jawed fish exhibited a cartilaginous endoskeleton [94]. (**c**) An axial and appendicular endoskeleton formed by endochondral ossification is the evolutionarily novel identifying feature of the osteichthyes [95]. Many of these fish also exhibit both dermal and perichondral/intramembranous bones. In contrast to osteichthyans, chondrichthyans fail to develop bone and their endoskeleton is cartilaginous. Asterisks represent paraphyletic groups. (Schematic adapted from Hirasawa and Kuratani [82] with permission)

modern-day eels, lampreys, and hagfish [99]. Remains of these creatures are found in the fossil record 540 million years ago in the Carboniferous rocks around Edinburgh, Scotland. Conodonts have been pictured as a group of not-so-handsome jawless and bulbous-eyed organisms which had tooth-like structures (odontodes), arranged around the openings of the mouth and gills. The odontodes contained apatitic mineral, and their structural and chemical analysis revealed the presence of zones of enamel-like tissue (ganoine) and dentin [100]. Thus, as early as 500 million years ago, mechanisms were in place which led to the formation of mineralized dentin, enamel, and the basal bone which connected the teeth to their underlying tissues [101].

How odontodes evolved into teeth located in the oral cavity is not understood. It was initially thought that odontodes were anchored to, and the product of, a dermal epithelium, which migrated with time from the skin into the mouth and pharynx of the organism (the "outside-in" theory) [102]. Remnants of the epithelium, the "dental lamina," regulated by chemical signals from the local underlying mesenchyme, induced tooth formation. An alternative hypothesis was that teeth arose from odontodes already present within the oropharyngeal cavity, and they then migrated from the mouth region to the thorax and abdomen of the animals to form denticles associated with their dermal skeleton (the "inside-out" theory) [103]. Fraser and colleagues [104] proposed an "inside and out" theory in which one set of denticles originated in the dermal ectoderm (out) and one set of odontodes was positioned in the endoderm of the mouth region (in). Implicit here is the suggestion that the two types of structures evolved separately and thus represented a type of convergent evolution.

It should be mentioned that tooth formation is regulated by chemical signals from both the covering ectoderm and underlying mesenchyme. Moreover, there is overwhelming evidence that neural crest cells (a group of specialized cells in the developing cranium) participate in tooth formation (see Chap. 3). The "inside-out" concept would imply that the signals for tooth development are generated by the oral (endodermal) mesenchyme and that signals for odontode migration are mediated through the skin ectodermal mesenchyme [102]. In support of this proposal, the extant Mexican salamander (axolotl) indicates that both ectodermal and endodermal cells contribute to the formation of a tooth [105].

With respect to true bone, probably the most exciting insight into the evolution of the skeleton has been attained through studies of fossil members of a group of heavily armored fish-like sea creatures called ostracoderms (Greek, "ostracon" or "shell" and "derma" or "skin") (Fig. 1.9). Ostracoderms were gnathostomes and likely survived by sucking in and ingesting nutrients from the sea floor. In this way, the animal would have been able to access adequate quantities of particulate phosphorus for mineralization purposes. Members of the ostracoderm family (osteostracans) exhibited a mineralized exoskeleton, while their cranium itself was enclosed in a bone-like exoskeleton overlaid with tubercles of dentin having a cap of an enamel-like tissue (Fig. 1.10) [106]. In his book, *The Ancestor's Tale: A Pilgrimage*

Fig. 1.9 Artist view of an ostracoderm. Ostracoderms were a clade of jawless fish (superclass Agnatha, phylum Chordata) within the heterostracan and osteostracan classes which lived 500,000 years ago and became extinct at the end of the Devonian period. Ostracoderms had no lungs or air bladder and probably survived by feeding on detritus from the ocean floor. Ostracoderms were protected by an exoskeleton of dermal bone. The cephalic portion of the fish was covered with a dense armor of dermal bone embedded with tooth-like structures consisting of dentin-like and enamel-like tissues (image based on studies of Janvier [96] and modeled from the archived collection of ostracoderms at the Museum of National History, New York). The size of a mature ostracoderm ranged up to 15 cm in length, based on the collection. (The figure was drawn by Benjamin Edlitz and modified by Bradley Snyder, M.S.)

to the Dawn of Evolution, Dawkins conjectured that it was this dermal exoskeleton which protected creatures from large scorpion-like arthropods that roamed the seas in that era [107]. With the demise of such predators, the possible need for a dermal skeleton was lost and the jawed descendants from this group evolved an internal skeleton.

Not surprisingly since the fossil record is incomplete, there is still some dispute concerning the evolutionary development of endochondral bone, that is, bone deposited within calcified cartilage. Certain evidence documents that calcified cartilage represented a separate stage of evolution: One tiny, fossilized fish bearing the unlikely name of *Palaeospondylus gunni* appears to contain a skeleton of mineralized cartilage. Johanson et al. [108] reported that the tissue was comprised of hypertrophic chondrocytes in an apatitic matrix but there was a total absence of bone. In other words, this little fish denoted an intermediate species in which the endoskeleton consisted entirely of mineralized cartilage (Fig. 1.11).

From an evolutionary perspective, the concept that calcified cartilage formation preceded bone formation has not been settled. In a discussion of fossil tissue

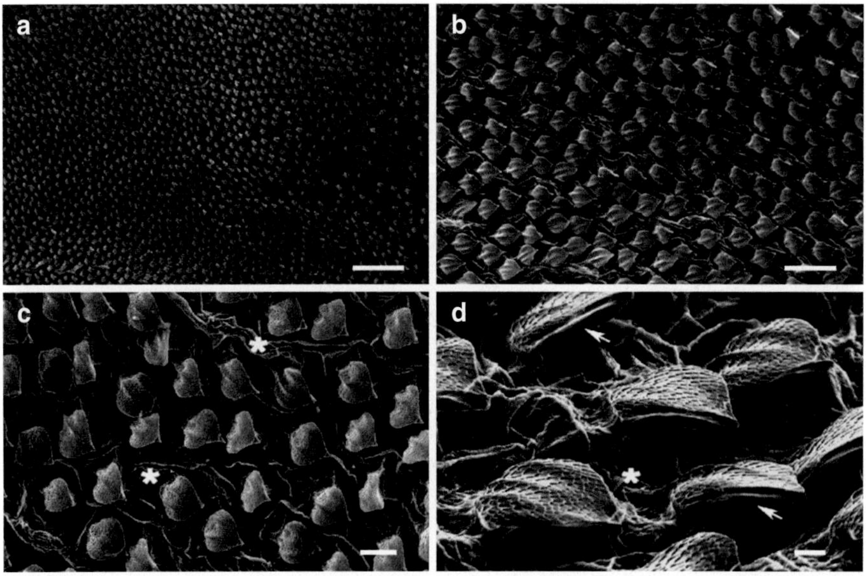

Fig. 1.10 Scanning electron micrographs of oral denticles (small, tooth-like structures) of the shark, *Rhizoprionodon lalandii*. (**a**) Low power view of a field of oral denticles. (**b**) Higher power view of denticles from (**a**) showing they are of uniform sizes with each individual apex facing the caudal direction. (**c**) View showing the presence of bundles of collagen fibers (*) between the denticles. (**d**) The oral denticles present a smooth surface (arrows) on their caudal surface and bundles of collagen fibers (*) lie between the denticles. Scale bars: 1 mm (**a**), 300 μm (**b**), 100 μm (**c**), 30 μm (**d**). (Images from Ciena et al. [101] with permission)

succession, Smith and Hall have suggested that the primary tissue of the endoskeleton is cartilage-based, while bone and possibly dentin represent the principal tissues of the exoskeleton [109]. Romer made an attempt to address this important question in his treatise, "The 'Ancient History' of Bone" [110]. He wrote, "Cartilage, as mentioned previously, is apparently an ancient type of skeletal material. But, in the vertebrates, it is essentially an embryonic adaptation, and its prominence in the adult skeleton of many fishes is not a primitive feature but an evidence of neoteny" [110]. Dare it be stated that Romer may have been incorrect in his reasoning and that bone and cartilage co-evolved as did bone and tooth. In subsequent chapters of this book, similar themes are explored but, instead of using information from the fossil record, questions are evaluated from the viewpoint of the cell biology of mineralizing systems, DNA-based taxonomy, and molecular pathways of mineral accretion and transport.

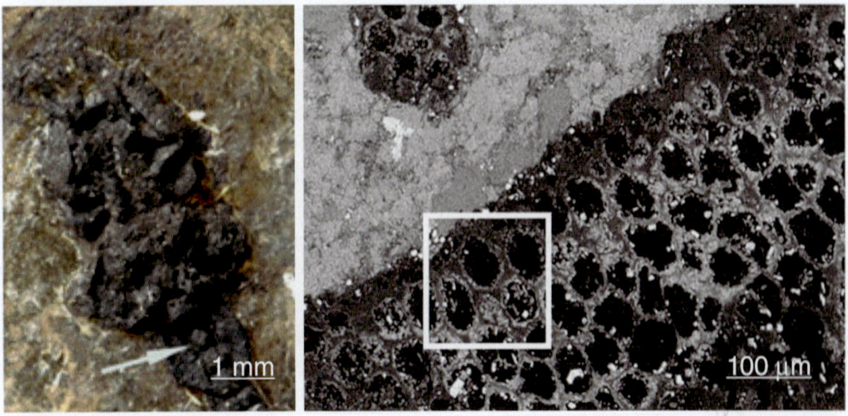

Fig. 1.11 Structure and chemical composition of the skeleton of *Palaeospondylus gunni*, a fossilized fish which lived approximately 385 million years ago. The left panel presents a dorsal view of the fossil showing the chondrocranium and vertebral region (arrow). The right panel illustrates a section through the fossil revealing the matrix surrounding the empty locations of each of the skeletal cells. The size and shape of the cells are very similar to those of chondrocytes which populate hypertrophic cartilage. The square outlined in white in the right panel defines a region of the lacunar space and the perilacunar and interlacunar matrix in the section from which X-ray emission maps were obtained. Resulting data showed calcium and phosphorus principally localized in the matrix around the lacunae of the fossil and confirming that the tissue was mineralized. (Images and spectra adapted from Johanson et al. [108] with permission)

1.7 Perspective: In the Beginning

Before proceeding with a discussion concerning the energetics of mineralization (Chap. 2), it is worthwhile reflecting on events occurring in the several-billion-year time interval which linked apatite geochemistry with the formation of molecules critical to the emergence of life. Despite numerous fascinating research ideas, the hypotheses surrounding the prebiotic–biotic transition remain exceptionally vague and uncertain scientifically. However, what the current narrative lacks in specificity, it more than makes up for in assumptions: DNA or RNA did not hitchhike to Earth clinging to a meteor, there was sufficient soluble phosphate in the Hadean Sea or in freshwater on land to permit nucleotide synthesis, and a lipid bilayer separating the abiotic-synthesized organic molecules from their external environment was an autocatalytic event. Notwithstanding these and many other conjectures, the most likely scenario is that, whether mediated by a beach catalyst, crystal-tide mineral, or an underwater thermal vent, life began with the appearance of RNA.

While this chapter has reviewed the tenets of the RNA World hypothesis to some extent, other ideas for the origin of life have been presented. Very recently, Chatterjee and Yadav proposed a further modification of the RNA World hypothesis: That the system would be more functional if it were not totally reliant on RNA [1]. The essence of this suggestion was that peptides together with ribozymes are needed to generate an active protocell [1]. As noted earlier, peptides were critical since amino acids could be synthesized from very basic molecules which were readily available in the prebiotic world and peptide bonds could be generated. Within the protocell, RNA, peptides, and other organic molecules would interact with each other co-dependently [111]. Whether peptides were introduced early or were later additions to the system, there can be little doubt that RNA in a variety of molecular configurations provides information storage properties and serves as an enzyme(s) involved with basic biosynthetic processes. Indeed, these two activities of RNA would meet the first part of a definition propounded by NASA that "Life is a self-sustaining chemical system capable of Darwinian evolution" [112].

Where does apatite fit into the origin of life scenario? Does existing evidence support the Neumans' rather naive statement that life began with apatite? [3]. While the largest context of experimental evidence suggests that minerals are involved in many prebiotic chemical reactions and related physical events such as those linked to molecule binding and concentration, such a role for apatite specifically seems unlikely. Nevertheless, in a very detailed paper based on apatitic lattice parameters and other structural features, Kostetsky presented the opinion that apatite and co-crystallizing minerals were associated with the synthesis of protocell components which included polynucleotides, proteins, and even lipids [113]. This study [113] supports the concept that apatite may have exerted considerable effects on the origin of life in addition to the recognized importance of clays, silica, calcium carbonate, and many other minerals. Of course, whether these minerals participated in abiotic processes as a catalyst or a reactive surface or were just the precipitated product of a chemical reaction is difficult to ascertain.

While the surface character of limited mineral species may have been relevant to certain defined aspects of abiotic chemistry, the wide diversity of evolved mineralized structures associated with living organisms provides the animal with a biological advantage. Whereas intracellular mineral in some organisms is used to detect changes in gravitational forces, for example, it forms in others a mechanically strong extracellular matrix which can be used for functions as disparate as vision, predation, protection, and locomotion. Moreover, controlled release of the ions contained within the mineral plays a major role in systems of enzyme regulation, energetics, signal transduction, and information storage.

Finally, in addressing the second part of the NASA definition of life, that a need exists for expressed chemical activities to support Darwinian evolution [112], it is likely that the mechanical properties of cartilage and cartilage-like matrices promoted the evolution of a defined shape of a species (compare, for instance, the unshapely sea-dwelling octopus with the solidly modeled land-dwelling tortoise). For many animals, acquisition of semi-firm and hard tissues enabled migration from the aquatic environment to dry land or the change from movement on land to that of flight among several additional functions. Thus, cartilage, a tissue which can remain stiff and non-mineralized, is found in many forms in numerous different phyla across the animal kingdom. In contrast to cartilage, bone appeared as a protective external dermal armor of vertebrates, possibly derived from a dentin-like tissue.

The presence of a mechanically strong tissue, whatever its origin, containing an almost insoluble apatitic mineral afforded many biological advantages to its inorganically invested species. A critical advantage in this context may be exemplified by the passage from sea to land established by certain ancient fish. In this fundamental developmental step in skeletonization, it might be predicted that evolutionary changes in which there is a transition of fish fin to tetrapod limb would require activation of a multigene circuit with numerous feedback systems (see Chap. 4). Surprisingly, this situation does not appear to be the case. In a recent study, Hawkins et al. reported that just two genes were required to induce such a change [114]. Using gene editing techniques, these investigators found that an activating mutation in appendage-patterning genes, *VAV2* (a member of the diffuse B-cell lymphoma family of genes/proteins, so named for the sixth letter of the Hebrew alphabet) and Wiskott-Aldrich syndrome (*WASL)b*, caused a fin-to-limb alteration characterized by the appearance of supernumerary long bones in zebrafish [114]. These new bones, which integrated with the musculature and formed articulated joints [114], underpin and demonstrate the striking evolutionary value for a species possessing mineralized tissues. Further details on how such tissues are formed, including the genetic regulation of cartilage and bone biogenesis and possible mechanisms by which inorganic ions may be deposited within the extracellular, commonly collagenous, matrices of vertebrates, are topics discussed in considerable detail in later chapters of this book.

References

1. Chatterjee S, Yadav S. The origin of prebiotic information system in the peptide/RNA world: a simulation model of the evolution of translation and the genetic code. Life (Basel). 2019;9:25.
2. Kasting JF. When methane made climate. Sci Am. 2004;291:78–85.
3. Neuman WF, Neuman MW. The chemical dynamics of bone mineral. Chicago, IL: University of Chicago Press; 1958.
4. Miller SL. A production of amino acids under possible primitive earth conditions. Science. 1953;117:528–9.
5. James Cleaves H II, Michalkova Scott A, Hill FC, Leszczynski J, Sahai N, Hazen R. Mineral–organic interfacial processes: potential roles in the origins of life. Chem Soc Rev. 2012;41:5502–25.
6. Cummings LJ, Snyder MA, Brisack K. Protein chromatography on hydroxyapatite columns. Methods Enzymol. 2009;463:387–404.
7. Israelachvili JN, Mitchell DJ, Ninham BW. Theory of self-assembly of lipid bilayers and vesicles. Biochim Biophys Acta. 1977;470:185–201.
8. Kitt JP, Bryce DA, Minteer SD, Harris JM. Confocal Raman microscopy investigation of self-assembly of hybrid phospholipid bilayers within individual porous silica chromatographic particles. Anal Chem. 2019;91:7790–7.

9. Greenwell HC, Coveney PV. Layered double hydroxide minerals as possible prebiotic information storage and transfer compounds. Orig Life Evol Biosph. 2006;36:13–37.
10. Moreno EC, Kresak M, Hay DI. Adsorption of molecules of biological interest onto hydroxyapatite. Calcif Tissue Int. 1984;36:48–59.
11. Shaw WJ. Solid-state NMR studies of proteins immobilized on inorganic surfaces. Solid State Nucl Magn Reson. 2015;70:1–14.
12. Ferris JP, Hill AR Jr, Liu R, Orgel LE. Synthesis of long prebiotic oligomers on mineral surfaces. Nature. 1996;381:59–61.
13. Gibbs D, Lohrmann R, Orgel LE. Template-directed synthesis and selective adsorption of oligoadenylates on hydroxyapatite. J Mol Evol. 1980;15:347–54.
14. De Sanctis MC, Ammannito E, McSween HY, Raponi A, Marchi S, Capaccioni F, et al. Localized aliphatic organic material on the surface of Ceres. Science. 2017;355:719–22.
15. Cronin JR, Cooper GW, Pizzarello S. Characteristics and formation of amino acids and hydroxy acids of the Murchison meteorite. Adv Space Res. 1995;15:91–7.
16. Cooper GW, Cronin JR. Linear and cyclic aliphatic carboxamides of the Murchison meteorite: hydrolyzable derivatives of amino acids and other carboxylic acids. Geochim Cosmochim Acta. 1995;59:1003–15.
17. Bernstein MP, Dworkin JP, Sandford SA, Cooper GW, Allamandola LJ. Racemic amino acids from the ultraviolet photolysis of interstellar ice analogues. Nature. 2002;416:401–3.
18. Koga T, Naraoka H. A new family of extraterrestrial amino acids in the Murchison meteorite. Sci Rep. 2017;7:1–8.
19. Hazen RM, Sholl DS. Chiral selection on inorganic crystalline surfaces. Nat Mater. 2003;2:367–74.
20. Hazen RM, Sverjensky DA. Mineral surfaces, geochemical complexities, and the origins of life. Cold Spring Harb Perspect Biol. 2010;2:1–21.
21. Kasting JF, Howard MT. Atmospheric composition and climate on the early Earth. Philos Trans R Soc Lond Ser B Biol Sci. 2006;361:1733–41.
22. Oro J, Guidry CL. A novel synthesis of polypeptides. Nature. 1960;186:156–7.
23. Miller SL, Urey HC. Organic compound synthesis on the primitive earth. Science. 1959;130:245–51.
24. Adande GR, Woolf NJ, Ziurys LM. Observations of interstellar formamide: availability of a prebiotic precursor in the galactic habitable zone. Astrobiology. 2013;13:439–53.
25. Pino S, Sponer JE, Costanzo G, Saladino R, Mauro ED. From formamide to RNA, the path is tenuous but continuous. Life (Basel). 2015;5:372–84.
26. Costanzo G, Saladino R, Crestini C, Ciciriello F, Di Mauro E. Formamide as the main building block in the origin of nucleic acids. BMC Evol Biol. 2007;7:1–8.
27. Powner MW, Gerland B, Sutherland JD. Synthesis of activated pyrimidine ribonucleotides in prebiotically plausible conditions. Nature. 2009;459:239–42.
28. Powner MW, Sutherland JD, Szostak JW. Chemoselective multicomponent one-pot assembly of purine precursors in water. J Am Chem Soc. 2010;132:16677–88.
29. Patel BH, Percivalle C, Ritson DJ, Duffy CD, Sutherland JD. Common origins of RNA, protein and lipid precursors in a cyanosulfidic protometabolism. Nat Chem. 2015;7:301–7.
30. Sutherland JD. The origin of life--out of the blue. Angew Chem Int Ed Eng. 2016;55:104–21.
31. Badu-Tawiah AK, Campbell DI, Cooks RG. Reactions of microsolvated organic compounds at ambient surfaces: droplet velocity, charge state, and solvent effects. J Am Soc Mass Spectrom. 2012;23:1077–184.
32. Lee JK, Banerjee S, Nam HG, Zare RN. Acceleration of reaction in charged microdroplets. Q Rev Biophys. 2015;48:437–44.
33. Nam I, Lee JK, Nam HG, Zare RN. Abiotic production of sugar phosphates and uridine ribonucleoside in aqueous microdroplets. Proc Natl Acad Sci U S A. 2017;114:12396–400.
34. Kelley DS, Karson JA, Blackman DK, Fruh-Green GL, Butterfield DA, Lilley MD, et al. An off-axis hydrothermal vent field near the Mid-Atlantic Ridge at 30 degrees N. Nature. 2001;412:145–9.

35. Kelley DS, Karson JA, Fruh-Green GL, Yoerger DR, Shank TM, Butterfield DA, et al. A serpentinite-hosted ecosystem: the Lost City hydrothermal field. Science. 2005;307:1428–34.
36. Herschy B, Whicher A, Camprubi E, Watson C, Dartnell L, Ward J, et al. An origin-of-life reactor to simulate alkaline hydrothermal vents. J Mol Evol. 2014;79:213–27.
37. Bassez MP. Anoxic and oxic oxidation of rocks containing Fe(II)Mg-silicates and Fe(II)-monosulfides as source of Fe(III)-minerals and hydrogen. Geobiotropy. Orig Life Evol Biosph. 2017;47:453–80.
38. Eschenmoser A. On a hypothetical generational relationship between HCN and constituents of the reductive citric acid cycle. Chem Biodivers. 2007;4:554–73.
39. Sahai N, Kaddour H, Dalai P, Wang Z, Bass G, Gao M. Mineral surface chemistry and nanoparticle-aggregation control membrane self-assembly. Sci Rep. 2017;7:1–13.
40. Branscomb E, Russell MJ. Frankenstein or a submarine alkaline vent: who is responsible for abiogenesis?: Part 2: As life is now, so it must have been in the beginning. Bioessays. 2018;40:e1700182.
41. Ourisson G, Nakatani Y. The terpenoid theory of the origin of cellular life: the evolution of terpenoids to cholesterol. Chem Biol. 1994;1:11–23.
42. Albertsen AN, Duffy CD, Sutherland JD, Monnard PA. Self-assembly of phosphate amphiphiles in mixtures of prebiotically plausible surfactants. Astrobiology. 2014;14:462–72.
43. Cooper GW, Onwo WM, Cronin JR. Alkyl phosphonic acids and sulfonic acids in the Murchison meteorite. Geochim Cosmochim Acta. 1992;56:4109–15.
44. Hanczyc MM, Fujikawa SM, Szostak JW. Experimental models of primitive cellular compartments: encapsulation, growth, and division. Science. 2003;302:618–22.
45. Pizzarello S, Shock E. Carbonaceous chondrite meteorites: the chronicle of a potential evolutionary path between stars and life. Orig Life Evol Biosph. 2017;47:249–60.
46. Huber C, Kraus F, Hanzlik M, Eisenreich W, Wächtershäuser G. Elements of metabolic evolution. Chemistry. 2012;18:2063–80.
47. Bada JL. New insights into prebiotic chemistry from Stanley Miller's spark discharge experiments. Chem Soc Rev. 2013;42:2186–96.
48. Niether D, Afanasenkau D, Dhont JK, Wiegand S. Accumulation of formamide in hydrothermal pores to form prebiotic nucleobases. Proc Natl Acad Sci U S A. 2016;113:4272–7.
49. Shapiro R. A simpler origin for life. Sci Am. 2007;296:46–53.
50. Crick FH. The origin of the genetic code. J Mol Biol. 1968;38:367–79.
51. Orgel LE. Evolution of the genetic apparatus. J Mol Biol. 1968;38:381–93.
52. Gilbert W. Evolution of antibodies. The road not taken. Nature. 1986;320:485–6.
53. Guerrier-Takada C, Altman S. Catalytic activity of an RNA molecule prepared by transcription in vitro. Science. 1984;223:285–6.
54. Chen IA, Salehi-Ashtiani K, Szostak JW. RNA catalysis in model protocell vesicles. J Am Chem Soc. 2005;127:13213–9.
55. Cech TR. The ribosome is a ribozyme. Science. 2000;289:878–9.
56. Ekland EH, Szostak JW, Bartel DP. Structurally complex and highly active RNA ligases derived from random RNA sequences. Science. 1995;269:364–70.
57. Muller S. Engineering of ribozymes with useful activities in the ancient RNA world. Ann N Y Acad Sci. 2015;1341:54–60.
58. Ekland EH, Bartel DP. RNA-catalysed RNA polymerization using nucleoside triphosphates. Nature. 1996;382:373–6.
59. Orgel LE. Prebiotic chemistry and the origin of the RNA world. Crit Rev Biochem Mol Biol. 2004;39:99–123.
60. Cairns-Smith AG. Chemistry and the missing era of evolution. Chemistry. 2008;14:3830–9.
61. Joshi PC, Aldersley MF, Delano JW, Ferris JP. Mechanism of montmorillonite catalysis in the formation of RNA oligomers. J Am Chem Soc. 2009;131:13369–74.
62. Jheeta S, Joshi PC. Prebiotic RNA synthesis by montmorillonite catalysis. Life (Basel). 2014;4:318–30.

63. Saladino R, Carota E, Botta G, Kapralov M, Timoshenko GN, Rozanov AY, et al. Meteorite-catalyzed syntheses of nucleosides and of other prebiotic compounds from formamide under proton irradiation. Proc Natl Acad Sci U S A. 2015;112:2746–55.
64. Acevedo OL, Orgel LE. Template-directed oligonucleotide ligation on hydroxylapatite. Nature. 1986;321:790–2.
65. Stephenson JD, Popovic M, Bristow TF, Ditzler MA. Evolution of ribozymes in the presence of a mineral surface. RNA. 2016;22:1893–901.
66. Hathaway JC, Degens ET. Methane-derived marine carbonates of pleistocene age. Science. 1969;165:690–2.
67. Grotzinger JP, Kasting JF. New constraints on Precambrian ocean composition. J Geol. 1993;101:235–43.
68. Pasek MA, Gull M, Herschy B. Phosphorylation on the early earth. Chem Geol. 2017;475:149–70.
69. Handschuh GJ, Orgel LE. Struvite and prebiotic phosphorylation. Science. 1973;179:483–4.
70. Hazen RM. Paleomineralogy of the Hadean Eon: a preliminary species list. Am J Sci. 2013;313:807–43.
71. Yamagata Y, Watanabe H, Saitoh M, Namba T. Volcanic production of polyphosphates and its relevance to prebiotic evolution. Nature. 1991;352:516–9.
72. Pasek MA, Lauretta DS. Aqueous corrosion of phosphide minerals from iron meteorites: a highly reactive source of prebiotic phosphorus on the surface of the early Earth. Astrobiology. 2005;5:515–35.
73. Bryant DE, Kee TP. Direct evidence for the availability of reactive, water soluble phosphorus on the early Earth. H-phosphinic acid from the Nantan meteorite. Chem Commun (Camb). 2006;14:2344–6.
74. Pasek M, Herschy B, Kee TP. Phosphorus: a case for mineral-organic reactions in prebiotic chemistry. Orig Life Evol Biosph. 2015;45:207–18.
75. Bowler FR, Chan CK, Duffy CD, Gerland B, Islam S, Powner MW, et al. Prebiotically plausible oligoribonucleotide ligation facilitated by chemoselective acetylation. Nat Chem. 2013;5:383–9.
76. Cassone G, Sponer J, Saija F, Di Mauro E, Marco Saitta A, Sponer JE. Stability of 2′,3′ and 3′,5′ cyclic nucleotides in formamide and in water: a theoretical insight into the factors controlling the accumulation of nucleic acid building blocks in a prebiotic pool. Phys Chem Chem Phys. 2017;19:1817–25.
77. Sponer JE, Sponer J, Mauro ED. New evolutionary insights into the non-enzymatic origin of RNA oligomers. Wiley Interdiscip Rev RNA. 2017;8:1–15.
78. Choimet M, Tourrette A, Drouet C. Adsorption of nucleotides on biomimetic apatite: the case of cytidine 5' monophosphate (CMP). J Colloid Interface Sci. 2015;456:132–7.
79. Schopf JW. The early evolution of life: solution to Darwin's dilemma. Trends Ecol Evol. 1994;9:375–7.
80. Murdock DJ, Donoghue PC. Evolutionary origins of animal skeletal biomineralization. Cells Tissues Organs. 2011;194:98–102.
81. Briggs DE. The Cambrian explosion. Curr Biol. 2015;25:864–8.
82. Hirasawa T, Kuratani S. Evolution of the vertebrate skeleton: morphology, embryology, and development. Zoological Lett. 2015;1:1–17.
83. Donoghue PC, Sansom IJ. Origin and early evolution of vertebrate skeletonization. Microsc Res Tech. 2002;59:352–72.
84. Donoghue PC, Sansom IJ, Downs JP. Early evolution of vertebrate skeletal tissues and cellular interactions, and the canalization of skeletal development. J Exp Zool B Mol Dev Evol. 2006;306:278–94.
85. Janvier P. Facts and fancies about early fossil chordates and vertebrates. Nature. 2015;520:483–9.
86. Vynios DH, Papageorgakopoulou N, Sazakli H, Tsiganos CP. The interactions of cartilage proteoglycans with collagens are determined by their structures. Biochimie. 2001;83:899–906.

87. Hall BK. Bones and cartilage: developmental and evolutionary skeletal biology. 2nd ed. London; San Diego, CA; Waltham, MA; Oxford: Academic Press; 2015.
88. Sebe-Pedros A, Ballare C, Parra-Acero H, Chiva C, Tena JJ, Sabido E, et al. The dynamic regulatory genome of Capsaspora and the origin of animal multicellularity. Cell. 2016;165:1224–37.
89. Revert-Ros F, Lopez-Pascual E, Granero-Molto F, Macias J, Breyer R, Zent R, et al. Goodpasture antigen-binding protein (GPBP) directs myofibril formation: identification of intracellular downstream effector 130-kDa GPBP-interacting protein (GIP130). J Biol Chem. 2011;286:35030–43.
90. Rodriguez-Pascual F. How evolution made the matrix punch at the multicellularity party. J Biol Chem. 2019;294:770–1.
91. Kumar S, Hedges SB. A molecular timescale for vertebrate evolution. Nature. 1998;392:917–20.
92. Kutschera U, Niklas KJ. The modern theory of biological evolution: an expanded synthesis. Naturwissenschaften. 2004;91:255–76.
93. Atake OJ, Eames BF. Mineralized cartilage and bone-like tissues in chondrichthyans offer potential insights into the evolution and development of mineralized tissues in the vertebrate endoskeleton. Front Genet. 2021;12:762042.
94. Keating JN, Marquart CL, Marone F, Donoghue PCJ. The nature of aspidin and the evolutionary origin of bone. Nat Ecol Evol. 2018;2:1501–6.
95. Brazeau MD, Giles S, Dearden RP, Jerve A, Ariunchimeg Y, Zorig E, et al. Endochondral bone in an early Devonian 'placoderm' from Mongolia. Nat Ecol Evol. 2020;4:1477–84.
96. Janvier P. Early jawless vertebrates and cyclostome origins. Zool Sci. 2008;25:1045–56.
97. Sallan LC, Coates MI. End-Devonian extinction and a bottleneck in the early evolution of modern jawed vertebrates. Proc Natl Acad Sci U S A. 2010;107:10131–5.
98. Dean MN, Summers AP. Mineralized cartilage in the skeleton of chondrichthyan fishes. Zoology (Jena). 2006;109:164–8.
99. Murdock DJ, Dong XP, Repetski JE, Marone F, Stampanoni M, Donoghue PC. The origin of conodonts and of vertebrate mineralized skeletons. Nature. 2013;502:546–9.
100. Sansom IJ, Smith MP, Armstrong HA, Smith MM. Presence of the earliest vertebrate hard tissue in conodonts. Science. 1992;256:1308–11.
101. Ciena AP, Rangel BS, Bruno CE, Miglino MA, de Amorim AF, Rici RE, et al. Morphological aspects of oral denticles in the sharpnose shark rhizoprionodon lalandii (Muller and Henle, 1839) (Elasmobranchii, Carcharhinidae). Anat Histol Embryol. 2016;45:109–14.
102. Donoghue PC, Rucklin M. The ins and outs of the evolutionary origin of teeth. Evol Dev. 2016;18:19–30.
103. Smith MM, Coates MI. Evolutionary origins of the vertebrate dentition: phylogenetic patterns and developmental evolution. Eur J Oral Sci. 1998;106:482–500.
104. Fraser GJ, Cerny R, Soukup V, Bronner-Fraser M, Streelman JT. The odontode explosion: the origin of tooth-like structures in vertebrates. Bioessays. 2010;32:808–17.
105. Soukup V, Epperlein HH, Horacek I, Cerny R. Dual epithelial origin of vertebrate oral teeth. Nature. 2008;455:795–8.
106. Keating JN, Marquart CL, Donoghue PC. Histology of the heterostracan dermal skeleton: insight into the origin of the vertebrate mineralised skeleton. J Morphol. 2015;276:657–80.
107. Dawkins R. Ancestor's tale: a pilgrimage to the dawn of evolution. Boston, MA: Houghton Mifflin Harcourt; 2004.
108. Johanson Z, Kearsley A, den Blaauwen J, Newman M, Smith MM. No bones about it: an enigmatic Devonian fossil reveals a new skeletal framework--a potential role of loss of gene regulation. Semin Cell Dev Biol. 2010;21:414–23.
109. Smith MM, Hall BK. Development and evolutionary origins of vertebrate skeletogenic and odontogenic tissues. Biol Rev Camb Philos Soc. 1990;65:277–373.
110. Romer AS. The "ancient history" of bone. Ann N Y Acad Sci. 1963;31:168–76.

111. Kunnev D, Gospodinov A. Possible emergence of sequence specific RNA aminoacylation via peptide intermediary to initiate Darwinian evolution and code through origin of life. Life (Basel). 2018;8:8040044.
112. Deamer DW, Fleischaker GR. Origins of life: the central concepts. Boston, MA: Jones and Bartlett Publishers; 1994.
113. Kostetsky EY. The possibility of the formation of protocells and their structural components on the basis of the apatite matrix and cocrystallizing minerals. J Biol Phys. 2005;31:607–38.
114. Hawkins MB, Henke K, Harris MP. Latent developmental potential to form limb-like skeletal structures in zebrafish. Cell. 2021;184:899–911.

The Enigma of Edith: Free Energy, Nucleation, and the Formation of Mineral

2

Imagine for a moment that we are nothing but the product of billions of years of molecules coming together and ratcheting up through natural selection, that we are composed only of highways of fluids and chemicals sliding along roadways within billions of dancing cells, … To me, that understanding would be a numinous experience, better than anything ever proposed in anyone's holy text. [1]

Glossary

Amorphous mineral A mineral having only short-range structural order and which may serve as a precursor to hydroxyapatite or isomorphs of calcium carbonate.

Enthalpy change A thermodynamic concept which assesses the internal energy of a system. In chemistry, a change in enthalpy is the heat released/absorbed during a chemical reaction.

Entropy A thermodynamic measure of the disorder or random nature of a macroscopic system. It is a measure of the energy change among atoms or molecules in a closed system.

Gibbs free energy (ΔG) The energy associated with a system which can be used to do work. The free energy is equivalent to the enthalpy plus the product of the entropy times the temperature.

Interfacial energy The surface energy which determines the interaction of two phases (solid, liquid, gas) at their interface.

Ion cluster An association of ions of opposing charges in solution.

Ion pair In the context of vertebrate mineralization chemistry, the association of hydrated calcium with hydrated phosphate ions.

Liquid–liquid phase A multicomponent system in which there is a condensation of hydrated ions or molecules in a liquid. The term may be used to describe prenucleation clusters of hydrated and aggregated ion pairs (for example, calcium and phosphate) which exist in a bulk liquid phase.

I. M. Shapiro, W. J. Landis, *Mechanisms of Mineralization of Vertebrate Skeletal and Dental Tissues*, https://doi.org/10.1007/978-3-031-34304-9_2

Metastable solution A non-equilibrium (unstable) state of a solution such as that exhibited by a supersaturated fluid in which ions may cluster or remain in solution depending on the energy of the system.

Nucleating agent An agent which promotes the formation of a crystal from an aqueous solution.

Posner cluster An aggregate of calcium and phosphate ions forming a particle with the stoichiometric formula $Ca_9(PO_4)_6$ and considered to be a basic component of amorphous calcium phosphate.

Prenucleation and critical cluster A small aggregate of ions which remains stable in solution while having the potential to nucleate and form a definable mineral phase.

Spinodal decomposition A process by which a thermodynamically unstable but apparently homogeneous liquid phase can separate to form two or more individual phases.

2.1 Setting the Scene

Leaving events which took place during the Hadean period when single-celled organisms first appeared in a calcium-rich carbonated sea, time jumps forward four billion years to the historical era. A mere 2000 years ago, during the Biblical period, an unusual form of mineralization was reported in the book of Genesis. Two Angels advised Lot that God was about to destroy the city of Sodom. Responding to this warning, Lot and his family decided to flee the accursed city, but Edith, the wife of Lot, chanced to look back at the destruction and was turned into a pillar of salt. While neither X-ray diffraction analysis nor Fourier transform infrared (FTIR) spectroscopy was available at that time to confirm the cubic crystal structure of salt, it was inferred that Edith's final state was precipitated sodium chloride. This type of salt nucleation was not unreasonable as Sodom was positioned by the Dead Sea, a huge lake almost saturated with respect to salt (29% vs. 9% in the sea) together with magnesium, calcium, potassium, bromine, sulfur, and iodine.

Although there are numerous commentaries and songs on the ethics of looking backward, even the most liberal interpretation would find it unlikely that the cubic array of salt could replace the hexagonal close-packed structure of the bone mineral, hydroxyapatite. Of course, there is the possibility that unlike many other regions of the world, the Jordan Valley is rich in phosphate, and thus Edith may have crystallized out as a block of hydroxyapatite. In that case, mineralization may have begun in her bones and teeth and extended outward (vide infra for a discussion of the "outside and in" theory), and some type of inhibitor must have come into play to limit mineral growth and prevent Edith from spreading to the rest of the Valley. These comments concerning saturation, nucleation, mixed ions, inhibitors, and types of minerals provide a backdrop for considering how nanocrystals of apatite can form using changes in free energy and without the necessary participation of a higher being.

As a first approach to understanding how minerals can form in situ, it is important to move away from the murky depths of Sodom and the Dead Sea and consider a more serene and clarified setting in which the environmental constituents and conditions are fairly well demarcated, in more scientific language, research in vitro. Under these circumstances, events resulting in mineral deposition can be delineated as follows: In an aqueous environment, apatite formation is the resultant of a free energy change causing ions to aggregate, cluster, clump, and then combine prior to undergoing one or more phase transitions to form a final mineral species. To understand how these behaviors are mediated and controlled, several factors may be considered:

- The availability of calcium and phosphate ion levels in serum/extracellular fluid in relation to mineralization.
- The chemical potential of the ions in the aqueous environment.
- Saturation status of tissue extracellular fluid with respect to apatite.
- Classical nucleation theory—homogeneous nucleation.
- The heterogeneous nucleation of hydroxyapatite.
- Heterogeneous nucleation: Identification of clusters and intermediate products.
- Structural aspects of bone and tooth mineral.

The effects of protein regulators of mineralization are discussed in more detail in Chap. 9.

2.2 The Availability of Calcium and Phosphate Ion Levels in Serum/Extracellular Fluid in Relation to Mineralization

A simplistic explanation for mineral formation is that the concentrations of mineral ions in the extracellular fluid of bone and other mineralizing tissues are sufficient to cause deposition/precipitation of hydroxyapatite, $Ca_{10}(PO_4)_6(OH)_2$ [2]. Arguably, the ion levels at mineralizing sites are so high (supersaturated) that powerful circulating inhibitors exist to regulate their deposition in bone matrices as well as to prevent the crystallization of inorganic salts in the interstices of soft tissues. In this and later chapters, the validity of the points of this statement will be explored. Determinations will be made, first, as to the levels of calcium and phosphate ions in human serum and extracellular fluid; the relation of these ion levels to the formation and solubility characteristics of hydroxyapatite; and, finally, the free energy requirements for mineral deposition.

To focus first on calcium, the total serum calcium concentration is 2.1–2.6 mM [3]. Extrapolating results from a beating heart removed from a frog (which is exquisitely sensitive to the external calcium concentration), McLean and Hastings [4] reported that about 44% of the total calcium in the human is ionized and, for a given individual, this value is quite constant. Another 44% of total calcium complexes with the protein albumin (the major protein of serum) and some binds to the

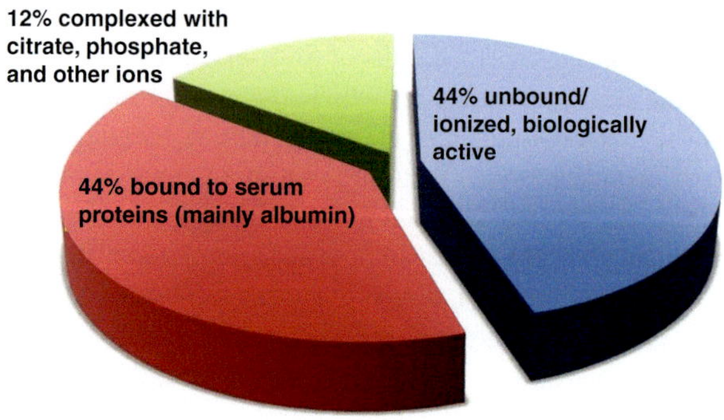

Fig. 2.1 A schematic of the molecular state of calcium in serum. Serum calcium may occur in a number of independent forms: free or ionic, bound to serum proteins, or complexed with small molecules like citrate and phosphate. From a mineralization standpoint, the most active fraction is that of unbound calcium. Since equilibria exist between each of the different calcium species, if the free calcium falls then less calcium will remain bound to ligands in the other two compartments. (Schematic based on data from McLean and Hastings [4] with permission)

globulins (many of which participate in acquired immunity). The remaining 12% of calcium pairs with bicarbonate, lactate, citrate, and phosphate (Fig. 2.1) [4–6].

When the ionized fraction of calcium in the serum is elevated, dystrophic mineral is often present in blood vessels, kidney, and liver (for a discussion of dystrophic mineralization, see Chap. 10). As an aside, in most hospital laboratories, ionic calcium values are now determined using an ion-specific electrode [7]. Further, in light of earlier discussions concerning interactions with other chemical species, the term, "ionized," may be inappropriate as none of the calcium is "free." Perhaps a more proper wording should be "freely diffusible."

Whether calcium is ionic or freely diffusible, the emphasis of this discussion concerns the role of ionic calcium in the formation of bone mineral. As such, it is important to note that organisms evolved in a sea enriched in ionic or freely diffusible calcium, a factor which favored calcium involvement in multiple physiological processes. Indeed, while calcium levels are three to five times lower in serum than they are in seawater [8], human blood calcium levels are highly and carefully regulated physiologically with minimum daily variation. That nature chose the calcium ion for so many physiologic functions is not surprising: the outermost orbital of the ion contains two valence electrons. This configuration prompts the formation of a stable oxidation state (2+) and the ability of calcium to form two covalent bonds with anionic partners. Moreover, the ionic radius of calcium is larger than that of other alkaline earth metals. The size is sufficient to balance effects attributable to steric hindrance with electrostatic field intensity, thereby enhancing calcium binding to many biological ligands, including phosphate ions and nucleic acids [9].

Calcium homeostasis is strictly maintained by hormones, especially vitamin D, parathyroid hormone, and fibroblast growth factor 23. Other aspects of calcium control involve small molecules and the pH of the blood itself (a topic discussed in subsequent chapters of this book). If the level of any of these components or pH is altered or impaired, significant changes occur in serum calcium ion concentrations, which in turn influence mineral formation and dissolution. If calcium ion levels decrease, they usually modify the phosphate ion status in serum and ultimately affect new bone formation, resorption, and density. This type of interaction implies a direct relation between the levels of calcium in the serum and biological mineralization.

Since many early invertebrates evolved in aquatic environments rich in both calcium and carbonate ions, these organisms were comprised of calcium carbonate as their predominant mineral phase. With respect to phosphate-containing apatitic tissues principally of vertebrates, from the discussion of the geobiology of phosphorites (Chap. 1), it is clear that phosphorus appeared late in the evolution of the biosphere and, compared with calcium, was of low abundance. The most basic and stable form of the element in biological tissues is founded on a chemical structure in which pentavalent (5+) phosphorus is accommodated in the center of a tetrahedron formed by four oxygen atoms, $(PO_4)^{3-}$. This arrangement is established by orbital electrons in the phosphorus outer shell forming a double bond with one oxygen atom while the remaining three oxygen atoms individually bond to the phosphorus. Such a structure maintains the stability of the phosphate tetrahedron and prevents nucleophilic attack especially when phosphate esters are formed. The free energy change associated with cleavage of the ester bond is sufficient to drive many chemical reactions, and the high negative charge of the ion delays or prevents its movement through lipid bilayers [10].

The critical mineral and structural anion in skeletal and dental tissues is phosphate. While this anion is of low abundance compared with calcium, as just noted, the actual concentration of phosphate in the blood is not that different from calcium. The anion phosphate concentration in plasma is about 1 mM [3]. In diseases in which there is a disturbance of phosphate metabolism or simply a low metabolic intake of the ion, bone and tooth mineralization undergo severe defects, as detailed in later chapters of this book. Thus, like calcium, changes in circulating phosphate ion levels exert profound consequences on mineral deposition.

With respect to the relation between phosphate ion concentration and apatite deposition in humans, the question of bound versus free phosphate levels in the circulation has been examined [11]. In this case, most of the serum phosphate appears to exist in a freely diffusible or ionized state (about 53%). Of the remaining phosphate, some ions pair with sodium (29%), calcium (3%), magnesium (3%), and protein (12%) [11].

Unlike the divalent calcium ion, the trivalent phosphate ion can occur in the circulation as three separate species: $H_2PO_4^-$, HPO_4^{2-}, and PO_4^{3-}. Ionization (and speciation) is dependent on pH and, hence, mineral formation and composition will vary with the pH of the medium. Thus, when titrated with base, phosphoric acid dissociates as follows:

$$H_3PO_4 \rightleftharpoons H^+ + H_2PO_4^- \left(pK_a = 2.1\right).$$

In other words, the phosphoric acid loses one proton and forms the $H_2PO_4^-$ ion. With further addition of a base, the pH increases toward neutral, and there is loss of a second proton:

$$H_2PO_4^- \rightleftharpoons H^+ + HPO_4^{2-}\left(pK_b = 6.7\right).$$

At a high pH, the HPO_4^{2-} discharges its last proton:

$$HPO_4^{2-} \rightleftharpoons H^+ + PO_4^{3-}\left(pK_c = 11.4\right),$$

where $pK_{a\text{–}c}$ is the dissociation constant for each of the reactions and is equal to the negative logarithm of the undissociated anion (left) divided by the product of the concentration of the dissociated products (right).

Since most metabolic activities are optimum in a neutral pH range (7.0–7.6), these equations inform the reader that the two active species in serum are a mixture of HPO_4^{2-} and $H_2PO_4^-$. Is there a predominant phosphate ion species in human serum? This question may be addressed using the equation developed by Henderson [12]:

$$pH = pK_a + \log_{10}\frac{\left[A^-\right]}{\left[HA\right]},$$

where pK_a is the log of the dissociation constant for an acid [HA] and [A^-] is the anion. From this equation, the ratio of HPO_4^{2-} to $H_2PO_4^-$ at pH 7.4 is about 4.0. In other words, the predominant phosphate ion in serum is HPO_4^{2-}. However, when expressed in activities (see below), values for these two ions (HPO_4^{2-} and $H_2PO_4^-$) are very similar (about 0.12 and 0.19 mM, respectively).

To continue the discussion of serum ions and mineralization, a firm basis has now been established for stating that there is a direct relation between specific ions of serum or tissue extracellular fluid (calcium and HPO_4^{2-} to $H_2PO_4^-$) and mineral deposition; indeed, the results of hundreds of animal experiments and numerous clinical observations of human disease lend direct support for this association. However, these studies do not address the central question as to whether mineral formation is just chemical precipitation from concentrated non-ideal solutions and, if it is, how it is controlled. To begin to examine these questions, a closer understanding of the investing fluid is needed together with a consideration of the interactions of populations of ions which generate much of the potential energy required to form the mineral.

2.3 The Chemical Potential of Ions in an Aqueous Environment

Prior to any discussion of the worthiness of the ideas mentioned above, it is important to comment that the quantitative measure of concentrations, whether mol/L or fluid ounces/bottle, does not provide insight into the chemical "energy" of ions in a solution. Even in very dilute solutions, electrostatic and other interactions take place between positive and negative ions; as a result, there is a decrease in the number of free (unattached) ions. These interactions are the effect of the differing electric charges on each of the ions and reflect in part the valency of the ionic species, the number and concentration of cations and anions, dipole moments of charged proteins, and atomic radii among other considerations. Thus, for a biological medium like serum or extracellular fluid containing appreciable quantities of monovalent, divalent, and trivalent ions, the actual ionic or chemical potential is highly complicated.

As just noted, interactions which influence the chemical potential of ions in solution include weak ionic and electrostatic forces and also the presence of water molecules. Water has a dipole moment—a term used to indicate a separation of positive from negative charges in ionic or atomic bonding. The oxygen atom induces the water dipole by attracting electrons from the hydrogen atom and thus accumulating a partial negative charge, leaving a slightly positive hydrogen atom. The resulting asymmetrical charge distribution allows one of the poles of the water molecule to react with both anions and cations. Accordingly, a positively charged calcium ion binds to the negative charge of the oxygen atom of water. The interaction is more complicated within a phosphate ion; the multiple bonds between phosphorus and oxygen are covalent, but electrons of the oxyanion (oxygen atoms shared with phosphorus) bind those of the hydrogen atom of water. Thus, water molecules help to form quasi-stable hydration shells around each phosphate ion. In less than ideal solutions, positive and negative ions interact and their hydration shells re-orient to produce a new and discrete species, an ion pair (Fig. 2.2).

Why is ion pairing important? Its formation complicates the measurement of both ionic concentration and activity of a solution, and ion pairing also decreases the entropy of a system. With regard to the latter, ion pairing allows water to become less ordered and is a major driving force for precipitation events. The loss of entropy by ion association is more than offset by the gain in randomness of water and hence an overall increase in entropy occurs for the system. The value of these changes is discussed in more detail in subsequent sections of this chapter concerning homogeneous and heterogeneous nucleation. Additionally, when ion pairs form, they change the effective physical properties of a solution, such as those associated with osmotic pressure. To summarize these preceding comments, because of the complexity of ionic interactions between water and anionic and cationic elements, rules governing dilute or "ideal" solutions cannot be applied to physiological media like serum and interstitial fluid.

With these caveats in mind, what measures are used to ascertain the concentration of ions in complex non-ideal physiological fluids? A first approximation would be to consider the individual concentrations of each of the solutes; however, as indicated above, this approach is of limited value because of interionic interactions. A more realistic approach is to consider the concentration in terms of ion activity (a), where:

$$a = \gamma c,$$

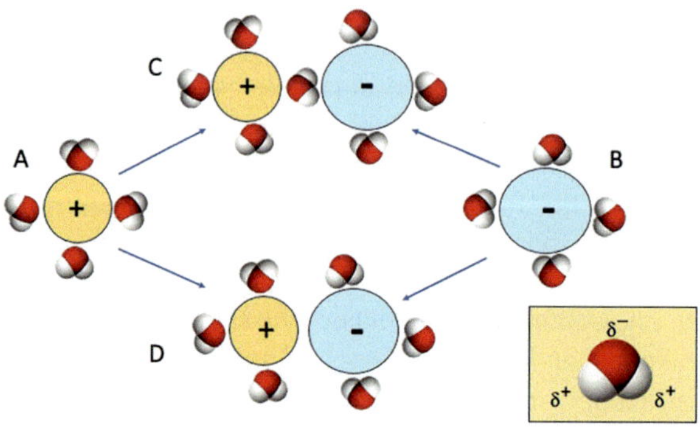

Fig. 2.2 Formation of ion pairs by hydrated and oppositely charged ions. Ion pairing results from the interaction of positive (gold, A) and negative (blue, B) ions such as calcium and phosphate, respectively, and the overlapping of their hydration shells (C and D). Weak electrostatic bonds are formed between each of the charged ions and polarized asymmetric water solvent molecules. Polarization of water occurs as the consequence of a slight negative charge (δ^-) of its oxygen atom (red) and a slight positive charge (δ^+) of each of its two hydrogen atoms (grey-white) (inset). The dipole produced governs the orientation of water molecules around the charged ions. Partial overlapping of the electron orbitals of the two ions leads to the elimination of solvent (water) with the enhancement of non-covalent interactions between the oppositely charged ions (D). If solvent molecules are not eliminated and are shared by the two charged ions, then the attraction between the charged ions is decreased and the concentration of ion pairs is reduced (C). In the case of phosphate, the anion exists principally as HPO_4^{2-} and $H_2PO_4^-$ in serum and hence water bonding is complex, involving associations between both the H^+ and the negatively charged oxyanions of HPO_4^{2-} and $H_2PO_4^-$. These additional interactions promote ion pairing and complexation

with γ the activity coefficient and c the ion concentration. In "ideal" solutions in which the ion concentration is very low and there is little interionic activity, then $\gamma = 1$. Of course, for serum and other physiological fluids, the solute is complex, and the number of charged species is high. As a result, the activity coefficient is much greater than 1.

Determination of the activity coefficient of physiological solutions is dependent on knowing the ionic strength (I), defined as:

$$I = \frac{1}{2}\sum_{n}^{i} c_i z_i^2,$$

where c_i is the concentration of the ith ion species and z_i is the ion charge. Debye, the 1936 recipient of the Nobel Prize in Chemistry, and the physicist Hückel used information about the ionic strength to generate a formula which describes the activity coefficient (γ):

$$\log(\gamma_i) = \frac{-Az_i^2\sqrt{I}}{1 + \mathrm{Ba}_i\sqrt{I}},$$

where the parameter z_i is the valence of ion i, a_i is the effective ion diameter, I the total ionic strength of the solution, and A and B are the temperature and dielectric constant, respectively. While this equation is appropriate for extracellular fluid and serum (ionic strength 0.15 mmol/L), for solutions with higher ionic strength (up to 0.5 mmol/L) a further term needs to be added:

$$\log(\gamma_i) = -Az_i^2\left(\frac{\sqrt{I}}{1+\sqrt{I}} - 0.3I\right).$$

For ionic strengths up to 1 mol/L, one should consider the Truesdell-Jones equation:

$$\log(\gamma_i) = -Az_i^2\left(\frac{\sqrt{I}}{1+\mathrm{Ba}_i\sqrt{I}}\right) + b_i I$$

where parameters are defined as above and b_i is a constant specific to the ion species.

In summary, this section first considered the possibility that vertebrate mineralization results from the precipitation of mineral from elevated concentrations of calcium and phosphate ions in the tissue extracellular fluid. Profound interionic effects preclude the use of conventional concentration values of the solution ions. Instead, ion activity measurements provide a more robust descriptor of solute molecules in a physiological fluid. In the following section, consideration is given to the importance of using calcium and phosphate ion activities rather than concentration values in relation to mineral deposition.

2.4 Saturation Status of Tissue Extracellular Fluid

So far, the proposition has been made implicitly that serum is supersaturated with the ions which form apatite. If this is the case, a major function of all vertebrate soft tissues is to prevent aberrant precipitation, which would interfere with normal cell function. One way to consider this concept is to conduct a thought experiment in which a small fragment of Lot's wife (after she were mineralized, of course) is treated with a fixed volume of serum at a specific (neutral) pH, and the release of calcium and/or phosphate ions measured. More than likely, it would be found that only a small number of her ions would go into solution. On the other hand, despite the high activities of mineral ions contained in serum, little or no new Edith mineral would be formed.

This thought experiment tells us something about the physical characteristics of the mineral and the tissue fluid. Thus, while the medium contains high levels of calcium and phosphate ions, they are insufficient to precipitate new mineral spontaneously; also, the mineral is insoluble or sparsely soluble in tissue fluids. In other words, the ion content of physiological fluids is insufficient to promote the spontaneous precipitation of mineral. This type of supersaturated solution is described as being metastable with respect to the ions which form apatite.

The terms saturation, supersaturation, and metastability are used to describe solute–solvent relationships. Being saturated implies that the solute concentration is maximum. However, utilizing heat or other physical processes, it is possible for a saturated solution to accommodate even further quantities of solute to form a supersaturated solution. Then, by cooling or by introducing a "nucleating" agent, the solute can be "coaxed" to come out of solution to form a solid, often crystalline phase. Applied to the mineralization process, if serum or extracellular fluid is supersaturated or metastable, then nucleation should yield hydroxyapatite or a precursor mineral. As this phenomenon is directly relevant to the mineralization mechanism, it raises the questions as to whether the chemical potential (ion activity product) of calcium and phosphate in the extracellular fluid is sufficient to drive mineral formation and, if it is, what the nature of the resulting mineral may be in this system.

The solubility of apatite in an aqueous phase can be determined by equilibrating a fixed volume of serum with hydroxyapatite and measuring the release of calcium and phosphate ions (and protons) at a selected pH. The value of the ion activity product (K_{ap}) can then be compared with the solubility product (K_{sp}) of the mineral using the following formula:

$$S = \left(K_{ap} / K_{sp}\right)^{1/v},$$

where v is the number of ions. The calculation of the actual solubility of apatite has been a fertile ground for scientific dispute (probably because it is sensitive to the conditions of a given experiment). For hydroxyapatite in an aqueous medium, the mineral dissociates as follows:

$$Ca_{10}\left(PO_4\right)_6\left(OH\right)_2 \rightarrow \left(aCa^{2+}\right)_{10}\left(aPO_4^{3-}\right)_6\left(aOH^-\right)_2,$$

where a is the ion activity. Building on the early work of Moreno et al. [13], Sohnel and Grases [14] reported that the K_{ap} is 1.93×10^{-49} mol/L and the solubility (activity) product, K_{sp}, for hydroxyapatite is 2.03×10^{-59} mol/L at 37°C. From the formula above, the tendency for these ions to crystallize can then be expressed by (S). In this case,

$$S = \left(1.93 \times 10^{-49} / 2.03 \times 10^{-59}\right)^{1/9} = 12.4 - 13.5.$$

The value of S depends on a number of factors including the chemical composition of the mineral, the dimensions of the first formed mineral particles, whether the phosphate term is PO_4^{3-} or HPO_4^{2-}, and the status of the initial mineral phase (amorphous calcium phosphate, octacalcium phosphate, or another species). For example, Sohnel and Grases [14] note that, if the size of a carbonated apatitic particle of diameter 0.7 nm is increased to 1 nm, then the tissue fluid supersaturation with respect to apatite decreases from S = 12.4–13.5 to 2.1–3.7. Regardless of this approximation, the K_{sp} value indicates that hydroxyapatite in an aqueous medium releases an extremely small number of calcium and phosphate ions. If the stoichiometric formula $(Ca^{2+})_5(PO_4^{3-})_3(OH^-)$ is used for hydroxyapatite, then $K_{sp} = 1 \times 10^{-57}$ [15].

That serum is less saturated with amorphous calcium phosphate and octacalcium phosphate is surprising as they are both considered precursors of hydroxyapatite. Thus, if precipitation chemistry plays a role in the mineralization process, it might be supposed that both these mineral species would form after apatite. However, as will be discussed in detail later, the observations that these apatitic precursor phases form before apatite [16] fit with the concepts that, in a metastable solution, the initial phase is thermodynamically unstable and forms ion clusters; cluster interactions with each other form a mineral. The first mineral phase subsequently undergoes a number of phase transformations as it adopts a more stable configuration and matures into hydroxyapatite. These changes are in line with the proposal by Ostwald in 1897 that the first solid to appear in a solution is the least stable phase, which undergoes transformation and recrystallizes to establish the most thermodynamically stable state (Ostwald's Rule of Stages) [17].

These observations in vitro are to be taken with some caution. First, the activity coefficients are based on calculated values and thus may not be representative of systems in vivo. Second, while the mathematics employed accommodate ion pairing such as $Ca\text{-}HPO_4^0$, $Ca\text{-}H_2PO_4^+$, and $Ca\text{-}OH^+$ (0 and + indicate the zero and positive charge on the ion pair), the solubility measurements in vitro often utilize saline or a synthetic tissue fluid and certain assumptions are made concerning oxygen and CO_2 tension. In addition, associations with organic molecules, which can foster and inhibit mineral formation and dissociation, are often ignored. Moreover, aside from calcium and phosphate, other ions such as carbonate and fluoride impact mineral solubility. Thus, while fluoride stabilizes an apatite lattice and decreases its solubility, carbonate promotes mineral dissolution and lowers the ion activity product of the medium. All these factors influence the saturation status of the medium and the formation of the first mineral phase and its transition to the final mature mineral.

One further comment is relevant to the metastability of the medium and its relation to the formation of an initial mineral phase. In the presence of proteins with an appropriate structure, the medium does not need to be supersaturated to form these mineral particles. Indeed, in fluids under-saturated with respect to both amorphous calcium phosphate and octacalcium phosphate, ions aggregate to form clusters through a process described as nucleation. Many authors have commented on the importance of this type of nucleation process (homogeneous nucleation) as a mechanism for deposition of mineral. Further development of this concept involves the participation of a nucleating agent, such as an extracellular matrix protein(s), to induce mineral deposition (heterogeneous nucleation). In the next two sections of this chapter, consideration is made of each of the nucleation processes in relation to the formation of the mineral.

2.5 Classical Nucleation Theory: Homogeneous Nucleation

Most readers will be familiar with particular examples of the chemical aspect of nucleation: a mass of crystals appearing in a supersaturated solution of copper sulfate, for instance, or the release of CO_2 bubbles from champagne. During the transformation of solvated ions or dissolved molecules to a final gaseous or crystalline state, there is a dramatic alteration in their structure. Why do these changes in state occur and what environmental factors govern the nucleation process?

Whether a fluid is blue copper sulfate, pink champagne, or straw-colored serum, nucleation is the result of minute fluctuations in the density of the medium caused by changes in thermal energy. These fluctuations influence the kinetic energy of the ions of a system and lead to enhanced molecular interactions and the clustering of positive and negative molecules. Most of these clusters are unstable and they dissociate into the bulk liquid phase of the system. However, when there is bonding between hydrated ions and molecules, more stable clusters may form. This stochastic process is dependent on the kinetic energy of the particles and hence directly responsive to the temperature and the concentration (activity) of ions in the medium. As will be discussed later, the ionic and molecular movements relate to the change in the free energy landscape of the bulk fluid phase and the surface energy of the developing ion clusters. For metastable fluids like serum, the details of how ions in the bulk fluid phase aggregate with each other to form clusters are not understood. In a physiological context, the intracellular or extracellular medium is complex and contains more than just calcium and phosphate or carbonate ions, as noted above.

Since ions which form the mineral in a medium complex with water molecules as well as H^+ and OH^- ions, there are profound changes in thermodynamic enthalpic and entropic values during clustering and subsequent crystal formation. The enthalpy term, a measure of the internal energy of a system, includes the energy of ion and water bonding in forming clusters and eventually in generating crystals. Entropy is different from enthalpy as it reflects the disorder of a system. Thus, when cluster formation takes place, the increase in molecular organization causes a net change in entropy. In other words, the disordered medium itself has high entropy while the more ordered clustering and crystal formation are indicative of a lower entropy status. The pH of the medium influences both entropy and enthalpy, and indeed even small changes in pH can serve to modify nucleation and clustering by affecting ionic interactions, the overall charge characteristics of clusters, and formation of interatomic bonds within crystals [18].

To this point of discussion, the energy requirements for mineral formation from metastable serum or extracellular fluid have been described in limited fashion. The driving force for the nucleation of the mineral is the change in free energy in the energetically metastable bulk extracellular fluid or serum [19]. Expressed mathematically by Gibbs [20]:

$$\Delta G = \Delta H - T\Delta S,$$

where ΔG is the free energy, ΔH is the change in the internal energy or the enthalpy of the system (in biochemical terms this would be equivalent to bond energy), ΔS is the change in entropy in the system, and T is the temperature.

The importance of the equation was underscored by the English inorganic chemist Joseph Mellor, who stated in 1922 that this "relationship was a fundamental driving force in nature" [21]. Thus, for an unstable supersaturated medium like serum or extracellular fluid, which would eventually form mineral and thereby become less saturated and more ordered, a phase transition (liquid to solid) would occur, resulting in the development of nucleation centers. During this transition, there would be an abrupt change in the free energy of the system. For a cluster to reach a size which promotes phase transition, a difference would be needed in the free energy gain of the cluster and the free energy loss in the medium, aided by a decrease in the saturation of the medium.

Regarding the Gibbs equation itself, the larger the value of $-\Delta G$ and the higher the ion activity product, the greater the propensity of a solution to undergo a phase change and generate clusters of ions or grow mineral nuclei. Concerning $T\Delta S$, when this value is greater than ΔH, the right-hand side of the Gibbs equation becomes negative and indicates that a reaction is spontaneous. Alternatively, when ΔH is greater than $T\Delta S$, the resultant is positive and suggests that nuclei will not form unless the free energy barrier is lowered. While the sign of ΔG dictates whether a reaction is favorable, it does not indicate reaction speed. In most enzymatically driven reactions, the rate is very high, while for mineral formation, the rate is much lower.

The following equation represents a more detailed mathematical description of the Gibbs free energy change (ΔG) in mineral nucleation [22]:

$$\Delta G = \frac{4}{3}\pi r^3 \sigma_s \Delta\mu + 4\pi r^2 \gamma,$$

where r = the radius of a spherical cluster, σ_s = the number-density of the cluster, $\Delta\mu$ = the change in chemical potential (activity) of the supersaturated solution, and γ = the interfacial (or surface) energy of the crystal embryo.

The equation above relates energy changes in clustering, nucleation, and crystal growth. It consists of two terms: the first (longer) term is negative and refers to the bulk energy change caused by the alteration in chemical potential as ions cluster; the second (shorter) term is the interfacial energy deficit at the nucleated mineral–medium interface (Fig. 2.3) [23]. Notably, in a growing mineral particle, the newly acquired ions present at the particle periphery are packed differently (more loosely) from those at the particle center. The former molecules are at a greater free energy state because of their low level of bonding and high mobility. When the cluster is small (r less than critical), the interfacial free energy term is greater than the bulk free energy. However, as the radius of the cluster enlarges, the bulk energy phase surpasses the interfacial energy requirements, and there is a fall in the free energy of the system; as the cluster grows into a defined mineral phase, the free energy of the system is lowered further [24]. Navrotsky saw cluster formation as a competition between "positive surface free energy of the particle" (proportional to the surface

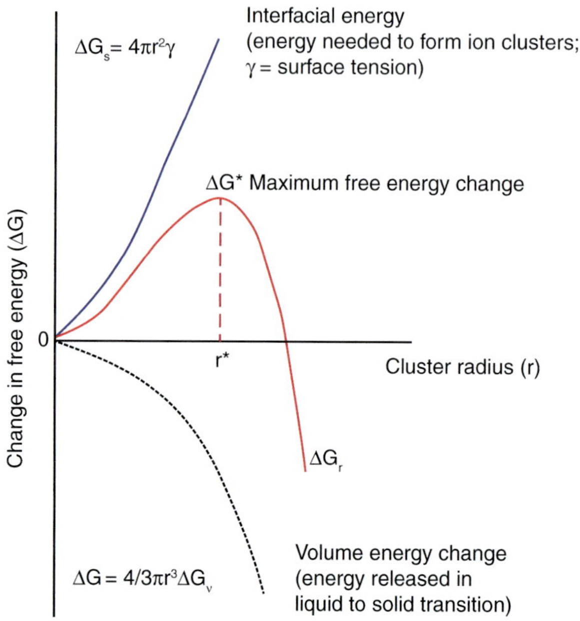

Fig. 2.3 Schematic of thermodynamic changes in Gibbs free energy during particle nucleation in solution. The *x*-axis indicates the radius (*r*) of the developing particle; the *y*-axis shows the change in Gibbs free energy (ΔG). At r^*, the radius of the growing cluster of ions forms a critical-sized nucleus, at which point the overall change in the free energy of the system is zero. Accordingly, particles with radii greater than r^* will serve to lower the overall free energy of the system while particles with radii smaller than r^* will increase the overall free energy. The upper blue line in the schematic is a measure of the surface free (interfacial) energy of the growing particle. Energy is required to form the new particle, which is destabilized by the surface tension (γ) at the particle–medium interface. The energy expended at the particle surface is $\Delta G_s = 4\pi r^2\gamma$. The lower dashed line indicates the volume free energy change or, stated otherwise, the energy which is generated ($-\Delta G_v$) as the particle is formed in the medium. The notation is written as a minus value to indicate the release or gain of energy per unit volume. Since the particle is expected to be spherical in shape with a radius *r*, the free energy volume change = $4/3\pi r^3\Delta G_v$. The red line shows the change in total (bulk) free energy of the system ΔG_r. At r^*, ΔG_r becomes negative with particle stabilization and growth and is the sum of ΔG_v and ΔG_s. In other words, as the nucleus increases in radius beyond r^*, the free energy change is favorable to further growth and the particle increases in size. (Schematic based on Ragone [23] and other papers)

area divided by the square of the particle radius) and "the negative bulk free-energy change" (proportional to the cube volume of the particle radius) [25]. A slightly different view of clustering was provided by ten Wolde et al., who commented that, if the distance between two particles is less than 1.5 × diameter, "they are considered to be connected and belong to the same cluster" [26]. Further incorporation of ions into the cluster-forming layered structures fosters the generation of a nucleus (Fig. 2.4).

Accompanying the changes in free energy in a supersaturated solution, a series of intermediate metastable phases is formed before the formation of a final

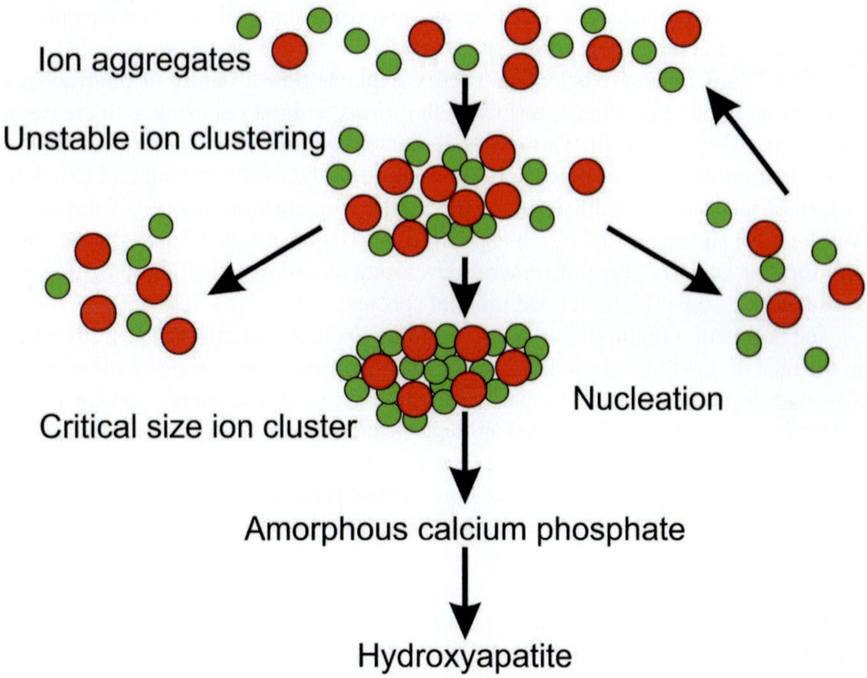

Fig. 2.4 A schematic of events which lead to homogeneous nucleation of mineral particles in a bulk fluid. The change in free energy associated with high local concentrations of calcium (green) and phosphate (red) ion aggregates results in the formation of elemental ion clusters or embryos. The first formed clusters are relatively unstable because of the destabilizing effect of high surface energy, and the ions comprising such clusters may return to a non-aggregated state. When the rate of ion accumulation is greater than the rate of release of ions into the bulk fluid, a stable ion cluster of critical size or nucleus of calcium phosphate forms. The change in free energy of the system enhances further ion accumulation, and an initial rudimentary ordering or organization of the ions into a mineral phase occurs. In the case of bone, an amorphous calcium phosphate phase is first formed in which the ions possess a short-range, but no long-range, order. In a series of steps involving intermediate mineral phases, amorphous calcium phosphate is ultimately transformed to a poorly crystalline hydroxyapatite (apatite) having long-range ordering of its calcium and phosphate ions. From a thermodynamic viewpoint, crystal nucleation, crystal growth, and crystal development are distinct processes with significantly different rate constants

thermodynamically stable phase. During formation of these intermediate and unstable phases, there is a return of weakly bound ions to the bulk aqueous phase. These phase transformations can be regarded as a series of precipitation, dissolution, and re-precipitation events. When an intermediate species forms, which is sufficiently large and stable in the mother medium, it acts as a nucleus for the formation of the subsequent species and ultimately the final crystal product. Once a large crystal has formed, it will grow and deplete ions from the solution at the expense of growth of smaller, more soluble crystals and ion clusters, a phenomenon called Ostwald's ripening. The large crystal grows stepwise by incorporation of the smaller clusters into surface defects (dislocations) and so-called kink sites [27, 28].

Over the past two decades, experimental and computational studies have pointed to a need to amend critical tenets of the classical nucleation theory [25, 29]. The resulting Non-Classical Nucleation Theory explains how a family of intermediate (precursor) mineral products (mesocrystals) forms without requiring an increase in free energy (Fig. 2.5) [30]. This two-step nucleation process stipulates that two or more intermediate phases precede the formation of the final bone mineral [31, 32]. A further theoretical modification of the classical nucleation theory developed by Vekilov [19] and supported by Zhang and Liu [33] indicates that liquid–liquid clusters form at an early stage of mineral development and that such ionic aggregates subsequently convert into defined mineral species.

The notion of a liquid–liquid phase is an intellectually exciting concept. It refers to the idea that, within a bulk liquid phase, segregation of ions or molecules creates a second, liquid phase [34]. In the dense liquid phase, a free energy change would foster the development of hydrated aggregates and prenucleation clusters of calcium and phosphate ions. Veis and Dorvee took this idea one step further to show that the change in free energy within a dense liquid phase promotes the growth of the earliest mineral nuclei [35]. A subsequent change in the free energy landscape of the system would accompany the conversion of this first solid phase into hydroxyapatite.

How and why clustering occurs is not understood. Wallace et al. [36] and Smeets et al. [37] used molecular dynamics simulation (computational resolution of molecular movements) to probe the stability of liquid–liquid clusters formed in calcium carbonate-containing media. The clusters were reported to be mobile structures contained within the bulk aqueous phase [38] and they may exist as polymers

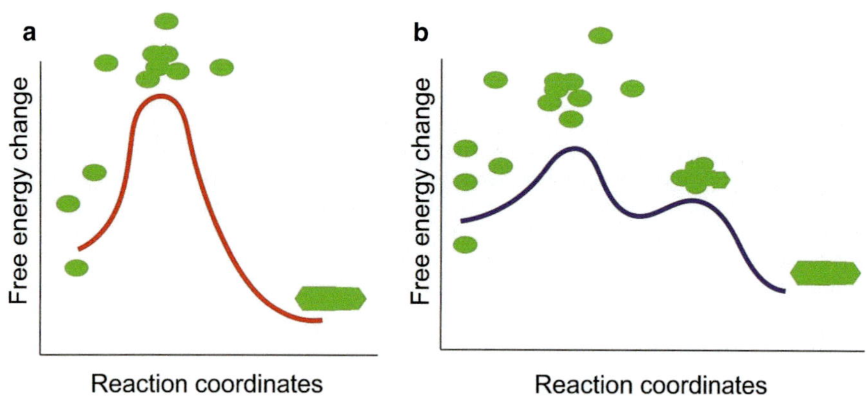

Fig. 2.5 Free energy changes accompanying single and multistep pathways for crystal nucleation, growth, and development. (**a**) Classical one-step pathway showing the free energy change for individual ions and ion aggregates (green circles) and the subsequent formation of product (green mineral crystal). (**b**) Two-step pathway typically involving the presence of a nucleating heterogeneous surface and a metastable intermediate (green solid aggregate of ions), which is formed prior to a phase transition(s) yielding the final product. The less acute energy change which accompanies the formation of the metastable intermediate allows more regulated ion binding and therefore more control over the final mineral product. (Schematic based on Myerson and Trout [30] with permission)

forming branched chains or rings of calcium and carbonate ions (Fig. 2.6) [39]. Once clusters have aggregated, free energy changes facilitate their conversion to an amorphous state. These ideas concerning an initial dense liquid–liquid phase are supported by studies, for example, of Gebauer et al. [38], Gower [40], and Rodriguez-Navarro et al. [41] More than likely, a similar liquid–liquid paradigm applies to the nucleation of calcium phosphate-containing minerals.

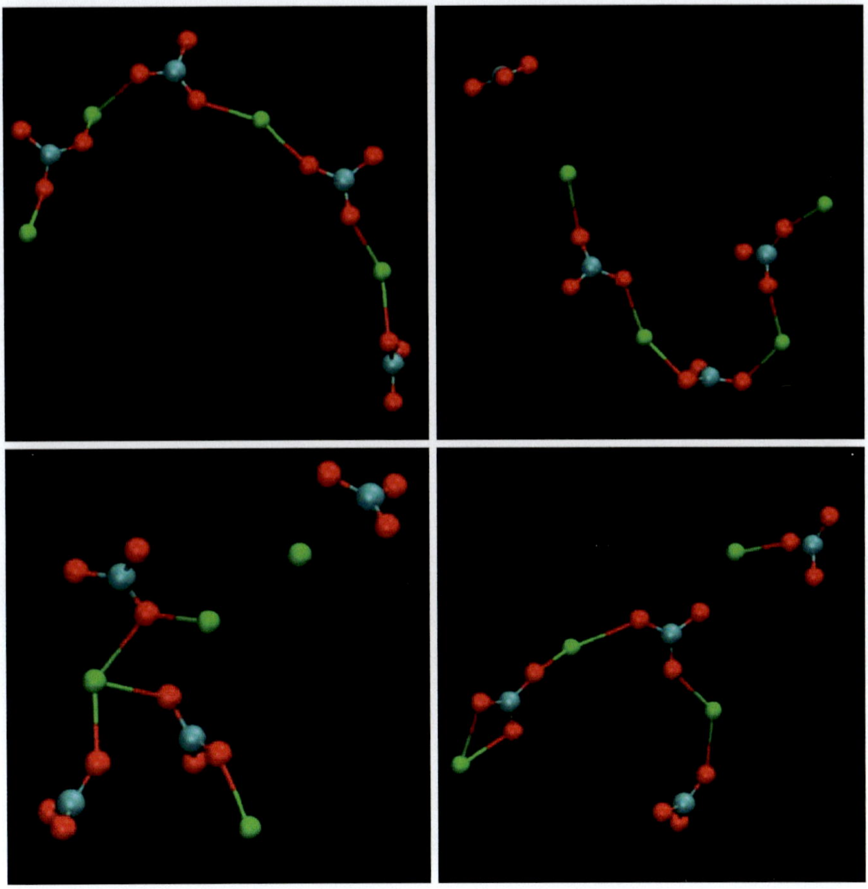

Fig. 2.6 Predicted structures of prenucleation clusters. Prenucleation clusters of calcium carbonate have been visualized here using molecular dynamics computer simulations. The simulations presented required that the structures be hydrated although simulations may be obtained without hydration. In this example, stable prenucleation clusters exist as ionic polymers of calcium and carbonate held together by ionic bonds. The structures assume both linear and branched shapes and in dynamic motion they may break, reform, re-branch, and even form ring-like conformations. Structures shown represent the configurations of four separate clusters after 1 ns of simulation at experimental conditions ($[Ca^{2+}] = 0.4$ mM, $[HCO_3^-] = 10$ mM, pH = 10). Atoms are colored green, blue, and red for calcium, carbon, and oxygen, respectively. (Simulated calcium carbonate structures from Demichelis et al. [39] with permission)

In a recent report, Veis and Dorvee drew attention to the importance of water molecules in mineral formation [35]. Water molecules not only form the environment in which cluster formation takes place but also incorporate into the mineral products. Moreover, interactions between hydrated calcium and phosphate ions directly influence the entropic state of the bulk water phase. Accordingly, entropic changes, as well as other measurements of the energy landscape, need to accommodate the influence of hydrated ions or particles in the formation of prenucleation clusters, nucleated particles, and even the transition of amorphous to crystalline states. Vekilov et al. [42] and Yau et al. [43] noted the critical significance of the water-dependent positive change in free energy during mineral formation and the negative change resulting from the liberation of water molecules during formation of ion clusters and crystal growth, respectively.

Before completing this section of the chapter, it should be emphasized that neither extracellular fluid nor serum are "ideal liquids." These fluids contain many different types of anionic and cationic electrolytes as well as macromolecules like albumin. Also, in the bone or cartilage microenvironment, there are fibril-forming proteins, polyanions, calcium-binding proteins, phosphorylated proteins, and cellular structures which, like the electrolytes, exhibit high affinities for charged molecules. Thus, while the classical, as well as non-classical, nucleation concepts are appropriate when dealing with relatively simple solutions, for mineral forming in biological niches, these proposals must accommodate the impact of the other populating ionic or molecular species. This consideration has prompted a second look at the mathematical underpinnings of the nucleation concept when applied to complex biological systems. Indeed, Vekilov [44] showed on theoretical grounds that values of $-\Delta G$ grossly underestimated the actual free energy changes. In fact, classical nucleation theory overestimates the rate of apatite formation by ten orders of magnitude. As a result, to account for the free energy needs of mineralizing systems and the "non-ideal" state of tissue fluids and the influence of their numerous composite ionic and molecular species, the concept of homogeneous nucleation has been superseded by a more inclusive concept which is heterogeneous nucleation, described immediately below.

2.6 Application of the Theory of Heterogeneous Nucleation to the Process of Biological Mineralization

Few scholars would deny the importance of the Gibbs equation and the tenets of classical- and non-classical homogeneous nucleation in guiding understanding of the process of mineral formation. However, as indicated above, there are theoretical limits to the value of homogeneous nucleation in explaining the process by which mineral is deposited in a complex extracellular matrix. These practical bounds were first heard at international conferences in the 1970s as the rumblings and arguments of a cadre of brilliant investigators which included A. Posner, W. Neuman, and M.J. Glimcher. In particular, Neuman and Glimcher presented evidence that the chemical nature of the organic phase of a mineralizing tissue promoted and guided mineral deposition. Together with other collaborators, Neuman with Glimcher examined the role of the extracellular matrix of bone in facilitating mineralization. Experiments were conducted to evaluate the impact of collagen and possibly other components of the matrix in both directing the placement and location of the mineral and lowering the free energy required for mineral deposition [45]. The concept that there existed a "magic bullet" in the organic matrix which enhanced mineralization led to multiple efforts to identify and find this molecule and to assess the mechanism by which apatite could be nucleated in developing bone and tooth.

Parenthetically, the concept of heterogeneous nucleation promoted ideas concerning a branch of chemistry called biomimetics, in which systems used in nature to form bone (and other tissues) are "hijacked" to create new mineralized and mineralizable materials. Accordingly, proteins, peptides, and amino acids which are part of the mineralization process are used to build mineralized matrices [46]. Such a biomimetic concept found widespread application in orthopedics and dentistry where there is an emphasis on forming new mineralized tissue as well as enhancing osseointegration of metallic implants with the surrounding bone [47].

How does heterogeneous nucleation work in mineralization? The most immediate explanation is that specific matrix proteins lower the Gibbs free energy requirements for mineral formation [48]. Figure 2.7 shows that the free energy required for the formation of a critical nucleation cluster under homogeneous nucleation conditions decreases in the presence of a nucleating surface provided by specific extracellular molecules and leading to heterogeneous nucleation events [49]. Such a change is not unlike that discussed in fundamental enzyme kinetics where the formation of an enzyme (Enz)-substrate (S) complex (S-Enz) lowers the free energy of formation of the product (P):

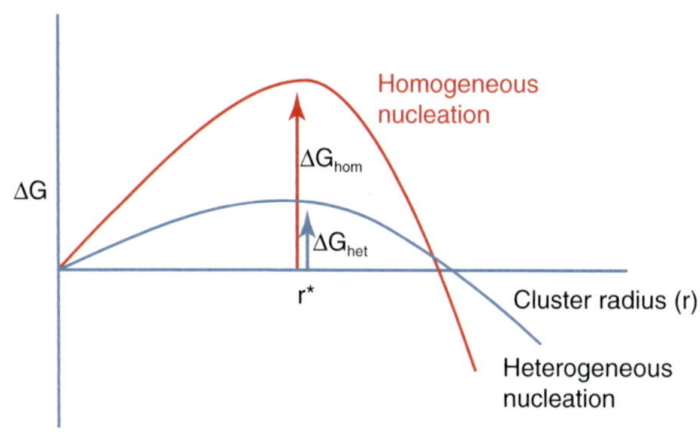

Fig. 2.7 Comparison of free energy changes (ΔG) during homogeneous versus heterogeneous nucleation of ion clusters in solution. As discussed in Fig. 2.3, r^* represents the critical radius of a cluster which must be reached before nuclei of calcium phosphate can be formed. For both types of nucleation, ion clusters less than r^* dissociate because of high interfacial forces, while radii greater than r^* result in a decrease in ΔG with formation of stable nuclei and the eventual formation of hydroxyapatite. For both homogeneous and heterogeneous nucleation, the critical radius size of an ion cluster is the same. However, the presence of a nucleating surface (heterogeneous nucleation) stabilizes the ion clusters and lowers the surface free energy of nuclei formation compared to that for homogeneous nucleation (ΔG_{het} compared to ΔG_{hom} at r^*). (Schematic adapted from Kamano et al. [49] with permission)

$$S + Enz \rightarrow S - Enz \rightarrow Enz + P.$$

Mullin [50] took the heterogeneous nucleation concept one stage further, showing that the interactions of charged sites between a forming crystal and those of its matrix are greater than those between the mineral and its aqueous medium. In other words, the stronger the matrix–mineral bond, the lower the interfacial free energy requirements of the system. This consideration differs from that of homogeneous nucleation, where the interfacial energy term refers only to the interface between the crystal and the medium [51]. Moreover, the bonds formed between the crystal and its matrix influence the size of the crystal and its orientation [52, 53]. This relation has biomechanical implications, namely that, in tissue regions subjected to the greatest stresses, a crystal is optimally oriented to accept the applied load [54]. Indeed, in such an optimal orientation, the crystal lattice strain is low and the crystal solubility is minimal. From this perspective, the presence of molecules mediating heterogeneous nucleation not only promotes mineralization but also positively impacts major physical characteristics of the mineral phase and the tissue itself.

To summarize the discussion to this point, it is quite clear that heterogeneous nucleation provides a much more credible description of events which drive mineralization compared to homogeneous nucleation. Heterogeneous nucleation also encompasses a role for charged groups of extracellular matrix proteins in the regulation of mineralization events. However, heterogeneous nucleation does not answer questions concerning the composition and structure of nucleating ion species or possible ion clusters, and it does not address the issue of a potential transition of these aggregates into crystals with defined chemistry or long-range structure. Such questions are considered further in the following section of this chapter.

2.7 Heterogeneous Nucleation: Identification of Clusters and Intermediate Mineral Products

If the concept of heterogeneous nucleation is valid, then it should be possible to isolate and characterize mineral cluster aggregates and identify each of the intermediates of the mineral-forming process. It turns out that identifying new, unstable, and transient mineral phases is extremely difficult. A major problem is the retrieval of samples representative of the individual stages in the mineralization process without inducing artifacts in the cluster itself and the transient mineral phases. For example, during the evaluation of an aqueous sample by light or electron microscopy, photons or electrons or simply heating alone may likely provoke artifactual changes in cluster composition, critical early events in mineral phase transformations, and final crystal structure [55]. Several analytical techniques have attempted to circumvent or overcome these limitations, including the use of specialized protocols for specimen preparation, anhydrous fixation techniques, and freeze-substitution and quench-freezing methods [56]. The development of cryoelectron microscopy has had a positive impact on mineral phase analysis as the technique maintains a frozen sample near liquid nitrogen temperature. With this system, low dose electron imaging can document mineral and cluster characteristics with minimal alterations in sample structure and hydration state [57].

Spinodal decomposition chemistry provided further insights into mechanisms by which mineral forms in relatively simple solutions. The term refers to the process by which a thermodynamically unstable but apparently homogeneous liquid phase can separate out to form two or more individual phases. When applied to a supersaturated or saturated solution along with ion clustering, there are corresponding changes in the free energy status which provide a low-barrier path to a new metastable phase. These and related concepts attracted the attention of scientists studying invertebrate (seashell) calcification where the mineral formed from a metastable solution in the mature shell was a calcium carbonate, not the calcium phosphate characteristic of vertebrate mineralization. Prior to the formation of the final and most stable crystalline component of the shell, the presence was proposed of an amorphous calcium carbonate phase along with crystalline calcium carbonate [58, 59]. In this context, the term, "amorphous," described a mineral having short-range structural order and yielding a poorly defined X-ray diffraction pattern. Further work with calcium carbonate systems demonstrated both amorphous calcium carbonate and vaterite, an isomorphic crystalline phase of calcium carbonate [31, 60–62], and also showed that such an amorphous form of calcium carbonate preceded development of calcite or aragonite, two other isomorphic crystalline calcium carbonate phases [61]. In addition, utilizing calcium carbonate chemistry, Gower [40] and Wolf et al. [63] made the seminal observation that densification of an aqueous phase proceeds in the absence of a nucleating moiety. These data, then, presented evidence for a multi-stage nucleation process beginning with an amorphous phase of calcium carbonate.

With regard to calcium phosphates, Betts and Posner first suggested that calcium and phosphate ion clusters may occur in solution as precursors to other mineral phases [64]. Using several different approaches, they described these clusters with an assigned stoichiometric formula, $Ca_9(PO_4)_6$ (Fig. 2.8). Such Posner clusters were thought to pack together to form spheres of about 300–1000 nm in diameter [65]. In a further development, Posner proposed that the aggregated spheres generated a mineral phase identified as amorphous calcium phosphate [64]. As a related note, quantum chemistry calculations suggest that pairs of Posner molecules may chemically bind together as the first step toward larger-scale aggregation, an event which may lead to the formation of amorphous calcium phosphate and eventually hydroxyapatite [66]. Moreover, in a simple aqueous medium, the presence of a templating surface such as a membrane lipid promotes amorphous calcium phosphate cluster formation and, from an energetic perspective during the formation of small crystal nuclei, the unfavorable interfacial contribution dominated the balance between bulk and surface energy [38].

Investigators in the Sommerdijk laboratory provided the first experimental evidence that a series of intermediate phases preceded apatite formation. They reported that, prior to the formation of apatite in solution, there was an association of nanometer-sized structures into prenucleation clusters which formed an amorphous calcium phosphate [67]. Subsequently, several steps were defined in the transition ultimately to apatite, including an early phase identified as hydrated prenucleation ion pairs with a low calcium:phosphate ratio (essentially $[Ca(HPO_4)_3]^{4-}$), of which over half the pairs were protonated $[Ca(HPO_4)_2(H_2PO_4)]^{3-}$ [68]. These species existed as disk-like structures with a central calcium ion associated with phosphate

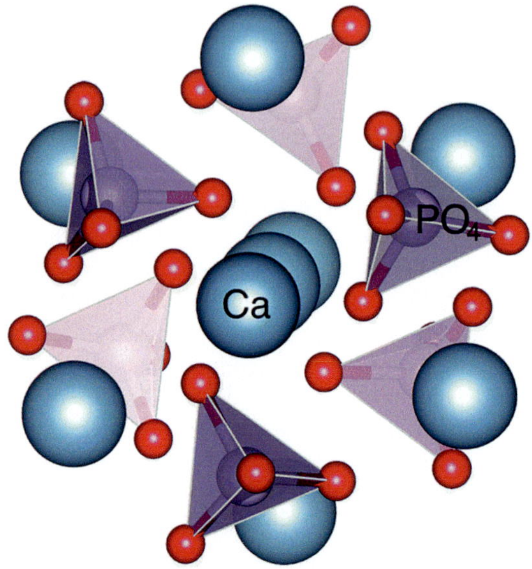

Fig. 2.8 Ionic structure of a conceptual Posner cluster, stoichiometrically $Ca_9(PO_4)_6$. A single unit cell of hydroxyapatite can accommodate two Posner clusters, one of which is shown in the schematic with its constituent calcium (blue), phosphate tetrahedra (purple), and oxygen (red) ions. Oxygen ions lie at the vertices of the phosphate tetrahedra in this slightly tilted view along the central calcium ion axis. Each tetrahedral phosphate group associates with both a peripheral calcium ion and a central calcium ion. A cluster is about 1 nm in diameter and is proposed to form at low levels of solution supersaturation or at mineralizing sites in vivo [68]. The appearance of Posner clusters is thought to be brief and transient as it is governed by the Ostwald-Lussac Principle (Rule of Stages), which dictates that the least thermodynamically stable phase of a mineral species (that is, the most soluble and having the lowest interfacial energy) undergoing crystallization is the first to develop before transforming sequentially to intermediate phases and ultimately to the most stable phase (the least soluble and having the highest interfacial energy). Quantum chemistry calculations suggest that pairs of Posner molecules can chemically bind together as the first step toward larger-scale aggregation, which may lead to the formation of amorphous calcium phosphate and eventually hydroxyapatite [64]. (Schematic from Swift et al. [65] with credit to The Royal Society of Chemistry)

ions and with an excess of free energy [68]. Additional discrete temporal, chemical, and morphological steps were determined in this landmark study. Thus, spherical amorphous calcium phosphate aggregates with the formula $[Ca_2(HPO_4)_3]^{2-}$ transitioned into a ribbon-like $[Ca_6(HPO_4)_4(PO_4)_2]^{2-}$ and then formed plates of an octacalcium phosphate-like mineral, $Ca_8(HPO_4)_2(PO_4)_4$ (Fig. 2.9) [68]. During this transformation process, there was densification of the particles, accumulation of calcium ions, and a loss of water, all leading to the final mineral, apatite, in the system.

Of course, as the calcium phosphate intermediate species described above were isolated using model systems in vitro, it was still necessary to confirm these findings with both cells and animal tissues which undergo mineralization to form the mineral

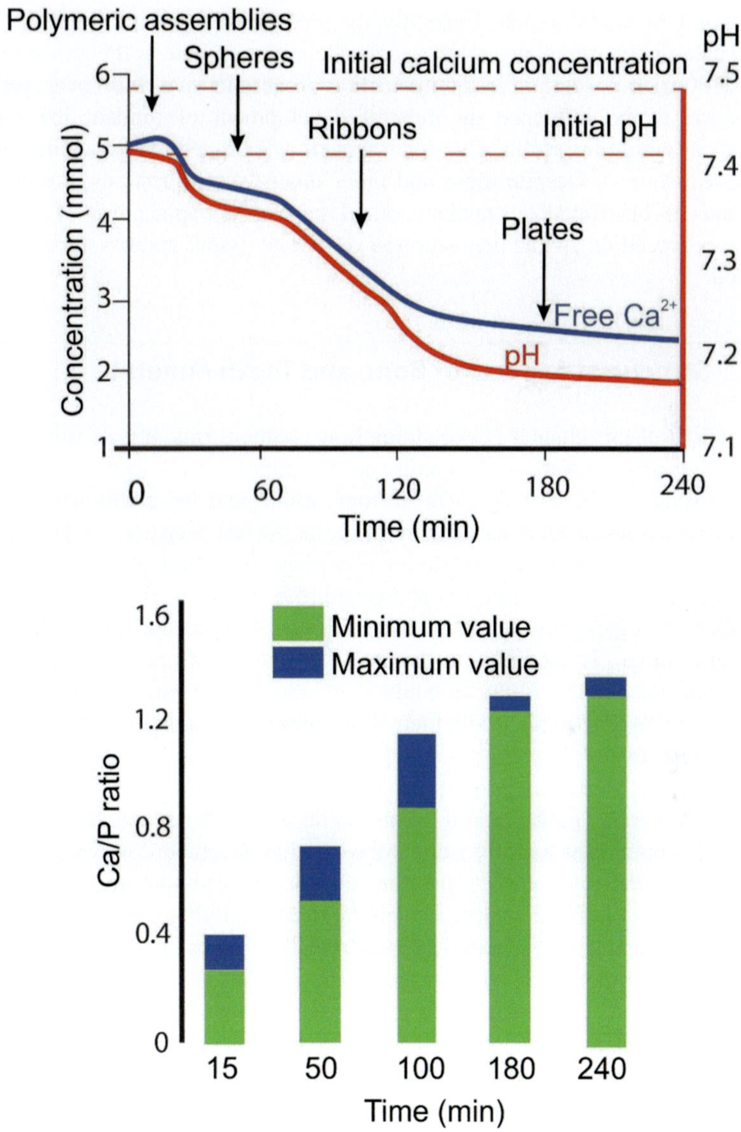

Fig. 2.9 Mineral complexes formed in a metastable solution of calcium phosphate and the corresponding change in Ca/P ratio with time. The reactions were initiated by adding phosphate ions (10 mM/L) to a buffered saline solution containing calcium ions (10 mM/L) and maintained at constant ionic strength with stirring. Calcium concentration and pH were monitored throughout the study using ion-selective electrodes. During the reaction, samples were collected and vitrified on electron microscope grids and subjected to high-resolution cryotransmission electron microscopy [68]. Other samples were examined by FTIR spectroscopy [68]. By combining cryotransmission electron microscopy of solution samples with measurements of pH and free calcium and phosphate concentrations, the form and Ca/P ratio of a series of calcium phosphate intermediates (polymeric aggregates, spheres, ribbons, and plates) were determined as a function of time in vitro. The final product of the study (plates) had a structure which was similar to octacalcium phosphate [$Ca_8H_2(PO_4)_6{\cdot}5H_2O$]. (Image modified from Habraken et al. [68] with permission)

of bone and the dental tissues. Currently, the presence and potential significance of several calcium intermediates, such as octacalcium phosphate or the mineral brushite ($CaHPO_4$), are uncertain in the nucleation process in vivo. Moreover, yet to be determined is the influence on mineral development of unidentified proteins which exist within intracellular vesicles as well as extracellular organelles (matrix vesicles/exosomes). Despite these and other unanswered questions, however, the importance of heterogeneous nucleation and the transient appearance of intermediate phases preceding apatite deposition in vertebrate tissues appears to be very well accepted.

2.8 Structural Aspects of Bone and Tooth Mineral

As the intent of this chapter is explaining how apatite forms, it is of value to comment further on the nature of the mature apatite mineral itself, as well as other calcium phosphates which are present in bone, tooth, and the additional vertebrate mineralized tissues under discussion in this book. As has been noted earlier, precursor phases to apatite include octacalcium phosphate, dicalcium phosphate dihydrate (brushite), and amorphous calcium phosphate [69]. A listing of calcium phosphate minerals in biological tissues is presented in Table 2.1. As mentioned previously, the term "amorphous" is used here to denote a mineral having no long-range atomic order when evaluated by X-ray techniques. In contrast, "crystalline" minerals have long-range ordered structures with their atoms arranged in a regular or quasi-regular repeating pattern.

The name of the mineral present in mammalian tissues, "apatite," is derived from the Greek "απατεíν" (apatein) and means "to deceive." Thus, "apatite" would seem to be a very appropriate term for a mineral which has an extraordinary and deceptive chemistry. As indicated earlier, the formula for the geologic apatite mineral is $Ca_{10}(PO_4)_6(OH)_2$. In bone, dentin, cementum, and calcifying cartilage and tendon, the biological apatite mineral exists as nanocrystalline plates (or platelets), 30–50 nm long, 15–30 nm wide, and 2–10 nm thick [16, 69]. The structure of the biological mineral is not homogeneous as each platelet has an outer (1–2 nm) amorphous or non-structured surface layer. Moreover, the stoichiometry of the individual constituents which comprise the underlying mineral is variable and, for that reason, biological apatite is described as a calcium-deficient mineral, enriched in magnesium, acid phosphate (HPO_4^{2-}), and carbonate with variable amounts of sodium and several other elements. Within the unit cell (Fig. 2.10), carbonate ions comprise 3–9 wt% and can replace orthophosphate (PO_4^{3-}) ions [70, 71]. Accordingly, apatite has been assigned a generalized formula $Ca_{8.1}Mg_{0.2}(PO_4)_{4.3}(HPO_4)_{0.5}(CO_3)_{1.2}(OH)_{0.3}$ [72]. Based on infrared analyses of bone and enamel, von Euw et al. suggest that the apatite formula should be amended to $Ca_{7.5}(PO_4)_{2.8}(HPO_4)_{2.6}(CO_3)_{0.6}(OH)_{0.2}$ [73].

Because of its ability to bind extraneous ions, apatite readily exchanges many other ions from the biological medium bathing it and substitutes them in its crystal lattice for calcium, phosphate, and carbonate. Therefore, the apatitic formula is not fixed but is rather quite variable and dependent on the composition of the local

Table 2.1 Possible calcium phosphate phases present in mineralizing systems

Compound	Formula	Ca/P molar ratio	Occurrence
Monocalcium phosphate monohydrate	$Ca(H_2PO_4)_2 \cdot H_2O$	0.5	
Dicalcium phosphate dihydrate (DCPD), (brushite)	$CaHPO_4 \cdot 2H_2O$	1.0	Dental calculus; pathological calcifications; intermediate in the formation of hydroxyapatite; chick and bovine bone, following bone resorption
Dicalcium phosphate anhydrous (DCPA or DCP), mineral monetite	$CaHPO_4$	1.0	
Octacalcium phosphate (OCP)	$Ca_8H_2(PO_4)_6 \cdot 5H_2O$	1.33	Dental and urinary calculi; bone and enamel mineralization
β-Tricalcium phosphate (β-TCP)	β-$Ca_3(PO_4)_2$	1.5	Dystrophic mineralization; arthritic tissue; dental and urinary calculi
Amorphous calcium phosphate (ACP)	$Ca_xH_y(PO_4)_z \cdot nH_2O$, $n = 3$–4.5; 15–20% H_2O	1.2–2.2	Intermediate in newly forming bone and enamel
Calcium-deficient hydroxyapatite (CDHA)	$Ca_x(HPO_4)_x(PO_4)_{6-x}(OH)_{2-x}$ $(0 < x < 1)$	1.5–1.67	Bone; calcified cartilage; dentin; enamel
Hydroxyapatite [apatite] (HA, Hap, or OHAp)	$Ca_{10}(PO_4)_6(OH)_2$	1.67	

Table adapted from Dorozhkin [69] with permission

aqueous environment. While water molecules present at the crystal surface are distinct from those of the hydration layer and the bulk aqueous phase, hydrated ions move easily into the crystal structure. As a result, minute changes occur in the structure and composition of the surface layers of apatite crystals, and they can lead to rapid mineral phase transitions, allowing both dissolution and re-precipitation reactions. These reactions can release calcium into the general circulation or give rise to a defective mineral which is characteristic of dental caries, for example.

The presence of exogenous or “foreign” inorganic ions which are different from those of calcium, phosphate, oxygen, and carbonate distorts the apatite lattice and contributes to changes in the biochemistry, biology, and biomechanical properties of apatite crystals and mineral. Thus, elements like fluoride and lead, from drinking water and pollutants in the air, respectively, may be incorporated into the crystal lattice structure [74], especially in enamel, and alter the nature of apatite. Plutonium disseminated from nuclear weapons or as fallout from accidents at nuclear reactors

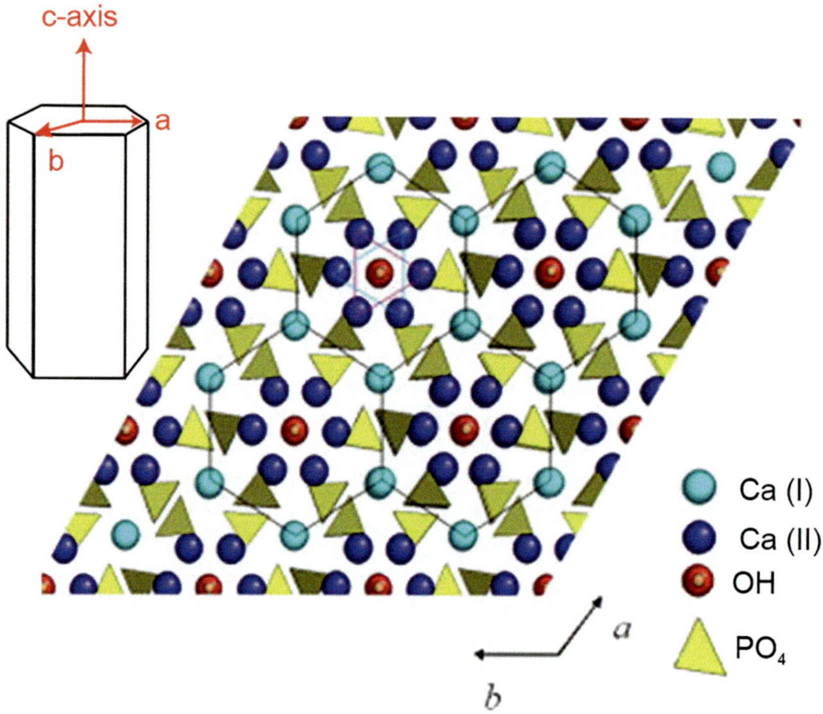

Fig. 2.10 The structure of a unit cell of hydroxyapatite [$Ca_{10}(PO_4)_6(OH)_2$] as viewed in the *c*-plane of the crystal (perpendicular to its crystallographic *c*-axis). Directions for the three mutually orthogonal axes of a hydroxyapatite crystal are shown in the upper left diagram. Four complete hexagonal structures within the unit cell are defined by lines connecting six calcium ions [Ca (I), green] in-plane and located at the periphery of each hexagon. There are six other calcium ions [Ca (II), blue] internal to each hexagon. The Ca (I) ions are aligned three-dimensionally in columns along the *c*-axis. Ca (II) ions are arranged in-plane as two triangles (thin green and red lines shown in the upper left hexagon) rotated 60° with respect to each other. The Ca (I) and (II) ions at each of their locations have a certain mobility and Ca (II) ions are staggered three-dimensionally in consecutive planes along the crystallographic *c*-axis. Calcium mobility and shifts in location create voids or vacancies in the hydroxyapatite structure which can be substituted and filled by many other cations. Phosphate ions exist in the form of tetrahedrons (one central phosphorus atom surrounded by four oxygen atoms; grey-green and yellow, inverted by 180° with respect to each other) and may be substituted by several other anions. These tetrahedrons occupy much of the space of the unit cell. Three of these tetrahedral structures are positioned close to the Ca (II) planes. While the oxygen ions from these three tetrahedrons are strongly bound to phosphorus, they also interact with calcium ions. Thus, four Ca (I) ions are tightly bound to six oxygen ions and less tightly associated with three other oxygen ions. In contrast, six Ca (II) ions are associated with seven oxygen ions. Unlike the phosphate groups comprising the hexagonal structures, other phosphate groups located toward the outside of the unit cell are protonated as HPO_4^{2-}. In contrast to calcium, the tetrahedral phosphate groups, despite their possible substitution, provide a greater degree of stability within the unit cell. Therefore, the overall stability of the hydroxyapatite lattice structure is more accurately related to the presence and location of the tetrahedral phosphate ions. Parallel channels of OH ions (red) are positioned in the center of each of the hexagons and extend throughout the *c*-axis of the unit cell. These OH channels provide pores for entry and exit of ions of the mineral and allow the mineral to reflect the local fluid environment and its composition. (Schematic adapted from Boanini et al. [82] with permission)

like Chernobyl is a bone-seeking element which localizes to apatite and is retained in the skeleton for long periods [75]. The apatite crystal lattice can likely accommodate one-half to two-thirds of the elements listed in the Periodic Table [76]. The wide range of substitutions enhances the highly complex and deceptive chemistry of apatite.

Aside from those ions and elements which constitute the unit cell of apatite, the high surface energy of the mineral promotes adsorption of many metabolites. Among these is citrate, a negatively charged ion which is a product of the tricarboxylic acid cycle in the mitochondria of cells. This organic acid is taken up by apatite such that its concentration is 100–400-fold higher than that of blood. In fact, 60–70% of the total citrate concentration is in bones and teeth [77]. Unlike carbonate, fluoride, and other ion or elemental species, however, citrate is not incorporated into the lattice structure of apatite but instead is thought to act as a cement which binds the individual apatite crystallites together [78].

Thus far, several aspects of the external structure of the biological nanoscale apatite platelets have been described. However, as Fig. 2.10 shows, the unit cell of apatite, much like a protein, has a complex inner (tertiary) structure. Understanding its atomic arrangement helps explain the properties of the mineral. Within the interior of the hydroxyapatite unit cell, oxygen atoms associate with phosphorus to form tetrahedral phosphate ions which stabilize the crystal lattice and maintain the position of ten surrounding calcium ions. Apatite contains two groups of calcium atoms: columnar triangles of calcium atoms which traverse the crystallographic c-axis of the unit cell, and an inner group of calcium ions which forms overlapping and shared hexagons of the unit cell when projected on its basal a-plane. Along the c-axis, channels exist within the mineral that can be occupied by hydroxyl ions or other ions such as fluoride, carbonate, or even water [70, 79–82]. A case in point is carbonate, which is incorporated into the central channel of the apatite lattice and causes an increase in the crystal a-axis length and a decrease in the c-axis length, changes which then impact apatite stability and solubility [80, 82]. In contrast, fluoride ions form strong electronegative bonds with adjacent ions, markedly influencing the dimensions of the unit cell and decreasing the solubility of the bone or tooth mineral [83]. How the ion clusters or crystals of apatite mineral interact with proteins of the extracellular matrix is considered later in this volume.

To summarize the discussions so far, this chapter was begun by reflecting on the mineralization of Lot's wife, Edith, and it was ended with the current section considering the "deceitful" properties of the mineral itself. Perhaps "deceitful" is an incorrect description of bone and tooth mineral, and mysterious is a better depiction of Edith's fate. Although the mineral is extraordinarily complex, a great deal more is now known about the physical events which promote its formation: the importance of the free energy landscape, the crowded status of the ions, molecules, and metabolites in the fluid media, delineation of precursor mineral phases, and involvement of extraneous proteins leading to heterogeneous nucleation. In the following chapters of this book, the discussions move from the physical forces influencing mineral formation to the role of cells and extracellular matrix constituents directing the accumulation and location of the ions which comprise the mineral phase of the skeletal tissues. It is this process which generates the uniqueness of each of the structures which characterize bone, calcifying cartilage and tendon, dentin, cementum, and enamel of vertebrate tissues.

2.9 Perspective: The Enigma of Edith: Free Energy, Nucleation, and the Formation of Mineral

For those readers whose interests lie more with cellular events which regulate the mineralization process in vivo, and less with Biblical stories concerning Lot and Edith, some of the major themes previously addressed in this chapter may be considered too theoretical and possibly only distantly related to the realities of bone, tooth, and other vertebrate hard tissues. With this possibility in mind, the focus of the concluding section of this chapter is to summarize some of its major ideas, thereby providing the reader with a snapshot of critical physical–chemical events governing hydroxyapatite formation whether in a collagen- or an amelogenin-containing matrix. However, if this book had been written 10 or 20 years earlier, it might have ended here, as the work of many researchers was concerned with defining which specific phosphoprotein in the matrix was the "nucleator." Most scholars would now agree that no single molecule serves as a "magic bullet" and that the formation of apatite in an extracellular matrix is not the product of a precipitation reaction. In other words, the mineralization process in a tissue is far more complex than the crystallization of a mineral like copper sulfate from a supersaturated medium.

This chapter was begun by considering the status of calcium and phosphate ions in serum and their impact on mineral deposition. While both ions are present at approximately similar concentrations (mM), their values are an incomplete assessment of their chemical potential in the tissue fluid or serum. Reasons for this limitation include the properties of mono-, di-, and multi-valent ions, solvent effects, and the formation of ion pairs. However, if expressed as ion activities, these concentration values are no longer confounding. With ion activity as an appropriate measure of chemical potential, the serum is metastable in relation to the ions which are present in apatite. Moreover, the calcium and phosphate ion activity product is sufficiently high to cause spontaneous mineral deposition.

The mechanism for the formation of a solid phase within a supersaturated or metastable solution continues to be developed both from a theoretical and a practical viewpoint. The Gibbs concept of crystal formation is the underlying precept for understanding how crystals form in supersaturated fluids. Nucleation of crystals occurs when ion clusters reach a critical size. During this process, depletion of ions from the mother solution causes the clusters to become unstable and dissociate. However, when the cluster of ions reaches a critical size, the energy barrier is overcome and more stable aggregates form, which eventually become the mature mineral. The introduction of spinodal decomposition concepts modified the classical homogeneous nucleation theory: during spinodal decomposition, the theory recognizes that some ion clusters form intermediate phases which are more stable than the ionic clusters [67]. Also, because of the lower interfacial energy, there is preferential formation of intermediate mineral phases [84]. Eventually, nuclei of intermediate stability undergo further phase transitions generating the final mature and stable mineral [31, 62, 68]. While not intuitive, such a model has found wide appeal, not only for mineral formation but also as a mechanism by which macromolecular cellular proteins change their state, aggregate, and form compartments within the cytosol [85].

In biological systems, mineral formation is more complex than in simple supersaturated solutions and is dependent on factors which include the metastability of the aqueous solution, the interfacial energy barrier, and the presence of associated proteins or lipids. The formation of a mineral crystal in concert with a molecular template (heterogeneous nucleation) relies on the interaction between charged groups on the template and the ions required for mineral formation [86]. When these two parameters are stereochemically "matched," the interfacial energy barrier is lowered, leading to a more rapid formation of the mineral phase. By controlling each of these parameters, Li et al. were able to generate an extraordinary group of mineralized structures, each with its own morphology and composition [87]. The importance of templating in relation to clustering will be developed in later chapters and, in particular, mechanisms will be discussed concerning the regulation of ion clustering within subcellular compartments as well as the templating of mineral by proteins of the extracellular matrix. In this regard, the direction and extent of mineral growth are likely dependent on the interactions between charged groups of specific extracellular proteins (such as collagen and phosphoproteins) of bone, dentin, and other vertebrate mineralized tissues and the lattice characteristics of the crystallographic faces of apatite.

Considerable information now exists on the thermodynamics of clustering, much of which has emerged from studies of calcium carbonate-based systems. When applied to apatite-forming systems, there is strong support for the notion that, before the appearance of a stable, mature mineral, there is the formation of prenucleation clusters consisting of about 70 ions with an average diameter of about 2 nm. According to the seminal studies of Posner and Betts [64], ion clusters exist as roughly spherical molecules with the stoichiometric formula, $Ca_9(PO_4)_6$. Moreover, nanometer-sized aggregates of these clusters form larger spherical particles with water molecules in their interstices. Subsequently, maturation of these particles results in the formation of amorphous calcium phosphate. It should be noted that the structure of Posner clusters is still debated, and Kanzaki et al., for example, have calculated that a number of different stable calcium phosphate ion clusters could exist in which a calcium atom forms six Ca–O bonds with six different PO_4 groups [88].

Aside from the change in the free energy, the conversion of an amorphous mineral to its final mature form elicits a change in hydration status. With reference again to the calcium carbonate system, the transition of an amorphous calcium carbonate phase to the final crystalized state with the loss of water triggers solid-state transformation and crystallization. Ihli et al. showed that this phase transition probably involves dissolution/recrystallization events, a "ripening effect" initiated by changes in the hydration status of the amorphous mineral [89]. Similar dissolution and recrystallization events probably are involved with the conversion of amorphous calcium phosphate to hydroxyapatite. As the transition to higher-ordered structures takes place, some of the more stable intermediate phases are retained and may even accumulate in the mature tissue. Not surprisingly, minor quantities of amorphous calcium phosphate, brushite, and octacalcium phosphate minerals are present in bone and the dental tissues (see Table 2.1). These metastable phases may also incorporate traces of elements such as fluoride, strontium, and lead. If, for some reason, ion aggregation or cluster formation slows, there will be profound influences on mineral maturation and crystallinity, and such effects may account for some of the observed variations in the chemistry, structure, and morphology of the mineral phases of bone, dentin, enamel, and other vertebrate hard tissues.

The mechanism by which apatite crystals of the skeletal and dental tissues attain their final platelet-like dimensions requires the stepwise addition of calcium phosphate clusters or amorphous calcium phosphate particles to the mineral, a process described as nanotextural or nanogranular [90]. These sequential events are consistent with the Ostwald Rule of Stages, which predicts that a thermodynamically unstable mineral phase is formed before the appearance of a thermodynamically more stable phase. In this regard, another consideration is the influence of small crystal size (less than 100 nm) which characterizes vertebrate mineral, a factor leading to a possibility that the free energy of the enthalpy difference between phases is sufficiently low that the less stable mineral would be formed first and before the development of apatite [25].

During final stages of mineral deposition, the platelets of apatite may become organized in a "hierarchical" structural fashion (this topic is discussed in more detail in Chap. 7). In other words, when deposited within an organic matrix, apatite crystals may be oriented and aligned in a direction which delineates the characteristic architecture and provides the mechanical properties appropriate for the functional needs of the tissue. In bone, dentin, and cementum extracellular matrix, for instance, the mineral crystals associated with collagen are oriented with their crystallographic *c*-axes effectively parallel to the long axis of the collagen fibers. How this specific structural feature is achieved is still not clearly understood, but certain possible mechanisms have been proposed, including the effects of electric fields [91], directed crystal growth mediated by phosphoproteins or other molecules in the vicinity of collagen [92], and orientation and alignment dictated by the stereochemical nature of collagen, particularly the presence of charged amino acids comprising the neighborhood of the gap and overlap regions of the protein [52, 53].

While the basic chemical structure of apatite is similar in all vertebrate mineralized tissues, it is interesting to discuss enamel in the context of its morphological character. The crystals of enamel are significantly different in size and shape from those of bone, dentin, or the other normally mineralizing vertebrate tissues. Enamel crystals form in a matrix rich in the protein, amelogenin, rather than collagen which comprises the matrices of bone, dentin, and the like. Unlike bone and its collagen-based counterparts, enamel crystals are also exceptionally long and ribbon-like, and they typically extend for microns in length from the dentin surface to the exterior of the tooth. On the other hand and similar to bone, the enamel mineral phase is reported to appear first as prenucleation clusters of amorphous calcium phosphate [93], which is transiently stabilized by the matrix proteins, amelogenin and ameloblastin [94]. As will be discussed in Chap. 3, these proteins are intrinsically disordered and, as such, both maintain enamel mineral and promote mineralization [95]. Interaction with the mineral is entropy-driven, causing a decrease in tissue hydration and ultimate cleavage of the associated matrix proteins. The proteins permit development of highly organized, parallel linear arrays of mineral structures, which subsequently fuse to form bundles of elongated mineral particles [94]. This type of self-assembly generates the unique mineral phase of dental enamel [96]. Further details of the structure of enamel and the role of major protein constituents in the development of enamel tissue are considered in Chaps. 3 and 4 of this book.

To conclude this chapter, certain cautions must be noted in a discussion of free energy, interface energy barriers, nucleation, and the formation of mineral. Many of the studies described herein were performed in vitro, in other words in glass or plastic using synthetic solutions. In doing so, these studies sidestep the influence of cells and systemic tissue factors on mineral formation. Even more importantly, if mineral nucleation is initiated in intracellular locales, then, in line with the ideas expressed in this book, many new factors need to be taken into consideration. These factors include information about intracellular compartments and volumes, ion activities, competing signaling pathways, and the chemical nature of nucleating molecules and surfaces. Of course, these considerations do not rule out the importance of collagen and the intrinsically disordered proteins in participating in the nucleation process as these molecules are present within subcellular compartments. Nevertheless, whether there is extracellular or intracellular mineralization or both, the principles of heterogeneous multistep nucleation provide a powerful insight into the generation of an inorganic phase which characterizes multiple mineralization processes.

References

1. Eagleman D. Incognito: the secret lives of the brain. New York, NY: Pantheon Books; 2011.
2. Wang L, Nancollas G. Calcium orthophosphates: crystallization and dissolution. Chem Rev. 2008;108:4628–69.
3. Bazydlo LAL, Needham M, Harris NS. Calcium, magnesium, and phosphate. Lab Med. 2014;45:e44–50.
4. McLean F, Hastings A. The state of calcium in the fluids of the body: I. The conditions affecting the ionization of calcium. J Biol Chem. 1935;108:285–321.
5. Walser M. Ion association. VI. Interactions between calcium, magnesium, inorganic phosphate, citrate and protein in normal human plasma. J Clin Invest. 1961;40:723–30.
6. Bilezikian JP. Primer on the metabolic bone diseases and disorders of mineral metabolism. 9th ed. Hoboken, NJ: Wiley-Blackwell; 2019.
7. Burnett R, Christiansen T, Covington A, Fogh-Andersen N, Kulpmann W, Lewenstam A, et al. IFCC recommended reference method for the determination of the substance concentration of ionized calcium in undiluted serum, plasma or whole blood. Clin Chem Lab Med. 2000;38:1301–14.

8. Cuyler E, Byrne R. Spectrophotometric calibration procedures to enable calibration-free measurements of seawater calcium carbonate saturation states. Anal Chim Acta. 2018;1020:95–103.
9. Kamerlin SC, Sharma PK, Prasad RB, Warshel A. Why nature really chose phosphate. Q Rev Biophys. 2013;46:1–132.
10. Kinjo TG, Schnetkamp PPM. Ca2+ chemistry, storage and transport in biologic systems. Voltage-gated calcium channels. Boston, MA: Springer US; 2005. p. 1–11.
11. Marshall RW. Plasma fractions. In: Nordin BEC, editor. Calcium, phosphate and magnesium metabolism. Edinburgh: Churchill Livingstone; 1976. p. 162–85.
12. Henderson L. Concerning the relationship between the strength of acids and their capacity to preserve neutrality. Am J Physiol. 1908;21:173–9.
13. Moreno EC, Gregory TM, Brown WE. Preparation and solubility of hydroxyapatite. J Res Natl Bur Stand A Phys Chem. 1968;72A:773–82.
14. Sohnel O, Grases F. Supersaturation of body fluids, plasma and urine, with respect to biological hydroxyapatite. Urol Res. 2011;39:429–36.
15. Moreno E, Aoba T. Comparative solubility study of human dental enamel, dentin, and hydroxyapatite. Calcif Tissue Int. 1991;49:6–13.
16. Dorozhkin S. Amorphous calcium (ortho)phosphates. Acta Biomater. 2010;6:4457–75.
17. Ostwald W. Studien über die Bildung und Umwandlung fester Körper. Z Phys Chem. 1897;22:289–330.
18. Rodriguez-Navarro C, Jroundi F, Schiro M, Ruiz-Agudo E, Gonzalez-Munoz M. Influence of substrate mineralogy on bacterial mineralization of calcium carbonate: implications for stone conservation. Appl Environ Microbiol. 2012;78:4017–29.
19. Vekilov P. The two-step mechanism of nucleation of crystals in solution. Nanoscale. 2010;2:2346–57.
20. Gibbs J. On an alleged exception to the second law of thermodynamics. Science. 1883;1:160.
21. Mellor J. A comprehensive treatise on inorganic and theoretical chemistry. London: Longmans, Green and Co.; 1924.
22. Auer S, Frenkel D. Prediction of absolute crystal-nucleation rate in hard-sphere colloids. Nature. 2001;409:1020–3.
23. Ragone DV. Thermodynamics of materials. New York, NY: John Wiley & Sons, Inc.; 1995.
24. Abraham F. Homogeneous nucleation theory. 1st ed. New York, NY: Academic Press; 1974.
25. Navrotsky A. Energetic clues to pathways to biomineralization: precursors, clusters, and nanoparticles. Proc Natl Acad Sci U S A. 2004;101:12096–101.
26. ten Wolde PR, Frenkel D. Enhancement of protein crystal nucleation by critical density fluctuations. Science. 1997;277:1975–8.
27. Teng H, Dove P, Orme C, De Yoreo J. Thermodynamics of calcite growth: baseline for understanding biomineral formation. Science. 1998;282:724–7.
28. Wallace A, DeYoreo J, Dove P. Kinetics of silica nucleation on carboxyl- and amine-terminated surfaces: insights for biomineralization. J Am Chem Soc. 2009;131:5244–50.
29. De Yoreo J, Gilbert P, Sommerdijk N, Penn R, Whitelam S, Joester D, et al. Crystallization by particle attachment in synthetic, biogenic, and geologic environments. Science. 2015;349:aaa6760.
30. Myerson AS, Trout BL. Nucleation from solution. Science. 2013;341:855–6.
31. Gebauer D, Volkel A, Colfen H. Stable prenucleation calcium carbonate clusters. Science. 2008;322:1819–22.
32. Lutsko J, Duran-Olivencia M. Classical nucleation theory from a dynamical approach to nucleation. J Chem Phys. 2013;138:1–14.
33. Zhang T, Liu X. Nucleation: what happens at the initial stage? Angew Chem Int Ed. 2009;48:1308–12.
34. Luft J, Wolfley J, Snell E. What's in a drop? Correlating observations and outcomes to guide macromolecular crystallization experiments. Cryst Growth Des. 2011;11:651–63.
35. Veis A, Dorvee J. Biomineralization mechanisms: a new paradigm for crystal nucleation in organic matrices. Calcif Tissue Int. 2013;93:307–15.

36. Wallace A, Hedges L, Fernandez-Martinez A, Raiteri P, Gale J, Waychunas G, et al. Microscopic evidence for liquid-liquid separation in supersaturated CaCO3 solutions. Science. 2013;341:885–9.
37. Smeets PJM, Finney AR, Habraken WJ, Nudelman F, Friedrich H, Laven J, et al. A classical view on nonclassical nucleation. Proc Natl Acad Sci U S A. 2017;114:7882–90.
38. Gebauer D, Kellermeier M, Gale JD, Bergstrom L, Colfen H. Pre-nucleation clusters as solute precursors in crystallisation. Chem Soc Rev. 2014;43:2348–71.
39. Demichelis R, Raiteri P, Gale JD, Quigley D, Gebauer D. Stable prenucleation mineral clusters are liquid-like ionic polymers. Nat Commun. 2011;2:1–8.
40. Gower L. Biomimetic model systems for investigating the amorphous precursor pathway and its role in biomineralization. Chem Rev. 2008;108:4551–627.
41. Rodriguez-Navarro C, Ruiz-Agudo E, Harris J, Wolf S. Nonclassical crystallization in vivo et in vitro (II): Nanogranular features in biomimetic minerals disclose a general colloid-mediated crystal growth mechanism. J Struct Biol. 2016;196:260–87.
42. Vekilov P, Feeling-Taylor A, Petsev D, Galkin O, Nagel R, Hirsch R. Intermolecular interactions, nucleation, and thermodynamics of crystallization of hemoglobin C. Biophys J. 2002;83:1147–56.
43. Yau S, Thomas B, Vekilov P. Molecular mechanisms of crystallization and defect formation. Phys Rev Lett. 2000;85:353–6.
44. Vekilov P. Nucleus in a droplet. Nat Mater. 2012;11:838–40.
45. Glimcher M. The possible role of collagen fibrils and collagen-phosphoprotein complexes in the calcification of bone in vitro and in vivo. Biomaterials. 1990;11:7–10.
46. Oliveira I, Carvalho A, Radhouani H, Goncalves C, Oliveira J, Reis R. Promising biomolecules. Adv Exp Med Biol. 2018;1059:189–205.
47. Coelho P, Jimbo R, Tovar N, Bonfante E. Osseointegration: hierarchical designing encompassing the macrometer, micrometer, and nanometer length scales. Dent Mater. 2015;31:37–52.
48. Sommerdijk N, de With G. Biomimetic CaCO3 mineralization using designer molecules and interfaces. Chem Rev. 2008;108:4499–550.
49. Kamano Y, Kadota K, Shimosaka A, Shirakawa Y, Hidaka J. Quantitative evaluation on the heterogeneous nucleation of amino acid by a thermodynamic analysis. J Mol Liq. 2014;200:474–9.
50. Mullin JW. Crystallization. 4th ed. Oxford: Butterworth-Heinemann; 2001.
51. Travaille A, Steijven E, Meekes H, van Kempen H. Thermodynamics of epitaxial calcite nucleation on self-assembled monolayers. J Phys Chem B. 2005;109:5618–26.
52. Landis W, Silver F. Mineral deposition in the extracellular matrices of vertebrate tissues: identification of possible apatite nucleation sites on type I collagen. Cells Tissues Organs. 2009;189:20–4.
53. Silver FH, Landis WJ. Deposition of apatite in mineralizing vertebrate extracellular matrices: a model of possible nucleation sites on type I collagen. Connect Tissue Res. 2011;52:242–54.
54. Jimbo R, Nakano T, Sawase T. Alignment of biological apatite c-axis under functional loading: a preliminary report. Implant Dent. 2016;25:594–8.
55. Egerton RF. Radiation damage to organic and inorganic specimens in the TEM. Micron. 2019;119:72–87.
56. Manning L, Richmond J. High-pressure freeze and freeze substitution electron microscopy in C. elegans. Methods Mol Biol. 2015;1327:121–40.
57. Quentin D, Raunser S. Electron cryomicroscopy as a powerful tool in biomedical research. J Mol Med (Berl). 2018;96:483–93.
58. Aizenberg J, Weiner S, Addadi L. Coexistence of amorphous and crystalline calcium carbonate in skeletal tissues. Connect Tissue Res. 2003;44:20–5.
59. Weiner S, Sagi I, Addadi L. Choosing the crystallization path less traveled. Science. 2005;309:1027–8.
60. Zhang T, Liu X. Multistep crystal nucleation: a kinetic study based on colloidal crystallization. J Phys Chem B. 2007;111:14001–5.

61. Pichon B, Bomans P, Frederik P, Sommerdijk N. A quasi-time-resolved CryoTEM study of the nucleation of CaCO3 under langmuir monolayers. J Am Chem Soc. 2008;130:4034–40.
62. Pouget E, Bomans P, Goos J, Frederik P, de With G, Sommerdijk N. The initial stages of template-controlled CaCO3 formation revealed by cryo-TEM. Science. 2009;323:1455–8.
63. Wolf S, Müller L, Barrea R, Kampf C, Leiterer J, Panne U, et al. Carbonate-coordinated metal complexes precede the formation of liquid amorphous mineral emulsions of divalent metal carbonates. Nanoscale. 2011;3:1158–65.
64. Posner A, Betts F. Synthetic amorphous calcium phosphate and its relation to bone mineral structure. Acc Chem Res. 1975;8:273–81.
65. Swift M, Van de Walle C, Fisher M. Posner molecules: from atomic structure to nuclear spins. Phys Chem Chem Phys. 2018;20:12373–80.
66. Yunger Halpern N, Crosson E. Quantum information in the Posner model of quantum cognition. Ann Phys. 2019;407:92–147.
67. Dey A, Bomans P, Muller F, Will J, Frederik P, de With G, et al. The role of prenucleation clusters in surface-induced calcium phosphate crystallization. Nat Mater. 2010;9:1010–4.
68. Habraken W, Tao J, Brylka L, Friedrich H, Bertinetti L, Schenk A, et al. Ion-association complexes unite classical and non-classical theories for the biomimetic nucleation of calcium phosphate. Nat Commun. 2013;4:1–12.
69. Dorozhkin SV. Calcium orthophosphate-containing biocomposites and hybrid biomaterials for biomedical applications. J Funct Biomater. 2015;6:708–832.
70. Elliott JC. Structure, crystal chemistry and density of enamel apatites. Ciba Found Symp. 1997;205:67–72.
71. Dorozhkin S. Calcium orthophosphates: occurrence, properties, biomineralization, pathological calcification and biomimetic applications. Biomatter. 2011;1:121–64.
72. Legros R, Balmain N, Bonel G. Age-related changes in mineral of rat and bovine cortical bone. Calcif Tissue Int. 1987;41:137–44.
73. Von Euw S, Wang Y, Laurent G, Drouet C, Babonneau F, Nassif N, et al. Bone mineral: new insights into its chemical composition. Sci Rep. 2019;9:1–11.
74. Shapiro I, Dobkin B, Tuncay O, Needleman H. Lead levels in dentine and circumpulpal dentine of deciduous teeth of normal and lead poisoned children. Clin Chim Acta. 1973;46:119–23.
75. Froidevaux P, Haldimann M. Plutonium from above-ground nuclear tests in milk teeth: investigation of placental transfer in children born between 1951 and 1995 in Switzerland. Environ Health Perspect. 2008;116:1731–4.
76. Hughes J, Rakovan J. The crystal structure of apatite, Ca5(PO4)3(F,OH,Cl). Rev Mineral Geochem. 2002;48:1–12.
77. Costello L, Franklin RB. Plasma citrate homeostasis: how it is regulated; and its physiological and clinical implications. An important, but neglected, relationship in medicine. HSOA J Hum Endocrinol. 2016;1:1–19.
78. Davies E, Muller K, Wong W, Pickard C, Reid D, Skepper J, et al. Citrate bridges between mineral platelets in bone. Proc Natl Acad Sci U S A. 2014;111:E1354–63.
79. LeGeros R, Trautz O, Klein E, LeGeros J. Two types of carbonate substitution in the apatite structure. Experientia. 1969;25:5–7.
80. Elliott JC. The problems of the composition and structure of the mineral components of the hard tissues. Clin Orthop Relat Res. 1973;93:313–45.
81. Elliott JC. Structure and chemistry of the apatites and other calcium orthophosphates. Amsterdam: Elsevier; 1994.
82. Boanini E, Gazzano M, Bigi A. Ionic substitutions in calcium phosphates synthesized at low temperature. Acta Biomater. 2010;6:1882–94.
83. Uskokovic V. The role of hydroxyl channel in defining selected physicochemical peculiarities exhibited by hydroxyapatite. RSC Adv. 2015;5:36614–33.
84. Raiteri P, Gale J. Water is the key to nonclassical nucleation of amorphous calcium carbonate. J Am Chem Soc. 2010;132:17623–34.
85. Bernard A, Klionsky D. Defining the membrane precursor supporting the nucleation of the phagophore. Autophagy. 2014;10:1–2.

86. Hamm LM, Giuffre AJ, Han N, Tao J, Wang D, De Yoreo JJ, et al. Reconciling disparate views of template-directed nucleation through measurement of calcite nucleation kinetics and binding energies. Proc Natl Acad Sci U S A. 2014;111:1304–9.
87. Li L, Fijneman A, Kaandorp J, Aizenberg J, Noorduin W. Directed nucleation and growth by balancing local supersaturation and substrate/nucleus lattice mismatch. Proc Natl Acad Sci U S A. 2018;115:3575–80.
88. Kanzaki N, Treboux G, Onuma K, Tsutsumi S, Ito A. Calcium phosphate clusters. Biomaterials. 2001;22:2921–9.
89. Ihli J, Wong WC, Noel EH, Kim YY, Kulak AN, Christenson HK, et al. Dehydration and crystallization of amorphous calcium carbonate in solution and in air. Nat Commun. 2014;5:1–10.
90. Drouet C. Apatite formation: why it may not work as planned, and how to conclusively identify apatite compounds. Biomed Res Int. 2013;2013:1–12.
91. Guler G, Atalay NS, Ozogul C, Erdogan D. Biochemical and structural approach to collagen synthesis under electric fields. Gen Physiol Biophys. 1996;15:429–40.
92. Baht GS, O'Young J, Borovina A, Chen H, Tye CE, Karttunen M, et al. Phosphorylation of Ser136 is critical for potent bone sialoprotein-mediated nucleation of hydroxyapatite crystals. Biochem J. 2010;428:385–95.
93. Jokisaari JR, Wang C, Qiao Q, Hu X, Reed DA, Bleher R, et al. Particle-attachment-mediated and matrix/lattice-guided enamel apatite crystal growth. ACS Nano. 2019;13:3151–61.
94. Fang PA, Conway JF, Margolis HC, Simmer JP, Beniash E. Hierarchical self-assembly of amelogenin and the regulation of biomineralization at the nanoscale. Proc Natl Acad Sci U S A. 2011;108:14097–102.
95. Wald T, Spoutil F, Osickova A, Prochazkova M, Benada O, Kasparek P, et al. Intrinsically disordered proteins drive enamel formation via an evolutionarily conserved self-assembly motif. Proc Natl Acad Sci. 2017;114:E1641–50.
96. Beniash E, Simmer J, Margolis H. Structural changes in amelogenin upon self-assembly and mineral interactions. J Dent Res. 2012;91:967–72.

3 Form and Function of Tissues which Undergo Mineralization

... in most of the Mammalian class, the ends of the long cylindrical bones, which support the articular cartilages, are distinct in the growing bone from the shaft and are termed "epiphyses," the shaft being the "diaphysis:" the seat of the active growth of the bone is in a cartilaginous crust at the ends of the diaphysis. [1]

Glossary

Amelogenesis The process of forming dental enamel.

Anlage The earliest developmental stage of an organ, beginning with a condensation of embryonic cells; a provisional cartilaginous structure forms the anlage for a long bone.

Appendicular skeleton Endochondral bones of the limbs together with the bones of the pelvis and shoulder.

Axial skeleton Bones of the vertebrae, sacrum, coccyx, ribs, and sternum.

Cementum A bone-like mineralized tissue present typically on the dentin surface which anchors the tooth to alveolar bone through the periodontal ligament.

Dental follicle Mesenchymal tissue forming a sac around the developing tooth.

Endochondral bone Bone which is formed in and replaces a cartilaginous matrix.

Enthesis The site of insertion or attachment of a tendon, ligament, fascia, or joint capsule to bone. At the bone interface, an enthesis exhibits a zone of mineralizing fibrocartilage.

Epithelial sheath (of Hertwig) Fusion of inner and outer enamel epithelia to form a bilayered sheath required for cementum and root formation of a tooth.

Growth plate/cartilage A strip of cartilage which serves as a growth center for bone elongation.

Membrane bone Bone which is formed by intramembranous ossification. These bones are present in the skull, clavicle, and jaw.

I. M. Shapiro, W. J. Landis, *Mechanisms of Mineralization of Vertebrate Skeletal and Dental Tissues*, https://doi.org/10.1007/978-3-031-34304-9_3

Neural crest ectoderm Ectodermal cells originating from the neural fold in an embryo and contributing to the formation of many tissues, including the enamel organ and bones of the skull.

Sclerotome The region of a somite which will form osteogenic tissues of the skeleton.

SIBLING A small integrin-binding ligand, N-linked glycoprotein.

Somatopleure The lateral and ventral walls of an embryo which give rise to the lateral plate mesoderm.

Stem cell Undifferentiated cells which have the capacity to commit to a specific lineage while retaining the ability to self-renew.

Tomes' process The distal aspect of a secretory and maturing ameloblast.

Gene/Protein Nomenclature

Symbols for genes are written in uppercase and italicized, whereas symbols for proteins are given in uppercase but not italicized. For simplicity, the chapter does not distinguish between the same genes/proteins found in animals and humans.

Abbreviations and Proper Names for Genes/Proteins and Other Molecules

αvβ3 and αvβ5	Integrin receptors
AMBN	Ameloblastin
AMELX	Amelogenin
AMTN	Amelotin
bFGF2	Basic fibroblast growth factor 2
BMP	Bone morphogenetic protein
BSP	Bone sialoprotein
c-MYC	Avian myelocytomatosis viral oncogene
CD34	Cluster of differentiation 34
COLI	Type I collagen
COLII	Type II collagen
COLX	Type X collagen
DMP	Dentin matrix protein
DSP	Dentin sialoprotein
DSPP	Dentin sialophosphoprotein
ENAM	Enamelin
FGF	Fibroblast growth factor
FGF(R)	Fibroblast growth factor (receptor)
HOX	Homeobox
IHH	Indian hedgehog
IHH(R)	Indian hedgehog (receptor)
KLF	Krüppel-like factor
KLK	Kallikrein-related peptidase
MMP	Matrix metalloproteinase
MSX	Muscle segment homeobox

NANOG	Homeobox transcription factor Nanog
OCT4A	Octamer-binding transcription factor 4a
ODAM	Odontogenic ameloblast-associated protein
PITX	Paired-like homeodomain transcription factor
PPRγ2	Peroxisome proliferator-activated receptor γ2
PTHrP	Parathyroid hormone related peptide
PTPN1	Tyrosine-protein phosphatase non-receptor type 1
RUNX	Runt-related transcription factor
SCA1	Spinocerebellar ataxia type 1
SHH	Sonic hedgehog
SOX	Sex determining region Y (SRY)-box
TBX5	T-box transcription factor 5
TGF	Transforming growth factor
TWIST	Twist family bHLH transcription factor
VEGF	Vascular endothelial growth factor
WNT	Wingless-related integration site

3.1 Commonalities and Differences in Structural Aspects of Mineralized Tissues

Just as Lot turned his back on Sodom, this chapter turns its back on Edith and the change in free energy which brought about her crystallization and moves on to consider the structure and some of the cellular characteristics of tissues which undergo mineralization. The chapter goal is to provide the reader with a detailed understanding of the structure of mineralized tissues and to identify commonalities in form and function which relate to events which lead to mineral deposition. The chapter focuses extensively on cartilage (elaborated by chondroblasts) and bone (osteoblasts) together with dentin (odontoblasts), cementum (cementoblasts/cementocytes), and enamel (ameloblasts).

It is important to acknowledge that there are profound differences in the developmental characteristics of each of the tissue and cell types mentioned above. For example, in the tooth, odontoblasts and ameloblasts follow a temporal and interdependent differentiation cascade. This maturation process is substantially different from endochondral mineralization of the axial (vertebral) and appendicular (limb) skeleton, a process characterized by the formation of a cartilage "anlage" which is then replaced by bone. "Anlage" is a term used in embryology to describe a condensation of mesenchymal cells forming a provisional morphological replica of an organ. Remarkably, the growth of the axial and appendicular skeleton is dependent on the activities of cells populating a strip of specialized cartilage, the epiphyseal growth plate. Chondrocytes comprising this tissue exhibit a well-ordered maturation process, resulting first in the formation of calcified cartilage and then the induction of bone. This "cartilaginous crust" is present through adolescence and is responsible for longitudinal bone growth.

It would be remiss to overlook a second form of osteogenesis which originates in a dense connective tissue membrane on the outside of bone. This type of intramembranous mineralization is responsible for the development of the bone of the scapula and the bones of the skull, the lower jaw, and the clavicle of various vertebrates. For completion, some bones form in investing muscle and tendon. The largest of these is the patella; the smooth surface of this bone enhances the sliding movement of the quadriceps femoris tendon. This sliding action allowed Lot to escape Sodom, although it did not prevent Edith from swiveling around on her axial skeleton, thereby actuating her demise.

As a parenthetical consideration, many cartilaginous tissues are not replaced by bone. Instead, they form structural components which are the bronchial, articular, laryngeal, nasal, and ear cartilages. The resiliency and elastic properties of these cartilages are mechanically suited for their function such as breathing, sound generation, hearing, and locomotion.

From a structural viewpoint, endochondral and membrane bones, as well as dentin and cementum, are biphasic structures and, like rebars used in building construction, consist of a brittle, hard mineral (apatite) enmeshed in a softer resilient (collagenous) extracellular matrix. The presence of the embedded mineral precludes the use of conventional technical approaches developed for studying soft tissues like heart and lung. Thus, of necessity, specialized methodologies have been devised to probe these tissues to assess the nature of the protein matrix without being confounded by the presence of mineral. Details of some of these techniques are described in Chap. 4.

3.2 A Brief Introduction to the Developmental Biology of Bone

From a developmental viewpoint, both axial and appendicular bones arise from the interaction between two embryonic structures, the notochord and the lateral plate mesoderm. Like the earliest chordates, the developing human fetus in the first trimester has no spinal column or indeed no visible skeleton. During this period, the notochord forms in the midline of the embryo located below the neural tube where it establishes the longitudinal axis of the fetus. The notochord acts as a primitive endoskeleton extending like a flexible rod from the fetal head (cephalic) to the tail (caudal) region and, because of its structure and biomechanical features, it keeps the embryo in an extended configuration. Indeed, histology of the notochord shows that its strength and flexibility are likely a result of its vacuolated cells which generate an outward osmotic pressure which is resisted by a tough connective tissue sheath [2, 3]. Aside from its skeletal function, the notochord is an important signaling tissue which promotes the commitment of cells to the osteogenic and chondrogenic lineages [3]. Cells located around the notochord are called paraxial mesoderm and those farther away are lateral plate mesoderm (Fig. 3.1). It is these cells which will form the provisional skeleton.

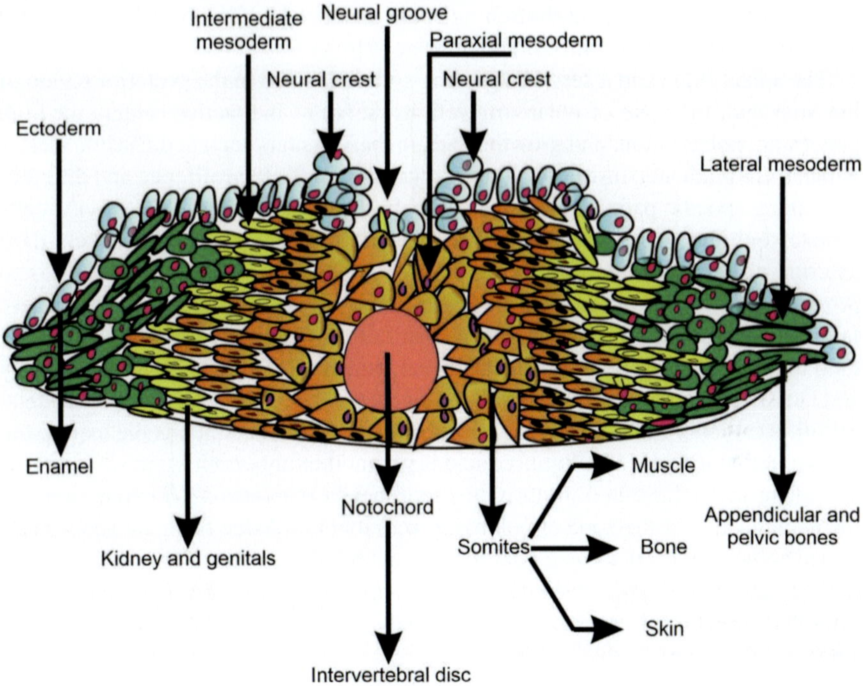

Fig. 3.1 Diagram of a transverse section through a developing human embryo viewed along the cranial-to-caudal axis. After approximately 3 weeks of development, midline mesenchymal cells form a rod of tissue called the notochord. Lateral to this structure, a condensation of cells, the paraxial mesoderm, becomes segmented to form somites. Subsequent differentiation of paraxial mesoderm generates three layers of cells, the myotome (muscle), dermatome (skin), and sclerotome. Cells of the sclerotome form connective tissues, in particular cartilage and bone of the axial skeleton. An additional condensation of cells, the intermediate mesoderm, forms genitourinary structures, while cells of the lateral plate mesoderm contribute to the appendicular, pelvic, and shoulder skeleton as well as the circulatory system, the body cavity, and extra-embryonic tissues. The ectoderm-derived neural crest cells contribute to tissues which form dentin (odontoblasts), enamel (ameloblasts), and many of the bones of the skull (osteoblasts). Finally, the intervertebral disk is formed from cellular remnants of the notochord following vertebrae formation and mineralization

While each vertebrate limb develops independently, formation and growth are exquisitely controlled, such that the ipsilateral and contralateral characteristics of each of the bones are almost identical [4]. Moreover, the growth activities between bones are tightly coordinated to ensure equivalent bilateral development. Limb formation commences in specific locales of the lateral plate mesoderm in response to chemical signals from the notochord, paraxial mesoderm, and possibly the endoderm [5]. Proliferation of mesenchymal cells in the vicinity of the putative cervical-thoracic and lumbosacral region generates limb buds. Mesenchymal cells contained within the buds condense and proliferate, forming the cartilaginous anlage of the limb bones [6, 7]. Epithelial thickening at the distal edge of the limb bud generates a structure called the apical ectodermal ridge [8].

The apical ridge and a second signaling system located in the posterior region of the limb bud, the zone of polarizing activity, serve as interactive centers for limb patterning, polarization, and growth. Cells in these centers secrete diffusible factors which form gradients, instructing the mesenchymal cells to proliferate and differentiate along specific pathways. For example, in the zone of polarizing activity, cells secrete sonic hedgehog (SHH), a protein which regulates digit number and their anterior–posterior positioning in the embryo [9]. Another function of SHH is to promote the expression of genes encoding fibroblast growth factors (FGFs). These factors, together with co-receptor heparan sulfate proteoglycans (such as syndecan), bind to cognate receptors during early development of the limb bud and both early and late stages of chondrogenesis and osteogenesis [10]. Besides promoting skeletal cell differentiation, FGF is part of a feedback loop which maintains the expression and secretion of SHH [11]. Retinoic acid is yet another important agent which forms a gradient in the limb bud. In turn, this morphogen regulates *SHH* expression and the maintenance of the zone of polarizing activity [12]. Aside from the genes mentioned above, *wingless-related integration site* (*WNT*)*5a*, *T-box transcription factor* (*TBX*)*5*, and *paired-like homeodomain transcription factor* (*PITX*)*1* are required for limb bud growth, differentiation, and polarity [13, 14]. The roles of *WNT5a* and *TBX5* are discussed further in Chap. 4.

Through a combination of cell proliferation and matrix synthesis, cartilages of the appendicular skeleton enlarge and form a rudimentary limb. Mesenchymal cells outside the condensate of the early limb proliferate and fashion a fibrous perichondrium [15]. By the eighth week of human development, cells in the mid-diaphysis of each of the tissue anlage become hypertrophic, and many of its constituent chondrocytes secrete a new collagen (type X collagen, COLX) within a calcifying matrix. Endothelial capillaries invade the calcified cartilage and begin a process of erosion and the creation of endochondral bone. The stimulus for vascularization and erosion is not precisely known, but it is likely related to events which govern the terminal differentiation of the hypertrophic chondrocytes (discussed in more detail in the next section of this chapter). With vascularization, incoming phagocytic cells (chondroclasts and possibly macrophages) remove the very dense cartilage. In addition to bringing oxygen and nutrients to the tissue regions, the invading capillaries remodel the cartilage and develop a new tissue, the bone marrow, which contains osteogenic stem cells. These cells are anchored to cartilage remnants and secrete a matrix comprised of type I collagen (COLI) and a number of non-collagenous proteins [16, 17]. The spread of primary ossification then continues from the mid-diaphyseal region up and down the bone shaft (Fig. 3.2) [18].

At birth in the human, the cartilages at the proximal and distal ends of the bones of the appendicular skeleton (epiphyses) initiate activities which result in longitudinal growth just described and at the same time serve as secondary centers of ossification. Such secondary centers begin as an avascular cartilage permeated by blood

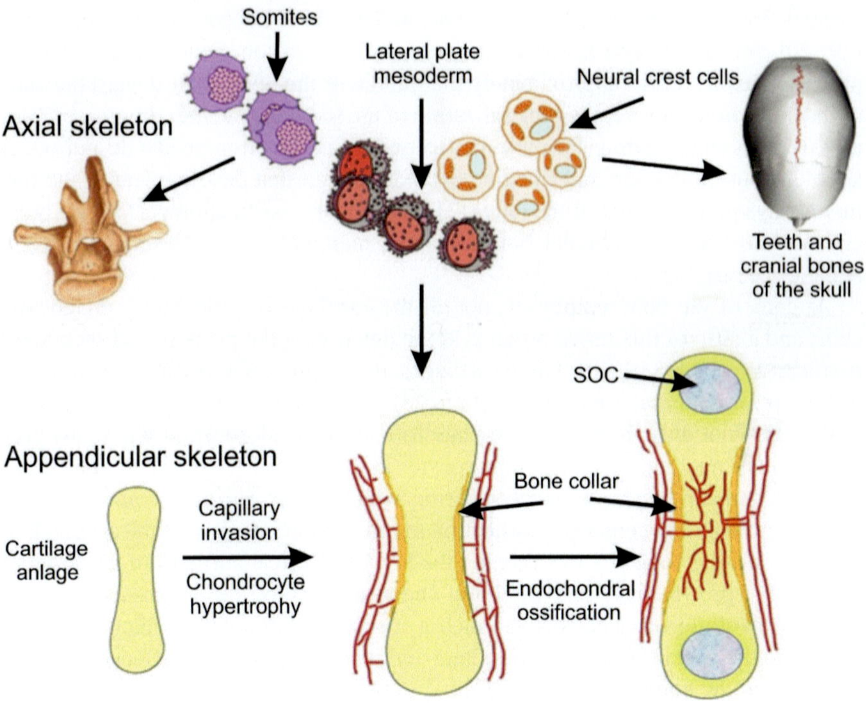

Fig. 3.2 Fate of cells derived from the neural crest, lateral plate mesoderm, and somites in a developing human embryo. These cells form much of the cranial skeleton and the teeth, as well as the appendicular and axial skeleton. By 6 weeks of development, lateral plate mesodermal cells differentiate into chondrocytes which condense and form a cartilaginous anlage. The chondrocytes which populate the anlage then undergo a series of changes described as endochondral ossification. An early event in the ossification process is the formation of a fibrous sheath, the bone bark or bone collar, around the middle of the anlage shaft. Osteoblasts within the collar form a tube of compact bone which encloses the chondrocytes. Although initially avascular, the shaft becomes surrounded by blood vessels, which invade the cartilage and cause the chondrocytes to become hypertrophic (enlarge) and eventually die through the induction of apoptosis. In concert with the loss of chondrocytes, the cartilaginous tissue is populated by osteoprogenitor cells which initiate extracellular matrix synthesis and mineralization of the midshaft (diaphysis) of the anlage. Following further invasion of blood vessels, a primary ossification center is formed by the spread of trabeculated bone up and down the diaphysis. Subsequently, a secondary ossification center (SOC) develops at the two ends of the bone. A strip of cartilage lying between the primary and secondary centers of ossification forms the growth cartilage while the central region of the diaphysis becomes the site of provisional bone marrow

vessels contained within cartilage canals, delivering oxygen, nutrients, and possibly osteoprogenitor cells [19]. Further development of the secondary ossification centers is similar to that of primary ossification, characterized by hypertrophy of centrally located chondrocytes, invasion of vascular elements, digestion of cartilage remnants by phagocytic cells, replacement of the canals by a vascular plexus and stromal tissue, secretion of bone proteins, and eventual trabecular bone formation [19, 20]. As the mineralization of the primary ossification center of a long bone proceeds linearly from approximately the middle of the bone shaft toward the ends of the bone, noted above, the mineralization of the secondary center progresses radially from its mid-portion. Primary and secondary ossification events do not occur simultaneously, and Xie and Chagin have pointed out that development of the former always precedes that of the latter [21]. Secondary ossification centers are present in almost all appendicular bones of most mammals while absent from many aquatic species.

At each of the bone epiphyses, not all the cartilage is removed by phagocytic cells, and a strip of this tissue remains at the junction of the primary and secondary ossification centers [19, 20]. These cartilages, the epiphyseal growth plates, are sites of further longitudinal bone growth (Fig. 3.3). Interestingly, the secondary ossification centers not only induce growth plate formation developmentally in many taxa but also continually protect the chondrocytes in the plate of terrestrial animals from high mechanical forces of weight-bearing which occur during locomotion [24]. Thus, the secondary center of ossification maintains the integrity of the articulating complex of long bones by overcoming the low mechanical stiffness of the resident cells of the growth plate and shielding and strengthening the epiphysis [21, 24]. From an evolutionary perspective, such a mechanical adaptation likely occurred when aquatic species transitioned to land and lost the modulating effects on forces provided by the buoyancy of water [21].

Aside from forming the limb bones, the paraxial mesoderm is also involved with development of the axial skeleton. During embryogenesis, mesoderm condenses on either side of the notochord. The process is followed by segmentation and the formation of structures called somites (Fig. 3.1) [25]. Microscopically, individual somites are comprised of spherical groups of epithelial cells arranged around a central lumen. Somites bud sequentially and pair-wise from the paraxial mesoderm in a cranial-to-caudal direction [25]. Somite formation and activity are mediated by the expression of a number of growth factors as well as several homeobox (*HOX*) genes along the longitudinal axis of the embryo. The *HOX* genes specify the overall body plan, ensuring that tissues and organs are organized and located in their appropriate developmental position. The genes serve as transcription factors which regulate the expression of networks of possibly hundreds of target genes. The expression of these genes is in turn regulated by retinoic acid as well as other gene products [26].

Mesodermal cells of the somites differentiate into distinct sub-compartments (Fig. 3.1). These include the dermatome, whose cells will form the dermis (skin); the myotome, whose cells will form muscle; and the sclerotome, whose cells generate the bone of the cartilaginous end plates and the vertebral column [27]. In the latter instance, the paired somites present in the embryo are the forerunners of the 33 vertebrae of the human skeleton [28]. Following the development of the vertebrae, which protect the nascent nervous system, cells of notochordal origin form the intervertebral disks. The disks mimic one aspect of the notochord in that they provide flexibility to the vertebral column.

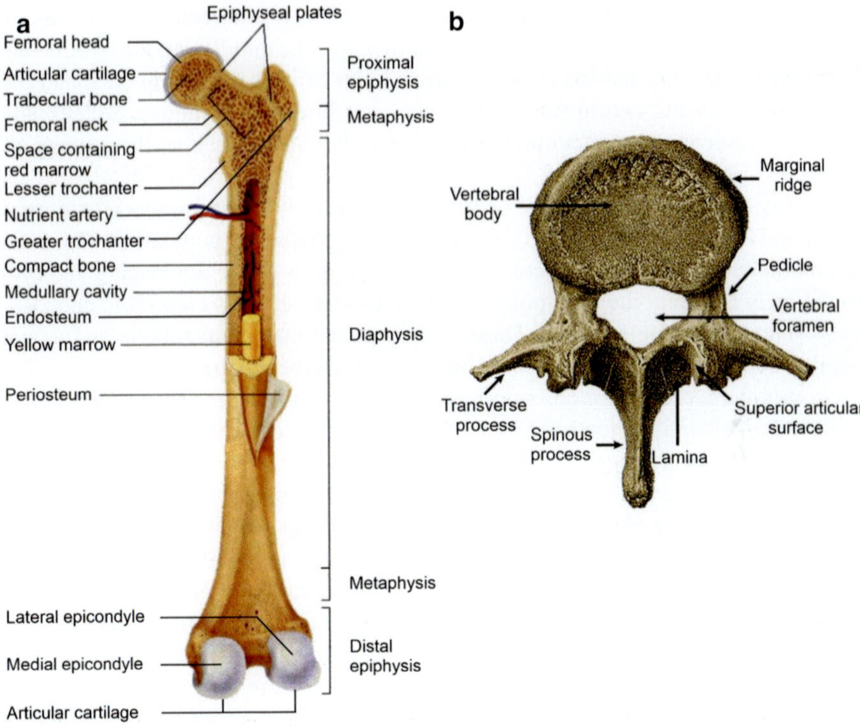

Fig. 3.3 Schematic and histological features of a human long bone (femur) and a vertebra (lumbar region), representing elements of the appendicular and axial skeleton, respectively. (**a**) Long bones may be separated into three regions along the length of the tissue: the diaphysis, metaphysis, and epiphysis. The shaft of the bone, the diaphysis, is comprised principally of an outer region of dense compact bone and an inner region of trabecular or spongy bone. The two ends of the bone are the epiphyses and the region between the diaphysis and epiphysis is the metaphysis. Epiphyses are covered with cartilage to allow the long bone to articulate smoothly with neighboring bones in a joint capsule. For example, the femoral head and condyles articulate with bones of the hip and the knee. In growing individuals, epiphyseal growth plates are present at both ends of a long bone as well as the trochanters of a femur. When long bone growth ceases, the growth plate closes and forms an epiphyseal line. The long bone is covered along its diaphysis and metaphysis with a thin, cellular membrane, the periosteum. To puberty and the point of growth plate closure, periosteum lies adjacent and joins to the perichondrium, another thin, cellular membrane which envelopes the epiphysis to the point of the appearance of articular cartilage. After closure, the perichondrium forms the periosteum, which subsequently surrounds the epiphysis in the adult [18]. The periosteum and perichondrium carry arterial blood to supply nutrients to the bone, and red and yellow marrow develops within the internal cavities and spaces of this tissue. Muscle and tendon attachment sites are prominent at the ends of long bones. (**b**) Superior view of a lumbar spinal vertebra, showing its essential features of a central body and vertebral arch, comprised of the pedicle and lamina, with associated bony processes. A typical vertebra is an irregularly shaped complex of bone and hyaline cartilage. An intervertebral disk, a unique non-mineralized tissue derived from the notochord, separates each of the spinal vertebrae. The vertebral body consists of spongy or cancellous bone covered by a thin layer of cortical bone. Thicker cortical bone covers the vertebral body and processes. The junction between the vertebral body and arch creates the vertebral foramen, a space which accommodates the nerves of the spinal cord together with blood vessels. (Schematics adapted from (**a**) Dorozhkin [22] and (**b**) Shapiro and Risbud [23] with permission)

While the above discussion focuses on the axial and appendicular skeleton, many of the bones of the head and neck (skull and bones of the middle ear, the hyoid bone in the neck, the ribs, and the sternum) originate in cells of the neural crest and the dorsal aspect of the neural tube [29]. Using cell lineage techniques, Selleck and Bronner-Fraser showed that surface ectoderm also contributes to the genesis of these cells [30].

During development, the crest cells undergo an epithelial-to-mesenchymal transition and exhibit characteristics of a multipotent progenitor population. The cells are motile and migrate to and colonize specific regions of the developing embryo. In the early cranium, the multipotent neural crest-derived cells commit to chondrocytic and osteoblastic lineages. These cells populate the dental follicle and dental papilla; dentin-secreting odontoblasts originate in the neural crest [31].

In summary, the neural crest-derived cells form a major portion of the craniofacial complex and the teeth. Those cells which migrate to the trunk region of the embryo fail to form bone. Thus, as far as osteogenesis is concerned, neural crest osteogenic activity is confined to the rostral regions of the developing embryo. Notably, while most skull bones are formed from these cells, some of these bones (the parietal bones, for example) arise from paraxial-derived tissue [32]. Thus, the skull is comprised of cells derived from both the lateral plate mesoderm and the neural crest. Not surprisingly, disruption of neural crest formation or prevention of migration leads to diseases of both the nervous system and the skull. For instance, abnormal patterning resulting from mutations in genes required for the synthesis of ribosomal RNA causes Treacher Collins syndrome, a congenital disorder of the craniofacial complex [33]. Premature fusion of cranial sutures, craniosynostosis, is linked to mutations in at least 20 genes, including *fibroblast growth factor receptor* (*FGF*(*R*)), *Indian hedgehog* (*IHH*) and its receptor (*IHH*(*R*)), *muscle segment homeobox* (*MSX*)*2*, and *twist family bHLH transcription factor* (*TWIST*)*1* [34].

The bones comprising the axial and appendicular skeleton of vertebrates develop, grow, and mature over the lifetime of the species. These stages of progressive bone formation lead to several characteristic structural features distinguishing axial and appendicular components. In the case of a long bone representing a typical appendicular element, such features include the bone epiphysis, metaphysis, and diaphysis, periosteum, perichondrium, and articular and growth plate cartilages (Fig. 3.3) [22]. For bones representative of the axial skeleton such as a vertebra, anatomical landmarks include transverse processes, the vertebral body, spinous process, and lamina, for example (Fig. 3.3) [23, 35].

Whether axial or appendicular skeletal components, the developing, growing, and maturing bones undergo a series of remodeling events, in which older tissue, including both organic and inorganic material, is constantly replaced and renewed during the life of the animal. The human skeleton consists of two types of bone tissue: compact or cortical/lamellar bone, principally forming the walls of the midshaft (diaphysis) of long bones, for example, and trabecular or cancellous/spongy bone, generally found at the ends (metaphysis and epiphysis) of long bones (Fig. 3.3) [36]. With regard to bone remodeling in humans, cortical bone comprises 75% of the entire skeleton with a turnover rate of 4%/year while trabecular bone makes up 25% of the skeleton with a turnover rate of 28%/year (turnover rate of the whole skeleton is ~10%/year) [37]. Such replacement and renewal events are regulated by communication and interaction between osteoprogenitor cells and osteoblasts, osteoclasts, and osteocytes, the three principal types of cells comprising bone and whose microstructure has been characterized innumerable times. In this regard, these cells are commonly observed microscopically near one another such that multinucleated osteoclasts involved in enzymatically degrading and resorbing bone tissue recruit osteoprogenitor cells and osteoblasts to begin and complete new bone formation (Fig. 3.4).

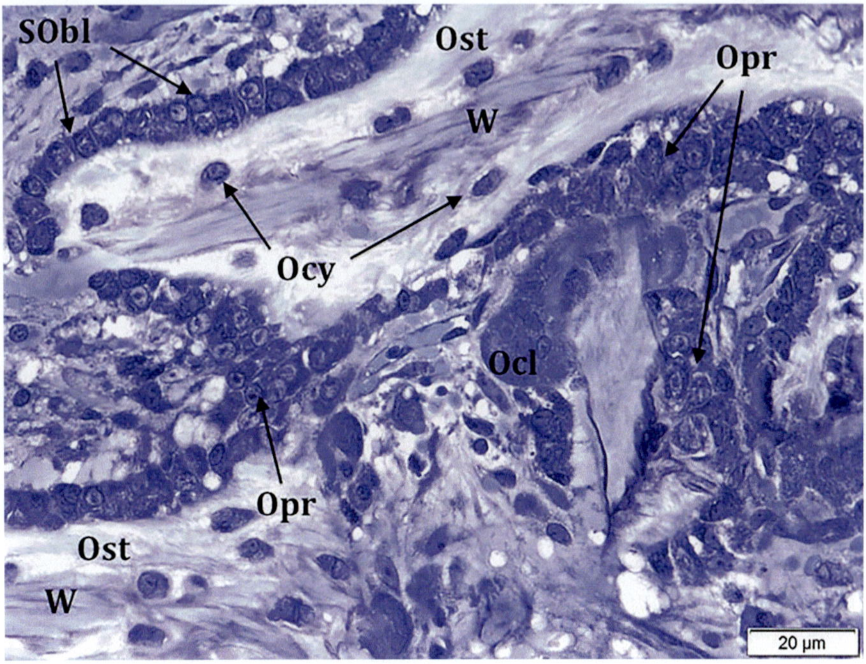

Fig. 3.4 Light photomicrograph of a region of repair tissue from a stable rabbit cortical bone fracture site. The section of the tissue has been stained with 1% toluidine blue and shows woven bone (W), numerous osteoblasts (SObl) covering the surface of osteoid (Ost; unmineralized bone matrix), osteocytes (Ocy), an osteoclast (Ocl), and osteoprogenitor cells (Opr). An interesting structural feature illustrated in this image is the organization of osteoblasts adjacent to each other to form a palisade of cells. This characteristic arrangement of cells is also present in dentin, cementum, and enamel among the vertebrate mineralized tissues and likely represents an optimization of form and function by the cells. (Image courtesy of Frederic Shapiro, M.D., Department of Bioengineering, Northeastern University, Boston, MA)

A defining characteristic of such osteoblasts is that they may be found disposed in extensive palisades lining the surfaces of bone which have been digested by osteoclast activity (Fig. 3.4). Osteoblasts forming palisades secrete seams of an organic extracellular complex, the osteoid, ~2–5 μm in width comprised principally of COLI and necessary for the subsequent deposition of mineral. Mineralization is reported to occur in two temporal waves. The first wave is deposited rapidly (days) and accounts for 70–75% of the total mineral; the second, is slower (months) and results in an increase in mineral density which serves to achieve its physiological limit [38]. Parfitt pointed out that osteoid seam thickness remains constant for only about 20% of the bone-forming surfaces [39]. The remaining bone-forming surfaces are lined by a single layer of thin, flattened osteoblasts forming a "leaky envelope" which is permeable to large molecules, and these cells elaborate an unmineralized "lamina limitans" [40–42]. Once an osteoblast becomes embedded in the mineralized osteoid matrix, it becomes terminally differentiated as an osteocyte, which maintains an extensive signaling network with other osteocytes and osteoblasts as well as osteoclasts; the osteocytes serve to orchestrate the overall remodeling process (Fig. 3.4) [43, 44]. Further details related to the interactions and functions of these three types of cells are presented in Chaps. 4 and 9.

3.3 The Cartilaginous Growth Plate and the Initiation of Bone Mineralization

As noted above, limb formation and growth in vertebrates are tightly controlled such that particular characteristics of each developing long bone are nearly identical [4]. Thus, centers of growth (growth plates) are present in the proximal and distal epiphyses of growing long bones (Fig. 3.3). While these plates are the major cartilaginous growth centers, similar structures are evident in other regions of developing long bone, for example, in the greater and lesser trochanter of the femur (Fig. 3.3). The trochanter serves as a site for muscle or ligament attachment, where mechanical forces promote additional bone growth [45]. The activities of cells in each of the epiphyseal growth plates, responding in part to the mechanical forces applied to the bone, modulate and control overall bone length and thereby regulate overall height of the growing child or other vertebrate species.

The contribution of individual growth plates to the overall growth of bone in children has been estimated. In this instance, the proximal plates of the human fibula and tibia are reported to be similar, 61% and 57%, respectively, in total bone growth [46]. In contrast, Pritchett determined that 70% of growth in the human femur occurred at the distal growth plate [46]. Growth plates are also associated with permanent cartilages. One such cartilage, the nasal septum, contains a growth plate at its caudal and dorsal borders. Growth at these sites pushes the facial skeleton away from the cranium, a movement which serves to enlarge or extend the skull [47]. In most bones, once their genetically predetermined size has been reached, the cartilage band is lost, growth ceases, and a bone bridge unites the epiphysis with the diaphysis (Fig. 3.3). The concurrent change in cartilage function probably reflects the loss of stem cells to feed into the epiphysis and replace the terminally differentiated chondrocytes together with changes in the level of hormones regulating childhood and adolescent growth. Morita et al. have shown that there is also decreased telomerase activity [48].

The growth plate is contained within the epiphysis and is separated from the secondary ossification center of a long bone (Fig. 3.2). Early in growth, the growth cartilage is enclosed in a tube of membrane bone, the "bone bark or bone collar" or the Ring of Lacroix (Fig. 3.2) [49]. The Ring prevents shearing dislocation of the epiphysis from the metaphysis and provides mechanical rigidity to these regions of the developing bone [49]. There is also some evidence to indicate that the attached periosteum may be involved with the post-natal growth process [50]. Indeed, Warrell and Taylor reported that surgical incision and elevation of the periosteum probably release pressure on the plate and enhance ipsilateral bone growth [51]. One other anatomical structure of importance in the epiphysis is a circumferential groove, the Node of Ranvier (not to be confused with the Node of Ranvier which limits neural axon myelination). Cells within the node associated with the epiphysis are thought to serve as a reservoir of stem cells for the growth cartilage [52].

A characteristic feature of the cartilaginous growth plate is its avascular nature. Since there is limited oxygen delivery to cells in this tissue region, its composite chondrocytes generate much of their metabolic energy through glycolysis [53]. It is of particular interest to note that the thick cartilage of rapidly growing birds is extensively permeated by vascular canals; the cartilage of even dinosaurs was likewise replete with such canals [54]. Canals are also found in the growth plates of most other species where they function to maintain chondrocyte viability by accessing nutrients and some oxygen to the chondrocytes [55]. The channels in the cartilage matrix are probably excavated by macrophages as these cells secrete a number of matrix metalloproteinase (MMP) enzymes [20, 56]. Other roles of the channels are to provide stem cells to the growing bone and to serve as centers for mineral deposition [57, 58].

To summarize these brief comments about the cartilaginous epiphysis and bone formation, it is clear, first, that the mineralization of cartilage represents a critical stage in the formation of endochondral bone, while long bone growth requires the formation of a transient calcified cartilage phase. Thus, the formation of cartilage is obligatory for long bone development and its subsequent growth. In this context, as indicated earlier, there is the likelihood that calcified cartilage itself existed as a skeletal structure in early vertebrate evolution. Second, epiphyseal chondrocytes are easily harvested in vitro and have been documented to define events which lead to cartilage mineralization. Studies of these cells have revealed pathways and mechanisms which are relevant to almost all tissues which undergo mineralization. Third, investigations of the growth plate have yielded new information about cell–cell interactions and communication. In other words, the growth plate provides a foundation and platform for understanding unique principles of intercellular communication in an avascular field. Finally, since the cartilaginous plate is the organ which forms trabecular (endochondral) bone and is involved with membrane (intramembranous) bone formation, it has enhanced the knowledge of crosstalk between chondrocytes and osteoblast progenitors in the investing membrane of the bone. Each of these roles for cells of the epiphyseal growth plate is discussed further in the next section of this chapter and in additional detail in later chapters of this book.

3.4 Microscopic Architecture of the Growth Plate

This section of the chapter intends to familiarize the reader with structural aspects of the growth plate and associations between each part of the plate and its surrounding and adjacent tissues. Control of chondrocyte function and elaboration of principles and mechanisms of mineralization are considered in other book chapters. Beginning with the cellular structure of the plate, the epiphyseal growth plate of a long bone in the appendicular skeleton of a vertebrate can be divided into four distinct zones based on morphological, compositional, and functional considerations: a resting or reserve zone, and proliferating, hypertrophic, and degenerative regions (Fig. 3.5) [59]. The major characteristics of each of the zones are as follows.

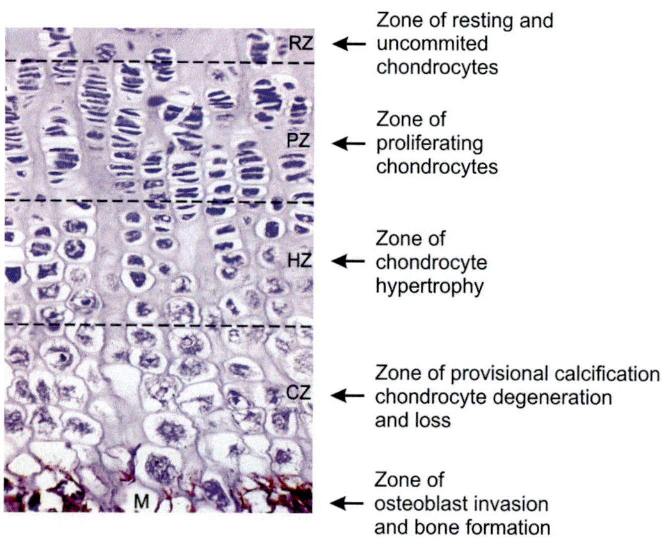

Fig. 3.5 Histological organization of the epiphyseal growth plate in a toluidine blue-stained section from a rat femur. This cartilaginous structure is populated by chondrocytes which undergo a maturation sequence. Cells in the resting zone (RZ) become activated and undergo rounds of proliferation (proliferating zone, PZ). Once proliferation stops, the post-mitotic chondrocytes enlarge (hypertrophic zone, HZ) and initiate events which lead to mineralization of the cartilage extracellular matrix (calcifying zone, CZ). The mineralized matrix is subsequently removed by invading metaphyseal blood vessels. In the late hypertrophic zone, it is generally thought that many of the chondrocytes die (degenerative region) and the cartilage is replaced by osteoblasts, which lay down metaphyseal bone. The metaphysis is designated as (M). (Image adapted from Anderson and Shapiro [59] with permission)

The resting zone is most distant from the site of bone formation (metaphysis). Resting zone activity reflects the potential of cells to enter a proliferative phase and become mature chondrocytes. Some investigators have suggested that the reserve zone is fueled by mesenchymal stem cells, possibly originating in the Node of Ranvier, which feed into this cartilage layer and are committed to the chondrocyte phenotype [60, 61]. Located below the zone of resting chondrocytes is a region containing stacks of chondrocytes which form columns of proliferating cells. Each column, or chondron, in mammalian tissues may contain as few as five cells but the numbers vary; in the very rapidly growing epiphysis of the young chick, the proliferating zone may hold as many as 50 cells in a column [62]. Use of live imaging techniques which visualize chondrocytes as they transition from a resting to a proliferative zone shows that, after a small, round resting cell [63] divides, the flattened discoid progeny [64] shimmy into position while undergoing a tipping process. As a result of this movement, the daughter cell or cells are now orthogonal (perpendicular) to the direction of bone growth (Fig. 3.6) [65, 66]. In this way, the groups of cells become nested in columns in the proliferative zone and generate the characteristic zonal architecture of the growth plate (Fig. 3.5).

Proteins which are secreted during the formation of columns of proliferative cells include N-cadherins and WNT [67]. Indeed, WNT ligands serve as molecular cues which activate cell movements; this observation has been interpreted as involvement of a more generalized biological system, termed the planar cell polarity pathway [65, 68]. Li and Dudley [67] found that inhibition of the protein, frizzled, a receptor involved with WNT signaling, influences chondrocyte orientation and chondron formation in the growth plate. N-cadherins, robustly expressed by chondrocytes, enhance cell–cell adhesion and are secreted prominently at the membrane interface between the chondrocytes [65]. Moreover, along other regions of the membrane, the cell attaches to proteins of the extracellular matrix to initiate and produce chondrocyte elongation and spreading [67].

Hypertrophy or swelling, the third stage of chondrocyte maturation, begins when cell proliferation stops. In a study of the growth plate, Stump reported the use of the descriptor, "the hypertrophic zone of Retterer" [69], to label the terminally differentiated cells of the epiphysis [70]. The concept of a hypertrophic cell is an interesting one, especially as swelling is often associated with tissue disease (for example, cardiac hypertrophy), metabolic dysfunction with accumulation of fat and glycogen, or osmotic dysregulation and fluid increase [71]. However, in terms of growth plate physiology, swelling is a normal maturation event for chondrocytes and not indicative of cellular pathology. During this stage in the life cycle of the epiphyseal cell, large post-mitotic chondrocytes become invested in a complex extracellular matrix. Compared to the more flattened, discoid neighboring proliferative cells, the hypertrophic cells are oval-to-polygonal in shape with greatly increased circumference and diameter.

The degree of swelling is considerable in the cells of the deeper and more distal aspects of the growth plate. Buckwalter et al. showed that, between the upper proliferative zone and the lower hypertrophic zone of a growth plate, the cells increased their mean volume more than 500% [72]. More importantly, from a functional viewpoint, the absolute volume of endoplasmic reticulum, Golgi membranes, and mitochondria in each chondrocyte rose by over 120%, while the volume of cytoplasm

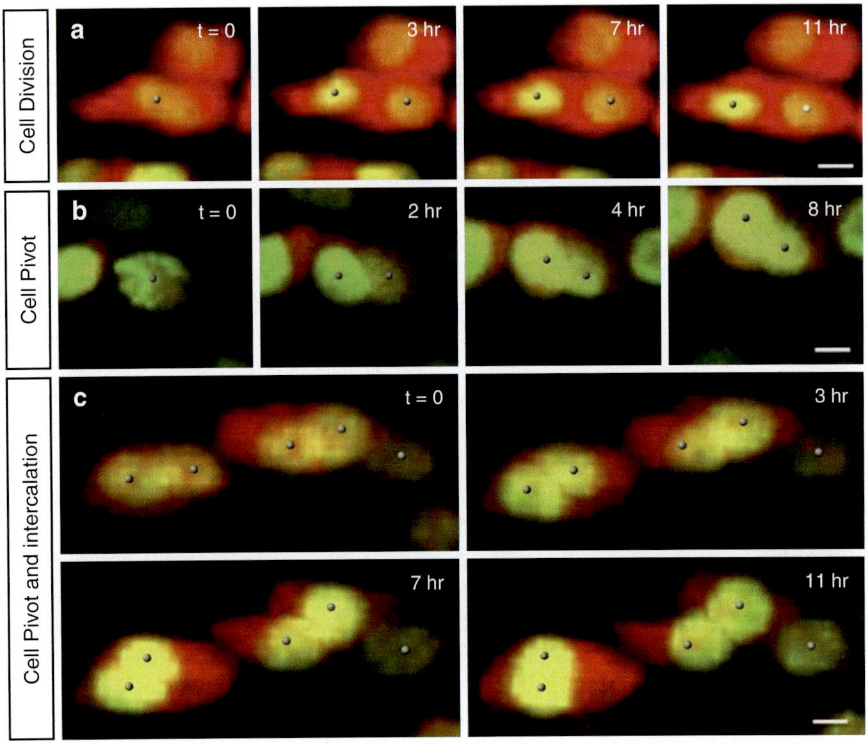

Fig. 3.6 Oriented division of proliferating chondrocytes shown by light photomicroscopy. Chick chondrocytes in forming metacarpal bone labeled with green and red dyes using a replication-competent avian retroviral (RCAS) infection procedure were evaluated by live imaging microscopy over time (0–11 h). (**a**) Cell division: Daughter cells appeared to be positioned orthogonal to the tissue proximodistal axis and physically coupled after cytokinesis. (**b**) Cell pivot and mediolateral intercalation in the proliferative zone: The mother cell divides to generate a pair of daughter cells which have modified their starting positions after 8 h. (**c**) Initiation of columniation: At $t = 0$, on the left, two laterally aligned daughter cells (cell doublets) reorganize their orientations and stack into a single column by 11 h. Scale bars = 4 μm. (Images adapted from Li et al. [65] with permission)

and nucleoplasm increased by nearly 800% [72]. These findings suggested that these enlarged cells maintained their biosynthetic activity and, by expanding, promoted bone elongation [72]. In a much later study using a technique called diffraction phase microscopy, Cooper et al. measured the dry mass and mean cell height of individual murine growth plate chondrocytes (Fig. 3.7) [73]. These investigators showed that initially there were marked increases in both cell volume and dry mass, evidence of true hypertrophy in the growth plates of the tibia and metatarsus of the animals [73]. As the cells continued to mature, there were additional but smaller increases in cell height, and cell volume became greater while dry mass remained constant [73]. These observations directly supported the notion that the maturation-dependent enlargement in cell size was the result of enhanced functional activity, possibly because of hormonal or local growth factor stimulation. This type of change in cell size is not without precedent as it is also observed in uterine smooth muscle cells caused by the stimulatory effect of estrogen during pregnancy [74].

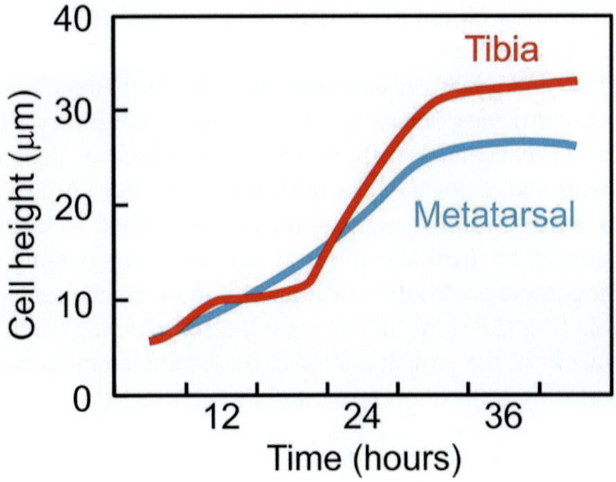

Fig. 3.7 Increase in average cell height of post-mitotic (hypertrophic) chondrocytes in the proximal tibia and distal metatarsal of mice. The plot shows that the mean height of the cells was elevated three- to fourfold over a 36-h time period. Experimental measurements were made with proliferating cells labeled with a nuclear stain which defines their last mitotic cycle while the change in cell height was determined by the technique of diffraction phase microscopy. In the same study, the authors showed that cell shape changes were an inherent property of the cells and not caused by osmotic pressure changes in the fluid of the extracellular matrix. (Plot adapted from Cooper et al. [73] with permission)

During the period of proliferation and hypertrophy, several critical events occur which relate to the function of chondrocytes as mediators of bone growth and mineralization. First, the hypertrophic cells secrete a unique protein, COLX, as noted above. This protein, together with neighboring proteoglycan subunits, may provide considerable mechanical support to the tissue [75]; and there is some indication that its presence is linked to the conversion of chondrocytes to osteoblasts, a topic which is discussed in more detail later in this chapter. Second, these chondrocytes secrete vascular endothelial growth factor (VEGF), which serves to promote the invasion of vascular elements into the cartilage, leading to degradation of the tissue and its mineralization [76]. In the chick femur, for example, cartilage mineralization takes place close to the edges of the penetrating vascular canals. Mineralization is incomplete with small islands of unmineralized matrix connecting chondrocyte lacunae [77].

While much of the cartilage is removed during the processes of vascular incursion and tissue degradation, residual cartilage struts remain as scaffolds for attachment of invading osteogenic cells from the blood vessels [78]. Bound to the struts and often linked by tight junctions, these cells secrete collagen and other proteins to form a contiguous organic layer (osteoid) while osteoclasts initiate a bone remodeling cycle, mentioned previously, in which bone is renewed and the first poorly organized bone is replaced by more ordered structures. With respect to the deposition of mineral, hypertrophic chondrocytes release small membranous organelles or blebs, called matrix vesicles, which may be intimately involved in nascent apatite formation in the cartilage. The majority of such vesicles is found within the longitudinal septae between hypertrophic cells. The vesicle blebbing phenomenon and the mechanism of formation of mineral putatively associated with the vesicles are discussed in some detail in Chap. 6.

As the cells bleb and release vesicles into the cartilage extracellular matrix, most hypertrophic chondrocytes become autophagic. Autophagy is a process which enhances survival of the cells through degradation of their own lipid and protein components and removal of dysfunctional organelles [79]. Eventually, when these molecular stores are consumed and the cell is assailed by calcium and phosphate ions generated by cartilage removal and bone remodeling activities, the hypertrophic chondrocyte activates a death program in which many of the cells die through apoptosis (Fig. 3.8) [80]. In this regard, certain findings indicate that, while cells at the periphery of the growth plate die, considerable numbers of cells appear to survive and possibly transdifferentiate into osteoblasts [81].

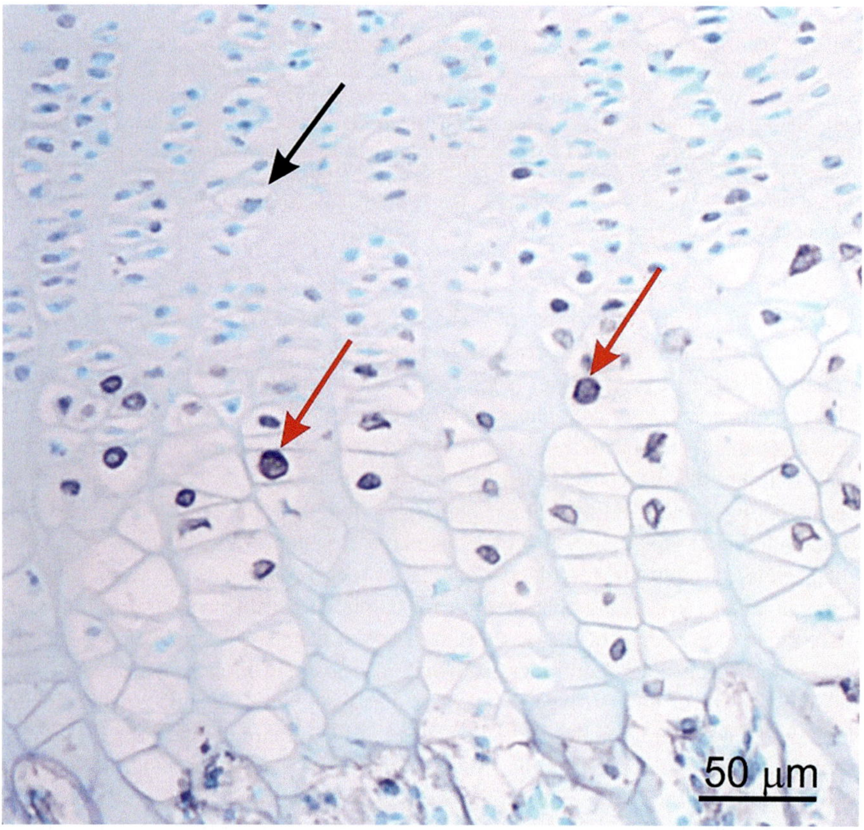

Fig. 3.8 Light photomicrograph of a longitudinal section through a mouse growth plate. The section shows proliferating, maturing (black arrow), and terminally differentiated hypertrophic chondrocytes. The section has been stained using the TUNEL procedure to label and identify fragmented cell nuclei. TUNEL-positive (apoptotic) nuclei (red arrows) are observed in chondrocytes present in the hypertrophic region of the growth plate. (Image from Shapiro et al. [80] with permission)

3.5 Functional Importance of Chondrocyte Hypertrophy, Apoptosis, and Transdifferentiation

The fate of the hypertrophic chondrocyte is a topic relevant to endochondral bone formation in the neo-natal and post-natal state and possibly to intramembranous ossification as well. As proposed many years ago, these cells comprising the epiphyseal growth plates of long bones were suggested to change lineages and transform (transdifferentiate) into osteoblasts [82, 83]. Transdifferentiation is a term used to describe the process by which a differentiated cell changes its identity and assumes the characteristics of another separate and distinct cell type.

An alternative concept for these cells is that they mineralize their matrix and die, after which they are replaced by osteoblasts of stromal origin. Support for this hypothesis has been principally observational, by which several microscopic studies investigating hypertrophic chondrocytes at the metaphyses, the chondro-osseous junction of a long bone, reported that the cells were lost [84–86]. Indeed, Brighton labeled the region as the zone of degeneration [87].

A caveat to these findings is that the "degenerative zone" is difficult to examine histologically and interpret unequivocally. However, when tissue is optimally prepared for microscopy, the morphology of many of the hypertrophic cells is consistent with cells dying through apoptosis [88]. This mode of cell death was confirmed using a nuclear labeling technique (TUNEL) showing that the DNA of the chondrocytes was fragmented, a critical indication of apoptosis [89–91]. That death may not be the sole fate of all the chondrocytes was the observation by Erenpreisa and Roach, who reported that the cells were "in limbo, unable to live or die" but temporarily "stable" while their chromatin was condensed [92, 93].

Present understanding of the fate of hypertrophic chondrocytes is that many of these cells in the growth plate die through apoptosis, but some of them survive and transdifferentiate into osteoblasts, as mentioned above. Transdifferentiation occurs without the differentiated cell first exhibiting the genetic profile characteristic of a dedifferentiated state. While relatively uncommon in higher organisms, such an identity switch has been observed in frogs [94], flies [95], and worms [96]. There is a growing awareness that the transdifferentiation process is regulated by the expression of a number of regulatory genes and responsive to environmental factors. In terms of specific pathway involvement, autophagy appears to play an important role in nuclear reprogramming and subsequent differentiation [97]. Thus, it is not unreasonable to assume that changes in transcription factor activity and environmental status at the mineralization front would promote the transdifferentiation of hypertrophic chondrocytes into metaphyseal osteoblasts during endochondral ossification.

The idea that chondrocytes could transform into osteoblasts received early support from the noted biologist, Dame Honor Fell. In 1933, she reported that cartilage tissue in organ culture formed both bone and osteoid [98]. These findings were reinforced by morphological evidence of transdifferentiation of chondrocytes in Meckel's cartilage, chondrosarcomas, and epiphyseal cartilage [99–101]. Biochemical analyses also showed that hypertrophic chondrocytes could synthesize extracellular matrix proteins and proteoglycans [102, 103]. Further, Oshima et al. demonstrated that hypertrophic chondrocytes secreted type II collagen (COLII), osteopontin, osteocalcin, osteonectin, and aggrecan core-binding protein transcripts [104].

By the 1990s, the intrinsic plasticity of what was earlier considered to be a terminally differentiated chondrocyte was well established [105–109]. Using a cell culture system to assess possible fates of the hypertrophic chondrocyte, Bianco et al. proposed that "all early hypertrophic chondrocytes have the inherent potential to differentiate to osteoblast-like cells and to contribute to initial bone formation" [110]. The pathway taken by a cell would depend on local conditions: Those chondrocytes located toward the peripheral perichondrial tissues would be exposed to a microenvironment which potentiates osteogenesis, while those cells located in the avascular center of the cartilage would undergo apoptosis. In this context, ultrastructural studies suggested that the last identifiable hypertrophic chondrocyte could become mitotic and divide asymmetrically; it was hypothesized that one of the resulting cells would die by apoptosis while the other cell would be free to transdifferentiate into an osteoblast [60, 111]. Thus, while transdifferentiation was accepted as a plausible explanation for endochondral osteogenesis, a number of local possibilities provided avenues for further research.

The optimal experimental approach to address the transdifferentiation question is through lineage tracing. The technique follows the fate of labeled cells to learn if they die or, in the case of epiphyseal chondrocytes, are "resurrected" as osteoblasts. Indeed, recombination lineage tracing methods support the notion that chondrocyte-to-osteoblast transdifferentiation takes place in the growth plate [112]. Another related and particularly insightful study was reported by Park et al. [113]. Using COLX as a marker for hypertrophic chondrocytes, these investigators showed that many osteoblasts of the subchondral metaphysis and, surprisingly, those in the endosteal and cortical bone originated from hypertrophic chondrocytes (Fig. 3.9); there was also labeling of osteocytes [113]. Moreover, these workers demonstrated that, within chondrocyte lacunae, there were small condensed cells which probably were associated with hypertrophic chondrocytes [113], a result which added strength to the earlier findings reported by Roach et al. [111]. A more practical value of such a chondrocyte–osteoblast relationship indicated that bone repair could be facilitated using chondrocyte graft material [114].

The route taken by cells to become a putative osteoblast is not clear. As just discussed, there is the possibility that chondrocytes directly transdifferentiate into the osteoblast phenotype [105, 107, 112]. More likely, prior to becoming osteogenic, hypertrophic chondrocytes revert to a more primitive stem cell-like state. This idea has some support as the cells express stem cell marker genes such as *cluster of differentiation* (*CD*)*34*, *avian myelocytomatosis viral oncogene* (*c-MYC*), *spinocerebellar ataxia type* (*SCA*)*1*, and *sex determining region Y* (*SRY*)*-box* (*SOX*)*2* [113, 115]. Also, during repair of a fracture, where there is a chondrocyte-to-osteoblast transition, chondrocytes in the cartilaginous callus express the stem cell markers, *octamer-binding transcription factor* (*OCT*)*4A*, *homeobox transcription factor Nanog* (*NANOG*), and *SOX2* [116]. Based on these observations, it might be

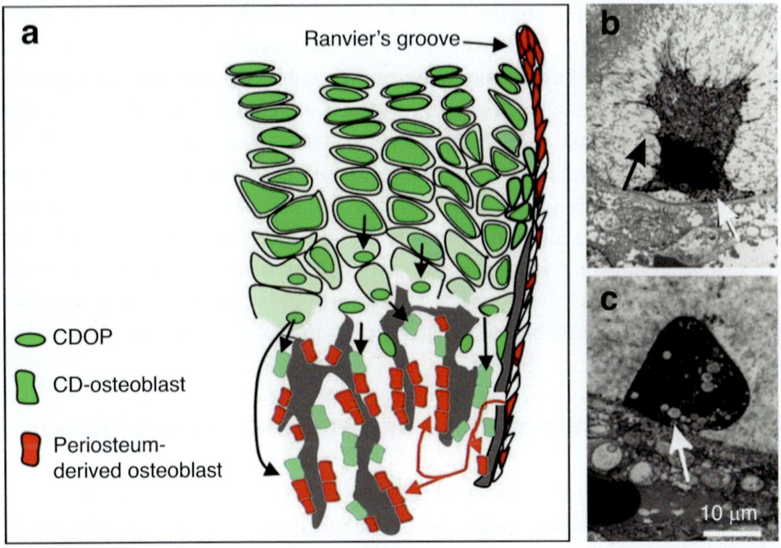

Fig. 3.9 Differentiation of populations of murine epiphyseal chondrocytes into metaphyseal osteoblasts. Lineage tracing was performed using dyes which label epiphyseal chondrocytes. Panel (**a**) indicates that osteoblasts arise from at least two distinct pools of cells: Some hypertrophic chondrocytes proliferate and generate small condensed cells which then assume the characteristics of chondrocyte-derived osteoprogenitor cells (CDOPs, marked in green with pathways denoted with black arrows), secreting osteogenic markers (CD-osteoblast) which include COLI and osterix. A second pool arises from cells in the periosteum (marked in red), possibly from proliferating cells located in the Groove or Node of Ranvier (pathways denoted by red arrows). Panels (**b**) and (**c**) show electron micrographic images of small, contracted cells located in the hypertrophic region of the growth plate. These cells appear to occupy only a portion of their chondrocyte lacuna, exhibit a condensed nucleus and extensive vacuolization of their cytoplasm, and may have plasma membrane protrusions and small processes (black arrow). Their plasma membrane is attached asymmetrically to the transverse septum (white arrows) but not to the extracellular matrix. Scale bar estimated and is the same for both (**b**) and (**c**). (Images adapted from Park et al. [113] with permission)

expected that some chondrocytes return to a more pluripotential state before assuming their new osteogenic character.

Once a primitive stem cell status is achieved, the process of transdifferentiation described above needs careful regulation since subsequent differentiation can lead to commitment to a number of non-bone phenotypes. Thus, regulatory control is necessarily complex and numerous transcription factors are required for chondrocytes to become bone cells. This putative change involves β-catenin, a ligand in the WNT/β-catenin pathway [117], the tyrosine-protein phosphatase non-receptor type (PTPN)1 [118], and SOX9 [119]. The role of *SOX9* in transdifferentiation events is yet unclear and complicated. Lui et al. [120] showed that, unless this transcription factor were down-regulated, chondrocyte transdifferentiation into osteoblasts did not take place. All the findings concerning the fate of the terminal chondrocyte are at odds with the older and commonly held assumption that every late-stage hypertrophic cell is terminally differentiated. Contrary to conventional thought, then, these cells appear to be capable of undergoing further division. To summarize these and other investigations, it is highly likely that, while many of the centrally located hypertrophic chondrocytes are removed from the growth plate by apoptosis, a population of mature, supposedly hypertrophic cells undergoes a subsequent round of proliferation which provides a secondary cohort of osteogenic cells. Such cells synthesize and secrete an organic matrix (osteoid) which then mineralizes to form new metaphyseal bone. An interesting speculation from this present discussion may be made if chondrocytes in the epiphysis undergo transdifferentiation to newly forming osteoblasts. In this instance, both sets of cells may generate mineral by a common mechanism, regardless of the differentiation stage of the cells in this tissue region.

3.6 Origin of Endochondral and Intramembranous Bones

In general, the ability to form bone in the late-embryonic and post-natal state lies in the multitude of cells which populate the bone marrow. In a compartment of the marrow known as the bone stroma, populations of mesenchymal stem cells proliferate and commit to different lineages (osteoblastic, chondrocytic, adipogenic, myogenic, and fibroblastic) while retaining the ability to maintain their own stem cell phenotype. Lineage commitment is regulated by transcription factors, the most important of which are the runt-related transcription factor (RUNX)2 and transcription factor (SP)7 (osterix) (bone lineage), SOX9 (cartilage lineage), and peroxisome proliferator-activated receptor (PPR)γ2 (fat lineage) [121]. Removed from the marrow and grown under appropriate culture conditions, the stromal cells form nodules of bone or cartilage or fat [122]. It is interesting to note that other populations of stem cells in the hematopoietic portion of the marrow have the capacity to commit to the osteoclastic lineage [123]. Accordingly, the bone marrow contains all the cell types critical for bone mineralization as well as for bone turnover and remodeling.

Based on the studies described above, it may be suggested that, while there is an evolutionary separation between cartilage and bone, there appears to be a commonality between chondrocyte and osteoblast lineages. Indeed, as will be discussed further in Chap. 4, many of the genes which were considered to be specific for osteoblasts are now thought to be expressed in chondrocytes [124, 125]. Thus, rather than being viewed as very separate cell lineages, chondrocytes and osteoblasts may or are likely to represent variations of a common cell theme [126]. Likewise, lineage studies reveal similarities between the cells of the tooth (odontoblasts and cementoblasts) and bone. Stem cell lineage in the dental tissues will be discussed in more detail later in this chapter.

If chondrocytes and osteoblasts share a common lineage, then tissues should exist in which there is expression of the characteristics of both cell types. Indeed, in a study of bone repair in zebrafish, cells co-expressed genes of both chondrocytes and osteoblasts [127]. Moreover, the cells were not unlike those found in avascular chondroid bone, a tissue with chondrocytes embedded in a bone-like matrix [127].

The architecture of this peculiar tissue and the types of genes expressed add strength to the notion that cartilage and bone have overlapping origins [128, 129].

The development of endochondral and intramembranous bones differs substantially. Rather than beginning from a cartilage anlage which is ultimately replaced by bone as described above for endochondral bone, intramembranous bone forms in the investing perichondrial/periosteal membrane of the metaphysis and diaphysis. The membrane itself is complex; its outer layer provides biomechanical support particularly during embryonic development, while its inner cambium layer contains progenitor cells having osteogenic potential [130]. Intramembranous bone is formed from such mesenchymal cells populating the dense and nearly avascular collagenous matrix of the cambium layer. Intramembranous bone formation begins when the mesenchymal cells aggregate, proliferate, and commit to an osteoprogenitor phenotype. These cells, osteoblasts, secrete COLI and other molecules to the extracellular matrix to constitute an osteoid, which then becomes mineralized. As osteoid progressively mineralizes, osteoblasts become surrounded and entrapped by mineral and they differentiate into osteocytes (Fig. 3.10) [131]. This form of mineralization is regulated initially by the hypoxic environment of the tissue and probably the presence of anti-angiogenic factors. Transcription factors regulating osteoprogenitor cell proliferation and differentiation include VEGF, IHH, bone morphogenetic protein (BMP), and parathyroid hormone related protein (PTHrP).

The source of intramembranous bone cells has been examined by Colnot et al., who used the capsule of the kidney as a tissue "container" to evaluate bone formation [132]. Very unexpectedly, they found that the perichondrium provided osteogenic cells for both cortical and trabecular bone formation [132]. Very recently, Debnath et al. identified a periosteal stem cell population which formed intramembranous bone [133]. The stem cells were multipotent such that they could commit to other lineages, and they were self-renewing, a basic stem cell property. Moreover, the cells exhibited a distinct transcriptional signature, separate from those of cells which form endochondral bone. Genes expressed by these cells included *RUNX2* and *SOX9* together with stemness-associated genes such as *MYC*, *NANOG*, *WNT5a*, and *Krüppel-like factor* (*KLF*)*4* [133]. It should be noted additionally that a molecular profiling study reported by Duchamp de Lageneste et al. indicated that cells of the periosteum contained periostin, a protein which binds cell membranes through their integrin receptors (αvβ3 and αvβ5) and extracellular proteins like fibronectin and tenascin [134]. Based on these results, Debnath et al. considered that multiple distinct pools of stem cell progenitors existed and that "these periosteal stem cells provide a cellular basis for long-observed differences between intramembranous and endochondral bone formation" [133]. Thus, there is the possibility that osteoblasts are derived from several different sources. During development, they may originate from stem cells resident within blood vessels invading newly forming bone tissue, transdifferentiating hypertrophic chondrocytes, and the perichondrium; during the post-natal state, such osteoblasts may arise from stromal cells in the bone marrow and cells comprising the investing membranes of bone.

As a conclusion to this section of the chapter, it is appropriate to comment briefly on human mature bone. The microstructure of such mature tissue consists in part of

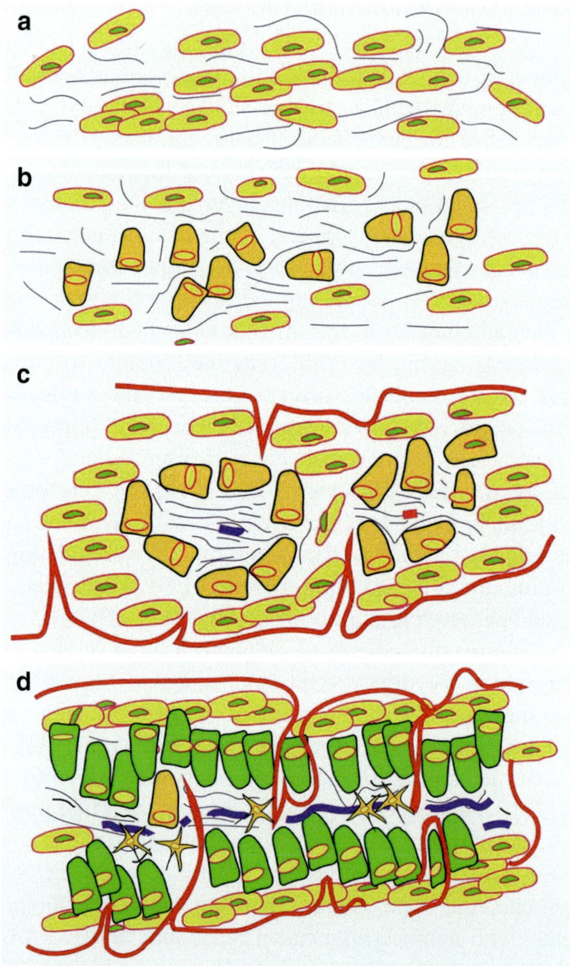

Fig. 3.10 A schematic showing major steps in the formation of intramembranous bone. (**a**) Aggregation and condensation of cells derived from the paraxial mesoderm. (**b**) Differentiation of clones of osteogenic stem cells (brown) which secrete COLI. (**c**) Invasion of blood capillaries (red) into the tissue and localization of osteogenic cells into nests; initiation of secretion of bone matrix components. (**d**) Further invasion of blood capillaries into the tissue and formation of palisades of osteoblasts (green); additional osteoblast secretion of the organic matrix of bone; initial deposition of bone mineral (blue) accompanied by the appearance of osteocytes (stellate cells) and development of a periosteum

a lattice of branching mineralized trabeculae enclosing marrow elements and blood vessels. In long bones, these components are found in the metaphysis (a region lying between the epiphysis and the diaphysis; see Fig. 3.3) as well as along the inner aspect of the diaphysis. Contained within the bone trabeculae are terminally differentiated osteocytes and, decorating the periphery of the trabeculae, are osteoblasts which actively synthesize proteins and other molecules responsible for osteoid formation and deposition of new bone. In some areas of the tissue, large multinucleated osteoclasts are present on the trabecular surfaces. These cells often produce pitting and scalloping of the surfaces, an appearance indicative of osteoclast resorption of the underlying bone tissue.

As noted previously, trabecular bone (also known as cancellous or spongy bone) is distinct from compact bone (cortical or lamellar bone), and this latter type of tissue comprises the outer periphery of the diaphysis and much of the bone mass in mature long bones. Compact bone is the product of a disordered initial bone tissue derived from cells of the perichondrium and subsequently replaced. During the replacement process, concentric rings of lamellae become organized around a central vascular canal to form primary osteons, which in turn are remodeled to yield secondary osteons and Haversian bone [135, 136]. Osteons are approximately cylindrical structures ranging from millimeters to about a centimeter in length in humans and they are arranged in columns with their long axes oriented in parallel fashion with the long axis of their associated compact bone [137]. Osteons vary in their diameter (0.25–0.35 mm in humans) depending on the number and thickness of their constituent lamellae (Fig. 3.11) [137–140].

Within the multiple concentric lamellae of an osteon, osteocytes are embedded in their resident tissue spaces or lacunae (Fig. 3.11). The cells develop long cytoplasmic dendritic processes which are located in channels or canaliculi extended from the lacunae to form an interconnected lacunocanalicular network which traverses the osteons. A quantitative estimate of the network in a cubic centimeter of human osteonal bone has been reported as an astonishing 74 km in length with a total surface area of 215 m^2 in the human body [141]. By accommodating the cells and their finger-like dendrites, the lacunocanalicular network allows the osteocytes to join with one another and with osteoblasts and osteoclasts to facilitate intracellular communication through their processes coupled by gap junctions. In this manner, an enormous volume of bone comes into contact with the syncytium of cells which can rapidly mobilize and transport growth factors, proteins, and other molecules within and about the milieu. Blood vessels comprising the central Haversian canals of osteons (Fig. 3.11) are interconnected by lateral or transverse channels called Volkmann's canals. Further details of osteocyte-osteoblast-osteoclast function are described in Chaps. 4 and 9.

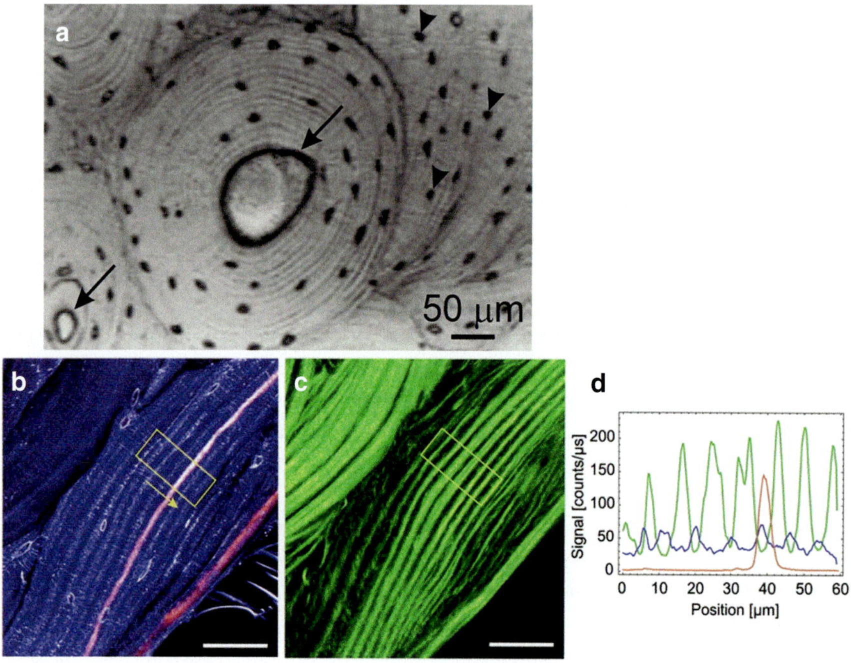

Fig. 3.11 Histological structure of lamellar bone. (**a**) Cross-sectional profile of decalcified, plastic-embedded femoral/tibial bone viewed by polarized light microscopy. The transverse section shows the structure of a number of osteons of varying diameter, each consisting of concentric bone lamellae deposited cylindrically around a central Haversian canal (arrows). Numerous osteocytes (small, black structures; arrowheads) may be found disposed within and following the concentric lamellae. The specimen was obtained from a human osteogenesis imperfecta bone biopsy. (**b–d**) Images of individual lamellae of a bone specimen identified principally by collagen orientation under third- (**b**) and second- (**c**) harmonic multiphoton microscopy together with a plot (**d**) of the signal profiles from a selected sample region. The bone specimen was obtained from the iliac crest of a 6-year-old ewe, injected with calcein during bone growth [138]. The tissue was fixed in 70% ethanol, embedded in polymethylmethacrylate, and examined by third harmonic generation (THG), second harmonic generation (SHG), and fluorescence detection [138]. (**b**) THG (blue) demonstration of bone lamellae accompanied by calcein fluorescence (pink) revealed along specific lamellae. (**c**) SHG (green) image of bone lamellae. (**d**) THG (blue), SHG (green), and calcein (orange) signals plotted simultaneously following the direction of the arrow across the bone tissue (yellow-outlined box in **b** and **c**). THG and SHG signal profiles are anticorrelative and of different amplitude, results which are ascribed to the varying orientation and other particular properties and features of the collagen fibrils comprising individual lamellae [138]. Scale bars: (**b**) and (**c**) 50 µm. (Images adapted from (**a**) Shapiro and Wu [139] and (**b–d**) Doube [140] after Genthial et al. [138] with permission)

3.7 Introductory Comments on the Structure and Formation of the Vertebrate Dental Tissues

Phylogenetic analysis indicates that, in concert with bone, teeth have a long evolutionary history marked by the development of a great degree of structural complexity and a high level of biomechanical strength to accommodate the need for lifetime durability. The anatomical form of mammalian teeth varies from the smooth, permanently erupting incisors of many rodents to the grooved, pitted, and cusp-like occlusal surfaces of ungulates. In all cases, the fundamental mineralized tooth structure is comprised of three tissues, enamel, dentin, and cementum, the latter connecting the tooth to the surrounding "alveolar" bone through a periodontal ligament (Fig. 3.12) [142]. Dentin forms the core tissue of both the tooth crown and root and their respective attachment tissues of enamel and cementum. Contained within dentin is the dental pulp, a soft, vascular, and innervated tissue (Fig. 3.12) [142]. The marked structural, functional, and developmental differences between each of these three mineralized tissues (dentin, cementum, and enamel) provide overlapping insights into the diverse biological processes which regulate vertebrate hard tissue formation and mineralization. With this concept in mind, this section of the chapter reviews pertinent aspects of the architecture and development of the dental tissues in relation to their mineral deposition. The regulatory, mechanistic, and phenotypic aspects of tooth and bone mineralization are discussed in later chapters of this book.

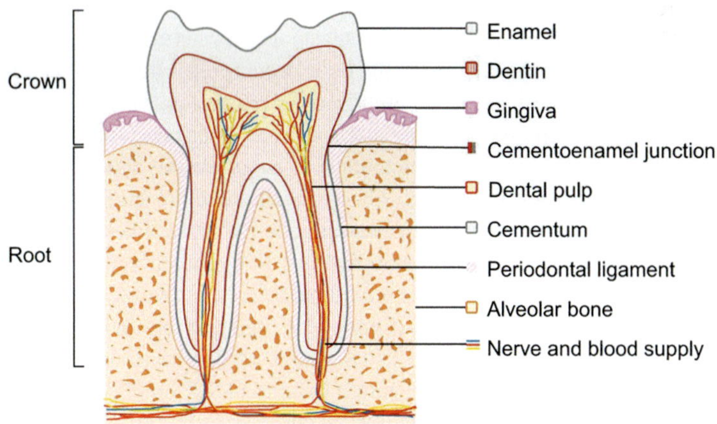

Fig. 3.12 Schematic diagram of a human tooth molar illustrating major features and the location of four composite mineralized tissues: enamel, dentin, cementum, and bone. The periodontal ligament attaches the cementum to the bone. (Schematic from Li et al. [142] based on a figure (Figure 4) created by Jonathan Dimes for the NIH Report, *Oral Health in America: Advances and Challenges* [Internet]. Bethesda (MD): National Institute of Dental and Craniofacial Research (US); 2021 Dec. https://www.ncbi.nlm.nih.gov/books/NBK578299/figure/ch3.fig4/)

3.8 Dentin and Odontogenesis

Dentin is a complex tissue whose morphological appearance and function have undergone considerable evolutionary changes. In Mammalia, the protective neuro-ectodermal odontomes (see Chap. 2) have evolved into dentin and its associated structures of enamel, cementum, and basal bone. These teeth are incorporated into the masticatory apparatus. Even its name has evolved: Rather than Ivory or Tooth Bone, Sir Richard Owen specified that this structure should be "dentine" [143], and the term remained so for about a century until the American literature removed the final "e" and "dentine" became "dentin."

While not as hard as enamel or as mineralized as calcified cartilage, dentin is more mineralized than either endochondral or membrane bone [144]. In humans, other than for individuals who aggressively grind their teeth (bruxism), a cap of enamel protects the dentin layer. In many species, the tooth anatomy is different from that of the human; for example, plates of dental tissues (comprised of dentin, coronal cementum, and enamel) form the occlusal surface of many ruminants. Differential wear on these three tissues produces a serrated surface, which can be used to shear or shred vegetation.

Three types of human dentin have been described: Mantle dentin at the junction of enamel and cementum, primary dentin or circumpulpal dentin which makes up the bulk of the tooth extending from mantle dentin to the central dental pulp, and a relatively narrow region of secondary dentin in the inner (pulpal) aspects of the tooth which reflects a low level of odontoblastic activity following tooth eruption. As a means of introduction to dentin structure, it may be noted that mantle dentin is a layer of tissue ~15–30 μm thick, consisting principally of collagen with additional proteins and other molecules, deposited by odontoblasts as they retreat into the tooth pulp (Fig. 3.13) [146]. Mantle dentin covers the bulk circumpulpal dentin, which is a wide layer around the pulpal wall and also comprised of collagen and associated molecules. Circumpulpal dentin consists of peritubular (or intratubular) and intertubular dentin. Narrow channels or tubules permeate mantle and circumpulpal dentin, accommodating long, extended odontoblast processes, extracellular fluid, and nerve fibers and connecting the pulp with the dentin–cementum junction and the dentin–enamel junction. Tubules in mantle dentin are few in number, thin, and curved compared to the much more numerous, wider, and straighter tubules in circumpulpal dentin [146].

In the pulpal region of the tooth, odontoblasts organize into palisades of cells as they form an initially unmineralized extracellular predentin around the tubules of the tissue (Fig. 3.13). Tubule mineralization is rapid (~5–6 μm/day) [147]. In many mammals, the earliest circumpulpal mineralization is found around collagen fiber bundles (von Korff fibers), which originate in the developing dental pulp and interdigitate between the migrating odontoblasts [148]. The peritubular dentin directly surrounds and forms the walls of dental tubules. Such dentin is densely packed and hypermineralized (~40% greater mineral content than that of intertubular dentin) and the tubule walls have a thin inner organic lining (lamina limitans) with a high glycosaminoglycan content [146]. Intertubular dentin lies loosely packed between tubules, forming the majority of dentin. The initial mineralization of dentin appears

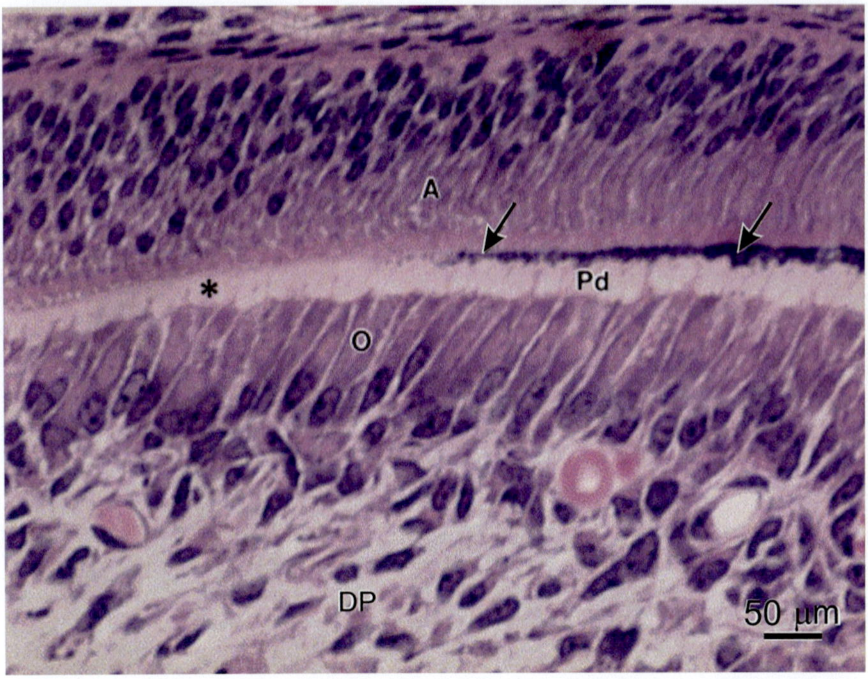

Fig. 3.13 Structural relationship between ameloblasts and odontoblasts in the developing mammalian tooth. Odontoblasts (O), columnar in shape and polarized in the distribution of their components, form a palisade of cells which faces ameloblasts (A) and cells of the dental pulp (DP). As the odontoblasts elongate and retreat into the pulp, they secrete predentin (Pd), which is initially unmineralized (*) and then progressively mineralizes (arrows). Like odontoblasts, ameloblasts form a palisade of long, polarized cells (A). As these cells move away from the odontoblasts, they secrete enamel matrix proteins. (Image adapted from Arana-Chavez and Maasa [145] with permission)

related to the presence of significant numbers of matrix vesicles, especially in mantle dentin, mentioned previously in the context of epiphyseal cartilage mineral formation. Subsequently, mineral is deposited elsewhere in the extracellular collagenous dentin matrix [149, 150]. It may be noted that matrix vesicles are considered to be one of several different mechanisms responsible for mineral formation. Not all tissues rely on a single mechanism and matrix vesicle-mediated mineralization appears to be tissue- and maturation/age-dependent. Matrix vesicles are discussed in greater detail in other chapters of this book. The third dentinal zone borders the pulpal region of the tooth and is evident following tooth eruption into the oral cavity. Odontoblasts in this tissue zone continue a slow (~4 µm/day) secretion of a narrow zone of "secondary dentin" throughout the life of the tooth [151].

To continue, the following defines the formation of the principal dentin type, circumpulpal dentin, and the deposition of apatite into its extracellular matrix comprised largely of COLI. Within the developing jaw, interactions between epithelial cells and an investing neural crest-derived mesenchyme govern and reciprocally induce the appearance of dentin and tooth placement. Signaling molecules like SHH, FGF, and BMP cause differential proliferation of the epithelial cells to form placodes which invaginate the underlying mesenchyme to create a provisional tissue structure, the enamel organ [152, 153]. Maturation of the enamel organ leads to morphological changes in tooth structure, commonly described as progressing through bud-, cap-, and bell-like stages. During these developmental events, the mesenchyme becomes organized to form the dental papilla, the precursor of the dental pulp, and stem cells in the forming pulp commit to the odontoblastic lineage to generate dentin [154]. In this regard, stem cells isolated from the periodontal ligament, dental follicle, alveolar bone, apical papilla, and gingiva, seeded on a mineralized surface in vitro, produce long cytoplasmic processes which resemble those of normal dentin [155]. Further, on a hydrogel scaffold, these cells secrete dentin sialophosphoprotein (DSPP) and dentin matrix protein (DMP)1 as well as other associated molecules [156]. Details of the roles of these proteins in the initiation of mineralization of dentin and bone are discussed elsewhere (Chap. 8). Of the dentin extracellular matrix molecules, COLI is the most abundant while DSPP is the most common non-collagenous component. Changes in any dentin constituent, regardless of its abundance, can lead to profound alterations in the tissue. For example, genetic studies in this respect show that loss of DMP1 causes a more severe tooth phenotype compared to a loss of DSPP [157].

During odontogenesis, molecular signals promote the differentiation of enamel-forming cells (ameloblasts) in the inner enamel epithelium and such cells then secrete enamel proteins. These two processes, amelogenesis and odontogenesis, are dependent on a series of complex feed-forward and feedback regulatory signaling loops [158], involving molecules released from the inner enamel epithelium as well as signals originating from the neural crest-derived mesenchymal cells [159].

The committed post-mitotic mesenchymal cells (odontoblasts) which line the inner surface of the papilla become polarized with the nucleus and many of their secretory organelles, including endosomes and multivesicular bodies, accommodated at the basal end of the cells (Fig. 3.13). The cells elongate and produce highly extended cytoplasmic processes, which are aligned roughly parallel to each other and orthogonal to the dentin–enamel or the dentin–cementum junction (Fig. 3.13). Such processes curve once they leave the basal area of the cell and then undergo a sigmoid-shaped turn which points toward the exterior surface of the developing tooth [145]. Like osteocyte processes, odontoblast processes anastomose with adjacent cells. At their basal ends, they communicate with other odontoblasts and, at their coronal ends, their terminal processes make connections within the enamel-forming layer of the tooth. Surrounding each cytoplasmic process of the odontoblast is a cone-shaped tubule of (peritubular) matrix, noted above, which is of maximum thickness at the cell body and narrows as it reaches the junction with enamel. In these tubules, collagen fibrils are tightly packed and arranged orthogonally to the dentinal processes [160].

The peritubular matrix and the matrix between each of the dentinal tubules become mineralized [161]. Smooth surface matrix vesicle-like structures have been observed as individual organelles or as chains decorating the odontoblastic process (Fig. 3.14). Similar organelles have been reported in the dentin intratubular space [145].

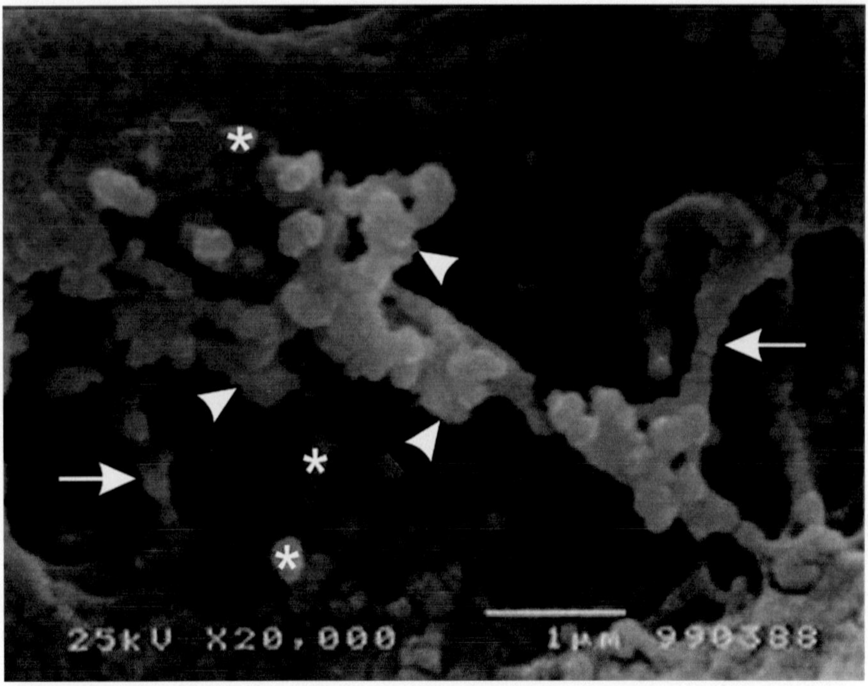

Fig. 3.14 A scanning electron micrograph of human dentin showing an odontoblast process covered with small round particles or vesicles. Such organelles have a smooth surface and appear attached to the wall of the process (arrowheads) or form chains which extend into dentin (arrows). In addition to the attached bodies, clumps of particles are found in the intratubular space (*). The structural features suggest that these organelles are equivalent to matrix vesicles present in other tissues. Scale bar = 1 μm. (Image adapted from Garcés-Ortíz et al. [161] with permission)

Like bone, where secretion of an unmineralized osteoid precedes mineral deposition, the basal end of the odontoblast synthesizes a zone or layer of unmineralized dentin (predentin), which is subsequently mineralized (Fig. 3.15) [146]. As also mentioned previously, the layer of unmineralized predentin is present at the pulpal surface of the forming tooth. The degree of mineralization of the matrix between the tubules is not as great as that of the tubule walls themselves, and peritubular dentin comprises the hardest aspect of dentin (Fig. 3.16) [160].

In summary, dentinogenesis is mediated by interaction of the oral mesenchyme with epithelial cells of the neuroectoderm. Under the control of numerous signaling molecules, the cells involved and the tissue layers they elaborate not only mediate the formation of dentin but also activate events which lead to the development of enamel, the dental pulp, the investing layers of cementum, and eventually the periodontium itself. The formation of enamel and cementum is discussed below while the mechanism of apatite deposition in their extracellular matrices is detailed in Chaps. 5 and 8.

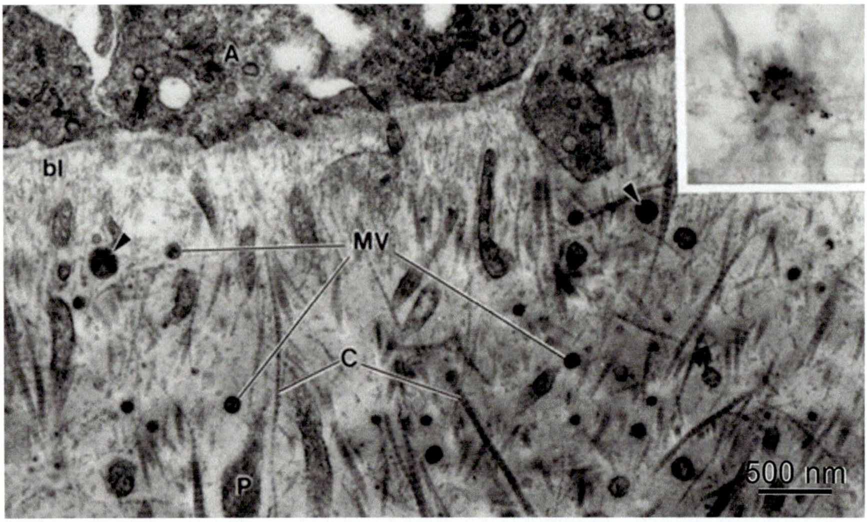

Fig. 3.15 Transmission electron micrograph of mammalian mineralizing dentin matrix. Numerous small vesicles (MV) populate the collagen-rich (**c**) extracellular dentin matrix. Several of such vesicles contain mineral particles (arrowheads). Cytoplasmic processes (P) of the parent odontoblast are shown along the lower border of the photomicrograph and a portion of an ameloblast (A) lies adjacent to degraded basement lamina (bl). Inset: Enlargement of a single matrix vesicle which has been immunostained with an antibody against dentin matrix protein 1 (DMP1). The stain is co-localized with mineral, a result suggesting that DMP1 is associated with the mineral. (Image adapted from Arana-Chavez and Massa [145] with permission)

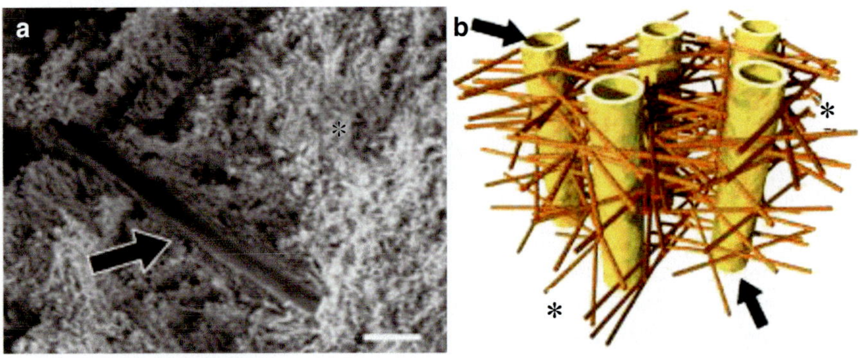

Fig. 3.16 Microstructure of human dentin. (**a**) Scanning electron microscopic image of dentin at a fractured root surface and showing a matrix of mineralized collagen fibers (*) and a portion of a dentin tubule in longitudinal profile (black arrow). The tubule (black) is surrounded by a ring of hypermineralized peritubular dentin, which appears as a white line along the dentin periphery in the image. The matrix of mineralized collagen fibers (*) consists of less mineralized intertubular dentin. Scale bar = 2 μm. (**b**) Schematic diagram of the conceptual microstructure of dentin. Tubules (yellow hollow cylinders; black arrows) are enveloped by mineralized collagen fibers (brown rods, *), oriented principally orthogonally to the tubule axis. The apatite mineral comprising peritubular dentin about the tubules is not crystallographically oriented while the apatite mineral comprising collagen has its crystallographic *c*-axis oriented along the long axis of the fibers [160]. (Images from Forien et al. [160] with permission)

3.9 Cementum, Cementogenesis, and the Periodontal Membrane

Unlike dentin, which forms the bulk of the mineral phase of the tooth, and enamel, which protects much of the tooth crown, cementum comprises a layer of mineralized tissue as thin as 50 μm which is deposited on the cervical region of root dentin in human teeth (Fig. 3.12). This layer of primary cementum is acellular; secondary cementum covers the apical portion of the root dentin and often the primary cementum itself. This more expansive cementum layer is cellular. In mammals, cementum functions to anchor the tooth to the alveolar bone of the jaw through an intermediate, soft, highly vascularized tissue, the periodontal ligament (Fig. 3.12), whose terminal ends of collagen fibers (Sharpey's fibers) insert into the cementum and the alveolar bone periosteum [162]. In other taxa, cementum is absent and the root dentin is directly fused to the bone [163]. From the perspective of joining root dentin and alveolar bone, cementum serves as an enthesis (the attachment site typically connecting ligament or tendon to bone) and thus, from a functional viewpoint, cementum differs from bone, calcified cartilage, calcified tendon, dentin, and enamel among the vertebrate mineralized tissues.

Like dentin, cementum is a product of surrounding mesenchymal cells and the enamel organ. During tooth development, the double-layered epithelium of the enamel organ grows into the neural crest-derived ectomesenchyme to enclose the dental papilla and form the epithelial root sheath (of Hertwig) (Fig. 3.17). The

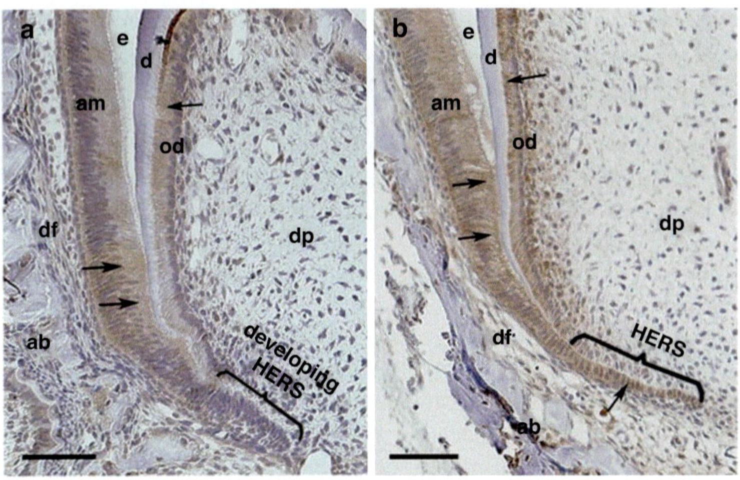

Fig. 3.17 Light microscopy of sections of the developing molar teeth from a mouse and stained with an antibody against insulin growth factor-1 (IGF1). (**a**) Image at 3 days of post-natal tooth development. The Hertwig epithelial root sheath (HERS) is formed when the ameloblast cell layer (am) and outer layers of the enamel organ grow into the dental pulp (dp). Odontoblasts (od), dentin (d), enamel (e), the dental follicle (df) and alveolar bone (ab) are shown. (**b**) Image at 5 days of post-natal tooth development. The bilayered epithelial sheath is fully formed and invaginated into the pulp. The appearance of the sheath is coincident with the start of root formation. Arrows show positive IGF1 staining. Scale bar = 50 μm. (Images adapted from Fujiwara et al. [164] with permission)

growth of the sheath in an apical direction serves as a guide for subsequent root dentin formation [165]. At the end of the growth period, the epithelium becomes fenestrated and signals cementum formation while the remains of the sheath are incorporated into the periodontal membrane [166].

As with many other areas of vertebrate hard tissue research, there has been considerable debate concerning the origin of cementoblasts and cementum. Over 50 years ago and based on histologic analysis, Paynter and Pudy wrote that cementoblasts, the constituent cells of cementum, were derived from the dental follicle, a structure which also provides cells for alveolar bone and the periodontal ligament [167]. Others felt that root sheath cells differentiated into cementoblasts following an epithelial–mesenchymal transition [168, 169]. There was also evidence that sheath cells regulated the differentiation of dental follicle cells [164, 170]. Present considerations suggest that reciprocal interactions between cells of the epithelium itself and the mesenchyme of the follicle more than likely play a critical role in the formation of cementum as well as the periodontal tissues.

Like cementoblast origin, regulation of cementoblast differentiation is also unclear and complex. Chen et al. reported that transforming growth factor (TGF)-β1 and basic fibroblast growth factor 2 (bFGF2) promoted an epithelial–mesenchyme transition of root sheath cells [171]. It is also known that FGF2 targets the secretion of both periostin and fibronectin [172] and that both BMP and dentin sialoprotein (DSP) enhance cementogenesis [173]. Finally, a number of enamel proteins, including ameloblastin (AMBN) and amelogenin (AMELX), participate in the regulation of the phenotype of dental follicle cells to form periodontal structures and cementum [174].

The information above is interesting despite its current inability to provide an integrated concept(s) to explain the origin and differentiation of cementoblasts. In any case, there is general agreement regarding three points of issue: First, the source of cementoblasts lies with both the epithelium of the root sheath and the investing mesenchyme. Second, regulation of the transition between these tissues and cementoblasts and control of cementoblast differentiation are complicated events which involve many genes concerned with dentin and bone formation as well as amelogenesis. Finally, the root sheath needs to be disrupted prior to cementogenesis and the subsequent mineralization of the cementum extracellular matrix [164].

The conjecture over the origin and differentiation of cementoblasts has also given rise to the questions of whether cementum is simply another type of bone or is indistinguishable from dentin. From a morphological viewpoint, like osteocytes, cementoblasts are trapped in lacunae within their own mineralized extracellular matrix. These embedded cementoblasts (or more properly cementocytes) are closer in shape to osteocytes in that they form numerous cell processes but, unlike osteocytes, cementoblasts have few intracellular connections (Fig. 3.18) [175]. From a biochemical viewpoint, there is certainly some similarity between cementum, bone, and dentin. For instance, like bone and dentin, cementum is a collagenous (principally COLI) tissue and cementoblasts, like osteoblasts, secrete a spectrum of non-collagenous proteins, including fibronectin, DMP1, DSPP, and bone sialoprotein

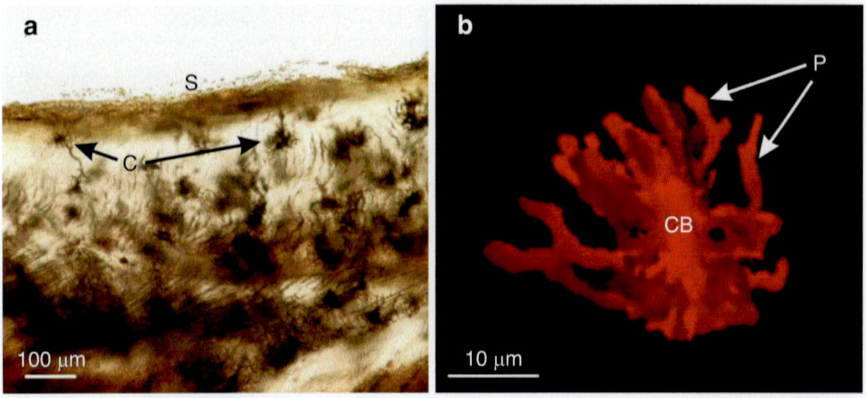

Fig. 3.18 (**a**) Light photomicrograph of a 100 μm thick section of human cellular cementum cut transversely to the cementum surface (S) with a diamond disk. The cells (cementocytes, C) are characterized by numerous processes but few apparent intracellular connections. (**b**) Three-dimensional reconstruction of a cementocyte showing several well-defined cell processes (P) extending from the cell body (CB). (Images adapted from Scivetti et al. [175] with permission)

(BSP) [168]. On the other hand, certain compositional differences exist between cementoblasts and odontoblasts as the former cells secrete their own profile of mitochondrial proteins as well as cellular (non-mitochondrial) proteins involved with various functionalities, such as calcium transport [176]. In addition, the tissue-specific protein, cementum-derived growth factor, an insulin-like growth protein, promotes cell division and enhances the differentiation of progenitor cells into cementoblasts [177]. Another protein, cementum attachment protein, shares some features with collagen, but it binds tightly to apatite and activates progenitor cells to secrete a mineralizing phenotype [178].

A detailed proteomic comparison of cementum and alveolar bone confirms that, while differences exist, their respective extracellular proteins are similar and, in fact, the protein profiles of these two tissues are closer than those of cementum and dentin (Fig. 3.19) [176]. Besides COLI, cementum also contains several proteoglycans (decorin, biglycan, fibromodulin, asporin, lumican, and osteomodulin/osteoadherin) as well as periostin, vimentin, tenascin, and vitronectin [176]. These molecular analyses [168, 175–178] indicate, then, that cementum shares many structural and biochemical attributes with bone and dentin, but that cementum maintains its own particular set of proteins and differs developmentally as well as metabolically from these two counterpart vertebrate mineralized tissues.

As an added note regarding the mineralization of cementum, cementoblasts are reported to produce extracellular matrix vesicles which contain electron-dense apatitic deposits which grow and form calcospherites [179]. The presence and structural nature of such vesicles and mineral deposits resemble those features reported in early bone and dentin (mantle) formation [179]. Apatite is also deposited in a collagenous matrix intrinsic to cementum as it is with bone and dentin collagen [179]. Together, these factors suggest that, while aspects of the biochemistry,

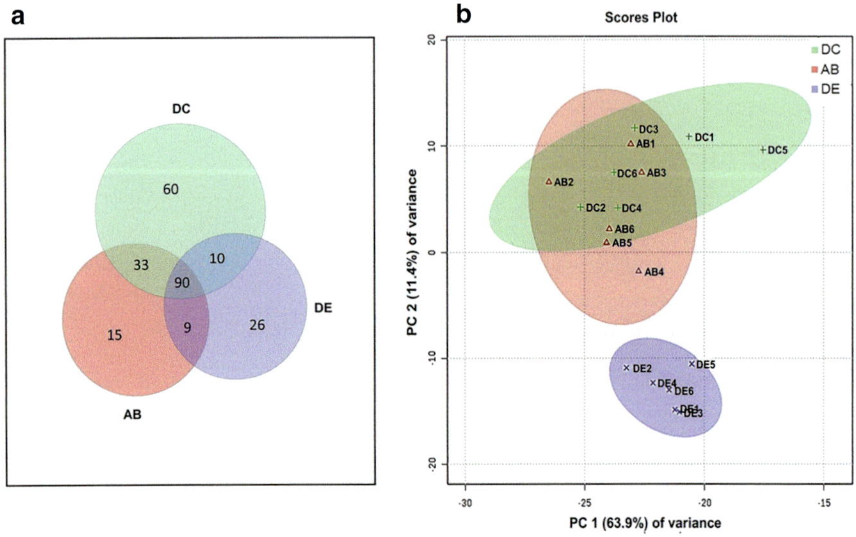

Fig. 3.19 Proteomic profile of cementum (DC, green), alveolar bone (AB, red), and dental enamel (DE, blue) determined by liquid chromatography tandem mass spectrometry following laser capture microdissection of first mandibular molars from a mouse. (**a**) A Venn diagram of 243 identified proteins. Of these, 193, 147, and 135 proteins were associated with DC, AB, and DE, respectively. Ninety proteins (37% of total) were common to all three tissues; enamel contained 26 unique proteins, bone contained 15, and cementum 60. (**b**) The proteomic data transformed using principal component (PC) analysis. This approach reduces the complexity of data sets for each of the three tissues while emphasizing variation. The axes used here are based on the two largest variations in data, 63.9% (PC1) and 11.4% (PC2), respectively. The analysis showed that, while dental cementum and alveolar bone clustered together, a result suggesting that the proteomes of cementum and bone were similar, the enamel proteome segregated into a separate and different domain. (Diagrams from Salmon et al. [176] with permission)

structure, and developmental and metabolic activities differ among cementum, bone, and dentin as just mentioned, the means by which these three tissues elaborate mineral through matrix vesicles and/or collagen appear to be common.

3.10 Enamel Mineral, Amelogenesis, and Proteins of the Enamel Matrix

In contrast to bone, dentin, and cementum, mature enamel is an acellular neuroectodermal tissue which provides teeth with a protective barrier against attack by physical and chemical agents. From a mechanical viewpoint, enamel is by far the hardest and most densely mineralized of the vertebrate tissues [180]. Moreover, despite considerable speculation by patients and dentists in the earlier part of the last century that this tissue contains nerve cells, mineralized enamel is a non-vital structure. In stark contrast to bone, dentin, or cementum, there are no living cells in enamel. Indeed, enamel consists of apatite (~95%), organic constituents (~4%), and water (~1%) [180]. The hardness of this rock-like tissue is in part attributable to its extremely high mineral content and unique organization. Measured on a scale which compares the ability of one material to scratch another (the Mohs Scale), the hardness of enamel is greater than bone and even steel, but less than carborundum or diamond. Thus, to cut through this highly mineralized tissue, dentists employ diamond-encrusted drill bits. In engineering terms, enamel is a bioceramic and it functions in the tooth to protect its underlying dentin from the weak acids generated by the oral flora and from mechanical damage caused by occlusal (biting) forces.

The structure of enamel is complex and its microanatomy has eluded scientists for more than two centuries. However, from seminal work of Boyde, Reith, and Slavkin and their collaborators, there is now a detailed understanding of enamel structure and biophysical properties, although the mechanism(s) by which the tissue becomes mineralized is incompletely understood [181–183]. A notable feature of enamel is that its composite apatite mineral crystals are very large. In contrast to the apatite crystals of other vertebrate mineralized tissues which are discrete, thin platelets about 40–60 × 20 × 2 nm in length, width, and thickness, respectively, enamel crystals form long, thin, somewhat flattened ribbons [181, 184, 185]. As enamel matures, its crystals increase in width and thickness and assume an irregular hexagonal shape when viewed end-on by electron microscopy [185]. The thickness of crystals varies with their distance from the dentin–enamel junction. With greater distance, Boyde and Bromage estimate a doubling of crystal thickness from approximately 50 nm to 100 nm [150]. The continuously lengthening crystals may eventually span the full breadth of the enamel layer of the tooth (Fig. 3.17).

Enamel crystals are aligned with their crystallographic *c*-axes (along their length) generally parallel to one another, and together the crystals cluster to form extensive three-dimensional cylindrical rods, referred to as prismatic enamel. Each rod is surrounded by tightly packed interrod crystals of interprismatic enamel; the *c*-axes of these crystals are also in parallel alignment but oriented at approximately 60° with respect to the *c*-axes of the rods, an angular direction which is clearly distinct from the rod direction. An interface between adjacent, packed rods and interrods is reported to exist as a narrow organic sheath found in early forming enamel of most vertebrates [181]. The sheath encompasses the rods of enamel crystals over most of the rod periphery with the remaining portion at the apical ends of the rods consisting of an intermingled continuum of rod and interrod crystals. In mature enamel, a well defined interface is evident between adjacent, packed rods and interrods, but there is little evidence for any proteinaceous, organic, sheath-like material [150]. A recent study by Koldehoff et al. failed to identify an organic layer and showed that, within the boundary region between rods and interrods, the enamel crystals were in direct contact with each other, fused together, or separated by gaps [186].

Enamel is elaborated from its origins at the dentin–enamel junction and extends to the outer surface of the tooth. Unlike bone (osteoid) or dentin (predentin), enamel produces an extracellular matrix which begins to mineralize immediately as it is secreted by ameloblasts. Thus, there is no layer of unmineralized enamel. As ameloblasts move toward the crown of the tooth, the cells appear to act in tandem as sheets to regulate the synthesis of alternating rows of rod and interrod mineral bundles and establish a unique pattern of mineral deposition. In this instance, cells comprising the ameloblast sheets move in both the direction in which they propel themselves by secreting enamel and a different direction relative to their neighbors within the plane of the sheets. The dual cellular movement gives rise to a complex, albeit highly organized, crisscrossing of mineral bundles. This regularly alternating orientation of rod and interrod bundles is described as decussation (after the Roman numeral, "decus," X). The pattern of decussation is species-specific and provides enamel with a hierarchical microstructure which is extraordinarily strong, hard, and crack-resistant.

The formation of the complex structure of enamel rods and interrods begins with the synthesis and secretion of specific proteins implicated in enamel development. This cell-mediated activity is known as amelogenesis (Old French, "amelo" or "enamel") and has been defined principally by studies of the rodent incisor. While amelogenesis in the rodent tooth is limited in the sense that it does not necessarily define or precisely represent the identical events in humans, selected and critical aspects of the process in rodents recapitulate steps in enamel formation in the human tooth. Thus, in both rodents and humans, ameloblasts are responsible for the synthesis and secretion of enamel matrix proteins and the maintenance of the enamel extracellular milieu conducive to mineralization.

Enamel formation takes place in the enamel organ and is a classic example of interaction and crosstalk between mesenchymal cells of the developing tooth and the neural crest ectoderm. Indeed, without instructional signals from the mesenchyme, enamel formation cannot take place. Once initiated, the process of amelogenesis can be divided functionally into three major stages: presecretory, secretory, and maturation. The presecretory stage of amelogenesis includes cellular morphogenetic and differentiation sub-stages. The secretory phase involves the development of ameloblast cytoplasmic extensions, the Tomes' processes, through which extracellular matrix proteins are transported and secreted, leading to the deposition of mineral. The maturation stage is characterized by changes in cell morphology including the appearance of long, polarized ruffle-ended and smooth-ended ameloblasts and the elaboration of a mineralized extracellular matrix. Details of each of these stages are presented below.

Within the inner enamel epithelium of the developing tooth, the presecretory ameloblasts appear as cells with a prominent nucleus located centrally in the cytoplasm. These cells are separated from the adjacent dentin matrix by a basal lamina containing abundant collagen fibrils and odontoblastic processes. On their differentiation, ameloblasts elongate into a columnar cell shape which exhibits a proximal shift in the location of the nucleus and a distal movement of an enlarged complex of Golgi saccules and other organelles. Following degradation of the basal lamina, the formation of a junctional complex occurs and defines a distal projection of the cell called the Tomes' process. At this projection, there is exocytosis of vesicle-containing proteins, as noted above. As differentiation proceeds, ameloblasts become polarized in terms of the locations of their nuclei and cytoplasmic constituents. The cells are no longer capable of dividing, and they sparingly secrete proteins such as AMELX (described below) as well as small amounts of DSPP (also secreted by odontoblasts). The ameloblasts tightly associate with each other to form a palisade of cells interconnected through their junctional complexes. Electron microscopy of the distal aspects of the Tomes' processes provides the first evidence for the deposition of enamel as dense patches formed within the adjacent collagenous dentin matrix and at the sites of basal lamina degradation and infolded ameloblast membranes [159, 187, 188].

The initial secretory stage of amelogenesis is marked by increases in the volume of ameloblast cytoplasmic organelles, notably the endoplasmic reticulum and Golgi. Concurrently, a similar elevation occurs in the number of membrane-invested secretory vesicles which traffic to both proximal and distal domains of the Tomes' process. These two Tomes' regions release their contents into the extracellular matrix and separately regulate the deposition of enamel interrods and rods, respectively.

The proteins secreted by ameloblasts and comprising the enamel matrix are surprisingly limited in their number. Amelogenin is the principal enamel protein, encoded by two genes. The allele on the X chromosome is strongly expressed (~90%) and required for enamel mineral formation while the allele on the Y chromosome is expressed at a very low level (~10%) [189]. Hence, the amelogenin gene is conventionally written as *AMELX* [190]. The enamel matrix also contains lesser amounts of AMBN [191, 192], enamelin (ENAM) [192, 193], amelotin (AMTN) [194, 195], and odontogenic ameloblast-associated protein (ODAM) [196, 197]. These proteins belong to the secretory calcium-binding phosphoprotein gene family [198, 199]. An additional protein found during the secretory stage of amelogenesis is a tooth-specific matrix metalloproteinase (MMP20) and maturing ameloblasts secrete kallikrein-related peptidase (KLK)4 [200]. These two enzymes cleave other enamel proteins [192, 200]. Each of these proteins appears to be highly regulated with regard to its location in enamel and its duration in amelogenesis, and each is associated functionally with aspects of mineral formation, growth, and organization [201]. Keratins, proteins intimately associated with the skin, are also secreted by ameloblasts and their presence lends considerable strength to developmental studies which indicate that enamel is an ectodermal tissue [202].

The relationship between secretion of the proteins above and enamel formation at rod and interrod sites is characterized by several features. First, there is a continual production of secretory granules which traffic from the ameloblast body to the Tomes' process with constant exocytosis and release of granule (protein) contents into the extracellular milieu. Second, the formation of interrod enamel precedes that of rod enamel and creates an initial cavity site for the developing rod crystals. Third, the movement of each ameloblast occurs synchronously with its neighbor and ameloblasts migrate successively farther from the dentin–enamel junction, a process which causes a constant lengthening of interrod and rod components and the formation of characteristically long ribbons of the mineral. Lastly, as ameloblasts retract from the dentin surface, their distal Tomes' regions narrow while they simultaneously lengthen, a change which leaves the Tomes' infolded membranes compressed within mineral which has already been elaborated. In early enamel, the result of these transitions in structure ultimately yields rod crystals which are enclosed in membrane sheaths. This type of mineral deposition does not appear to involve matrix vesicles.

By the end of the secretory phase, the enamel crown of the tooth has been formed although it is only partially mineralized and, in bovine species, has the consistency of hard cheese at this stage. The events which occur during the maturation stage of amelogenesis involve growth in thickness of the enamel crystals, mineral ripening, and protein removal from the extracellular spaces of the tissue. To accommodate mineral growth, proteases degrade the extracellular organic matrix and water is removed from this tissue region. In terms of cellular changes, ameloblasts become shorter in length, devoid of Tomes' processes, and reduced in the number and volume of their constituent organelles. The cells also slow their biosynthetic activity, a decrease which leads to a lowered expression of their unique enamel genes and protein products. In contrast, expression of genes involved with activities which include ion transport and proteolysis is elevated. At the end of this period, many of the ameloblasts are lost, probably through activation of apoptosis, a change mediated by high levels of environmental calcium and phosphate ions [203].

During the maturation stage of amelogenesis, ameloblasts undergo relatively rapid synchronous cyclical changes in which their apical surface transitions between highly infolded or ruffled and smooth cell membranes. These transitions are thought to enhance the influx of mineral ions to interrod and rod enamel necessary for further crystal growth (ruffle-ended ameloblasts) and/or provide a mechanism for removal of organic material and water from the enamel matrix. This latter process is required for progressive crystal development. The cyclical and repetitive modulations of maturation-stage ameloblasts are accompanied by changes in pH and the action of MMPs which degrade enamel proteins and mobilize calcium and other ions. During this stage, the cells are loaded with intracellular vesicles, many of which are positive for calcium [204]. Variations in the tightness of the sealing zones have been linked to ameloblast metabolism and water and ion transport [201]. For ruffle-ended ameloblasts, their distal tight junctions would be expected to limit access to ions in the tissue fluid whereas the junctions would be leaky or disappear for smooth-ended ameloblasts. This change would likely facilitate movement of ions from the tissue fluid to the forming enamel and from the mineralizing enamel back into the systemic circulation. Both ruffle- and smooth-ended ameloblasts allow the passage of acid-degraded or enzymatically digested cellular components as well as water into or out of the cells [201]. With the culmination of the maturation stage, the ameloblasts enter a so-called protective stage of amelogenesis in which they transition to cells of a smaller and more compact nature. During this phase, the cells display a centrally located nucleus and maintain a level of cytoplasmic organelle activity which supports the existing highly mineralized enamel matrix.

A further discussion of the form, function, and mineralization of enamel involves a basic understanding of its component genes and proteins. In this regard, certain aspects of its principal constituent will be described below and additional details of other important molecules mediating amelogenesis will be presented in subsequent chapters of the book. As was noted above, the major encoded protein of enamel is AMELX, which accounts for 80–90% of the secreted proteins of this tissue [205]. AMELX is critically involved in the growth and development of enamel although its precise role remains incompletely understood. Depending on the species examined, AMELX consists of 170–180 amino acid residues, of which 13 are charged and found within the peptide sequences comprising the amino- and carboxy-terminal ends of the intact protein [206]. Native AMELX has a single phosphorylated residue, serine, at position 16 of the sequence. The protein is predominantly hydrophobic and maintains a high degree of homology across species [206]. The central region of AMELX contains elevated histidine, glutamine, and proline residues. AMELX may interact with calcium and phosphate ions or apatite, itself, through its charged residues at the termini of the molecule as well as its histidine residues (when they are protonated) in the central portion of the molecule [207].

AMELX is reported to self-assemble to form higher-order structures which appear microscopically as nanospheres, small, rounded particles ~20–70 nm in diameter; such nanospheres are suggested to assemble in arrays which define the space for the formation and subsequent growth of the long linearly arranged rods of enamel (Fig. 3.20) [206, 208–214]. While the interior of the nanospheres is thought to be hydrophobic, their exterior surface is considered hydrophilic and acidic, consisting of the carboxy-terminal end of AMELX with its charged amino acids suggested to function in controlling nanosphere aggregation [215]. When studied in solution under normal physiological conditions, the single phosphorylation of native porcine AMELX appears to modulate a transition from the nanosphere structure of the protein to one of an extended chain and, in addition, stabilizes an early amorphous calcium phosphate phase of mineral [216]. Porcine AMELX which is unphosphorylated in related solution experiments leads to the formation of crystalline apatite [217].

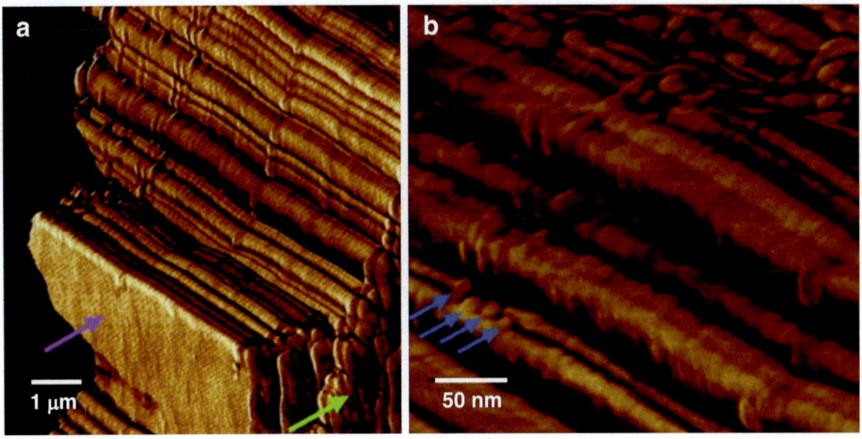

Fig. 3.20 Developing enamel from a pig mandibular molar examined in the tapping mode of an atomic force microscope. (**a**) A phase image (1 × 1 μm scan field) showing several plate-shaped crystallites of enamel apatite in parallel arrangement (purple arrow) and associated organic matrix (green arrow). The (100) face of apatite is thought to be most prominently observed in the image [185]. (**b**) An enlargement of a phase image (500 × 500 nm scan field) shows presumptive amelogenin nanospheres (blue arrows) aligned along the side faces of and between enamel crystallites. Such nanospheres are suggested to provide the organized microstructure for the initiation and oriented growth of enamel crystals. During early stages of enamel formation, the amelogenin is also thought to envelope the crystals to inhibit premature crystal–crystal fusion [185]. (Images adapted from Moradian-Oldak [185] with permission)

A structural model of AMELX different from that of nanospheres has been presented relatively recently in which the protein is proposed to self-assemble into thin nanoribbons, ~17 nm wide and ~3–4 nm thick (Fig. 3.21) [201, 218]. The origin of such nanoribbons is suggested to reside in AMELX dimers stabilized by calcium and phosphate ions to form anti-parallel β-sheets which resemble amyloid fibers [218–221]. Based on studies in vitro, the nanoribbons are thought to serve as organic templates binding calcium and phosphate ions to develop mineral in the form of narrow threads which mimic nascent enamel crystals [201, 218]. Possibly, vesicles at the Tomes' process of ameloblasts putatively secrete AMELX as dimers which immediately self-assemble into initial nanoribbons in the extracellular milieu. In the presence of calcium and phosphate ions, nanoribbons accrete additional dimers as AMELX secretion continues, and the ribbons lengthen while mineral begins to develop along them and within very close distances to the Tomes' membranes (Fig. 3.21) [201]. Moreover, native AMELX is insufficient to lead to direct apatite deposition and may require the presence of other enamel proteins, including MMP20, to assemble and organize mineral in association with the AMELX nanoribbons [201]. In a related context and as mentioned briefly above, it is important to emphasize that enamel has effectively little or no extracellular space which remains unmineralized. This feature of enamel is in marked contrast to the unmineralized osteoid of bone and the predentin region of dentin. Mineral deposition in enamel, then, occurs rapidly at the ameloblast–extracellular matrix interface, a notable distinction between this tissue and osteoid and predentin. The transport of calcium and phosphate ions, critical in mediating the formation of nanoribbons and mineral in the events just described, is a topic discussed in more detail in later chapters of this book.

Native full-length AMELX may be experimentally cleaved and spliced to generate various molecules which can be tested for their resulting functional properties. One such common splice variant is leucine-rich amelogenin peptide (LRAP), which consists of the 59 amino acid residues of the charged amino and carboxy-termini of AMELX [222–224]. Discovery of this transcript led to considerable excitement, particularly in terms of its possible role in mediating mineral deposition in the enamel extracellular matrix, inducing osteogenesis, and directing cementoblast activity [225, 226]. On the other hand, an additional study of potential LRAP function was made by Chen et al., who developed a transgenic mouse which secreted the protein in the absence of *AMELX* [227]. In this *AMELX* knockout model, LRAP did not restore the mineralization defects caused by the loss of AMELX. The investigators concluded that, "LRAP does not appear to have a substantial role in mineral formation," although the study hinted that there was "the intriguing possibility of signaling or other function to be determined" [227]. However, such putative roles of LRAP have not yet been documented. Indeed, subsequent work has reported that LRAP up-regulates *AMELX* transcription and, when it is phosphorylated, LRAP both stabilizes amorphous calcium phosphate and enhances mineral deposition [228, 229]. At the present time, these investigations, together with those of another splice variant, M180, the most abundant AMELX isoform in mice consisting of 180 amino acids [230–232], demonstrate several different effects concerning the restoration of defects in enamel, including aspects correcting apatite crystal and prism structure [231, 233].

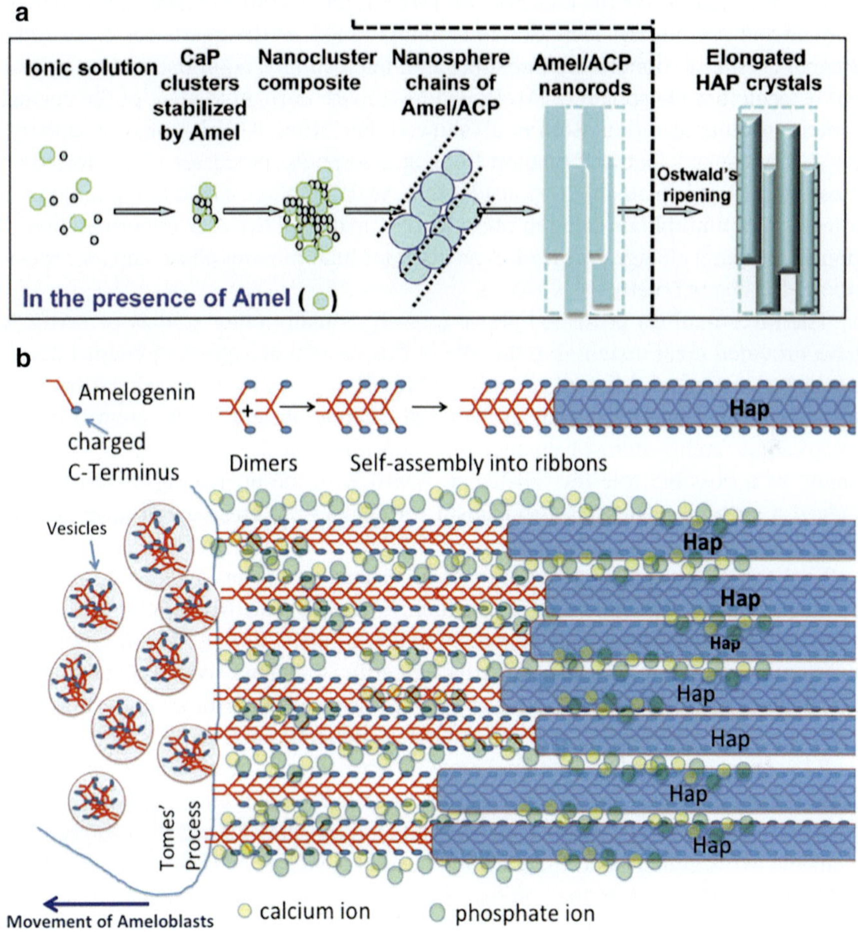

Fig. 3.21 Schematic diagrams, as conceptualized by Lacruz et al. [201], comparing the current principal models of apatite formation of enamel during the secretory stage of amelogenesis. (**a**) In the model of apatite deposition directed by amelogenin nanospheres, amelogenin (Amel/AMELX) is initially secreted from ameloblasts to the extracellular matrix where it forms nanospheres. Nanospheres in the presence of extracellular calcium and phosphate ions in the extracellular milieu interact and stabilize the ions as prenucleation clusters (nanocluster composites). Alignment of nanospheres into chains subsequently occurs while the ions transition into amorphous calcium phosphate (ACP) in the shape of nanorods. Through Ostwald ripening, the nanorods become crystalline hydroxyapatite (HAP). In this model, the nanospheres of AMELX serve as spacers to separate the ACP nanorods in early events of mineral formation [210, 214]. (**b**) In the model of apatite deposition directed by AMELX nanoribbons, AMELX is initially secreted through vesicles from ameloblasts to the extracellular matrix. Rather than nanospheres, AMELX takes the form of dimers which, on secretion, self-assemble in an anti-parallel fashion as nanoribbons bridged by calcium and phosphate ions. Additional secreted dimers extend the length of individual nanoribbons, whose molecular structure acts as a template for calcium phosphate growth and rod-like development ultimately as hydroxyapatite [218, 219]. (Schematic from Lacruz et al. [201] with permission)

In other examples of the knockout of *AMELX*, the resulting enamel is found to be thinned and profoundly decreased in hardness [234]. At the microscopic level, the enamel crystallite ribbons are indistinct and, interestingly, comprised not of apatite but octacalcium phosphate [234]. Providing a more normative view of the enamel protein in mineralization, Shin et al. showed that, when AMELX was phosphorylated, it enhanced apatite formation from an amorphous precursor [235]. Based on these studies, it is plausible to consider that AMELX plays a role in regulating the transition of amorphous calcium phosphate to apatite [236]. Any abnormal disruption with such a change will lead to an aberrant final mineral phase and its deposition in the enamel matrix.

The investigations presented above as well as many other studies of AMELX have provided great insight into the role of this protein in aspects of enamel development. A principal function of native AMELX appears to be its critical involvement in normal apatite crystal growth and orientation and in the control of the microscopic architecture of enamel rods and interrods [231]. However, both the nature of a possible role for full-length AMELX in the nucleation of apatite and indeed a specific nucleation mechanism for enamel are elusive and have not yet been determined.

Like the poorly understood mechanism of enamel crystal nucleation, how the enamel matrix may regulate mineralization is also still far from clear. Mazumder et al. suggested that fragments of enamel matrix proteins form a "mesh" around the developing mineral which "supports the integrity of enamel rods, maintaining the hierarchical microstructure of enamel" [237]. Besides establishing an environment promoting the organization and maturation of mineral, such protein fragments as well as another co-localizing enamel protein (AMBN) have been considered to modulate and enhance the transition of nascent nucleated mineral into apatite crystallites [216, 228]. As enamel crystallites develop and mature, they fill and then outgrow the available extracellular matrix space comprised of AMELX and other proteins. With continued crystal growth, the volume of these proteins must be reduced to accommodate ever-increasing crystallite size. In this context, ameloblasts at the secretory stage of amelogenesis secrete MMP20 to cleave AMELX, LRAP, AMBN, and ENAM [238, 239]. Additionally, during the maturation stage of amelogenesis, KLK4 hydrolyzes any partially degraded AMELX, AMBN, and ENAM [240]. These two enzymes, MMP20 and KLK4, are responsible, then, for creating the space required for further enamel crystal development through degradation of extracellular matrix proteins. As would be expected, deletion or mutation of the genes which encode these two enzymes causes severe enamel defects [200, 241, 242].

From all these perspectives, it is clear that enamel formation is completely different from that of bone or the other vertebrate hard tissues which are elaborated when mineral is deposited in association with matrix vesicles and collagen fibers. Indeed, the only evidence of collagen participation in the mineralization of enamel is at a very early stage when mineralized collagen fibers are present at the dentin–enamel interface [243]. The structure of a number of the enamel-specific and non-collagenous proteins noted above and their roles in amelogenesis and enamel formation are discussed in detail in Chap. 8.

3.11 Perspective: Form and Function of Tissues which Undergo Mineralization

Numerous studies now confirm that bone is formed developmentally in three ways: first, by an endochondral process wherein bone cells replace a cartilage anlage; second, by osteoblasts originating in the perichondrial membrane and producing intramembranous bone; and third, through the activities of neural crest-derived mesenchyme cells. These latter cells are responsible for bone located mainly in the head and neck regions of the skeleton as well as tissues of the tooth. These three principal means of bone development are supported by a number of different stem cell pools in the growing embryo which can commit to the osteogenic phenotype and form the 270 bones which comprise the human skeleton [133]. Interestingly, from investigations of bone repair, it is likely that these stem cells exhibit some plasticity in that they can co-repress genes for either chondrocytes or osteoblasts [114, 244].

The process of bone formation of a limb, for instance, depends on the invasion of blood vessel sprouts into the cartilaginous anlage of the tissue and subsequent removal of hypertrophic chondrocytes. Associated with such vessel invasion, perichondrial osteogenic precursors migrate into and populate the anlage. Within the regions of bone which will become its primary ossification center, these perichondrial cells commit to the osteogenic phenotype and replace the cartilage with trabecular bone while at the same time generating the presumptive structure of the bone marrow.

At a later time in bone development, vessel invasion is also important but, in this instance, not with the cartilage anlage but with the epiphyseal growth plate of the tissue. Such a circumstance elicits the controversial question of the fate of hypertrophic chondrocytes during formation of the limb bones. The two opposing concepts of the disposition of these cells pit the view of the hypertrophic chondrocyte as a mediator of osteogenesis (pro-life) against the view that the hypertrophic chondrocyte dies and is replaced by bone cells (pro-death). Lineage tracing experiments support the latter notion, that a portion of these cells in bone epiphyses survives and transdifferentiates into osteoblasts. In this case, the common usage which describes the hypertrophic chondrocyte as a terminally differentiated cell would need to be revised. Tsang et al. have commented that, regarding hypertrophic chondrocytes, “the hypertrophic state is just their last stop along the chondrocytic lineage and represents a turning point to begin a new journey within the growing bone. They may then re-enter the cell cycle and/or become osteo-progenitors before adopting another “terminal” state as osteoblasts/osteocytes or even adipocytes” [245].

As just noted, the idea that cells transdifferentiate clashes with the long-standing pro-death fate of the terminally differentiated cell. Indeed, as a counterpoint to transdifferentiation, a very recent morphological study by Pazzaglia et al. shows that hypertrophic chondrocytes in rabbits die and trigger invasion of hematopoietic cells [246]. Osteoblasts appear within the cartilage 70 μm below the vascular invasion line, and no evidence of mitoses or commitment to a particular lineage is found in the late hypertrophic cells [246].

The conflict between these two concepts may reach some resolution through the following consideration: The recent major approaches in lineage tracing and morphological analyses used to investigate hypertrophy of growth plate cartilage may represent sampling of different populations of hypertrophic cells. Lineage tracing provides a kinetic view of events in the growth plate while morphology offers a detailed image snapshot of the one-time status of the hypertrophic cells of the plate. The respective data gathered are necessarily different and perhaps complementary in that some cells may undergo transdifferentiation and others suffer death. Thus, lineage tracing gives evidence of a transformative event which affects a defined population of chondrocytes in growing bone. Yang et al., for instance, estimated that transdifferentiated cells numbered approximately 60% of all mature osteogenic cells [112]. The structural studies would reflect the conditions of the remaining 40% of non-transdifferentiated cells, which die and become replaced by osteogenic cells associated with the invading vasculature. In this regard, Maes et al. suggested that the late-stage hypertrophic chondrocytes may secrete chemoattractants which promote the movement of osteoblast precursors and endothelial cells to the bone [247].

Aside from its role of delivering stem cells to bone, vascularization provides nutrients and oxygen to the metaphysis, the endosteum, and the growth plate of this tissue. The anatomical distribution of blood vessels, capillaries, and sinusoids establishes areas of varying oxygen tension which would be expected to govern the direct metabolic and anabolic activities of resident cells or cells moving to bone. For example, the low oxygen tension in the bone marrow cavity, growth plate, and diaphysis would enhance glycolysis and inhibit respiration of resident populations of osteoblasts and chondrocytes [53]. It is likely that hypoxia would also influence the commitment of osteoprogenitor cells to a defined phenotype and promote terminal differentiation of osteoblasts to osteocytes [248]. In later chapters of this book, these vascular effects are discussed with regard to regulation of chondrocyte death and matrix vesicle synthesis.

The presence of an initial cartilaginous phase in bone raises the question as to why this transient tissue is needed for vertebrate bone deposition. One possibility is that cartilage and bone evolved separately. Studies of the fossil record provide some support for this concept in that a calcified cartilage phase, rather than bone, may have existed as a separate skeletal phase (see Chap. 1). Thus, an evolutionary stage may have occurred during which the skeleton of some animals was comprised entirely of calcified cartilage. Since this tissue has few of the mechanical advantages of bone, it may have been modified over time to form the anlage, which was then replaced by bone as a means to achieve a greater mechanical response. An additional explanation to the question is that the cartilage anlage affords the developing bones of the fetus with a structure which can withstand the biomechanical forces applied to the tissue complex during embryonic life. In post-embryonic life, osseous tissue is better suited to the mechanical forces experienced by the skeleton.

While endochondral bone predominates in the vertebrate appendicular skeleton, cells within the inner aspect of the perichondrial/periosteal membrane form intramembranous bone [131]. This form of mineralization is especially common in the flat bones of the skeleton. In the post-natal state, the periosteum preserves its osteogenic activity, and such activity is evident in bone fracture repair. During such healing, a cartilaginous "callus" recapitulates some of the osteogenic events occurring in endochondral bone formation, while periosteal osteogenic cells complete fracture repair by promoting the synthesis of compact bone.

Besides the cells of the lateral and paraxial mesoderm which form most of the bones of the axial and appendicular system, the other source of osteogenic cells in vertebrates resides in the neural crest. From this neural tissue, osteogenic cells migrate to generate both bone and dentin and, from an evolutionary perspective, provide the "armor" of dermal denticles for both extinct and extant organisms [249]. In this regard, Cebra-Thomas et al. have commented that trunk neural crest cells contribute to the ventral plastron (underside shell) bones of the turtle [250]. Unlike turtles and other diapsids, osteogenic activities of these neural crest-derived cells are restricted in mammals to the bones of the head and neck and tissues of the tooth as noted previously. Not surprisingly, defects in neural crest development lead to diseases of the skeleton, many of which cause craniofacial deficiencies as well as malformations of dentin and enamel [251].

With respect to dentin formation, the process of odontogenesis appears to be quite different from that of endochondral or intramembranous bone formation. Since teeth develop from ectodermal structures or placodes, an initial or early mesenchymal anlage is absent. Further, odontogenesis is dependent on enamel formation; there is no turnover or remodeling of dentin, and post-natal growth is limited to a region of secondary dentin formation located in the inner aspect of the tooth. Despite the developmental differences just presented and the fact that the highly organized tubular structure of dentin is morphologically distinct from that of endochondral or intramembranous bone, there are several similarities between dentin and bone and also between dentin and calcifying cartilage. Like bone and its osteoid, there is an organic matrix in dentin, the predentin, comprised of secreted proteins and other molecules necessary for mineral formation. The mineral phase of dentin, like bone and calcifying cartilage, is apatitic and the organic matrix of both bone and dentin consists predominantly of COLI together with a number of non-collagenous, less abundant constituents, including SIBLING (a small integrin-binding ligand, N-linked glycoprotein) proteins. Finally, like bone and calcifying cartilage, dentin mineralization is collagen-dependent and mediated in part by the early secretion of extracellular matrix vesicles.

Cementum bears numerous similarities to both bone and dentin. From a biochemical viewpoint, cementum is comprised of COLI complemented by several non-collagenous proteins, including the SIBLINGs, tissue non-specific alkaline phosphatase, and a host of proteoglycans as noted above [252, 253]. Cementoblasts in cementum are the cellular equivalent to osteoblasts in bone or odontoblasts in dentin in that they synthesize and secrete these proteins and other molecules to an extracellular matrix which mineralizes. Cementoblasts also secrete extracellular matrix vesicles in early cementogenesis, an event which is comparable to the secretion of matrix vesicles by osteoblasts in early osteogenesis or by chondrocytes in calcifying cartilage [254]. Cementum thickens by appositional growth in the absence of a cartilage anlage or a growth plate [255]. Cementoblasts become embedded in the forming extracellular cementum milieu, modulate their function as cementocytes residing in individual lacunae, and populate the now mineralized tissue much like osteocytes in bone. However, cementocytes occupy cementum with many fewer cells per unit volume than osteocytes in bone, and some regions of cementum are almost entirely acellular as well as avascular [256]. While their respective cell density is relatively low with respect to that of osteocytes in bone, cementocytes develop extended cell processes (Fig. 3.18), structures whose presence suggests cell–cell signaling and communication [175], like that found among osteocytes in bone. However, cementocytes have few apparent intracellular connections [175].

While the commonalities noted here exist between cementum, bone, and dentin, the principal function of cementum is very different from that of bone and dentin. Cementum serves as an attachment tissue as it bridges alveolar bone and dentin by means of the periodontal ligament. In this respect, cementum appears to fulfill a role like that of an enthesis, typically the attachment site joining tendons or ligaments to bone. Although cementum and bone share numerous biochemical and structural features, their major functions are distinct and the basis for attachment by cementum is unclear. Possibly the large number of proteins revealed by proteomic analysis to be exclusive to cementum compared to both bone and dentin (Fig. 3.19) [176] provides the foundation by which cementum performs its effective attachment between bone and dentin.

Unlike the other vertebrate mineralized tissues discussed above, which exhibit several overlapping characteristics, enamel is unique in many ways. It is different in its degree of mineralization (in excess of 95% versus 65% in bone, 71% in dentin, and 45–50% in cementum); absence of a collagenous matrix; reliance on dentinogenesis; architecture of its mineral phase; and expression and secretion of unusual enamel matrix proteins during development, including AMELX, ENAM, AMBN, AMTN, ODAM, MMP20, and KLK4. Additionally, maturation of enamel is distinct from that of bone, dentin, or other vertebrate hard tissues in that the protein-rich enamel matrix is almost completely removed by proteases which cleave the enamel proteins into lower molecular weight fragments. Following such cleavage and protein degradation, mineralization occurs such that mature enamel contains only traces (~1% or less) of its original organic constituents.

There is limited information on how the ameloblast synthesizes apatitic precursors and uses these proteins to form enamel as the external mineral "crust" of the teeth. No evidence of the presence and involvement of matrix vesicles has been reported in enamel deposition, although intracellular vesicles which contain mineral (as organelles staining for calcium with potassium pyroantimonate) have been found [204]. Possibly the type of extracellular mineralization which characterizes enamel is similar to that of intramembranous bone formation in which the mineral is formed directly in association with collagen fibrils. Indeed, as mentioned previously, mineralized collagen fibers are evident at the dentin–enamel interface at the inception of amelogenesis [243].

Regarding the evolutionary history of enamel and the relationship between enamel and teeth, odontodes, skin, and scales, the anthropologist, Per Erik Ahlberg, noted that scales present on some extinct fish contained an enamel-like homologue, ganoine [257]. Based on both morphological and phylogenetic studies, he and colleagues suggested that enamel originated in the skin before its appearance in teeth [258]. This hypothesis was supported by a number of observations. First, the teeth of the fossil bony fish, *Psarolepis romeri*, were found to be comprised only of dentin, while its dermal odontodes were comprised of both dentin and ganoine [258]. Second, the scales of another fossil bony fish, *Andreolepis*, contained ganoine while the teeth and even the odontodes were comprised only of dentin [258]. Third, the spotted gar, *Lepisosteus oculatus*, secreted both ENAM and AMBN in its skin [258]. These results underscore the concepts that enamel is an ectodermal tissue, that skin is the initial site of enamel formation, and that possibly keratins are involved with enamel formation and maturation [202, 259]. As such, these findings lead to the conclusions that dentin and enamel evolved separately, and that dentin signaling became tightly linked to amelogenesis when dentin and enamel came together during evolution to form the composite tooth.

References

1. Owen R. Lectures on the comparative anatomy and physiology of the vertebrate animals. London: Royal College of Surgeons of England; Longman, Brown, Green and Longmans; 1846. Delivered at the Royal College of Surgeons of England, 1844.
2. Adams DS, Keller R, Koehl MA. The mechanics of notochord elongation, straightening and stiffening in the embryo of Xenopus laevis. Development. 1990;110:115–30.
3. Stemple DL. Structure and function of the notochord: an essential organ for chordate development. Development. 2005;132:2503–12.
4. Summerbell D. Evidence for regulation of growth, size and pattern in the developing chick limb bud. J Embryol Exp Morphol. 1981;65:129–50.
5. Ohuchi H, Nakagawa T, Yamamoto A, Araga A, Ohata T, Ishimaru Y, et al. The mesenchymal factor, FGF10, initiates and maintains the outgrowth of the chick limb bud through interaction with FGF8, an apical ectodermal factor. Development. 1997;124:2235–44.
6. Martin P. Tissue patterning in the developing mouse limb. Int J Dev Biol. 1990;34:323–36.
7. Johnson RL, Tabin CJ. Molecular models for vertebrate limb development. Cell. 1997;90:979–90.
8. Saunders JW Jr. The proximo-distal sequence of origin of the parts of the chick wing and the role of the ectoderm. J Exp Zool. 1948;108:363–403.
9. Litingtung Y, Dahn RD, Li Y, Fallon JF, Chiang C. Shh and Gli3 are dispensable for limb skeleton formation but regulate digit number and identity. Nature. 2002;418:979–83.
10. Ornitz DM, Marie PJ. Fibroblast growth factor signaling in skeletal development and disease. Genes Dev. 2015;29:1463–86.
11. Niswander L, Jeffrey S, Martin GR, Tickle C. A positive feedback loop coordinates growth and patterning in the vertebrate limb. Nature. 1994;371:609–12.
12. Bastien J, Rochette-Egly C. Nuclear retinoid receptors and the transcription of retinoid-target genes. Gene. 2004;328:1–16.
13. Ng JK, Kawakami Y, Buscher D, Raya A, Itoh T, Koth CM, et al. The limb identity gene Tbx5 promotes limb initiation by interacting with Wnt2b and Fgf10. Development. 2002;129:5161–70.
14. Sun X, Mariani FV, Martin GR. Functions of FGF signalling from the apical ectodermal ridge in limb development. Nature. 2002;418:501–8.
15. Eames BF, de la Fuente L, Helms JA. Molecular ontogeny of the skeleton. Birth Defects Res C Embryo Today. 2003;69:93–101.

16. Farnum CE, Wilsman NJ. Ultrastructural histochemical evaluation of growth plate cartilage matrix from healthy and osteochondritic swine. Am J Vet Res. 1986;47:1105–15.
17. Hunziker EB. Mechanism of longitudinal bone growth and its regulation by growth plate chondrocytes. Microsc Res Tech. 1994;28:505–19.
18. Di Nino DL, Long F, Linsenmayer TF. Regulation of endochondral cartilage growth in the developing avian limb: cooperative involvement of perichondrium and periosteum. Dev Biol. 2001;240:433–42.
19. Ganey TM, Love SM, Ogden JA. Development of vascularization in the chondroepiphysis of the rabbit. J Orthop Res. 1992;10:496–510.
20. Alvarez J, Costales L, Serra R, Balbin M, Lopez JM. Expression patterns of matrix metalloproteinases and vascular endothelial growth factor during epiphyseal ossification. J Bone Miner Res. 2005;20:1011–21.
21. Xie M, Chagin AS. The epiphyseal secondary ossification center: evolution, development and function. Bone. 2021;142:115701.
22. Dorozhkin SV. Calcium orthophosphates in nature, biology and medicine. Materials (Basel). 2009;2:399–498.
23. Shapiro IM, Risbud MV. Introduction to the structure, function, and comparative anatomy of the vertebrae and the intervertebral disc. In: Shapiro IM, Risbud MV, editors. The intervertebral disc: molecular and structural studies of the disc in health and disease. Vienna: Springer; 2014. p. 3–15.
24. Xie M, Gol'din P, Herdina AN, Estefa J, Medvedeva EV, Li L, et al. Secondary ossification center induces and protects growth plate structure. elife. 2020;9:e55212.
25. Tam PP, Trainor PA. Specification and segmentation of the paraxial mesoderm. Anat Embryol (Berl). 1994;189:275–305.
26. Duester G. Retinoic acid synthesis and signaling during early organogenesis. Cell. 2008;134:921–31.
27. Bruggeman BJ, Maier JA, Mohiuddin YS, Powers R, Lo Y, Guimaraes-Camboa N, et al. Avian intervertebral disc arises from rostral sclerotome and lacks a nucleus pulposus: implications for evolution of the vertebrate disc. Dev Dyn. 2012;241:675–83.
28. Haque M, Ohata K, Takami T, Soares SB Jr, Aree SN, Hakuba A, et al. Development of lumbosacral spina bifida: three-dimensional computer graphic study of human embryos at Carnegie stage twelve. Pediatr Neurosurg. 2001;35:247–52.
29. White TD, Black MT, Folkens PA. Human osteology. 3rd ed. San Diego, CA: Academic Press; 2012.
30. Selleck MA, Bronner-Fraser M. Origins of the avian neural crest: the role of neural plate-epidermal interactions. Development. 1995;121:525–38.
31. Hall BK, Gillis JA. Incremental evolution of the neural crest, neural crest cells and neural crest-derived skeletal tissues. J Anat. 2013;222:19–31.
32. Jiang X, Iseki S, Maxson RE, Sucov HM, Morriss-Kay GM. Tissue origins and interactions in the mammalian skull vault. Dev Biol. 2002;241:106–16.
33. van Gijn DR, Tucker AS, Cobourne MT. Craniofacial development: current concepts in the molecular basis of Treacher Collins syndrome. Br J Oral Maxillofac Surg. 2013;51:384–8.
34. Lattanzi W, Barba M, Di Pietro L, Boyadjiev SA. Genetic advances in craniosynostosis. Am J Med Genet A. 2017;173:1406–29.
35. Bougery JM, Jacob NH. Atlas of complete treatise on human anatomy comprising operative medicine. Paris: CA Delaunay; 1833.
36. Akbarzadeh Khorshidi M. Length scale parameter of single trabecula in cancellous bone. Biomech Model Mechanobiol. 2020;19:1917–23.
37. Manolagas SC. Birth and death of bone cells: basic regulatory mechanisms and implications for the pathogenesis and treatment of osteoporosis. Endocr Rev. 2000;21:115–37.
38. Fuchs RK, Allen MR, Ruppel ME, Diab T, Phipps RJ, Miller LM, et al. In situ examination of the time-course for secondary mineralization of Haversian bone using synchrotron Fourier transform infrared microspectroscopy. Matrix Biol. 2008;27:34–41.

39. Parfitt AM. The actions of parathyroid hormone on bone: relation to bone remodeling and turnover, calcium homeostasis, and metabolic bone disease. Part I: Mechanisms of calcium transfer between blood and bone and their cellular basis: morphological and kinetic approaches to bone turnover. Metabolism. 1976;25:809–44.
40. Talmage RV, Cooper CW, Park HZ. Regulation of calcium transport in bone by parathyroid hormone. Vitam Horm. 1970;28:103–40.
41. Scherft JP. The lamina limitans of the organic matrix of calcified cartilage and bone. J Ultrastruct Res. 1972;38:318–31.
42. Luk SC, Nopajaroonsri C, Simon GT. The ultrastructure of endosteum: a topographic study in young adult rabbits. J Ultrastruct Res. 1974;46:165–83.
43. Bonewald LF. The amazing osteocyte. J Bone Miner Res. 2011;26:229–38.
44. Robling AG, Bonewald LF. The osteocyte: new insights. Annu Rev Physiol. 2020;82:485–506.
45. Sissons HA, Kember NF. Longitudinal bone growth of the human femur. Postgrad Med J. 1977;53:433–7.
46. Pritchett JW. Growth and growth prediction of the fibula. Clin Orthop Relat Res. 1997;334:251–6.
47. Wealthall RJ, Herring SW. Endochondral ossification of the mouse nasal septum. Anat Rec A Discov Mol Cell Evol Biol. 2006;288:1163–72.
48. Morita M, Nakanishi K, Kawai T, Fujikawa K. Telomere length, telomerase activity, and expression of human telomerase reverse transcriptase mRNA in growth plate of epiphyseal articular cartilage in femoral head during normal human development and in thanatophoric dysplasia. Hum Pathol. 2004;35:403–11.
49. Fenichel I, Evron Z, Nevo Z. The perichondrial ring as a reservoir for precartilaginous cells. In vivo model in young chicks' epiphysis. Int Orthop. 2006;30:353–6.
50. Jenkins DH, Cheng DH, Hodgson AR. Stimulation of bone growth by periosteal stripping. A clinical study. J Bone Joint Surg (Br). 1975;57:482–4.
51. Warrell E, Taylor JF. The role of periosteal tension in the growth of long bones. J Anat. 1979;128:179–84.
52. Shapiro F, Holtrop ME, Glimcher MJ. Organization and cellular biology of the perichondrial ossification groove of Ranvier: a morphological study in rabbits. J Bone Joint Surg Am. 1977;59:703–23.
53. Shapiro IM, Srinivas V. Metabolic consideration of epiphyseal growth: survival responses in a taxing environment. Bone. 2007;40:561–7.
54. Holliday CM, Ridgely RC, Sedlmayr JC, Witmer LM. Cartilaginous epiphyses in extant archosaurs and their implications for reconstructing limb function in dinosaurs. PLoS One. 2010;5:e13120.
55. Wilsman NJ, Van Sickle DC. Cartilage canals, their morphology and distribution. Anat Rec. 1972;173:79–93.
56. Melton JT, Clarke NM, Roach HI. Matrix metalloproteinase-9 induces the formation of cartilage canals in the chondroepiphysis of the neonatal rabbit. J Bone Joint Surg Am. 2006;88:155–61.
57. Kugler JH, Tomlinson A, Wagstaff A, Ward SM. The role of cartilage canals in the formation of secondary centres of ossification. J Anat. 1979;129:493–506.
58. Blumer MJ, Longato S, Richter E, Perez MT, Konakci KZ, Fritsch H. The role of cartilage canals in endochondral and perichondral bone formation: are there similarities between these two processes? J Anat. 2005;206:359–72.
59. Anderson HC, Shapiro IM. The epiphyseal growth plate. In: Bronner F, Farach-Carson MC, Roach HI, editors. Bone and development. London: Springer; 2010. p. 39–64.
60. Erenpreisa J, Roach HI. Aberrations of cell cycle and cell death in normal development of the chick embryo growth plate. Mech Ageing Dev. 1999;108:227–38.
61. DeLise AM, Fischer L, Tuan RS. Cellular interactions and signaling in cartilage development. Osteoarthr Cartil. 2000;8:309–34.

62. Linsenmayer TF, Long F, Nurminskaya M, Chen Q, Schmid TM. Type X collagen and other up-regulated components of the avian hypertrophic cartilage program. Prog Nucleic Acid Res Mol Biol. 1998;60:79–109.
63. Abad V, Meyers JL, Weise M, Gafni RI, Barnes KM, Nilsson O, et al. The role of the resting zone in growth plate chondrogenesis. Endocrinology. 2002;143:1851–7.
64. van der Eerden BC, Karperien M, Wit JM. Systemic and local regulation of the growth plate. Endocr Rev. 2003;24:782–801.
65. Li Y, Li A, Junge J, Bronner M. Planar cell polarity signaling coordinates oriented cell division and cell rearrangement in clonally expanding growth plate cartilage. elife. 2017;6:e23279.
66. Romereim SM, Conoan NH, Chen B, Dudley AT. A dynamic cell adhesion surface regulates tissue architecture in growth plate cartilage. Development. 2014;141:2085–95.
67. Li Y, Dudley AT. Noncanonical frizzled signaling regulates cell polarity of growth plate chondrocytes. Development. 2009;136:1083–92.
68. Davey CF, Moens CB. Planar cell polarity in moving cells: think globally, act locally. Development. 2017;144:187–200.
69. Retterer E. Evolution du cartilage transitoire. J Anat Physiol. 1900;36:467–565.
70. Stump CW. The histogenesis of bone. J Anat. 1925;59:136–54.
71. Shao D, Tian R. Glucose transporters in cardiac metabolism and hypertrophy. Compr Physiol. 2015;6:331–51.
72. Buckwalter JA, Mower D, Ungar R, Schaeffer J, Ginsberg B. Morphometric analysis of chondrocyte hypertrophy. J Bone Joint Surg Am. 1986;68:243–55.
73. Cooper KL, Oh S, Sung Y, Dasari RR, Kirschner MW, Tabin CJ. Multiple phases of chondrocyte enlargement underlie differences in skeletal proportions. Nature. 2013;495:375–8.
74. Chang KC, Barth AS, Sasano T, Kizana E, Kashiwakura Y, Zhang Y, et al. CAPON modulates cardiac repolarization via neuronal nitric oxide synthase signaling in the heart. Proc Natl Acad Sci U S A. 2008;105:4477–82.
75. Gibson G, Lin DL, Francki K, Caterson B, Foster B. Type X collagen is colocalized with a proteoglycan epitope to form distinct morphological structures in bovine growth cartilage. Bone. 1996;19:307–15.
76. Gerber HP, Vu TH, Ryan AM, Kowalski J, Werb Z, Ferrara N. VEGF couples hypertrophic cartilage remodeling, ossification and angiogenesis during endochondral bone formation. Nat Med. 1999;5:623–8.
77. Shapiro IM, Boyde A. Mineralization of normal and rachitic chick growth cartilage: vascular canals, cartilage calcification and osteogenesis. Scanning Microsc. 1987;1:599–606.
78. Odgren PR, Witwicka H, Reyes-Gutierrez P. The cast of clasts: catabolism and vascular invasion during bone growth, repair, and disease by osteoclasts, chondroclasts, and septoclasts. Connect Tissue Res. 2016;57:161–74.
79. Bohensky J, Leshinsky S, Srinivas V, Shapiro IM. Chondrocyte autophagy is stimulated by HIF-1 dependent AMPK activation and mTOR suppression. Pediatr Nephrol. 2010;25:633–42.
80. Shapiro IM, Adams CS, Freeman T, Srinivas V. Fate of the hypertrophic chondrocyte: microenvironmental perspectives on apoptosis and survival in the epiphyseal growth plate. Birth Defects Res C Embryo Today. 2005;75:330–9.
81. Boyde A, Shapiro IM. Morphological observations concerning the pattern of mineralization of the normal and the rachitic chick growth cartilage. Anat Embryol (Berl). 1987;175:457–66.
82. van der Stricht O. Recherches sur le cartilage articulaire des oiseaux. Arch Biol. 1890;10:1–41.
83. Brachet A. Etude sur la resorption du cartilage et le developpement des os longs chez les oiseaux. J Anat Physiol. 1893;10:391–417.
84. Ham AW. Some histophysiological problems peculiar to calcified tissues. J Bone Joint Surg Am. 1952;24A:701–28.
85. Cameron DA. The fine structure of bone and calcified cartilage. A critical review of the contribution of electron microscopy to the understanding of osteogenesis. Clin Orthop Relat Res. 1963;26:199–228.

86. Anderson HC. Electron microscopic studies of induced cartilage development and calcification. J Cell Biol. 1967;35:81–101.
87. Brighton CT. The growth plate. Orthop Clin North Am. 1984;15:571–95.
88. Adams CS, Shapiro IM. The fate of the terminally differentiated chondrocyte: evidence for microenvironmental regulation of chondrocyte apoptosis. Crit Rev Oral Biol Med. 2002;13:465–73.
89. Hatori M, Klatte KJ, Teixeira CC, Shapiro IM. End labeling studies of fragmented DNA in the Avian growth plate: evidence of apoptosis in terminally differentiated chondrocytes. J Bone Miner Res. 1995;10:1960–8.
90. Bronckers AL, Goei W, Luo G, Karsenty G, D'Souza RN, Lyaruu DM, et al. DNA fragmentation during bone formation in neonatal rodents assessed by transferase-mediated end labeling. J Bone Miner Res. 1996;11:1281–91.
91. Zenmyo M, Komiya S, Kawabata R, Sasaguri Y, Inoue A, Morimatsu M. Morphological and biochemical evidence for apoptosis in the terminal hypertrophic chondrocytes of the growth plate. J Pathol. 1996;180:430–3.
92. Roach HI, Erenpreisa J. The phenotypic switch from chondrocytes to bone-forming cells involves asymmetric cell division and apoptosis. Connect Tissue Res. 1996;35:85–91.
93. Erenpreisa J, Roach H. Aberrant death in dark chondrocytes of the avian growth plate. Cell Death Differ. 1998;5:60–6.
94. Mukhi S, Brown DD. Transdifferentiation of tadpole pancreatic acinar cells to duct cells mediated by Notch and stromelysin-3. Dev Biol. 2011;351:311–7.
95. Schaub C, Marz J, Reim I, Frasch M. Org-1-dependent lineage reprogramming generates the ventral longitudinal musculature of the Drosophila heart. Curr Biol. 2015;25:488–94.
96. Zuryn S, Ahier A, Portoso M, White ER, Morin MC, Margueron R, et al. Transdifferentiation. Sequential histone-modifying activities determine the robustness of transdifferentiation. Science. 2014;345:826–9.
97. Rothman J, Jarriault S. Developmental plasticity and cellular reprogramming in Caenorhabditis elegans. Genetics. 2019;213:723–57.
98. Fell HB. The osteogenic capacity in vitro of periosteum and endosteum isolated from the limb skeleton of fowl embryos and young chicks. J Anat. 1932;66:157–80.
99. Lutfi AM. The fate of chondrocytes during cartilage erosion in the growing tibia in the domestic fowl (Gallus domesticus). Acta Anat (Basel). 1971;79:27–35.
100. Silbermann M, Frommer J. The nature of endochondral ossification in the mandibular condyle of the mouse. Anat Rec. 1972;172:659–67.
101. Aigner T, Loos S, Müller S, Sandell LJ, Unni KK, Kirchner T. Cell differentiation and matrix gene expression in mesenchymal chondrosarcomas. Am J Pathol. 2000;156:1327–35.
102. von der Mark K, von der Mark H. The role of three genetically distinct collagen types in endochondral ossification and calcification of cartilage. J Bone Joint Surg (Br). 1977;59B:458–64.
103. Horton WA, Dwyer C, Goering R, Dean DC. Immunohistochemistry of types I and II collagen in undecalcified skeletal tissues. J Histochem Cytochem. 1983;31:417–25.
104. Oshima O, Leboy PS, McDonald SA, Tuan RS, Shapiro IM. Developmental expression of genes in chick growth cartilage detected by in situ hybridization. Calcif Tissue Int. 1989;45:182–92.
105. Moskalewski S, Malejczyk J. Bone formation following intrarenal transplantation of isolated murine chondrocytes: chondrocyte-bone cell transdifferentiation? Development. 1989;107:473–80.
106. Descalzi Cancedda F, Gentili C, Manduca P, Cancedda R. Hypertrophic chondrocytes undergo further differentiation in culture. J Cell Biol. 1992;117:427–35.
107. Roach HI. Trans-differentiation of hypertrophic chondrocytes into cells capable of producing a mineralized bone matrix. Bone Miner. 1992;19:1–20.
108. Ishizeki K, Takigawa M, Nawa T, Suzuki F. Mouse Meckel's cartilage chondrocytes evoke bone-like matrix and further transform into osteocyte-like cells in culture. Anat Rec. 1996;245:25–35.

109. Zerega B, Cermelli S, Bianco P, Cancedda R, Cancedda FD. Parathyroid hormone [PTH(1-34)] and parathyroid hormone-related protein [PTHrP(1-34)] promote reversion of hypertrophic chondrocytes to a prehypertrophic proliferating phenotype and prevent terminal differentiation of osteoblast-like cells. J Bone Miner Res. 1999;14:1281–9.
110. Bianco P, Cancedda FD, Riminucci M, Cancedda R. Bone formation via cartilage models: the "borderline" chondrocyte. Matrix Biol. 1998;17:185–92.
111. Roach HI, Erenpreisa J, Aigner T. Osteogenic differentiation of hypertrophic chondrocytes involves asymmetric cell divisions and apoptosis. J Cell Biol. 1995;131:483–94.
112. Yang G, Zhu L, Hou N, Lan Y, Wu XM, Zhou B, et al. Osteogenic fate of hypertrophic chondrocytes. Cell Res. 2014;24:1266–9.
113. Park J, Gebhardt M, Golovchenko S, Perez-Branguli F, Hattori T, Hartmann C, et al. Dual pathways to endochondral osteoblasts: a novel chondrocyte-derived osteoprogenitor cell identified in hypertrophic cartilage. Biol Open. 2015;4:608–21.
114. Bahney CS, Hu DP, Taylor AJ, Ferro F, Britz HM, Hallgrimsson B, et al. Stem cell-derived endochondral cartilage stimulates bone healing by tissue transformation. J Bone Miner Res. 2014;29:1269–82.
115. Hu DP, Ferro F, Yang F, Taylor AJ, Chang W, Miclau T, et al. Cartilage to bone transformation during fracture healing is coordinated by the invading vasculature and induction of the core pluripotency genes. Development. 2017;144:221–34.
116. Bais M, McLean J, Sebastiani P, Young M, Wigner N, Smith T, et al. Transcriptional analysis of fracture healing and the induction of embryonic stem cell-related genes. PLoS One. 2009;4:e5393.
117. Jing Y, Jing J, Wang K, Chan K, Harris SE, Hinton RJ, et al. Vital roles of beta-catenin in trans-differentiation of chondrocytes to bone cells. Int J Biol Sci. 2018;14:1–9.
118. Wang L, Huang J, Moore DC, Song Y, Ehrlich MG, Yang W. SHP2 regulates intramembranous ossification by modifying the TGFbeta and BMP2 signaling pathway. Bone. 2019;120:327–35.
119. Zuo C, Wang L, Kamalesh RM, Bowen ME, Moore DC, Dooner MS, et al. SHP2 regulates skeletal cell fate by modifying SOX9 expression and transcriptional activity. Bone Res. 2018;6:12.
120. Lui JC, Yue S, Lee A, Kikani B, Temnycky A, Barnes KM, et al. Persistent Sox9 expression in hypertrophic chondrocytes suppresses transdifferentiation into osteoblasts. Bone. 2019;125:169–77.
121. Bianco P, Robey PG. Skeletal stem cells. Development. 2015;142:1023–7.
122. Owen M, Friedenstein AJ. Stromal stem cells: marrow-derived osteogenic precursors. CIBA Found Symp. 1988;136:42–60.
123. Xiao W, Wang Y, Pacios S, Li S, Graves DT. Cellular and molecular aspects of bone remodeling. Front Oral Biol. 2016;18:9–16.
124. Abzhanov A, Rodda SJ, McMahon AP, Tabin CJ. Regulation of skeletogenic differentiation in cranial dermal bone. Development. 2007;134:3133–44.
125. Eames BF, Amores A, Yan YL, Postlethwait JH. Evolution of the osteoblast: skeletogenesis in gar and zebrafish. BMC Evol Biol. 2012;12:27.
126. Apschner A, Schulte-Merker S, Witten PE. Not all bones are created equal - using zebrafish and other teleost species in osteogenesis research. Methods Cell Biol. 2011;105:239–55.
127. Paul S, Schindler S, Giovannone D, de Millo Terrazzani A, Mariani FV, Crump JG. Ihha induces hybrid cartilage-bone cells during zebrafish jawbone regeneration. Development. 2016;143:2066–76.
128. Benjamin M, Ralphs JR, Eberewariye OS. Cartilage and related tissues in the trunk and fins of teleosts. J Anat. 1992;181:113–8.
129. Vickaryous MK, Hall BK. Development of the dermal skeleton in Alligator mississippiensis (Archosauria, Crocodylia) with comments on the homology of osteoderms. J Morphol. 2008;269:398–422.
130. Kronenberg HM. The role of the perichondrium in fetal bone development. Ann N Y Acad Sci. 2007;1116:59–64.

131. Erlebacher A, Filvaroff EH, Gitelman SE, Derynck R. Toward a molecular understanding of skeletal development. Cell. 1995;80:371–8.
132. Colnot C, de la Fuente L, Huang S, Hu D, Lu C, St-Jacques B, et al. Indian hedgehog synchronizes skeletal angiogenesis and perichondrial maturation with cartilage development. Development. 2005;132:1057–67.
133. Debnath S, Yallowitz AR, McCormick J, Lalani S, Zhang T, Xu R, et al. Discovery of a periosteal stem cell mediating intramembranous bone formation. Nature. 2018;562:133–9.
134. Duchamp de Lageneste O, Julien A, Abou-Khalil R, Frangi G, Carvalho C, Cagnard N, et al. Periosteum contains skeletal stem cells with high bone regenerative potential controlled by Periostin. Nat Commun. 2018;9:773.
135. Boyde A, Hordell MH. Scanning electron microscopy of lamellar bone. Z Zellforsch Mikrosk Anat. 1969;93:213–31.
136. Locke M. Structure of long bones in mammals. J Morphol. 2004;262:546–65.
137. Schünke M, Schumacher U, Schulte E. General anatomy and musculoskeletal system. In: Ross LM, Lamperti ED, editors. Thieme atlas of anatomy. Stuttgart: Thieme; 2007.
138. Genthial R, Beaurepaire E, Schanne-Klein M-C, Peyrin F, Farlay D, Olivier C, et al. Label-free imaging of bone multiscale porosity and interfaces using third-harmonic generation microscopy. Sci Rep. 2017;7:3419.
139. Shapiro F, Wu JY. Woven bone overview: structural classification based on its integral role in developmental, repair and pathological bone formation throughout vertebrate groups. Eur Cell Mater. 2019;38:137–67.
140. Doube M. Closing cones create conical lamellae in secondary osteonal bone. R Soc Open Sci. 2022;9:220712.
141. Repp F, Kollmannsberger P, Roschger A, Kerschnitzki M, Berzlanovich A, Gruber GM, et al. Spatial heterogeneity in the canalicular density of the osteocyte network in human osteons. Bone Rep. 2017;6:101–8.
142. Li J, Parada C, Chai Y. Cellular and molecular mechanisms of tooth root development. Development. 2017;144:374–84.
143. Owen R. Odontography. London: Hippolyte Bailliere; 1840.
144. Dubey DK, Tomar V. Role of molecular level interfacial forces in hard biomaterial mechanics: a review. Ann Biomed Eng. 2010;38:2040–55.
145. Arana-Chavez VE, Massa LF. Odontoblasts: the cells forming and maintaining dentine. Int J Biochem Cell Biol. 2004;36:1367–73.
146. Goldberg M, Kulkarni AB, Young M, Boskey A. Dentin: structure, composition and mineralization. Front Biosci (Elite Ed). 2011;3:711–35.
147. Dean MC, Scandrett AE. Rates of dentine mineralization in permanent human teeth. Int J Osteoarchaeol. 1995;5:349–58.
148. Shroff B, Thomas HF. Investigation of the role of Von Korff fibers during murine dentinogenesis. J Biol Buccale. 1992;20:139–44.
149. Stratmann U, Schaarschmidt K, Wiesmann HP, Plate U, Höhling HJ. Mineralization during matrix-vesicle-mediated mantle dentine formation in molars of albino rats: a microanalytical and ultrastructural study. Cell Tissue Res. 1996;284:223–30.
150. Boyde A, Brommage TG. Histology of dental hard tissues. In: de Buffrenil V, de Ricqles AJ, Zylberberg L, Padian K, editors. Vertebrate skeletal histology and paleohistology. Boca Raton, FL: CRC Press; 2021. p. 260–8.
151. Massler M, Schour I. The appositional life span of the enamel and dentin-forming cells; human deciduous teeth and first permanent molars; introduction. J Dent Res. 1946;25:145–50.
152. Mina M, Kollar EJ. The induction of odontogenesis in non-dental mesenchyme combined with early murine mandibular arch epithelium. Arch Oral Biol. 1987;32:123–7.
153. Lumsden AG. Spatial organization of the epithelium and the role of neural crest cells in the initiation of the mammalian tooth germ. Development. 1988;103:155–69.
154. Gronthos S, Mankani M, Brahim J, Robey PG, Shi S. Postnatal human dental pulp stem cells (DPSCs) in vitro and in vivo. Proc Natl Acad Sci U S A. 2000;97:13625–30.

155. Nuti N, Corallo C, Chan BM, Ferrari M, Gerami-Naini B. Multipotent differentiation of human dental pulp stem cells: a literature review. Stem Cell Rev. 2016;12:511–23.
156. Cavalcanti BN, Zeitlin BD, Nor JE. A hydrogel scaffold that maintains viability and supports differentiation of dental pulp stem cells. Dent Mater. 2013;29:97–102.
157. Ye L, MacDougall M, Zhang S, Xie Y, Zhang J, Li Z, et al. Deletion of dentin matrix protein-1 leads to a partial failure of maturation of predentin into dentin, hypomineralization, and expanded cavities of pulp and root canal during postnatal tooth development. J Biol Chem. 2004;279:19141–8.
158. Jussila M, Thesleff I. Signaling networks regulating tooth organogenesis and regeneration, and the specification of dental mesenchymal and epithelial cell lineages. Cold Spring Harb Perspect Biol. 2012;4:a008425.
159. Balic A, Thesleff I. Tissue interactions regulating tooth development and renewal. Curr Top Dev Biol. 2015;115:157–86.
160. Forien JB, Fleck C, Cloetens P, Duda G, Fratzl P, Zolotoyabko E, et al. Compressive residual strains in mineral nanoparticles as a possible origin of enhanced crack resistance in human tooth dentin. Nano Lett. 2015;15:3729–34.
161. Garces-Ortiz M, Ledesma-Montes C, Reyes-Gasga J. Scanning electron microscopic study on the fibrillar structures within dentinal tubules of human dentin. J Endod. 2015;41:1510–4.
162. Kuroiwa M, Chihara K, Higashi S. Electron microscopic studies on Sharpey's fibers in the alveolar bone of rat molars. Kaibogaku Zasshi. 1994;69:776–82.
163. Luan X, Walker C, Dangaria S, Ito Y, Druzinsky R, Jarosius K, et al. The mosasaur tooth attachment apparatus as paradigm for the evolution of the gnathostome periodontium. Evol Dev. 2009;11:247–59.
164. Fujiwara N, Tabata MJ, Endoh M, Ishizeki K, Nawa T. Insulin-like growth factor-1 stimulates cell proliferation in the outer layer of Hertwig's epithelial root sheath and elongation of the tooth root in mouse molars in vitro. Cell Tissue Res. 2005;320:69–75.
165. Ten Cate AR. The role of epithelium in the development, structure and function of the tissues of tooth support. Oral Dis. 1996;2:55–62.
166. Huang X, Bringas P Jr, Slavkin HC, Chai Y. Fate of HERS during tooth root development. Dev Biol. 2009;334:22–30.
167. Paynter KJ, Pudy G. A study of the structure, chemical nature, and development of cementum in the rat. Anat Rec. 1958;131:233–51.
168. Bosshardt DD. Are cementoblasts a subpopulation of osteoblasts or a unique phenotype? J Dent Res. 2005;84:390–406.
169. Huang XF, Chai Y. Molecular regulatory mechanism of tooth root development. Int J Oral Sci. 2012;4:177–81.
170. Zeichner-David M. Regeneration of periodontal tissues: cementogenesis revisited. Periodontol. 2000;2006(41):196–217.
171. Chen J, Chen G, Yan Z, Guo Y, Yu M, Feng L, et al. TGF-beta1 and FGF2 stimulate the epithelial-mesenchymal transition of HERS cells through a MEK-dependent mechanism. J Cell Physiol. 2014;229:1647–59.
172. Itaya S, Oka K, Ogata K, Tamura S, Kira-Tatsuoka M, Fujiwara N, et al. Hertwig's epithelial root sheath cells contribute to formation of periodontal ligament through epithelial-mesenchymal transition by TGF-beta. Biomed Res. 2017;38:61–9.
173. Lee SY, Auh QS, Kang SK, Kim HJ, Lee JW, Noh K, et al. Combined effects of dentin sialoprotein and bone morphogenetic protein-2 on differentiation in human cementoblasts. Cell Tissue Res. 2014;357:119–32.
174. Thesleff I. Epithelial-mesenchymal signalling regulating tooth morphogenesis. J Cell Sci. 2003;116:1647–8.
175. Scivetti M, Pilolli GP, Corsalini M, Lucchese A, Favia G. Confocal laser scanning microscopy of human cementocytes: analysis of three-dimensional image reconstruction. Ann Anat. 2007;189:169–74.
176. Salmon CR, Giorgetti APO, Paes Leme AF, Domingues RR, Kolli TN, Foster BL, et al. Microproteome of dentoalveolar tissues. Bone. 2017;101:219–29.

177. Arzate H, Zeichner-David M, Mercado-Celis G. Cementum proteins: role in cementogenesis, biomineralization, periodontium formation and regeneration. Periodontol. 2000;2015(67):211–33.
178. Liu HW, Yacobi R, Savion N, Narayanan AS, Pitaru S. A collagenous cementum-derived attachment protein is a marker for progenitors of the mineralized tissue-forming cell lineage of the periodontal ligament. J Bone Miner Res. 1997;12:1691–9.
179. Takano Y, Sakai H, Baba O, Terashima T. Differential involvement of matrix vesicles during the initial and appositional mineralization processes in bone, dentin, and cementum. Bone. 2000;26:333–9.
180. Elliott JC. Structure, crystal chemistry and density of enamel apatites. CIBA Found Symp. 1997;205:54–67; discussion 67–72.
181. Boyde A. The development of enamel structure. Proc R Soc Med. 1967;60:923–8.
182. Reith EJ, Boyde A. A correlated scanning and transmission electron microscopic study of maturation ameloblasts in developing molar teeth of rats. Cell Tissue Res. 1979;197:421–31.
183. Slavkin HC, Hu CC, Sakakura Y, Diekwisch T, Chai Y, Mayo M, et al. Gene expression, signal transduction and tissue-specific biomineralization during mammalian tooth development. Crit Rev Eukaryot Gene Expr. 1992;2:315–29.
184. Glimcher MJ, Daniel EJ, Travis DF, Kamhi S. Electron optical and X-ray diffraction studies of the organization of the inorganic crystals in embryonic bovine enamel. J Ultrastruct Res. 1965;12:1–77.
185. Moradian-Oldak J. Amelogenins: assembly, processing and control of crystal morphology. Matrix Biol. 2001;20:293–305.
186. Koldehoff J, Swain MV, Schneider GA. The geometrical structure of interfaces in dental enamel: a FIB-STEM investigation. Acta Biomater. 2020;104:17–27.
187. Smith CE, Nanci A. Overview of morphological changes in enamel organ cells associated with major events in amelogenesis. Int J Dev Biol. 1995;39:153–61.
188. Paine ML, Snead ML. Tooth developmental biology: disruptions to enamel-matrix assembly and its impact on biomineralization. Orthod Craniofacial Res. 2005;8:239–51.
189. Chen E, Piddington R, Decker S, Park J, Yuan ZA, Abrams WR, et al. Regulation of amelogenin gene expression during tooth development. Dev Dyn. 1994;199:189–98.
190. Lattanzi W, Di Giacomo MC, Lenato GM, Chimienti G, Voglino G, Resta N, et al. A large interstitial deletion encompassing the amelogenin gene on the short arm of the Y chromosome. Hum Genet. 2005;116:395–401.
191. MacDougall M, DuPont BR, Simmons D, Reus B, Krebsbach P, Kärrman C, et al. Ameloblastin gene (AMBN) maps within the critical region for autosomal dominant amelogenesis imperfecta at chromosome 4q21. Genomics. 1997;41:115–8.
192. Bartlett JD. Dental enamel development: proteinases and their enamel matrix substrates. ISRN Dent. 2013;2013:684607.
193. Al-Hashimi N, Lafont AG, Delgado S, Kawasaki K, Sire JY. The enamelin genes in lizard, crocodile, and frog and the pseudogene in the chicken provide new insights on enamelin evolution in tetrapods. Mol Biol Evol. 2010;27:2078–94.
194. Iwasaki K, Bajenova E, Somogyi-Ganss E, Miller M, Nguyen V, Nourkeyhani H, et al. Amelotin--a novel secreted, ameloblast-specific protein. J Dent Res. 2005;84:1127–32.
195. Somogyi-Ganss E, Nakayama Y, Iwasaki K, Nakano Y, Stolf D, McKee MD, et al. Comparative temporospatial expression profiling of murine amelotin protein during amelogenesis. Cells Tissues Organs. 2012;195:535–49.
196. Kestler DP, Foster JS, Macy SD, Murphy CL, Weiss DT, Solomon A. Expression of odontogenic ameloblast-associated protein (ODAM) in dental and other epithelial neoplasms. Mol Med. 2008;14:318–26.
197. Lee HK, Park SJ, Oh HJ, Kim JW, Bae HS, Park JC. Expression pattern, subcellular localization, and functional implications of ODAM in ameloblasts, odontoblasts, osteoblasts, and various cancer cells. Gene Expr Patterns. 2012;12:102–8.
198. Hu JC-C, Lertlam R, Richardson AS, Smith CE, McKee MD, Simmer JP. Cell proliferation and apoptosis in enamelin null mice. Eur J Oral Sci. 2011;119:329–37.

199. Rosset EM, Bradshaw AD. SPARC/osteonectin in mineralized tissue. Matrix Biol. 2016;52–54:78–87.
200. Hu Y, Smith CE, Richardson AS, Bartlett JD, Hu JC, Simmer JP. MMP20, KLK4, and MMP20/KLK4 double null mice define roles for matrix proteases during dental enamel formation. Mol Genet Genomic Med. 2015;4:178–96.
201. Lacruz RS, Habelitz S, Wright JT, Paine ML. Dental enamel formation and implications for oral health and disease. Physiol Rev. 2017;97:939–93.
202. Chiba R, Okubo M, Yamamoto R, Saito MM, Kobayashi S, Beniash E, et al. Porcine keratin 75 in developing enamel. J Oral Biosci. 2019;61:163–72.
203. Hubbard MJ. Calcium transport across the dental enamel epithelium. Crit Rev Oral Biol Med. 2000;11:437–66.
204. Deporter DA. The early mineralization of enamel. Fine structural observations on the cellular localization of calcium with the potassium pyroantimonate technique. Calcif Tissue Res. 1977;24:271–4.
205. Moradian-Oldak J, Lakshminarayanan R. Intrinsic disorder in amelogenin. In: Goldberg M, editor. Amelogenins: multifaceted proteins for dental and bone formation and repair. Paris: Bentham Science Publishers Ltd.; 2012. p. 106–32.
206. Shaw WJ, Tarasevich BJ, Buchko GW, Arachchige RMJ, Burton SD. Controls of nature: secondary, tertiary, and quaternary structure of the enamel protein amelogenin in solution and on hydroxyapatite. J Struct Biol. 2020;212:107630.
207. Gungormus M, Oren EE, Horst JA, Fong H, Hnilova M, Somerman MJ, et al. Cementomimetics-constructing a cementum-like biomineralized microlayer via amelogenin-derived peptides. Int J Oral Sci. 2012;4:69–77.
208. Fincham AG, Moradian-Oldak J, Simmer JP, Sarte P, Lau EC, Diekwisch T, et al. Self-assembly of a recombinant amelogenin protein generates supramolecular structures. J Struct Biol. 1994;112:103–9.
209. Fincham AG, Moradian-Oldak J, Diekwisch TG, Lyaruu DM, Wright JT, Bringas P Jr, et al. Evidence for amelogenin "nanospheres" as functional components of secretory-stage enamel matrix. J Struct Biol. 1995;115:50–9.
210. Fincham AG, Moradian-Oldak J, Simmer JP. The structural biology of the developing dental enamel matrix. J Struct Biol. 1999;126:270–99.
211. Moradian-Oldak J, Paine ML, Lei YP, Fincham AG, Snead ML. Self-assembly properties of recombinant engineered amelogenin proteins analyzed by dynamic light scattering and atomic force microscopy. J Struct Biol. 2000;131:27–37.
212. Bartlett JD, Ganss B, Goldberg M, Moradian-Oldak J, Paine ML, Snead ML, et al. Protein-protein interactions of the developing enamel matrix. Curr Top Dev Biol. 2006;74:57–115.
213. Wang L, Guan X, Du C, Moradian-Oldak J, Nancollas GH. Amelogenin promotes the formation of elongated apatite microstructures in a controlled crystallization system. J Phys Chem C Nanomater Interf. 2007;111:6398–404.
214. Yang X, Wang L, Qin Y, Sun Z, Henneman ZJ, Moradian-Oldak J, et al. How amelogenin orchestrates the organization of hierarchical elongated microstructures of apatite. J Phys Chem B. 2010;114:2293–300.
215. Aichmayer B, Margolis HC, Sigel R, Yamakoshi Y, Simmer JP, Fratzl P. The onset of amelogenin nanosphere aggregation studied by small-angle X-ray scattering and dynamic light scattering. J Struct Biol. 2005;151:239–49.
216. Wiedemann-Bidlack FB, Kwak SY, Beniash E, Yamakoshi Y, Simmer JP, Margolis HC. Effects of phosphorylation on the self-assembly of native full-length porcine amelogenin and its regulation of calcium phosphate formation in vitro. J Struct Biol. 2011;173: 250–60.
217. Green DR, Schulte F, Lee K-H, Pugach MK, Hardt M, Bidlack FB. Mapping the tooth enamel proteome and amelogenin phosphorylation onto mineralizing porcine tooth crowns. Front Physiol. 2019;10:925.
218. Carneiro KMM, Zhai H, Zhu L, Horst JA, Sitlin M, Nguyen M, et al. Amyloid-like ribbons of amelogenins in enamel mineralization. Sci Rep. 2016;6:23105.

219. Martinez-Avila O, Wu S, Kim SJ, Cheng Y, Khan F, Samudrala R, et al. Self-assembly of filamentous amelogenin requires calcium and phosphate: from dimers via nanoribbons to fibrils. Biomacromolecules. 2012;13:3494–502.
220. Habelitz S. Materials engineering by ameloblasts. J Dent Res. 2015;94:759–67.
221. Akkineni S, Zhu C, Chen J, Song M, Hoff SE, Bonde J, et al. Amyloid-like amelogenin nanoribbons template mineralization via a low-energy interface of ion binding sites. Proc Natl Acad Sci U S A. 2022;119:e2106965119.
222. Fincham AG, Belcourt AB, Termine JD, Butler WT, Cothran WC. Dental enamel matrix: sequences of two amelogenin polypeptides. Biosci Rep. 1981;1:771–8.
223. Fincham AG, Belcourt AB, Termine JD, Butler WT, Cothran WC. Amelogenins. Sequence homologies in enamel-matrix proteins from three mammalian species. Biochem J. 1983;211:149–54.
224. Gibson CW, Golub E, Ding WD, Shimokawa H, Young M, Termine J, et al. Identification of the leucine-rich amelogenin peptide (LRAP) as the translation product of an alternatively spliced transcript. Biochem Biophys Res Commun. 1991;174:1306–12.
225. Boabaid F, Gibson CW, Kuehl MA, Berry JE, Snead ML, Nociti FH Jr, et al. Leucine-rich amelogenin peptide: a candidate signaling molecule during cementogenesis. J Periodontol. 2004;75:1126–36.
226. Warotayanont R, Frenkel B, Snead ML, Zhou Y. Leucine-rich amelogenin peptide induces osteogenesis by activation of the Wnt pathway. Biochem Biophys Res Commun. 2009;387:558–63.
227. Chen E, Yuan ZA, Wright JT, Hong SP, Li Y, Collier PM, et al. The small bovine amelogenin LRAP fails to rescue the amelogenin null phenotype. Calcif Tissue Int. 2003;73:487–95.
228. Le Norcy E, Kwak SY, Wiedemann-Bidlack FB, Beniash E, Yamakoshi Y, Simmer JP, et al. Leucine-rich amelogenin peptides regulate mineralization in vitro. J Dent Res. 2011;90:1091–7.
229. Le Norcy E, Lesieur J, Sadoine J, Rochefort GY, Chaussain C, Poliard A. Phosphorylated and non-phosphorylated leucine rich amelogenin peptide differentially affect ameloblast mineralization. Front Physiol. 2018;9:55.
230. Haruyama N, Hatakeyama J, Moriyama K, Kulkarni AB. Amelogenins: multi-functional enamel matrix proteins and their binding partners. J Oral Biosci. 2011;53:257–66.
231. Wright JT, Li Y, Suggs C, Kuehl MA, Kulkarni AB, Gibson CW. The role of amelogenin during enamel-crystallite growth and organization in vivo. Eur J Oral Sci. 2011;119:65–9.
232. Pugach MK, Suggs C, Li Y, Wright JT, Kulkarni AB, Bartlett JD, et al. M180 amelogenin processed by MMP20 is sufficient for decussating murine enamel. J Dent Res. 2013;92:1118–22.
233. Gibson CW, Li Y, Suggs C, Kuehl MA, Pugach MK, Kulkarni AB, et al. Rescue of the murine amelogenin null phenotype with two amelogenin transgenes. Eur J Oral Sci. 2011;119:70–4.
234. Hu Y, Smith CE, Cai Z, Donnelly LA, Yang J, Hu JC, et al. Enamel ribbons, surface nodules, and octacalcium phosphate in C57BL/6 Amelx(-/-) mice and Amelx(+/-) lyonization. Mol Genet Genomic Med. 2016;4:641–61.
235. Shin NY, Yamazaki H, Beniash E, Yang X, Margolis SS, Pugach MK, et al. Amelogenin phosphorylation regulates tooth enamel formation by stabilizing a transient amorphous mineral precursor. J Biol Chem. 2020;295:1943–59.
236. Beniash E, Metzler RA, Lam RS, Gilbert PU. Transient amorphous calcium phosphate in forming enamel. J Struct Biol. 2009;166:133–43.
237. Mazumder P, Prajapati S, Bapat R, Moradian-Oldak J. Amelogenin-ameloblastin spatial interaction around maturing enamel rods. J Dent Res. 2016;95:1042–8.
238. Chun YH, Yamakoshi Y, Yamakoshi F, Fukae M, Hu JC, Bartlett JD, et al. Cleavage site specificity of MMP-20 for secretory-stage ameloblastin. J Dent Res. 2010;89:785–90.
239. Yamakoshi Y, Richardson AS, Nunez SM, Yamakoshi F, Milkovich RN, Hu JC, et al. Enamel proteins and proteases in Mmp20 and Klk4 null and double-null mice. Eur J Oral Sci. 2011;119:206–16.
240. Lu Y, Papagerakis P, Yamakoshi Y, Hu JC, Bartlett JD, Simmer JP. Functions of KLK4 and MMP-20 in dental enamel formation. Biol Chem. 2008;389:695–700.

241. Bartlett JD, Simmer JP. New perspectives on amelotin and amelogenesis. J Dent Res. 2015;94:642–4.
242. Prajapati S, Tao J, Ruan Q, De Yoreo JJ, Moradian-Oldak J. Matrix metalloproteinase-20 mediates dental enamel biomineralization by preventing protein occlusion inside apatite crystals. Biomaterials. 2016;75:260–70.
243. Smith CE, Hu Y, Hu JC, Simmer JP. Ultrastructure of early amelogenesis in wild-type, Amelx(-/-), and Enam(-/-) mice: enamel ribbon initiation on dentin mineral and ribbon orientation by ameloblasts. Mol Genet Genomic Med. 2016;4:662–83.
244. Bahney CS, Zondervan RL, Allison P, Theologis A, Ashley JW, Ahn J, et al. Cellular biology of fracture healing. J Orthop Res. 2019;37:35–50.
245. Tsang KY, Chan D, Cheah KS. Fate of growth plate hypertrophic chondrocytes: death or lineage extension? Develop Growth Differ. 2015;57:179–92.
246. Pazzaglia UE, Reguzzoni M, Casati L, Sibilia V, Zarattini G, Raspanti M. New morphological evidence of the 'fate' of growth plate hypertrophic chondrocytes in the general context of endochondral ossification. J Anat. 2020;236:305–16.
247. Maes C, Kobayashi T, Selig MK, Torrekens S, Roth SI, Mackem S, et al. Osteoblast precursors, but not mature osteoblasts, move into developing and fractured bones along with invading blood vessels. Dev Cell. 2010;19:329–44.
248. Araldi E, Schipani E. Hypoxia, HIFs and bone development. Bone. 2010;47:190–6.
249. Gillis JA, Alsema EC, Criswell KE. Trunk neural crest origin of dermal denticles in a cartilaginous fish. Proc Natl Acad Sci U S A. 2017;114:13200–5.
250. Cebra-Thomas JA, Betters E, Yin M, Plafkin C, McDow K, Gilbert SF. Evidence that a late-emerging population of trunk neural crest cells forms the plastron bones in the turtle Trachemys scripta. Evol Dev. 2007;9:267–77.
251. Johnston MC, Bronsky PT. Prenatal craniofacial development: new insights on normal and abnormal mechanisms. Crit Rev Oral Biol Med. 1995;6:25–79.
252. McKee MD, Zalzal S, Nanci A. Extracellular matrix in tooth cementum and mantle dentin: localization of osteopontin and other noncollagenous proteins, plasma proteins, and glycoconjugates by electron microscopy. Anat Rec. 1996;245:293–312.
253. Beertsen W, VandenBos T, Everts V. Root development in mice lacking functional tissue non-specific alkaline phosphatase gene: inhibition of acellular cementum formation. J Dent Res. 1999;78:1221–9.
254. Hayashi Y. Ultrastructural characterization of extracellular matrix vesicles in the mineralizing fronts of apical cementum in cats. Arch Oral Biol. 1985;30:445–9.
255. Bosshardt DD, Selvig KA. Dental cementum: the dynamic tissue covering of the root. Periodontol. 2000;1997(13):41–75.
256. Foster BL, Nagatomo KJ, Nociti FH Jr, Fong H, Dunn D, Tran AB, et al. Central role of pyrophosphate in acellular cementum formation. PLoS One. 2012;7:e38393.
257. Ahlberg PE. Palaeontology: birth of the jawed vertebrates. Nature. 2009;457:1094–5.
258. Qu Q, Haitina T, Zhu M, Ahlberg PE. New genomic and fossil data illuminate the origin of enamel. Nature. 2015;526:108–11.
259. Beniash E. Co-option of hair follicle keratins into amelogenesis is associated with the evolution of prismatic enamel: a hypothesis. Front Physiol. 2017;8:823.

Genes and Gene Networks Regulating Mineralization

4

Complex organisms cannot be construed as the sum of their genes, nor do genes alone build particular items of anatomy or behavior by themselves. Most genes influence several aspects of anatomy and behavior – as they operate through complex interactions with other genes and their products, and with environmental factors both within and outside the developing organism. We fall into a deep error, not just a harmful oversimplification, when we speak of genes 'for' particular items of anatomy or behavior. [1]

Glossary

Basic helix-loop-helix An evolutionarily conserved protein motif comprising many dimeric transcription factors which mediate DNA binding and regulate gene expression.

Chromatin Chromosomal DNA (and proteins) packaged in the nucleus by interacting with nucleosomes.

Chromosome A thread-like structure in the nucleus of most eukaryotic cells consisting of condensed DNA and small proteins (chromatin) and carrying genomic information of an organism.

Cis-acting elements Non-coding DNA sequences which regulate the expression of mRNA.

Cre-recombinase An enzyme which facilitates the recombination of DNA between loxP sites and commonly used in studies of the mouse genome.

CRISPR/Cas9 A genome-editing technology applied to modify specific regions of chromatin; the acronym represents "*c*lustered *r*egularly *i*nterspaced *s*hort *p*alindromic *r*epeats/*C*RISPR-*as*sociated protein *9*."

Enhancer Cis-regulatory DNA sequences which regulate mRNA transcription by binding transactivating proteins.

Environmental niche A potential tissue space in which cells receive and integrate signals promoting functional activities.

Epigenetic Local or environmental factors which alter gene transcription.

I. M. Shapiro, W. J. Landis, *Mechanisms of Mineralization of Vertebrate Skeletal and Dental Tissues*, https://doi.org/10.1007/978-3-031-34304-9_4

Gene regulatory network An interaction of genes which govern the expression of mRNA.

Histone A family of five types of proteins which serve to interact with DNA to form a nucleosome; histones H2A, H2B, H3 and H4 form two doublets (an octamer) comprising the nucleosome core, and histone H1, located outside the nucleosome, binds to linker DNA joining neighboring nucleosomes.

Interactome Term used to describe the interaction of proteins with transcription factors regulating gene transcription.

Kinase An enzyme which transfers phosphate groups from ATP and other energy-rich molecules to proteins and other substrates.

MicroRNA Small non-coding RNA species which regulate gene expression at the post-translational level.

Nucleosome The basic subunit of chromatin, comprised of eight histone proteins, a histone octamer, about which a short (146 base pairs) span of DNA wraps.

Pericytes Fibroblastic cells which are bound to endothelial cells lining blood vessels.

"Pioneer" transcription factor Proteins which promote gene transcription by directly interacting with and displacing condensed chromatin to allow the subsequent recruitment of other transcription factors and proteins.

Polyubiquitination The process of linking multiple ubiquitin molecules onto a protein, causing it to be degraded in the proteasome or in autophagic vacuoles.

Proteolysis The process of protein degradation.

SMADs A family of proteins of similar structure which are the principal transducers of signals for receptors of the TGF-β superfamily.

Stem cell An undifferentiated cell which undergoes asymmetric division to produce a self-renewing daughter cell and a cell which can commit to a differentiated pathway.

Super-enhancer Many clusters of enhancer elements on DNA which bind proteins required for gene transcription.

Topologically associating domain (TAD) A dense cluster of DNA sequences which interact with each other.

TRAF adaptor proteins A protein family of seven members (in mammalia) which activate latent transcription factors involved in signal transduction.

Trans-acting factors Proteins which bind to cis-acting elements and regulate gene expression.

Transcription factor A protein generated by a non-coding gene which controls the transcription of DNA to mRNA.

TWIST proteins A family of basic helix-loop-helix transcription factors important in embryonic development and thought to regulate certain genes mediating bone formation.

Abbreviations and Proper Names for Genes/Proteins and Other Molecules

ACAN	Aggrecan
ALP	Alkaline phosphatase
AMBN	Ameloblastin
AMELX	Amelogenin
AMTN	Amelotin
APC	Adenomatous polyposis coli
ATF4	Activating transcription factor 4
AXN	AXIN
BMP	Bone morphogenetic protein
BMP(*R*)	Bone morphogenetic protein (receptor)
BSP	Bone sialoprotein
CA	Carbonic anhydrase
CBFA1	Core-binding factor subunit alpha 1
CBFB	Core-binding factor subunit beta
CD	Cluster of differentiation
CDH5	Cadherin 5
CK	Casein kinase
COLI/COL1	Type I collagen
COLII/COL2	Type II collagen
COLX/COL10	Type X collagen
CTNNB1	β-catenin
DES	Desmin
Dll	Delta-like
DLX5	Distal-less homeobox 5
DMP1	Dentin matrix protein 1
DSPP	Dentin sialophosphoprotein
DVL1	Dishevelled (1) or segment polarity protein dishevelled homologue (1)
EGFL1	Epidermal growth factor-like domain 1
ENAM	Enamelin
EPFN	Epiprofin
ERK1/2	Extracellular signal-regulated protein kinases 1 and 2
FGF	Fibroblast growth factor
FGF(*R*)	Fibroblast growth factor (receptor)
FLT1	FMS-related tyrosine kinase 1
FN	Fibronectin
FOSL1	Fos-like antigen 1
FOX(*C1*)	Forkhead box (C1)
FZD	Frizzled
GDF	Growth differentiation factor
GLI	Glioma-associated oncogene homologue
GSK3	Glycogen synthase kinase 3
HAPLN3	Hyaluronan and proteoglycan link protein 3
HAT	Histone acetyltransferase

HDAC	Histone deacetylase
HEY	Hairy ear, Y-linked
HH	Hedgehog
HIF	Hypoxia-inducible factor
IHH	Indian hedgehog
KLK4	Kallikrein-related peptidase 4
LRP	Low-density lipoprotein receptor-related protein
MAPK	Mitogen-activated protein kinase
MEF	Myocyte enhancer factor
MEPE	Matrix extracellular phosphoglycoprotein
MMP	Matrix metalloproteinase
MSX2	Muscle segment homeobox 2
mTOR	Mechanistic target of rapamycin
NELL1	Neural EGFL-like 1
NF-kB	Nuclear factor kappa-light-chain-enhancer of activated B cells
NFAT	Nuclear factor of activated T-cells
NFIC	Nuclear factor IC
NG2	Neural/glial antigen 2
NOTCH	Neurogenic locus notch homologue
OCN	Osteocalcin
OPG	Osteoprotegerin
OPN	Osteopontin
OSX	Osterix
p38	p38 mitogen-activated protein kinase
p300	Transcriptional co-activator protein
PANX	Pannexin
PDGF	Platelet-derived growth factor
PDGF(R)	Platelet-derived growth factor (receptor)
PECAM	Platelet endothelial cell adhesion molecule
PHD	Prolyl hydroxylase
PHEX	Phosphate-regulating neutral endopeptidase, X-linked
POL II	RNA polymerase II
PTH	Parathyroid hormone
PTH(R)	Parathyroid hormone (receptor)
PTHrP	Parathyroid hormone-related protein
RANK	Receptor activator of nuclear factor-kappa B
RANKL	Receptor activator of nuclear factor-kappa B ligand
RSK	Ribosomal S6 kinase
RUNX	Runt-related transcription factor
SATB	Special AT (adenine, thymine)-rich sequence-binding protein
SHH	Sonic hedgehog
SLC	Solute carrier
SMA	Smooth muscle actin
SMAD	Mothers against decapentaplegic, homolog of Drosophila gene
SOST	Sclerostin

SOX	Sex determining region Y (SRY)-box
STAT	Signal transducer and activator of transcription
TBX	T-box transcription factor
TCF7	Transcription factor 7
TCF/LEF	T-cell specific transcription factor/lymphoid enhancer-binding factor
TGF-β	Transforming growth factor-β
TMSB4	Thymosin beta 4
TNAP	Tissue non-specific alkaline phosphatase
TNF	Tumor necrosis factor
TRAF	Tumor necrosis factor receptor-associated factor
VEGF	Vascular endothelial growth factor
VIM	Vimentin
vWF	von Willebrand factor
WNT	Wingless-related integration site
WWP2	WW domain-containing protein 2

4.1 Hierarchy and the Microenvironment

The story of Lot and his family illustrates that, if a command from *on-high* is unheeded or flouted, there are unforeseen human problems: For non-compliant Edith, her defiant behavior resulted in salt encrustation. The Biblical text indicates that the fate of the Lot family was supervised by the Deity at the top of a hierarchical regulatory ladder; below him/her were the Angels, and toward the bottom of the ladder was Lot, who conveyed instructions in a patriarchal manner to his wife and children. Despite the warning that the family should flee Sodom and the "epigenetic" urging of the accompanying angels, Edith was unable to transduce these signals into the appropriate phenotypic response and, as a result, she was cast into a pillar of salt.

The Biblical narrative may be considered as a metaphor which can be applied to the hierarchical control of gene expression, leading to the question of how the process is regulated. To prevent the appearance of a disordered phenotype and responses which could be fatal, such as that which befell Edith in the Family Lot, gene control is multilayered and systematic. Perched high on the control ladder in the cell nucleus is chromatin, whose physical state exerts a powerful effect on expressed genes. Transcription begins when regions of otherwise non-accessible nucleosomal DNA are unmasked by proteins called "pioneer transcription factors." Bound to DNA, such proteins induce chromatin remodeling and recruitment of non-pioneer transcription factors together with enzymes which provide epigenetic stability. As a result of these interactions, groups of genes become organized within larger structures or topologically associating domains (TADs), as great as 1 Mb in length [2].

Within an active TAD (also called euchromatin, as distinct from heterochromatin), considerable numbers of encoded proteins bind each other and DNA and function as repressors or activators of transcription. Binding results in chromatin looping as TAD and non-coding enhancer and promoter sequences of DNA are brought close enough to influence transcription. Transcription, the process by which information contained within a specific DNA sequence is used to generate a complementary strand of RNA (specifically, messenger RNA or mRNA), is mediated by enzymes, in particular RNA polymerases, which bind promoter sequences, localized upstream of the transcription start site. Individual steps in the transcription mechanism are described extensively in other texts and will not be recapitulated here, but aspects of transcription and certain transcription factors which regulate the mineralization process in vertebrate tissues are presented in this chapter.

Farther down the hierarchical ladder, tissue-specific changes in secreted proteins are transduced by alterations in the tight interaction between genomic DNA and nucleosomal proteins (nucleosomes are regions of DNA tightly wound around particles comprised of histone and non-histone proteins). It should be emphasized that the mechanisms by which chromatin regulatory proteins control chromosomal architecture and eventually gene transcription are substantially different from those which affect transcription factor–DNA interaction [3]. Such "epigenetic" interaction is mediated by enzymes (histone deacetylase, acetyltransferases, histone kinases, and methyltransferases), which modify the charge on histone proteins and influence nucleosomal binding to the negatively charged acidic groups of DNA [3]. In addition, enzymes of the switch/sucrose-nonfermentable ATPase family (SWI/SNF) may remodel chromatin by moving or repositioning the nucleosome so as to change the interaction of specific DNA-binding regulatory proteins of the nucleosome octamer with chromatin strands. In this instance, when chromatin unwinds, a more open architecture appears and presents a landscape in which the bound regulatory proteins interact with promoter or enhancer regions of a specific gene to alter transcriptional activity [4]. Thus, the relative inaccessibility of DNA which limits the rate of gene transcription is overcome by epigenetic reactions remodeling chromatin. Each of the hierarchical factors discussed above, then, exerts a coordinated response which impacts target gene expression. Moreover, the system provides a mechanism by which the genome senses and responds to the epigenetic environment through changes in metabolic intermediates and cellular energetics.

How do these hierarchical ideas concerning chromatin structure relate to the mineralization process? In the previous chapter, it was pointed out that mineralization is a stepwise process: Cells assume a specific differentiated state and synthesize and secrete a complex tissue-specific extracellular organic matrix; calcium and phosphate ions are accumulated in the cell and transferred to the matrix where the mineral is subsequently deposited in a defined configuration. While this sequence indicates a hierarchical and carefully controlled and ordered set of events, it also suggests that there is a coupling between the differentiation status of a cell and mineral formation. It is also clear that the system can be uncoupled and, depending on local environmental signals, cellular activity can lead to the mineral deposition which may be dystrophic or aberrant (Such mineralization is discussed in more detail in Chap. 10). The goal of the present chapter is to build on these basic understandings of gene regulation and define hierarchical regulatory circuits that mediate mineral deposition in bone, teeth, and other vertebrate tissues. In this context, aspects of the roles of selected transcription factors as well as target genes affecting mineralization are addressed.

4.2 Gene Regulatory Networks Involved with the Induction of Mineralization

The functional blueprint of a cell is dependent on the transcription of specific genes from selected regions of its nuclear chromatin. Besides the obvious interest in these protein-coding genes in mediating events linked to mineral deposition, it should be mentioned that the greater portion of the mammalian genome contains sequences that are non-coding. Many of these nucleotide sequences, together with proteins called transcription factors and influenced by the structural status of the coiled nucleosomal DNA, play an important role in regulating the expression of protein clusters which characterize a specific cell type. From these macromolecular interactions and associations, the "interactome" [5], a complex web, governs every aspect of cell function including, of course, the mineralization process. In this regard, mechanistic studies of genes and the interactome have led to a new branch of science, genomic analysis, and with it, informatics. Advances in these fields have relied on the development of computational tools involving more and more refined algorithms to probe hierarchical functions. High throughput genomics and gain and loss of function techniques are also now in use to explore the role of regulatory genes responsive to transcriptional control. The detailed information being obtained in this context connects many different biological systems and provides a new understanding of evolutionary genomics [6]. Whether knowledge of the interactome would have predicted that Edith would peek at Sodom as it burned is doubtful; whether curiosity was built into her personality is probable; whether personality is a gene trait is likely! Transcriptional traits such as curiosity and personality are complex genetic issues which will take years of sequencing and analysis to untangle. In contrast, more discrete functions such as the control of events which lead to mineral deposition are now yielding many new insights.

As indicated above, transcription is exquisitely regulated by groups or clusters of transcription factors and other proteins known as enhancers, which bind to stretches of DNA often located at distances hundreds of kilobases from the start site of a gene or even within the gene itself [7]. For example, the enhancer which regulates sonic hedgehog (*SHH*) is found 1 Mb away from the gene start site and is actually present in the intron of the limb development membrane protein 1 gene in the mouse and human [8]. If the enhancer is mutated, then the phenotype expresses a form of polydactyly [8].

In addition to enhancer elements, there is evidence of "super-enhancer" motifs, that is, a collection or group of enhancers which drive cell-type-specific expression programs defining a particular tissue type [9]. Compared to enhancers, such super-enhancer elements are larger in size, include a much higher number and occupancy of transcriptional regulatory proteins, and function to transcribe DNA to RNA more rapidly [10]. In skeletal biology, super-enhancers have been identified in chondroprogenitor cells which are responsive to sex-determining region Y-box (SOX)9 as well as cells of the dental epithelium [11]. Embryonic stem cells utilize a combination of super-enhancers and a protein, Mediator, which is itself a complex of more than 20 proteins [12], to maintain transcriptional activity. Together with other enhancer elements, Mediator regulates the transcription of pluripotency and self-renewal genes, *Oct3/4*, *SOX2*, *Nanog*, and *KLF4* [12]. Additionally and in an extremely important functional context, Mediator binding with RNA polymerase II (Pol II) activates gene transcription [13].

Besides these regulatory elements described above, there are the "pioneer" factors, mentioned earlier, which mediate transcription by opening up or "remodeling" chromatin [14]. Several transcription factors affecting the phenotype of vertebrate mineralized tissues are listed in Table 4.1 [15–46].

In all cases in Table 4.1, regulation is achieved by a number of transcriptional proteins binding to enhancer and promoter elements of chromatin. Unless indicated otherwise in Table 4.1, all syndromes are related to gene mutation or deletion.

In eukaryotic cells, DNA is wound tightly around nucleosomes, as noted previously, and such interaction restricts transcription factor access to the genomic DNA. Pioneer transcription factors recognize specific sequences in regions of compacted DNA about the nucleosomes and allow the recruitment of other proteins, particularly histone-modifying enzymes [14]. Chromatin is structurally remodeled as a result of the activity of such enzymes, allowing enhancer regulatory elements to bind to DNA sequences which have thereby become accessible. Transcription follows. Examples of pioneer factors are SOX2, paxillin 7 (PAX7), KLF4, and special AT-rich sequence-binding protein-1 (SATB1). Like SWI/SNF, which controls embryonic cell lineage by modulating DNA–histone protein interactions [47], SATB1 alters chromatin architecture by folding chromatin into loop domains to facilitate transcription [48]. SATB1-dependent transcriptional activities are expressed by presecretory ameloblasts and promote nuclear polarization, Tomes' process formation, and ultimately secretion of proteins which form the enamel matrix [49]. In growth plate chondrocytes, forkhead box (FOX)C, meets the definition of a pioneer factor [40, 50, 51]. This protein disturbs nucleosomal–DNA interaction, permitting the binding of the transcription factor, glioma-associated oncogene homolog (GLI)2, and provoking the expression of Indian hedgehog (*IHH*), parathyroid hormone-related protein (*PTHrP*), and type X collagen (*COLX*) target genes [51]. As will be discussed later, proteins encoded by these genes play major roles in endochondral mineralization.

The expressed genes, associated proteins, and other cis-regulatory elements together form a gene regulatory network (Fig. 4.1). Although such a network is a mathematical construct, it provides a basis for understanding and defining the complexity of gene transcription. Its role is to control the expression of an unspecified number of genes: Depending on the developmental state of the tissue or metabolic function, these genes can vary in number from tens to many hundreds. A characteristic of the network is the presence of specific transcription factors, some of which

Table 4.1 Partial listing of important transcription factors regulating the phenotype of vertebrate skeletal and dental tissues

Transcription factor	Bone and cartilage formation and mineralization	Dental tissue formation and mineralization	Abnormality associated with gene mutation and reference
Distal-less homeobox (DLX)	Induced by BMP; with MSX regulates osteogenesis in the jaw; promotes chondrocyte differentiation (DLX-2, -3, -5 and -6)	Promotes organization of dental epithelial cells and required for odontogenic patterning; regulates enamel morphogenesis and *AMELX* expression (DLX-1 and -2)	Tricho-Dento-Osseous syndrome (enamel hypoplasia, taurodontism, and increased density of intramembranous and endochondral bone) [15, 16]
Epiprofin (EPFN/SP6)		Expressed by secretory ameloblasts and odontoblasts; together with MSX-2 controls ameloblast cell fate in the dental epithelium	Amelogenesis imperfecta [17]
Forkhead box (FOX and FOX-C)	Promotes osteogenesis and endochondral ossification; required for craniofacial development; interacts with RUNX2 and ATF4; inhibits RANKL expression (FOX-1, -3 and -4; FOX-C1)	Promotes enamel maturation (FOX-1)	May play a role in pathogenesis of osteoporosis and amelogenesis imperfecta (see Chap. 10) [18, 19]
Hypoxia-inducible factor (HIF)	Positive regulator of osteogenesis; increases osteoblast number and activity; couples angiogenesis to osteogenesis during endochondral ossification; required for extracellular protein synthesis; influences osteoblast-osteoclast coupling (HIF-1α)	Enhances expression of enamel matrix genes (*AMELX*, *AMBN*, *ENAM*, *MMP20* and *TUFTELIN*) as well as *HIF-1α* and possibly hypoxia	Metabolic bone diseases, osteoporosis, degenerative disc disease, bone fracture, osteonecrosis of the jaw [20–22]
	Inhibits osteoblast proliferation and stimulates RANK-induced osteoclastogenesis (HIF-2α)		

(continued)

Table 4.1 (continued)

Transcription factor	Bone and cartilage formation and mineralization	Dental tissue formation and mineralization	Abnormality associated with gene mutation and reference
Muscle segment homeobox (MSX)	With BMP, required for skeletogenesis; promotes chondrogenic differentiation (MSX-1 and -2)	Required during tooth germ development; prevents odontoblast differentiation by inhibiting expression of BMP at the late bell stage; promotes ameloblast survival and expression of secreted proteins and cell adhesion molecules	Affects tooth agenesis and cleft palate, parietal foramina 1 and type 2 craniosynostosis, and possibly bisphosphonate-induced osteonecrosis of the jaw [23–25]
Nuclear factor of activated T-cells (NFAT)	Required for osteoclast formation; inhibits osteoblast differentiation; with NOTCH, blocks osteoblast maturation and mineralization (NFATc1, -2, and -3)		[26, 27]
Nuclear factor I (NFI)	Regulates postnatal proliferation of chondrocytes and endochondral bone formation in the growth plate (NFI-C)	Regulates odontoblast and ameloblast differentiation; required for root formation with OSX, RUNX2 (NFI-C2 and -4)	[28, 29]
Neurogenic locus notch homologue (NOTCH)	Promotes chondrogenic differentiation; suppresses early and late stages of osteoblastogenesis; modulates osteocyte differentiation and mineralization; induces osteoclastogenesis	Required for generation of the enamel organ; promotes differentiation of ameloblasts and odontoblasts; enhances enamel and dentin matrix secretion and formation of the tooth root	Hajdu-Cheney syndrome (spondylocostal dysostosis, kyphosis, osteoporosis, multiple dental abnormalities) [30, 31]
Osterix (OSX/SP7)	Controls osteoblast commitment and osteogenesis; required for intramembranous bone formation; inhibits chondrocyte maturation	Required for odontoblast and cementoblast differentiation and tooth and root development; with RUNX2 and NFI, regulates ameloblast differentiation	Loss of bone density (osteoporosis) and possibly congenital hip dysplasia; recessive form of osteogenesis imperfecta [32–34]

Table 4.1 (continued)

Transcription factor	Bone and cartilage formation and mineralization	Dental tissue formation and mineralization	Abnormality associated with gene mutation and reference
Runt-related transcription factor (RUNX)	Promotes osteogenesis, osteoblast differentiation, and bone formation (RUNX2, -1 and -3)	Promotes dentin formation and mineralization as well as enamel maturation (RUNX2, -1 and -3)	Cleidocranial dysostosis (delayed closure of fontanels, hypoplastic clavicles, supernumerary teeth, retained primary teeth, and delayed or arrested eruption of permanent teeth) [35–37]
Mothers against decapentaplegic, homolog (SMAD)	Responds to TGF-β/BMP and interacts with other transcription factors (RUNX2) to promote osteogenesis and bone formation (SMAD1, -4, -5, and -8)	Regulates crown dentinogenesis through BMP-WNT-SMAD signaling in odontoblasts	Myhre syndrome (SMAD4 gain of function mutation, causing short stature, brachydactyly, facial dysmorphism, deafness, and developmental delay) [38, 39]
Sex-determining region Y (SRY)-box (SOX)	Promotes chondrogenic differentiation and cartilage formation; involved with early stages of osteogenesis (SOX5, -6, and -9)	Promotes odontogenesis and amelogenesis (SOX5, -8, -9, -11, -13, -14, -17, -18, and -21)	Campomelic dysplasia (shortening and bowing of long bones, development of clubfeet, loss of ribs, cleft palate, scoliosis, and micrognathia) [40–42]
TWIST	With FGFR2, required for maturation of osteoprogenitor cells; blocks chondrogenesis (decreases ability of RUNX to bind DNA)	With FGF10, required for tooth morphogenesis and odontoblast differentiation; promotes ameloblast differentiation and expression of *AMBN*	Saethre-Chotzen syndrome (craniosynostosis, premature closure of craniofacial sutures, and limb abnormalities) [43, 44]
β-catenin and transcription factor 7 (TCF7)	Affects WNT signaling; promotes osteogenic differentiation of mesenchymal stem cells and intramembranous and endochondral ossification; required for transdifferentiation of epiphyseal chondrocytes	Required for odontoblast maturation, cementogenesis, and root formation and elongation; promotes differentiation of ameloblasts	Disrupted WNT signaling associated with osteoporosis and sclerosteosis [45, 46]

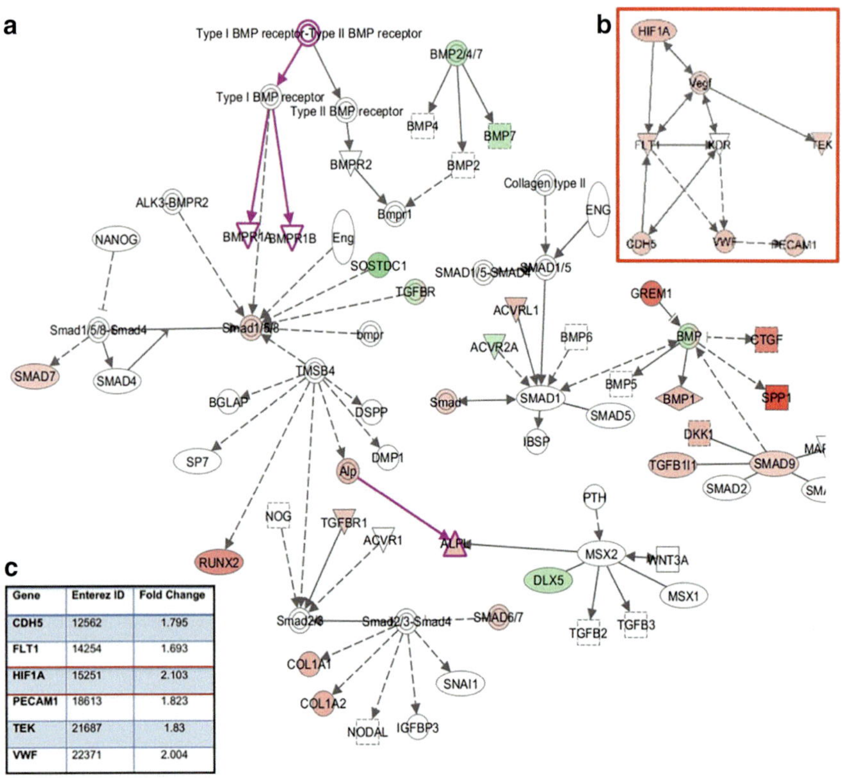

Gene	Enterez ID	Fold Change
CDH5	12562	1.795
FLT1	14254	1.693
HIF1A	15251	2.103
PECAM1	18613	1.823
TEK	21687	1.83
VWF	22371	2.004

Fig. 4.1 Informatic analysis (Ingenuity Pathway Analysis) of gene regulatory networks activated in patients with heterotopic ossification caused by large surface area burns. Over 25,000 genes in the adipose tissue of more than 200 patients with burns together with a control population were evaluated. These genes were interrogated using (**a**) an interactive network of mRNA transcripts of burned or unburned "control" patients. Transcripts (3500 in this study) were directly (solid arrow) or indirectly (broken lines) influenced by burns. Arrow colors indicate the strength of interactions between genes: solid purple denotes strongest, black weaker, and broken arrow weakest relationships. Transcripts heavily involved with the induction of heterotopic ossification include *SMADs*, *BMPs*, *BMPR*, *MSX*, *ALP*, *DLX5*, *WNT3A*, and the cytoskeleton gene, *TMSB4*. (**b**) Significant upregulation of *HIF1α* determined to be influenced by burns in this study and its downstream network of genes, including *vWF*, *PECAM1*, *FLT1*, *CDH5*, and *VEGF*. (**c**) Fold increase in gene expression of burn patients compared with control subjects. Entrez ID is a system used to categorize specific genes. (Two transcripts linked to SMAD9 are incompletely illustrated as in the original schematic from Agarwal et al. [52], reprinted with permission)

act as "genetic nodes." Such nodes interconnect with each other and integrate feedback and feedforward signals. Accordingly, each node or group of nodes receives signals from numerous inputs which amplify or dampen nodal activity, and the integrated output is then represented mathematically as logic functions. This interconnected nodal system (a node cluster) provides measured and robust control of the mRNA ultimately transcribed. Clusters often function together and, in turn, regulate the fidelity of the development of a specific cell type or a tissue.

Gene regulatory networks are being delineated for each of the defined stages of skeletogenesis. Networks (or circuits) which have been examined include neural crest induction [53], stem cell differentiation [54], osteogenesis [55], and tooth development [56]. Farrell et al. [57] examined the earliest stages of embryogenesis using a zebra fish model of development and profiled the transcriptomes of more than 38,000 individual embryonic cells. From this information, they traced the changes of cells, initially defined as originating in the axial or paraxial mesoderm of the zebra fish, as they transitioned into a more defined skeletal phenotype. Interestingly, Farrell et al. found that the fate of a specific embryonic cell type was not fixed: At an early developmental stage, embryonic cells responded to local cues and could assume more than one phenotype (Fig. 4.2) [57].

In the context of gene regulatory networks related to mineralizing tissues, Li et al. analyzed data from RNA-seq and chromatin immunoprecipitation followed by DNA sequencing (ChIP-seq) of human fetal long bone chondrocytes together with other tissue samples [59]. They reported a circuit comprised of 332 protein-coding genes and 34 long non-coding RNA (lncRNA) genes. Their results revealed gene activation of 34 transcription factors including runt-related transcription factor (*RUNX*)*2*, (*SOX*)*8/9*, mothers against decapentaplegic (*SMAD*)*3*, signal transducer and activator of transcription (*STAT*)*1/4*, fos-like antigen (*FOSL*)*1*, and *GLI*. In addition, 11,334 distal enhancers were found to function as transcriptional regulators of cartilage-selective genes with signaling mediated by fibroblast growth factor (*FGF*), hedgehog (*HH*), bone morphogenetic protein (*BMP*), transforming growth factor-β (*TGF-β*), and wingless-related integration site (*WNT*)/*β-catenin* (*CTNNB1*). Predictably, this extremely complex network involved many hierarchical levels of control and its signals were considered to be iterative in nature with built-in redundancy of function [59].

Gao and Davidson [60] have defined the network of regulatory as well as mineralization-related genes in the sea urchin (*Strongylocentrotus purpuratus*). During urchin development, they found that a small group of cells secreted the calcium carbonate skeleton of the organism. Over 30 genes were reported as comprising the network needed for the formation of the sea urchin skeleton and, of these, 23 were regulatory (transcription factors) and signaling genes and the rest were linked to genes required for the synthesis/mineralization of the skeleton. During gastrulation and the initiation of skeletal morphogenesis, there was differential expression of over 400 RNA transcripts [61]. Those transcripts identified in the skeletogenic phase led to six spicule matrix proteins involved with amorphous-crystalline transitions and otopetrin, a protein associated with the formation of otoliths, the small calcium carbonate deposits in the inner ear of vertebrates which function to maintain balance as well as sensitivity to gravity and acceleration [62]. Other genes found included carbonic anhydrase (*CA1*) and five solute carrier (*SLC*) genes which were thought likely to participate in calcium and carbonate ion transport into the sea urchin skeleton [63].

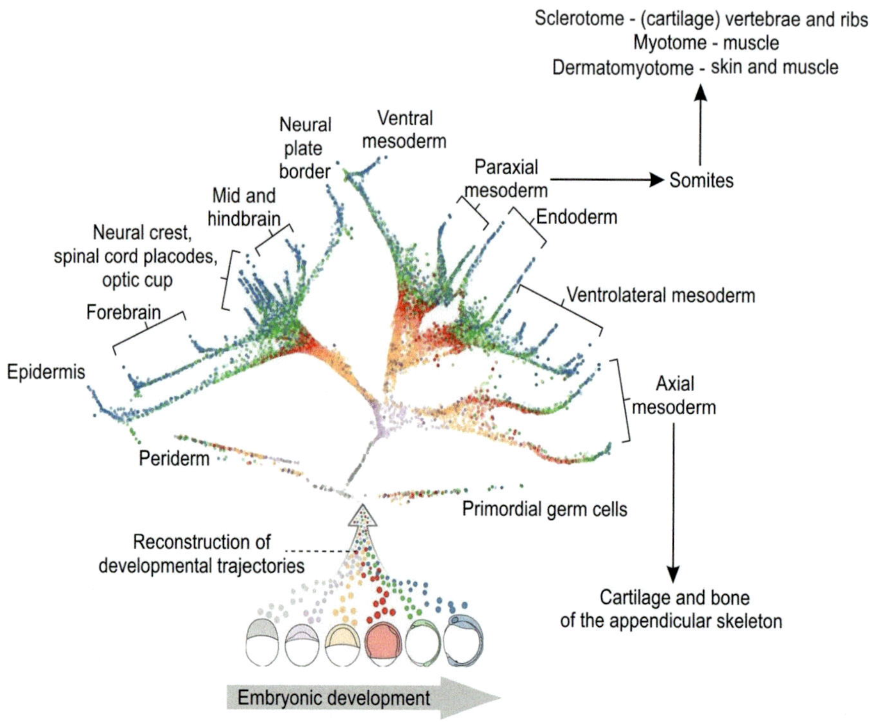

Fig. 4.2 Zebrafish transcriptome following the lineage of cells which commit to the skeleton as well as other organs and tissues of the body. During embryonic development, the transcriptome of over 38,000 individual cells was assessed by RNA-sequencing (RNA-seq) analysis. Cells were collected at 12 different stages of development, six of which are illustrated in the diagram for simplicity and depicted by color. Time points for the study were 3 h postfertilization, when most cells were pluripotent, to 12 h postfertilization, when cells had differentiated into specific cell types. Computational analysis of 25 pluripotent embryonic cells showed that they differentiated to form particular tissues: Axial mesodermal cells generated both bone and cartilage of the appendicular skeleton while somite formation from the paraxial mesoderm led to the vertebrae and their associated rib bones. The work also demonstrates that, despite such commitments, there was also considerable cell lineage plasticity. Tracking of the cells of the neuroectoderm, responsible for tooth formation, was not investigated. In addition, cells of the neural mesoderm give rise to many of the bones of the skull [58]. (Diagram modified from Farrell et al. [57] with permission)

Another example of this type of genomic analysis is its use in probing the causation of abnormal mineralization, in particular, the mechanism by which large surface area burns lead to regions of heterotopic ossification in the soft tissues of a human population (Fig. 4.1) [52]. When genomic information of the burn patients in this study was compared to a control population, the risk of heterotopic ossification was linked to aberrant responses of a number of genes known to be involved directly with regulating the mineralization process. Among those genes most affected by the burns was hypoxia-inducible factor (*HIF*)-*1*, and inhibition of *HIF-1* expression by drugs appeared to mitigate the ossification response. Other genes affected by burns included *SMADs* and the *BMP* receptors (I and II). The role of these and other genes which mediate physiological tissue mineralization is discussed in later sections of this and other book chapters. In this and the preceding illustrations, analysis of gene regulatory networks has provided insight and understanding of gene transcription and helped identify genes and proteins and their interrelated pathways critical to tissue developmental processes, mineralization in these instances.

As an added note regarding gene networks, Davidson and colleagues speculated that network variation might drive evolutionary changes at cis-regulatory genomic nodes [64]. These nodes bind transcription factors and other proteins usually found on the same DNA strand as the transcribed gene. Evolutionary changes in gene sequences were suggested to alter relationships between nodes, possibly causing functional modifications [64]. As more of these networks are mapped and nodal structure identified, detailed information on genomic architecture is expected to provide new perceptions into the genomic control of mineralization, the mechanics of species evolution, and even species survival and extinction.

There are few studies of gene regulatory networks in organisms which generate hydroxyapatite. In the bones of developing zebrafish, Fisher and Franz-Odendaal identified at least three gene networks: Fibrillar collagens, calcium-binding phosphoproteins, and vitamin K-dependent proteins, with *RUNX2*, the master regulator of osteogenesis, serving as the critical regulatory gene [57, 65].

To conclude this section of the chapter, it is important to emphasize that, while detailed studies of regulatory networks of bone and teeth are sparse, considerable numbers of pioneer and other transcription factors participate in both hard tissue formation and mineralization. Among the most extensively studied transcription factors are those which comprise the SOX family, WNT/β-catenin, RUNX2/3, and osterix (OSX). As will be subsequently discussed, the cell phenotype reflects cross-talk between each of the genes, mRNA, and secreted proteins such as BMP and FGF. Other important factors governing gene expression and cell fate are molecular signals originating from secreted proteins and microRNAs as well as environmental cues released by surrounding tissues. The mechanisms by which these niche signals influence gene expression are discussed in the next chapter section, followed by a consideration of the roles of the transcription factors mentioned above.

4.3 The Environmental Niche

From the earliest zygote to the fully mature organism, individual tissues form or exist in protective and complex microenvironmental niches. The concept of a tissue niche was first mentioned by Schofield in 1978 in relation to spleen colony-forming cells and hematopoietic stem cells [66]. Since then, the term, "niche," has been used to indicate a potential space, bounded by a number of different tissue types, in which cells receive and integrate a multitude of signals which guide and enhance their metabolic, genetic, and functional activities [67].

In biomedical research, "niche" has been used rather loosely, often referring to a monolayer of cells in which the effect of a selected environmental factor, such as a specific surface, nutrient, gas, or force application, can be studied. While this approach has yielded a considerable amount of data concerning almost every aspect of cell function, its value is limited as it does not provide a platform for the integration of the multitude of signals generated by the tissues which comprise the niche. In the organism, these signals include the oxemic state of the tissue, endocrinal secretions from the systemic circulation, chemical signals from surrounding cells (paracrine control), chemical mediators synthesized by the cells within the tissue (autocrine control), microRNAs released from exosomes, and biomechanical forces. If a single signal is corrupted or absent, or the local environment is disturbed, mineralization may be incomplete or defective, and as a result, the tissue will be unable to fulfill its intended function.

What types of microenvironments exist in mineralizing tissues? Probably the best-studied example is the endosteal surface of the bone marrow cavity. Marrow itself is a complex fatty tissue containing a myriad of cell types resident within a three-dimensional, poorly organized "stroma." The adult bone marrow stroma sustains stem cells which can differentiate into osteoblasts, chondrocytes, adipocytes, and other cell types, while hematopoietic stem cells within the marrow can commit to the osteoclast lineage [68].

Aside from the bone niche, several other microenvironments are present within hard tissues. For example, nucleus pulposus cells of the intervertebral disc, a remnant of the notochord, populate a niche with boundaries formed by end plate cartilages of the adjacent vertebrae and the annulus fibrosus. In healing fractures, the niche is a cartilaginous callus formed at the fracture site and enclosed by a perichondrium and the marrow cavity; in synovial joints, niche boundaries exist between cartilage and the synovial membranes; and in the growth plate, the tissue limits are the perichondrium and the metaphyseal bone. The developing tooth also exhibits a number of complex microenvironments. For dentin, odontoblasts are confined between a region of enamel and secreting ameloblasts and a region of pulpal tissue. For mineralizing enamel, cells of the stellate reticulum form one boundary tissue while dentin and secreting odontoblasts form another at the tooth inner margin.

Not surprisingly, disruption of any of the niches mentioned above by trauma, infection, changes in applied biomechanical forces or disturbances in oxemic status, for example, results in disease and loss of function [69]. The reason for mentioning oxemic status is that a common feature of most connective tissues is their poor vascularity: In the growth plate, the central hypertrophic zone is almost avascular [70] and, in bone, the oxygen tension in osteocyte lacunae is below 10% [71]. These observations have raised questions as to how the cells populating such tissues generate energy for both survival and anabolic activities and, relevant to the principal theme of this book, initiate events which lead to mineralization [72]. A partial answer to these questions is that niche survival is mediated by the activation of hypoxia-inducible factor-1 (HIF-1), considered as the master transcription factor responsible for regulating cellular energy metabolism (Fig. 4.3) [69]. HIF-1

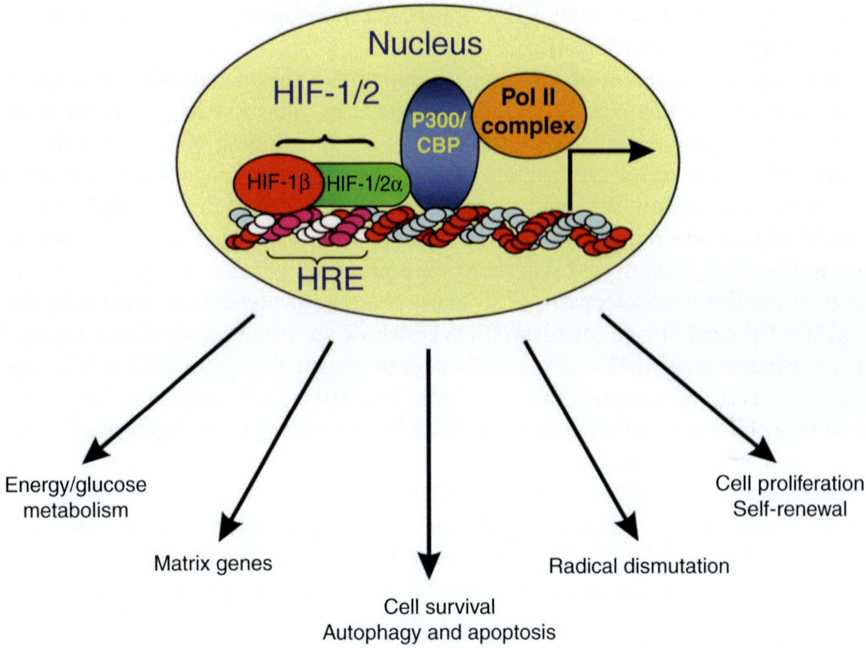

Fig. 4.3 Schematic of the role of the transcription factor HIF in the regulation of cell metabolism and function. Low oxygen tension (hypoxia), such as that experienced by cells of cartilage, cortical bone, or the intervertebral disc, results in the expression and stabilization of the members of the HIF family of transcription factors. Hypoxia stabilizes HIF-1α and it then forms a heterodimer with HIF-1β. These proteins associate with coactivators which include p300/CBP and DNA polymerase II (Pol II) complex and bind to hypoxia-responsive elements (HRE) in the enhancer region of a large number of target genes. The genes are required for critical cell and tissue functions, such as energy metabolism, cell survival, cell proliferation, and several others shown schematically. (Schematic adapted from Risbud et al. [69] with permission)

promotes metabolic adaptation to hypoxia by switching energy pathways from oxidation to glycolysis [72].

HIF-1 is encoded as two separate isoforms, α and β. HIF-1α is sensitive to oxygen tension and its expression is increased in hypoxia. In contrast, HIF-1β is stable in both normoxia and hypoxia [69]. Sensitivity to oxygen tension is mediated by a group of enzymes, oxygen-activated prolyl hydroxylases (PHDs). Under normoxia, these enzymes hydroxylate HIF-1α and thereby render it degradable through the ubiquitin-proteasome pathway. In hypoxia, the activity of the PHD system is inhibited, allowing HIF-1α to accumulate in cells. The protein heterodimerizes with its counterpart HIF-1β and the heterodimer affects the activity of several downstream target genes (Fig. 4.3) [69]. In concert with earlier comments concerning the complexities of transcription, it is not surprising that HIF interacts with a number of transcription factors, including SOX9, RUNX2, and β-catenin [26, 27]. Since each of these proteins is required for the elaboration of a mineralized structure, the hypoxic response mediated by HIF is integrated with functions linked to the induction of mineralization [73, 74].

Genetic manipulation of the mouse genome has shown that HIF-1α supports chondrocyte and osteoblast differentiation, function, and survival [75]. In the avascular growth plate, stabilization of HIF leads to the secretion of a number of isoforms of vascular endothelial growth factor (VEGF), which enhance the invasion of vascular elements into the cartilage [76]. The dependence of the bone HIF system on the oxygen-sensing activity of PHDs has also been investigated. Merceron et al. silenced *PHD2* and showed that there were profound changes in the activities of both osteoblasts and osteoclasts [77]. These changes involved the transcription factor, WNT/β-catenin, and sclerostin (SOST), whose reciprocal expression was found to be dependent on SIRT-1, a key transducer of epigenetic signals [77]. Aside from HIF-1α, PHDs regulate the HIF-2α isoform. This HIF species promotes the interaction between bone formation and resorption by influencing the expression of receptor activator of nuclear factor kappa-B ligand (*RANKL*) and thereby enhancing osteoclastic activity and bone turnover [78].

In summary, the characteristically low oxygen tension in cartilage and bone decreases the activity of PHDs and leads to HIF stabilization and the induction of genes affecting a myriad of cell and tissue functions which include tissue mineralization and bone turnover. These HIF-related changes involve numerous additional transcription factors which are closely related to bone formation. Together with epigenetic factors, HIF-associated proteins form a complex circuit which adapts the cells of cartilage and bone to the challenge of function and survival in a hypoxic environment.

The components of the niche cooperate to permit normal development and mineralization. For many of these niches, the internal hypoxic conditions of a cell promote metabolic coupling, that is, communication between groups of cells which serve to nourish and direct energy generation. That metabolic coupling may exist between cartilage, bone, ligament, and other connective tissues are discussed in more detail in Chap. 9. In recent years, many other niches have been described, including those in the intestine, skin, spinal cord, and brain, as well as those related to the cancer microenvironment [79]. In these examples, it is clear that neighboring cells impact the niche and affect its cellular development. Moreover, as the niche abounds with stem cells which have yet to commit to a defined phenotype, the effects of local conditions on the fate and function of these cells need to be considered. Such local factors favoring stem cell commitment and a number of key transcription factors regulating the mineralization activities of cells of bone and teeth are described in the following sections of this chapter.

4.4 Stem Cells and Mineralization

In a landmark paper, Caplan proposed that the adult bone marrow niche contained stem cells which, under the appropriate conditions, commit to one of a number of different mesenchymal lineages [80]. Experimentally, adherent marrow cells cultured on selected substrates in the presence of complex media caused the commitment to chondrogenic, osteogenic, myogenic, fibrogenic, tenogenic, and adipogenic lineages. The cells were self-renewing and, since they exhibited low immunogenicity, they were regarded as an attractive modality for use in cell therapy and regeneration [81]. The importance of these multilineage marrow stem cells cannot be overstated: There is a large consensus that the regenerative power of these cells, easily harvested from marrow, can be harnessed to foster tissue repair and healing [82], and the field of regenerative medicine has been built around this function.

While marrow stem cells may be readily accessed and utilized in a host of basic and clinical science applications, their consistency of responses and commitment may be variable in terms of mineral deposition and bone formation [83, 84]. In this regard, these observations have led to studies concerned with identifying the true stem cells from among the entirety of cells present in marrow extracts. Initial studies focused on the cellular structure of the marrow itself, in particular vascular tissue [85]. Penetrating the endosteal niche from the marrow cavity are vessels which range in size from microscopic blood sinusoids to larger collecting sinuses emptying into macroscopic sinuses or veins. Surrounding the small caliber vessels are specialized cells, pericytes (Greek, "peri" or "around"), discovered in 1873 by the French scientist, Rouget [86]. These cells, embedded in the basement membrane, are closely associated with the abluminal surface of the marrow vascular elements (Fig. 4.4) [87, 88]. Work in vitro showed that these cells form adherent colonies which can be induced to differentiate along osteogenic, chondrogenic, or adipogenic lineages [89]. Other investigations demonstrated that pericytes commit to neurons and related cells [90], that pericyte cultures of human dental pulp generate odontoblast-like cells [91], and that implanted retinal pericytes form both chondrocytes and adipocytes [92]. These and other results have led some investigators to the conclusion that pericytes are the true mesenchymal stem cell [93]. The assumption that pericytes represent the marrow stem cell remains equivocal but, if it is valid, then it could explain why blood vessels and heart tissue become mineralized (see Chap. 10). Moreover since the number of these cells in marrow would be expected to differ from donor to donor, that variation could help clarify understanding of the inconsistencies in mineralization response noted previously.

An approach to establishing cell lineage is by determining whether a specific gene, protein, or group of proteins are common to both the stem cell, in this case, the pericyte, and the differentiated osteoblast or chondrocyte. In early experiments in this regard, pericytes were found to secrete an abundant number of proteins, including smooth muscle actin (SMA), desmin (DES), vimentin (VIM), neural/glial antigen 2 (NG2), a cluster of differentiation 13 and 146 (CD13, CD146), and platelet-derived growth factor receptor (PDGF(R)) [94]. Initially, it was thought that PDGFR was a specific marker of stem cells and hence it was used to explore the origin of marrow stem cells [95]. However, the protein was detected in stromal cells, perivascular fibroblasts, and vascular smooth muscle cells [96]. Utilization of PDGFR in defining whether pericytes commit to a bone or cartilage phenotype was of limited value, therefore, and yielded to a possibly more accurate methodology.

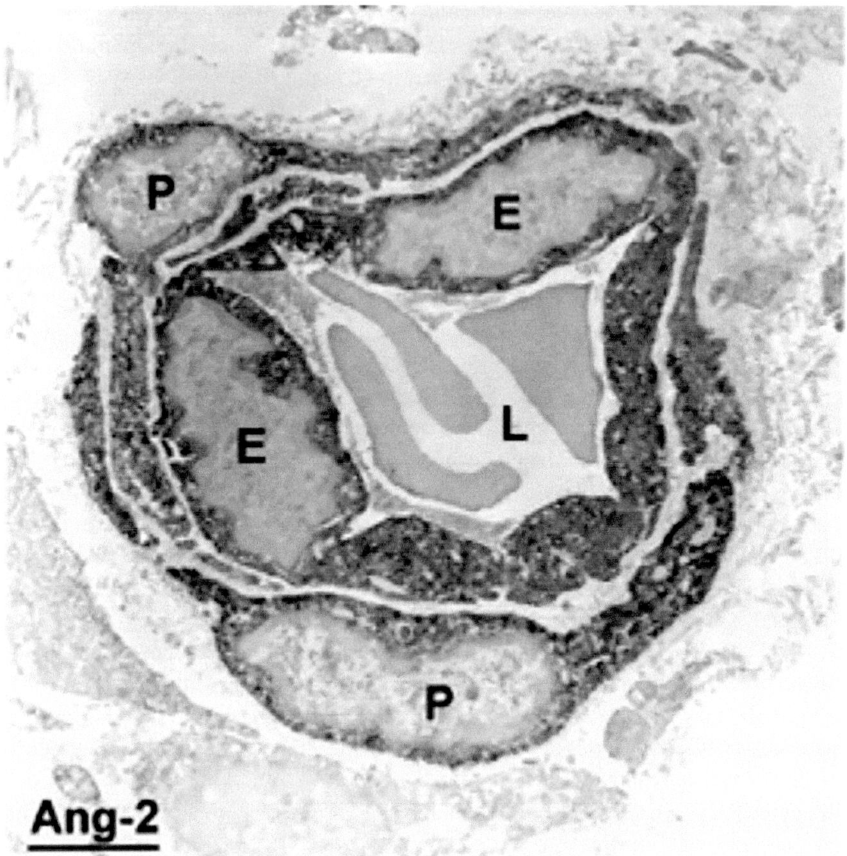

Fig. 4.4 Immunocytochemistry of pericytes associated with a blood capillary. In this experiment, a thermoreversible gelation polymer disc was implanted subcutaneously in a rat to stimulate new capillary growth. After 10 days, discs were retrieved and frozen sections were prepared and treated with an antibody against angiopoietin (Ang). Sections were fixed, stained with osmium tetroxide, and imaged by transmission electron microscopy. Pericytes (P) labeled by Ang-2 were found disposed about a capillary lumen (L) lined by endothelium cells (E). Scale bar = 2.5 μm. (Image adapted from Wakui et al. [87] with permission)

An alternative technique to assessing the origin and fate of a particular cell type during development is that of lineage tracing. This method relies on the fact that cells can be labeled with a gene expressing a multicolor fluorescent protein. For pericytes so labeled and putatively committed to cells of skeletal origin, detection of the labeled gene would then be evident in the stem cell itself (the pericyte), all daughter cells, and the differentiated cells of mineralized tissues. T-box transcription factor (*TBX*)*4* is known to be expressed by pericytes in multiple organs of the mouse and, like bone marrow stromal cells, TBX-labeled murine cells differentiate in culture into fat, bone, and cartilage [97]. TBX4, then, was considered a useful lineage marker and it was evaluated in lineage-tracing experiments of pericytes

[97]. These cells were found to maintain their identity and did not commit to other cell lineages, a result which was inconsistent with the concept that pericytes are multipotent tissue-resident progenitors which can contribute to multiple cell types [27]. Thus, the question of the origin of cells which eventually become bone or cartilage remains unresolved.

The lack of clarity in understanding the ontology of cells of bone or cartilage impacts the new field of regenerative medicine, where so many of its clinical approaches rely on isolating populations of adult stem cells from bone marrow. This shortcoming notwithstanding, what is known with certainty is that, as far as bone and cartilage formation and mineralization are concerned, the cells which commit to these lineages derive from embryonic cells originating during the development of an organism in the somites, the lateral plate mesoderm and the cranial mesoderm [58] (Fig. 4.2), and that a hierarchy of transcription factors carefully controls differentiation along specific and complex pathways. Four major transcription factors, RUNX2, OSX, SOX9, and β-catenin, among several others, play significant roles in regulating the expression of genes which commit cells to a specific mineralization phenotype. Their activities and mechanisms by which many of the secreted proteins of mineralized tissues like BMP interrogate transcription factors and influence both osteogenesis and chondrogenesis are described in the following sections of this chapter.

4.5 *RUNX2* and *OSX*: The "Uber" Genes of Osteogenesis and Dentinogenesis

The *RUNX* family of genes and transcription factors is required for bone formation (Fig. 4.5). The name, *RUNX*, reflects a common family feature: An α-subunit sharing a stretch of DNA 120–130 nucleotides in length labeled the *RUNT* domain [98]. *RUNX* regulates the commitment of mesenchymal cells to the osteogenic and odontogenic lineages and participates in the control of endochondral and intramembranous bone mineralization. In humans, a heterozygous mutation in *RUNX2* (also called core binding factor a [*CBFA*]1) is a cause of cleidocranial dysplasia syndrome (see Table 4.1). This disorder is characterized by anomalies in membrane bone formation, absence of bone growth, open skull fontanels in children, and dental abnormalities, the most common of which is supernumerary teeth [99]. In addition, *RUNX2* regulates the expression of many of the genes linked directly to enamel and dentin mineralization. Since the absence of *RUNX2* prevents bone formation in mice, this transcription factor is regarded as the master (German, "uber" or "over") gene of osteoblast function.

There are three members of the *RUNX* family, *RUNX1*, *RUNX2*, and *RUNX3*. While some redundancy of function exists among them, it is thought that *RUNX1* and *RUNX3* associate with *RUNX2* to regulate chondro- and osteogenesis [100, 101]. In order for RUNX2 to bind DNA in normal function, a second protein is required, core-binding factor b (CBFB)1 [102]. The formation of the RUNX2-CBFB1 heterodimer promotes the interaction of RUNX2 with chromatin and

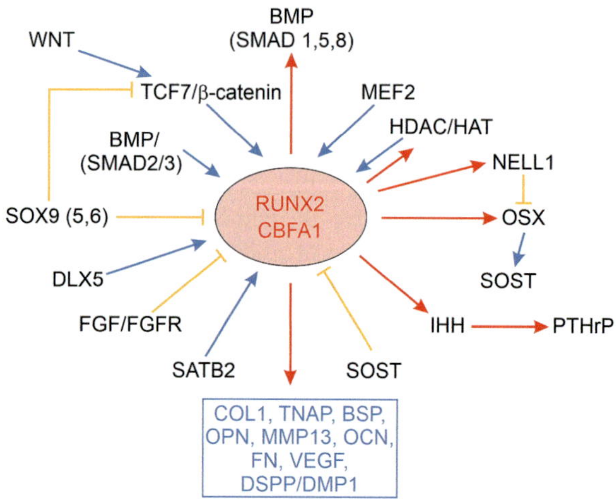

Fig. 4.5 Schematic of transcriptional regulation of *RUNX2/CBFA1* expression and target gene regulation in osteoblasts. Many of the proteins secreted during the differentiation of osteoblasts and their subsequent maturation to osteocytes are influenced by and interact with *RUNX2*. While the interactions are complex, blue arrows indicate proteins which promote *RUNX2* expression, yellow lines indicate proteins which tend to block or inhibit *RUNX2* activity, and red arrows show *RUNX2* target genes and proteins. Those target genes expressed and their proteins subsequently secreted during osteoblast maturation are boxed in blue and include *COL1* (*type 1 collagen*), *TNAP* (*tissue non-specific alkaline phosphatase*), *BSP* (*bone sialoprotein*), *OPN* (*osteopontin*), *MMP* (*matrix metalloproteinase*)*13*, *OCN* (*osteocalcin*), *FN* (*fibronectin*), *VEGF* (*vascular endothelial growth factor*), and *DSPP/DMP1* (*dentin sialophosphoprotein/dentin matrix protein 1*). Other genes activated by *RUNX2* are *BMP* (*bone morphogenetic protein*) *through SMAD1, 5* and *8*; *HDAC/HAT* (*histone deacetylase/histone acetyltransferase*), *NELL1* (*neural EGFL-like 1*), *OSX* (*osterix*), and *IHH* (*Indian hedgehog*), which in turn activates *PTHrP* (*parathyroid hormone-related protein*). Proteins which promote *RUNX2* expression include TCF7 (lymphoid enhancer-binding factor/T-cell specific transcription factor7) in conjunction with β-catenin, BMP through SMAD 2/3, DLX5 (distal-less homeobox)5, SATB2 (special AT-rich sequence-binding protein)2, HDAC/HAT, and MEF2 (myocyte enhancer factor)2. Inhibitors of *RUNX2* include SOX9, 5 and 6, FGF/FGFR (fibroblast growth factor and its receptor), and SOST (sclerostin). NELL1 inhibits *OSX* and OSX activates *SOST*. Only protein notation (non-italicized) is shown in the schematic

reduces RUNX2 ubiquitination (thereby increasing its longevity and prolonging its transcriptional activity) [103]. Absence of either *RUNX2* or *CBFB1* severely and negatively impacts bone formation and mineralization [102, 104]. These and other results indicate that *RUNX2* and *CBFB1* are each required for both endochondral and intramembranous ossification.

Transcription of *RUNX2* is controlled by a phalanx of proteins, many of which have been mentioned previously. Together, the proteins form an enhanceosome; components of this complex include distal-less homeobox (DLX)5, SMAD1, β-catenin (CTNNB1), transcription factor 7 (TCF7), myocyte enhancer factor (MEF)2, SOX6, and OSX [105, 106]. Another transcription factor, SATB2, binds RUNX2 at nuclear scaffolding sites and mediates chromatin-loop remodeling, a process noted earlier which allows histone deacetylases (HDACs) and acetyltransferases (HATs) access to the highly coiled and packaged nucleosomal DNA, thereby providing non-genomic control of cellular activity [107]. Neural EGFL-like 1 (*NELL1*) is a gene regulated by *RUNX2* [108]. *NELL1* encodes an osteoinductive secreted protein which promotes both endochondral as well as membrane bone formation [109]. Thus, *RUNX2* functions to influence the maturation of chondrocytes while enhancing bone formation (Fig. 4.6) [110, 111].

In the growth plate, *RUNX2* plays a key role in controlling the fate of maturing chondrocytes. The gene is expressed in chondrocytes at high levels in proliferating and pre-hypertrophic stages of growth plate development, where it regulates the expression of *COLX*, bone sialoprotein (*BSP*), matrix metalloproteinase (*MMP*)*13*, and *VEGF* [112]. In addition, by regulating the expression of *IHH*, which itself promotes the expression of the secreted protein, PTHrP, *RUNX2* slows chondrocyte terminal differentiation and in a feedback loop inhibits its own further expression [110]. If *RUNX2* is deleted experimentally in an animal model, there is a lack of endochondral bone formation linked to a failure of chondrocytes to assume the hypertrophic state [111]. Accompanying the absence of hypertrophy, cells express low levels of *COLX*, osteopontin (*OPN*), *BSP*, and *MMP13*, and the activity of the enzyme, tissue non-specific alkaline phosphatase (TNAP), is reduced (Fig. 4.6) [111].

With respect to osteogenesis, *RUNX2* regulates the secretion of many proteins which are required for the deposition of endochondral bone. Aside from type I collagen (COLI), such proteins include BSP, osteocalcin (OCN), fibronectin (FN)1, and dentin matrix protein (DMP)1 [113]. In the process of intramembranous mineralization, the commitment of stem cells to the osteogenic lineage is also mediated by *RUNX2*, likely on promoting the expression of *IHH*, *FGF*, and *FGFR2*. *RUNX2-CBFB1* does not promote the transition of osteoblasts to osteocytes, a step which is dependent on *OSX* [114]. Deletion of *RUNX2* decreases the expression of *DMP1* and *SOST*, as well as osteoprotegerin (*OPG*) and *RANKL* [115, 116]. When RANKL binds to its cognate receptor, RANK, in cells, accumulation of tumor necrosis factor receptor-associated factor (TRAF) adaptor proteins and induction of NFATc1 occur [26]. Together with neurogenic locus notch homolog (NOTCH), NFATc1 forms a circuit which prevents osteoblast maturation [26]. Associated with other transcriptional proteins, NFAT activates genes required for the initiation of a resorptive response [117]. Thus, the deletion of *RUNX2* negatively influences osteoclast activity, thereby decreasing the rate of bone removal [115]. In this situation, bone turnover and subsequent bone formation and replacement by activation of resorption are impaired.

As indicated earlier, in addition to *RUNX2*, the *RUNX1* isoform is involved with chondro- and osteogenesis. *RUNX1* is expressed in both cartilaginous growth plate and bone formation [118]. Tang et al. showed that *RUNX1* upregulated the expression of both *RUNX2* and *OCN* and its deletion caused defective endochondral and intramembranous bone ossification [101]. Moreover, when *RUNX1* was deleted from chondrocytes, there was a loss of chondrocyte maturation in the growth plate with low levels of expression of *IHH* and *PTHrP* [36]. Such results, as mentioned above, indicate that *RUNX1* with *RUNX2* mediate chondrogenesis and endochondral bone formation. An overview of *RUNX2* control of chondrocyte terminal differentiation and osteogenesis is shown in Fig. 4.6 [119].

In common with osseous tissue derived from the lateral plate mesoderm, *RUNX2* is also expressed in mineralizing tissues originating from the neural crest [120].

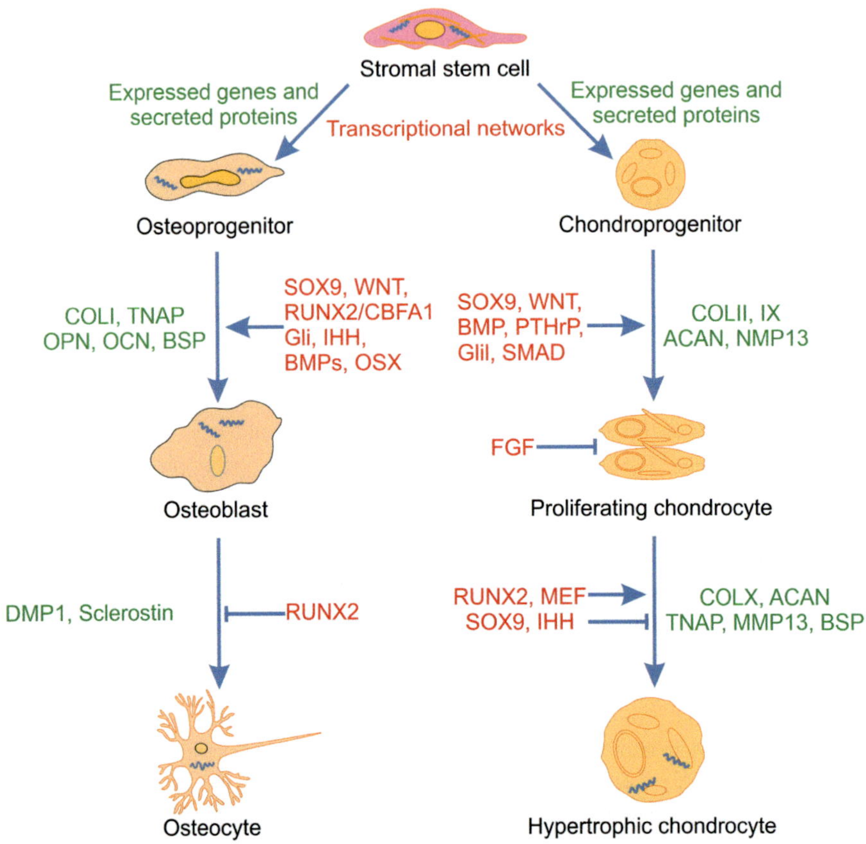

Fig. 4.6 Schematic representation of the roles of proteins in transcription networks (red) and the expression of genes and secretion of their proteins (green) during chondrogenic and osteogenic differentiation. SOX9 expression/secretion is key to promoting multipotent mesenchymal stromal cell commitment to osteoblastic and chondrocytic lineages. While transcriptional circuits differ during the maturation of progenitors into chondrocytes and osteoblasts, cells are activated by a number of common transcriptional proteins including SOX9 and WNT. RUNX2 regulates the maturation of osteoprogenitor cells into osteoblasts while slowing the differentiation of osteoblasts into osteocytes. SOX9 together with FGF slows the differentiation of chondroprogenitor cells and proliferating chondrocytes into the terminal hypertrophic state. Further control is mediated by IHH which, together with PTHrP, forms a negative feedback loop regulating chondrocyte hypertrophy. Thus, IHH expression/secretion in pre-hypertrophic chondrocytes induces PTHrP expression/secretion in immature chondrocytes while PTHrP inhibits hypertrophic differentiation. Gene expression and protein secretion by chondrocytes and osteoblasts depend on the stage of maturation of these cells. The early bone mineralizing phenotype is characterized by the secretion of COLI, TNAP, OPN, OCN, and BSP. Osteocytes also secrete a number of proteins, including PHEX, MEPE, DMP1, FGF23, and sclerostin. During chondrogenesis, proteins secreted by proliferating chondrocytes include COLII and IX, ACAN, and BSP; mature hypertrophic cells secrete COLX, TNAP, and MMP13 among others

RUNX2 is required for the very early stages of enamel and dentin formation [121]. The transcription factor is highly expressed in pre-odontoblasts, and its level of expression decreases significantly with cell maturation. *RUNX2* promotes the expression of a number of genes which encode proteins of the extracellular dentin matrix, including the major non-collagenous protein, dentin sialophosphoprotein (DSPP), together with OCN, OPN, and DMP [122]. If *RUNX2* is deleted in an animal model, the pulp chamber of a tooth is enlarged, odontoblasts lose their columnar shape, dentinal tubules fail to form and cell polarization is lost [123]. Moreover, the tissue which is developed resembles bone more than dentin [123]. This gene defect also influences the development of enamel in several ways: enamel maturation is delayed and enamel mineral density is low, enamel rods and interrod substance are disorganized, and there is an accumulation of enamel proteins, possibly caused by a decrease in kallikrein-related peptidase (KLK)4 activity [124]. Thus, *RUNX2* is required for odontoblasts and ameloblasts to develop their full phenotype and, without this transcription factor, odontogenesis, amelogenesis, and dentin and enamel mineralization cannot be completed [125].

As an additional note, differentiation of RUNX2-positive osteoprogenitor cells into functioning osteoblasts and eventually osteocytes is dependent on the synthesis and secretion of the transcription factor, OSX [126]. The *OSX* gene is expressed by many, if not all, bone cells (osteoblasts, osteocytes) as well as pre-hypertrophic and hypertrophic chondrocytes [127]. Target genes for OSX associated with the development of the osteocyte phenotype involve X-linked phosphate-regulating neutral endopeptidase (*PHEX*), *SOST,* and *DMP1* [128, 129]. Following the deletion of *OSX*, animal models show a lower rate of both chondrogenesis and osteogenesis (endochondral and intramembranous bone formation) and a decrease in matrix vesicle biogenesis [128, 130]. Several proteins regulate OSX transcription. Besides RUNX2 (Fig. 4.5), DLX5, BMP2, and mitogen-activated protein kinase (MAPK) promote *OSX* expression, and epigenetic factors and tumor necrosis factor (TNF) negatively control *OSX* expression [131].

OSX is active in tooth formation and mineralization. An *OSX* animal knockout model showed that, while the gene did not influence tooth morphogenesis, it dramatically reduced the content of enamel proteins, including enamelin (ENAM), the amelogenin X isoform (AMELX), MMP20, amelotin (AMTN), and KLK4 [132]. *OSX* facilitates the differentiation of odontoblasts and regulates odontoblast proliferation in root dentin and *DMP1* and *DSPP* expression [133, 134]. When *OSX* is overexpressed, cementum formation increases [135]. Together, the results above support the concept that *OSX* is required for and profoundly influences bone, cartilage, dentin, enamel, and cementum formation and mineralization.

4.6 The *SOX* Family: Major Regulators of Chondrogenesis

The *SOX* or *SRY* family consists of more than 20 genes and counterpart transcription factors which regulate and promote chondrogenesis. Several examples and their known functions are the following: Together with *SOX5* and *6*, *SOX9* maintains the proliferative state of chondrocytes in the growth plate, delays early hypertrophy, and counteracts the activity of other transcription factors which favor osteogenesis [127, 136]. During tooth formation, *SOX9* and related family members exert a similar critical level of control: *SOX6, 9, 13, 14,* and *21* are expressed in the dental epithelium, *SOX5* and *18* are expressed in odontoblasts, *SOX8, 11, 17,* and *21* are expressed in preameloblasts; and *SOX13* is expressed in the primary enamel knot [41]. Heterozygous mutations within the *SOX9* gene cause campomelic dysplasia, a form of dwarfism, and Pierre Robin sequence, a condition in which there is defective growth of the lower jaw and other cranial and facial abnormalities [137].

SOX9 has been studied extensively. During skeletogenesis, this transcription factor is intimately involved with the generation of a cartilaginous anlage while it suppresses the commitment of progenitor cells to other lineages, in particular, those which would foster fat and bone formation [40]. This process occurs hand-in-hand with signals mediated through the WNT/β-catenin pathway, which serves as a molecular brake on *SOX9* expression to prevent stem cells from becoming osteogenic (Additional details of this pathway are discussed later in this chapter) [138]. Among the cartilage genes targeted by *SOX9* are those which encode type II collagen (*COLII*), aggrecan (*ACAN*) and hyaluronan, and proteoglycan link protein (*HAPLN*)*3* [11]. As the cartilaginous skeleton transits to the osteogenic phase, *SOX9* enhances the orderly expression of *COLX* in association with proteins which include WNT and RUNX2 [136, 139, 140].

In the growth plate, *SOX9* promotes the proliferation and maturation of chondrocytes while controlling the expression of the hypertrophic phenotype, as noted above [136]. With SOX5 and 6, SOX9 binds to regulatory elements of genes mediating cartilage formation. Here, SOX9 downregulates expression of *TNAP*, *COLX,* and *MMP13*, and it interacts with β-catenin and blocks WNT signaling, a requirement for chondrocyte maturation and hypertrophy [141]. Moreover, SOX9 transcriptionally increases the expression of *PTHrP*, a key regulator of chondrocyte proliferation and maturation which serves to limit chondrocyte hypertrophy [142].

While SOX9 regulates the expression of many target genes, its own expression is under tight transcriptional control. *SOX9* expression is increased by HIF-1α, WNT (non-canonical), FGF, BMP, and PTHrP while canonical WNT and NOTCH decrease expression [143, 144]. As an aside with respect to NOTCH, hypoxia/HIF-1α signals upregulate the NOTCH ligand, delta-like (Dll)4, resulting in an increase in osteogenesis, while the NOTCH-regulated gene, hairy ear, Y-linked (*HEY*)*1*, inhibits osteogenesis [145]. Structural analysis of SOX9 shows that transactivation domains exist in both the C-terminus and the central region of the protein [146]. With coactivator proteins, these domains interact with the promoter region of chondrocytic genes as well as other major transcriptional proteins required for bone and tooth formation, including β-catenin [146]. A schematic of the transcriptional complex formed by SOX9 and coactivators bound to the COLII enhancer region is shown in Fig. 4.7 [147].

SOX9 causes a rapid and robust expression of mRNA from many target genes linked to the expression of the chondrocyte phenotype [148]. This response is attributable to the presence and activity of clusters of super-enhancers [147, 149], which were described earlier in this chapter. Such super-enhancer clusters are often located at a distance from the promoter regions of DNA but, because of chromatin looping, they are spatially close to protein-coding genes (Fig. 4.7) [149, 150].

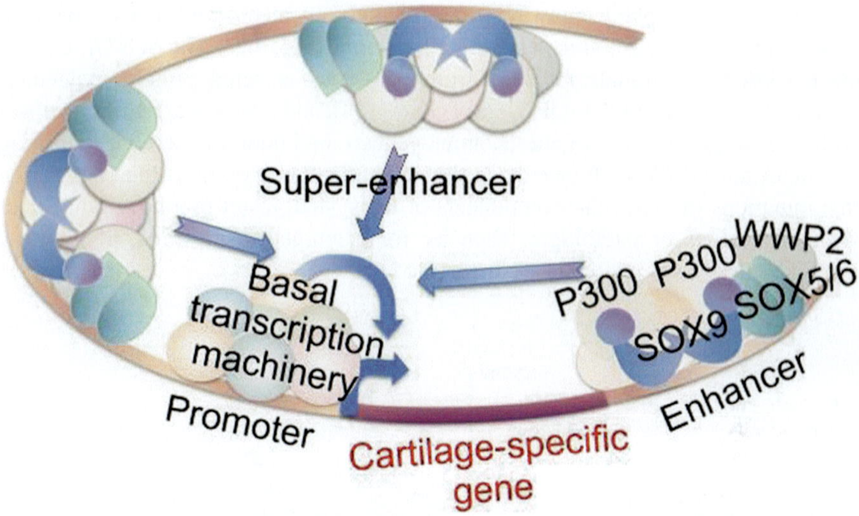

Fig. 4.7 Schematic of regulation and super-enhancement of DNA transcription. In this example, a super-enhancer cluster is shown as a group of three enhancer loci, bound up- and downstream of a cartilage-specific gene, such as *COLIIA1*. Each enhancer consists of several transcription factors, including SOX9, SOX5/6, WWP2, and P300. The clusters alter the three-dimensional structure of DNA and thereby lead to multiple interactions with the basal transcription machinery in the promoter region of the DNA. The resulting effect of such interactions is a high level of transcription of cartilage-specific genes along with an elevation in the activity of RNA polymerase. (Schematic adapted from Lefebvre and Dvir-Ginzberg [147] with permission)

Despite the substantial information now known about *SOX9* structure and function, Lefebvre recently stated that "the SOX molecular network is likely far more sophisticated than currently appreciated" [150]. Nevertheless, the widespread importance of this transcription factor is evident since its loss results in changes in over a thousand other genes, many of which are downregulated [151]. Such a result clearly identifies *SOX9* as a critically important activator of cartilage and bone formation.

4.7 The *WNT/β-Catenin* System: From Wingless Fruit Flies to Spina Bifida

In his treatise, "The Guide for the Perplexed," Maimonides in 1190 wrote that "Your purpose … should always be to know … the whole that was intended to be known." In terms of the mineralization process, the "whole" is enormous, and only now are the major gene families being identified which contribute to the processes of the formation of bone, teeth, and other vertebrate mineralized tissues. Of the genes involved in these events, another of the most critical is Wingless (*WNT*). This gene took its name from mutations in the gene which result in wingless fruit flies. WNT is a ligand and a member of a family of lipoproteins [152]. There are 19 genes in the WNT family in vertebrates [152]. WNTs evoke multiple functions mediated through a common or canonical pathway which involves β-catenin and a non-canonical WNT pathway which is independent of β-catenin (Fig. 4.8). The WNT/β-catenin system integrates signals received from numerous secreted protein molecules, including FGF, IHH, SOST, OPG, and TGF-β [154], and it directs activities involved with chondrogenesis, osteogenesis, amelogenesis, and odontogenesis. Underlining the importance of WNT/β-catenin in the mineralization process is the observation that mutations in molecular components of the system result in a large number of skeletal diseases or pathologies such as, for example, spina bifida, a condition caused by deficits in neural tube closure [155].

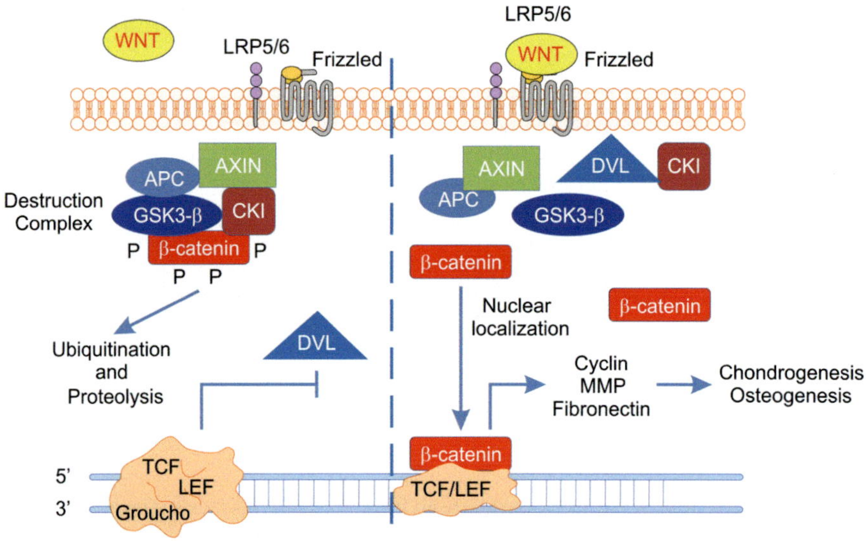

Fig. 4.8 Schematic of the WNT canonical signaling pathway illustrating WNT-β-catenin regulation of signaling in cells committed to the processes of osteogenesis and chondrogenesis. WNT protein is a robust activator of both processes and is also intimately involved with odontogenesis. In addition, β-catenin represses osteoclastogenesis. Interference with the secretion of β-catenin or its interaction with its receptor (Frizzled) and co-receptor low-density lipoprotein receptor-related protein 5 (LRP5) or LRP6 inhibits biological mineralization. In the quiescent state (left aspect of the schematic), WNT signaling is prevented or dampened by the formation of a protein complex comprised of axin, casein kinase 1 (CK1), adenomatous polyposis coli (APC), and glycogen synthase kinase 3-β (GSK3-β). Together these proteins bind and phosphorylate β-catenin. The phosphorylated form of β-catenin undergoes polyubiquitination and proteolysis, thereby preventing it from activating target genes. In the active state (right aspect of the schematic), WNT binding to receptors and its association with LRP5/6 promote migration of the protein, Dishevelled (DVL), to the plasma membrane and subsequent dissociation of the protein complex. In the absence of the complex, β-catenin is not phosphorylated and can translocate to the nucleus. It activates transcription of WNT target genes by first displacing the protein inhibitor, Groucho, from the transcription complex and then forming a complex with a number of proteins including T-cell specific transcription factor/lymphoid enhancer-binding factor (TCF/LEF). The transcription complex promotes the expression of target genes which include *cyclins*, *fibronectin,* and *matrix metalloproteinases* (*MMPs*) and leads to osteogenesis or chondrogenesis. It is interesting that WNT can serve as a ligand which is independent of β-catenin and the basis of the WNT non-signaling pathway. (Schematic adapted from Lerner et al. [153] with permission)

Unlike the other transcription factors discussed previously, WNT/β-catenin is different in that WNT is a ligand, secreted by osteoblasts, osteocytes, chondrocytes, and bone marrow cells. Further complicating the understanding of how this family of transcription factors influences bone cell function is that the ligand is bound to and stored with extracellular matrix molecules like heparan sulfate proteoglycans [156]. These interactions reduce the available concentration and signaling activity of WNT [156].

While the secretion of a number of WNT ligands increases during osteoblastic differentiation, one such ligand (WNT10B) is most strongly related to the development of mineralization [157]. *WNT10B* overexpression is associated with elevated osteoblastogenesis and increased bone mass and thickness [158]. Deletion of *WNT10B* causes a reduction of osteoblastic progenitor cells and loss of bone density [159]. The other highly secreted WNT protein, WNT5A, also promotes osteoblast survival and osteogenesis [160].

The canonical signal is activated when the WNT protein binds to the cell membrane receptor protein, frizzled (FZD) (Fig. 4.8). On binding, a conformational change occurs which promotes a complex between WNT and FZD with members of a low-density lipoprotein receptor-related protein family (LRP5/6). This activity triggers the dissociation of β-catenin from a cytoplasmic grouping of proteins characterized as a "destruction complex," comprised of the molecules axin (AXN), adenomatous polyposis coli (APC), casein kinase (CK), protein phosphatase, and glycogen synthase kinase 3-β (Fig. 4.8). The function of the complex is to phosphorylate β-catenin, rendering it bound and inactive. Once released from the complex, free β-catenin translocates to the cell nucleus, where it interacts with other proteins serving as transcriptional coactivators. There are considerable numbers of such DNA-binding proteins, and among the most common are members of the T-cell specific transcription factor/lymphoid enhancer-binding factor (TCF/LEF) family (Fig. 4.8). The β-catenin–TCF/LEF complex activates transcription of many bone target genes, especially those required for stem cell proliferation and commitment to the osteoblastic phenotype, an action which thereby leads to trabecular bone formation and mineralization (Fig. 4.8) [153, 161].

Animals with a *β-catenin* mutation exhibit skeletal deformities and aberrant growth. Experimental deletion of *β-catenin* at an early developmental stage in animal models promotes chondrogenesis and delays osteogenesis [162]. On the other hand, if *β-catenin* is stabilized, osteogenesis is enhanced and is accompanied by a significant increase in bone mass [163]. In the post-natal state, WNT/β-catenin signaling plays a major role in regulating chondrocyte terminal differentiation by reducing hypertrophy and increasing chondrocyte apoptosis [164]. There are also changes in secondary bone ossification centers with loss of expression of genes linked to vascularization (*VEGF*) and matrix turnover (*MMP13*) [164]. In mature osteoblasts, the bone protein, OPG, is a direct target of *β-catenin* expression and changes in *β-catenin* activity affect OPG modulation of osteoclast function [165].

As noted above, besides regulating the activities of osteoblasts and chondrocytes, the WNT/β-catenin system also influences the activity of osteoclasts (Fig. 4.9). Mature osteocytes secrete the WNT ligand as well as SOST, an inhibitor of osteogenesis (Fig. 4.9) [166]. By binding to LRP5/6, SOST blocks canonical WNT/β--catenin-signaling activity (Fig. 4.8), an event which lowers osteoblast number and mineralization status and decreases trabecular and periosteal bone formation [167]. In humans, heterozygous mutation of *SOST* leading to low levels of its secretion causes sclerosteosis (van Buchem disease), characterized by elevated and dense bone formation [168]. As a result of this finding, neutralizing monoclonal antibodies raised against SOST are now being evaluated as possible treatments for osteoporosis and fracture repair [169].

The control of osteoclasts by WNT signaling is interesting and significant (Fig. 4.9). These cells resorb or remove bone, a process which is linked with new bone formation by osteoblasts. Osteoblasts and osteocytes regulate osteoclast function through the secretion of a number of molecules, the most important of which is the protein, RANKL (Fig. 4.9) [170]. There is strong evidence to suggest that canonical WNT signaling through β-catenin promotes RANKL secretion by osteoblasts, and this action in turn enhances osteoclast formation and activity [165]. Osteoblasts and osteocytes also generate the soluble protein, OPG, noted above, which serves as a "decoy" receptor for RANKL (Fig. 4.9). The term, "decoy," implies that OPG will compete with the RANK receptor for the available RANKL. As a result, the quantity of RANKL free to bind to RANK at the surface of osteoclast precursors is lowered. Accordingly, stimulation of fully mature osteoclasts is reduced and bone resorption is inhibited (Fig. 4.9) [170]. These complex actions of WNT β-catenin provide a mechanism to fine-tune the resorptive process while integrating osteoblastic and osteoclastic activities during bone remodeling (Fig. 4.9). Details of the RANK, RANKL, and OPG system are described elsewhere in this book.

With correlates to its roles in chondrogenesis and osteogenesis, the WNT/β--catenin system participates in the morphogenesis of teeth. Multiple WNT isoforms and β-catenin are present during odontogenesis and, when mutations occur in the genes which encode these proteins, tooth development is blocked [171, 172]. Remarkably, when WNT/β-catenin signaling is prolonged, there is ectopic tooth formation [173]. Moreover, if the cytoplasmic protein, AXN2, of the "destruction complex" mentioned previously (Fig. 4.8) is deleted, then permanent teeth do not form [174]. All of these observations lend support to the notion that WNT/β-catenin is an early signaling molecule required for odontogenesis.

While each of the studies discussed above focuses on signaling through the WNT/β-catenin system, numerous secreted molecules are also required for dental morphogenesis and mineralization. Such secreted proteins include FGF, BMP, IHH, and OSX [175]. How information from these secreted molecules affects the development program of the tooth is not clear. A very intriguing consideration in this context is that the process is mediated by a cellular organelle called the primary cilium [176, 177]. This structure is found along the surface of all cells and it detects, integrates, and transduces chemical, physical, and biomechanical information to the

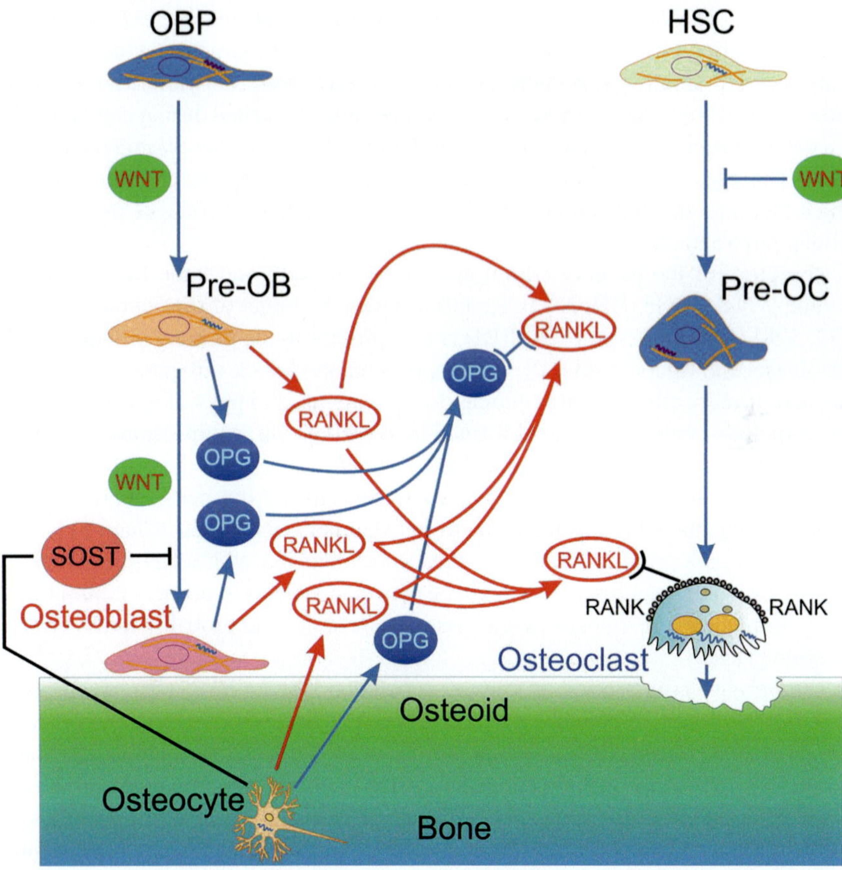

Fig. 4.9 WNT regulation of bone remodeling. WNT promotes osteogenesis by stimulating the maturation of osteoblast progenitor cells (OBP) to preosteoblasts (Pre-OB) and then to osteoblasts. In contrast, WNT blocks the development of hematogenous stem cells (HSC) to preosteoclasts (Pre-OC) in their maturation to osteoclasts. WNT also promotes the expression and secretion of osteoprotegerin (OPG) by preosteoblasts and osteoblasts. Thus, WNT acts to control osteoblastic bone formation in several critical ways. Besides the activity of WNT, the activity of the RANKL/RANK/OPG signaling pathway regulates bone mass. In this instance, preosteoblasts and osteoblasts secrete RANKL in addition to OPG. RANK, the receptor for RANKL (ligand), resides in the osteoclast cell membrane. On the interaction between RANKL and RANK, osteoclasts are activated for resorption. To prevent excess osteoclastic resorption, OPG acts as a decoy receptor for RANKL. On binding between OPG and RANKL, the available amount of RANKL needed for activation of osteoclasts is decreased and hence bone resorption and the remodeling process are inhibited. Osteocytes also express and secrete OPG and RANKL to contribute to the pools of these proteins in further controlling bone mass. Osteocytes express and secrete SOST (sclerostin) as well, which blocks the commitment of preosteoblasts to osteoblasts through WNT suppression as another means of regulating bone remodeling

cell nucleus, where such signals ultimately lead to changes in gene expression and protein secretion [178]. The organelle is a thin, thread-like structure which binds to many secreted proteins. It is microtubule-based and extends outward from the basal body (derived from the centriole, the protein complex inherited during cell division) of a cell and is encased in an extension of the lipid bilayer of the plasma membrane [179]. The organelle is enriched in specific signaling receptors and ion channels which facilitate the detection of chemical and biomechanical changes in the extracellular environment.

Receptors of the primary cilium are critical to tooth and bone formation and include WNT, PDGF, HH, NOTCH, and mechanistic target of rapamycin (mTOR) [172, 178]. A PTH receptor (PTH(R)) is also present in the primary cilium of both osteoblasts and osteocytes [180]. Overexpression and knock-out gene studies indicate that primary cilia regulate cellular, developmental, and physiological processes which include dentin and enamel formation as well as intramembranous and endochondral bone formation [181, 182]. Besides chemical signals, the primary cilium detects and transmits mechanical signals which change the permeability of cell membrane channels such as the calcium ion channel, TRPV4 (see Chap. 5) [183]. Accordingly, the primary cilium is likely to be involved with adapting tissues to applied environmental forces. As might be expected, defects in primary cilium function (ciliopathies) are associated with diseases of bone (and probably tooth) mineralization [184].

The primary cilium aside, it may be concluded that *RUNX2*, *OSX*, *SOX9, and WNT/β-catenin* serve a common purpose to promote the specific phenotypes of cells of each of the various vertebrate mineralized tissues. Through their interactions, individually or commonly with others in this group of four, these important genes and their counterpart transcription factors enhance the smooth development, maturation, and function of osteogenic, chondrogenic, tenogenic, odontogenic, amelogenic, and cementogenic cells. Moreover, these factors are responsive to, and controlled by, external signals mediated by both secreted and local molecules. In later chapters of this book, evidence is presented that the four factors also critically regulate intracellular events which lead to calcium accumulation, the formation of matrix vesicles, and the generation of inhibitory factors which prevent dystrophic mineralization.

4.8 Bone Morphogenetic Protein (BMP): A Family of Endogenous Bone Factors in Clinical Use

Of the considerable number of molecules required for the formation and mineralization of bone, teeth, and other vertebrate connective tissues, the most important clinically comprise a class described as bone morphogenetic proteins (BMPs). The possible presence of these proteins in bone was first investigated and reported by Urist, who suspected their existence as a result of his observation that demineralized bone particles actively promoted bone healing [185]. He surmised that this property was attributable to unidentified bone matrix molecules which mediated osteogenesis and chondrogenesis [185]. While Urist was unable to characterize these molecules completely, his idea was subsequently proven correct: Huggins et al. showed that bone and dentin contained osteogenic agents [186] and Wozney et al. sequenced, cloned, and described several of such osteogenic proteins from bone [187].

It is now known that the BMPs are members of the TGF-β family of secreted extracellular matrix-associated glycosylated proteins, and phylogenetic studies suggest that BMP signaling is at least 700 million years old [188]. While the actual name given to the proteins indicates a principal effect on osteogenesis, BMPs have many other functions. These include specifying body axes, homeotic gene expression, and the formation of specialized tissues [189]. The ligands themselves, as well as their membrane receptors and SMADs (components of the BMP signal transduction system), are present in primitive organisms such as cnidaria and sponges as well as mineralizing corals [190, 191]. These ancient molecules are highly conserved: Sampath et al. showed that endochondral bone formation was induced in mammalian cells treated with fly (Drosophila) orthologues of BMP [192].

Not surprisingly, a molecule as old as BMP has generated many (at least) 20 isoforms. One group of BMP molecules includes the growth and differentiation factors (GDFs) and a second group is comprised of the transforming growth factors, nodal and activins [193]. The biological activities of BMPs or GDFs are buffered by tissue levels of the proteins, gremlin, chordin, and noggin; these are BMP antagonists which bind BMP and consequently reduce its effective concentration and influence on cellular activity [194]. The level of the BMP antagonist, noggin (which binds both BMP2 and 7), is in turn dependent on the secretion of heparan sulfate proteoglycans, cell surface binding molecules [195]. Heparan sulfate proteoglycans also bind chordin, another BMP antagonist as just noted [195]. Since downstream BMP signaling events are also sensitive to the transcription factor, RUNX2, and the secreted protein, FGF, cellular mechanisms exist to integrate BMP pathways with signals from other gene and protein systems [196, 197]. For example, in the dental epithelium, WNT/β-catenin suppresses odontogenic gene expression by inducing both WNT and BMP antagonists [198].

By binding to transmembrane cell surface receptors, BMPs activate a complex downstream signaling system (Fig. 4.10). The receptor proteins in the system are categorized according to their phosphorylation potential. The type II receptor (BMPR2) is constitutively active, while the type I receptor (BMPR1, also known as active-like kinase or ALK) is activated by a phosphorylation reaction mediated by the type II receptor. Thus, once a ligand is bound to the BMPR2 receptor, it phosphorylates BMPR1, which then recruits and phosphorylates a group of intracellular proteins or R-SMADs (SMAD1, SMAD5, and SMAD8). Binding of yet another SMAD, SMAD4, to the R-SMADs permits the entire SMAD complex to migrate into the cell nucleus, associate with *cis*-acting DNA regulatory motifs, and cause chromatin remodeling and transcription of target genes such as *OCN*, *OPN*, and *COLX* in skeletal cells (Fig. 4.10) [199]. The SMADs also receive inhibitory signals from cellular kinases (MAPK and glycogen synthase kinase 3-β) [200].

Two types of BMPs (human recombinant rhBMP2 or InFuse, manufactured by Medtronic, and rhBMP7 or OP1 [osteogenic protein-1], manufactured by

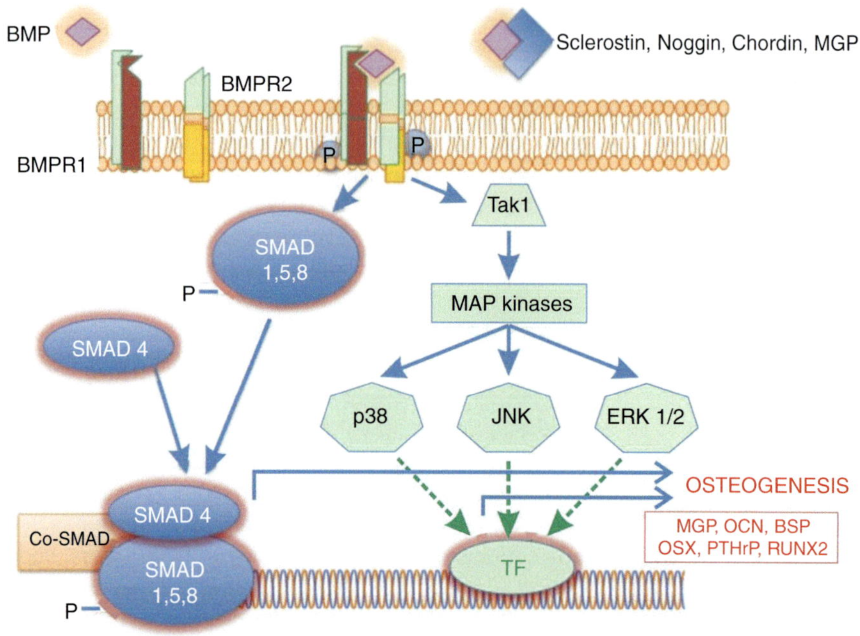

Fig. 4.10 Schematic of bone morphogenetic protein (BMP) signaling pathways in cells of mineralizing tissues. Binding of BMP2 or BMP4 to their receptors (BMPR) on the cell membrane results in the formation of a receptor complex (BMPR1 and 2). Complex formation is regulated by the extracellular protein antagonists, noggin and chordin, and by proteins resident in the extracellular bone matrix such as matrix gla protein (MGP) and sclerostin. Intracellular inhibitors exist, including SMADs and coreceptors, which can also downregulate BMP activity. The BMP-activated receptor complex initiates phosphorylation of SMAD1, 5, and 8 and subsequent association with SMAD4. In the cell nucleus, these SMADs interact with other proteins (Co-SMADs) to activate the transcription of osteogenic genes (*OCN*, *OPN*, *COLX*), *distal-less homeobox 5* (*DLX5*), *sclerostin*, and *RUNX2*. BMP can also signal through non-SMAD pathways in which there is phosphorylation of transforming growth factor activated kinase-1 (TAK1) by the BMP-ligand receptor complex. Phosphorylation leads to the activation of MAP kinases, which in turn activate other kinases (p38, JNK, and ERK 1/2). These kinases translocate to the nucleus to phosphorylate transcription factors (TF) which induce events leading to tissue mineralization. OCN = osteocalcin, OSX = osterix, BSP = bone sialoprotein, PTHrP = parathyroid hormone-related peptide

Olympus) are used clinically for the treatment of non-union fractures and vertebral bone repair. In addition, BMP2 is now utilized for maxillary sinus grafting and some dental applications [201]. Since bone normally contains very low levels of BMP, clinical applications of these proteins must be regarded with caution as prescribed amounts can be 100–1000 times the bone endogenous protein content. Such excess physiological levels of BMP are of concern: Poynton and Lane commented that the use of BMP in the spine may promote bony overgrowths, interactions with exposed dura, cancer risk, systemic toxicity, reproductive toxicity, immunogenicity, local toxicity, osteoclastic activation, and untoward effects on distal organs [202].

4.9 Perspective: Genes and Gene Networks Regulating Mineralization

RNA sequencing (RNA-seq) is a method commonly employed to analyze genes expressed by cells. When RNA-seq is applied to cells concerned with matrix synthesis and mineralization, marked changes are observed in the expression of a great number of transcription factors and coding and non-coding RNA transcripts. For example, as noted earlier, in the cartilage of the mineralizing human growth plate, there is an expression of several hundred protein-coding genes and many lncRNA genes [59]. Of the protein-coding genes, 32 genes are linked with over 60 human skeletal diseases [59]. RNA-seq analysis of rapidly growing deer antler bone indicates a regulatory network of over 300 differentially expressed genes associated with calcium signaling, chondrocyte proliferation and differentiation, and matrix homeostasis, together with a cohort of microRNAs [203, 204]. Deletion of a single transcription factor (SOX9) caused changes in over a thousand genes, transcription factors, enhancers, microRNAs, and lncRNAs [205]. These and other examples underscore the complexity of the myriad temporal, developmental, and tissue-specific interactions among transcription factors and the numerous gene products which comprise a gene regulatory network.

While work is in progress to untangle the interactions among genes which lead to the differentiation and subsequent maturation of an osteoprogenitor, chondroprogenitor, or other progenitor cells of vertebrate mineralized tissues, there remains considerable uncertainty regarding the origin of stem cells in adult mineralized tissues. It is clear from lineage tracing experiments that pericytes are not synonymous with the stromal stem cells which can commit to the osteogenic or chondrogenic phenotype. Accordingly, the question remains as to which stromal cell fulfills the role of an adult osteoblastic stem cell. In this context, the term "adult" is used guardedly and should not be confused with "embryonic;" during development, bone stem cells originate from cells of the axial and paraxial mesoderm and the neuroectoderm (Fig. 4.2).

In a complex tissue such as bone or cartilage, there are numerous niches which may contain stem cells. If pericytes do not function as stem cells, which of these cells or groups of cells participate in the formation of both cancellous and trabecular bone? In contrast to current views concerning the importance of specific populations of marrow stromal cells serving as stem cells, there is some evidence that small "spore-like" cells exist in most human tissues. These cells undergo asymmetric division and exhibit pluripotency [206, 207], and the possibility has been suggested that this small overlooked population of cells, when activated, could provide precursor cells for bone and cartilage formation [208].

A second possibility is that stromal marrow cells in the environmental niche release extracellular vesicles, organelles which contain mRNAs, microRNAs, and proteins [209]. The contents of these structures promote pluripotential behavior by target cells in bones and teeth [209, 210]. In other words, such vesicles activate target cells and promote their commitment to an osteogenic or odontogenic lineage. In fact, when these organelles are injected into critical-sized or osteochondral defects, they promote healing [211].

Key transcription factors which regulate bone and tooth formation and tissue mineralization include RUNX2, OSX, SOX9, and WNT/β-catenin. In a tissue like the bony epiphysis, SOX9 together with co-regulators RUNX2/3 and CBFB modulates the expression of many target genes in the growth plate and responds to secreted proteins such as IHH and PTHrP [127]. Interestingly, SOX9 is present early in the development of the cartilaginous plate and this transcription factor remains active even while its expression level decreases dramatically during the cartilage-to-bone transition of the tissue [212]. To add yet one more level of complexity, the transcription factor, TWIST1, is required for the development of both the appendicular and craniofacial skeleton of vertebrates. The protein can exist as a homo- and heterodimer: As a heterodimer, it maintains mesenchymal cells in a pre-osteoblastic state and with tissue maturation TWIST1 homodimers are formed, which promote osteogenesis [213]. Accordingly, gene transcription is dependent on both the presence of specific proteins as well as their conformational state.

SOX9 prevents the activation of *RUNX2*, curbs the hypertrophic response of proliferating epiphyseal chondrocytes, and blocks WNT signaling [141]. The interactions among SOX9, RUNX2, and other transcription factors ensure the smooth change of growth plate chondrocytes into their terminally differentiated state, while preventing their differentiation into articular chondrocytes or other cell types. In addition, there is strong evidence in the growing limb that a reciprocal inhibitory control exists between SOX9 and RUNX2: SOX9 inhibits the transcriptional activity of *RUNX2*, while RUNX2 decreases *SOX9* expression [40]. While acknowledging the importance of SOX9, Lefebvre et al. noted that "more work is needed to identify the chondrogenic pioneers that act before or with SOX9 and thereby answer the important question of whether SOX9 is a master chondrogenic factor" [40]. Indeed, the question has been raised as to whether this putative SOX9 role could be attributed to other transcription factors such as those of the FOX family [214]. Clearly, there is room for much more work in this arena.

In a related context, while it is known that the terminal differentiation of osteoblasts into osteocytes is regulated by RUNX2 together with OSX [128], other transcription factors are likely involved in this process. For example, activating transcription factor 4 (ATF4) is of some importance since its gene deletion results in a dramatic alteration in collagen synthesis [215]. Yang et al. reported that ATF4 activity was dependent on a kinase (ribosomal S6 kinase, RSK) [216]. When *ATF4* was inactivated, there were marked changes in growth, stature, and other human skeletal anomalies (Coffin-Lowry syndrome) [217].

As detailed above, WNT/β-catenin is another critically important transcription factor which directly mediates osteogenesis, bone growth, and homeostasis. The ligand WNT binds to the receptor FZD, which then associates with lipoproteins LRP5/6 in the plasma membrane of osteoblasts and other mineralizing connective tissue cells. Mutation of the LRP5 receptor causes a profound increase in bone mass, directly linked to the disorder, osteoporosis pseudoglioma syndrome [218]. Triggering of the canonical WNT signaling pathway releases β-catenin from the cytoplasmic protein "destruction complex" and stimulates transcription of many of the genes required for the expression of the osteoblastic phenotype.

Complementing the canonical pathway, a non-canonical WNT pathway mediates changes in calcium ion translocation from stores in the endoplasmic reticulum of cells. The effect of the non-canonical pathway on calcium transport raises questions concerning the role of WNT and other transcription factors on events linked to intracellular calcium-phosphate ion cluster generation and mineral formation. De reported that, in response to WNT-5A binding to FZD, there was activation of phospholipase C and the formation of the calcium ionophore, inositol 1,4,5-trisphosphate [219]. This molecule causes the release of calcium from intracellular stores in the endoplasmic reticulum with the activation of calcium-sensitive proteins like protein kinase C and calcium-calmodulin-dependent kinase II [220]. These findings lend strength to the notion that the non-canonical WNT pathway is involved with intracellular calcium ion homeostasis and suggest that the canonical WNT pathway may be relegated to the synthesis of the proteins of the extracellular matrix.

Regarding the control of events associated with the earliest stages of tooth formation, as was discussed earlier, induction of the enamel organ requires reciprocal interactions between two separate cell types, those of the neural crest-derived ectoderm located in the mid-brain region of the embryo and those of the associated mesenchyme [221]. In this context, in the provisional jaws of the developing embryo, cells lining the inner surface of the dental epithelium condense and induce the proliferation of the mesenchyme. Cells at the periphery of the condensate (odontoblasts) become columnar and secrete their extracellular collagenous predentin matrix, noted previously in Chap. 3. When the predentin matrix is about 5 μm thick, its mineralization begins as a series of calcospherites which fuse and form dentin. The presence of predentin induces differentiation of ameloblasts from the inner layer of the dental epithelium, and these cells secrete the enamel matrix proteins which subsequently orchestrate enamel mineralization [222]. Interactions between the ectodermal (dental epithelium) and mesenchymal components of the developing tooth are described as "reciprocal" since the dental epithelium provokes dentin formation and odontoblasts and predentin induces ameloblast differentiation [223, 224].

The reciprocal control of tooth morphogenesis is mediated by groups of cells located in the embryonic mesenchyme and characterized as the "primary enamel knot" [224]. This signaling center appears early in development (at the cap stage of the tooth) as a condensed mass of non-proliferating cells at the tip of the enamel organ. Enamel knot cells secrete molecules which organize surrounding tissues and initiate the events which eventually lead to tooth formation [224]. Secreted proteins which include BMP, SHH, FGF, HH, and WNT stimulate cells of ectodermal origin to form the dental epithelium as well as cells of mesenchymal origin to aggregate and form the dental papilla [224, 225]. Crosstalk between these two cell types is responsible for enamel and dentin formation and tooth morphogenesis, in particular cusp arrangement and number [224].

Other genes maintaining the ectodermal phenotype of the dental epithelium and the activity of the enamel knot include muscle segment homeobox (*MSX*)*2* and *Mediator*, the latter gene mentioned previously, which also enhances the transcription of TNAP in mineralizing enamel [226]. MSX2 is a transcriptional repressor and controls the expression of *AMELX* [227]. Lallemand et al. [228] and Vieux-Rochas et al. [229] have shown that, in the absence of *MSX*, the resulting limb phenotype was similar to that of a mouse *BMP* knockout, a finding suggesting that *MSX* was downstream of *BMP*. Paradoxically, in the tooth germ, *MSX* was required for the expression of *BMP*, and hence it appeared that the transcription factor lay upstream of *BMP* [24]. Not surprisingly, when *MSX* genes are mutated, then both skeletal and dental deformities occur, including cleft palate and tooth agenesis [230].

Details of the control of epithelial–mesenchymal cell interactions in the dental epithelium are very limited. Two master transcription factors, epiprofin (EPFN/SP6) and SOX21, are thought to be required for the early development of the enamel organ [231]. The ameloblast phenotype is supported by *SOX21*, a gene expressed in the proliferating neural progenitor population in the brain [232]. In the dental lamina, which contains cells of the neural crest origin, *SOX21* helps maintain the ameloblast phenotype [231]. *EPFN/SP6* is widely expressed in the enamel organ. It plays a role in controlling epithelial–mesenchymal cell interactions and stimulates the expression of genes involved in the synthesis of the enamel matrix proteins, AMELX, and ameloblastin (AMBN) [231]. At the later stages of tooth development, *EPFN/SP6* promotes dentin formation by stimulating the expression of *DSPP*, it regulates root formation, and it controls other morphogenetic features of the dentition including tooth number [233].

Parenthetically, recent studies have shown that the expression of *EPFN/SP6* is regulated by Pannexin (*PANX*)*3* [234]. Hexamers of PANX3 form channels in the plasma membrane and the endoplasmic reticulum which regulate calcium ion transport. The protein releases calcium ions from the cisternae of the endoplasmic reticulum into the cytosol and facilitates the transfer of calcium ions from cell to cell. Notably, this protein forms similar channels in odontoblasts, as well as osteoblasts and osteocytes [235, 236]. PANX3 also promotes OSX transcription, thereby enhancing the differentiation of osteoblasts to osteocytes [235]. Additional details of calcium ion transport are discussed in Chaps. 5 and 6.

Before concluding the topic of epithelial–mesenchymal cell interactions in the dental epithelium, it is worthwhile considering the possibility that the reciprocal exchange of information between odontoblasts and ameloblasts may be mediated by the release of extracellular particles. In the mid-70s, a study of the developing tooth by Slavkin and colleagues reported that vesicles were associated with cells of the proliferating inner enamel epithelium [237]. Such vesicles contained RNA and they were proposed to "mediate developmental information from mesenchyme to epithelia during "instructive" stages of tooth development" [237]. Available technology at the time did not allow precise determination of the cells of origin or the possible release, movement, or target of the extracellular vesicles, but the vesicles were considered important to cell–cell information transfer [237]. Possibly these vesicles are analogous to those now described as exosomes/matrix vesicles, which have been suggested to regulate aspects of cell function by transferring microRNA from cell to cell to help modulate bone, cartilage, and dentin mineralization [238].

Finally, the emphasis on a small number of major transcription factors in hard tissue formation should not obscure the importance of signals of the FOX regulatory networks, especially in regard to tissue specificity. For example, the nuclear factor I family of transcription proteins (NFIC) is required for tooth development [239]. When NFIC is deleted, odontoblasts lose their characteristic shape and polarity, and they secrete a low level of DSPP. In addition, the cells express reduced levels of several enamel genes, including *AMELX* and *AMBN* [240]. Most likely, a regulatory network interconnects and communicates between NFIC with other transcription factors, including *OSX* [241]. Signals from *OSX* cause a decrease in the secretion of two important dentin matrix proteins, DMP1 and DSPP [133]. These and other findings thus link together NFIC with OSX in dentin, while OSX interacts with RUNX2 and possibly SOX9 in developing enamel [34].

What is very surprising about the relationship between NFIC and OSX is that OSX appears to influence the development of root dentin, rather than crown dentin. In the tooth, then, the NFIC-OSX axis is responsible for root-polarized odontoblast activity, involving both DMP1 and DSPP [241]. Therefore, in common with the interactions between SOX9 and other transcription factors, NFIC-OSX functions to promote regional odontoblast differentiation and root formation and elongation. This finding suggests that variations in common gene regulatory systems are in play to control tissue development. In the next chapter, this concept is expanded by an assessment of the regional development of tooth and bone and a consideration of the roles of many of the secreted proteins of these tissues in regulating components of the mineralization process.

As one further consideration of this Perspective and the topic of genes and gene networks which regulate mineralization, there is relevance to a brief comment about the use of certain recombinant molecular biological techniques applied to investigate the role of a specific gene in the formation or function of cells of bone, cartilage, mineralizing tendon, and the tooth. Very importantly, these techniques allow the investigator to establish a causal, as distinct from an associative, link between a particular gene and tissue or animal character. Results of the application of the methods in many instances have provided some exceptionally powerful insights into gene function. The most common approach to studying the function of an individual gene and its encoded protein is to generate an animal model, often a mouse, which lacks or overexpresses a specific gene. Analysis of the model utilizing a wide spectrum of techniques (microscopic, immunohistochemical or immunocytochemical, and microcomputed tomographic, for example, among a host of others) yields intimate details of the role of the gene in directing the anatomical, histological, developmental, biochemical, and behavioral characteristics of the animal, or its phenotype.

A second common approach to understanding how the genome regulates the phenotype is through RNA sequencing (RNA-seq) analysis. This type of investigation leads to a unique global description of the transcriptome during development or following experimental manipulation. The method details not only expressed genes, but also non-coding RNAs and small RNAs, 5′ and 3′ ends of genes, and splicing patterns. The procedure relies on reverse-transcribing RNA to cDNA, which is then examined using very high-throughput quantitative "next-generation" sequencing systems [242]. Another exciting development is the CRISPR (clustered regularly interspaced short palindromic repeats)/Cas (CRISPR-associated protein)9 gene editing technique, which is now used to revise or delete regions of

DNA. The CRISPR RNA base pairs with genomic DNA are cut by the Cas9 protein. Repair can then take place following the removal of the nucleotides of interest or identifying a site for the insertion of a new gene [243]. These different technical approaches permit the establishment of gene maps or protein circuits which explain how cells like chondrocytes, for instance, become terminally differentiated and induce mineralization. Likewise, these methods are used to probe cell lineage, survival, apoptosis, metabolism, and ion transport in tissues of the skeleton and the tooth.

References

1. Gould SJ. The without and within of smart mice, 1999; p. 234 (originally appearing in Time magazine, 9-13-1999).
2. Szabo Q, Bantignies F, Cavalli G. Principles of genome folding into topologically associating domains. Sci Adv. 2019;5:eaaw1668.
3. Bates SE. Epigenetic therapies for cancer. N Engl J Med. 2020;383:650–63.
4. Centore RC, Sandoval GJ, Soares LMM, Kadoch C, Chan HM. Mammalian SWI/SNF chromatin remodeling complexes: emerging mechanisms and therapeutic strategies. Trends Genet. 2020;36:936–50.
5. Baudot A, Angelelli JB, Guenoche A, Jacq B, Brun C. Defining a modular signalling network from the fly interactome. BMC Syst Biol. 2008;2:1–17.
6. Hawe JS, Theis FJ, Heinig M. Inferring interaction networks from multi-omics data. Front Genet. 2019;10:1–13.
7. Kleinjan DA, van Heyningen V. Long-range control of gene expression: emerging mechanisms and disruption in disease. Am J Hum Genet. 2005;76:8–32.
8. Lettice LA, Heaney SJ, Purdie LA, Li L, de Beer P, Oostra BA, et al. A long-range Shh enhancer regulates expression in the developing limb and fin and is associated with preaxial polydactyly. Hum Mol Genet. 2003;12:1725–35.
9. Whyte WA, Orlando DA, Hnisz D, Abraham BJ, Lin CY, Kagey MH, et al. Master transcription factors and mediator establish super-enhancers at key cell identity genes. Cell. 2013;153:307–19.
10. Loven J, Hoke HA, Lin CY, Lau A, Orlando DA, Vakoc CR, et al. Selective inhibition of tumor oncogenes by disruption of super-enhancers. Cell. 2013;153:320–34.
11. Liu CF, Lefebvre V. The transcription factors SOX9 and SOX5/SOX6 cooperate genome-wide through super-enhancers to drive chondrogenesis. Nucleic Acids Res. 2015;43:8183–203.
12. Yamanaka S. Pluripotency and nuclear reprogramming. Philos Trans R Soc Lond Ser B Biol Sci. 2008;363:2079–87.
13. Kornberg RD. Mediator and the mechanism of transcriptional activation. Trends Biochem Sci. 2005;30:235–9.
14. Mayran A, Drouin J. Pioneer transcription factors shape the epigenetic landscape. J Biol Chem. 2018;293:13795–804.
15. Thomas BL, Tucker AS, Qui M, Ferguson CA, Hardcastle Z, Rubenstein JL, et al. Role of Dlx-1 and Dlx-2 genes in patterning of the murine dentition. Development. 1997;124: 4811–8.
16. Zhang J, Zhang W, Dai J, Wang X, Shen G. Overexpression of Dlx2 enhances osteogenic differentiation of BMSCs and MC3T3-E1 cells via direct upregulation of Osteocalcin and Alp. Int J Oral Sci. 2019;11:12.
17. Nakamura T, de Vega S, Fukumoto S, Jimenez L, Unda F, Yamada Y. Transcription factor epiprofin is essential for tooth morphogenesis by regulating epithelial cell fate and tooth number. J Biol Chem. 2008;283:4825–33.
18. Ma X, Su P, Yin C, Lin X, Wang X, Gao Y, et al. The roles of FoxO transcription factors in regulation of bone cells function. Int J Mol Sci. 2020;21:692.

19. Poché RA, Sharma R, Garcia MD, Wada AM, Nolte MJ, Udan RS, et al. Transcription factor FoxO1 is essential for enamel biomineralization. PLoS One. 2012;7:e30357.
20. Stegen S, Carmeliet G. Hypoxia, hypoxia-inducible transcription factors and oxygen-sensing prolyl hydroxylases in bone development and homeostasis. Curr Opin Nephrol Hypertens. 2019;28:328–35.
21. Schipani E, Mangiavini L, Merceron C. HIF-1α and growth plate development: what we really know. Bonekey Rep. 2015;4:730.
22. Sidaly R, Landin MA, Suo Z, Snead ML, Lyngstadaas SP, Reseland JE. Hypoxia increases the expression of enamel genes and cytokines in an ameloblast-derived cell line. Eur J Oral Sci. 2015;123:335–40.
23. Kero D, Vukojevic K, Stazic P, Sundov D, Mardesic Brakus S, Saraga-Babic M. Regulation of proliferation in developing human tooth germs by MSX homeodomain proteins and cyclin-dependent kinase inhibitor p19(INK4d). Organogenesis. 2017;13:141–55.
24. Feng XY, Wu XS, Wang JS, Zhang CM, Wang SL. Homeobox protein MSX-1 inhibits expression of bone morphogenetic protein 2, bone morphogenetic protein 4, and lymphoid enhancer-binding factor 1 via Wnt/β-catenin signaling to prevent differentiation of dental mesenchymal cells during the late bell stage. Eur J Oral Sci. 2018;126:1–12.
25. Wehrhan F, Hyckel P, Amann K, Ries J, Stockmann P, Schlegel K, et al. Msx-1 is suppressed in bisphosphonate-exposed jaw bone analysis of bone turnover-related cell signalling after bisphosphonate treatment. Oral Dis. 2011;17:433–42.
26. Zanotti S, Smerdel-Ramoya A, Canalis E. Reciprocal regulation of Notch and nuclear factor of activated T-cells (NFAT) c1 transactivation in osteoblasts. J Biol Chem. 2011;286:4576–88.
27. Takayanagi H, Kim S, Koga T, Nishina H, Isshiki M, Yoshida H, et al. Induction and activation of the transcription factor NFATc1 (NFAT2) integrate RANKL signaling in terminal differentiation of osteoclasts. Dev Cell. 2002;3:889–901.
28. Lee DS, Roh SY, Choi H, Park JC. NFI-C is required for epiphyseal chondrocyte proliferation during postnatal cartilage development. Mol Cell. 2020;43:739–48.
29. Lamani E, Gluhak-Heinrich J, MacDougall M. NFI-C2 temporal-spatial expression and cellular localization pattern during tooth formation. Develop Growth Differ. 2015;57:625–38.
30. Canalis E. Notch in skeletal physiology and disease. Osteoporos Int. 2018;29:2611–21.
31. Cai X, Gong P, Huang Y, Lin Y. Notch signalling pathway in tooth development and adult dental cells. Cell Prolif. 2011;44:495–507.
32. Cao Z, Liu R, Zhang H, Liao H, Zhang Y, Hinton RJ, et al. Osterix controls cementoblast differentiation through downregulation of Wnt-signaling via enhancing DKK1 expression. Int J Biol Sci. 2015;11:335–44.
33. Liu Q, Li M, Wang S, Xiao Z, Xiong Y, Wang G. Recent advances of osterix transcription factor in osteoblast differentiation and bone formation. Front Cell Dev Biol. 2020;8:601224.
34. Lee DS, Roh SY, Park JC. The Nfic-osterix pathway regulates ameloblast differentiation and enamel formation. Cell Tissue Res. 2018;374:531–40.
35. Komori T. Regulation of proliferation, differentiation and functions of osteoblasts by Runx2. Int J Mol Sci. 2019;20:1694.
36. Tang CY, Chen W, Luo Y, Wu J, Zhang Y, McVicar A, et al. Runx1 up-regulates chondrocyte to osteoblast lineage commitment and promotes bone formation by enhancing both chondrogenesis and osteogenesis. Biochem J. 2020;477:2421–38.
37. Yamashiro T, Aberg T, Levanon D, Groner Y, Thesleff I. Expression of Runx1, -2 and -3 during tooth, palate and craniofacial bone development. Mech Dev. 2002;119:S107–10.
38. Zou ML, Chen ZH, Teng YY, Liu SY, Jia Y, Zhang KW, et al. The Smad dependent TGF-β and BMP signaling pathway in bone remodeling and therapies. Front Mol Biosci. 2021;8:593310.
39. Lin AE, Alali A, Starr LJ, Shah N, Beavis A, Pereira EM, et al. Gain-of-function pathogenic variants in SMAD4 are associated with neoplasia in Myhre syndrome. Am J Med Genet A. 2020;182:328–37.
40. Lefebvre V, Angelozzi M, Haseeb A. SOX9 in cartilage development and disease. Curr Opin Cell Biol. 2019;61:39–47.
41. Kawasaki K, Kawasaki M, Watanabe M, Idrus E, Nagai T, Oommen S, et al. Expression of Sox genes in tooth development. Int J Dev Biol. 2015;59:471–8.

42. Saito K, Michon F, Yamada A, Inuzuka H, Yamaguchi S, Fukumoto E, et al. Sox21 regulates anapc10 expression and determines the fate of ectodermal organ. iScience. 2020;23:101329.
43. Hirsch N, Eshel R, Bar Yaacov R, Shahar T, Shmulevich F, Dahan I, et al. Unraveling the transcriptional regulation of TWIST1 in limb development. PLoS Genet. 2018;14:e1007738.
44. Meng T, Huang Y, Wang S, Zhang H, Dechow PC, Wang X, et al. Twist1 is essential for tooth morphogenesis and odontoblast differentiation. J Biol Chem. 2015;290:29593–602.
45. Duan P, Bonewald LF. The role of the wnt/β-catenin signaling pathway in formation and maintenance of bone and teeth. Int J Biochem Cell Biol. 2016;77:23–9.
46. Jing Y, Jing J, Wang K, Chan K, Harris SE, Hinton RJ, et al. Vital roles of β-catenin in trans-differentiation of chondrocytes to bone cells. Int J Biol Sci. 2018;14:1–9.
47. Alver BH, Kim KH, Lu P, Wang X, Manchester HE, Wang W, et al. The SWI/SNF chromatin remodelling complex is required for maintenance of lineage specific enhancers. Nat Commun. 2017;8:14648.
48. Cai S, Han HJ, Kohwi-Shigematsu T. Tissue-specific nuclear architecture and gene expression regulated by SATB1. Nat Genet. 2003;34:42–51.
49. Zhang Y, Zheng L, Le M, Nakano Y, Chan B, Huang Y, et al. SATB1 establishes ameloblast cell polarity and regulates directional amelogenin secretion for enamel formation. BMC Biol. 2019;17:104.
50. Cirillo LA, Lin FR, Cuesta I, Friedman D, Jarnik M, Zaret KS. Opening of compacted chromatin by early developmental transcription factors HNF3 (FoxA) and GATA-4. Mol Cell. 2002;9:279–89.
51. Yoshida M, Hata K, Takashima R, Ono K, Nakamura E, Takahata Y, et al. The transcription factor Foxc1 is necessary for Ihh-Gli2-regulated endochondral ossification. Nat Commun. 2015;6:6653.
52. Agarwal S, Loder S, Brownley C, Cholok D, Mangiavini L, Li J, et al. Inhibition of Hif1alpha prevents both trauma-induced and genetic heterotopic ossification. Proc Natl Acad Sci U S A. 2016;113:E338–47.
53. Prasad MS, Charney RM, Garcia-Castro MI. Specification and formation of the neural crest: perspectives on lineage segregation. Genesis. 2019;57:e23276.
54. Ha M, Hong S. Gene-regulatory interactions in embryonic stem cells represent cell-type specific gene regulatory programs. Nucleic Acids Res. 2017;45:10428–35.
55. Kang X, Sun Y, Zhang Z. Identification of key transcription factors - gene regulatory network related with osteogenic differentiation of human mesenchymal stem cells based on transcription factor prognosis system. Exp Ther Med. 2019;17:2113–22.
56. Balic A, Thesleff I. Tissue interactions regulating tooth development and renewal. Curr Top Dev Biol. 2015;115:157–86.
57. Farrell JA, Wang Y, Riesenfeld SJ, Shekhar K, Regev A, Schier AF. Single-cell reconstruction of developmental trajectories during zebrafish embryogenesis. Science. 2018;360:eaar3131.
58. Galea GL, Zein MR, Allen S, Francis-West P. Making and shaping endochondral and intramembranous bones. Dev Dyn. 2021;250:414–49.
59. Li B, Balasubramanian K, Krakow D, Cohn DH. Genes uniquely expressed in human growth plate chondrocytes uncover a distinct regulatory network. BMC Genomics. 2017;18:983.
60. Gao F, Davidson EH. Transfer of a large gene regulatory apparatus to a new developmental address in echinoid evolution. Proc Natl Acad Sci U S A. 2008;105:6091–6.
61. Rafiq K, Shashikant T, McManus CJ, Ettensohn CA. Genome-wide analysis of the skeletogenic gene regulatory network of sea urchins. Development. 2014;141:950–61.
62. Hughes I, Binkley J, Hurle B, Green ED, Program NCS, Sidow A, et al. Identification of the Otopetrin Domain, a conserved domain in vertebrate otopetrins and invertebrate otopetrin-like family members. BMC Evol Biol. 2008;8:41.
63. Piacentino ML, Zuch DT, Fishman J, Rose S, Speranza EE, Li C, et al. RNA-Seq identifies SPGs as a ventral skeletal patterning cue in sea urchins. Development. 2016;143:703–14.
64. Davidson EH, Rast JP, Oliveri P, Ransick A, Calestani C, Yuh CH, et al. A genomic regulatory network for development. Science. 2002;295:1669–78.
65. Fisher S, Franz-Odendaal T. Evolution of the bone gene regulatory network. Curr Opin Genet Dev. 2012;22:390–7.

66. Schofield R. The relationship between the spleen colony-forming cell and the haemopoietic stem cell. Blood Cells. 1978;4:7–25.
67. Fuchs E, Tumbar T, Guasch G. Socializing with the neighbors: stem cells and their niche. Cell. 2004;116:769–78.
68. Väänänen HK, Laitala-Leinonen T. Osteoclast lineage and function. Arch Biochem Biophys. 2008;473:132–8.
69. Risbud MV, Schipani E, Shapiro IM. Hypoxic regulation of nucleus pulposus cell survival: from niche to notch. Am J Pathol. 2010;176:1577–83.
70. Brighton CT, Heppenstall RB. Oxygen tension in zones of the epiphyseal plate, the metaphysis and diaphysis. An in vitro and in vivo study in rats and rabbits. J Bone Joint Surg Am. 1971;53:719–28.
71. Spencer JA, Ferraro F, Roussakis E, Klein A, Wu J, Runnels JM, et al. Direct measurement of local oxygen concentration in the bone marrow of live animals. Nature. 2014;508:269–73.
72. Shapiro IM, Srinivas V. Metabolic consideration of epiphyseal growth: survival responses in a taxing environment. Bone. 2007;40:561–7.
73. Kaidi A, Williams AC, Paraskeva C. Interaction between beta-catenin and HIF-1 promotes cellular adaptation to hypoxia. Nat Cell Biol. 2007;9:210–7.
74. Lee SH, Che X, Jeong JH, Choi JY, Lee YJ, Lee YH, et al. Runx2 protein stabilizes hypoxia-inducible factor-1alpha through competition with von Hippel-Lindau protein (pVHL) and stimulates angiogenesis in growth plate hypertrophic chondrocytes. J Biol Chem. 2012;287:14760–71.
75. Wan C, Shao J, Gilbert SR, Riddle RC, Long F, Johnson RS, et al. Role of HIF-1alpha in skeletal development. Ann N Y Acad Sci. 2010;1192:322–6.
76. Maes C, Araldi E, Haigh K, Khatri R, Van Looveren R, Giaccia AJ, et al. VEGF-independent cell-autonomous functions of HIF-1α regulating oxygen consumption in fetal cartilage are critical for chondrocyte survival. J Bone Miner Res. 2012;27:596–609.
77. Merceron C, Ranganathan K, Wang E, Tata Z, Makkapati S, Khan MP, et al. Hypoxia-inducible factor 2α is a negative regulator of osteoblastogenesis and bone mass accrual. Bone Res. 2019;7:7.
78. Lee SY, Park KH, Yu HG, Kook E, Song WH, Lee G, et al. Controlling hypoxia-inducible factor-2α is critical for maintaining bone homeostasis in mice. Bone Res. 2019;7:14.
79. Scadden DT. The stem-cell niche as an entity of action. Nature. 2006;441:1075–9.
80. Caplan AI. Mesenchymal stem cells. J Orthop Res. 1991;9:641–50.
81. Lohan P, Treacy O, Lynch K, Barry F, Murphy M, Griffin MD, et al. Culture expanded primary chondrocytes have potent immunomodulatory properties and do not induce an allogeneic immune response. Osteoarthr Cartil. 2016;24:521–33.
82. Tolar J, Le Blanc K, Keating A, Blazar BR. Concise review: hitting the right spot with mesenchymal stromal cells. Stem Cells. 2010;28:1446–55.
83. Diefenderfer DL, Osyczka AM, Reilly GC, Leboy PS. BMP responsiveness in human mesenchymal stem cells. Connect Tissue Res. 2003;44:305–11.
84. Mastrolia I, Foppiani EM, Murgia A, Candini O, Samarelli AV, Grisendi G, et al. Concise review: challenges in clinical development of mesenchymal stromal/stem cells. Stem Cells Transl Med. 2019;8:1135–48.
85. Ramasamy SK, Kusumbe AP, Itkin T, Gur-Cohen S, Lapidot T, Adams RH. Regulation of hematopoiesis and osteogenesis by blood vessel-derived signals. Annu Rev Cell Dev Biol. 2016;32:649–75.
86. Rouget C. Mémoire sur le développement, la structure et les propriétés physiologique des cappilaires sanguins et lymphatiques. Arch Phys. 1873;5:603.
87. Wakui S, Yokoo K, Muto T, Suzuki Y, Takahashi H, Furusato M, et al. Localization of Ang-1, -2, Tie-2, and VEGF expression at endothelial-pericyte interdigitation in rat angiogenesis. Lab Investig. 2006;86:1172–84.
88. Braun A, Xu H, Hu F, Kocherlakota P, Siegel D, Chander P, et al. Paucity of pericytes in germinal matrix vasculature of premature infants. J Neurosci. 2007;27:12012–24.
89. Corselli M, Chen CW, Crisan M, Lazzari L, Péault B. Perivascular ancestors of adult multipotent stem cells. Arterioscler Thromb Vasc Biol. 2010;30:1104–9.

90. Dore-Duffy P, Katychev A, Wang X, Van Buren E. CNS microvascular pericytes exhibit multipotential stem cell activity. J Cereb Blood Flow Metab. 2006;26:613–24.
91. Alliot-Licht B, Bluteau G, Magne D, Lopez-Cazaux S, Lieubeau B, Daculsi G, et al. Dexamethasone stimulates differentiation of odontoblast-like cells in human dental pulp cultures. Cell Tissue Res. 2005;321:391–400.
92. Farrington-Rock C, Crofts NJ, Doherty MJ, Ashton BA, Griffin-Jones C, Canfield AE. Chondrogenic and adipogenic potential of microvascular pericytes. Circulation. 2004;110:2226–32.
93. Crisan M, Yap S, Casteilla L, Chen CW, Corselli M, Park TS, et al. A perivascular origin for mesenchymal stem cells in multiple human organs. Cell Stem Cell. 2008;3:301–13.
94. Gerhardt H, Betsholtz C. Endothelial-pericyte interactions in angiogenesis. Cell Tissue Res. 2003;314:15–23.
95. Armulik A, Abramsson A, Betsholtz C. Endothelial/pericyte interactions. Circ Res. 2005;97:512–23.
96. Krautler NJ, Kana V, Kranich J, Tian Y, Perera D, Lemm D, et al. Follicular dendritic cells emerge from ubiquitous perivascular precursors. Cell. 2012;150:194–206.
97. Buckingham ME, Meilhac SM. Tracing cells for tracking cell lineage and clonal behavior. Dev Cell. 2011;21:394–409.
98. Ducy P, Zhang R, Geoffroy V, Ridall AL, Karsenty G. Osf2/Cbfa1: a transcriptional activator of osteoblast differentiation. Cell. 1997;89:747–54.
99. Otto F, Kanegane H, Mundlos S. Mutations in the RUNX2 gene in patients with cleidocranial dysplasia. Hum Mutat. 2002;19:209–16.
100. Yoshida CA, Yamamoto H, Fujita T, Furuichi T, Ito K, Inoue K, et al. Runx2 and Runx3 are essential for chondrocyte maturation, and Runx2 regulates limb growth through induction of Indian hedgehog. Genes Dev. 2004;18:952–63.
101. Tang J, Xie J, Chen W, Tang C, Wu J, Wang Y, et al. Runt-related transcription factor 1 is required for murine osteoblast differentiation and bone formation. J Biol Chem. 2020;295:11669–81.
102. Lim KE, Park NR, Che X, Han MS, Jeong JH, Kim SY, et al. Core binding factor beta of osteoblasts maintains cortical bone mass via stabilization of Runx2 in mice. J Bone Miner Res. 2016;31:245.
103. Qin X, Jiang Q, Matsuo Y, Kawane T, Komori H, Moriishi T, et al. Cbfb regulates bone development by stabilizing Runx family proteins. J Bone Miner Res. 2015;30:706–14.
104. Park J, Gebhardt M, Golovchenko S, Perez-Branguli F, Hattori T, Hartmann C, et al. Dual pathways to endochondral osteoblasts: a novel chondrocyte-derived osteoprogenitor cell identified in hypertrophic cartilage. Biol Open. 2015;4:608–21.
105. Nishio Y, Dong Y, Paris M, O'Keefe RJ, Schwarz EM, Drissi H. Runx2-mediated regulation of the zinc finger Osterix/Sp7 gene. Gene. 2006;372:62–70.
106. Kawane T, Komori H, Liu W, Moriishi T, Miyazaki T, Mori M, et al. Dlx5 and mef2 regulate a novel runx2 enhancer for osteoblast-specific expression. J Bone Miner Res. 2014;29:1960–9.
107. Jensen ED, Schroeder TM, Bailey J, Gopalakrishnan R, Westendorf JJ. Histone deacetylase 7 associates with Runx2 and represses its activity during osteoblast maturation in a deacetylation-independent manner. J Bone Miner Res. 2008;23:361–72.
108. Truong T, Zhang X, Pathmanathan D, Soo C, Ting K. Craniosynostosis-associated gene nell-1 is regulated by runx2. J Bone Miner Res. 2007;22:7–18.
109. Chen F, Zhang X, Sun S, Zara JN, Zou X, Chiu R, et al. NELL-1, an osteoinductive factor, is a direct transcriptional target of Osterix. PLoS One. 2011;6:e24638.
110. Iwamoto M, Kitagaki J, Tamamura Y, Gentili C, Koyama E, Enomoto H, et al. Runx2 expression and action in chondrocytes are regulated by retinoid signaling and parathyroid hormone-related peptide (PTHrP). Osteoarthr Cartil. 2003;11:6–15.
111. Komori T. Roles of Runx2 in skeletal development. Adv Exp Med Biol. 2017;962:83–93.
112. Komori T. Runx2, an inducer of osteoblast and chondrocyte differentiation. Histochem Cell Biol. 2018;149:313–23.

113. Chen H, Ghori-Javed FY, Rashid H, Adhami MD, Serra R, Gutierrez SE, et al. Runx2 regulates endochondral ossification through control of chondrocyte proliferation and differentiation. J Bone Miner Res. 2014;29:2653–65.
114. Liu W, Toyosawa S, Furuichi T, Kanatani N, Yoshida C, Liu Y, et al. Overexpression of Cbfa1 in osteoblasts inhibits osteoblast maturation and causes osteopenia with multiple fractures. J Cell Biol. 2001;155:157–66.
115. Enomoto H, Shiojiri S, Hoshi K, Furuichi T, Fukuyama R, Yoshida CA, et al. Induction of osteoclast differentiation by Runx2 through receptor activator of nuclear factor-kappa B ligand (RANKL) and osteoprotegerin regulation and partial rescue of osteoclastogenesis in Runx2-/- mice by RANKL transgene. J Biol Chem. 2003;278:23971–7.
116. Bae JS, Gutierrez S, Narla R, Pratap J, Devados R, van Wijnen AJ, et al. Reconstitution of Runx2/Cbfa1-null cells identifies a requirement for BMP2 signaling through a Runx2 functional domain during osteoblast differentiation. J Cell Biochem. 2007;100:434–49.
117. Nakashima T, Takayanagi H. New regulation mechanisms of osteoclast differentiation. Ann N Y Acad Sci. 2011;1240:E13–8.
118. Lian JB, Balint E, Javed A, Drissi H, Vitti R, Quinlan EJ, et al. Runx1/AML1 hematopoietic transcription factor contributes to skeletal development in vivo. J Cell Physiol. 2003;196:301–11.
119. Capulli M, Paone R, Rucci N. Osteoblast and osteocyte: games without frontiers. Arch Biochem Biophys. 2014;561:3–12.
120. Shirai Y, Kawabe K, Tosa I, Tsukamoto S, Yamada D, Takarada T. Runx2 function in cells of neural crest origin during intramembranous ossification. Biochem Biophys Res Commun. 2019;509:1028–33.
121. D'Souza RN, Aberg T, Gaikwad J, Cavender A, Owen M, Karsenty G, et al. Cbfa1 is required for epithelial-mesenchymal interactions regulating tooth development in mice. Development. 1999;126:2911–20.
122. Chen S, Rani S, Wu Y, Unterbrink A, Gu TT, Gluhak-Heinrich J, et al. Differential regulation of dentin sialophosphoprotein expression by Runx2 during odontoblast cytodifferentiation. J Biol Chem. 2005;280:29717–27.
123. Li S, Kong H, Yao N, Yu Q, Wang P, Lin Y, et al. The role of runt-related transcription factor 2 (Runx2) in the late stage of odontoblast differentiation and dentin formation. Biochem Biophys Res Commun. 2011;410:698–704.
124. Athanassiou-Papaefthymiou M, Kim D, Harbron L, Papagerakis S, Schnell S, Harada H, et al. Molecular and circadian controls of ameloblasts. Eur J Oral Sci. 2011;119:35–40.
125. Miyazaki T, Kanatani N, Rokutanda S, Yoshida C, Toyosawa S, Nakamura R, et al. Inhibition of the terminal differentiation of odontoblasts and their transdifferentiation into osteoblasts in Runx2 transgenic mice. Arch Histol Cytol. 2008;71:131–46.
126. Nakashima K, Zhou X, Kunkel G, Zhang Z, Deng JM, Behringer RR, et al. The novel zinc finger-containing transcription factor osterix is required for osteoblast differentiation and bone formation. Cell. 2002;108:17–29.
127. Nishimura R, Hata K, Nakamura E, Murakami T, Takahata Y. Transcriptional network systems in cartilage development and disease. Histochem Cell Biol. 2018;149:353–63.
128. Zhou X, Zhang Z, Feng JQ, Dusevich VM, Sinha K, Zhang H, et al. Multiple functions of Osterix are required for bone growth and homeostasis in postnatal mice. Proc Natl Acad Sci U S A. 2010;107:12919–24.
129. Plotkin LI, Bellido T. Osteocytic signalling pathways as therapeutic targets for bone fragility. Nat Rev Endocrinol. 2016;12:593–605.
130. Nishimura R, Hata K, Matsubara T, Wakabayashi M, Yoneda T. Regulation of bone and cartilage development by network between BMP signalling and transcription factors. J Biochem. 2012;151:247–54.
131. Sinha KM, Zhou X. Genetic and molecular control of osterix in skeletal formation. J Cell Biochem. 2013;114:975–84.

132. Bae JM, Clarke JC, Rashid H, Adhami MD, McCullough K, Scott JS, et al. Specificity protein 7 is required for proliferation and differentiation of ameloblasts and odontoblasts. J Bone Miner Res. 2018;33:1126–40.
133. Feng JQ, Huang H, Lu Y, Ye L, Xie Y, Tsutsui TW, et al. The Dentin matrix protein 1 (Dmp1) is specifically expressed in mineralized, but not soft, tissues during development. J Dent Res. 2003;82:776–80.
134. Kim TH, Bae CH, Lee JC, Kim JE, Yang X, de Crombrugghe B, et al. Osterix regulates tooth root formation in a site-specific manner. J Dent Res. 2015;94:430–8.
135. Cao Z, Zhang H, Zhou X, Han X, Ren Y, Gao T, et al. Genetic evidence for the vital function of Osterix in cementogenesis. J Bone Miner Res. 2012;27:1080–92.
136. Dy P, Wang W, Bhattaram P, Wang Q, Wang L, Ballock RT, et al. Sox9 directs hypertrophic maturation and blocks osteoblast differentiation of growth plate chondrocytes. Dev Cell. 2012;22:597–609.
137. Antwi P, Hong CS, Duran D, Jin SC, Dong W, DiLuna M, et al. A novel association of campomelic dysplasia and hydrocephalus with an unbalanced chromosomal translocation upstream of SOX9. Cold Spring Harb Mol Case Stud. 2018;4:a002766.
138. Bhattaram P, Penzo-Méndez A, Kato K, Bandyopadhyay K, Gadi A, Taketo MM, et al. SOXC proteins amplify canonical WNT signaling to secure nonchondrocytic fates in skeletogenesis. J Cell Biol. 2014;207:657–71.
139. Amano K, Hata K, Sugita A, Takigawa Y, Ono K, Wakabayashi M, et al. Sox9 family members negatively regulate maturation and calcification of chondrocytes through up-regulation of parathyroid hormone-related protein. Mol Biol Cell. 2009;20:4541–51.
140. Szklarczyk D, Franceschini A, Kuhn M, Simonovic M, Roth A, Minguez P, et al. The STRING database in 2011: functional interaction networks of proteins, globally integrated and scored. Nucleic Acids Res. 2011;39:D561–8.
141. Topol L, Chen W, Song H, Day TF, Yang Y. Sox9 inhibits Wnt signaling by promoting beta-catenin phosphorylation in the nucleus. J Biol Chem. 2009;284:3323–33.
142. Li J, Dong S. The signaling pathways involved in chondrocyte differentiation and hypertrophic differentiation. Stem Cells Int. 2016;2016:2470351.
143. Rutkovskiy A, Stensløkken KO, Vaage IJ. Osteoblast differentiation at a glance. Med Sci Monit Basic Res. 2016;22:95–106.
144. Kovács B, Vajda E, Nagy EE. Regulatory effects and interactions of the Wnt and OPG-RANKL-RANK signaling at the bone-cartilage interface in osteoarthritis. Int J Mol Sci. 2019;20:4653.
145. Ballhause TM, Jiang S, Baranowsky A, Brandt S, Mertens PR, Frosch KH, et al. Relevance of notch signaling for bone metabolism and regeneration. Int J Mol Sci. 2021;22:1325.
146. Haseeb A, Lefebvre V. The SOXE transcription factors-SOX8, SOX9 and SOX10-share a bipartite transactivation mechanism. Nucleic Acids Res. 2019;47:6917–31.
147. Lefebvre V, Dvir-Ginzberg M. SOX9 and the many facets of its regulation in the chondrocyte lineage. Connect Tissue Res. 2017;58:2–14.
148. Hata K, Takahata Y, Murakami T, Nishimura R. Transcriptional network controlling endochondral ossification. J Bone Metab. 2017;24:75–82.
149. Peng Y, Zhang Y. Enhancer and super-enhancer: positive regulators in gene transcription. Anim Model Exp Med. 2018;1:169–79.
150. Lefebvre V. Roles and regulation of SOX transcription factors in skeletogenesis. Curr Top Dev Biol. 2019;133:171–93.
151. Ohba S, He X, Hojo H, McMahon AP. Distinct transcriptional programs underlie Sox9 regulation of the mammalian chondrocyte. Cell Rep. 2015;12:229–43.
152. Miller JR. The Wnts. Genome Biol. 2002;3:3001–15.
153. Lerner UH, Ohlsson C. The WNT system: background and its role in bone. J Intern Med. 2015;277:630–49.
154. Chen G, Deng C, Li YP. TGF-β and BMP signaling in osteoblast differentiation and bone formation. Int J Biol Sci. 2012;8:272–88.

155. Logan CY, Nusse R. The Wnt signaling pathway in development and disease. Annu Rev Cell Dev Biol. 2004;20:781–810.
156. Lin X. Functions of heparan sulfate proteoglycans in cell signaling during development. Development. 2004;131:6009–21.
157. Zhong Z, Zylstra-Diegel CR, Schumacher CA, Baker JJ, Carpenter AC, Rao S, et al. Wntless functions in mature osteoblasts to regulate bone mass. Proc Natl Acad Sci U S A. 2012;109:E2197–204.
158. Bennett CN, Longo KA, Wright WS, Suva LJ, Lane TF, Hankenson KD, et al. Regulation of osteoblastogenesis and bone mass by Wnt10b. Proc Natl Acad Sci U S A. 2005;102:3324–9.
159. Stevens JR, Miranda-Carboni GA, Singer MA, Brugger SM, Lyons KM, Lane TF. Wnt10b deficiency results in age-dependent loss of bone mass and progressive reduction of mesenchymal progenitor cells. J Bone Miner Res. 2010;25:2138–47.
160. Kobayashi Y, Uehara S, Udagawa N, Takahashi N. Regulation of bone metabolism by Wnt signals. J Biochem. 2016;159:387–92.
161. Rao TP, Kuhl M. An updated overview on Wnt signaling pathways: a prelude for more. Circ Res. 2010;106:1798–806.
162. Hill TP, Spater D, Taketo MM, Birchmeier W, Hartmann C. Canonical Wnt/beta-catenin signaling prevents osteoblasts from differentiating into chondrocytes. Dev Cell. 2005;8:727–38.
163. Rodda SJ, McMahon AP. Distinct roles for Hedgehog and canonical Wnt signaling in specification, differentiation and maintenance of osteoblast progenitors. Development. 2006;133:3231–44.
164. Chen M, Zhu M, Awad H, Li TF, Sheu TJ, Boyce BF, et al. Inhibition of beta-catenin signaling causes defects in postnatal cartilage development. J Cell Sci. 2008;121:1455–65.
165. Glass DA II, Bialek P, Ahn JD, Starbuck M, Patel MS, Clevers H, et al. Canonical Wnt signaling in differentiated osteoblasts controls osteoclast differentiation. Dev Cell. 2005;8:751–64.
166. Collette NM, Genetos DC, Economides AN, Xie L, Shahnazari M, Yao W, et al. Targeted deletion of Sost distal enhancer increases bone formation and bone mass. Proc Natl Acad Sci U S A. 2012;109:14092–7.
167. Semenov M, Tamai K, He X. SOST is a ligand for LRP5/LRP6 and a Wnt signaling inhibitor. J Biol Chem. 2005;280:26770–5.
168. Sebastian A, Loots GG. Genetics of Sost/SOST in sclerosteosis and van Buchem disease animal models. Metabolism. 2018;80:38–47.
169. Kobza AO, Papaioannou A, Lau AN, Adachi JD. Romosozumab in the treatment of osteoporosis. Immunotherapy. 2020;12:965–81.
170. Yasuda H, Shima N, Nakagawa N, Yamaguchi K, Kinosaki M, Mochizuki S, et al. Osteoclast differentiation factor is a ligand for osteoprotegerin/osteoclastogenesis-inhibitory factor and is identical to TRANCE/RANKL. Proc Natl Acad Sci U S A. 1998;95:3597–602.
171. Fjeld K, Kettunen P, Furmanek T, Kvinnsland IH, Luukko K. Dynamic expression of Wnt signaling-related Dickkopf1, -2, and -3 mRNAs in the developing mouse tooth. Dev Dyn. 2005;233:161–6.
172. Liu F, Chu EY, Watt B, Zhang Y, Gallant NM, Andl T, et al. Wnt/beta-catenin signaling directs multiple stages of tooth morphogenesis. Dev Biol. 2008;313:210–24.
173. Wang XP, O'Connell DJ, Lund JJ, Saadi I, Kuraguchi M, Turbe-Doan A, et al. Apc inhibition of Wnt signaling regulates supernumerary tooth formation during embryogenesis and throughout adulthood. Development. 2009;136:1939–49.
174. Lammi L, Arte S, Somer M, Jarvinen H, Lahermo P, Thesleff I, et al. Mutations in AXIN2 cause familial tooth agenesis and predispose to colorectal cancer. Am J Hum Genet. 2004;74:1043–50.
175. Jussila M, Thesleff I. Signaling networks regulating tooth organogenesis and regeneration, and the specification of dental mesenchymal and epithelial cell lineages. Cold Spring Harb Perspect Biol. 2012;4:a008425.
176. Nakatomi M, Hovorakova M, Gritli-Linde A, Blair HJ, MacArthur K, Peterka M, et al. Evc regulates a symmetrical response to Shh signaling in molar development. J Dent Res. 2013;92:222–8.

177. Liu B, Chen S, Cheng D, Jing W, Helms JA. Primary cilia integrate hedgehog and Wnt signaling during tooth development. J Dent Res. 2014;93:475–82.
178. DeRouen MC, Oro AE. The primary cilium: a small yet mighty organelle. J Invest Dermatol. 2009;129:264–5.
179. Joukov V, De Nicolo A. The centrosome and the primary cilium: the yin and yang of a hybrid organelle. Cells. 2019;8:701.
180. Martín-Guerrero E, Tirado-Cabrera I, Buendía I, Alonso V, Gortázar AR, Ardura JA. Primary cilia mediate parathyroid hormone receptor type 1 osteogenic actions in osteocytes and osteoblasts via Gli activation. J Cell Physiol. 2020;235:7356–69.
181. Whitfield JF. The solitary (primary) cilium--a mechanosensory toggle switch in bone and cartilage cells. Cell Signal. 2008;20:1019–24.
182. Hampl M, Cela P, Szabo-Rogers HL, Kunova Bosakova M, Dosedelova H, Krejci P, et al. Role of primary cilia in odontogenesis. J Dent Res. 2017;96:965–74.
183. Lee KL, Guevarra MD, Nguyen AM, Chua MC, Wang Y, Jacobs CR. The primary cilium functions as a mechanical and calcium signaling nexus. Cilia. 2015;4:7.
184. Brugmann SA, Cordero DR, Helms JA. Craniofacial ciliopathies: a new classification for craniofacial disorders. Am J Med Genet A. 2010;152a:2995–3006.
185. Urist MR. Bone: formation by autoinduction. Science. 1965;150:893–9.
186. Huggins C, Wiseman S, Reddi AH. Transformation of fibroblasts by allogeneic and xenogeneic transplants of demineralized tooth and bone. J Exp Med. 1970;132:1250–8.
187. Wozney JM, Rosen V, Celeste AJ, Mitsock LM, Whitters MJ, Kriz RW, et al. Novel regulators of bone formation: molecular clones and activities. Science. 1988;242:1528–34.
188. Marques CL, Fernández I, Viegas MN, Cox CJ, Martel P, Rosa J, et al. Comparative analysis of zebrafish bone morphogenetic proteins 2, 4 and 16: molecular and evolutionary perspectives. Cell Mol Life Sci. 2016;73:841–57.
189. Lowery JW, Rosen V. The BMP pathway and its inhibitors in the skeleton. Physiol Rev. 2018;98:2431–52.
190. Lelong C, Mathieu M, Favrel P. Identification of new bone morphogenetic protein-related members in invertebrates. Biochimie. 2001;83:423–6.
191. Zoccola D, Moya A, Beranger GE, Tambutte E, Allemand D, Carle GF, et al. Specific expression of BMP2/4 ortholog in biomineralizing tissues of corals and action on mouse BMP receptor. Mar Biotechnol (NY). 2009;11:260–9.
192. Sampath TK, Rashka KE, Doctor JS, Tucker RF, Hoffmann FM. Drosophila transforming growth factor beta superfamily proteins induce endochondral bone formation in mammals. Proc Natl Acad Sci U S A. 1993;90:6004–8.
193. Weiss A, Attisano L. The TGFbeta superfamily signaling pathway. Wiley Interdiscip Rev Dev Biol. 2013;2:47–63.
194. Brazil DP, Church RH, Surae S, Godson C, Martin F. BMP signalling: agony and antagony in the family. Trends Cell Biol. 2015;25:249–64.
195. Jasuja R, Allen BL, Pappano WN, Rapraeger AC, Greenspan DS. Cell-surface heparan sulfate proteoglycans potentiate chordin antagonism of bone morphogenetic protein signaling and are necessary for cellular uptake of chordin. J Biol Chem. 2004;279:51289–97.
196. Bae SC, Lee KS, Zhang YW, Ito Y. Intimate relationship between TGF-beta/BMP signaling and runt domain transcription factor, PEBP2/CBF. J Bone Joint Surg Am. 2001;83-A:S48–55.
197. Naganawa T, Xiao L, Coffin JD, Doetschman T, Sabbieti MG, Agas D, et al. Reduced expression and function of bone morphogenetic protein-2 in bones of Fgf2 null mice. J Cell Biochem. 2008;103:1975–88.
198. Chen X, Liu J, Li N, Wang Y, Zhou N, Zhu L, et al. Mesenchymal Wnt/beta-catenin signaling induces Wnt and BMP antagonists in dental epithelium. Organogenesis. 2019;15:55–67.
199. Javed A, Bae JS, Afzal F, Gutierrez S, Pratap J, Zaidi SK, et al. Structural coupling of Smad and Runx2 for execution of the BMP2 osteogenic signal. J Biol Chem. 2008;283:8412–22.
200. Shimmi O, Newfeld SJ. New insights into extracellular and post-translational regulation of TGF-beta family signalling pathways. J Biochem. 2013;154:11–9.

201. James AW, LaChaud G, Shen J, Asatrian G, Nguyen V, Zhang X, et al. A review of the clinical side effects of bone morphogenetic protein-2. Tissue Eng Part B Rev. 2016;22:284–97.
202. Poynton AR, Lane JM. Safety profile for the clinical use of bone morphogenetic proteins in the spine. Spine (Phila Pa 1976). 2002;27:S40–8.
203. Yao B, Zhang M, Liu M, Liu Y, Hu Y, Zhao Y. Transcriptomic characterization elucidates a signaling network that controls antler growth. Genome. 2018;61:829–41.
204. Hu P, Wang T, Liu H, Xu J, Wang L, Zhao P, et al. Full-length transcriptome and microRNA sequencing reveal the specific gene-regulation network of velvet antler in sika deer with extremely different velvet antler weight. Mol Gen Genomics. 2019;294:431–43.
205. Yao B, Zhang M, Liu M, Wang Q, Liu M, Zhao Y. Sox9 functions as a master regulator of antler growth by controlling multiple cell lineages. DNA Cell Biol. 2018;37:15–22.
206. Vacanti MP, Roy A, Cortiella J, Bonassar L, Vacanti CA. Identification and initial characterization of spore-like cells in adult mammals. J Cell Biochem. 2001;80:455–60.
207. Taichman RS, Wang Z, Shiozawa Y, Jung Y, Song J, Balduino A, et al. Prospective identification and skeletal localization of cells capable of multilineage differentiation in vivo. Stem Cells Dev. 2010;19:1557–70.
208. Ratajczak MZ, Ratajczak J, Suszynska M, Miller DM, Kucia M, Shin DM. A novel view of the adult stem cell compartment from the perspective of a quiescent population of very small embryonic-like stem cells. Circ Res. 2017;120:166–78.
209. Gangaraju VK, Lin H. MicroRNAs: key regulators of stem cells. Nat Rev Mol Cell Biol. 2009;10:116–25.
210. Hu X, Zhong Y, Kong Y, Chen Y, Feng J, Zheng J. Lineage-specific exosomes promote the odontogenic differentiation of human dental pulp stem cells (DPSCs) through TGFβ1/smads signaling pathway via transfer of microRNAs. Stem Cell Res Ther. 2019;10:170.
211. Zhang S, Chu WC, Lai RC, Lim SK, Hui JH, Toh WS. Exosomes derived from human embryonic mesenchymal stem cells promote osteochondral regeneration. Osteoarthr Cartil. 2016;24:2135–40.
212. Kozhemyakina E, Lassar AB, Zelzer E. A pathway to bone: signaling molecules and transcription factors involved in chondrocyte development and maturation. Development. 2015;142:817–31.
213. Franco HL, Casasnovas J, Rodríguez-Medina JR, Cadilla CL. Redundant or separate entities?--roles of Twist1 and Twist2 as molecular switches during gene transcription. Nucleic Acids Res. 2011;39:1177–86.
214. Kurakazu I, Akasaki Y, Hayashida M, Tsushima H, Goto N, Sueishi T, et al. FOXO1 transcription factor regulates chondrogenic differentiation through transforming growth factor β1 signaling. J Biol Chem. 2019;294:17555–69.
215. Elefteriou F, Benson MD, Sowa H, Starbuck M, Liu X, Ron D, et al. ATF4 mediation of NF1 functions in osteoblast reveals a nutritional basis for congenital skeletal dysplasiae. Cell Metab. 2006;4:441–51.
216. Yang X, Matsuda K, Bialek P, Jacquot S, Masuoka HC, Schinke T, et al. ATF4 is a substrate of RSK2 and an essential regulator of osteoblast biology; implication for Coffin-Lowry Syndrome. Cell. 2004;117:387–98.
217. Rogers RC, Abidi FE. Coffin-Lowry syndrome. In: Adam MP, Ardinger HH, Pagon RA, Wallace SE, Bean LJH, Stephens K, et al., editors. GeneReviews. Seattle, WA: University of Washington; 1993.
218. Gong Y, Slee RB, Fukai N, Rawadi G, Roman-Roman S, Reginato AM, et al. LDL receptor-related protein 5 (LRP5) affects bone accrual and eye development. Cell. 2001;107:513–23.
219. De A. Wnt/Ca2+ signaling pathway: a brief overview. Acta Biochim Biophys Sin Shanghai. 2011;43:745–56.
220. Kuhl M, Sheldahl LC, Malbon CC, Moon RT. Ca(2+)/calmodulin-dependent protein kinase II is stimulated by Wnt and Frizzled homologs and promotes ventral cell fates in Xenopus. J Biol Chem. 2000;275:12701–11.
221. Köntges G, Lumsden A. Rhombencephalic neural crest segmentation is preserved throughout craniofacial ontogeny. Development. 1996;122:3229–42.

222. Nanci A. Ten Cate's oral histology. 9th ed. St. Louis, MO: Elsevier; 2018.
223. Mina M, Kollar EJ. The induction of odontogenesis in non-dental mesenchyme combined with early murine mandibular arch epithelium. Arch Oral Biol. 1987;32:123–7.
224. Thesleff I, Vaahtokari A, Partanen AM. Regulation of organogenesis. Common molecular mechanisms regulating the development of teeth and other organs. Int J Dev Biol. 1995;39:35–50.
225. Du W, Hu JK, Du W, Klein OD. Lineage tracing of epithelial cells in developing teeth reveals two strategies for building signaling centers. J Biol Chem. 2017;292:15062–9.
226. Yoshizaki K, Hu L, Nguyen T, Sakai K, Ishikawa M, Takahashi I, et al. Mediator 1 contributes to enamel mineralization as a coactivator for Notch1 signaling and stimulates transcription of the alkaline phosphatase gene. J Biol Chem. 2017;292:13531–40.
227. Babajko S, de La Dure-Molla M, Jedeon K, Berdal A. MSX2 in ameloblast cell fate and activity. Front Physiol. 2015;5:510.
228. Lallemand Y, Nicola MA, Ramos C, Bach A, Cloment CS, Robert B. Analysis of Msx1; Msx2 double mutants reveals multiple roles for Msx genes in limb development. Development. 2005;132:3003–14.
229. Vieux-Rochas M, Bouhali K, Mantero S, Garaffo G, Provero P, Astigiano S, et al. BMP-mediated functional cooperation between Dlx5;Dlx6 and Msx1;Msx2 during mammalian limb development. PLoS One. 2013;8:e51700.
230. Alappat S, Zhang Z-Y, Chen Y-P. Msx homeobox gene family and craniofacial development. Cell Res. 2003;13:429–42.
231. Yoshizaki K, Fukumoto S, Bikle DD, Oda Y. Transcriptional regulation of dental epithelial cell fate. Int J Mol Sci. 2020;21:8952.
232. Whittington N, Cunningham D, Le TK, De Maria D, Silva EM. Sox21 regulates the progression of neuronal differentiation in a dose-dependent manner. Dev Biol. 2015;397:237–47.
233. Nakamura T, Jimenez-Rojo L, Koyama E, Pacifici M, de Vega S, Iwamoto M, et al. Epiprofin regulates enamel formation and tooth morphogenesis by controlling epithelial-mesenchymal interactions during tooth development. J Bone Miner Res. 2017;32:601–10.
234. Zhang P, Ishikawa M, Doyle A, Nakamura T, He B, Yamada Y. Pannexin 3 regulates skin development via Epiprofin. Sci Rep. 2021;11:1779.
235. Ishikawa M, Yamada Y. The role of pannexin 3 in bone biology. J Dent Res. 2017;96:372–9.
236. Ishikawa M, Williams G, Forcinito P, Ishikawa M, Petrie R-J, Saito K, et al. Pannexin 3 ER Ca2+ channel gating is regulated by phosphorylation at the Serine 68 residue in osteoblast differentiation. Sci Rep. 2019;9:18759.
237. Slavkin HC, Croissant RD, Bringas P, Matosian P, Wilson P, Mino W, et al. Matrix vesicle heterogeneity: possible morphogenetic functions for matrix vesicles. Fed Proc. 1976;35:127–34.
238. Shapiro IM, Landis WJ, Risbud MV. Matrix vesicles: are they anchored exosomes? Bone. 2015;79:29–36.
239. Lee HK, Lee DS, Park SJ, Cho KH, Bae HS, Park JC. Nuclear factor I-C (NFIC) regulates dentin sialophosphoprotein (DSPP) and E-cadherin via control of Kruppel-like factor 4 (KLF4) during dentinogenesis. J Biol Chem. 2014;289:28225–36.
240. Steele-Perkins G, Butz KG, Lyons GE, Zeichner-David M, Kim HJ, Cho MI, et al. Essential role for NFI-C/CTF transcription-replication factor in tooth root development. Mol Cell Biol. 2003;23:1075–84.
241. Zhang H, Jiang Y, Qin C, Liu Y, Ho SP, Feng JQ. Essential role of osterix for tooth root but not crown dentin formation. J Bone Miner Res. 2015;30:742–6.
242. Wang Z, Gerstein M, Snyder M. RNA-Seq: a revolutionary tool for transcriptomics. Nat Rev Genet. 2009;10:57–63.
243. Doudna JA, Charpentier E. Genome editing. The new frontier of genome engineering with CRISPR-Cas9. Science. 2014;346:1258096.

5 Calcium and Phosphate Ion Uptake, Distribution, and Homeostasis in Cells of Vertebrate Mineralized Tissues

It is believed that a relatively early step of great consequence in evolution was the development of the nucleated eukaryotic cell[1]*. Our first interest is not so much in the nucleus itself but in the variety of vesicles and filamentous proteins which also appeared and in particular the reticula and then the organelles .... In effect, there were generated new aqueous solution compartments, in addition to the cytoplasm and the sea and not in equilibrium with either of them. These compartments, vesicle solutions, served as chemical depositories, as buffers for internal solutions, as zones for isolated activities such as decorative digestion locked away from the synthesis in the cytoplasm. They became useful for endocytosis and exocytosis of wanted and unwanted small or large molecules. Since they could be at a different pH or a different redox potential from the cytoplasm, different reactions were carried out in each one of them including synthesis steps such as protein glycosylation. The vesicles were also used to prepare materials for export – Biominerals, wall, wool or periplasmic proteins and polysaccharides.* [1]

Glossary

Annexins A family of calcium-dependent, membrane-binding proteins.

Apoptosis A highly controlled and evolutionarily conserved form of cell death which occurs commonly during development and aging.

Calbindins A family of calcium-binding proteins which act as calcium buffers and calcium sensors and originally described as vitamin D-dependent calcium-binding proteins.

Calcium oscillations Intracellular increases or fluctuations in cytosolic calcium levels resulting from periodic discharges of calcium ions from organelles or plasma membrane channel activation.

Calcium release-activated channels (CRAC) Specialized plasma membrane calcium ion channels which are activated on depletion of intracellular calcium and which reestablish calcium ion levels in the endoplasmic reticulum of a cell.

Calmodulin A multifunctional protein found in all eukaryotic cells which senses and binds calcium ions. It regulates the activities of other calcium-binding proteins, ion channels, and signaling pathways.

[1] Margulis L., Schwartz KV. Five kingdoms: an illustrated guide to the phyla of life on earth. 2nd ed. New York, WH Freeman, 1988.

I. M. Shapiro, W. J. Landis, *Mechanisms of Mineralization of Vertebrate Skeletal and Dental Tissues*, https://doi.org/10.1007/978-3-031-34304-9_5

Calnexin A calcium-binding chaperone protein integral to the endoplasmic reticulum of a cell which helps direct, together with calreticulin, folding and assembly of proteins and glycoproteins, and also regulates calcium levels in the cytosol and endoplasmic reticulum.

Calreticulin A calcium-binding chaperone protein resident in the endoplasmic reticulum of a cell which helps direct, together with calnexin, folding and assembly of proteins and glycoproteins and also regulates calcium levels in the cytosol and endoplasmic reticulum.

Chaperone protein A molecule which promotes the folding and assembly of proteins while shielding them from other proteins which might obstruct the processes.

Chondrodysplasia A family of rare inherited disorders in which there is aberrant development and growth of endochondral bone characterized by widening of the epiphyses and short thickened bones. The most common clinical manifestations of chondrodysplasia are multiple epiphyseal dysplasia and pseudoachondroplasia.

Claudins A family of proteins which participate in the structure of tight junctions.

Cooperativity An act of binding a molecule to an acceptor protein which in turn promotes the binding of a second molecule.

Cytochrome C and SMAC/DIABLO Protein components contained within mitochondria and which induce apoptosis when they are released.

EF-hand sequence A 40-amino acid residue motif which comprises the helix-loop-helix region of a significant number of proteins and which binds calcium ions with high affinity.

Inositol 1,4,5-trisphosphate A signaling molecule synthesized from membrane lipids which causes the release of calcium ions from intracellular organelles.

Inositol hexakisphosphate A signaling molecule concerned with phosphate ion efflux from cells.

K_m The affinity constant for a reaction, commonly measuring the ability between molecules or other reaction components to bind to one another, as an enzyme and its substrate. A high K_m indicates a low affinity and, for optimum function, K_m is close to the substrate concentration.

Leucine zipper, EF-hand-containing transmembrane protein 1 (LETM1) A transmembrane protein of the mitochondrial inner membrane leaflet which mediates proton-dependent calcium ion efflux from the mitochondrion.

Mitochondrial calcium uniporters Transmembrane proteins which allow the passage of calcium ions into mitochondria of cells.

Mitochondrial-associated endoplasmic reticulum membranes Sites of attachment of mitochondria to the endoplasmic reticulum of a cell.

ORAI A family of plasma membrane pore-forming proteins of the calcium release-activated calcium (CRAC) channel which associate with STIM and mediate calcium ion influx into a cell.

Paracellular transport An epithelial pathway which fosters the transport of ions, water, and other molecules between cells.

Phosphate transporters Proteins which transport inorganic phosphate ions across the cell membrane. Major phosphate transporters are encoded by *SLC20A1*, *SLC20A2*, and *SLC34A1*.

PIEZO1/2 Homologous pressure-sensitive channels in the plasma membrane of a cell which regulate calcium ion influx.

Potassium-dependent sodium/calcium exchanger A protein complex which utilizes the extracellular sodium ion gradient to co-transport calcium and potassium ions out of a cell.

Primary familial brain calcification A genetic disease characterized by the mineralization of blood vessels of the brain.

Purinergic receptor-operated channels (P2X and Y) Cation channels which are preferentially stimulated by adenosine or ATP.

Reduced co-enzymes Molecules such as NADH and FADH2 which are used to generate ATP.

Ryanodine receptor A receptor molecule present in the membranes of the endoplasmic reticulum activated by an increase in calcium ions as well as ligands derived from NADP.

Single nucleotide polymorphisms (SNPs) A variation in DNA caused by a difference in a single nucleotide base-pair.

SPX domain of XPR1 The peptide domain within the receptor XPR1 responsible for phosphate ion efflux from cells.

STIM1/2 Sensor proteins resident in the endoplasmic reticulum which regulate calcium ion homeostasis in a cell by controlling store-operated calcium channels in the plasma membrane.

Store-operated calcium entry (SOCE) Following calcium ion depletion, the process of refilling intracellular (endoplasmic reticulum) calcium stores.

Tight junction A complex of proteins which joins together epithelial cells to limit the movement of ions and water between cells.

Transcellular transport The movement of molecules through cells.

Transient receptor potential voltage (TRPV) channels Nonselective cation channels which enhance the passage of calcium ions into a cell. TRPV channel activity is dependent on calcium ion store depletion, as well as ligands such as inositol 1,4,5-trisphosphate.

Unfolded protein response A signaling mechanism which orchestrates the recovery of endoplasmic reticulum function following stress in which chaperones, foldases, and protein disulfide isomerases fold and package proteins such that they can be trafficked for further processing in the Golgi apparatus of a cell.

Uniporter An integral membrane protein responsible for the transport of a single species of ion or molecule across a cell membrane in one direction.

Voltage-gated calcium channels Ion channels of a cell membrane which promote calcium entry into cells when membrane depolarization occurs.

Abbreviations and Proper Names for Genes/Proteins and Other Molecules

AMEL	Amelogenin
BMP	Bone morphogenetic protein
CA	Carbonic anhydrase
COLI	Type I collagen
CRAC	Calcium release-activated channel

DMP	Dentin matrix protein
FGF23	Fibroblast growth factor 23
IP3	Inositol 1,4,5-trisphosphate
IP6	Inositol hexakisphosphate
LETM1	Leucine zipper, EF-hand containing transmembrane protein 1
MAM	Mitochondrial-associated endoplasmic reticulum membrane
MARC	Mitochondrial-associated endoplasmic reticulum contact
NCX(K)	Sodium/calcium (potassium) exchanger
OCN	Osteocalcin
ORAI	Olfactory receptor class A related 1
P2, P2X, and P2Y	Two families of purinergic G protein-coupled receptors
PMCA	Plasma membrane calcium ATPase
RUNX	Runt-related transcription factor
SERCA	Sarcoendoplasmic reticulum calcium-ATPase
SLC20A1/PiT1	Solute carrier family 20 member A1; also called sodium-dependent phosphate transporter 1
SLC20A2/PiT2	Solute carrier family 20 member A2; also called sodium-dependent phosphate transporter 2
SLC24A4	Solute carrier family 24 member A4; also called potassium-dependent calcium-sodium transporter
SLC34A1	Solute carrier family 34 member A1; also called sodium/phosphate cotransporter, *NaPi-IIa* or *NPT2a*
SLC34A2	Solute carrier family 34 member A2; also called sodium/phosphate cotransporter, *NaPi-IIb* or *NPT2b*
SMAC/DIABLO	Diablo IAP-binding protein
SOCE	Store-operated calcium entry
STIM	Stromal interacting molecule
TNAP	Tissue non-specific alkaline phosphatase
TRPV	Transient receptor potential voltage channel
VDCC	Voltage-gated calcium channel
VEGF	Vascular endothelial growth factor

5.1 Snapshots in Time

A microscopic view of the mineralization front of a vertebrate calcifying tissue shows a confusing complex of round, ovoid, and dendritic cells, located amid a landscape of apatitic mineral with crystallites enfolded within an extracellular matrix. The chemical nature of the matrix depends on the tissue type, age and stage of maturation, and the phenotype of the cells which populate the forming bone, tooth, or other mineralizing tissue. For example, in developing enamel, its cheese-like matrix contains a number of unique proteins, in particular, the hydrophobic protein, amelogenin (AMEL). This protein presumably modulates thin and very long enamel crystallites, which form rods winding their way, like partially cooked spaghetti, from a pre-mineralized region closest to the ameloblast to the most distal surface at the periphery of the dental sac. The earliest crystallites of bone are evident within and about collagen fibers; they are external to the osteoblasts and shorter than those of enamel. In mineralizing cartilage and early dentin, calcospherites (a word meaning spherical mineralized particles) are found together with matrix vesicles, anchored to the organic matrix secreted by and surrounding maturing hypertrophic chondrocytes and odontoblasts. Morphologic evidence suggests that such vesicles sequester calcium and phosphate ions and produce foci of mineral; with time, the vesicle membranes may become leaky and possibly rupture (see below). Vesicle contents then escape to extravesicular spaces where additional crystallites may nucleate, fuse, and grow in all dimensions as they develop a more permanent and better organized mineral phase in the tissue extracellular milieu.

Although light and conventional electron microscopy as well as other sophisticated forms of imaging yield insights into the mineralization process, the information which they provide represents only static snapshots in time and does not present a temporal view of events leading to the deposition of mineral. In other words, these types of microscopic techniques do not afford a complete frame-by-frame explanation of how mineralization occurs. Despite such a shortcoming, many investigators have made assumptions and drawn conclusions about mineral formation from the still images. Like the discussion of the saga of Lot's wife, against the commands of the Lord, Edith looked back at the destruction of Sodom (Fig. 5.1). Her action has prompted two markedly different interpretations of this singular incident. Indeed, commentaries on the Biblical mineralization event chronicle that two of Edith's daughters remained in the doomed city, as did her two future sons-in-law (Gen.

Fig. 5.1 A view of the terrible event in which the biblical cities of Sodom and Gomorrah are destroyed. The only survivors are Lot and his family, and this medieval painting shows the Lot family leaving the doomed city. Note the sad backward-looking figure of Edith, who was subsequently turned to a salt pillar. (Woodcut from the Nuremberg Chronicle, compiled and written by Hartmann Schedel and illustrated by Michael Wolgemut and Wilhelm Pleydenwurff, 1493)

Rabbah 50:9; see also "Lot's daughters"). With close family members remaining in Sodom, it would not be unreasonable to conjecture that the backward glance was an expression of concern and worry about her relatives. This explanation is markedly different from the more conventional one which views her actions as a case of willful disobedience. These are two very different explanations for a sad event which ends with her mineralization. In a similar manner, the presence of pre-formed mineral, like a mineralized Edith, provides little information as to the why and wherefore of events which promote mineral nucleation, growth and development leading to the formation of apatite in vertebrate tissues.

As will be elaborated in more detail later in this chapter, studies over many years by numerous research groups have led to two principal mechanisms by which apatitic crystallites form in bone, dentin, cementum, and mineralizing cartilage and tendon (possible mechanisms of enamel mineralization are discussed elsewhere in this book). The first suggested mechanism proposes that vesicles loaded with calcium and phosphate ions are shuttled through the cytosol to the plasma membrane [2] and exported from the cells to the extracellular matrix as exosomes or matrix vesicles [3]. In the growth plate, for example, hypertrophic chondrocytes exocytose matrix vesicles into the cartilaginous matrix [4, 5]; in newly forming bone and dentin, osteoblasts, and odontoblasts, respectively, secrete matrix vesicles into the collagen-rich extracellular matrix of these tissues [6, 7]. Following rupture or loss of the membrane envelope of these secreted vesicles, calcium and phosphate ions released or exposed to the extracellular milieu associate as mineral foci which coalesce and spread to neighboring proteins and other molecules.

The second mechanism of mineralization involves the biogenesis and secretion of calcium phosphate particles. Unlike the membrane-bound organelles just noted above, osteoblasts and chondrocytes secrete these "naked" particles during both endochondral and intramembranous bone formation and replacement [8]. One clue to their origin is the observation that particles of this type have been visualized on the inner walls of blood vessels in developing bone, a result suggesting a possible vascular source [9]. Presumably, these particles find their way to collagen molecules, which then become mineralized. A more complete discussion of the nature of the particles is presented in the next chapter of this book.

Whether the source for mineral formation is calcium and phosphate particles enclosed in matrix vesicles or vascular-derived calcium and phosphate particles, local and systemic molecules play a major role in controlling subsequent aspects of the mineralization process. To accommodate apatite nucleation and mineral crystal growth and development, these molecules influence the transition of precursor phases to crystalline apatite and promote the turnover of matrix proteins. In addition, they limit the propagation of mineral and thereby inhibit dystrophic mineralization in resident soft tissues. Both of the mechanisms described above for vertebrate tissue mineralization are energy-dependent and take place in a relatively hypoxic tissue environment. From this perspective, apatite deposition impacts both metabolic and energy conserving pathways of the cell and would be expected to generate the unique characteristics of each of the specialized mineralized tissues.

5.2 Some General Comments on Calcium Homeostasis in Cells of Mineralizing Tissues

One of the themes explored in this book is that mineralization is a cell-directed process in which calcium and phosphate ions are partitioned into intracellular organelles such as vesicles and exported or secreted across the plasma membrane of a cell as ion complexes or clusters. In this regard, a preponderance of evidence supports the concept that one of the mechanisms leading to early mineralization of extracellular matrices involves exocytosis of matrix vesicles from chondrocytes, osteoblasts and odontoblasts [4–7]. In the extracellular milieu, the release of packaged ions from these organelles or the free ion complexes themselves serve as foci for mineral formation. For this purpose, cellular pathways evolved which facilitated calcium ion accumulation, trafficking, and discharge without triggering metabolic changes which precipitated cell death and tissue failure.

As the cell membrane forms an almost impermeable barrier to calcium ions, an initial question concerns the mechanism(s) by which cells accumulate or concentrate cations and anions. To begin with, the term, "impermeable," is somewhat misleading as limited amounts of calcium ions diffuse through the membrane. Aside from this avenue, there exist specific transport systems which provide controlled cation passage into the cell: Indeed, carrier proteins, electrochemical channels, and ATP-driven pumps in the plasma membrane can now be delineated which transport calcium ions into subcellular compartments while maintaining the ionic composition of the cytosol. Notably, while the actual cytosolic calcium concentration remains low, in the micromolar range, the cell experiences substantive and rapid calcium oscillatory changes. These oscillations in calcium ion levels regulate an ever-growing list of intracellular signaling functions, especially those concerned with anabolic and catabolic processes. It is for this reason, as noted earlier, that cation transport linked to mineral deposition must take place in such a way that it does not interfere with metabolic signaling activities requisite for cell survival and function in a specialized and often hostile metabolic niche.

How a small amplitude spike in the cytosolic calcium concentration is accommodated by the cell has received intense study [10]. Since flooding the cytosol with calcium ions interferes with energy metabolism, the cell has developed intricate means of maintaining homeostasis. Protein buffering systems and storage organelles protect the cell against high temporal and spatial oscillatory fluxes. For example, the Golgi cisternae, endosomes, and lysosomes have a high capacity for calcium [11, 12]. These organelles accumulate and concentrate calcium and phosphate ions while activation of membrane pumps and channels and sequestration by calcium buffering proteins smooth out localized shifts in cytosolic calcium levels. Following an oscillatory spike and the restoration of the cytosolic calcium ion concentration, levels of key molecules like ATP and NADH are re-established and cell function is maintained. However, if the increase in cytosolic calcium is extended in time, then irreversible deficits occur. Such deficits include damaged contractile proteins, altered signaling pathways, changes in the reduction-oxidation (redox) state of the cell, and eventual death (apoptosis). Each of these functional deficits can precipitate irreparable events which lead to tissue damage and often dystrophic mineralization.

In the context of calcium regulation by cells, many investigations have focused on the role of calcium-binding proteins of mineralizing tissues. Several of such proteins, especially those directly related to mineralization phenomena (annexins, calcineurin, calmodulin, calreticulin, fetuin and matrix Gla protein and other vitamin K-dependent proteins, for example), have attracted considerable attention. Annexins, calcineurin, and calreticulin are discussed later in this chapter and fetuin and vitamin K-dependent proteins are described in Chap. 9. Besides buffering the calcium ion level to preserve cell function, some of these proteins may serve as co-factors for a number of enzymes and act as signaling molecules through numerous pathways. The function of certain cytosolic proteins such as parvalbumin and the calbindins (vitamin D-dependent proteins required for the intestinal absorption of calcium) is dependent upon the presentation of common structural motifs which endow the protein with a high affinity (or strength of binding) for calcium ions. The affinity of these calcium-binding proteins, measured as a binding constant, is usually one or more orders of magnitude greater than the steady-state cytosolic calcium ion level. In other words, such proteins will bind or take up calcium ions (buffering) in response to even minor changes in calcium ion concentration. Ligand binding by these proteins is reversible, and the kinetics of binding are rapid. Thus, when the ion flux is high and rapid and leads to elevations in intracellular calcium ion levels, these proteins return the calcium ion concentration in the cytosol to the physiologic "set-point." From a functional perspective, this series of events means that the cell can respond to calcium challenges with time periods which range from microseconds (when activating membrane fusion) to hours (when promoting gene transcription or protein processing, for example).

Protein databases which are currently available indicate that nearly 300 proteins, including parvalbumin, calmodulin, and the calbindins, have one or more calcium-binding domains. All of these proteins contain a highly conserved peptide region comprising a pair of helices (actually a helix and then a loop in the protein backbone followed by another helix) called an EF-hand domain which binds calcium ions (Fig. 5.2) [13–16]. The loop is rich in acidic residues and forms coordinate linkages with calcium ions. Within the loop or "palm" region of the E-F hand structure, then, positively charged hydrated calcium ions bind preferentially to negatively charged amino acid residues, such as aspartic and glutamic acid (Fig. 5.2) [16]. Uncharged amino acids form a localized protective hydrophobic umbrella around the calcium ions which serves to stabilize the protein–calcium complex [17]. Some closely related proteins vary in the number of EF-hand regions they contain. The vitamin D-dependent protein, calbindin D9k, expresses a single EF-hand pair, for instance, while calbindin D28k contains three paired EF-hand motifs. An added advantage of the latter type of helix pairing is that ligand binding to one region of the protein complex causes a structural change which promotes calcium ion binding at a second site, a phenomenon termed "cooperativity." This conformational event promotes interactions with other proteins, as well as with the long carbon chains of fatty acids and even charged sites on RNA and DNA, while at the same time buffering the cytosol against a rise in free calcium [18]. As indicated above, a variety of calcium-binding proteins with differing calcium ion affinities exists within the cell. Bagur and

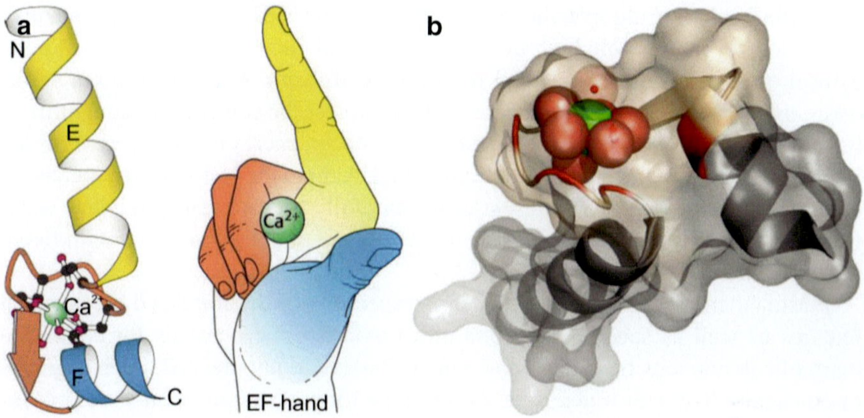

Fig. 5.2 Topology of the helix-loop-helix structure of many calcium ion-binding proteins and its analogy with fingers of the hand. (**a**) The index finger and thumb of the right hand are used to model the EF-hand calcium-binding motif. The yellow (E) and blue (F) helical portions of the motif correspond to the first finger and thumb, respectively, and the calcium-binding region is contained within the curvature of the middle finger. Coordinate bonds are formed between the calcium ion (green) and the charged amino acids (black) of the protein. Calcium ion movement either "on" or "off" the protein is facilitated by the variability coordination chemistry of calcium. N and C refer to the N- and C-termini of the peptide chain. (**b**) Schematic of the highly conserved region of an EF-hand protein, calmodulin, comprising a pair of helices (actually an α-helix and then a loop in the protein backbone followed by another α-helix). The loop region (tan) in an EF-hand peptide motif is typically rich in acidic residues and oxygen atoms (red) are found closest to the bound calcium. The schematic also shows a calcium ion coordinately bound to water. α-Helices of the protein are depicted in grey. (Panel **a** adapted from Kazakov et al. [13]; Panel **b** reproduced from Halling et al. [14] with permission)

Hajnóczky commented that those proteins with high affinity will buffer calcium ions and preserve ion homeostasis in the cell, while those proteins with affinities closer to the cytosolic calcium concentration (10^{-7} M) "act as calcium sensors and as such modify their conformational state and interact with downstream targets" [19].

Against this background of ion buffering and signaling, mechanisms by which both cells and certain proteins regulate the activities of calcium ions will next be briefly described. This discussion will include consideration of the means by which osteoblasts and other cells of vertebrate mineralized tissues maintain high calcium ion loads in specific organelles without triggering noxious events and mechanisms of the release of accumulated stores of calcium ions from cells. Chaps. 6 and 7 contain a more detailed treatment of calcium ion release mechanisms related to mineral deposition.

5.3 The Role of Membrane Channels in the Regulation of Calcium Ion Entry into Cells

Cells which populate bone, calcifying cartilage and tendon (in some avians), dentin, cementum, and enamel in vertebrates must survive in an environment which is constantly battered with a torrential flow of calcium ions. As the first line of defense in controlling calcium, the plasma membrane of these cells serves as a substantive "dam" to prevent possible high tides of external calcium ions from swamping the cytosol, precipitating proteins, and triggering cell death. And, like the sea walls, dikes, and sea gates of the Netherlands which protect the country from engulfment by the North Sea, when porosities and breaches occur, energy-requiring systems are in place to maintain the land dry and functional. Akin to the Netherlands, metabolic energy-requiring systems are present which carefully and exquisitely regulate water flow and ion transport across the plasma membrane of cells concerned with tissue mineralization.

Plasma membrane channels with a broad spectrum of structural and biochemical features as well as specific functionalities serve as critical portals for controlled entry of calcium ions from the extracellular tissue fluid into the cell. Such channels or pores are filled with water, often occur in the form of multimolecular complexes which traverse the lipid bilayer of the membrane, and permit diffusion of ions along electrochemical gradients from outside to inside the cell.

Calcium ions are especially important with respect to cellular processes as they are ubiquitous signaling molecules among their several functions and even slight variations in their concentration influence many signaling pathways in the cell. Hence, it is of considerable physiological consequence that the entry of hydrated calcium ions into the cytosol is precisely maintained so as to keep the ion activity product low and constant. Because of this limitation, cells which generate mineralized tissues are suggested here to accumulate calcium ions by a process which requires strict compartmentalization. In this manner, such a localized environment would permit ion packaging and storage and foster ion cluster formation so that subsequent free energy changes would be sufficient to promote mineral nucleation.

The most common transmembrane channels are transient receptor potential voltage channels (TRPV), store-operated calcium entry channels (SOCE), voltage-gated calcium channels (VDCC), channels for Na^+/Ca^{2+} exchange (NCX), purinergic receptors (P2), and receptor-operated channels (ROC; P2X and Y) (Fig. 5.3) [20]. SOCE channels are considered in greater detail later in this chapter in terms of their relation with the endoplasmic reticulum in the cell cytoplasm. A plasma membrane calcium ATPase (PMCA) is also involved in calcium ion homeostasis (discussed in more detail in the following chapter).

Following the opening of an ion channel to allow calcium ions to enter and move down their concentration gradient into the cell, a spike or oscillations occur in the cytosolic calcium ion load. Before buffering systems might respond to such changes and mediate a return to baseline levels, the increase or shift in calcium ion activity may induce alterations of cell structure and function. In this instance, the cell can employ a second line of defense involving membrane systems of intracellular organelles, particularly those of the endoplasmic reticulum and mitochondria. Within the endoplasmic reticulum, for example, chaperone proteins, such as calreticulin and calnexin, which are responsible for protein folding, are sensitive to calcium ion levels and can buffer and control them. The endoplasmic reticulum of the cell, then, functions not only in the synthesis of molecules critical to the formation of the

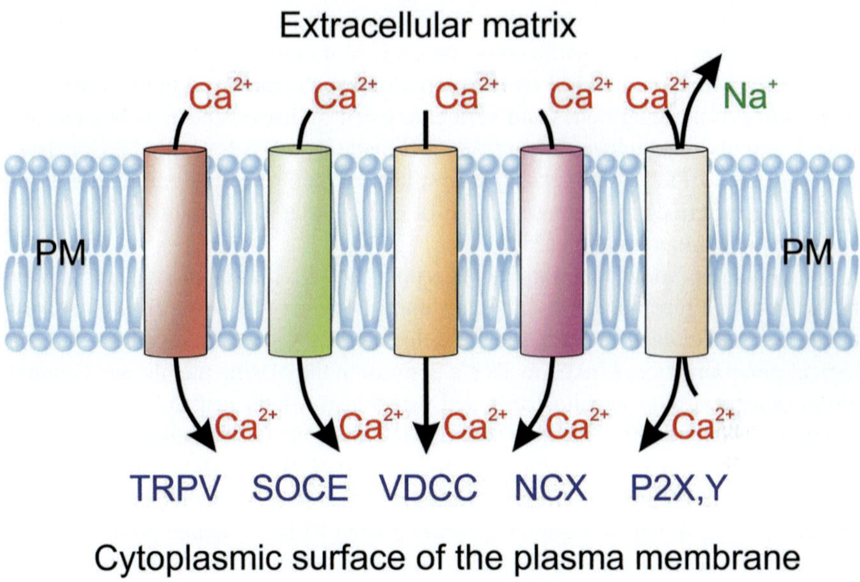

Fig. 5.3 Schematic showing plasma membrane (PM) channels regulating calcium ion entry into a cell. The most common pathways for calcium ion entry and relevance to the mineralization process include transient receptor potential voltage channels (TRPV), store-operated calcium entry channels (SOCE), voltage-dependent calcium channels (VDCC), channels for Na^+-Ca^{2+} exchange (NCX), and receptor-operated purinergic (P) channels (P2X and Y). There are also plasma membrane calcium ATPase channels (not shown). All these channels are active in osteoblasts. It is likely that a receptor type of channel is species-specific, redundancies between channels exist, and groups of calcium transporters work coordinately to regulate specific functions which include calcium sensing, signal transduction, and organelle calcium loading. (Schematic after Parekh and Putney, Jr. [20] with permission)

complex intracellular and extracellular matrices of hard tissues but also in the regulation of calcium and other ions for mineral formation [21, 22]. In its latter role, the endoplasmic reticulum, as well as the mitochondria, relies indispensably on proteins which are associated specifically with its membrane and channel system.

Of the channels illustrated in Fig. 5.3, the TRPV channel in osteoblasts is responsive to the vanilloid compound, capsaicin. TRPVs are representative of a large family of channels and their composite proteins, of which at least three are highly selective for calcium (TRPV3, 6 and 7). It is noteworthy that at least two variants in TRPV structure are present in osteoblasts, and four members of the TRPV family are active in odontoblasts [23]. There is some apparent redundancy in the function of these channel proteins since inhibition of TRPV5 and 6 shows little change in expression of mineralization-related genes or the calcium uptake characteristics of osteoblasts [24]. Once the cation has traversed the TRPV channel and entered the cytoplasm, it immediately interacts with a calcium-binding protein, commonly a member of the calbindin family. These proteins are of varying molecular weights (9k and 28k as noted above) and are dependent on the presence of vitamin D [25, 26].

Human clinical studies show that TRPV channels are important in the initiation and progression of the mineralization process. Mutations in *TRPV4* cause striking irregularities in endochondral ossification along with decreased bone density and deformities of the long bones and vertebrae (seen in diseases such as brachyolmia type 3, spondylometaphyseal dysplasia, and metatropic dysplasia) [27]. Consistent with the role of *TRPV4* in metatropic dysplasia, a gain-of-function mutation in the gene is characterized by increased levels of intracellular calcium ions in osteoblasts and enhanced mineralization [28].

While there is some uncertainty concerning the activities of individual members of the vanilloid receptor family [24], it should be noted that currently a clear association exists between the activities of selected TRPV calcium channels and biological mineralization. Linked to TRPV activity in the plasma membrane, channels are present in osteoprogenitor cells and osteoblasts which respond to mechanical "stretch." These PIEZO channels (PIEZO 1 and 2) are stretch-sensitive ion transporters which are responsible for calcium ion influx [29–31]. The channels respond to hydrostatic pressure, in particular those forces exerted by blood vessels invading the tissues [32]. When activated, the force-sensing PEIZO channel proteins trigger a signal transduction pathway which promotes transcriptional changes enhancing osteogenesis and bone formation [33]. Loss of PIEZO transcription factor activity causes deficits in type II and IX collagen transcription and increases in osteoclast activity [34]. As a result of these changes, multiple spontaneous bone fractures occur [34]. It is possible that this channel is activated during bone and cartilage development, especially during the period when vascular tissue becomes involved with the induction of mineralization [34].

In contrast to TRPV channels which are principally non-gated, groups of voltage-dependent ion channels (VDCCs) control calcium ion transport across the plasma membrane in a different manner. VDCCs are ubiquitously present in excitable cells; high-voltage activated channels are present in osteoblasts [35]. VDCCs are typically in a closed configuration; a change in electrochemical potential causes channel opening and, in the presence of vitamin D, there is rapid transport of ions through the central pore of the channel along an electrochemical gradient [36]. Since VDCCs are unidirectional, they facilitate calcium ion entry into the cell with no co-incidental transport of a counterion. In this case, because positively charged calcium ions would outnumber negatively charged ions, there would be a loss of ionic equilibrium in the cells and functional deficits would ensue. To counteract this type of ionic disturbance, other transport proteins of the complex portal of gates and channels at the plasma membrane would be activated.

Consistent with the functional demands of cells of the vertebrate mineralized tissues, channel proteins are present in the plasma membrane which facilitate the transport of calcium and counterpart negatively charged ions into and out of the cell. Patch clamping, immunohistochemistry, and other techniques have shown that the cell can maintain charge neutrality by exchanging (X) sodium (N) for calcium (C) (sodium/calcium exchange protein, NCX) [37]. The mechanism of action of the NCX is regarded as one of the most critical in removing calcium from cells [37]. In addition, a potassium-dependent sodium/calcium exchanger (NCKX) may transport calcium from the cells to the extracellular environment [38]. For both NCX and NCKX exchangers, the direction of ion transport is dependent on the transmembrane ion gradient [39–41]. The importance of these NCX and NCKX channels in the physiology of the mineralization process is discussed further in Chap. 6.

The purinergic (P) channel proteins represent another family of exchangers linked to bone and cartilage mineralization. One class of these purinoceptors is a group of ligand-gated ion channels, P2X, which respond to nucleotides like ADP and ATP and shuttle calcium ions across the plasma membrane into the cell. As ligand-gated receptors for purines (P2X and P2Y receptors), the receptor pore in the membrane channel opens when ATP is bound and closes when ATP is unbound. Within this receptor family, ATP and ADP activate P2X and P2Y. A group of P1 receptors is sensitive to adenosine, a derivative of ATP [42]. In all cases, once the ligand is bound, calcium ions enter the cell.

Osteoblasts, like several types of neuronal cells, secrete ATP and the ATP receptor subtypes, P2X and P2Y, are present in the plasma membranes of bone and cartilage cells [43]. Activation of these receptors by purines and purine-based nucleotides causes an elevation in intracellular calcium levels: Binding of ATP to the P2X ion channel receptors leads to calcium influx. Activation of the P2Y receptors also causes a rise in intracellular calcium from both extracellular as well as internal cellular stores (Fig. 5.3).

While the role of nucleotides such as ADP and ATP is directly linked to metabolism, it is clear that they can also promote cell signaling activity. Thus, in the presence of ATP as just described, the receptor pore in the membrane channel opens, permitting the entry of calcium ions into the cell. Several sources of extracellular ATP have been delineated: those secreted by the cell itself (autocrine signaling); those secreted by neighboring cells (paracrine signaling); and those released from distant organs like the liver (endocrine signaling) [42]. When osteoblasts are treated with ATP, there is expression of tissue non-specific alkaline phosphatase (*TNAP*), runt-related transcription factor (*RUNX2*), and osteocalcin (*OCN*) [44]. These and other results indicate that purinergic signaling is required for the development and mineralization of the skeleton [45–47].

Since considerable numbers of P2X receptors exist, there has been discussion concerning the receptor subtype which mediates tissue formation and mineralization. In this regard, Ke et al. and Agrawal and Gartland reported that the receptor subtype, P2X7, is required for bone cell function [48, 49]. Activation of the P2X7 receptor has been found to promote osteoblast differentiation, plasma membrane blebbing, matrix mineralization, and increased expression of *WNT/β-catenin* transcriptional activity [50–52], while disruption of the gene encoding the P2X7 receptor reduces periosteal bone formation [48, 53]. From the observations just noted and others, Orriss et al. commented that the purinergic receptor in the differentiated and functional osteoblast could serve as an "off-switch" for bone formation [54].

As mentioned above, a great many membrane proteins participate in calcium ion loading of the cells of vertebrate mineralized tissues. It is of course very likely that redundancies in function and variations in responsiveness occur among such proteins and that the activities of these ion exchange and transporting systems reflect the function of cells at specific developmental phases of their life history. With respect to the latter point, it is probable that the P2X system is most active at early stages of osteo- and chondrogenesis when cells are in an anabolic phase and ATP generation and turnover are high. During later stages of the mineralization process, elevated cellular membrane phosphatase activity would decrease the levels of extracellular ATP. Other plasma membrane protein pumps and channels involved with calcium ion homeostasis are present and include a plasma membrane calcium ATPase and additional exchanger proteins. Such proteins are most active in osteoblasts during the phase of osteoid mineralization [40]. Thus, a wide diversity of calcium ion membrane exchanger/transporter systems is available and prepared to meet the demands of cells at specific stages of hard tissue development. At all times, once calcium ions have traversed the plasma membrane channel and been released into the cytosol, they interact with one or more calcium-binding proteins [55]. The bound proteins subsequently influence cell function at a number of different levels while protecting the cytosol from events which could promote cell toxicity.

5.4 The Role of Cytosolic Proteins in the Regulation of Calcium Ion Entry into Cells

As just noted, positively charged calcium ions which enter a cell through plasma membrane transporter systems discussed in the earlier part of this chapter bind a number of cytosolic proteins. In so doing, cellular ionic balance is preserved by the buffering action of the proteins. Such calcium-buffering proteins include those of the S100 family, calmodulin, calreticulin, and calbindins; these proteins bind calcium through a number of calcium ion-binding motifs, such as the EF-hand helix-loop-helix motif described above, and ion coordination by clustered acidic amino acids [56]. Regardless of the type of interaction, calcium ions induce a large conformational change in molecular structure of the binding protein, exposing motifs which permit the protein to activate downstream molecular targets. Such targets include membrane receptors, proteins involved with signal transduction, transcription factors, and nucleic acids [57]. In this way, the calcium-binding proteins also serve as both calcium sensors and trafficking agents, regulating the activities of enzymes as well as transcription factors [58].

In osteoblasts, a rise in ionic calcium levels is countered by enhanced interactions with proteins which include calmodulin-dependent protein kinases, calcineurin, calbindins, and transcriptional proteins such as nuclear factor of activated T-cells (NFAT); certain of these calcium-binding events lead to changes in osteoblast gene expression [59–61]. The conformational change in calmodulin caused by its binding calcium ions enhances the interaction of this complex with numerous proteins which contain calmodulin-sensitive motifs defined by basic and bulky hydrophobic amino acids bracketed by aromatic residues [62]. Calcium ions bound to calmodulin also activate protein kinases which promote the activity of the osteoblast-specific transcription factor, osterix (OSX) [63]. As noted previously, OSX in turn up-regulates the expression of a multitude of genes required for osteogenesis and development of the mature skeleton [63].

The observation that about 50% of the total calcium ion content of the endoplasmic reticulum is bound to calreticulin suggests that this protein is required for calcium accumulation and trafficking in cells [64]. Calcium–calreticulin complexes also modulate gene expression while playing a key role in maintaining calcium ion homeostasis in the endoplasmic reticulum [65]. Additionally, Pilquil and colleagues determined that calreticulin influenced the localization of NFAT: When NFAT nuclear trafficking was blocked, commitment of stems cells to chondrogenesis was favored over osteogenesis [66]. In a related context, the level of complexity in the control of osteogenesis as well as chondrogenesis may be amplified since activation of the NFAT pathway appears to decrease NF-κB, a potent regulator of osteoclastogenesis and bone remodeling [67]. Antagonism between ligand-dependent events which control bone and cartilage pathways of formation and resorption/loss can be viewed as providing another critical layer of control over cell commitment to a particular skeletal phenotype [68].

In developing dental tissues, there is increased expression of calcineurin by ameloblasts during the secretory stage of amelogenesis as well as in maturing odontoblasts during dentinogenesis; calreticulin is expressed in odontoblasts; calbindin-D9k and D28k are expressed in both ameloblasts and odontoblasts; and the calcium-binding protein, tuftelin, is highly expressed in ameloblasts [69–72]. While each of these proteins is associated with intracellular calcium ion homeostasis, Somogyi et al. reported that calreticulin was present in the predentin extracellular matrix, associated with both its collagen fibers and matrix vesicles [71]. These findings suggest that this protein is multifunctional in that it buffers intracellular changes in calcium ion concentrations and persists in the extracellular matrix, possibly stabilizing calcium phosphate ion clusters and mineral development [71].

The annexins comprise an additional group of calcium-binding proteins which is especially relevant and important to the mineralization process. The descriptor, annexin, is appropriate as these proteins annex or dock with charged anionic sites on membrane phospholipids, including those of the cell plasma membrane and subcellular organelles [73–75]. Of the 12 members of the annexin family of proteins, annexins A2, A5, and A6 are highly expressed in calcifying cartilage, bone, and matrix vesicles [76–79]. Other studies have shown that annexin A6 is present on the plasma membrane of cells and in autophagosomes as well as late endosomes and lysosomes, and it is thought that annexins function to form microdomains like with those of lipid rafts, while they enhance vesicle fusion and exocytosis [80]. Since annexins bind calcium ions and are present on intracellular organelles, the plasma membrane and exosomes, these proteins have also been suggested to affect mineralization processes by providing critical sites for intracellular calcium ion trafficking and cluster formation, possibly serving as nucleating centers for extracellular mineral deposition [81].

To conclude this section of the chapter, as has been pointed out earlier, a high level of control of calcium ionic activity is extremely consequential since many signal transduction pathways are sensitive to shifts in cytosolic calcium ions. Indeed, intracellular calcium ions have been perceived to act like hormones and have thus received the label of "second messengers" [82]. As described above, much of this control is achieved by the action of calcium-binding proteins in the cytosol, activation of membrane pumps and channels, and calcium ion capture by subcellular organelles, in particular those associated with the endoplasmic reticulum and mitochondria of cells. Together, these various cellular components are essential for maintaining the cytosolic calcium ion activity at low micromolar concentrations, thereby sparing cells from calcium overload, and they generate depots or reservoirs of high calcium content within the cell which may be utilized to initiate a mineralization response.

5.5 Metabolism of Calcium Ions by Mitochondria

Despite the buffering action of the calcium-binding proteins noted above, the distribution of calcium ions within a cell is not uniform; several studies show that there are micro- or nanodomains, enriched in calcium ions, in the vicinity of the endoplasmic reticulum [83], mitochondria [84, 85], Golgi apparatus [86], and the cell membrane [84, 85]. When plasma membrane-mediated fluxes in calcium ions occur in a cell, the calcium-binding activities of cytosolic proteins like calcineurin and calbindins are supported by other molecules to modulate cytoplasmic calcium concentration. One such protein is a sarcoendoplasmic reticulum calcium-ATPase (SERCA), which facilitates entry of calcium ions into the endoplasmic reticulum, and a related enzyme promotes calcium ion uptake into the mitochondria [62, 87]. Together, the calcium-binding proteins and these enzymes enhance the return of the cytoplasmic calcium ion concentration to baseline levels [88]. In this regard, Herrington et al. showed that, in response to an elevation in the external calcium ion concentration, there is increased mitochondrial uniporter expression and activity [89] whereby the channels respond rapidly to micromolar spikes in calcium ion levels [89]. More than likely, calcium ion accumulation in mitochondria is a physiological response directed toward maintaining calcium ion homeostasis.

The mitochondrial uniporter protein complex forms a transmembrane channel which transports calcium ions into the organelle matrix [90]. The uniporter straddles the inner mitochondrial membrane together with regulatory proteins which extend into the intramembranous space (Fig. 5.4). As the cytosolic calcium ion concentration rises, there is an increase in mitochondrial oxidative phosphorylation activity and an accompanying conformational change of the regulatory proteins which serves to open the membrane channel facilitating calcium entry and transport (Fig. 5.5) [94, 95]. Accordingly, for a cell in an actively respiring state, there is rapid clearance of calcium ions and return to homeostasis.

Aside from members of the uniporter protein family, many other calcium transporters exist in mitochondria (Fig. 5.6) [91]. One of these proteins with the unlikely name of leucine zipper, EF-hand containing transmembrane protein 1 (LETM1) allows the (electrogenic) transfer of one calcium ion into the organelle for the outward passage of one proton (Fig. 5.6) [96]. Another such protein is the sodium/calcium exchanger complex which utilizes the potential difference across the mitochondrial inner membrane to mobilize calcium in exchange for sodium ions (Fig. 5.6) [97]. A third protein is the highly abundant voltage-dependent anion

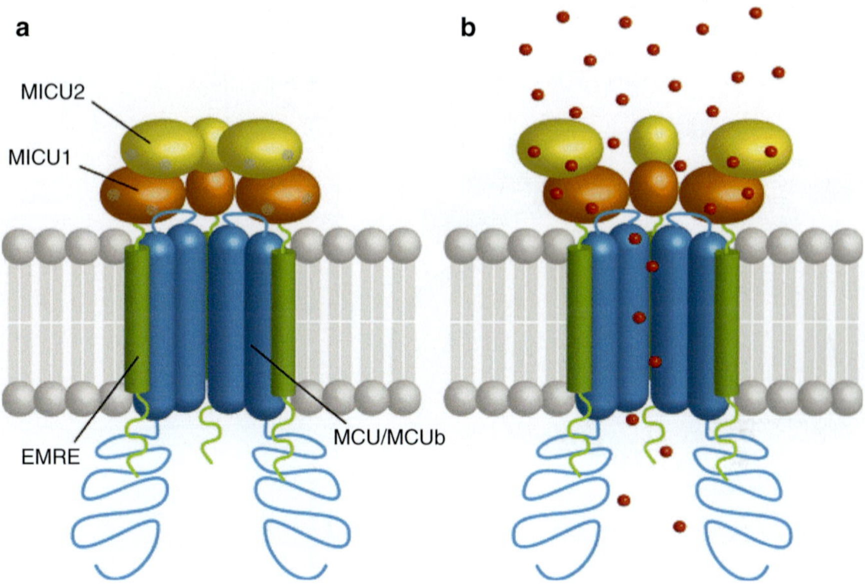

Fig. 5.4 Schematic of the mitochondrial calcium ion uniporter, MCU, which directs calcium ions into the mitochondria of a cell. The function of the MCU is to maintain calcium ion homeostasis within a cell by responding to fluxes in the cytosolic calcium ion concentration. The uniporter is situated on the inner mitochondrial membrane and consists of a transmembrane channel complex, MCU/MCUb, and EMRE (essential MCU regulator), together with regulatory proteins, MICU1 and MICU2 (mitochondrial calcium uptake 1 and 2, both EF-hand calcium-binding proteins). MICU1 and MICU2 extend into the intramembrane space between the outer (not shown) and inner mitochondrial membranes. (**a**) Under resting conditions of normal or low cytosolic calcium, MICU1 and MICU2 interact to prevent passage of calcium ions into the mitochondrion. (**b**) A rise in the cytosolic calcium ion concentration is reflected by an increase in calcium ion concentration in the mitochondrial intramembrane space (numerous small red spheres). The shift to a greater calcium ion concentration causes a conformational change in MICU1 and MICU2 which results in opening the MCU channel through the inner mitochondrial membrane, allowing calcium ion entry into the mitochondrial matrix (see also Fig. 5.6). MCUb is the paralog of MCU. (Image reproduced from Mammucari et al. [90] with permission)

channel (VDAC; Fig. 5.6) [98]. It is thought that this channel protein facilitates calcium ions as well as ATP transport across the outer membrane of the mitochondrion [98]. Finally, Sparagna et al. have described a very rapid calcium ion uptake mechanism, or RaM, which permits the mitochondrion to sequester a considerable number of calcium ions hundreds of times faster than the uniporter [99]. While these and several additional calcium transporters are active in tissues like the liver and heart [91, 100], their presence, activity, and physiological function in mineralized tissues have yet to be determined.

Why are there multiple types of channels for calcium uptake in mitochondria and what function(s) do they serve? One explanation may lie in the observation that, while the channels are selective for calcium, they vary in their sensitivity or affinity for the ion [101]. This feature possibly reflects the need for mitochondria to

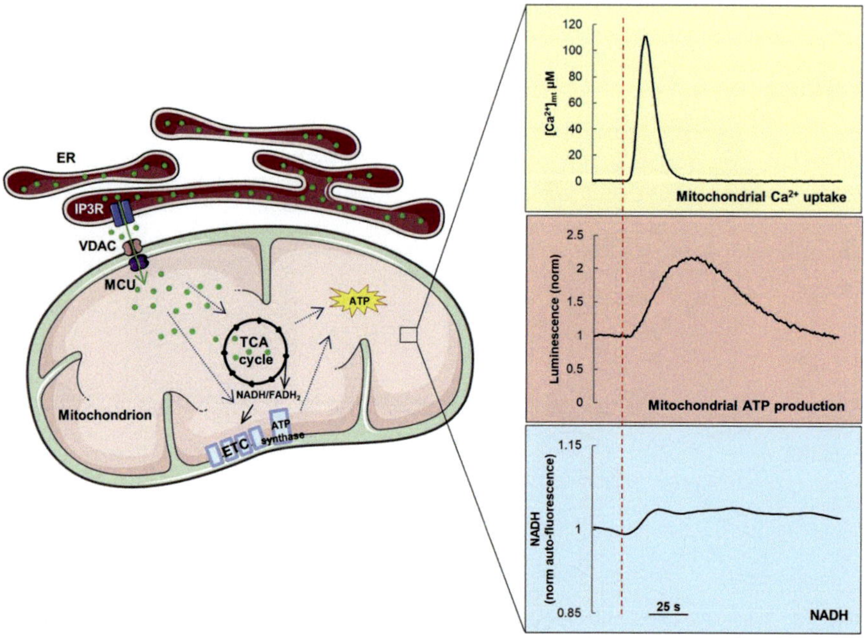

Fig. 5.5 Schematic of the transfer of calcium ions (small green circles) between the endoplasmic reticulum (ER) and mitochondria, mediated by inositol 1,4,5-trisphosphate (IP3) stimulation and sustaining the TCA cycle and ATP synthesis in the mitochondrial matrix. Mitochondria are structurally close or actually tethered to saccules of the ER [92, 93]. Calcium ions are released from the ER when IP3 binds to its cognate receptor (IP3R). Calcium ions are transported across the outer membrane of the mitochondrion by numerous voltage-dependent anion channels (VDAC), while the mitochondrial calcium uniporter (MCU) moves these ions into the mitochondrial matrix. Calcium ions influence the oxidative activity of the mitochondria as shown in plots of Ca^{2+}, ATP, and NADH measured in a representative system of SH-SY5Y cells on induction of IP3 with its agonist, bradykinin (BK) [94]. Red-dotted lines in each plot denote BK addition (100 nM). Ca^{2+} and ATP levels were determined by mitochondria-targeted aequorin and luciferase, respectively, while NADH was measured by auto-fluorescence [94]. BK causes marked increases in calcium ion influx into mitochondria (upper plot). Calcium ion increases stimulate the TCA cycle and lead to enhanced ATP synthesis (middle plot) and affect the electron transport chain (ETC) to generate NADH (lower plot). ATP and NADH have a longer duration compared to the extent of the transient calcium ion peak. Overall, calcium ions promote mitochondrial oxygen uptake, a requirement for electron transport/oxidative phosphorylation [95]. Adenosine triphosphate (ATP), nicotinamide adenine dinucleotide [reduced form] (NADH), flavin adenine dinucleotide [reduced form] ($FADH_2$), tricarboxylic acid (TCA). Changes in $FADH_2$ levels are not shown. (Schematic reproduced from Rossi et al. [94] with permission)

recognize and address a range of physiologic cation fluxes which are experienced by cells. For example, those mitochondria which are situated close to the endoplasmic reticulum network are likely to be exposed to high calcium ion fluxes and require transporters which respond to elevated levels of calcium, perhaps released at a slow rate. In contrast, those mitochondria bathed in low levels of cytoplasmic calcium ions may be best served by transporters which respond optimally to

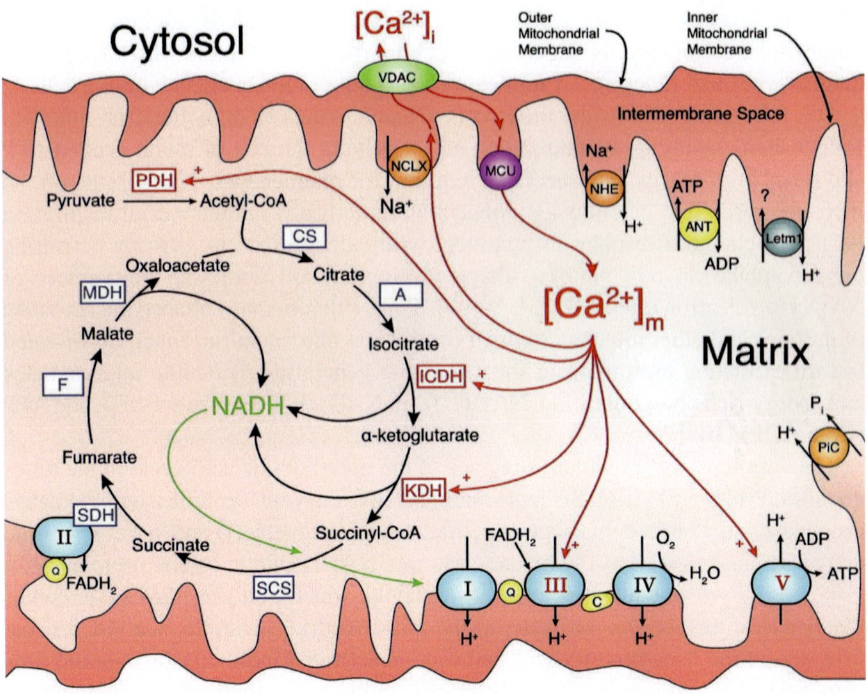

Fig. 5.6 Schematic of the effect of calcium ions ($[Ca^{2+}]_m$) on mitochondrial enzymes of the tricarboxylic acid (TCA) cycle and the cytochrome system (complexes I–V) of the electron transport chain. Calcium ions are transported from the cytosol into the mitochondrial intermembrane space by voltage-dependent anion channel proteins (VDAC) located in the outer mitochondrial membrane. Calcium ions enter the matrix of the mitochondrion by means of the mitochondrial calcium uniporter (MCU) integral to the inner mitochondrial membrane. With the uptake of calcium ions into the mitochondrion, ionic homeostasis is maintained by the activities of a number of transport systems which include the sodium/hydrogen exchanger (NHE), adenine nucleotide translocase (ANT), sodium calcium exchanger (NCLX), and the leucine zipper-EF-hand-containing transmembrane protein 1 (Letm1). Phosphate ion homeostasis is regulated by a mitochondrial proton/phosphate symporter (PiC). Enzymes of the TCA cycle which are calcium ion sensitive are indicated with a red arrow/plus sign. Of these, calcium ions increase α-ketoglutarate dehydrogenase activity by 100-fold. Calcium ions promote ATP synthesis by increasing the conductance of complexes III and V of the electron transport chain. Adenosine diphosphate (ADP), adenosine triphosphate (ATP), pyruvate dehydrogenase (PDH), flavin adenine dinucleotide [reduced] ($FADH_2$), citrate synthase (CS), aconitase (A), isocitrate dehydrogenase (ICDH), α-ketoglutarate dehydrogenase (KDH), succinyl CoA synthetase (SCS), succinate dehydrogenase (SDH), fumarase (F), malate dehydrogenase (MDH), cytochrome C (C), ubiquinone (Q). (Schematic reproduced from Williams et al. [91] with permission)

nanomolar fluxes in cation oscillations. In addition, distinct populations of mitochondria exist with regard to their location in a cell and proximity to their neighboring organelles. In chondrocytes, for instance, some mitochondria are localized at the cell periphery, whereas others occupy a perinuclear position [102]. As might be expected at these and other dissimilar sites, the mitochondria would exhibit modified functions and putatively express differing metabolic and respiratory properties.

Thus, the level of cytosolic calcium ions, the rate of release of calcium ions from intracellular stores, and the locale of mitochondria may be factors which lead to the diverse types and functions of mitochondrial calcium ion channels within a cell.

It is interesting to consider the effects of calcium ions on mitochondrial function. Rather than closing down respiration and depriving the cell of much needed ATP and other high energy intermediates required for maintenance of cell function, an increase in the free calcium ion concentration activates mitochondrial respiration [91]. Mitochondria respond to treatment with calcium ions in vitro by activating oxygen uptake (an indicator of oxidative respiration) and promoting the synthesis of NADH (an electron donor) (Fig. 5.5) [95]. Thus, calcium ions enhance the activities of proteins and other molecules which control key mitochondrial functions directed toward extracting protons from substrates and generating metabolic intermediates and energy-rich compounds like NADH (from NAD), $FADH_2$ (from FAD), and ATP (from ADP) [103].

Calcium ions also modulate enzymes of the tricarboxylic acid cycle in the mitochondrion (Fig. 5.6) [104]. Such calcium-sensitive enzymes include isocitrate dehydrogenase, succinate dehydrogenase, malate dehydrogenase, and α-ketoglutarate dehydrogenase as well as the malate-aspartate-NADH shuttle carrier protein, citrin [104, 105]. Pyruvate dehydrogenase, the enzyme which catalyzes the conversion of the cytosolic metabolite, pyruvate, to the mitochondrial substrate, acetyl CoA, the first step in the tricarboxylic acid pathway, is activated indirectly by calcium ions (Fig. 5.6). Calcium ions additionally activate individual cytochrome components of the electron transport chain; this process further enhances energy conservation by leading to an increase in ATP generation. Thus, elevations in mitochondrial calcium accumulation occur in concert with an increase in energy generation and utilization by the cell [106].

In the matrix of the mitochondrion, calcium ions rapidly dissociate from their binding proteins. Dissociation is likely the result of a lowered pH attributable to the formation of mitochondrial acids (malic, citric, oxaloacetic, succinic, glutaric, and fumaric) and accumulation of protons liberated from the electron transport chain within the organelle. As the pH changes, calcium ions bind to anionic groups such as phosphoryl groups on phospholipids. More central to the mineralization process, calcium is available to bind free phosphate ions to form a calcium phosphate ion pair.

In summary, the immediate effect of an increase in cytosolic calcium ion concentration is activation of the uniporter and other transporters in the mitochondrial membrane. As a result, calcium ions, either alone or possibly in tandem with phosphate anions accumulate in this organelle. These events are of considerable importance as they maintain ionic equilibrium and calcium ion homeostasis in the cytosol and avoid serious potential consequences such as cellular damage or apoptosis. The subsequent effect of calcium on mitochondrial function is to promote oxidative metabolism and the generation of energy for continued cellular anabolic and ion transport activities. All these activities are requisite for normal cell function during the expression of the mineralization phenotype and when there are substantive fluxes in calcium ion distribution within osteoblasts, odontoblasts, and other cells of vertebrate mineralized tissues.

5.6 Calcium Ion Storage in Mitochondria of Mineralizing Tissues

Not surprisingly, the distribution and compartmentalization of calcium ions in mitochondria of tissues undergoing mineralization have received intense study. Work performed by Lehninger and colleagues showed that isolated mitochondria had a high affinity for calcium ions [107–109]. In fact, these organelles could accumulate massive amounts of calcium (up to 3 mM/mg protein) in so-called mitochondrial granules (Fig. 5.7) [110, 111], and, consistent with comments presented above,

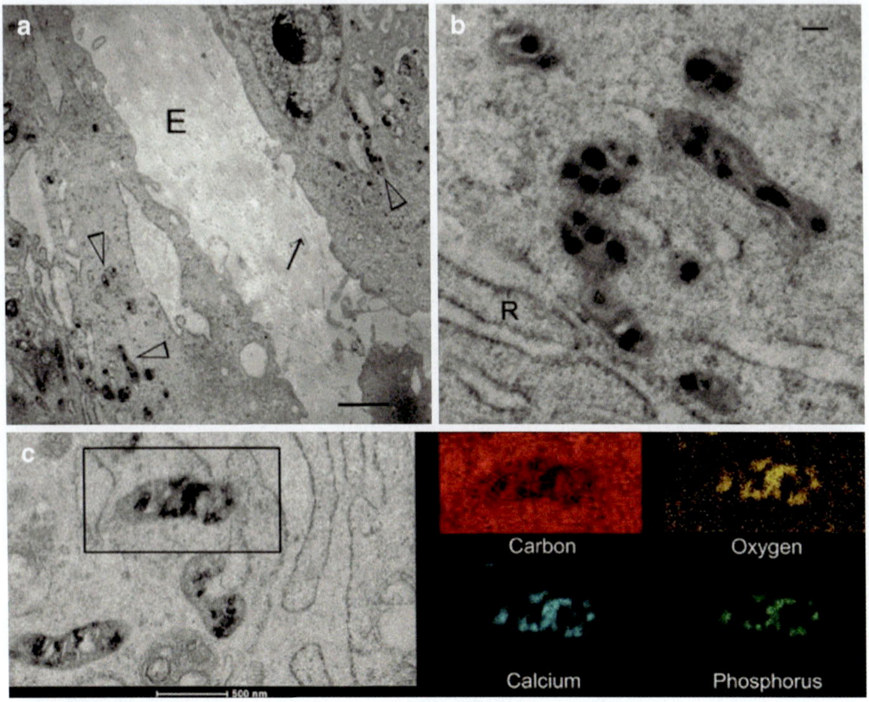

Fig. 5.7 Transmission electron microscopy (TEM) and energy dispersive X-ray spectroscopy (EDX) of mitochondrial granules in human dental pulp stem cells (hDPSCs). (**a**) A tissue section of differentiated hDPSCs which was fixed in glutaraldehyde and osmium tetroxide and then stained with uranyl acetate and lead citrate. TEM shows the cells contain electron dense mitochondrial granules (open arrowheads) and unmineralized collagen fibrils (thin arrow) located in extracellular matrices (E). (**b**) An enlarged view of a tissue section of hDPSCs illustrating dense mitochondrial granules close to endoplasmic reticulum (R). (**c**) A tissue section demonstrating a cluster of mitochondria with granules from hDPSCs (left panel). A single mitochondrion (enclosed box) was analyzed by EDX, where elemental mapping (right panel) revealed the presence of oxygen (yellow false color mapping), calcium (blue false color), and phosphorus (green false color) comprising granules. Carbon (red false color) was detected as the major background element in the specimen section. Tissue for EDX analysis was fixed in glutaraldehyde only and left unstained, examined by scanning transmission electron microscopy. Additional electron diffraction analysis of such granules indicated their composition was an amorphous calcium phosphate [110]. Scale bars = 200 nm (**a**), 100 nm (**b**), 500 nm (**c**). (Images adapted from Pei et al. [110] with permission)

calcium uptake stimulated respiration to a greater extent than that of even ADP [112]. The Lehninger group expanded its studies to examine the blue crab (*Callinectes sapidus*) and found that the mitochondria of its hepatopancreas accumulated calcium and, remarkably, in amounts which exceeded the solubility product of a calcium phosphate solution [113]. In this context, the crab hepatopancreas serves as both liver and pancreas. Prior to molting, calcium is removed from the exoskeleton and deposited in the hepatopancreas and, in the post-molt organism, calcium is removed from the hepatopancreas and presumably used to build a new exoskeleton [114]. Analysis of particulate deposits in the hepatopancreas mitochondria indicated that their calcium/phosphorus ratios and X-ray diffraction patterns were consistent with an amorphous calcium phosphate [115]. Based on these and other observations, it was suggested that "the specific property of mineralizing tissues is the ability of the cells concerned to produce a localized rise in the levels of mineral ions by means of mitochondrial activity, together with a matrix which is capable of being mineralized" [116].

While presented over 50 years ago, the ideas noted above had a considerable impact in focusing attention away from putative mineralization-specific properties of the extracellular matrix of a vertebrate tissue to an intrinsic metabolic function of the composite cells themselves. In this regard, many early studies were conducted by electron microscopic means to identify and localize calcium within cellular compartments or organelles, and a common approach utilized potassium pyroantimonate staining of a tissue for these purposes [111, 117–119]. However, pyroantimonate was experimentally limited since it was not strictly calcium-specific but also bound sodium and magnesium. Nonetheless, such staining under appropriate conditions provided useful insight into calcium loading, and the technique demonstrated calcium localization in the mitochondria of epiphyseal hypertrophic chondrocytes [117–119]. A similar observation was made by Morris and Appleton in an investigation of calcium transport in condylar growth cartilage [120]. Other studies examined pyroantimonate in the developing tooth to assess the role of mitochondria as a site for deposition of calcium. For example, Lyaruu et al. reported that, at the onset of mineralization of enamel, pyroantimonate was detected in the cytoplasmic side of the apical plasma membranes and in the mitochondria of ameloblasts [121]. These and additional analyses pointed to the particular importance of mitochondria in cells for mediating and accumulating calcium ions for tissue mineralization.

The focus on mitochondria as reservoirs for calcium ions has prompted questions concerning the cellular levels of free and total calcium ion concentration in relation to mineralization of the extracellular matrix of vertebrate hard tissues. Iannotti et al. showed, for instance, that the total intracellular calcium ion concentration increased during maturation of cells in the cartilaginous growth plate [122]. At the mineralization front of this tissue, the hypertrophic chondrocytes maintained the highest levels of both free and bound calcium ions within growth plate regions and the total endogenous calcium content of their mitochondria was well above that for the mitochondria of hepatocytes in liver, a non-mineralizing tissue, which also had a lower maximum calcium capacity [122]. In a study using radioactive calcium as a marker in osteoblasts, Imai et al. reported that, like epiphyseal chondrocytes, the calcium content was greatest in cells closest to forming bone [123]. While much of the calcium was localized to the cell membrane and mitochondria in chondrocytes [123], the cytosolic ionic calcium concentration was similar to that of non-mineralizing mammalian cell types [124]. This finding added strength to the concept that mitochondria accumulate calcium, thus maintaining the proper ionic composition of the cytosol and enhancing cell survival.

The observations above that mitochondria of cartilage, bone, and dental tissues are loaded with calcium ions can be related to calcium regulation in cells and the mineralization process in three different ways. The first is that mitochondria accumulate calcium and phosphate ions, and the subsequent export of these ions as ion pairs, clusters, and so forth from the cell into the extracellular matrix leads to mineral formation. Alternatively, since elevated levels of calcium ions are toxic to cells, the intracellular calcium overload is mitigated by mitochondria, which take up excess calcium ions to protect calcium-sensitive metabolic enzyme systems. A third concept is that mineralization is the result of accumulation and then release of calcium and phosphate ions from a yet undefined subcellular compartment and that mitochondria maintain calcium homeostasis during these events. All such studies point to a critical involvement of mitochondria in regulating calcium transport and underscore the importance of the cell, its subcellular compartments and particular component organelles in the initiation of the mineralization process in vertebrate hard tissues [3].

5.7 Storage of Calcium in the Cell: The Endoplasmic Reticulum, Golgi Apparatus, and Other Cellular Organelles

As described above, numerous studies of pyroantimonate-stained epiphyseal cartilage, developing dental tissues and bone cells provided evidence that mitochondria became calcium-loaded during the initiation of mineralization. Moreover, if the "static snapshots in time" of the microscopic images of the mitochondria were valid indicators of their function, then these organelles would be expected to play a critical role in mediating calcium ion transport from cells to extracellular matrices and subsequently to sites of matrix mineralization. Besides mitochondria, other cellular constituents were recognized to be associated with calcium and, in another set of experiments, Reith used pyroantimonate to examine cytoplasmic components of rat odontoblasts (Fig. 5.8) [111]. Among many investigators at the time of such work, Reith likely realized the full importance of additional intracellular membrane systems in calcium ion movement and extracellular mineral deposition. He wrote that, "... calcium is bound by Golgi elements of the odontoblast ... it is proposed that a collagen precursor is being seeded with calcium within the Golgi saccules and that these seeded sites will play a role in facilitating the extracellular mineralization which is to follow" [111]. Nearly 20 years later, the physiological importance of internal cellular membranes in calcium mediation and mineralization was further emphasized by Kogaya and Furuhashi, who determined in late-stage

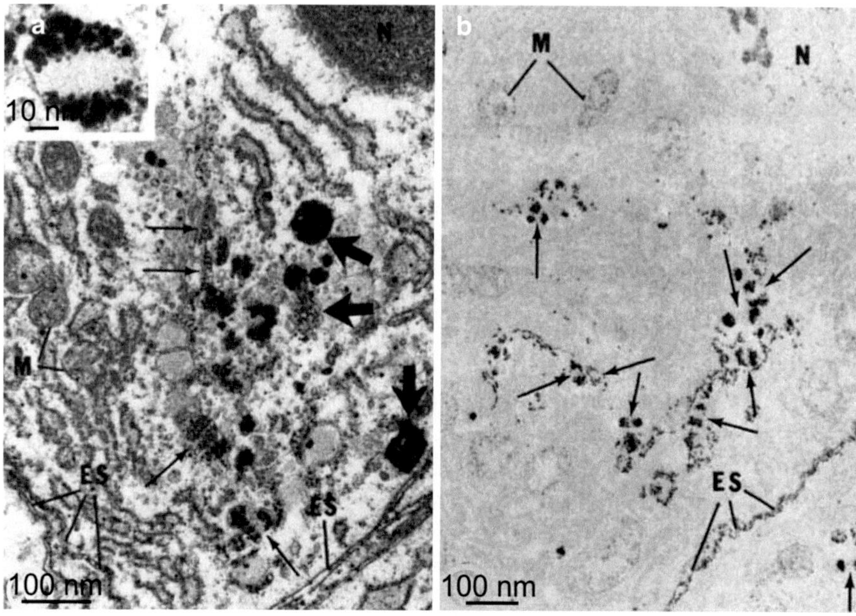

Fig. 5.8 (**a**) A photomicrograph obtained by transmission electron microscopy of a rat odontoblast treated with osmium tetroxide and potassium pyroantimonate to localize calcium. The tissue section shows calcium pyroantimonate reaction product present in Golgi saccules (thick arrows) of the cell. Thin arrows point to saccules containing filamentous material which is suggested to provide orientation of saccules. Calcium is also present in mitochondria (M) and the extracellular space (ES). The section was stained with uranyl acetate and lead nitrate. Inset. Enlargement of a Golgi saccule showing calcium pyroantimonate precipitate. (**b**) A photomicrograph of a Golgi region of an odontoblast treated with osmium tetroxide and potassium pyroantimonate but without uranyl acetate and lead citrate staining. Calcium is evident in Golgi saccules (thin arrows). The cell nucleus (N) may be identified in both images. Scale bars are approximate as based on original image magnifications. (Images from Reith [111] with permission)

preodontoblasts that pyroantimonate precipitate was mainly distributed in Golgi and numerous secretory granules (or vesicles) which were present in the developing dentin [125].

Calcium ion homeostasis is now known to involve all the cellular organelles discussed above to form an intracellular membrane network, and such a network also incorporates the endoplasmic reticulum [126, 127]. The tubular nature of this organelle with its many flattened saccules constitutes an internal transport system which contacts almost every compartment of the cell, including mitochondria, the Golgi apparatus, endosomal and lysosomal vesicles, and the plasma membrane with its constituent proteins (Fig. 5.9) [127, 128]. The endoplasmic reticulum is the major calcium ion storehouse of the cell. The free calcium concentration of the endoplasmic reticulum is reported to vary from 100 to 800 μM [129], whereas the cytosolic calcium ion concentration is given as 0.1 μM [130–133]. Within the endoplasmic reticulum, calcium ions are bound to EF-hand proteins like

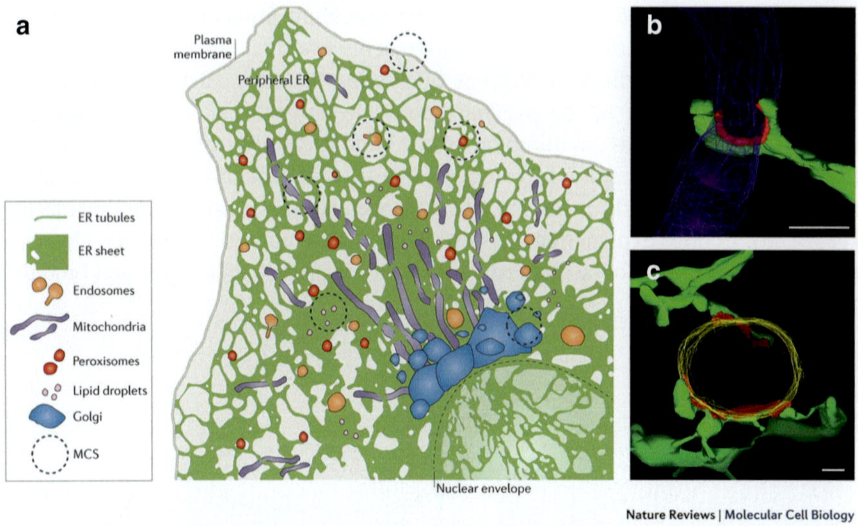

Fig. 5.9 Interaction of the endoplasmic reticulum with the plasma membrane and other organelles of a eukaryotic cell. (**a**) The endoplasmic reticulum (green), the largest membranous structure in a cell, associates with mitochondria (purple), the Golgi apparatus (blue), and endosomes (orange) as well as peroxisomes and lipid droplets. Tubules of the endoplasmic reticulum form dynamic contacts with sites on the plasma membrane (MCS). The figure shows that, while much of the endoplasmic reticulum is located around the nucleus (defined by the nuclear membrane), tubules and sheets of the endoplasmic reticulum extend outward and throughout the cytoplasm to contact the organelles noted above. (**b**, **c**) Interactions between tubules of the endoplasmic reticulum (green) and other cellular structures. The two panels illustrate by image reconstructions from electron tomography (a technique for generating three-dimensional structural information from cells, cellular constituents or other subjects) contact regions between (**b**) the endoplasmic reticulum and the plasma membrane of a yeast cell (red) and (**c**) an endosome of an animal cell (yellow). Scale bar = 200 nm for both (**b**) and (**c**). (Images from Phillips and Voeltz [128] with permission)

calbindins, calreticulin, and calmodulin, described previously in this chapter [16]. While there is well documented evidence that calcium ions are stored in the endoplasmic reticulum of cells of mineralizing tissues, the actual content has yet to be established.

Since the endoplasmic reticulum plays a key role in accumulating calcium ions for possible mineralization-related activities of a cell, certain salient features of its calcium uptake-release process are valuable to consider. Regarding first the mechanism of calcium ion release, a water-soluble agent, inositol 1,4,5-trisphosphate (IP3), is likely the most important ligand involved in the mechanism of calcium ion release [133]. By binding to its cognate receptor on the endoplasmic reticulum membrane, IP3 releases stored calcium ions [133]. Another metabolite, cADP ribose, a cyclic derivative of ADP, also releases calcium ions from the endoplasmic reticulum as well as from endocytic vesicles by binding to a specific receptor (the ryanodine receptor) [134].

Following their release, the endoplasmic reticulum immediately fills with calcium ions which enter the cell from the extracellular milieu. As might be predicted, this type of activity reflects a complex interaction between calcium ion transients occurring at the level of the plasma membrane and the endoplasmic reticulum (Fig. 5.10). The calcium ion filling process is known as store-operated calcium entry (SOCE) [88], and it is sensed by a group of EF-hand proteins, stromal interacting molecules (STIM 1 and 2), distributed across the endoplasmic reticulum membrane [135, 136]. In brief explanation, these proteins, on calcium ion depletion, cluster at the junction of the endoplasmic reticulum with the plasma membrane and bind to a

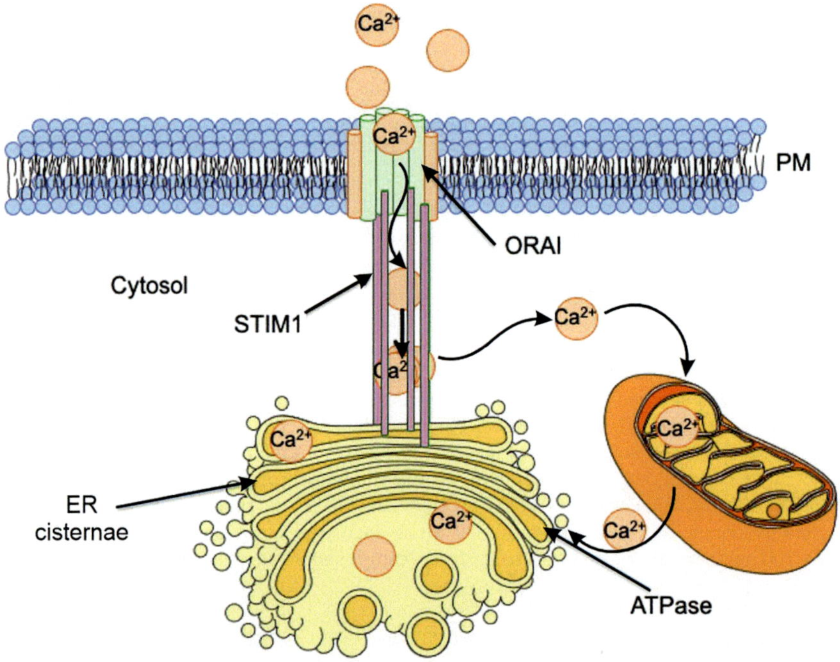

Fig. 5.10 Schematic diagram of the regulation of cytosolic calcium ion levels by endoplasmic reticulum (ER) and mitochondria in a cell. Passage of calcium ions into cells is mediated by store-operated calcium entry (SOCE) channels in the plasma membrane (PM) and occurs in response to depletion of calcium ion levels in the cytosol. Low cytosolic calcium ion levels cause stromal interaction molecules, STIM1, in the membranes of the endoplasmic reticulum to undergo a conformational change in which they aggregate to form a bridge which extends from the endoplasmic reticulum to the plasma membrane. The STIM1 calcium-sensing domain is located in the endoplasmic reticulum lumen and exhibits EF-hand calcium-binding characteristics. At the plasma membrane, STIM1 associates with ORAI proteins, and this interaction leads to membrane pore formation for the selective passage of calcium ions into the cell. Thus, when calcium stores are depleted in the cell, STIM1 molecules activate the ORAI channels to promote calcium influx into the cytosol. Low calcium ion influx is unlikely to perturb the cytosolic calcium ion level. However, when the influx is high and sustained, then it is likely to raise the cytosolic cation concentration. In this case, a significant fraction of the entering calcium ions is captured by mitochondria. This pool of calcium can subsequently be released to the endoplasmic reticulum through an ATPase pumping system and MAM-MARC attachment sites

second protein, olfactory receptor class A related 1 (ORAI1). The complex serves as a key component of calcium release-activated calcium (CRAC) channels which allow calcium ion entry into the organelle (Fig. 5.10) [137, 138].

Events leading to SOCE are likely to be more complicated as there is accumulating evidence that they are sensitive to the presence of microRNAs, together with other regulatory molecules and some agonists which cause calcium ion emptying of the endoplasmic reticulum [139]. In addition, it should be recognized that, although the cytosol calcium ion concentration is normally very low, an ATPase (P-type ATPase) in the wall of the endoplasmic reticulum serves to pump the calcium ions against a calcium ion gradient (low in the cytosol, high in the endoplasmic reticulum) into the cisternae and tubules of the endoplasmic reticulum network [140].

Is there a structural relation between the endoplasmic reticulum and mitochondria which facilitates their involvement in mineral deposition of vertebrate tissues? It is now known that the membranes of both the endoplasmic reticulum and mitochondria form very tight contact sites. Studies suggest that the gaps between these structures are of nanometer dimensions (~30 nm) and the tethering is robust so that the two organelles remain attached even when there is a rearrangement of the cytoskeleton [92, 93]. These attachment regions, mitochondrial-associated endoplasmic reticulum membranes (MAM), and mitochondrial-associated endoplasmic reticulum contacts (MARC), exhibit lipid raft-like properties and are thought to act as hubs for the regulation of activities which include autophagy, lipid metabolism, and apoptosis [141].

The MAM-MARC regions are very complex, sharing a considerable number of organellar proteins. Of these, the endoplasmic reticulum protein, mitofusin 2 (MFN2), interacts with a homologous protein in the outer membrane of the mitochondrion [142]. Also the endoplasmic reticulum protein, vesicle-associated membrane protein (VAMP)-associated protein B (VAPB), links with protein tyrosine phosphatase-interacting protein 51 (PTPIP51) of the mitochondrial membrane [143, 144]. The very proximity of the two membranes provides a functional connection facilitating the transfer of calcium ions between the two organelles [128]. Thus, a pore is formed between the endoplasmic reticulum glucose-regulated protein (GRP75) and mitochondrial voltage-dependent anion channel 1 (VDAC1) to create a complex together with the IP3 receptor (IP3R3) [145]. The close structural interaction between these organelles would certainly explain why mineral ions are found in both mitochondria and the endoplasmic reticulum of cells.

The role and interaction of organelles in regulating calcium ion storage and release in the cell, as discussed above, raises the question as to how they impact the mineralization-related activities of the cells of bone, dentin, and other vertebrate hard tissues. Osteoblasts as well as osteoprogenitor cells express both STIM and ORAI1 proteins [146, 147]. With respect to STIM proteins, their overexpression in osteoblast-like cells in culture leads to an increase in calcium deposition, accompanied by greater expression of key osteogenic markers, including the transcription factor, RUNX2, and two of its target genes, type I collagen (*COLI*) and bone morphogenetic protein (*BMP*)*4* [148].

When *ORAI1* is deleted in experimental animal models, conventional histological, biochemical, and microCT techniques indicate that there are profound changes in their bone architecture and mineral density [149]. Moreover, there is disruption of BMP signaling [150], a requirement for osteogenic differentiation, as well as a reduction in expression of the phosphate mobilizing hormone, fibroblast growth factor (FGF)23 [149]. Alterations also occur in the expression of *TNAP*, *COLI*, and *OCN* in both osteoblasts and osteocytes while osteoclast differentiation is impaired [146, 147, 149, 150].

Aside from bone cells, odontoblasts, as well as their progenitor cells, express both STIM and ORAI1 proteins [151]. When *STIM1* is silenced, odontoblast differentiation is blocked and dentin matrix synthesis is inhibited. Moreover, the extracellular dentin matrix contains fewer matrix vesicles and exhibits decreased deposition of mineral [151]. Together, these observations indicate that the SOCE system is active in bone and dentin, possibly influencing phosphate metabolism and enhancing osteoblast and odontoblast lineage commitment, differentiation, and the mineralization of their respective matrices [148]. In common with these two tissues, ameloblasts also express STIM1 and ORAI1 proteins (Fig. 5.10) [152, 153]. Interestingly, *STIM1* responds to dentin matrix protein (DMP), a molecule synthesized and secreted by both odontoblasts and bone cells, by releasing calcium ions from the endoplasmic reticulum [148]. The importance of both STIM1 and ORAI1 in relation to calcium ion transport and enamel mineralization is discussed in the next section of this chapter.

To summarize this section, it is clear that calcium ions accumulate in two subcellular compartments: the mitochondria and the saccules and tubules of the endoplasmic reticulum. Communication between the endoplasmic reticulum and the mitochondria, and the endoplasmic reticulum and the plasma membrane, permits the cell to accommodate calcium oscillations while at the same time providing sufficient levels of calcium for the maintenance of essential energetic, metabolic, and other processes in the cytosol. That these functions are mineralization-related is supported by studies which show that blockage or loss of individual proteins of the SOCE and CRAC channels causes aberrant mineralization. In addition, changes in calcium ion distribution in the cell influence expression of genes and secretion of proteins closely associated with the mineralization process. The accumulation of calcium ions together with phosphate ions will next be addressed as favoring intracellular calcium phosphate particle formation, primarily in compartments associated with the endoplasmic reticulum. Additional details concerning calcium/calcium phosphate efflux mechanisms at the onset of mineralization will be elaborated in Chap. 6.

5.8 Calcium Ion Transport Systems and Enamel Mineralization

As mentioned previously, enamel is distinct from bone and dentin in that it appears absent of an unmineralized zone of extracellular matrix equivalent to the seams of osteoid in bone or the predentin in dentin. How calcium ions are transferred across the enamel organ to generate such a matrix which is, then, almost completely devoid of protein is poorly understood. Most investigators would agree that calcium ions traverse the ameloblast layer from vascular supplies to the enamel organ rather than from developing dentin [154]. More uncertain and controversial is information on mechanisms by which the ameloblasts transport the calcium ions. From initial considerations, it would appear that ameloblasts mineralize their extracellular matrix using systems which are substantially different from those of osteoblasts or odontoblasts. As was noted earlier, studies performed by Reith and his collaborators [111, 154] suggested that calcium ions were transported by a paracellular route through the palisades of ameloblasts characterizing enamel (Fig. 5.11) [156]. Supporting this concept is the observation that calcium ions travel through paracellular spaces in other epithelial transport systems, such as those involved with intestinal calcium transport [157].

There is now accumulating evidence that a family of proteins called claudins, together with other tight junction proteins, form paracellular channels between cells which function to regulate permeability to small molecules and provide the driving force for calcium ion transport [158–160]. Similar tight junctional complexes are

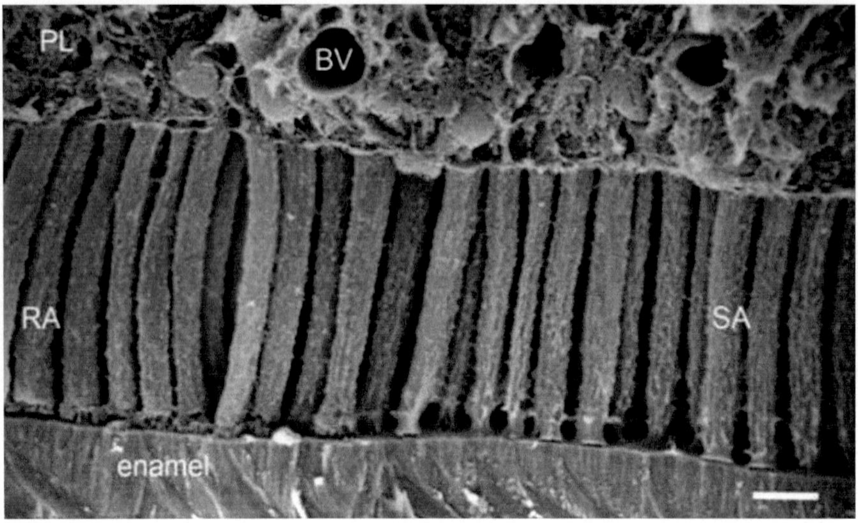

Fig. 5.11 A palisade of maturation stage ameloblasts in a rat lower incisor viewed by scanning electron microscopy. The tooth sample was fixed by glutaraldehyde perfusion, critical point dried, fractured to expose the inner tooth structure, and sputter-coated with gold. The papillary layer (PL) of the tissue is perforated with sites (empty) of blood vessels (BV). The palisade of numerous elongated cells consists of ruffle-ended (RA) and smooth-ended (SA) ameloblasts at the left and right, respectively, of the image. RAs are characterized by an invaginated membrane, the ruffled border, at the apical surface of the cells, adjacent to enamel, while SA are absent of such structures. Maturation stage ameloblasts regularly cycle between RA and SA forms to remove water and organic components from the enamel while enhancing enamel accretion of inorganic material [155]. Scale bar = 9.5 μm. (Original image kindly provided by Alan Boyde, B.D.S., Ph.D., St. Bartholomew's Hospital and Queen Mary University, London, UK, and reproduced from Paine et al. [156] with permission)

formed by ameloblasts; both claudin 16 and 19 are present together with claudin 3 at the apical end of secretory ameloblasts [161]. Moreover, when claudin 16 is deleted, there is impaired enamel formation [161]. Accordingly, paracellular transport may serve as one component of a complex system to ferry calcium ions to the mineralization front of enamel and other hard tissues. However, in line with the discussion at the beginning of this chapter and many other studies, it is clear that, besides paracellular transport, a cell relies on a number of different interactive pumps and channels to control its calcium ion influx and efflux [162]. Thus, a scenario more likely than that of paracellular transport is one in which the enamel epithelium mobilizes calcium ions by a transcellular route controlled by an integrated network of exchanger/transporter and buffering proteins which are responsive to both local and systemic signals. This type of calcium ion transport system is in keeping with those of most other cell types: It is highly specific; it is sensitive to changes in plasma membrane voltage and ligand-receptor signaling, organelle ion concentration, and the metabolic state of the cell; and it impacts many other cellular functions, including those which regulate gene expression and cell proliferation. These systems include the high-capacity, low-affinity sodium/calcium ion transport proteins at the plasma membrane, the high-affinity, low-capacity plasma membrane calcium ion pumps which drive calcium ions out of the cell or into the endoplasmic reticulum and associated organelles, and the SOCE system, each noted earlier.

Numerous proteins pump calcium ions from a cell, as described above. With respect to enamel, Robertson et al. have shown that the high-affinity PMCA proteins which transport calcium ions from a cell are present in ameloblasts [41]. Two of these proteins, PMCA1 and 4, are located in the basolateral membrane of both secretory- and maturation-stage ameloblasts [41]. At least two sodium/calcium transporters, NCX1 and NCX3, are also present in the plasma membrane of ameloblasts [163]. From considerations of the membrane potential and the relative intra- and extracellular ion concentrations, these proteins can promote calcium influx by enhancing the sodium efflux from the cell and vice versa. In addition, the sodium/calcium/potassium transporters, NCKX3 and NCKX4, are also localized to the ameloblast plasma membrane [164]. These molecules are similar in action to that of NCX as they enhance calcium ion entry into the cell while they transport sodium with potassium ions along a membrane ion gradient [164, 165]. Of these proteins, NCKX4 is most highly expressed during the maturation stage of amelogenesis, at both the apical poles of the ameloblast and the associated basolateral membranes [164]. In humans and mice, a mutation (*SLC24A4*) affecting this exchanger is linked to the pathogenesis of enamel [166]. Besides the membrane pumps which can serve to shuttle calcium ions into the cytosol of ameloblasts, enamel cells express a SERCA (SERCA2b) which traffics cytosolic calcium ions into the endoplasmic reticulum [167]. Like that of NCKX4 just noted, the expression of SERCA2b is greatest during ameloblast maturation [167].

In addition to the membrane pumps discussed above that transfer calcium ions directly into the endoplasmic reticulum, Lacruz and colleagues have shown that genes required to mediate SOCE activity (*STIM1*, *STIM2*, and *ORAI1*, *ORAI2*, and *ORAI3*) are expressed in ameloblasts [152, 156, 168]. ORAI1 is localized to the plasma membrane of the ameloblasts, and alignment of a diagnostic stain suggests that calcium influx into the cells occurs along the length of their plasma membrane (Fig. 5.12) [153]. Since STIM1 is diffused in the cytosol and co-localizes with the EF-hand protein, calnexin, the CRAC proteins are also in place for calcium ion uptake into the ameloblast [152]. It might be recalled that the SOCE system is triggered in response to a decrease in the level of calcium ions, possibly driven by IP3 [152]. The change in cation levels is sensed by STIM1, which clusters and opens CRAC channels together with ORAI. These channels allow calcium ions to enter the cell and refill the depleted cation stores of the endoplasmic reticulum [152, 153].

SOCE activities and the role of CRAC channels in ameloblasts are consistent with those in osteoblasts, where they likewise govern calcium storage and transport [169]. In this regard, when SOCE activity is reduced, there is impaired mineral deposition by human osteoprogenitor cells and osteoblasts [146, 147]. Additionally, utilizing chondrocyte cell lines, Inayama et al. showed that CRAC channels are dependent on both ORAI and STIM1, whose complexes sustain calcium ion entry into cells as just described above [170]. Thus, cells as unique as ameloblasts rely on the same network of membrane pumps, channels, and binding proteins which govern calcium ion homeostasis as those in other vertebrate mineralized tissues. As indicated earlier, among the better studied exchanger/transporter proteins, calcium ion release from the apical aspects of ameloblasts involves both sodium–calcium exchangers/transporters and calcium-ATPase while calcium ion entry at the basal-lateral region of the cell is governed by the activity of SOCE channels

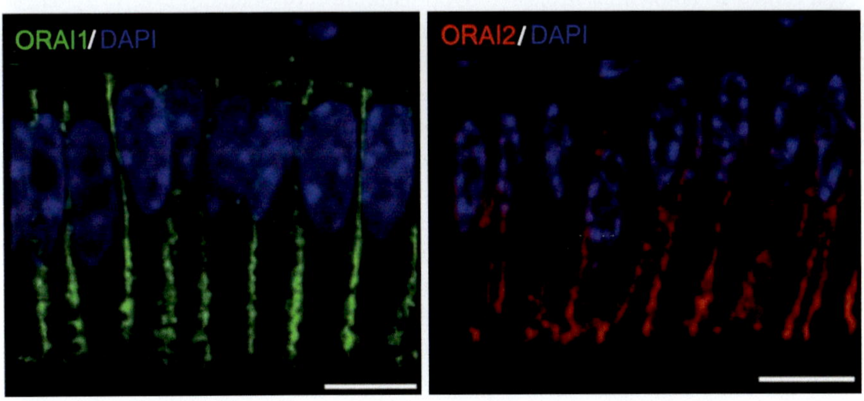

Fig. 5.12 Immunohistochemical localization of ORAI1 and ORAI2 homologs in normal murine ameloblasts. Histological sections through the enamel organ at the maturation stage show strong ORAI1 (green) and ORAI2 (red) staining in ameloblasts. Much of the staining is located in the baso-lateral plasma membrane of the cell. The nucleus of each cell is stained with DAPI (blue). Scale bars = 20 μm. (Images from Eckstein et al. [153] with permission)

together with the cation membrane channel transporter, TRPM7 [171, 172]. Recent studies indicate that TRPM7 is highly expressed in ameloblasts, especially during the maturation stage of amelogenesis when considerable quantities of mineral are formed [171, 172]. Moreover, Souza Bomfim and co-workers commented TRPM7 was a "positive modulator of SOCE by potentiating ORAI-dependent Ca^{2+} uptake" [173].

The combined activities of these exchanger/transporter proteins at the apical and basal-lateral aspects of the ameloblast during mineral formation cause shifts in intracellular calcium ion levels. Studies with calcium-sensitive dyes demonstrate that basal levels of calcium ions during the ameloblast maturation stage are considerably higher than at the secretory stage (230 nM versus 135 nM, respectively) [174]. These differences in calcium values support the idea that the maturation ameloblast provides much of the calcium ions needed to mineralize the AMEL-rich organic matrix of enamel [171, 172]. While these studies are relevant to intracellular calcium ion shifts, they do not address the possible role of paracellular calcium ion transport and release. The contribution of this pathway to enamel mineralization is currently unknown, and the relation, if any, between pericellular and transcellular calcium ion transport remains an unexplored avenue of dental research.

The question of the importance of the SOCE system in mediating enamel mineralization has been addressed directly by the deletion of two of the major genes required for CRAC formation. In mice, a mutation in *STIM1/STIM2* causes a CRAC channel deficit and extensive enamel hypomineralization [175, 176]. A mutation in ameloblast *ORAI1* is associated with a severe mineralization defect in human teeth [177], while a homozygous mutation in *STIM1* and *SLC24A4* has been linked to autosomal-recessive hypomaturation of enamel [178, 179]. These basic and clinical science observations strengthen the concept that enamel mineralization is mediated by a combination of pumps and channels which regulate the bulk transfer of calcium ions through palisades of ameloblasts to the mineralizing enamel. A schematic which relates changes in ameloblast morphology with calcium ion transport and enamel mineralization is shown in Fig. 5.13 [155, 162, 180–184].

What of the question as to whether calcium ions in the ameloblast are packaged for transport and discharge at the mineralization front like that in the osteoblast and other cells of vertebrate mineralized tissues? Although there is no evidence that ameloblasts secrete exosomes/matrix vesicles, it is clear that vesicles containing protein are released in great number from the Tomes' process of secretory ameloblasts [184, 185]. Whether these vesicles contain calcium ions or calcium phosphate ion clusters which are compartmentalized into exosome-type vesicles for exocytosis is unknown. New high resolution ultramicroscopic techniques will hopefully address this issue, while live cell imaging of ameloblasts at selected stages of maturation may provide more definitive information on the contribution of pericellular and transcellular calcium ion movements, mentioned above, in relation to the onset and progression of enamel mineralization.

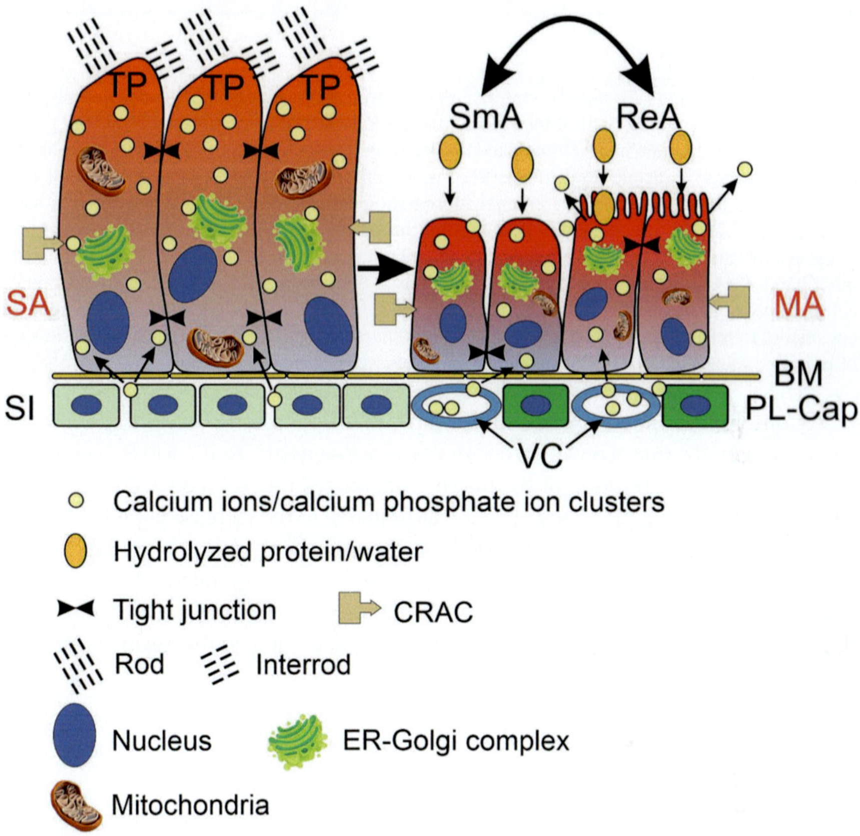

Fig. 5.13 Schematic diagram characterizing the enamel-forming activities of palisades of secretory phase ameloblasts (SA) and maturation phase ameloblasts (MA) and illustrating aspects of the structure of the cells and their local environment. MA may be either smooth-ended (SmA) or ruffle-ended (ReA) ameloblasts. During the cell transition between SA and MA (single-headed dark arrow), the stratum intermedium (SI) supporting the cells reorganizes to form a papillary layer (PL) accompanied by the presence of blood capillaries (CAP) or vascular channels (VC). A basement membrane (BM) separates the proximal/basal ends of the different cells from the underlying SI or PL-CAP. Calcium ions or calcium-phosphate ion clusters originating from the SI and/or vascular channels (VC) within the PL enter both SA and MA. One mechanism of ion uptake by the cells is mediated by calcium release-activated channels (CRAC) integral to the plasma membranes of the ameloblasts. In SA, the intracellular ions or ion clusters reach the extracellular matrices at the distal region of the cells, the Tomes' process (TP), along with secretory granules or vesicles (not shown) carrying proteins bound for export. In ReA, intracellular ions or mineral clusters enter the extracellular matrices through the distal/apical ruffled-border of these cells. Interrod and rod enamel are each secreted by SA at distinct sites of the TP. Secretion of interrod enamel at the proximal aspect of the TP precedes the secretion of rod enamel along one face of the TP, and the resulting interrod and rod enamel crystallites from the two sites of each ameloblast differ in their length-wise orientation. All ameloblasts are long, extended cells although SA are longer than MA. Both SA and MA maintain organelle polarity and MA decrease significantly in cell length and organelle content compared to SA. As amelogenesis progresses, changes in cell-cell interaction occur. SA and MA possess tight junctions along their proximal and/or distal aspects, serving in part to prevent paracellular transport of proteins and other molecules between neighboring ame-

Fig. 5.13 (continued)

loblasts. Such junctions are mainly comprised of the proteins, occludin and claudin(s) [180]. SAs have tight junctions along both of their distal and proximal plasma membranes [155, 162]. SmA and ReA, which alternate in developing and maintaining their two types of borders (double-headed dark arrow), are responsible for the removal of water, proteins and other organic constituents from enamel while ReA also promotes enamel inorganic accretion [155, 180]. While there is some uncertainty in details, it is generally thought that SmA tight junctions occur at the proximal ends of these cells and distal ends have incomplete or leaky junctional sites [155, 162, 181, 182]. On the other hand, ReA are characterized by tight junctions at their distal ends with relatively open intercellular spaces at their proximal ends [155, 162, 183]. The different tight junction locations are considered to reflect functional activity of the ameloblasts in regulating permeability and transport of proteins, water and ions during secretory and maturation stages of amelogenesis [180, 184].

As one further note pertinent to the topic of calcium ion transport systems and enamel mineralization, a downward shift in extracellular pH from 7.4 to 6.4 is reported to occur with the deposition of the highly mineralized enamel matrix during amelogenesis [186]. Such a pH change may be attributable to the increase in protons resulting from the conversion of acid phosphate groups, mainly HPO_4^{2-} available in the enamel extracellular matrix, to PO_4^{3-} groups which are incorporated into the growing apatite lattice. Restoration of neutral pH is achieved by the enzyme, carbonic anhydrase (CA), which scavenges the released protons according to the reactions below:

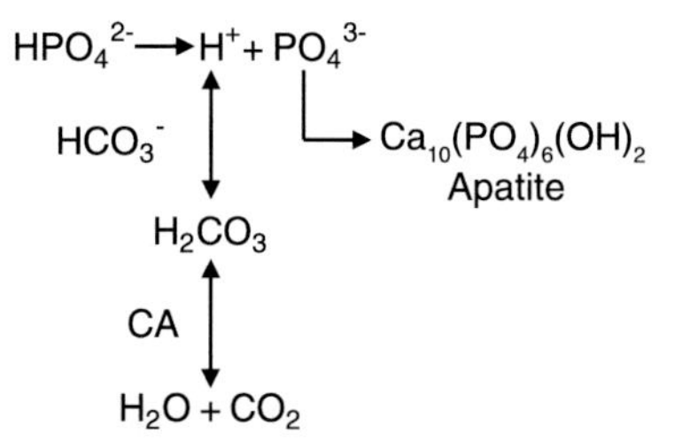

Of the many CA isoforms expressed by ameloblasts, CA9 and CA2 are particularly noteworthy [187]. CA9 is present in the enamel extracellular matrix and induced by HIF-1α. It may be recalled that this transcription factor responds to hypoxia by up-regulating a considerable number of genes linked to maintaining cell function, energy metabolism, and matrix synthesis (see Chap. 4). The second isoform, CA2, is intracellular and expressed by both secretory and maturing ameloblasts [188]. In bone, a loss of the CA2 isoform leads to osteopetrosis, a disease in which marked abnormalities in osteoclast function occur along with densification of bone tissue [189]. Together, both CA isoforms maintain cell function and mineral deposition in the face of growing tissue hypoxia and pH imbalance during amelogenesis.

5.9 Phosphate Uptake into the Endoplasmic Reticulum and Mitochondria

To this point of presentation in the chapter, nearly all discussion has focused on the uptake, accumulation and transport of calcium ions in cells with little comment about the other major and equally important ionic component of hydroxyapatite, phosphate. Besides being critical to extracellular mineral deposition, phosphate ions in the cytosol affect numerous cellular activities, including, for example, oxidative metabolism by mitochondria [190, 191] as well as transcription of certain proteins such as DMP1, OPN, ANK, and cyclin D1 which are linked to mineralization (Fig. 5.14) [193, 194]. Unlike the nanomolar-micromolar levels of intracellular calcium ions [131, 132], the levels of intracellular phosphate ions are likely to be considerably higher, in the millimolar range similar to that found in serum (~1–1.5 mM) [132, 196]. Phosphate ions, like calcium, need to be compartmentalized. In this instance, phosphate ions required for ATP synthesis and phosphorylation reactions must be separated from those utilized for building apatite. The reason for segregating these pools is that ATP generation in the mitochondria is sensed by the enzyme, 5′ AMP-activated protein kinase (AMP kinase), which responds to the ratios of AMP/ATP and ADP/ATP and concentrations of phosphate ions [190, 191]. During mineralization, when phosphate flux is expected to be elevated in the cell, adenine nucleotide sensing would become impossible and mitochondrial ATP synthesis would be severely impaired unless phosphate pools were separated.

Despite the critical role of phosphate ions in energy production and mineralization, present understanding of the mechanism(s) for control of cellular phosphate ion levels is somewhat limited. In this regard, there have been several studies assessing how phosphate ions, especially those associated with bones and teeth and their mineral deposition, enter cells. As with calcium ions in these tissues, cellular influx of phosphate ions is regulated by certain multimeric membrane exchanger/transporter proteins (Fig. 5.14). During development of the tooth germ, for example, specific phosphate exchanger/transporter genes are expressed which include three solute carrier members. These are family 20 member A1/sodium-dependent phosphate transporter 1 (*SLC20A1/PiT1*), family 20 member A2/sodium-phosphate transporter 2 (*SLC20A2/PiT2*), and family 34 member A2/sodium/phosphate co-transporter IIb (*SLC34A2/NaPi-IIb*) [192, 197]. These three genes are very highly expressed in the stratum intermedium and the sub-odontoblastic cell layer of the tooth although, somewhat surprisingly, the peak expression levels of these phosphate ion transporters do not occur in concert with events which lead to the commencement of dental tissue mineralization [192]. The major transporters of phosphate ions into the cells of bones and teeth are SLC20A1/PiT1, SLC20A2/PiT2, and another sodium-coupled phosphate anion transporter protein, SLC34A1/NaPi-IIa (sodium/phosphate co-transporter IIa) [192, 197]. Phosphate ion transport is linked to conformational changes in membrane protein structure which cause an energetically favorable movement of sodium ions down a membrane concentration gradient into the cell. Such sodium ion activity powers the otherwise energetically unfavorable movement of phosphate ions along the sodium gradient. Because sodium ions are brought into the cell by these integral membrane proteins, the ionic homeostasis of the cytosol is regulated by other types of exchanger proteins, functioning as sodium/hydrogen and sodium/potassium membrane pumps (Fig. 5.14) [164, 166].

As expected, alterations in the phosphate exchangers/transporters lead to an aberrant tissue phenotype. In a recent study, for instance, in which *SLC20A2/PiT2* was deleted in mice, Beck-Cormier et al. showed that the animals exhibited disorders of

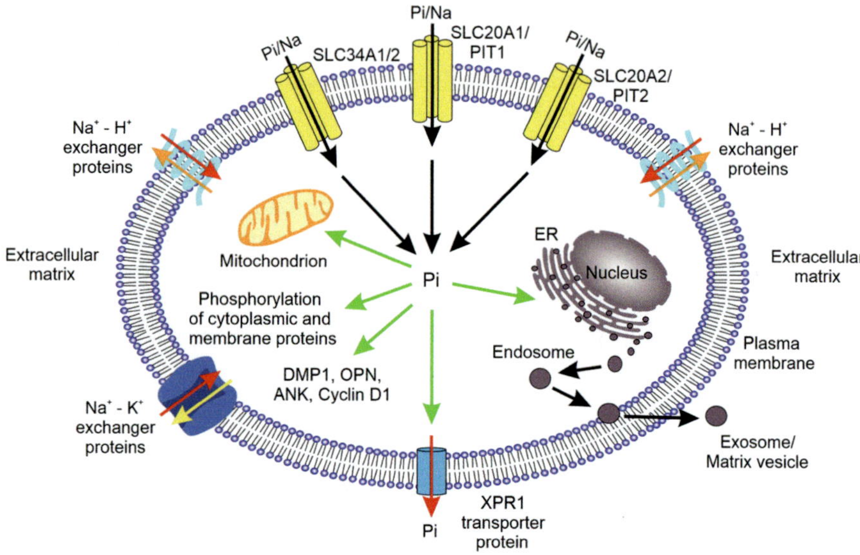

Fig. 5.14 Schematic diagram of possible phosphate metabolic pathways in osteoblasts. Three types of sodium-dependent phosphate exchangers/transporters have been identified in the plasma membranes of these cells: SLC20A1/PiT1, SLC20A2/PiT2, and SCL34A. Certain of these molecules have also been identified in odontoblasts and chondrocytes [192]. Although there is some uncertainty and confusion as to the role and importance of the PIT1 exchanger/transporter, the other two types of proteins have been linked with promoting the ingress of phosphate ions into the cells. Since all the membrane proteins co-transport sodium ions into the cell, cytosolic ionic stability is maintained by sodium/hydrogen and sodium/potassium membrane pumps [164, 166]. Phosphate ions and phosphorylated molecules are required for and regulate the activities of numerous cytoplasmic, organellar, and membrane protein phosphorylation reactions. Together with ATP and ADP, phosphate ions control the oxidative activities of osteoblast mitochondria; phosphate ions are also bound to the membranes and sequestered in small vesicles of the endoplasmic reticulum (ER) of these cells [192]. Such vesicles are shuttled as phosphate ion-laden endosomes and exosomes from the interior aspects to the periphery of the cell. Transport of endosomes and exosomes to the osteoblast plasma membrane occurs along microtubule tracks and in association with kinesin motor proteins carrying their cargo [2]. Changes in the level of cytoplasmic phosphate ions also influence the transcription of a number of mineralization-related gene products, including DMP1, OPN, ANK, and cyclin D1 [193, 194]. Efflux of phosphate ions from the cells is mediated by the transporter protein, XPR1, and through the release of phosphate-containing exosomes/matrix vesicles to the extracellular milieu [195]

their skeletal and dental tissues as well as diverse and unusual changes in their soft tissues such as the vasculature and brain (Fig. 5.15) [198]. Notably, the effects of the mutation were observed despite maintenance of calcium balance and phosphate ion homeostasis. Analysis of the mutant long bones in young animals indicated that there were disturbances in both endochondral and intramembranous mineralization; in adult mice, there was a major decrease in bone strength (Fig. 5.15) [198]. Despite all the pathological changes documented, the observation that both skeletal and dental tissues were mineralized led the authors to comment that, “Current data are thus not consistent with an essential role for SLC20A2 in providing Pi (phosphate) for

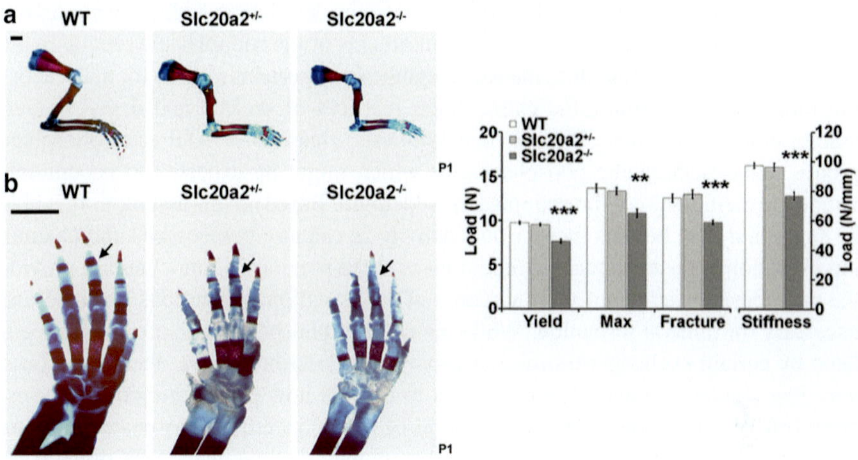

Fig. 5.15 Effects of *SLC20a2/PiT2* deletion on skeletal development in postnatal mice. (**a**) Forelimbs and (**b**) forelimb digits from wild type (WT), SLC20a2/PiT2$^{+/-}$ (heterozygous mutation), and *SLC20a2/PiT2*$^{-/-}$ (homozygous mutation) animals at postnatal day 1 (P1) showing staining for bone and cartilage with Alizarin red and Alcian blue, respectively. Compared to WT, formation of primary centers of ossification is delayed (black arrows) in both *SLC20a2/PiT2*$^{+/-}$ and *SLC20a2/PiT2*$^{-/-}$ mice. Scale bars = 1 mm (**a**) and 100 μm (**b**). Compared to WT and *SLC20a2/PiT2*$^{+/-}$ animals, *SLC20a2/PiT2*$^{-/-}$ mice also exhibit statistically significant decreased bone strength and stiffness as demonstrated by plots of yield, maximum load, fracture load and stiffness (Right panel; mean ± SE; $n = 6$ per genotype; $^{**}p < 0.01$ and $^{***}p < 0.001$ versus WT; ANOVA followed by Tukey's post hoc test). (Images and plots adapted from Beck-Cormier et al. [198] with permission)

initiation of the mineralization process" [198]. This conclusion was supported by the finding noted above that, in the developing tooth where both odontoblasts and ameloblasts are forming the mineralized extracellular matrix, *SLC20A2/PiT2* was more highly expressed in sub-odontoblastic cells rather than in odontoblasts or ameloblasts [192]. In this instance, the implication of the results recounted here is that some means other than the phosphate exchangers/transporters are also necessary for phosphate mobilization and matrix mineralization of bones and teeth. Further research is needed to identify and characterize such additional molecules.

The putative importance of the phosphate transporter, SLC20A1/PiT1, has also been evaluated using gene deletion technology. Conditional deletion of *SLC20A1/PiT1* in mice impaired endochondral ossification of the animals, causing endoplasmic reticulum stress with retention of aggrecan, extensive chondrocyte death, and abnormalities in mineralization [192]. This study pointed to the critical functioning of this phosphate ion transporter at the level of the endoplasmic reticulum with the suggestion that it acts as an endoplasmic reticulum stress-regulated gene rather than influencing mineralization events such as nucleation or crystal growth [192]. It was subsequently conjectured that the absence of *SLC20A1/PiT1* in phosphate ion transport with the activation of the stress response in the hypoxic environment of cartilage and an associated loss of vascular endothelial growth factor (*VEGF*) expression possibly led to the death of chondrocytes while triggering the observed disturbances in mineralization [199].

Phosphate ion transit intracellularly may be considered in an additional context. It is important to recognize that the ATP requirements of the endoplasmic reticulum are very high since this organelle is the site of synthesis of proteins needed for housekeeping functions and forming the extracellular matrices of skeletal and dental tissues, among other major roles. During such synthesis, cleavage of ATP ester phosphate groups releases inorganic phosphate ions which might be expected to accumulate within the membranes of the endoplasmic reticulum. Based on this assumption, it may be suggested that, besides their critical activity in calcium homeostasis and calcium ion liberation to the endoplasmic reticulum as noted previously, mitochondria provide the endoplasmic reticulum with a source of inorganic phosphate ions destined and necessary for mineral formation. While the entry of phosphate ions into cells is regulated by certain exchanger/transporter proteins as described above, there have been very few studies of mineralizing systems to explain how phosphate ions are transported out of cells. Nonetheless, from results of work concerning phosphate ion efflux in plants, yeast, and the fruit fly, it is clear that proteins containing the SPX domain are crucially important in maintaining phosphate ion homeostasis [200]. In human tissues, SPX is a multipass membrane receptor which resides in the plasma membrane receptor, XPR1 (xenotropic and polytropic murine leukemia retrovirus 1) (Fig. 5.14) [201]. Giovannini et al. demonstrated that, when XPR1 was non-functional, there was inhibition of phosphate ion efflux from metazoan cells [195]. Moreover, since the level of inhibition was profound, XPR1 (which subsumes SPX) was thought likely to be the sole protein which regulates phosphate ion transport from cells to their extracellular environment [195]. Indeed, it was concluded that, "Given the high conservation of XPR1 sequences (over 90% amino acid identity in mammals and 79% and 54% in zebrafish and fruit fly, respectively), it is likely that XPR1-mediated phosphate export is a universal feature in metazoans" [195].

The mechanism by which XPR1 mediates phosphate ion flux from a cell and interrogates phosphate ion homeostasis has focused on the inositol pyrophosphates. There is accumulating evidence that efflux of phosphate ions from cells appears to be mediated by a phosphorylated 6-carbon cyclic inositol containing a pyrophosphate group, inositol hexakisphosphate (IP6) [195]. As was noted earlier, another inositol derivative, IP3, regulates calcium ion flux in the endoplasmic reticulum [133]. IP6 is synthesized by plasma membrane kinases (inositol hexakisphosphate kinase 1 and 2), utilizing ATP as the phosphate donor [202]. When the genes encoding inositol hexakisphosphate kinases were deleted in human muscle cells, an increase was found in both the import and export of phosphate ions as well as an elevation in ATP synthesis in the treated cells [203]. Based on the results of this study [203], it is plausible that phosphate efflux from cells is mediated by signaling molecules such as IP6 binding receptors contained within the SPX domain of XPR1.

While all these studies above point to the potential importance of the IP6 system in mediating phosphate ion flux, there are few details describing its role in the mineralization of vertebrate hard tissues. As far as the transporter is concerned, inactivation of *XPR1* in mice is lethal [204]. When *XPR1* is conditionally ablated, an increase occurs in intracellular phosphate ion concentration [205]. An intracellular accumulation of phosphate ions in the presence of normal or elevated calcium ions would result in aberrant mineralization. As such, this event would explain the presence of mineral deposited in the basal ganglia of the brain (human familial brain calcification) in individuals with *XPR1* mutations [205]. A similar precipitation event is evident when *IP6 kinase* is ablated from osteoblast-like cells in vitro [206]. A decrease in kinase activity and a concomitant decrease in the level of IP6 cause an accumulation of stained mineral in the cultured cells [206] not unlike that reported in the basal ganglion [205]. Thus, it is plausible that the IP6 signaling system is active in bone cells and its disturbance in function may lead to dystrophic mineralization.

To summarize the findings above, it is clear that several plasma membrane exchanger/transporter proteins exist to support the movement of phosphate ions into a cell. Moreover, IP6 signaling to the SPX domain of XPR1 provides a coherent system for trafficking intracellular phosphate ions from a cell to the extracellular milieu and the mineralization front. A deeply grounded mechanistic understanding remains incomplete, however, concerning both cellular influx and efflux of phosphate regulated by exchanger/transporter proteins of this ion, and much is yet to be learned of the two systems as they relate to normal and aberrant mineralization of skeletal and dental tissues. Despite this shortcoming in knowledge, the studies noted above provide the basis for new insights into the control of phosphate ions for the deposition of extracellular mineral in vertebrate species.

5.10 Perspective: Calcium and Phosphate Ion Uptake, Distribution, and Homeostasis in Cells of Vertebrate Mineralized Tissues

The focus of this chapter has been the consideration of mechanisms by which cells regulate and control calcium and phosphate ions prior to the induction of extracellular matrix mineralization in vertebrate hard tissues. The central thesis is that osteoblasts, odontoblasts, and the cells of other vertebrate mineralized tissues sequester and transport these ions by means of their subcellular organelles. Ions are moved from the tissue fluid of the environmental niche into the cytosolic compartment of the cell and stored in the endoplasmic reticulum and mitochondria. The endoplasmic reticulum plays a central role in intracellular ion accumulation while mitochondria work to support endoplasmic reticulum function, calcium ion homeostasis and energy metabolism in the cell, and cell survival.

From a historical perspective, mitochondria were the first cellular organelles to be closely linked with calcium ion sequestration from the cytosol. Early studies in vitro of mitochondria isolated from cells showed they accumulated large quantities of calcium ions [112, 115]. Histochemical investigations of mineralizing tissues using the cation-sensitive probe, pyroantimonate, indicated that mitochondria sequestered calcium, which appeared in the form of granular material [117, 119]. Energy dispersive X-ray microanalysis and electron diffraction respectively demonstrated the presence of both calcium and phosphorus and an amorphous calcium phosphate phase of mineral in mitochondria [207, 208]. While there had been caution and criticism about the use of the pyroantimonate staining technique to identify calcium in biological tissues [207], subsequent reports confirmed that the organelle accumulated calcium (as well as phosphorus) ions in vivo [110, 123, 209]. Moreover, biochemical studies provided evidence that, accompanying its accumulation of calcium, there was up-regulation of ATP synthesis by mitochondria and hence the energy supply for anabolic reactions was maintained by the cell [95, 106]. From all these perspectives, the conclusion that mitochondria are involved with the mineralization process appears to be consistent and valid although the precise role of the organelle in this context has yet to be determined.

In common with mitochondrial involvement in the mineralization process, the endoplasmic reticulum system of a cell has been demonstrated to promote active calcium ion storage in both its extensive membranous reticulum and endosomal vacuoles [88, 129, 210]. That the endoplasmic reticulum can store calcium ions is underscored by the fact that, compared to the cytosolic calcium ion concentration (0.1 μM) [131, 132], its free calcium ion concentration can range from 100 to 800 μM [129]. Thus, it is likely that the calcium-loaded subcellular endoplasmic reticulum, which communicates with both the cell membrane and mitochondria, can generate vacuoles and vesicles which contain high levels of calcium. As such, the endoplasmic reticulum provides a means by which calcium ions may be accumulated intracellularly in vesicles or other organelles which could foster the possible formation of prenucleation ion clusters [211]. These organelles would follow tracks or pathways to the extracellular milieu defined by the cytoskeletal system of microtubules supporting kinesin motor proteins [212–214]. Kinesin proteins shuttle cargo (mitochondria and other organelles) unidirectionally to the periphery of a cell to dock at the plasma membrane for subsequent extracellular export [215, 216].

There is little doubt that the structural relation between mitochondria and the endoplasmic reticulum noted above enhances calcium trafficking activities between the two organelles and facilitates inter-organelle calcium ion transport. Additionally, Frieden et al. found that mitochondria located in close vicinity to plasma membranes generated calcium ion microdomains in the cytosol which enhanced store-operated calcium ion entry channels [85]. Opening of these channels in the endoplasmic reticulum promoted calcium ion re-filling, providing ions for both mineralization-related activities and preventing organelle stress (unfolded protein) response [85]. Likewise, VDAC channels and the mitochondrial uniporter coupled with IP3 and ryanadone receptor activity mediated entry of calcium ions into the mitochondrion [94, 217]. Frieden and colleagues commented that, "Mitochondria thus not only integrate intracellular calcium signals to produce the appropriate metabolic response but also control the fluxes of calcium inside cells to shape calcium signals" [85]. From this point of view, it may be suggested here that both mitochondria and the endoplasmic reticulum are active participants in processes which lead to the accumulation and subsequent transfer of calcium ions from the cell to the extracellular environment. Mechanisms of the export of calcium and phosphate ions from cells to extracellular matrixes are discussed further in Chap. 6.

The impact of calcium ions on mitochondrial function is complex, and high frequency physiological calcium ion oscillations, for example, promote mitochondrial oxidative phosphorylation, a requirement for the continued synthesis of proteins of the extracellular matrix [218]. However, if calcium ion loading is extensive, calcium phosphate particles are formed and become visible within the mitochondrial matrix [219]. Calcium ion loading by particle or granule formation ends with the opening of a large conductance pore, evoking a mitochondrial permeability transition [220]. This change uncouples electron transport from oxidative phosphorylation with a fall in energy production and eventual damage to the inner membrane structure of the organelle [219]. The mitochondrion then undergoes swelling, a decrease in the intramitochondrial pH, solubilization of the granules, and release of much of its accumulated calcium [221]. If the loss of membrane potential is extreme, there is concurrent release of proteins (cytochrome c and SMAC/DIABLO) from the mitochondrion and subsequent activation of events which lead to cell death (apoptosis) [220]. Relating this information to physiological mineral deposition, it is probable that calcium ion loading takes place within cells of bone, dentin, and other vertebrate mineralized tissues and enhances mitochondrial function while fueling the earliest steps in the mineralization process. In populations of hypertrophic chondrocytes of the growth plate, calcium ion loading may become extreme, leading first to the appearance of calcium phosphate granules in mitochondria and then, as loading continues, causing the release of proteins which activate cell death in certain of the hypertrophic cells.

Just as calcium ions enter cells through membrane channels and are accumulated in the endoplasmic reticulum and mitochondria, pools of phosphate ions exist in cells with evidence that inorganic phosphate ions and molecules such as β-glycerophosphate mediate SOCE channel activity and mineralization [222–225]. The phosphate donor, β-glycerophosphate, is a simple ester phosphate long associated with inducing mineralization in culture systems of osteoblasts, chondrocytes, odontoblasts, or their progenitors [222]. In this context, culture medium is very frequently supplemented with β-glycerophosphate so as to explore and putatively mimic mineralization events in vivo. The rationale for utilizing β-glycerophosphate in this way is that it is hydrolyzed by cell membrane-related TNAP to liberate phosphate ions for mineral formation. Moreover, there is some evidence that phosphate ions activate the cyclic-AMP/protein kinase A pathway to promote an ERK-dependent increase in the expression of the osteogenic protein, BMP [223]. However, the use of β-glycerophosphate is puzzling as this molecule is normally present in vanishingly small quantities in serum. Even more curious, β-glycerophosphate levels used in cell culture studies vary considerably, as high as 10 mM in some investigations [226]. Such concentrations are known to generate rapid mineral development, but the mineral is aberrant and can cause cell death [222, 224]. More recently, elevated levels of inorganic phosphate have been used instead of β-glycerophosphate to induce mineral deposition in cultured cells [227].

A recent study using megakaryocytes in culture demonstrated that β-glycerophosphate at low concentrations influenced mineral deposition by inducing a signaling cascade involving both ORAI1 and STIM to up-regulate SOCE

[228]. In a related investigation, Ma et al. showed that β-glycerophosphate in a mineralizing cell culture system promoted expression of both *ORAI1* and *STIM1* while increasing expression of transcription factors, *SOX9*, *RUNX2*, and *MSX2*, as well as transcripts of the target gene, *TNAP* [225]. Moreover, use of ORAI1 inhibitors and silencing of *ORAI1* gene expression grossly interfered with the action of β-glycerophosphate and resulted in decreased TNAP secretion and mineralizing activity [225]. While this latter work [225] was conducted with vascular smooth muscle cells which underwent mineralization, and not with cells of bone or the tooth, the findings are relevant to physiological mineralization. They, and the overall perspective from results of other cell culture work, provide new insights into the mechanism of action of β-glycerophosphate, including a novel role in promoting calcium ion accumulation in the endoplasmic reticulum, an event which leads to the trafficking of calcium-loaded endosomes and the induction of mineral formation [228]. Such investigations in vitro also point to the importance of SOCE and advance understanding of the regulation of several transcription factors critical to the initiation and progression of the process of mineralization in vertebrate tissues.

As for the cellular pools of inorganic phosphate ions are concerned, these are likely many-fold higher than the levels of cytoplasmic calcium ions as noted above. Direct measurements of phosphate ion concentrations in cells, as well as consideration of ubiquitous phosphorylation and dephosphorylation reactions and the presence of phosphate co-transporters, indicate that this anion is common and abundant in all major subcellular compartments. Additionally, phosphate ions utilized for mineralization may be derived, of course, from ATP, which is generated in great amounts in biosynthetically active skeletal and dental cells. What is not known is how phosphate ions are partitioned in a cell and the possible mechanisms of phosphate ion efflux. From studies of lower organisms [195] and a mammalian cell line [206], it is plausible to suggest that IP6 and its receptor provide a means for transporting intracellular phosphate ions to the extracellular milieu and the mineralization front. Such a conceptual release of phosphate from cells leads to other questions as to whether phosphate ions are packaged while in transit, whether phosphate ions are clustered with calcium ions during efflux, and whether phosphate ion trafficking promotes vesiculation, for example. These questions require yet further research regarding phosphate ion cellular influx, efflux, and control.

Regarding cellular phosphate ion influx, the K_m of the known membrane exchangers/transporters is approximately 0.1 mM [229, 230]. This value is quite low with respect to the serum phosphate ion concentration of ~1–1.5 mM reported in vertebrate mineralized tissues [132, 196]. In other words, these exchangers/transporters are most active at phosphate ion concentrations much lower than that of serum or the extracellular matrix [231]. Given such differences between intracellular and extracellular phosphate ion levels and the low capacity (binding affinity) of the known phosphate exchangers/transporters, this situation raises the questions as to whether other proteins are present in cells of mineralizing tissues which transport phosphate ions with affinities which are more in line with the phosphate ion concentration and whether phosphate is the rate-limiting factor in the formation of mineral in vertebrates. Once again, it is clear that more studies remain to define possible additional phosphate exchanger/transporter proteins and the ability of the cell to regulate and control the movement and access of phosphate ions to mineralizing sites of vertebrate tissues. The involvement of the currently identified transporters in the broader context of the regulation of phosphate and pyrophosphate metabolism is described further in Chap. 9.

From the summation of evidence at the present time, the role of the PiT exchangers/transporters in mineralization remains controversial and incompletely understood. Results from different research groups are inconsistent and confusing in defining precise functions for both PiT1 and PiT2. To expand on work mentioned above, a study, for example, with mice expressing *PiT1* at a reduced level (hypomorphic expression) demonstrated that the animals had unexpectedly normal bone mineralization [232]. It was speculated that the observed mineral deposition occurred because of a possible functional redundancy between *PiT1* and *PiT2* [232]. However, *PiT2*-deficient osteoblasts had reduced phosphate ion transport activity but no mineralization defects in vitro, and PiT2 was found to be instrumental in maintaining bone quality and strength [198]. Supporting the latter results, the loss of PiT2 appears to produce abnormal development of bone, decreased in its length, mineral density and biomechanical properties in mice [198, 233]. To confound these data, Yamazaki et al. recently found that the ablation of *Pit1* and *PiT2* actually enhanced mineralization in clonal osteoblastic cell lines [234].

With regard to putative Pit1 and PiT2 exchanger/transporter function in cartilage, Couasnay et al. reported that PiT1 resided principally in the endoplasmic reticulum of chondrocytes and not in their plasma membrane, the commonly accepted phosphate ion transport site in cells [199]. As a consequence of its specific interaction with an endoplasmic reticulum chaperone, protein disulfide isomerase, PiT1 was responsible for inducing an unfolded protein response which was independent of any phosphate ion transport function in these cells [199]. Additionally, the suggestion has been made that PiT1 may form a heterodimer with PiT2 to create a calcium sensor receptor [235, 236]. This idea was based on the inference that, in tissues like the parathyroid gland, the crystal structure of the calcium sensor receptor has multiple binding sites for phosphate ions [235]. There is some evidence to indicate that the PiT proteins as heterodimers can act as anion exchangers/transporters while also triggering downstream signaling events. Such heterodimers are described as transceptors [237].

On reflection of these various and quite different observations in bone and cartilage, ascribing a particular role to PiT1 and PiT2 in vertebrate mineralization is extremely challenging. Indeed, Beck-Cormier and Beck have proposed that mineralization may be a phenomenon which occurs through mechanisms independent of phosphate ion transport [238]. They have concluded that, "It is ... urgent to change the paradigm concerning the biology of PiTs, no longer considering them as Pi transporters only, but as multifunctional proteins" [238]. Such a consideration should lead to more attention and research concerning the involvement, for example, of the protein partners of the phosphate ion transporters in understanding the development of certain diseases such as primary familial brain calcification and could generate better explanations for the many differences reported between PiT1 and PiT2 in bone and cartilage. Further, a revision in thinking about PiT1 and PiT2 could help define more precise roles of these two molecules in ameloblasts and odontoblasts, where no one currently knows their possible effects in the mineralizing dental tissues. Far more work is certainly necessary to clarify the function(s) of PiT1, PiT2, and other putative phosphate exchangers/transporters as well as their regulation since phosphate is so critically important to the function and survival of both mineralizing and non-mineralizing cells of vertebrate tissues.

References

1. Williams RJR. Calcium: the developing role of its chemistry in biological evolution. In: Foley EC, Klee CB, editors. Calcium as a cellular regulator. London: Oxford University Press; 1999. p. 3–27.
2. Gennerich A, Vale RD. Walking the walk: how kinesin and dynein coordinate their steps. Curr Opin Cell Biol. 2009;21:59–67.
3. Shapiro IM, Landis WJ, Risbud MV. Matrix vesicles: are they anchored exosomes? Bone. 2015;79:29–36.
4. Bonucci E. Fine structure of early cartilage calcification. J Ultrastruct Res. 1967;20:33–50.
5. Anderson HC. Vesicles associated with calcification in the matrix of epiphyseal cartilage. J Cell Biol. 1969;41:59–72.
6. Bernard GW. Ultrastructural localization of alkaline phosphatase in initial intramembranous osteogenesis. Clin Orthop Relat Res. 1978;135:218–25.
7. Takano Y, Sakai H, Baba O, Terashima T. Differential involvement of matrix vesicles during the initial and appositional mineralization processes in bone, dentin, and cementum. Bone. 2000;26:333–9.
8. Mahamid J, Sharir A, Gur D, Zelzer E, Addadi L, Weiner S. Bone mineralization proceeds through intracellular calcium phosphate loaded vesicles: a cryo-electron microscopy study. J Struct Biol. 2011;174:527–35.
9. Boonrungsiman S, Gentleman E, Carzaniga R, Evans ND, McComb DW, Porter AE, et al. The role of intracellular calcium phosphate in osteoblast-mediated bone apatite formation. Proc Natl Acad Sci U S A. 2012;109:14170–5.
10. Pecze L, Blum W, Schwaller B. Routes of Ca2+ shuttling during Ca2+ oscillations: focus on the role of mitochondrial Ca2+ handling and cytosolic Ca2+ buffers. J Biol Chem. 2015;290:28214–30.
11. Pinton P, Pozzan T, Rizzuto R. The Golgi apparatus is an inositol 1,4,5-trisphosphate-sensitive Ca2+ store, with functional properties distinct from those of the endoplasmic reticulum. EMBO J. 1998;17:5298–308.
12. Calcraft PJ, Ruas M, Pan Z, Cheng X, Arredouani A, Hao X, et al. NAADP mobilizes calcium from acidic organelles through two-pore channels. Nature. 2009;459:596–600.
13. Kazakov AS, Sokolov AS, Vologzhannikova AA, Permyakova ME, Khorn PA, Ismailov RG, et al. Interleukin-11 binds specific EF-hand proteins via their conserved structural motifs. J Biomol Struct Dyn. 2017;35:78–91.
14. Halling DB, Liebeskind BJ, Hall AW, Aldrich RW. Conserved properties of individual Ca2+-binding sites in calmodulin. Proc Natl Acad Sci U S A. 2016;113:E1216–25.
15. Kawasaki H, Kretsinger RH. Calcium-binding proteins. 1: EF-hands. Prot Prof. 1994;1:343–517.
16. Gifford JL, Walsh MP, Vogel HJ. Structures and metal-ion-binding properties of the Ca2+-binding helix-loop-helix EF-hand motifs. Biochem J. 2007;405:199–221.

17. Yap KL, Ames JB, Swindells MB, Ikura M. Diversity of conformational states and changes within the EF-hand protein superfamily. Proteins. 1999;37:499–507.
18. Schwaller B. Cytosolic Ca2+ buffers. Cold Spring Harb Perspect Biol. 2010;2:a004051.
19. Bagur R, Hajnoczky G. Intracellular Ca(2+) sensing: its role in calcium homeostasis and signaling. Mol Cell. 2017;66:780–8.
20. Parekh AB, Putney JW Jr. Store-operated calcium channels. Physiol Rev. 2005;85:757–810.
21. Petersen OH, Burdakova N. The specificity of Ca2+ signalling. Acta Physiol Hung. 2002;89:439–50.
22. Dewenter M, von der Lieth A, Katus HA, Backs J. Calcium signaling and transcriptional regulation in cardiomyocytes. Circ Res. 2017;121:1000–20.
23. Sato M, Sobhan U, Tsumura M, Kuroda H, Soya M, Masamura A, et al. Hypotonic-induced stretching of plasma membrane activates transient receptor potential vanilloid channels and sodium-calcium exchangers in mouse odontoblasts. J Endod. 2013;39:779–87.
24. Little R, Muimo R, Robson L, Harris K, Grabowski PS. The transient receptor potential ion channel TRPV6 is expressed at low levels in osteoblasts and has little role in osteoblast calcium uptake. PLoS One. 2011;6:e28166.
25. Balmain N, Berdal A, Hotton D, Cuisinier-Gleizes P, Mathieu H. Calbindin-D9K immunolocalization and vitamin D-dependence in the bone of growing and adult rats. Histochemistry. 1989;92:359–65.
26. Faucheux C, Bareille R, Amedee J. Synthesis of calbindin-D28K during mineralization in human bone marrow stromal cells. Biochem J. 1998;333:817–23.
27. Nishimura G, Lausch E, Savarirayan R, Shiba M, Spranger J, Zabel B, et al. TRPV4-associated skeletal dysplasias. Am J Med Genet C: Semin Med Genet. 2012;160C:190–204.
28. Nonaka K, Han X, Kato H, Sato H, Yamaza H, Hirofuji Y, et al. Novel gain-of-function mutation of TRPV4 associated with accelerated chondrogenic differentiation of dental pulp stem cells derived from a patient with metatropic dysplasia. Biochem Biophys Rep. 2019;19:100648.
29. Lewis AH, Grandl J. Mechanical sensitivity of Piezo1 ion channels can be tuned by cellular membrane tension. elife. 2015;4:e12088.
30. Parpaite T, Coste B. Piezo channels. Curr Biol. 2017;27:R250–2.
31. Song Y, Li D, Farrelly O, Miles L, Li F, Kim SE, et al. The mechanosensitive ion channel Piezo inhibits axon regeneration. Neuron. 2019;102:e6373–89.
32. Zhang T, Chi S, Jiang F, Zhao Q, Xiao B. A protein interaction mechanism for suppressing the mechanosensitive Piezo channels. Nat Commun. 2017;8:1797.
33. Wang L, You X, Lotinun S, Zhang L, Wu N, Zou W. Mechanical sensing protein PIEZO1 regulates bone homeostasis via osteoblast-osteoclast crosstalk. Nat Commun. 2020;11:282.
34. Li J, Hou B, Tumova S, Muraki K, Bruns A, Ludlow MJ, et al. Piezo1 integration of vascular architecture with physiological force. Nature. 2014;515:279–82.
35. Rawlinson SC, Pitsillides AA, Lanyon LE. Involvement of different ion channels in osteoblasts' and osteocytes' early responses to mechanical strain. Bone. 1996;19:609–14.
36. van Abel M, Hoenderop JG, van der Kemp AW, van Leeuwen JP, Bindels RJ. Regulation of the epithelial Ca2+ channels in small intestine as studied by quantitative mRNA detection. Am J Physiol Gastrointest Liver Physiol. 2003;285:G78–85.
37. DiPolo R, Beaugé L. Sodium/calcium exchanger: influence of metabolic regulation on ion carrier interactions. Physiol Rev. 2006;86:155–203.
38. Tran DN, Jung EM, Yoo YM, Lee JH, Jeung EB. Potassium-dependent sodium/calcium exchanger 3 (Nckx3) depletion leads to abnormal motor function and social behavior in mice. J Physiol Pharmacol. 2020;71:1.
39. Stains JP, Gay CV. Inhibition of Na+/Ca2+ exchange with KB-R7943 or bepridil diminished mineral deposition by osteoblasts. J Bone Miner Res. 2001;16:1434–43.
40. Stains JP, Weber JA, Gay CV. Expression of Na(+)/Ca(2+) exchanger isoforms (NCX1 and NCX3) and plasma membrane Ca(2+) ATPase during osteoblast differentiation. J Cell Biochem. 2002;84:625–35.

41. Robertson SYT, Wen X, Yin K, Chen J, Smith CE, Paine ML. Multiple calcium export exchangers and pumps are a prominent feature of enamel organ cells. Front Physiol. 2017;8:336.
42. Burnstock G. Purine and pyrimidine receptors. Cell Mol Life Sci. 2007;64:1471–83.
43. Hoebertz A, Mahendran S, Burnstock G, Arnett TR. ATP and UTP at low concentrations strongly inhibit bone formation by osteoblasts: a novel role for the P2Y2 receptor in bone remodeling. J Cell Biochem. 2002;86:413–9.
44. Cutarelli A, Marini M, Tancredi V, D'Arcangelo G, Murdocca M, Frank C, et al. Adenosine Triphosphate stimulates differentiation and mineralization in human osteoblast-like Saos-2 cells. Develop Growth Differ. 2016;58:400–8.
45. Hatori M, Teixeira CC, Debolt K, Pacifici M, Shapiro IM. Adenine nucleotide metabolism by chondrocytes in vitro: role of ATP in chondrocyte maturation and matrix mineralization. J Cell Physiol. 1995;165:468–74.
46. Kaplan AD, Kilkenny DM, Hill DJ, Dixon SJ. Extracellular nucleotides act through P2U purinoceptors to elevate [Ca2+]i and enhance basic fibroblast growth factor-induced proliferation in sheep chondrocytes. Endocrinology. 1996;137:4757–66.
47. Hung CT, Allen FD, Mansfield KD, Shapiro IM. Extracellular ATP modulates [Ca2+]i in retinoic acid-treated embryonic chondrocytes. Am J Phys. 1997;272:C1611–7.
48. Ke HZ, Qi H, Weidema AF, Zhang Q, Panupinthu N, Crawford DT, et al. Deletion of the P2X7 nucleotide receptor reveals its regulatory roles in bone formation and resorption. Mol Endocrinol. 2003;17:1356–67.
49. Agrawal A, Gartland A. P2X7 receptors: role in bone cell formation and function. J Mol Endocrinol. 2015;54:R75–88.
50. Nakamura E, Uezono Y, Narusawa K, Shibuya I, Oishi Y, Tanaka M, et al. ATP activates DNA synthesis by acting on P2X receptors in human osteoblast-like MG-63 cells. Am J Physiol Cell Physiol. 2000;279:C510–9.
51. Panupinthu N, Rogers JT, Zhao L, Solano-Flores LP, Possmayer F, Sims SM, et al. P2X7 receptors on osteoblasts couple to production of lysophosphatidic acid: a signaling axis promoting osteogenesis. J Cell Biol. 2008;181:859–71.
52. Grol MW, Brooks PJ, Pereverzev A, Dixon SJ. P2X7 nucleotide receptor signaling potentiates the Wnt/beta-catenin pathway in cells of the osteoblast lineage. Purinergic Signal. 2016;12:509–20.
53. Li J, Liu D, Ke HZ, Duncan RL, Turner CH. The P2X7 nucleotide receptor mediates skeletal mechanotransduction. J Biol Chem. 2005;280:42952–9.
54. Orriss IR, Knight GE, Ranasinghe S, Burnstock G, Arnett TR. Osteoblast responses to nucleotides increase during differentiation. Bone. 2006;39:300–9.
55. Yanez M, Gil-Longo J, Campos-Toimil M. Calcium binding proteins. Adv Exp Med Biol. 2012;740:461–82.
56. Denessiouk K, Permyakov S, Denesyuk A, Permyakov E, Johnson MS. Two structural motifs within canonical EF-hand calcium-binding domains identify five different classes of calcium buffers and sensors. PLoS One. 2014;9:e109287.
57. Kolev SK, Petkov PS, Rangelov MA, Trifonov DV, Milenov TI, Vayssilov GN. Interaction of Na+, K+, Mg2+ and Ca2+ counter cations with RNA. Metallomics. 2018;10:659–78.
58. Zhou Y, Xue S, Yang JJ. Calciomics: integrative studies of Ca2+-binding proteins and their interactomes in biological systems. Metallomics. 2013;5:29–42.
59. Bellido T, Huening M, Raval-Pandya M, Manolagas SC, Christakos S. Calbindin-D28k is expressed in osteoblastic cells and suppresses their apoptosis by inhibiting caspase-3 activity. J Biol Chem. 2000;275:26328–32.
60. Zayzafoon M. Calcium/calmodulin signaling controls osteoblast growth and differentiation. J Cell Biochem. 2006;97:56–70.
61. Yeo H, Beck LH, Thompson SR, Farach-Carson MC, McDonald JM, Clemens TL, et al. Conditional disruption of calcineurin B1 in osteoblasts increases bone formation and reduces bone resorption. J Biol Chem. 2007;282:35318–27.

62. Clapham DE. Calcium signaling. Cell. 2007;131:1047–58.
63. Choi YH, Choi JH, Oh JW, Lee KY. Calmodulin-dependent kinase II regulates osteoblast differentiation through regulation of osterix. Biochem Biophys Res Commun. 2013;432:248–55.
64. Somlyo AP, Bond M, Somlyo AV. Calcium content of mitochondria and endoplasmic reticulum in liver frozen rapidly in vivo. Nature. 1985;314:622–5.
65. Dedhar S. Novel functions for calreticulin: interaction with integrins and modulation of gene expression? Trends Biochem Sci. 1994;19:269–71.
66. Pilquil C, Alvandi Z, Opas M. Calreticulin regulates a switch between osteoblast and chondrocyte lineages derived from murine embryonic stem cells. J Biol Chem. 2020;295:6861–75.
67. Boyce BF, Xiu Y, Li J, Xing L, Yao Z. NF-κB-mediated regulation of osteoclastogenesis. Endocrinol Metab (Seoul). 2015;30:35–44.
68. Bradley EW, Drissi MH. WNT5A regulates chondrocyte differentiation through differential use of the CaN/NFAT and IKK/NF-kappaB pathways. Mol Endocrinol. 2010;24:1581–93.
69. Berdal A, Hotton D, Saffar JL, Thomasset M, Nanci A. Calbindin-D9k and calbindin-D28k expression in rat mineralized tissues in vivo. J Bone Miner Res. 1996;11:768–79.
70. Deutsch D, Dafni L, Palmon A, Hekmati M, Young MF, Fisher LW. Tuftelin: enamel mineralization and amelogenesis imperfecta. CIBA Found Symp. 1997;205:135–55.
71. Somogyi E, Petersson U, Hultenby K, Wendel M. Calreticulin--an endoplasmic reticulum protein with calcium-binding activity is also found in the extracellular matrix. Matrix Biol. 2003;22:179–91.
72. Oshima S, Watanabe M. Elevated expression of calcineurin subunits during active mineralization of developing mouse molar teeth. Eur J Oral Sci. 2012;120:386–94.
73. Crumpton MJ, Dedman JR. Protein terminology tangle. Nature. 1990;345:212.
74. Swairjo MA, Concha NO, Kaetzel MA, Dedman JR, Seaton BA. Ca(2+)-bridging mechanism and phospholipid head group recognition in the membrane-binding protein annexin V. Nat Struct Biol. 1995;2:968–74.
75. Mayran N, Parton RG, Gruenberg J. Annexin II regulates multivesicular endosome biogenesis in the degradation pathway of animal cells. EMBO J. 2003;22:3242–53.
76. Genge BR, Wu LNY, Wuthier RE. Identification of phospholipid-dependent calcium-binding proteins as constituents of matrix vesicles. J Biol Chem. 1989;264:10917–21.
77. Kirsch T, Nah HD, Shapiro IM, Pacifici M. Regulated production of mineralization-competent matrix vesicles in hypertrophic chondrocytes. J Cell Biol. 1997;137:1149–60.
78. Thouverey C, Malinowska A, Balcerzak M, Strzelecka-Kiliszek A, Buchet R, Dadlez M, et al. Proteomic characterization of biogenesis and functions of matrix vesicles released from mineralizing human osteoblast-like cells. J Proteome. 2011;74:1123–34.
79. Veschi EA, Bolean M, Strzelecka-Kiliszek A, Bandorowicz-Pikula J, Pikula S, Granjon T, et al. Localization of annexin A6 in matrix vesicles during physiological mineralization. Int J Mol Sci. 2020;21:1367.
80. Enrich C, Rentero C, Grewal T. Annexin A6 in the liver: from the endocytic compartment to cellular physiology. Biochim Biophys Acta, Mol Cell Res. 2017;1864:933–46.
81. Kirsch T, Harrison G, Golub EE, Nah HD. The roles of annexins and types II and X collagen in matrix vesicle-mediated mineralization of growth plate cartilage. J Biol Chem. 2000;275:35577–83.
82. Hardingham GE, Bading H. Calcium as a versatile second messenger in the control of gene expression. Microsc Res Tech. 1999;46:348–55.
83. Rowland AA, Voeltz GK. Endoplasmic reticulum-mitochondria contacts: function of the junction. Nat Rev Mol Cell Biol. 2012;13:607–25.
84. Park MK, Ashby MC, Erdemli G, Petersen OH, Tepikin AV. Perinuclear, perigranular and sub-plasmalemmal mitochondria have distinct functions in the regulation of cellular calcium transport. EMBO J. 2001;20:1863–74.
85. Frieden M, Arnaudeau S, Castelbou C, Demaurex N. Subplasmalemmal mitochondria modulate the activity of plasma membrane Ca2+-ATPases. J Biol Chem. 2005;280: 43198–208.

86. Dolman NJ, Gerasimenko JV, Gerasimenko OV, Voronina SG, Petersen OH, Tepikin AV. Stable Golgi-mitochondria complexes and formation of Golgi Ca(2+) gradients in pancreatic acinar cells. J Biol Chem. 2005;280:15794–9.
87. Williams RJ. Calcium: outside/inside homeostasis and signalling. Biochim Biophys Acta. 1998;1448:153–65.
88. Krebs J, Agellon LB, Michalak M. Ca(2+) homeostasis and endoplasmic reticulum (ER) stress: an integrated view of calcium signaling. Biochem Biophys Res Commun. 2015;460: 114–21.
89. Herrington J, Park YB, Babcock DF, Hille B. Dominant role of mitochondria in clearance of large Ca2+ loads from rat adrenal chromaffin cells. Neuron. 1996;16:219–28.
90. Mammucari C, Raffaello A, Vecellio Reane D, Gherardi G, De Mario A, Rizzuto R. Mitochondrial calcium uptake in organ physiology: from molecular mechanism to animal models. Pflugers Arch. 2018;470:1165–79.
91. Williams GS, Boyman L, Lederer WJ. Mitochondrial calcium and the regulation of metabolism in the heart. J Mol Cell Cardiol. 2015;78:35–45.
92. Friedman JR, Lackner LL, West M, DiBenedetto JR, Nunnari J, Voeltz GK. ER tubules mark sites of mitochondrial division. Science. 2011;334:358–62.
93. West M, Zurek N, Hoenger A, Voeltz GK. A 3D analysis of yeast ER structure reveals how ER domains are organized by membrane curvature. J Cell Biol. 2011;193:333–46.
94. Rossi A, Pizzo P, Filadi R. Calcium, mitochondria and cell metabolism: a functional triangle in bioenergetics. Biochim Biophys Acta, Mol Cell Res. 2019;1866:1068–78.
95. Territo PR, Mootha VK, French SA, Balaban RS. Ca(2+) activation of heart mitochondrial oxidative phosphorylation: role of the F(0)/F(1)-ATPase. Am J Physiol Cell Physiol. 2000;278:C423–35.
96. Jiang D, Zhao L, Clapham DE. Genome-wide RNAi screen identifies Letm1 as a mitochondrial Ca2+/H+ antiporter. Science. 2009;326:144–7.
97. Tan W, Colombini M. VDAC closure increases calcium ion flux. Biochim Biophys Acta. 2007;1768:2510–5.
98. Bathori G, Csordas G, Garcia-Perez C, Davies E, Hajnoczky G. Ca2+-dependent control of the permeability properties of the mitochondrial outer membrane and voltage-dependent anion-selective channel (VDAC). J Biol Chem. 2006;281:17347–58.
99. Sparagna GC, Gunter KK, Sheu SS, Gunter TE. Mitochondrial calcium uptake from physiological-type pulses of calcium. A description of the rapid uptake mode. J Biol Chem. 1995;270:27510–5.
100. Buntinas L, Gunter KK, Sparagna GC, Gunter TE. The rapid mode of calcium uptake into heart mitochondria (RaM): comparison to RaM in liver mitochondria. Biochim Biophys Acta. 2001;1504:248–61.
101. Finkel T, Menazza S, Holmström KM, Parks RJ, Liu J, Sun J, et al. The ins and outs of mitochondrial calcium. Circ Res. 2015;116:1810–9.
102. Komarova SV, Ataullakhanov FI, Globus RK. Bioenergetics and mitochondrial transmembrane potential during differentiation of cultured osteoblasts. Am J Physiol Cell Physiol. 2000;279:C1220–9.
103. Lehninger AL, Vercesi A, Bababunmi EA. Regulation of Ca2+ release from mitochondria by the oxidation-reduction state of pyridine nucleotides. Proc Natl Acad Sci U S A. 1978;75:1690–4.
104. Denton RM, Rutter GA, Midgley PJ, McCormack JG. Effects of Ca2+ on the activities of the calcium-sensitive dehydrogenases within the mitochondria of mammalian tissues. J Cardiovasc Pharmacol. 1988;12:S69–72.
105. Numakura C, Tamiya G, Ueki M, Okada T, Maisawa SI, Kojima-Ishii K, et al. Growth impairment in individuals with citrin deficiency. J Inherit Metab Dis. 2019;42:501–8.
106. Szabadkai G, Pitter JG, Spat A. Cytoplasmic Ca2+ at low submicromolar concentration stimulates mitochondrial metabolism in rat luteal cells. Pflugers Arch. 2001;441:678–85.
107. Lehninger AL. Mitochondria and biological mineralization processes: an exploration. Horiz Biochem Biophys. 1977;4:1–30.

108. Lehninger AL, Reynafarje B, Vercesi A, Tew WP. Transport and accumulation of calcium in mitochondria. Ann N Y Acad Sci. 1978;307:160–76.
109. Lehninger A. The possible role of mitochondria and phosphocitrate in biological calcification. In: Westbroek P, de Jong EW, editors. Biomineralization and biological metal accumulation. Dordrecht: D. Reidel Publishing Co.; 1983. p. 107–21.
110. Pei D-D, Sun J-L, Zhu C-H, Tian F-C, Jiao K, Anderson MR, et al. Contribution of mitophagy to cell-mediated mineralization: revisiting a 50-year-old conundrum. Adv Sci. 2018;5:1800873.
111. Reith EJ. The binding of calcium within the Golgi saccules of the rat odontoblast. Am J Anat. 1976;147:267–70.
112. Greenawalt JW, Rossi CS, Lehninger AL. Effect of active accumulation of calcium and phosphate ions on the structure of rat liver mitochondria. J Cell Biol. 1964;23:21–38.
113. Chen CH, Greenawalt JW, Lehninger AL. Biochemical and ultrastructural aspects of Ca2+ transport by mitochondria of the hepatopancreas of the blue crab Callinectes sapidus. J Cell Biol. 1974;61:301–15.
114. Zilli L, Schiavone R, Storelli C, Vilella S. Analysis of calcium concentration fluctuations in hepatopancreatic R cells of Marsupenaeus japonicus during the molting cycle. Biol Bull. 2007;212:161–8.
115. Becker GL, Chen CH, Greenawalt JW, Lehninger AL. Calcium phosphate granules in the hepatopancreas of the blue crab Callinectes sapidus. J Cell Biol. 1974;61:316–26.
116. Shapiro IM, Greenspan JS. Are mitochondria directly involved in biological mineralisation? Calcif Tissue Res. 1969;3:100–2.
117. Brighton CT, Hunt RM. Mitochondrial calcium and its role in calcification. Histochemical localization of calcium in electron micrographs of the epiphyseal growth plate with K-pyroantimonate. Clin Orthop Relat Res. 1974;100:406–16.
118. Carson FL, Davis WL, Matthews JL, Martin JH. Calcium localization in normal rachitic, and D3-treated chicken epiphyseal chondrocytes utilizing potassium pyroantimonate-osmium tetroxide. Anat Rec. 1978;190:23–39.
119. Burger EH, de Bruijn WC. Mitochondrial calcium of intact and mechanically damaged bone and cartilage cells studied with K-pyroantimonate. Histochemistry. 1979;62:325–36.
120. Morris DC, Appleton J. Ultrastructural localization of calcium in the mandibular condylar growth cartilage of the rat. Calcif Tissue Int. 1980;30:27–34.
121. Lyaruu DM, Bronckers AL, Burger EH, Woltgens JH. Localization of calcium in differentiating odontoblasts and ameloblasts before and during early dentinogenesis and amelogenesis in hamster tooth germs. J Histochem Cytochem. 1985;33:595–603.
122. Iannotti JP, Brighton CT, Stambough JL, Storey BT. Calcium flux and endogenous calcium content in isolated mammalian growth-plate chondrocytes, hyaline-cartilage chondrocytes, and hepatocytes. J Bone Joint Surg Am. 1985;67:113–20.
123. Imai K, Neuman MW, Kawase T, Saito S. Calcium in osteoblast-enriched bone cells. Bone. 1992;13:217–23.
124. Star AM, Iannotti JP, Brighton CT, Armstrong PF. Cytosolic calcium concentration in bovine growth plate chondrocytes. J Orthop Res. 1987;5:122–7.
125. Kogaya Y, Furuhashi K. Ultrastructural localization of calcium in matrix vesicles and preodontoblasts of developing rat molar tooth germs during initial dentinogenesis. Acta Anat (Basel). 1988;132:100–8.
126. Corbett EF, Michalak M. Calcium, a signaling molecule in the endoplasmic reticulum? Trends Biochem Sci. 2000;25:307–11.
127. Wang WA, Agellon LB, Michalak M. Organellar calcium handling in the cellular reticular network. Cold Spring Harb Perspect Biol. 2019;11:a038265.
128. Phillips MJ, Voeltz GK. Structure and function of ER membrane contact sites with other organelles. Nat Rev Mol Cell Biol. 2016;17:69–82.
129. Burdakov D, Petersen OH, Verkhratsky A. Intraluminal calcium as a primary regulator of endoplasmic reticulum function. Cell Calcium. 2005;38:303–10.

130. Yu R, Hinkle PM. Rapid turnover of calcium in the endoplasmic reticulum during signaling. Studies with cameleon calcium indicators. J Biol Chem. 2000;275:23648–53.
131. Morgan AJ, Platt FM, Lloyd-Evans E, Galione A. Molecular mechanisms of endolysosomal Ca2+ signalling in health and disease. Biochem J. 2011;439:349–78.
132. Bazydlo LAL, Needham M, Harris NS. Calcium, magnesium, and phosphate. Lab Med. 2014;45:e44–50.
133. Berridge MJ. The inositol trisphosphate/calcium signaling pathway in health and disease. Physiol Rev. 2016;96:1261–96.
134. Ozawa T. Modulation of ryanodine receptor Ca2+ channels (Review). Mol Med Rep. 2010;3:199–204.
135. Liou J, Kim ML, Heo WD, Jones JT, Myers JW, Ferrell JE Jr, et al. STIM is a Ca2+ sensor essential for Ca2+-store-depletion-triggered Ca2+ influx. Curr Biol. 2005;15:1235–41.
136. Zhang SL, Yu Y, Roos J, Kozak JA, Deerinck TJ, Ellisman MH, et al. STIM1 is a Ca2+ sensor that activates CRAC channels and migrates from the Ca2+ store to the plasma membrane. Nature. 2005;437:902–5.
137. Penna A, Demuro A, Yeromin AV, Zhang SL, Safrina O, Parker I, et al. The CRAC channel consists of a tetramer formed by Stim-induced dimerization of Orai dimers. Nature. 2008;456:116–20.
138. Nwokonko RM, Cai X, Loktionova NA, Wang Y, Zhou Y, Gill DL. The STIM-Orai pathway: conformational coupling between STIM and Orai in the activation of store-operated Ca(2+) entry. Adv Exp Med Biol. 2017;993:83–98.
139. Pierro C, Sneyers F, Bultynck G, Roderick HL. ER Ca(2+) release and store-operated Ca(2+) entry - partners in crime or independent actors in oncogenic transformation? Cell Calcium. 2019;82:102061.
140. Feske S. ORAI1 and STIM1 deficiency in human and mice: roles of store-operated Ca2+ entry in the immune system and beyond. Immunol Rev. 2009;231:189–209.
141. Bravo-Sagua R, Parra V, López-Crisosto C, Díaz P, Quest AF, Lavandero S. Calcium transport and signaling in mitochondria. Compr Physiol. 2017;7:623–34.
142. de Brito OM, Scorrano L. Mitofusin 2 tethers endoplasmic reticulum to mitochondria. Nature. 2008;456:605–10.
143. Stenzinger A, Schreiner D, Koch P, Hofer HW, Wimmer M. Cell and molecular biology of the novel protein tyrosine-phosphatase-interacting protein 51. Int Rev Cell Mol Biol. 2009;275:183–246.
144. De Vos KJ, Morotz GM, Stoica R, Tudor EL, Lau KF, Ackerley S, et al. VAPB interacts with the mitochondrial protein PTPIP51 to regulate calcium homeostasis. Hum Mol Genet. 2012;21:1299–311.
145. Mendes CC, Gomes DA, Thompson M, Souto NC, Goes TS, Goes AM, et al. The type III inositol 1,4,5-trisphosphate receptor preferentially transmits apoptotic Ca2+ signals into mitochondria. J Biol Chem. 2005;280:40892–900.
146. Hwang SY, Foley J, Numaga-Tomita T, Petranka JG, Bird GS, Putney JW Jr. Deletion of Orai1 alters expression of multiple genes during osteoclast and osteoblast maturation. Cell Calcium. 2012;52:488–500.
147. Robinson LJ, Mancarella S, Songsawad D, Tourkova IL, Barnett JB, Gill DL, et al. Gene disruption of the calcium channel Orai1 results in inhibition of osteoclast and osteoblast differentiation and impairs skeletal development. Lab Investig. 2012;92:1071–83.
148. Chen Y, Ramachandran A, Zhang Y, Koshy R, George A. The ER Ca(2+) sensor STIM1 can activate osteoblast and odontoblast differentiation in mineralized tissues. Connect Tissue Res. 2018;59:6–12.
149. Choi H, Srikanth S, Atti E, Pirih FQ, Nervina JM, Gwack Y, et al. Deletion of Orai1 leads to bone loss aggravated with aging and impairs function of osteoblast lineage cells. Bone Rep. 2018;8:147–55.
150. Lee SH, Park Y, Song M, Srikanth S, Kim S, Kang MK, et al. Orai1 mediates osteogenic differentiation via BMP signaling pathway in bone marrow mesenchymal stem cells. Biochem Biophys Res Commun. 2016;473:1309–14.

151. Chen Y, Koshy R, Guirado E, George A. STIM1 a calcium sensor promotes the assembly of an ECM that contains extracellular vesicles and factors that modulate mineralization. Acta Biomater. 2021;120:224–39.
152. Nurbaeva MK, Eckstein M, Concepcion AR, Smith CE, Srikanth S, Paine ML, et al. Dental enamel cells express functional SOCE channels. Sci Rep. 2015;5:15803.
153. Eckstein M, Vaeth M, Aulestia FJ, Costiniti V, Kassam SN, Bromage TG, et al. Differential regulation of Ca(2+) influx by ORAI channels mediates enamel mineralization. Sci Signal. 2019;12:eaav4663.
154. Reith EJ, Boyde A. Histochemical and electron probe analysis of secretory ameloblasts of developing rat molar teeth. Histochemistry. 1978;55:17–26.
155. Nanci A. Ten Cate's oral histology: development, structure, and function. St. Louis, MO: Elsevier; 2018.
156. Paine ML, Boyde A, Lacruz RS. Transport functions of ectoderm epithelial cells forming dental enamel. In: Hamilton KL, Devor DC, editors. Ion transport across epithelial tissues and disease. Physiology in health and disease. London: Springer; 2020. p. 363–82.
157. Diaz de Barboza G, Guizzardi S, Tolosa de Talamoni N. Molecular aspects of intestinal calcium absorption. World J Gastroenterol. 2015;21:7142–54.
158. Bleich M, Shan Q, Himmerkus N. Calcium regulation of tight junction permeability. Ann N Y Acad Sci. 2012;1258:93–9.
159. Günzel D, Yu AS. Claudins and the modulation of tight junction permeability. Physiol Rev. 2013;93:525–69.
160. Zihni C, Mills C, Matter K, Balda MS. Tight junctions: from simple barriers to multifunctional molecular gates. Nat Rev Mol Cell Biol. 2016;17:564–80.
161. Bardet C, Ribes S, Wu Y, Diallo MT, Salmon B, Breiderhoff T, et al. Claudin loss-of-function disrupts tight junctions and impairs amelogenesis. Front Physiol. 2017;8:326.
162. Hubbard MJ. Calcium transport across the dental enamel epithelium. Crit Rev Oral Biol Med. 2000;11:437–66.
163. Okumura R, Shibukawa Y, Muramatsu T, Hashimoto S, Nakagawa K, Tazaki M, et al. Sodium-calcium exchangers in rat ameloblasts. J Pharmacol Sci. 2010;112:223–30.
164. Hu P, Lacruz RS, Smith CE, Smith SM, Kurtz I, Paine ML. Expression of the sodium/calcium/potassium exchanger, NCKX4, in ameloblasts. Cells Tissues Organs. 2012;196:501–9.
165. Thibodeau S, Yang W, Sharma S, Lytton J. Calmodulin binds and modulates K(+)-dependent Na(+)/Ca(2+)-exchanger isoform 4, NCKX4. J Biol Chem. 2021;296:100092.
166. Parry DA, Poulter JA, Logan CV, Brookes SJ, Jafri H, Ferguson CH, et al. Identification of mutations in SLC24A4, encoding a potassium-dependent sodium/calcium exchanger, as a cause of amelogenesis imperfecta. Am J Hum Genet. 2013;92:307–12.
167. Franklin IK, Winz RA, Hubbard MJ. Endoplasmic reticulum Ca2+-ATPase pump is up-regulated in calcium-transporting dental enamel cells: a non-housekeeping role for SERCA2b. Biochem J. 2001;358:217–24.
168. Lacruz RS, Smith CE, Bringas P Jr, Chen YB, Smith SM, Snead ML, et al. Identification of novel candidate genes involved in mineralization of dental enamel by genome-wide transcript profiling. J Cell Physiol. 2012;227:2264–75.
169. Wiemann M, Busselberg D, Schirrmacher K, Bingmann D. A calcium release activated calcium influx in primary cultures of rat osteoblast-like cells. Calcif Tissue Int. 1998;63:154–9.
170. Inayama M, Suzuki Y, Yamada S, Kurita T, Yamamura H, Ohya S, et al. Orai1-Orai2 complex is involved in store-operated calcium entry in chondrocyte cell lines. Cell Calcium. 2015;57:337–47.
171. Nakano Y, Le MH, Abduweli D, Ho SP, Ryazanova LV, Hu Z, et al. A critical role of TRPM7 as an ion channel protein in mediating the mineralization of the craniofacial hard tissues. Front Physiol. 2016;7:258.
172. Kádár K, Juhász V, Földes A, Rácz R, Zhang Y, Löchli H, et al. TRPM7-mediated calcium transport in HAT-7 ameloblasts. Int J Mol Sci. 2021;22:3992.
173. Souza Bomfim GH, Costiniti V, Li Y, Idaghdour Y, Lacruz RS. TRPM7 activation potentiates SOCE in enamel cells but requires ORAI. Cell Calcium. 2020;87:102187.

174. Nurbaeva MK, Eckstein M, Devotta A, Saint-Jeannet JP, Yule DI, Hubbard MJ, et al. Evidence that calcium entry into calcium-transporting dental enamel cells is regulated by cholecystokinin, acetylcholine and ATP. Front Physiol. 2018;9:801.
175. Eckstein M, Vaeth M, Fornai C, Vinu M, Bromage TG, Nurbaeva MK, et al. Store-operated Ca(2+) entry controls ameloblast cell function and enamel development. JCI Insight. 2017;2:e91166.
176. Eckstein M, Aulestia FJ, Nurbaeva MK, Lacruz RS. Altered Ca(2+) signaling in enamelopathies. Biochim Biophys Acta, Mol Cell Res. 2018;1865:1778–85.
177. McCarl CA, Picard C, Khalil S, Kawasaki T, Röther J, Papolos A, et al. ORAI1 deficiency and lack of store-operated Ca2+ entry cause immunodeficiency, myopathy, and ectodermal dysplasia. J Allergy Clin Immunol. 2009;124:1311–8.
178. Wang S, Choi M, Richardson AS, Reid BM, Seymen F, Yildirim M, et al. STIM1 and SLC24A4 are critical for enamel maturation. J Dent Res. 2014;93:94S–100S.
179. Wang SK, Choi M, Richardson AS, Reid BM, Lin BP, Wang SJ, et al. ITGB6 loss-of-function mutations cause autosomal recessive amelogenesis imperfecta. Hum Mol Genet. 2014;23:2157–63.
180. Hata M, Kawamoto T, Kawai M, Yamamoto T. Differential expression patterns of the tight junction-associated proteins occludin and claudins in secretory and mature ameloblasts in mouse incisor. Med Mol Morphol. 2010;43:102–6.
181. Smith CE. Cellular and chemical events during enamel maturation. Crit Rev Oral Biol Med. 1998;9:128–61.
182. Lacruz RS, Brookes SJ, Wen X, Jimenez JM, Vikman S, Hu P, et al. Adaptor protein complex 2-mediated, clathrin-dependent endocytosis, and related gene activities, are a prominent feature during maturation stage amelogenesis. J Bone Miner Res. 2013;28:672–87.
183. Garant PR, Nagy A, Cho MI. A freeze-fracture study of ruffle-ended post-secretory ameloblasts. J Dent Res. 1984;63:622–8.
184. Lacruz RS, Habelitz S, Wright JT, Paine ML. Dental enamel formation and implications for oral health and disease. Physiol Rev. 2017;97:939–93.
185. Pandya M, Rosene L, Farquharson C, Millán JL, Diekwisch TGH. Intravesicular phosphatase PHOSPHO1 function in enamel mineralization and prism formation. Front Physiol. 2017;8:805.
186. Sasaki S, Takagi T, Suzuki M. Cyclical changes in pH in bovine developing enamel as sequential bands. Arch Oral Biol. 1991;36:227–31.
187. Lacruz RS, Hilvo M, Kurtz I, Paine ML. A survey of carbonic anhydrase mRNA expression in enamel cells. Biochem Biophys Res Commun. 2010;393:883–7.
188. Wang X, Suzawa T, Ohtsuka H, Zhao B, Miyamoto Y, Miyauchi T, et al. Carbonic anhydrase II regulates differentiation of ameloblasts via intracellular pH-dependent JNK signaling pathway. J Cell Physiol. 2010;225:709–19.
189. Shah GN, Bonapace G, Hu PY, Strisciuglio P, Sly WS. Carbonic anhydrase II deficiency syndrome (osteopetrosis with renal tubular acidosis and brain calcification): novel mutations in CA2 identified by direct sequencing expand the opportunity for genotype-phenotype correlation. Hum Mutat. 2004;24:272.
190. Wilson DF, Owen C, Mela L, Weiner L. Control of mitochondrial respiration by the phosphate potential. Biochem Biophys Res Commun. 1973;53:326–33.
191. Garcia D, Shaw RJ. AMPK: mechanisms of cellular energy sensing and restoration of metabolic balance. Mol Cell. 2017;66:789–800.
192. Merametdjian L, Beck-Cormier S, Bon N, Couasnay G, Sourice S, Guicheux J, et al. Expression of phosphate transporters during dental mineralization. J Dent Res. 2018;97:209–17.
193. Julien M, Khoshniat S, Lacreusette A, Gatius M, Bozec A, Wagner EF, et al. Phosphate-dependent regulation of MGP in osteoblasts: role of ERK1/2 and Fra-1. J Bone Miner Res. 2009;24:1856–68.
194. Bon N, Couasnay G, Bourgine A, Sourice S, Beck-Cormier S, Guicheux J, et al. Phosphate (P(i))-regulated heterodimerization of the high-affinity sodium-dependent P(i) transporters

PiT1/Slc20a1 and PiT2/Slc20a2 underlies extracellular P(i) sensing independently of P(i) uptake. J Biol Chem. 2018;293:2102–14.
195. Giovannini D, Touhami J, Charnet P, Sitbon M, Battini JL. Inorganic phosphate export by the retrovirus receptor XPR1 in metazoans. Cell Rep. 2013;3:1866–73.
196. Risteli JP, Winter WE, Kleerekoper M, Risteli L. Bone and mineral metabolism. In: Burtis C, Ashwood E, Bruns D, editors. Tietz Textbook of clinical chemistry and molecular diagnostics. 5th ed. St. Louis, MO: Elsevier; 2012. p. 1733–802.
197. Adams CS, Mansfield K, Perlot RL, Shapiro IM. Matrix regulation of skeletal cell apoptosis: role of calcium and phosphate ions. J Biol Chem. 2001;276:20316–22.
198. Beck-Cormier S, Lelliott CJ, Logan JG, Lafont DT, Merametdjian L, Leitch VD, et al. Slc20a2, encoding the phosphate transporter PiT2, is an important genetic determinant of bone quality and strength. J Bone Miner Res. 2019;34:1101–14.
199. Couasnay G, Bon N, Devignes CS, Sourice S, Bianchi A, Veziers J, et al. PiT1/Slc20a1 is required for endoplasmic reticulum homeostasis, chondrocyte survival, and skeletal development. J Bone Miner Res. 2019;34:387–98.
200. Michigami T, Kawai M, Yamazaki M, Ozono K. Phosphate as a signaling molecule and its sensing mechanism. Physiol Rev. 2018;98:2317–48.
201. Battini J-L, Rasko JEJ, Miller AD. A human cell-surface receptor for xenotropic and polytropic murine leukemia viruses: possible role in G protein-coupled signal transduction. Proc Natl Acad Sci U S A. 1999;96:1385–90.
202. Voglmaier SM, Bembenek ME, Kaplin AI, Dorman G, Olszewski JD, Prestwich GD, et al. Purified inositol hexakisphosphate kinase is an ATP synthase: diphosphoinositol pentakisphosphate as a high-energy phosphate donor. Proc Natl Acad Sci U S A. 1996;93:4305–10.
203. Wilson MS, Jessen HJ, Saiardi A. The inositol hexakisphosphate kinases IP6K1 and -2 regulate human cellular phosphate homeostasis, including XPR1-mediated phosphate export. J Biol Chem. 2019;294:11597–608.
204. López-Sánchez U, Nicolas G, Richard A-C, Maltête D, Charif M, Ayrignac X, et al. Characterization of XPR1/SLC53A1 variants located outside of the SPX domain in patients with primary familial brain calcification. Sci Rep. 2019;9:6776.
205. Legati A, Giovannini D, Nicolas G, López-Sánchez U, Quintáns B, Oliveira JR, et al. Mutations in XPR1 cause primary familial brain calcification associated with altered phosphate export. Nat Genet. 2015;47:579–81.
206. Li X, Gu C, Hostachy S, Sahu S, Wittwer C, Jessen HJ, et al. Control of XPR1-dependent cellular phosphate efflux by InsP8 is an exemplar for functionally-exclusive inositol pyrophosphate signaling. Proc Natl Acad Sci U S A. 2020;117:3568–74.
207. Landis WJ, Glimcher MJ. Electron optical and analytical observations of rat growth plate cartilage prepared by ultracryomicrotomy: the failure to detect a mineral phase in matrix vesicles and the identification of heterodispersed particles as the initial solid phase of calcium phosphate deposited in the extracellular matrix. J Ultrastruct Res. 1982;78:227–68.
208. Appleton J, Lyon R, Swindin KJ, Chesters J. Ultrastructure and energy-dispersive x-ray microanalysis of cartilage after rapid freezing, low temperature freeze drying, and embedding in Spurr's resin. J Histochem Cytochem. 1985;33:1073–9.
209. Pozzan T, Rudolf R. Measurements of mitochondrial calcium in vivo. Biochim Biophys Acta. 2009;1787:1317–23.
210. Verkhratsky A. Physiology and pathophysiology of the calcium store in the endoplasmic reticulum of neurons. Physiol Rev. 2005;85:201–79.
211. Patel S, Cai X. Evolution of acidic Ca^{2+} stores and their resident Ca^{2+}-permeable channels. Cell Calcium. 2015;57:222–30.
212. Vale RD. The molecular motor toolbox for intracellular transport. Cell. 2003;112:467–80.
213. Chevalier-Larsen E, Holzbaur ELF. Axonal transport and neurodegenerative disease. Biochim Biophys Acta. 2006;1762:1094–108.
214. Leduc C, Padberg-Gehle K, Varga V, Helbing D, Diez S, Howard J. Molecular crowding creates traffic jams of kinesin motors on microtubules. Proc Natl Acad Sci U S A. 2012;109:6100–5.

215. Ross JL, Shuman H, Holzbaur EL, Goldman YE. Kinesin and dynein-dynactin at intersecting microtubules: motor density affects dynein function. Biophys J. 2008;94:3115–25.
216. Ross JL. The impacts of molecular motor traffic jams. Proc Natl Acad Sci U S A. 2012;109:5911–2.
217. De Stefani D, Bononi A, Romagnoli A, Messina A, De Pinto V, Pinton P, et al. VDAC1 selectively transfers apoptotic Ca2+ signals to mitochondria. Cell Death Differ. 2012;19:267–73.
218. Cardenas C, Miller RA, Smith I, Bui T, Molgo J, Muller M, et al. Essential regulation of cell bioenergetics by constitutive InsP3 receptor Ca2+ transfer to mitochondria. Cell. 2010;142:270–83.
219. Bernardi P. Mitochondria in muscle cell death. Ital J Neurol Sci. 1999;20:395–400.
220. Bonora M, Wieckowski MR, Chinopoulos C, Kepp O, Kroemer G, Galluzzi L, et al. Molecular mechanisms of cell death: central implication of ATP synthase in mitochondrial permeability transition. Oncogene. 2015;34:1475–86.
221. Chalmers S, Nicholls DG. The relationship between free and total calcium concentrations in the matrix of liver and brain mitochondria. J Biol Chem. 2003;278:19062–70.
222. Chung CH, Golub EE, Forbes E, Tokuoka T, Shapiro IM. Mechanism of action of beta-glycerophosphate on bone cell mineralization. Calcif Tissue Int. 1992;51:305–11.
223. Tada H, Nemoto E, Foster BL, Somerman MJ, Shimauchi H. Phosphate increases bone morphogenetic protein-2 expression through cAMP-dependent protein kinase and ERK1/2 pathways in human dental pulp cells. Bone. 2011;48:1409–16.
224. Orriss IR, Hajjawi MO, Huesa C, MacRae VE, Arnett TR. Optimisation of the differing conditions required for bone formation in vitro by primary osteoblasts from mice and rats. Int J Mol Med. 2014;34:1201–8.
225. Ma K, Liu P, Al-Maghout T, Sukkar B, Cao H, Voelkl J, et al. Phosphate-induced ORAI1 expression and store-operated Ca(2+) entry in aortic smooth muscle cells. J Mol Med (Berl). 2019;97:1465–75.
226. Boskey AL, Guidon P, Doty SB, Stiner D, Leboy P, Binderman I. The mechanism of beta-glycerophosphate action in mineralizing chick limb-bud mesenchymal cell cultures. J Bone Miner Res. 1996;11:1694–702.
227. Patel JJ, Bourne LE, Davies BK, Arnett TR, MacRae VE, Wheeler-Jones CP, et al. Differing calcification processes in cultured vascular smooth muscle cells and osteoblasts. Exp Cell Res. 2019;380:100–13.
228. Pelzl L, Sahu I, Ma K, Heinzmann D, Bhuyan AAM, Al-Maghout T, et al. Beta-glycerophosphate-induced ORAI1 expression and store operated Ca(2+) entry in megakaryocytes. Sci Rep. 2020;10:1728.
229. Biber J, Hernando N, Forster I. Phosphate transporters and their function. Annu Rev Physiol. 2013;75:535–50.
230. Beck L. Expression and function of Slc34 sodium-phosphate co-transporters in skeleton and teeth. Pflugers Arch. 2019;471:175–84.
231. Boyde A, Shapiro IM. Energy dispersive X-ray elemental analysis of isolated epiphyseal growth plate chondrocyte fragments. Histochemistry. 1980;69:85–94.
232. Bourgine A, Pilet P, Diouani S, Sourice S, Lesoeur J, Beck-Cormier S, et al. Mice with hypomorphic expression of the sodium-phosphate cotransporter PiT1/Slc20a1 have an unexpected normal bone mineralization. PLoS One. 2013;8:e65979.
233. Yamada S, Wallingford MC, Borgeia S, Cox TC, Giachelli CM. Loss of PiT-2 results in abnormal bone development and decreased bone mineral density and length in mice. Biochem Biophys Res Commun. 2018;495:553–9.
234. Yamazaki M, Kawai M, Kinoshita S, Tachikawa K, Nakanishi T, Ozono K, et al. Clonal osteoblastic cell lines with CRISPR/Cas9-mediated ablation of Pit1 or Pit2 show enhanced mineralization despite reduced osteogenic gene expression. Bone. 2021;151:116036.
235. Geng Y, Mosyak L, Kurinov I, Zuo H, Sturchler E, Cheng TC, et al. Structural mechanism of ligand activation in human calcium-sensing receptor. elife. 2016;5:e13662.

236. Figueres L, Beck-Cormier S, Beck L, Marks J. The complexities of organ crosstalk in phosphate homeostasis: time to put phosphate sensing back in the limelight. Int J Mol Sci. 2021;22:5701.
237. Kriel J, Haesendonckx S, Rubio-Texeira M, Van Zeebroeck G, Thevelein JM. From transporter to transceptor: signaling from transporters provokes re-evaluation of complex trafficking and regulatory controls: endocytic internalization and intracellular trafficking of nutrient transceptors may, at least in part, be governed by their signaling function. Bioessays. 2011;33:870–9.
238. Beck-Cormier S, Beck L. The need of a paradigm shift to better understand PiT1 and PiT2 biology: response to "Why is there no PiT1/SLC20A1 pathogenic variants yet linked to primary familial brain calcification?". J Bone Miner Res. 2020;35:825–6.

Calcium and Phosphate Ion Efflux from Cells: The Roles of Matrix Vesicles, Extracellular Vesicles, and Other Membrane-invested Transporters in Vertebrate Hard Tissue Mineralization

6

Secretion is widespread in all eukaryotic cells: all of us experience this in the course of daily life – saliva, mucus, sweat, tears, bile juice, adrenalin, etc. – the list is extremely long. How does a cell manage to repeatedly spit out some stuff without losing the rest? The answer is: through regulated vesicle trafficking within the cell. [1]

Glossary

Acidocalcisome A lysosome-like organelle rich in calcium ions and polyphosphates.

Amphisome A vesicle formed by fusion of an endosome and an autophagosome.

Annexin A family of calcium ion-regulated, phospholipid-binding proteins involved in exocytosis and stabilization of the plasma membrane of a cell.

Argonaute A protein family which functions in RNA silencing processes.

Autophagolysosome A vesicle formed by fusion of a lysosome with an autophagosome.

Autophagosome A double-membrane autophagic vesicle containing cytoplasmic damaged or non-functional proteins and organelles.

Autophagy The intracellular system for ensuring cell survival through removal of degenerate organelles and non-functional proteins.

Caveolin An integral membrane protein involved in receptor-independent endocytosis.

Class III phosphatidylinositol 3-kinase One of a number of enzymes which regulate autophagy, endosomal trafficking and vesicular transport by promoting the formation of the signaling lipid, phosphatidylinositol 3-phosphate.

Clathrin A protein involved with the formation of coated vesicles.

Coccolithophorid A unicellular eukaryotic phytoplankton which contains chloroplasts and generates a spherical-shaped exoskeleton (coccosphere) comprised of plates or scales of calcium carbonate (coccoliths).

Docking The process of vesicle delivery and vesicle membrane fusion and attachment to the plasma membrane of a cell.

I. M. Shapiro, W. J. Landis, *Mechanisms of Mineralization of Vertebrate Skeletal and Dental Tissues*, https://doi.org/10.1007/978-3-031-34304-9_6

EF-hand proteins (calbindin, nucleobindin, calreticulin) Proteins which have a helix-loop-helix structural binding motif and, for these proteins, a high affinity for calcium.

Endomembrane A membrane system comprising the endoplasmic reticulum, Golgi apparatus, endosomes, cell membrane and nuclear envelope, but not including the membranes of mitochondria.

Endoplasmic reticulum An organelle network of flattened membranes involved with protein and lipid synthesis and calcium storage in a cell.

Endosome A membrane-bound compartment comprising the endocytic vesicular transport system in the cell and derived from the trans-Golgi apparatus.

Exosomes Extracellular vesicles (~10 to 100 nm in diameter) derived from endosomes which are released following fusion of multivesicular bodies with the plasma membrane.

Extracellular vesicles Membrane-invested organelles, comprising both exosomes and microvesicles, released from cells into extracellular tissue spaces.

Flippase, floppase, scramblase Enzymes which move phospholipids between the outer and inner leaflets of the plasma membranes of a cell.

Fusion Melding of a vesicle membrane with the cell membrane.

Golgi apparatus A cellular organelle involved with transport and modification of proteins and the formation of a variety of cellular vesicles.

Intraluminal bodies Vesicles formed within a late-stage endosome.

Lipid raft Microdomains in a cell membrane which are enriched in cholesterol and glycosphingolipids.

Lysosome A vacuole involved with degradation of proteins, carbohydrates and lipids in a cell.

Lysosome-related organelle A specialized secretory organelle which shares many features of lysosomes.

Matrix vesicles Membrane-bound organelles present in the extracellular matrices of skeletal tissues and involved with mineralization.

MicroRNA A small non-coding RNA molecule.

Microvesicles Bioactive small vesicles (~100 to 1000 nm in diameter) released from cells and derived from the plasma membrane.

Multivesicular body An intracellular organelle formed from the aggregation of late-stage endosomes and which fuses with the plasma membrane to liberate intraluminal vesicles/exosomes to the extracellular matrix.

Phagophore A double-membrane structure formed in the cytosol that initiates autophagosome formation.

Phospholipases A class of enzymes which hydrolyze phospholipids, including phospholipase A, which removes acyl (fatty acid) groups to form lysophospholipids; phospholipase C, which generates diacylglycerol and inositol triphosphate from phosphatidylinositol; and phospholipase D, which hydrolyzes the phosphodiester bond of phosphatidylcholine to form phosphatidic acid.

Phospholipids (phosphatidylserine, phosphatidylethanolamine, phosphatidylcholine, sphingomyelin, phosphatidylinositol) A class of lipids present in the cell membrane, comprised of two fatty acid acyl groups and a phosphorylated hydrophilic (often charged) head group.

RAS A protein superfamily of small GTPases whose name originates from "*ra*t *s*arcoma virus" and consisting of five principal families of RAS, RHO, RAN, RAB and ARF GTPases.

SNARE A soluble NSF (*N*-ethylmaleimide-sensitive factor) accessory protein receptor which regulates endosome trafficking in the cell.

Syndecan A transmembrane heparan sulfate proteoglycan co-receptor for G-protein receptors involved with cell signaling.

Syntenin A molecule which binds to syndecan and is involved with exosome secretion.

Tethering A mechanism by which vesicles become linked to the cell membrane.

Tetraspanins Molecular scaffolding proteins which control receptor and integrin clustering and mediate linkages to the cytoskeleton.

Trafficking A process by which vesicles are transported to the cell membrane along microtubules of the cytoskeleton.

Abbreviations and Proper Names for Genes/Proteins and Other Molecules

ADP	Adenosine diphosphate
ALIX	ALG-2-interacting protein X
AMP	Adenosine monophosphate
ARF	ADP-ribosylation factor, a protein family within the RAS superfamily of small GTPases
ATG	Autophagy protein, such as ULK1 (ATG1), BECLIN-1 (ATG6), and LC3 (ATG8)
ATP	Adenosine triphosphate
BMP	Bone morphogenetic protein
ECM	Extracellular matrix
ENPP	Ectonucleotide triphosphate pyrophosphatase/phosphodiesterase
ERK	Extracellular signal-regulated kinase
ESCRT	Endosomal sorting complexes required for transport
FIP200	Focal adhesion kinase family interacting proteins
GTP	Guanosine triphosphate
GW182	A 182 Kd protein having numerous glycine (G) and tryptophan (W) residues and a component of GW/P-bodies involved in RNA processing and silencing
HIF	Hypoxia-inducible factor
LAMP	Lysosome-associated membrane protein
LC3	Microtubule-associated protein 1A/1B light chain 3/ATG8
LRP	Low-density lipoprotein receptor
miRNA/MIR	MicroRNA
MLCK	Myosin light-chain kinase
MMP	Matrix metalloproteinase

mRNA	Messenger RNA
MTOR	Mammalian target of rapamycin complex
NCX	Sodium-calcium exchange protein
ORAI1	Olfactory receptor class A related 1
P2, P2X and P2Y	Two families of purinergic G protein-coupled receptors
PHOSPHO1	Phosphoethanolamine/phosphocholine phosphatase 1
PMCA	Plasma membrane calcium ATPase
RAB	*Ra*s-related proteins in *b*rain; protein members of the RAS superfamily of small GTPases
RANK	Receptor activator of nuclear factor kappa-B
RANKL	Receptor activator of nuclear factor kappa-B ligand
RHO	*R*AS *ho*mologous protein; a protein family within the RAS superfamily of small GTPases
SNARE	Soluble *N*-ethylmaleimide-sensitive fusion protein-attachment protein receptor
SOCE	Store-operated calcium entry
STIM1	Stromal interaction molecule 1
TNAP	Tissue non-specific alkaline phosphatase
TRPV	Transient receptor potential cation channel subfamily V
ULK	Unc-51 like autophagy activating kinase
WNT	Wingless-related integration site

6.1 Some Thoughts on the Evolution of Biological Mineralization

At this juncture in the book, it is worth noting that many organisms generate mineralized structures. Studies of relatively simple invertebrate organisms such as calcifying bacteria, coccolithophores, calcareous sponges, corals, mollusks, and echinoderms provide a wide variety of unique insights into almost every aspect of their mineralization mechanisms. Since a great deal is known about these organisms and the interesting facets of their invertebrate biology, the question may be asked as to what can be learned from these creatures which would shed light on fundamental ideas pertaining directly to the mineralization of bone, cartilage, the tooth, and other vertebrate tissues.

The types of minerals formed by these simpler organisms are amazingly eclectic, and they range from complexes of iron and magnesium to silica and calcium (Table 6.1) [2]. The wide variation in mineral types formed by organisms across the evolutionary tree raises the question as to whether the creatures comprising each phylum have developed a unique mineralization mechanism (divergent evolution). Since such a situation is unlikely, a more reasonable assumption is that the mineralization of bones, teeth, shells, spines, and other types of calcified structures has evolved through a conserved pathway (convergent evolution). In support of this concept, studies of mineral formation in many organisms indicate that there is expression and secretion of common subsets of numerous molecules, including

Table 6.1 A few common forms of biominerals in the animal kingdom

Mineral	Formula	Organism
Calcite	$CaCO_3$	Coccolithophores
		Mollusks
		Foraminifera
Magnesium calcite	$(Mg,Ca)CO_3$	Echinoderms
		Corals
Aragonite	$CaCO_3$	Corals
		Mollusks
		Gastropods
		Cephalopods
Silica	SiO_2	Diatoms
		Radiolarians
		Sponges
Magnetite	Fe_3O_4	Bacteria
		Limpets
		Chitons

A more extensive list of the diversity of biominerals among animals may be found in Lowenstam and Weiner [2]

proteins, phosphorylated proteins, and polysaccharides. In addition, there is the utilization of similar intracellular ion trafficking pathways in the different organisms. Together, the spectrum of genomic, proteomic, and cellular observations suggests the existence of a shared transit system across the evolutionary tree leading to mineral deposition. Accordingly, information from lower organisms should provide insights into the mechanisms used by vertebrates to generate the mineral components of their skeletal and dental tissues.

With this comment about divergent and convergent evolution as a preamble, this chapter begins by asking what information characterizing invertebrate life promotes the view that the mineralization of bone, cartilage, the tooth, or other vertebrate tissues is an intracellular event. In other words, is there an evolutionary basis for the overarching hypothesis offered in this book that biological mineralization is produced and directed by the cell?

Far removed from the structure of bone, cartilage, and the teeth is the coccolithophorid or coccolithophore, a phytoplankton or alga which populates the oceans of the Earth. This tiny sea creature (of diameter less than ~100 μm) with its very long name generates an exoskeleton of plates or scales, coccoliths, on the surface of the cell, comprised of at least two isomorphic forms of calcite (calcium carbonate) together with an amorphous precursor phase of this mineral [3]. Several coccoliths form a coccosphere, as an organized, complex cluster of externally mineralized plates which encloses two chloroplasts, a nucleus, and other organelles of the coccolithophore (Fig. 6.1). Coccolithophores have generated great interest because of their environmental importance: During photosynthesis, they provide a mechanism to fix carbon dioxide, a gas accumulating in the atmosphere and widely recognized

Fig. 6.1 Scanning electron micrograph of the small, round coccolith, *Pleurochrysis,* aggregated to form individual coccospheres. Coccoliths represent invertebrate organisms which produce calcium carbonate (calcite) mineral mediated through activities of their Golgi apparatus. These organisms concentrate calcium and carbonate ions from their aquatic environment. The ions originate in an intracellular vesicle and are secreted to the cell surface. Scale bar = 1 μm. (Image from Hawkins et al. [5] with permission)

as a principal cause of global warming and weather change. In the sea, gigatons of coccolithophores and their mineralized coccoliths drop to the ocean bed and form chalk sediments. As the seabed is exposed and uplifted through geologic events, coccoliths become evident as white chalk cliffs. Indeed, the white cliffs of Dover, England, are the product of overwhelming numbers of promiscuous coccolithophores.

What clues can be extracted from coccolithophore biology which relate and help explain the mechanism of bone, tooth, and vertebrate mineralization in general? Like osteoblasts and odontoblasts, the coccolithophore plasma membrane provides an almost impermeable barrier to dissolved calcium ions. Calcium-hydrogen exchange proteins present in the plasma membrane promote the passage of calibrated quantities of calcium into these cells. The ions accumulate within the membrane cisternae of coccolith vesicles, Golgi-derived structures (also called coccolith vesicle-reticular bodies) [4, 5]. Mitochondria rapidly scavenge excess calcium ions which channel into the cell while bicarbonate also enters the cell through the plasma membrane by means of a bicarbonate-chloride (HCO_3^-/Cl^-) antiporter system [6]. Transporter proteins extract protons from the carbonate anions, thereby causing an increase in the pH of the coccolith vesicle and a condition which favors calcium carbonate particle formation on protein templates. Vesicles and their enclosed mineral particles are trafficked to the cell surface where they fuse to form the coccolith calcite plates surrounding and enveloping the organism [7]. While this description of coccolith mineralization is abbreviated, it is clear that mineralization starts as an intracellular event in the coccolithophore and involves the Golgi, mitochondria and other organelles, regulated calcium accumulation and sequestration, vesicle transport, and vesicle and cell membrane fusion.

A second example of invertebrate cell-mediated mineralization is the formation of the calcium carbonate skeleton of the sea urchin. In these prickly sea creatures, mineralized spicules develop rapidly in the embryo soon after fertilization [8]. Several different cell types appear to be involved with spicule formation, and Wilt emphasized the secretory nature and role of the primary mesenchymal cells in spicule development in sea urchin embryos [9]. He commented that these cells are extraordinary "factories" engaged in the secretion of not only mineralized spicules but also other extracellular components which include collagen and proteoglycans [9]. Thus, calcium carbonate granules (20–30 nm in diameter) are found inside vesicles of both mesenchymal and epithelial cells [10, 11]. Vidavsky et al. described these cells "as a highway for transporting vesicles, including mineral-bearing vesicles,…which may well be the source of mineral for the spicules." [12]. Close examination of the vesicles indicates that they are related to multivesicular bodies, which are characteristic of the endosomal-lysosome system of higher organisms (Fig. 6.2) [11, 13]. Mineral present in such vesicles has been characterized as an amorphous calcium carbonate or vaterite, both of which are highly unstable phases, and such particles were noted to function as ion storage sites, providing sources of ions for shell formation [12]. Together with coccolithophore biology, the morphologic analyses of the sea urchin and other invertebrates point to the importance of intravesicular bodies and endosomal trafficking as mechanisms for concentrating calcium ions for extracellular mineral formation.

This chapter began by suggesting an evolutionary approach to understanding the mechanisms of mineralization. While the organisms briefly presented above

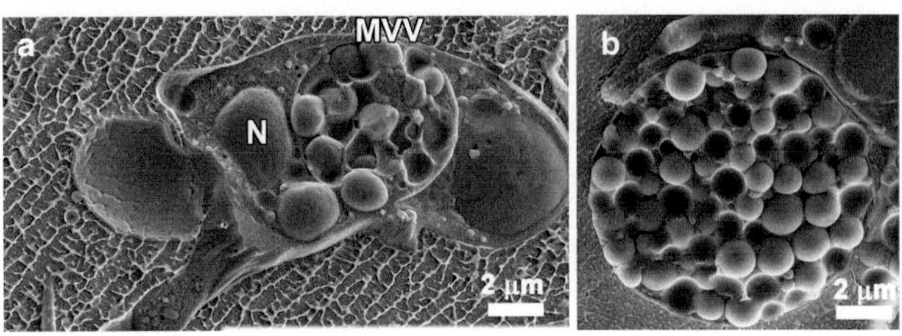

Fig. 6.2 Scanning electron microscopy of calcifying sea urchin embryos. These invertebrate organisms rapidly develop a calcite skeleton from calcium carbonate deposits evident within intracellular membranous vesicles. (**a**,**b**) Low temperature (−120 °C) scanning electron microscopy of a sea urchin embryo prepared by high freeze freezing and freeze fracture. The embryo reveals a nucleus (N) and an adjacent spherical body enclosing several small vesicles (multivesicular vesicles, MVV). These structures are labeled with calcein as a marker for calcium and yield peaks of calcium carbonate on Raman spectroscopy [12]. (Images from Vidavsky et al. [12] with permission)

(coccolithophores and sea urchins) are phylogenetically distant from mammalian skeletal and dental tissues, there is yet much to be learned from these lower-order invertebrates concerning the mechanisms of vertebrate mineral formation. First, although in most cases the type of mineral formed by these and other marine animals is distinct from apatite of vertebrates, it shares a critical common feature in that an amorphous phase precedes the appearance of intermediate phases and the final, mature crystalline mineral. Thus, both mineralized invertebrate and vertebrate tissues generate their respective ultimate crystalline mineral form through a series of phase transitions from an initial amorphous state. Second, from a mechanistic point of view, it is clear that both invertebrates and vertebrates maintain stringent control of cellular calcium levels. Like coccoliths as well as other marine species, the mineralized vertebrate tissues utilize a Golgi-endoplasmic reticulum internal membrane system, supported by mitochondria, to play a central role in calcium homeostasis. These activities are principally accomplished by accumulating, sequestering, and transporting calcium ions to prevent their overload in an otherwise calcium-rich intracellular environment. In this manner, both simpler and more well-developed species have evolved and incorporated systems which maintain low intracellular levels of the cation for signaling processes, while providing high extracellular levels of calcium ions for biomineralization. Third, the observation that invertebrates concentrate calcium ions in vesicles adds further strength to the idea that a similar packaging system is engaged with the initiation of the mineralization process throughout all species. Finally, in some cases, the mineral of invertebrates is deposited in an extracellular matrix, which is similar to the collagens and proteoglycans of vertebrate mineralized tissues. From all these perspectives (mineral maturation, intracellular calcium transport, vesicle function, and basic extracellular structure), then, findings from studies of these lower-order organisms provide an evolutionary basis for many of the key steps and events involved in the process of mineralization of bone, cartilage, teeth, and other hard tissues.

6.2 Efflux of Calcium and Phosphate Ions from Cells

Based on studies of ion transport in tissues in general and specific information gained from investigations of a range of mineralizing tissues, this chapter next addresses the trafficking of calcium and phosphate ions within a cell and from the cell to its extracellular milieu. An important focus of the chapter is the examination of the premise that the first observed nidus of mineral occurs in structures originally identified as extracellular matrix vesicles. In contrast to this supposition, consideration is also given to the concept that the discharge of calcium ions from intracellular storage organelles into a dense extracellular matrix causes mineral (apatite) formation in vertebrate tissues. In other words, rather than being an exclusive extracellular event mediated by matrix vesicles, vertebrate mineralization begins as an intracellular event marked by the accumulation of calcium and phosphate ions in vesicles which are then secreted into the extracellular matrix as ion clusters, ion pairs, or a putative early mineral phase of amorphous calcium phosphate.

One possible caveat to the proposal that intracellular organelles participate in the mineralization process is that mitochondria and the endoplasmic reticulum serve other functions which include calcium ion homeostasis in the cytosol. When cytosolic calcium ion levels are disturbed, whether through seismic shifts or just minor periodic oscillations, powerful and complex signaling systems are activated which impact many critical cellular functions. In cells of mineralizing tissues, these ion movements mediate changes in gene expression, metabolic activity, and cell survival and death. To avoid interfering with these other functions, Hubbard suggested that there was a division of calcium ions fated for mineral deposition from those linked to signaling activities [14]. A means of achieving such a division is to increase the abundance of calcium-binding proteins. Another possibility is to compartmentalize the calcium ions such that ions destined for mineralization are partitioned from those required for signaling and other activities. The latter is accomplished by anatomical separation of the two systems and by organelle-dependent removal of excess cytosolic calcium ions, thereby expediting the return to baseline calcium values. The initial accumulation of calcium ions is suggested here to promote energy-dependent anabolic reactions related to the synthesis of proteins trafficked or directed to the extracellular matrix while the release of sequestered calcium from intracellular storage organelles is the primary mediator of the mineralization process. In this manner, the cell may be involved in the synthesis and secretion of the proteins for both formation and subsequent mineralization of the extracellular matrix.

Mechanisms of control of the influx and intracellular storage of calcium ions have been discussed in some detail in Chap. 5. Ion efflux involves the trafficking and release of calcium ions from intracellular storage organelles as well as the activities of plasma membrane pumps and channels. As a component of the calcium ion efflux system, the endoplasmic reticulum performs a dual function: Refilling of the cytosol with calcium ions and trafficking of calcium ions from the endoplasmic reticulum to a host of different subcellular constituents, including mitochondria, lysosomes, endosomes, secretory vacuoles, and structures, identified as lysosome-related organelles [15]. Many of these intracellular components with their enclosed calcium ions are transported to the cell plasma membrane and, from there, the ions are translocated to the extracellular matrix. Additionally at the plasma membrane, protein pumps channel the cations outside the cell and into the extracellular matrix.

It has been pointed out in Chap. 5 that sodium-calcium exchange proteins (NCX/NCKX) and a family of plasma membrane calcium ATPases (PMCA) are the two most important membrane pumps which drive calcium ions into and from cells of vertebrate mineralized tissues [16, 17]. NCX proteins bind calcium ions even when the cytosolic calcium concentration is low, and they rapidly pump the ions from the cell to the extracellular milieu. This high-capacity system is capable of transporting 5000 calcium ions per second [18]. Pumping is not dependent on ATP but instead it piggybacks on the electrochemical gradient generated by the difference in the sodium ion concentration across membranes. Thus, as sodium ions slide down the gradient and into the cell, there is a counter transport of calcium ions. In other words, as sodium ions enter the cell, calcium ions are exported. From an energetic perspective, it takes three sodium ions to transport a single calcium ion.

Osteoblasts express both NCX and PMCA membrane pumps (Fig. 6.3) [17, 19]. The distribution and function of NCX were determined in cultured osteoblasts in critical studies which involved immunolabeling and inhibition of the NCX family member, NCX3 [17, 19]. Inhibition blocked the efflux of calcium ions from osteoblasts and, most importantly, diminished the capacity of these cells to mineralize

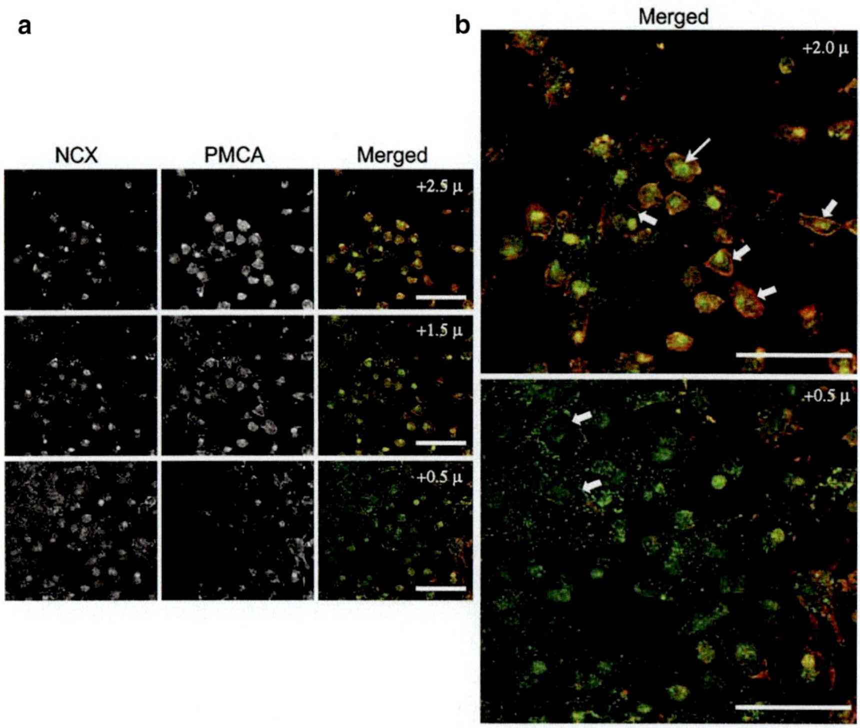

Fig. 6.3 Cellular immunolocalization by fluorescent confocal microscopy of NCX (sodium/calcium exchanger) and PMCA (plasma membrane calcium ATPase) transmembrane proteins in cultured primary chicken osteoblasts. The two calcium transporters were detected following specific antibody staining for NCX (green fluorescence) and PMCA (red fluorescence). (**a**) Detection of NCX, PMCA, and both proteins viewed as merged images (horizontal rows) within the same cells is shown in each horizontal row. Optical cross sections (*x*, *y* plane) of a field of cells presented at three different *z*-axis distances from a fixed reference point are denoted in microns in each vertical column of images. (**b**) Enlargements of merged images at 2.0 and 0.5 μm from the reference point show both NCX and PMCA at the plasma membrane of osteoblasts (thick arrows). Localization of the two proteins was found to be different in that NCX was associated with the plasma membrane abutting the osteoid secreted by the cells while PMCA was present on the plasma membrane facing the extracellular matrix (culture medium). The two transporters are then distributed in a polar manner, principally over surfaces of osteoblast plasma membranes which oppose each other. In this circumstance, osteoblasts express transporter proteins that promote calcium translocation from the cells to their immediate extracellular environment, either to the fluid milieu (PMCA) or osteoid (NCX). PMCA expression peaks early in the culture period while the level of expression of NCX is lower than that of PMCA and NCX is maintained throughout the osteoblast differentiation period [19]. Staining of one cell in the field (**b**, thin arrow) was identified as a fixation artifact [19]. Scale bars **a** = 50 μm, **b** = 25 μm. (Images from Stains et al. [19] with permission)

their bone matrix [19]. While these observations might be the result of disturbing other activities of the cultured cells, the loss of mineralization activity did not appear to alter osteoblast morphology, viability, or differentiation status [17, 19]. These reports [17, 19] have led to the conclusion that there is a direct linkage between NCX transporter-dependent calcium ion efflux and mineral deposition in bone and presumably other vertebrate mineralized tissues.

If NCX3 is the membrane pump required for mineral deposition in bone, then its location should be close to the mineralization front. Indeed, immunocytochemical staining with an antibody which recognizes NCX3 indicated that the protein was present in the osteoblast plasma membrane, which faces the bone mineralization front (Fig. 6.4) [19]. In the dental tissues, odontoblasts express both NCX3 and another NCX family member, NCX1 [20]. Immunolocalization showed that these proteins were positioned close to, or in, the membrane of odontoblastic processes which penetrated the mineralized matrix of the tooth [20]. Moreover, there is functional coupling between NCX and members of the transient receptor potential cation channel subfamily V (TRPV) channel proteins, TRPV1, TRPV2, and TRPV4 [21]. Accordingly, while some differences exist, there are similarities in the localization and ascribed importance of the NCX membrane pump in both odontoblasts and osteoblasts. Regarding enamel, calcium ion efflux in the formative stages of this tissue has received some study. NCX family members NCX1 and NCX3 are present in ameloblasts [22], and the sodium/calcium-potassium exchanger, NCX4, is

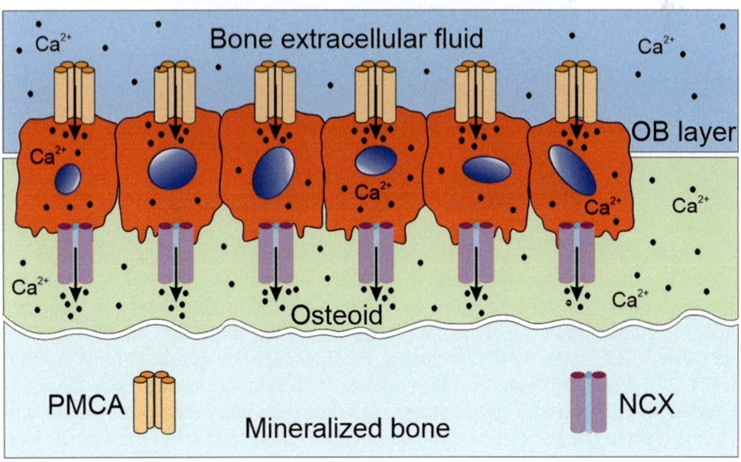

Fig. 6.4 Diagrammatic representation of PMCA and NCX protein pumps associated with osteoblasts (OB). Stains et al. [19] envision that the cell layer mediates mineralization by acting as an interface between the extracellular fluid and osteoid of the tissue. PMCA pumps bring calcium ions (Ca^{2+}, small black circles) into osteoblasts from the extracellular milieu (arrows) and maintain ion homeostasis in the cells. After processing, calcium ions are pumped out of the cells through NCX pumps located at the cell surface facing the osteoid (arrows). The darker blue structures in each osteoblast indicate cell nuclei. (Diagram adapted from Stains et al. [19] with permission)

expressed at moderately high levels and is upregulated during enamel maturation (Fig. 6.5) [23, 24].

From these analyses, there is convincing evidence that *NCX* family members are expressed in cells mediating the mineralization process in both bone and dental tissues. Based on their localization and other data, it is probable that NCX proteins have a significant role in mediating cell calcium ion efflux, generate high local levels of calcium ions in the extracellular matrices, and participate in the deposition of

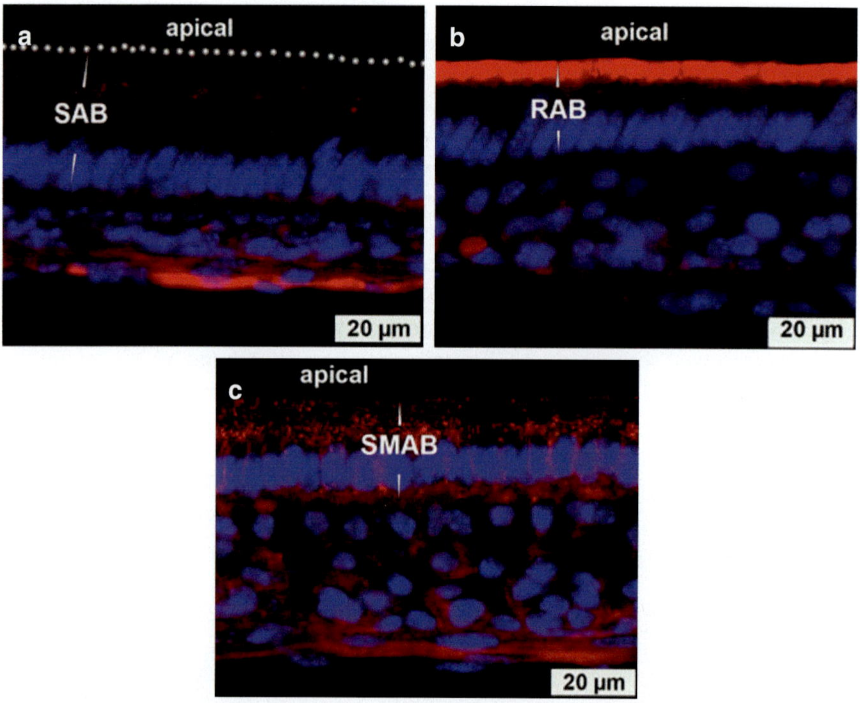

Fig. 6.5 Immunocytochemical localization of the sodium/calcium exchanger (NCX4) in the continously growing incisors from postnatal 7-week-old mice. NCX (or NCKX, sodium/calcium-potassium exchanger) proteins function as bidirectional transporters which regulate cellular transmembrane influx and efflux of calcium ions [23]. Of the six NCX(1-6) proteins encoded by the solute carrier (*SLC*) gene family in enamel, the most highly expressed are NCX3 and NCX4 isoforms [23, 24]. While substantially different from the developing human tooth, the continuously erupting rodent incisor provides temporal and spatial insights into events which occur when ameloblasts become polarized and secrete enamel proteins. Hemimandibles of the animals were decalcified, and sagittal sections of the tissues were treated for antigen retrieval, blocked, incubated with mouse anti-NCX4 antibody, and then exposed to a Fast Red complex. The latter demonstrates by red staining the level of *NCX4* expression in the sections. (**a**) The NCX4 signal was relatively weak in secretory ameloblasts (SAB), (**b**) concentrated principally at the apical surface of ruffle-ended maturation ameloblasts (RAB), and (**c**) distributed sparsely in the central cytosolic region of smooth-ended ameloblasts (SMAB). Cell nuclei are stained blue. The staining and additional data indicate that NCX4 serves an important regulatory role in calcium transport during amelogenesis and that ruffle-ended ameloblasts are particularly active in moving calcium ions into the enamel matrix [24]. (Images adapted from Chan et al. [24] with permission)

mineral. More than likely, they participate in the activities closely associated with the maintenance of cell calcium ion homeostasis, enabling essential signaling functions to be sustained. Through these different proteins and their specific functionalities, then, the cell can partition and compartmentalize cytosolic calcium so that it maintains calcium ion homeostasis and intracellular signaling activity while mediating calcium efflux and extracellular matrix mineralization.

Unlike many studies of NCX family members, information concerning the role of the PMCA family of proteins in mediating calcium ion transport in the cells of vertebrate hard tissues is relatively modest. However, a great deal is known of the function of these proteins in both excitable and non-excitable soft tissues. A priori, it should be acknowledged that the calcium ion transients in such soft tissues compared with osteoblasts, odontoblasts, or other hard tissues are limited in time, amplitude, and quantity. Nonetheless, mechanisms which have evolved for governing calcium ion movement are likely conserved and present in most tissue types, an assumption thereby suggesting that soft tissue investigations may provide valuable insight into the regulation of calcium ions in hard tissues.

In this context regarding membrane pumps mediating calcium ion transport, the hard tissue literature documents the immunolocalization of NCX3, as noted above, to the osteoblast plasma membrane facing the bone mineralization front [19]. In contrast, PMCA localizes to regions of the osteoblast plasma membrane which are distant from both NCX3 and the bone mineralization front (Figs. 6.3 and 6.4) [19]. These calcium ATPases (PMCA1, PMCA3, and PMCA4 isoforms) are also notably secreted during enamel mineralization [25], and PMCA is involved with dentin mineralization as well [26]. From the soft tissue literature, PMCA proteins are clustered within microdomains located within caveolae or lipid rafts of the plasma membrane [27] and at sites which knit together the plasma membrane and the endoplasmic reticulum [28]. PMCA proteins are also adapted to their membrane location and utilize acidic phospholipid molecules for activation [29] together with the EF-hand protein, calmodulin [30].

What is especially interesting is that the function of both PMCA and additional ATPases in soft tissues is closely bound to that of other membrane proteins, in particular stromal interaction molecule 1 (STIM1) [31] and the calcium ion channel protein, olfactory receptor class A related 1 (ORAI1) [32] From this perspective, the individual calcium ion membrane pumps exist as a component of a larger system involving store-operated calcium entry (SOCE) channels. Moreover, like the pumps of hard tissue cells, their role in calcium ion homeostasis occurs at the level of the plasma membrane and is directly related to calcium ion loading of membrane vesicles, lysosomes, autophagosomes, endosomes, and other organelles, especially at the junctional region encompassing the endoplasmic reticulum with the plasma membrane [33].

While these observations need to be confirmed using cells which manage large calcium ion transients, these findings indicate that PMCA functions by regulating calcium ion movement at microdomains in both the plasma membrane and the plasma membrane-endoplasmic reticulum junction. Indeed, Stains et al. suggested that PMCA fulfills a minor role in the mineralization process and functionally it is likely more closely linked to calcium ion uptake and homeostasis, rather than the induction of mineralization [19]. Compared with PMCA, the high expression level and spatial distribution of NCX3 in cells of hard tissues provide strong support for its role as the principal regulator of osteoblast calcium ion efflux, and as such it is required for the induction of mineralization (Fig. 6.4). Stains and Gay hypothesize that calcium "enters the cells and is then shuttled to the matrix-facing side of the cell, where it is in turn translocated out of the cell toward the site of mineralization by the Na-Ca exchanger" [17].

As an additional note regarding channels and pumps in vertebrate mineralized tissues, there is evidence of yet another transmembrane channel protein, anoctamin-6, in osteoblasts. The pump responds to an elevation in the cytosolic calcium and, in exchange for chloride ions, exports the cation from the cell (in more sophisticated terms, it generates an outwardly directed rectifying Cl^- current) [34]. This transport system interacts with NCX and P2Y receptors and, like NCX, is thought to play a significant role in the bone mineralization process [35, 36]. In a related context, a human *anoctamin-5* mutation leads to the clinical phenotype of dysplasia of the jaws and other osseous changes (gnathodiaphyseal dystrophy) [37]. Since anoctamin-6 exhibits calcium-dependent phospholipid scramblase activity, it is probable that this ion transport system is required for vesicle formation [38].

In summary, NCX alone or in partnership with other transporters releases calcium ions from cells of vertebrate mineralized tissues for passage into their extracellular matrices. By removing excess calcium ions from the cytosol, this efflux mechanism maintains the homeostatic activities of osteoblasts, odontoblasts, ameloblasts, and likely the cells of other vertebrate mineralized tissues. However, despite compelling research evidence of the importance of these transporter proteins, their activities in initiating mineralization have been underemphasized. In this regard, the perception that membrane pumps play a limited role in facilitating mineralization may perhaps be attributed to the numerous observations that cell-derived mineralizing organelles are immediately evident within the extracellular matrices of most hard tissues. These organelles, matrix vesicles, have dominated thoughts of the process of vertebrate mineralization for more than half a century. The occurrence and nature of such organelles and their relationship to calcium ion transporters like NCX and events leading to mineralization are considered below.

6.3 Matrix Vesicles: A Review of Their Morphological, Functional, and Developmental Characteristics

For many years, researchers assumed that the process of mineral accretion in vertebrate tissues was caused by the liberation of membrane-bound organelles or extracellular matrix vesicles from the cells of such tissues. Indeed, since their discovery in the 1970s investigators reported that these vesicles provided the initial nidus for apatite nucleation in almost all normally mineralized vertebrate tissues, as well as in dystrophic mineral accumulation (Fig. 6.6) [39–42]. The classic view was that the vesicles, once released from the mother cell and lodged in a dense extracellular collagenous matrix, possessed the molecular armament to initiate mineral nucleation

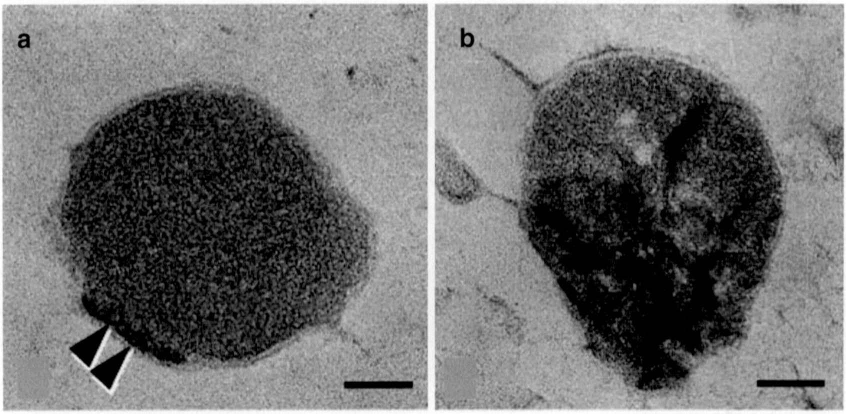

Fig. 6.6 Transmission electron micrographs of mineralizing matrix vesicles present in the extracellular matrix of the cartilaginous growth plate from a rat femur. Specimens were prepared by freeze substitution at liquid helium temperature to maintain a faithful structure. (**a**) Image of a vesicle filled with granular material. Dense mineral (arrowheads) appears within or close to the double membrane which encloses the vesicle. (**b**) Image of a vesicle in which much of the granular contents has been replaced by darkly staining mineral. Scale bars = 50 nm. (Images adapted from Amizuka et al. [42] with permission)

within their membrane-enclosed volumes. As the apatite crystals grew, they punctured the matrix vesicle "skin" to induce the formation of centers of mineralization in the extracellular spaces adjacent to the vesicles. The process triggered changes in the composition of the surrounding extracellular matrix and also influenced the cellular phenotype [43].

While not addressing the importance of matrix vesicles in mineral formation, recent studies indicate that mineral deposition begins as an intracellular, rather than an extracellular, event [44–50]. In other words, calcium ions and possibly prenucleation ion clusters or preapatitic salts are accumulated first within a small intracellular compartment(s), regulated by the energy status of the cell. Once secreted from the cell, these precursors to apatite become embedded or anchored in the dense extracellular tissue. In contact with ions of the extracellular fluid and in the absence of proteins inhibitory to mineralizing activity, the recruitment of additional calcium and phosphate ions leads to possible amorphous mineral phase formation, early- to late-stage mineral phase transformations, apatitic growth, and ultimately to tissue mineralization.

The following section of this chapter focuses on extracellular matrix vesicle biology in terms of the relationship of vesicles to events linked with the initiation of mineralization. Some detail of the varied types of vesicles which are formed will be presented, together with mechanisms by which these organelles exit the cell. At last, contemporary findings in cell biology are considered, including autophagy, mitochondrial function, and calcium secretion in potentially mediating the formation of the initial nidus of mineral which subsequently may lead to complete tissue mineralization.

6.4 Generalized Features of Matrix Vesicles

Extracellular matrix vesicle-mediated mineralization has been most thoroughly studied in the cartilaginous epiphyseal growth plate of several species [51, 52]. In this complex structure, zones of mineralization are distinct from those of growth, and the architecture of this tissue permits an analysis of the deposition of apatite in both cartilage and bone (endochondral bone formation). As was discussed in Chap. 3, terminally differentiated hypertrophic chondrocytes secrete matrix vesicles with the capacity to mineralize. While maturing chondrocytes also generate matrix vesicles, it is interesting that only those organelles resident in the hypertrophic growth plate regions lead to cartilage mineralization. Toward the end of this chapter, possible roles are reviewed of the non-calcifying vesicles in growth plate physiology.

From a historical perspective, the discovery of mineralizing vesicles in cartilage and other tissues was exciting and elicited much discussion [39–41, 51, 52]. Many scientists confirmed the initial observations about matrix vesicle mineralization in cartilage, bone, and dentin, and the scope of the vesicle phenomenon was extended to include pathological forms of mineralization such as those occurring in blood vessels [53]. By the mid-1980s, the study of vesicles reached a zenith with multiday conferences devoted to understanding how the organelles formed and induced mineralization. Experiments conducted in vivo and their published reports were enhanced by the development of systems to harvest vesicles from cells in culture. This methodological advancement provided scientists with a means of isolating vesicles, characterizing them, and evaluating factors influencing their formation and effects on mineralization [54].

In addition to chondrocytes in the growth plates of developing long bones [39–42, 55, 56], osteoblasts [56] and tendon cells from certain normally mineralizing avian species such as the turkey [57, 58] secrete vesicles (Fig. 6.7). Vesicles are also produced from the odontoblast cell body and cytoplasmic processes in the tooth [59]. Additionally, vesicles are secreted by cementoblasts and alveolar osteoblasts and they are evident during tissue repair (fracture callus) and reparative dentin formation [60–62]. However, it is unlikely that vesicles mediate mineral formation in tissues which undergo appositional growth [59]. Moreover, as discussed previously, while protein-filled vesicles are present in ameloblasts, such vesicles do not participate directly in the mineralization process of enamel.

Isolated from chondrocytes in culture or detected in growth plate cartilage or other tissues in vivo, matrix vesicles are observed by electron microscopy as generally oblate spheroids several hundreds of nanometers to a few microns in diameter [39–41, 43]. Such vesicles frequently contain small needle-like crystals of various sizes, on the order of a few tens of nanometers in their longest dimension [39–41, 56]. The manner of crystal formation within the vesicles is not entirely clear, but a common view is that the mineral occurs by association of calcium ions binding the protein, annexin, present in a lipid-protein membrane core complex of the vesicle, and/or binding charged phospholipids (phosphatidylserine or a calcium phosphate-phospholipid complex) in the vesicle membrane (Fig. 6.8) [43, 63]. The crystal nucleation process within vesicles is thought to be enhanced by the activity of tissue non-specific alkaline phosphatase (TNAP) and phosphoethanolamine/phosphocholine phosphatase 1 (PHOSPHO1), which increase local phosphate ion concentration, while a pyrophosphatase/phosphodiesterase (NPP1 or glycoprotein-1) simultaneously generates pyrophosphate which otherwise inhibits

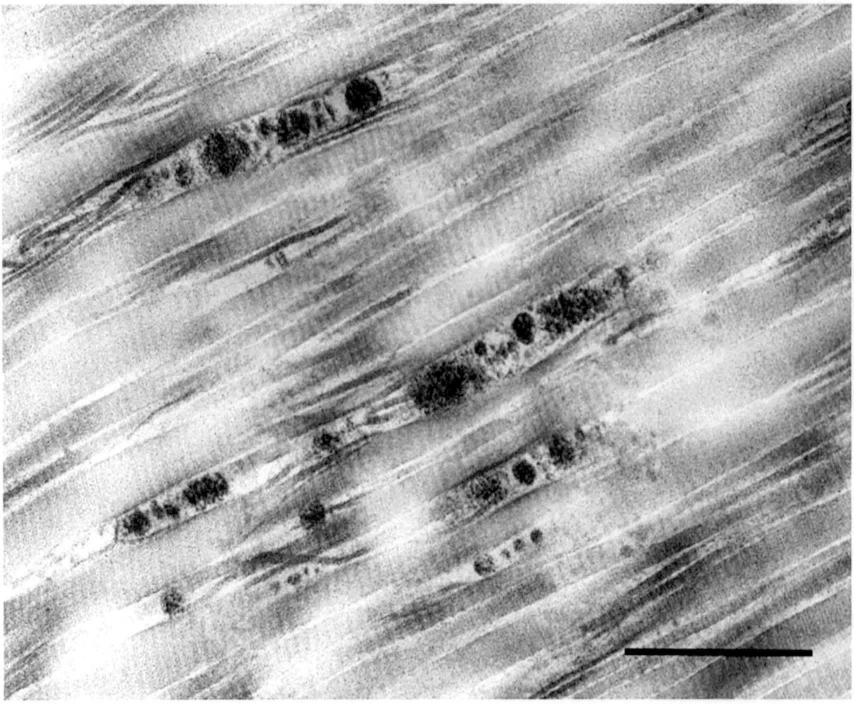

Fig. 6.7 Transmission electron micrograph of matrix vesicles appearing in the normally calcifying leg tendon of a 15-week-old turkey. The tendon was aldehyde-fixed and sections were stained with uranyl acetate and lead citrate. The tissue is composed of densely packed parallel arrays of type I collagen fibrils having the characteristic ~64 to 70 nm periodic banding pattern of the protein. The vesicles have been stained for alkaline phosphatase and exhibit increased electron density. Scale bar = 1 μm. (Image from Shapiro et al. [58] with permission)

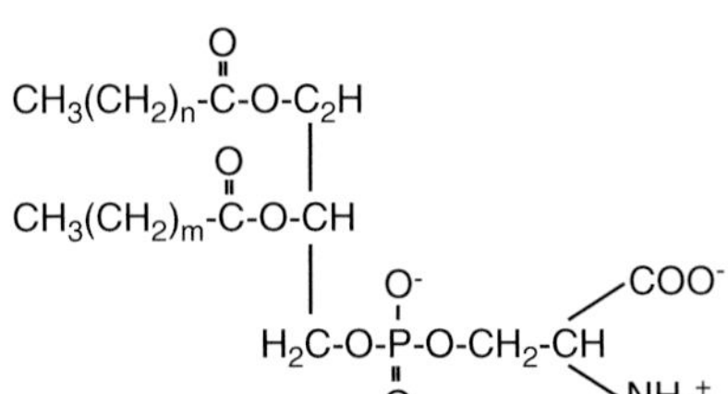

Fig. 6.8 Chemical structure of phosphatidylserine. This phospholipid molecule is highly polarized because of its two constituent long-chain fatty acid groups (lengths determined by the number of methylene (CH_2) groups, *n* and *m*) and a water-soluble and charged ester phosphoserine. Because of its excess negative charge (contributed by carboxyl and phosphate species), the molecule behaves as an anion. In comprising cellular and subcellular membranes, the fatty acid groups of phosphatidylserine face the lipid components of the membrane while the phosphoserine moiety is directed away from the membrane

mineral formation [43, 63]. E. Bonucci, one of the co-discoverers of matrix vesicles, termed these composite structures "calcifying globules." [39]. Immersed in cartilage, bone, or dentin extracellular tissue fluid containing an abundance of calcium and phosphate ions, the vesicles mineralize by recruitment of additional ions which bind to EF-hand proteins like calbindins. This mineral accretional process combined with activation of proteases and lipases leads to the disruption and loss of vesicle membrane integrity and promotes yet further crystal growth beyond the boundaries of the vesicles. Subsequently, the fusion of the numerous foci of mineralization provides much of the extracellular inorganic phase of cartilage, bone, dentin, and other vertebrate mineralized tissues.

6.5 Biogenesis of Matrix Vesicles and Other Forms of Extracellular Vesicles

The co-discoverer of matrix vesicles, H.C. Anderson, commented that vesiculation was caused by events which influenced the structure of localized regions of the plasma membrane of the mother cell [56]. In support of this view, the complement of proteins (ATPases, TNAP, PHOSPHO1, annexin V, and non-collagenous calcium-binding proteins) and the spectrum of phospholipids characterizing matrix vesicles were found to be different from those comprising the plasma membrane of the mother cell [43]. Moreover, early electron microscopic studies of cartilage confirmed that there was specificity to vesiculation in that blebs developed from cellular protrusions of the chondrocyte membrane (Fig. 6.9) [64]. The blebs originated at surface microvilli, interacted with proteins of the extracellular matrix, and were released from the cell when adhesion to matrix proteins was greater than that to the microfilament network of the cell [65].

While numerous reports [39–41, 51, 52] document observations that matrix vesicles are derived as protrusions or blebs or buds from cellular plasma membranes, there are other modes of formation and exit of vesicles to the extracellular milieu. For example, some vesicles appear to pass through the intact plasma membrane. Xiao et al. noted that matrix vesicles formed within cultured bone cells as clusters or aggregates enclosed in membranous sacs [66]. Such sacs separated from the plasma membrane to which they apparently attached, and they released the vesicles to the extracellular matrix. Liberated vesicles then led to mineral deposition (Fig. 6.10) [66]. Similarly, Thyberg and Friberg [67] showed that vesicles in rat growth plate cartilage were generated intracellularly from modified lysosomes, which were then released into extracellular matrices. The conclusion from these different investigations is that matrix vesicles have a mixed origin, some beginning within the internal membrane systems of cells and others forming de novo at the cell membrane. As will be discussed shortly, the intracellular origin of all these vesicles is associated with activities of the cellular endosomal-autophagic system.

All cells disgorge a variety of membrane-enveloped extracellular vesicles of diverse origin and size. Thus, it is not surprising that several different mechanisms exist for the formation and extracellular secretion of matrix vesicles, including membrane blebbing, described previously, multivesicular aggregates, and lysosome-like extrusions. A common finding is that vesicles are liberated from the cell membrane as microvesicles (~100 to 1000 nm in diameter) [68]. Other vesicular structures originating from the cellular endosomal compartment are described as exosomes (~10 to 100 nm in diameter), while tiny remnants of dying cells are designated as apoptotic bodies [68]. Many recent publications consider matrix vesicles as one form of exosomes, so these terms are used interchangeably. Given the interest in these various cellular components, publicly accessible databases (Exocarta, Expedia, and Vesiclepedia) provide information about almost every aspect of vesicle collection, purification, and content [69]. Frequently, since the source of the vesicles or vesicle-like organelles is unknown, exosomes and microvesicles are collectively known as extracellular vesicles.

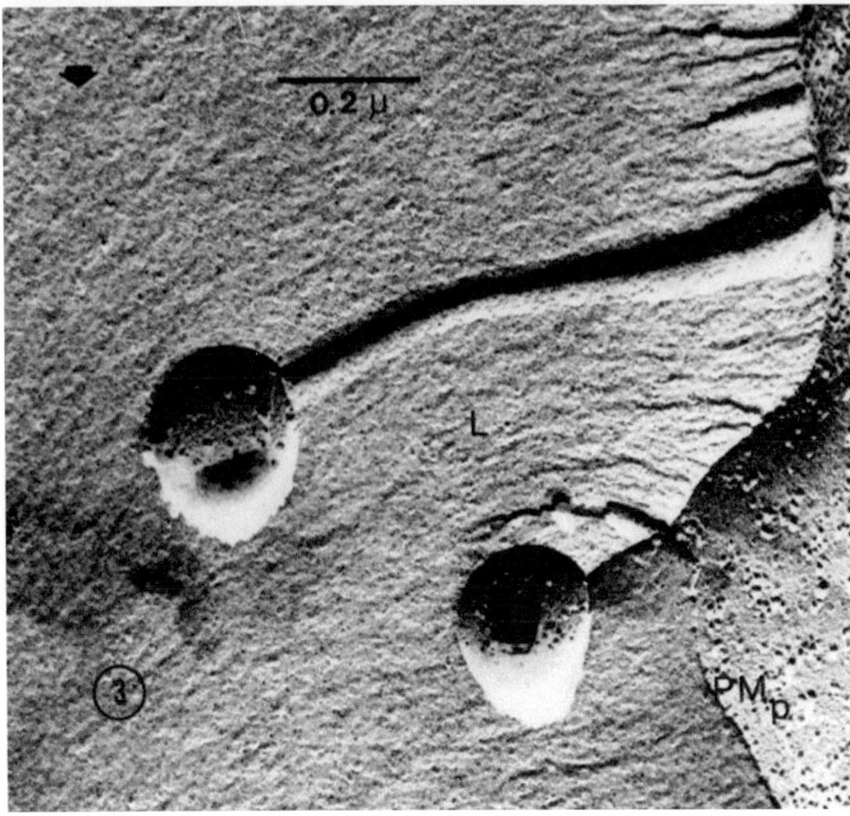

Fig. 6.9 Transmission electron micrograph of two matrix vesicles budding from the plasma membrane (PM) of hypertrophic chondrocytes from the cartilaginous growth plate of a tibia from a 6 to 10-week-old rat. Unfixed tissue was embedded in glycerol and freeze-fractured. Replicas of the ultrastructural surface of the fracture faces of samples were made by platinum coating and carbon reinforcement. Such replicas were carefully stripped from the sample surfaces and examined by microscopy. Vesicles appear in the lacunar matrix (L) of cells at the ends of microvillus projections from the chondrocyte plasma membrane (PM), forming then by exocytosis rather than by endocytosis or vesicle extrusion from invaginations of the plasma membrane [64]. Membrane-associated particles of unknown origin, possibly protein aggregates or proteoglycans, decorate the external, p-face of membranes (PM_p) [64]. Bar = 0.2 μm. (Image from Cecil and Anderson [64] with permission)

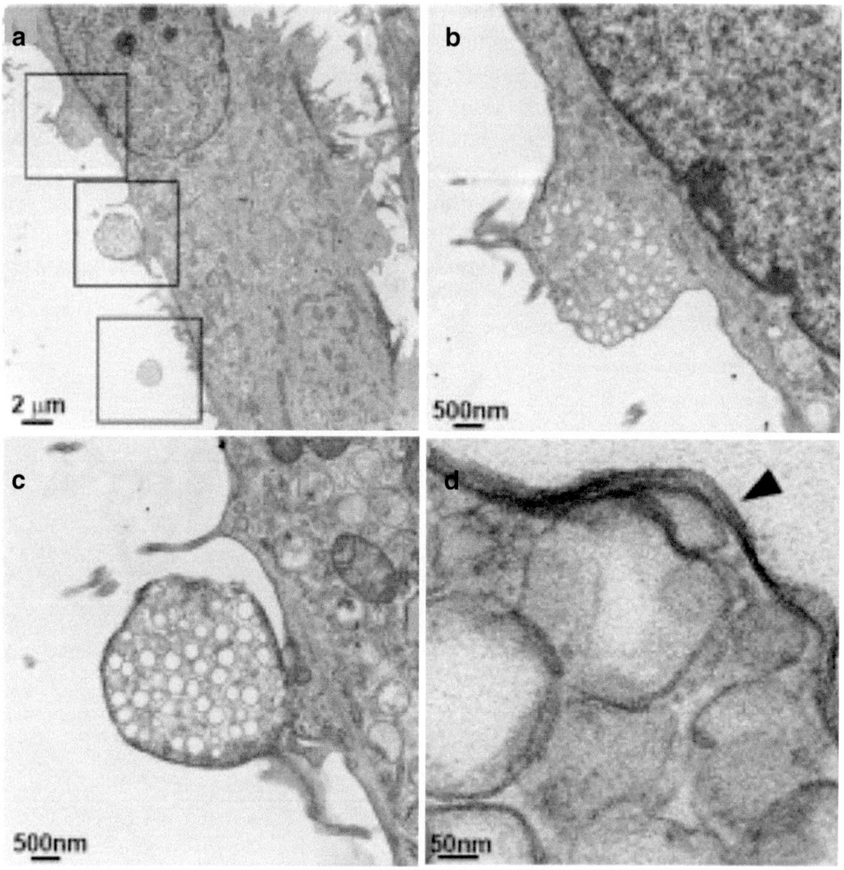

Fig. 6.10 Transmission electron microscopy of vesicle formation by MC3T3-E1 osteoblast-like cells in culture. (**a**) These cells appear to form clusters of vesicles enveloped in sacs continuous with the plasma membrane. (**b**, **c**) Enlargements of vesicle aggregates shown in (**a**) indicate a common double-membrane structure and the image in (**c**) suggests the separation of a sac from the plasma membrane. (**d**) The double membranes of a sac (arrowhead) and individual vesicles are apparent at high magnification. From the reported observations without details of specimen preparation, liberation of sacs and aggregated vesicles from the cells is thought to be followed by loss of sac membranes and release of separate (matrix) vesicles to the extracellular milieu where they then mineralize [66]. (Images adapted from Xiao et al. [66] with permission)

There is considerable variation in the contents of extracellular vesicles. The cargo can include a large number of both RNA and microRNA (miRNA or MIR) species, together with over 1000 different proteins and many neutral lipids and phospholipids [70]. Regarding their distribution, aside from those anchored in connective tissues, vesicles are present in body fluids, including serum, cerebrospinal fluid, urine, tears, and saliva. Each of these fluids can be collected and analyzed to obtain samples for vesicle analysis. From a functional viewpoint, extracellular vesicles participate in multiple physiological processes ranging from cell–cell signaling and immune regulation to tissue repair [71]. The contents of these organelles as well as the composition, organization, and membrane configuration delineate specific functional activities. Hence, from all these perspectives, a great deal of importance is attached to determining the origin, distribution, and molecular biological and structural nature of these cell-derived components. In the following sections of this chapter, the concept that matrix vesicles comprise a specific type or types of extracellular vesicle is discussed and, of direct importance to the overall theme of this book, the possibility that such vesicles accumulate calcium ions within the cell prior to being secreted in the extracellular matrix is addressed. How knowledge of extracellular vesicles and their potential for inducing mineralization correlates with emerging proposals concerning the role of the cell in partitioning calcium ions for mineral development is discussed later in the chapter.

6.6 Ion Loading, Compartmentalization, and Regulation of Extracellular Vesicle Formation

As a preface to the following discussion exploring the concept that cells generate vesicles which are trafficked from the cell to the mineralization front, no single experiment has been performed to follow the fate of calcium ions from the extracellular fluid into a cell, the processing of such ions within the cell, and the subsequent exit of the ions from the cell to the mineralization front of a tissue. However, as detailed in previous chapters of this book, numerous studies demonstrate calcium ion entry into a cell as well as intracellular sequestering and trafficking of these ions in organelles of vertebrate mineralized tissues. Furthermore, organelles which are expelled from the cell bear close molecular and morphological similarities to matrix vesicles found at the sites of mineral formation [39–41, 51, 52]. Such observations suggest that these various cellular activities may be linked together to promote both the initiation and propagation of mineralization within the extracellular matrix of vertebrate hard tissues.

As has been noted earlier, calcium ions are found in several subcellular compartments which include mitochondria, the Golgi complex, and the endoplasmic reticulum. Of these organelles, the endoplasmic reticulum is most closely associated with calcium ion accumulation while mitochondria sequester calcium to promote energetics and maintain charge neutrality and ionic homeostasis in the cell. In this manner, mitochondria protect cells from the cytotoxic effects of the cation. Mitochondria may also serve as a source of calcium ions for the endoplasmic reticulum as ultrastructural analysis indicates these two organelles are located within 30 nm of each other [72]. Consistent with the discussion noted in Chap. 5, the calcium ion concentration in the endoplasmic reticulum and vesicles derived from this organelle is very high, in some cases rivaling the levels found in the tissue extracellular fluid. These data, then, lend strength to the view that ion uptake, accumulation, and sequestration occur in several intracellular, structurally related organelles which may also translocate throughout the cell. The data also support the suggestion that interactions between these various organelles and their membrane systems mediate calcium ion loading, compartmentalization, and the trafficking of cargo to the extracellular milieu in the different cells of vertebrate mineralized tissues.

In osteoblasts, odontoblasts, and maturing ameloblasts, calcium ions contained within the endoplasmic reticulum and Golgi are associated with EF-hand calcium-binding proteins, in particular calbindins, calreticulins, and nucleobindins [73, 74]. Of these proteins, nucleobindins are especially interesting since they are both resident in several intracellular compartments, including the endoplasmic reticulum and mitochondria, and are secreted from cells [73–75]. Within the cells noted above, nucleobindins are localized to the cis-Golgi cisternae where they bind considerable numbers of free calcium ions; in the cytosol, nucleobindins are coupled with the Golgi surface as well as endosomal membranes [76]. Nucleobindins also function to transport and recycle the low-density lipoprotein receptor, LRP9, involved with the activation of wingless-related integration site, WNT, a transcription factor required for bone and tooth mineralization (see Chap. 4) [76].

Like the Angels which accompanied Lot and his family out of Sodom, calbindins, calreticulins, and nucleobindins chaperone calcium ions throughout cells and may facilitate the movement of endosomes and other organelles to the plasma membrane and onward to the extracellular matrix. This process, termed endosomal sorting, is initiated in a vesicle, the early endosome, and selectively separates ions and different molecules (the "cargo"), such as receptor proteins and lipids, for packaging and trafficking along a variety of distinct intracellular pathways, including the lysosomal pathway for cargo degradation and direct cargo transport pathways to the plasma membrane (Fig. 6.11) [77, 78].

The subsequent fate of the organelles just noted is complex, carefully regulated, and reflective of the functional and environmental needs of the cell, tissue, and organism in which they are found. Recycling early endosomes appear to originate at the plasma membrane or in the trans-Golgi network (Fig. 6.11) [78]. Electron microscopic images of early endosomes indicate they resemble irregular tubular structures which undergo further processing within the cells to form an organelle identified as a late endosome or multivesicular body (Fig. 6.11) [47, 78]. The structure of the multivesicular body is very different from that of an early endosome as invagination and then budding of its inner membrane generates intraluminal vesicles which are of the same range of sizes as secreted exosomes (Fig. 6.12) [78, 79]. Subsequent fusion of the multivesicular body with the plasma membrane allows the release of these intraluminal vesicles, which are now recognized as exosomes (Fig. 6.11). A principal function of endosomal vesicles is to sequester different ions and a variety of molecules, moving them throughout a cell (Fig. 6.11) [78].

On entry to cells, as noted above, calcium ions may accumulate in the endoplasmic reticulum and mitochondria (Fig. 6.11) [14–16, 44, 80]. As also mentioned earlier, calcium ions may interact with cytoplasmic calbindin, calreticulin, and nucleobindin molecules, and the three proteins may transit these ions to early-stage endosomes in addition to serving in effect as buffers to maintain intracellular calcium homeostasis and to prevent overloading of calcium ions in cells [73–75]. A mechanism(s), however, supporting the possible movement of calcium ions from the binding proteins to early endosomes has yet to be elaborated.

Besides the presence of calcium ions in endosomes formed at the plasma membrane, these ions may be accumulated in endocytic vesicles from several other possible sources. One is through the release of calcium ions into the cytosol from the

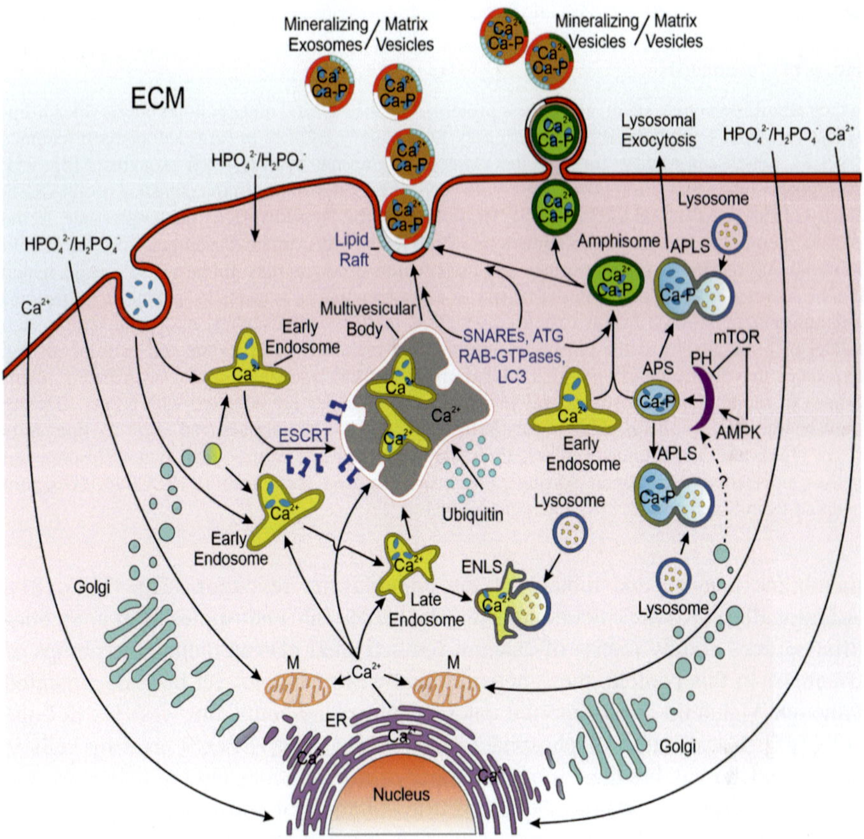

Fig. 6.11 Schematic view of intracellular calcium and phosphate ion trafficking and exosome release to the extracellular matrix (ECM) through autophagy-dependent secretory pathways and endosomal and lysosomal exocytosis. Recycling endosomes originate at the cellular plasma membrane (red) and internalize cargo of various ions, proteins, and other molecules (small light blue ellipsoids) by receptor-mediated endocytosis; endosomal vesicles also originate in the trans-Golgi network and ferry ions and molecules for processing to other cellular compartments [78]. Recycling endosomes and early/late endosomes can also form an intermediate structure, the multivesicular body, whose limiting membrane buds inwardly into its lumen to generate intraluminal vesicles. These vesicles are ultimately liberated to the ECM as exosomes. Calcium ions (Ca^{2+}) accumulate in the endoplasmic reticulum (ER); the ER subsequently loads the ions into mitochondria (M), endosomes, and multivesicular bodies [14–16, 44, 80]. Phosphate ions (HPO_4^{2-} and $H_2PO_4^-$) also enter the cells and appear in mitochondria. However, the trafficking of these ions to mitochondria and possibly other intracellular organelles is not clearly defined. The autophagic pathway provides yet another source of vesicles. In this aspect of intracellular calcium ion trafficking and exosome release, double membrane-invested phagophores (PH) of this pathway are thought to be generated from proteins of the trans-Golgi network (question mark and dashed arrow) and regulated by the activities of mTOR and AMP kinase (AMPK) [132]. These proteins are responsive to nutrient signaling and are coordinately related: During stress conditions, activation of AMPK promotes autophagy through the phosphorylation of two serine residues (317 and 777) of the mammalian autophagy-initiating kinase, ULK1; in contrast during homeostasis, mTOR phosphorylates ULK1 at serine residue 757, preventing AMPK-ULK1 interaction and the onset and progression of autophagy [132]. Phagophores transition to mature autophagic vesicles, autophagosomes (APS),

Fig. 6.11 (continued)

which engulf defective, spent, and excess proteins, organelles, and other cellular debris (blue ellipsoids) [83, 85]. Autophagosomes in turn may fuse with lysosomes to form autophagolysosomes (APLS), which can release cargo to the extracellular milieu by lysosomal exocytosis [83, 85]. Autophagosomes may alternatively fuse with early-stage endosomes in the presence of SNARES, RAB-GTPases, ATG, and LC3 [83–85]. Trafficking of this hybrid vesicle, the amphisome, to the plasma membrane provides a mechanism by which its contents can be discharged from the cell. In mineralizing tissues, autophagosomes and their fusion products may contain Ca-P nuclei which can be secreted into the extracellular matrix as exosomes/matrix vesicles [47]. Thus, amphisomes and autophagolysosome fusion vesicles, as well as multivesicular bodies, may each shuttle such nuclei or mineral ions to the plasma membrane and release them from the cell as mineralizing exosomes or vesicles/matrix vesicles, blebs, or exocytosed lysosomal products. Investing membranes of mineralizing exosomes and vesicles/matrix vesicles are depicted with colors showing their source from multivesicular bodies (white), the plasma membrane (red) with its lipid rafts (light blue), and amphisomes (green). Endolysosome (ENLS); cargo (small blue ellipsoids and yellow crescents); endosomal sorting complexes required for transport (ESCRT). (Diagram adapted from Shapiro et al. [58] with permission)

membrane lacunae and tubules of the endoplasmic reticulum (Fig. 6.11) [81]. Additionally, cytosolic calcium ions may be loaded into endosomes from mitochondrial sources and by means of calcium ion-activated plasma membrane pumps or channels. In this context, these potential pathways have not yet been documented although Mahamid et al. reported calcium phosphate-containing vesicles in bone cells [45]. Localization of phosphate ions (HPO_4^{2-} and $H_2PO_4^-$) entering cells is poorly understood, but these ions, like those of calcium, may also be sequestered in mitochondria (Fig. 6.11) [44, 80, 82]. Little is known of phosphate ion loading by endosomes. In work by Mahamid et al., just noted, a phosphate-enriched mineral was found in intracellular vesicles in bone [45]. Endosomes aggregate as multivesicular bodies, which now contain calcium ions, proteins, and other molecules. The endoplasmic reticulum may provide a further source of calcium ions to the multivesicular bodies and possibly phosphate ions from mitochondria may also be sequestered in these organelles. Multivesicular bodies may fuse with regions of the plasma membrane rich in raft lipids and proteins, a process mediated by other proteins such as *s*oluble *N*-ethylmaleimide-sensitive factor-*a*ttachment protein *r*eceptor*s* (SNARES), *Ra*s-related proteins members of the RAS superfamily of small GDP/GTPases (RAB-GDP/GTPases), autophagy proteins (ATG), and LCM3 (ATG8) and leading to the liberation at the plasma membrane of individual or groups of mineralizing exosomes/matrix vesicles into the extracellular matrix (ECM) (Fig. 6.11) [83–85]. In a separate but related series of events, late-stage endosomes may fuse with lysosomes to form endolysosomes, which promote the degradation of cargo and its ECM release (Fig. 6.11) [84].

An alternative fate of both early and late endosomes is fusion with an organelle derived from the autophagic pathway, an autophagosome (Fig. 6.11). Endosomes and autophagosomes form a hybrid structure, an amphisome. This organelle can fuse with a lysosome to generate a digestive vacuole, containing hydrolytic enzymes

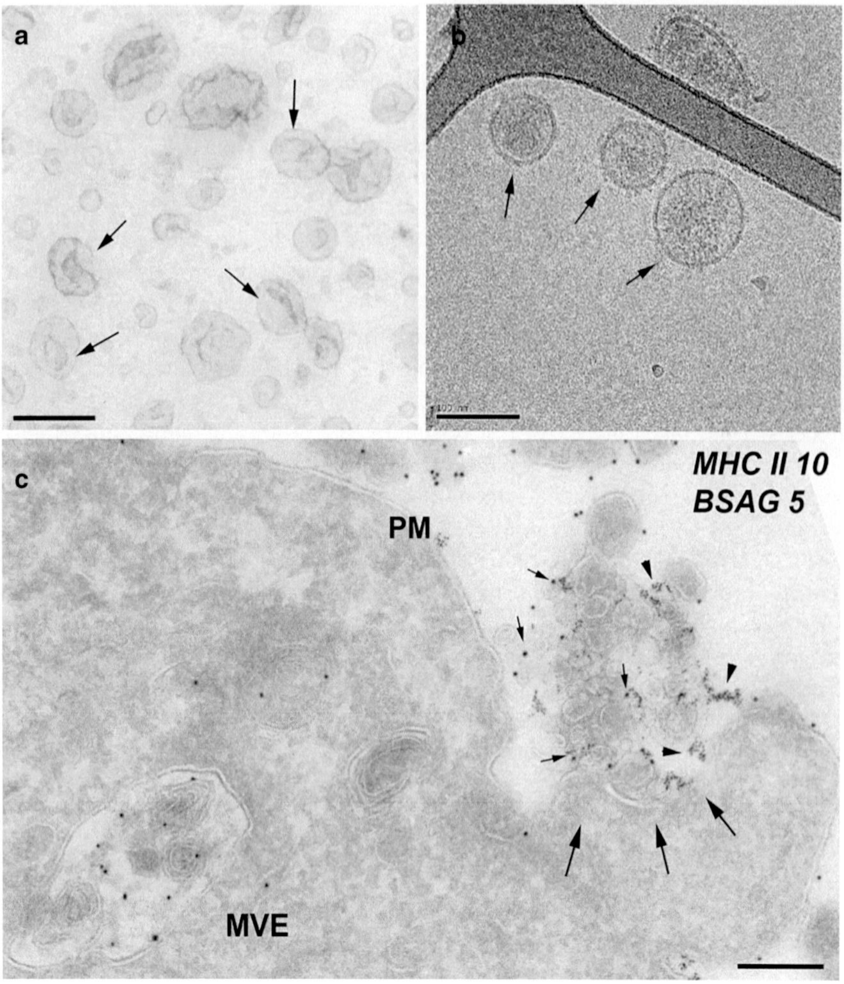

Fig. 6.12 Transmission electron microscopy of exosomes and late endosomes. (**a**) Exosomes from melanoma cells stained with uranyl acetate and embedded in methylcellulose. Exosomes (arrows) are generally round with diameters varying from 30 to 100 nm. (**b**) Exosomes from human prostate epithelial cells are directly frozen and imaged by cryo-electron microscopy (no chemical fixation or staining). Exosomes (arrows) appear round and heterogeneous in size. (**c**) A lymphocyte treated with bovine serum albumin coupled with 5-nm gold particles (BSAG5) and cryosectioned and immunolabeled for MHC class II (major histocompatibility complex molecules) with 10-nm protein-A gold (MHC II 10). The lymphocyte exhibits an intact multivesicular endosome or multivesicular body (MVE) and the same organelle in part fused to the lymphocyte plasma membrane (PM) and being exocytosed from the cell (large arrows). The fused MVE contains exosomes labeled with both 5- and 10-nm gold particles identifying BSAG5 (arrowheads) and MHC II 10 (small arrows), respectively. Scale bars for all panels = 100 nm. (Images adapted from Raposo and Stoorvogel [79] with permission)

required for lipolysis and proteolysis of trafficked lipids and proteins (Fig. 6.11) [86]. The amphisome can also fuse with the cellular plasma membrane, forming a secretory vesicle which expels its contents into the extracellular milieu. Thus, secretory autophagy represents a possible pathway for vesicle formation and secretion and calcium ion or calcium phosphate cluster secretion (Fig. 6.11) [87]. Details of the autophagic pathway are discussed in the next section of this chapter.

Exosome formation through the endocytic pathway and processing into multivesicular bodies involves the presence of ubiquitin and the participation of a protein family designated the endosomal sorting complex required for transport or ESCRT (Fig. 6.11) [78]. These latter proteins are also critically coupled to events which include endocytic vesicle navigation through the cell and the formation and budding of multivesicular bodies. Associated with the ESCRT family is vacuolar protein sorting (VPS)4, which removes ESCRT protein (usually ESCRT III) from a vesicle once processing has taken place. An ESCRT-interacting protein, apoptosis-linked gene-2-interacting protein X (ALIX), binds to a unique lipid, bis(monoacylglycero)phosphate on endosomal and cell surfaces and plays a role in packaging MIRs in the exosome [88]. Of principal importance in exosome biogenesis is a group of proteins, the small RAB-GTPases, which drive chemical reactions utilizing the high energy content of the nucleotide, GTP (Fig. 6.11). These proteins are involved with almost every aspect of exosome formation, including vesicle trafficking, tethering, docking, and fusion [89]. The RAB-GTPases are located on the cytoplasmic face of vesicles to ensure coordinated and faithful vesicle targeting to acceptor membranes [90].

Once vesicle docking to a membrane has been achieved, another group of proteins, the SNARES, enhances the fusion of vesicle membrane lipids with the target membrane by removal of bound water and changing the curvature of the plasma membrane, a process termed "bending" (Fig. 6.13) [91–94]. Bending causes membrane deformation and enhances vesicular trafficking to the extracellular matrix [91, 92]. In addition to the proteins mentioned above, many other molecules participate in vesicle docking and fusion and exosome biogenesis. For example, cells and developing vesicles both contain the cell surface heparan sulfate proteoglycan, syndecan, and cells also secrete the syndecan cytosolic interacting protein partner, syntenin [95]. Interaction between syndecan and syntenin links to ALIX and therefore to ESCRT, and these associations influence intraluminal vesicle formation and plasma membrane structure. In this way, syndecan-syntenin facilitates both membrane budding and exosome biogenesis [95]. Moreover, syndecan is responsive to a considerable number of extracellular ligands and hence provides a mechanism for the systemic control of exosome formation and release. Along with the syndecan-syntenin system, the tetraspanins, another group of transmembrane proteins which modify membrane structure, promote the formation of lipid domains that enhance exosome release by cells [96, 97]. Once exosomes are liberated from cells, ligands on the exosome surface bind to integrin receptors on extracellular matrix proteins which include fibronectin, collagen, vitronectin, and laminin [98].

While the ESCRT protein family, tetraspanins, SNARES, and a number of GTPases are highly important in both exosome and microvesicle formation, a major impetus for vesicle release is an increase in intracellular calcium ion concentration [79]. Vesicles are shed from the cell at lipid rafts in the plasma membrane,

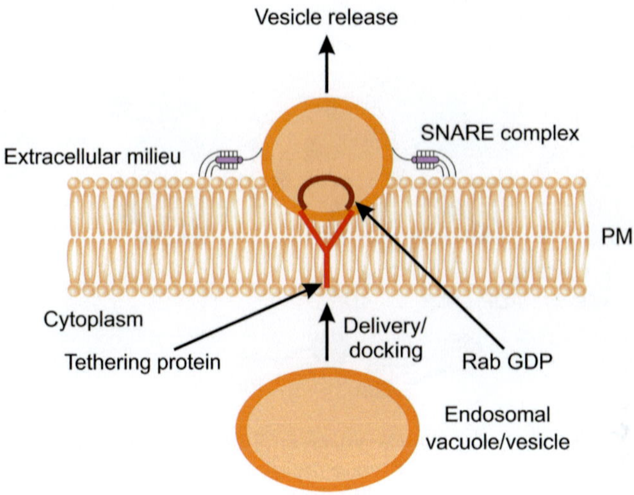

Fig. 6.13 Schematic of multiprotein complexes involved with vesicle exocytosis. While molecular interactions leading to matrix vesicle exocytosis in mineralizing cells are yet unknown, studies in non-mineralizing cells indicate that exocytosis is confined to specialized regions (rafts) of the plasma membrane (PM) enriched in cholesterol and sphingolipids [94]. Endosomal lipids in membranes enclosing vacuoles and vesicles dock and fuse with plasma membrane rafts in a process enhanced by SNARES (SNAP receptors), a family of proteins serving as molecular motors to deliver vacuoles/vesicles to target membranes and drive the docking and fusion events. The interaction between vacuoles/vesicles and plasma membranes is accompanied by the activity of Rab-GDP/GTPases, localized to the cytoplasmic face of the vacuole/vesicle, to ensure the fidelity of their targeting to the plasma membrane. Until they might be released, vesicles remain docked at the plasma membrane by tethering proteins. Separation of vesicles from plasma membranes into the extracellular milieu occurs by a blebbing phenomenon, possibly directed by several different factors such as the physical interaction between hydrophobic regions in the plasma membrane and the extracellular matrix; mechanical interactions between the cell surface heparan sulfate proteoglycan, syndecan, with the syndecan cytosolic-interacting protein partner, syntenin; and calcium-dependent movements of phospholipids from the outer to the inner leaflet of the plasma membrane [93]. (Schematic adapted from Astro and de Curtis [92] with permission)

comprised of specialized macromolecular assemblies enriched in cholesterol and sphingolipids. Such raft components mediate the formation of hydrophobic (water-hating) microdomains within the membrane, allowing the fat-enhanced regions to "float" in the lipid–protein structure of the plasma membrane. Possibly the physical interaction between these hydrophobic regions in the cell membrane and the extracellular matrix enhances the vesicle blebbing phenomena. A major factor causing blebbing is thought to be destabilizing changes in the plasma membrane mediated by a group of enzymes called flippases, floppases, and scramblases (Fig. 6.14) [99–103]. The "flip-flop" terminology is indicative of the movement of phospholipids between the inner and outer leaflets of the cell membrane [104, 105]. For example, phosphatidylethanolamine and phosphatidylserine are translocated across the membrane by a scramblase [103, 106]. This topological movement is affected by the influx of calcium ions, which both promotes the redistribution of membrane phospholipids and serves to enhance the release of vesicles from the cell [107].

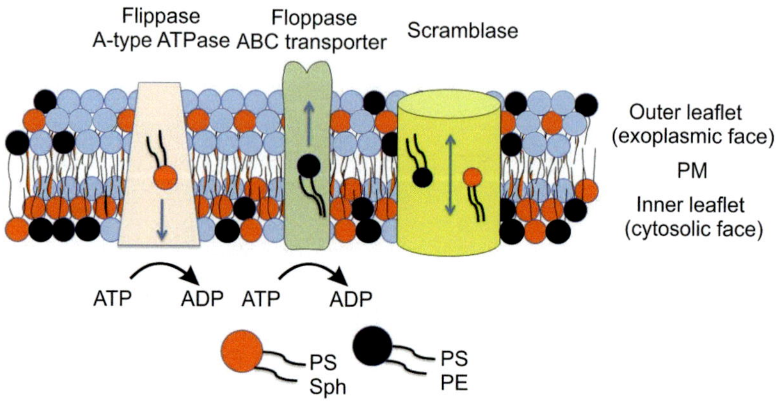

Fig. 6.14 Schematic diagram of enzyme systems involved in the generation and maintenance of membrane phospholipid structure. While most depictions of the plasma membrane represent its two opposing leaflets as the same, marked differences in composition exist between them and these variations influence functional activities [92]. For example, the known higher concentration of phosphatidylserine (PS) in the inner leaflet generates a negative charge which promotes interactions with the membrane as well as cytosolic proteins, and the presence of PS on the outer leaflet signals apoptosis [106]. While all phospholipids can diffuse and even rotate within the plane of the membrane, phospholipids also flip-flop between the leaflets in actions which require the input of chemical energy (ATP). Flippases are membrane translocase enzymes which move phospholipids (PS and sphingomyelin, Sph) inward toward the membrane center (exoplasmic to cytosolic faces of membranes), while floppases move phospholipids (PS and phosphatidylethanolamine, PE) outward from the membrane center (cytoplasmic to exoplasmic faces of membranes). The activity of these enzymes has been linked to the formation of vesicles [102]. Flippases are members of A-type ATPases and floppases are members of the ATP-binding cassette (ABC) superfamily of transporters (noted further in Chap. 9 in relation to the systemic release of ATP and mineralization). In contrast to the latter two translocases, scramblase promotes phospholipid transport between the two leaflets in an energy-independent manner. Scramblase activity is sensitive to and responds to the cytosolic calcium ion concentration [103]

Besides those enzymes directly involved with translocating phospholipids across the plasma membrane, another family of enzymes, the phospholipases, have been shown to influence membrane trafficking and stabilization as well as provide a source of phosphate ions for mineralization [108]. These enzymes sever the phosphodiester bond of membrane phospholipids to generate products such as long-chain diacylglycerol molecules, phosphatidic acid, and biologically active bases, which are commonly choline, ethanolamine, sphingosine, serine, and inositol [108]. In the case of the two most common membrane phospholipids, phosphatidylcholine and phosphatidylethanolamine, phospholipase A2 selectively removes an acyl group (fatty acid) from the phospholipid; if the severed fatty acid is arachidonic acid, then it, too, serves as a powerful signaling molecule [109]. The enzymatic action of phospholipase A2 on phosphatidylcholine and phosphatidylethanolamine produces the products, lysophosphatidylcholine and lysophosphatidylethanolamine, respectively, which can be further degraded to phosphocholine and phosphoethanolamine by lysophospholipase C [110]. Similar to phospholipase A2, an ectonucleotide pyrophosphatase/

phosphodiesterase (ENPP6), an enzyme which generates extracellular pyrophosphate, also hydrolyzes lysophospholipids to liberate the phosphorylated base [110, 111]. A possible fate of these phosphorylated bases is that they are cleaved by PHOSPHO1 to liberate free phosphate ions which may prime matrix vesicles for subsequent mineral formation [110]. As an added note in this context of membrane structure, sphingosine in the form of sphingosine-1 phosphate is required for the maintenance of membrane integrity and osteocyte function [112].

The blebbing phenomena and exosome/extracellular vesicle formation are influenced by yet another enzyme, phospholipase D2. This protein is regulated by small GTPases, adenine nucleotide ribosylation factor (ARF), and RAS homologous protein (RHO) and is physically associated with the actin cytoskeleton [86]. When phospholipase D2 is activated, it promotes the phosphorylation of extracellular signal-regulated kinase (ERK) and its recruitment to the plasma membrane where, in turn, ERK phosphorylates and activates myosin light-chain kinase (MLCK). This phosphorylation event leads to endosomal intraluminal budding and constriction of the plasma membrane at the base of forming exosomes/extracellular vesicles. The latter enhances the release of these organelles from the cell. In addition, phospholipase D2 induces membrane bending, as noted above [113–117].

Phospholipase D2 attacks and hydrolyzes phosphodiester bonds of phospholipids such as phosphatidylcholine, a major component of the plasma membrane, and this action generates the water-soluble product, phosphatidic acid, as mentioned previously [108]. Phosphatidic acid induces membrane bending, whose effect is linked to endosomal intraluminal budding and, when bound to annexin, phosphatidic acid promotes the release of vesicles from cells [116–119]. In addition, this phospholipid product can activate the ERK membrane budding system as described above, and influences autophagy through alterations in the activity of the mammalian target of rapamycin, mTOR [120, 121]. Based on these observations, phospholipase D2 and the products of phospholipid hydrolysis, in combination with other membrane enzymes, mediate multiple activities associated with the induction of autophagy, the formation of vesicles, the secretion of exosomes, and the availability of phosphate ions [115].

As noted previously, other types of vesicles are released from cells besides exosomes, the most common being microvesicles. Microvesicles are generally larger than exosomes, as mentioned earlier, and microvesicle formation, unlike that of exosomes, is independent of endosomal activity and reflects the submembrane accumulation of specific proteins and lipids [117]. Microvesicles and exosomes are similar, on the other hand, in that their interaction with the cell membrane is sensitive to many of the proteins active in cargo docking and exosome-membrane fusion. Such proteins include SNARES as well as small GTPases, RHO, RAS-related C3 botulinum toxin substrate (RAC)1, and cell division cell control protein (CDC)42 [122].

Like the blebbing or budding process which secretes exosomes to the extracellular milieu, the generation by blebbing or budding of microvesicles with their cytosolic contents to extracellular matrices also requires major changes in the plasma membrane structure of a cell. In this instance, phospholipases, sphingosine, phosphatidic acid, and other biologically active molecules are involved again in altering plasma membrane organization [108, 112]. Additionally, on the inner aspect of the plasma membrane, flippases, floppases, and scramblases cause the translocation of specific phospholipids to the external surface of the membrane [100, 101, 103, 123].

Both exosome and microvesicle liberation from cells respond to the activation of hypoxia-inducible factor (HIF). As noted previously, this transcription factor is secreted when the oxygen tension of a cell is low, a characteristic of most mineralizing tissues [124]. HIF-1α promotes exosome biogenesis and secretion, possibly by inducing the expression of RAB proteins [125, 126]. From this perspective, calcium ions, oxygen tension, and HIF-1α activity likely play a critical role in modulating pathways which generate and extrude vesicles from cells, key events in the mineralization of the extracellular matrices of vertebrate tissues.

6.7 Involvement of Autophagic Vacuoles, Lysosomes, and Acidocalcisomes in Intracellular Ion Transport

While the endosomal system may be considered a major source of extracellular organelles as described above, other circuits impact exosome biogenesis. Foremost among these is the "self-eating" or autophagic pathway (Fig. 6.11). This pathway provides a mechanism by which defective, spent, unneeded, or damaged cytosolic molecules and organelles are enveloped in double membrane-invested vesicles designated autophagosomes. These vesicles may fuse with lysosomes to form autophagolysosomes, which foster the degradation of their protein, lipid, carbohydrate, or organelle cargo (Fig. 6.11). The breakdown materials are released into the cytoplasm and recycled. Alternatively, the vesicle contents are exocytosed from the cell to the extracellular milieu [85, 127–129].

Although the process of autophagy was described over a hundred years ago, it was not until the early part of the twenty-first century that this novel pathway was recognized to be required for tissue function and cell survival. Indeed, the 2016 Nobel Prize was awarded to Y. Ohsumi for "his discoveries of mechanisms for autophagy." From the work of Ohsumi, Klionsky, and others, it is now known that autophagy can be activated in several different ways. For example, macroautophagy involves the envelopment and sequestration of organelles and cytoplasmic molecules in a double-membrane vesicle. In contrast, microautophagy is the wrapping and direct engulfment of cytoplasmic cargo along invaginating membranes of autophagic vesicles [127, 129, 130]. In each case, vesicle-enclosed cellular materials are hydrolyzed by lysosomal action and the products are reutilized, thereby supporting cell viability. The dysregulation of autophagy is now documented to influence diseases as varied as neurodegeneration, osteoporosis, and Paget's disease of bone [85].

Autophagy is very carefully controlled. In a healthy cell, the level of autophagic activity is normally low unless DNA damage occurs or there is nutritional, hypoxic, or endoplasmic reticulum stress on the cell [85, 131]. In response to such factors, the cell begins and continues scavenging and removing defective molecules and organelles as described above. Inhibition of autophagy is mediated by the protein complex, mTORC1, noted earlier, which phosphorylates and inactivates an autophagy signaling protein complex, a kinase, ULK1 (Fig. 6.11) [85, 132]. The enzyme, AMP kinase, promotes autophagy by interfering with mTORC1 function, thereby enhancing cellular processes which result in protein catabolism, a characteristic of autophagy (Fig. 6.11) [85, 132]. When signals for autophagy, such as altered DNA, cell starvation, hypoxia, or endoplasmic reticulum stress, are generated, inhibition of ULK1 is relieved [85, 132]. This event then initiates the formation of a double-membrane structure, the phagophore, possibly originating from a subdomain of the endoplasmic reticulum, the omegasome (Fig. 6.11) [133]. An additional protein complex, the class III phosphoinositide 3-kinase complex, next phosphorylates phosphatidylinositides to produce phosphatidylinositol 3-phosphate, a molecule involved in regulating the expansion of the membranes and maturation of the phagophore to an autophagic vacuole, the autophagosome (Fig. 6.11) [83]. Together with several autophagic (ATG) and vacuolar protein sorting 34 (VPS34) proteins, BECLIN-1 and microtubule-associated protein 1A/1B light chain 3 (LC3-II) [134], the phosphoinositide 3-kinase complex also promotes membrane trafficking (Fig. 6.15) [83, 131, 135].

Once formed, autophagosomes may fuse with lysosomes to generate digestive vacuoles, autophagolysosomes, as noted above, or with endosomes to produce amphisomes (Fig. 6.11) [83–85]. Amphisomes exhibiting the complex of proteins ATG3-12 interacting as mentioned earlier with ESCRT and ALIX [84] enhance exosome biogenesis or fusion with multivesicular bodies and participate in reactions which lead to the release of intraluminal bodies from the cell [84]. Accordingly, autophagy orchestrates crosstalk between vesicular trafficking pathways, involving endosomes, lysosomes, and autophagosomes, and favors exosome liberation to the

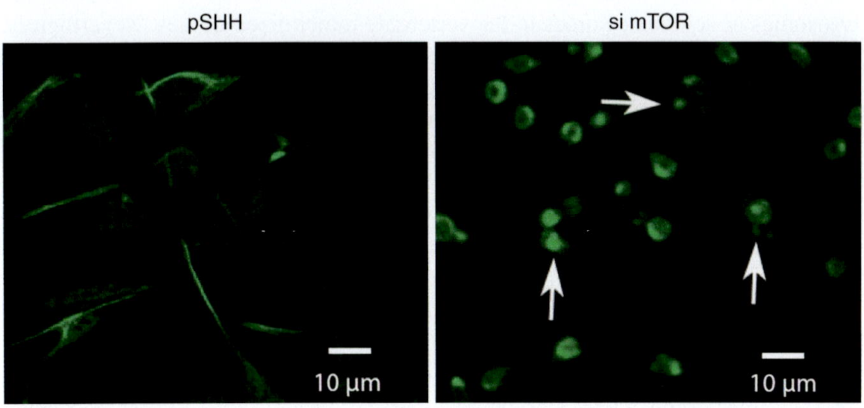

Fig. 6.15 Regulation of autophagy by mTOR as demonstrated by immunohistochemical staining of LC3, a marker of autophagic vacuole formation, in N1511 mouse chondrocytes cultured in fetal bovine serum. Left panel: Untreated (plasmid pSHH control) chondrocytes in the serum-containing medium were spread out with diffuse LC3 staining (green). Right panel: Following inhibition of mTOR by rapamycin (silencing [si] mTOR), chondrocytes condensed and exhibited punctate staining (arrows), a hallmark of autophagosome formation and autophagic activity. (Images from Srinivas et al. [135] and acknowledgment to Karger AG, Basel, Switzerland)

extracellular milieu [136, 137]. These events strengthen the hypothesis that interactions among exosomal, endosomal, and autophagic pathways are linked to activities associated with the initiation of tissue mineralization.

Lysosomes, or more likely a subgroup of these organelles called lysosome-related organelles, contain large reservoirs of calcium ions which can be released from the cell following fusion with the plasma membrane [138]. Like the activities of extracellular vesicles, trafficking and fusion of lysosomes involve SNARE proteins, RAB, syntaxin, and other proteins (Fig. 6.11) [139]. Experiments directly linking autophagic vesicles and lysosomes with apatite formation in mineralizing tissues are discussed in detail later in this chapter.

Another form of the lysosome is the acidocalcisome. Acidocalcisomes have been identified in trypanosomes and more recently in blood platelets and mast cells; they contain elevated concentrations of calcium and phosphate ions, calcium channel proteins (TRP family members, P2X receptors, and ryanodine receptors), and the calcium-mobilizing agent, nicotinic acid adenine dinucleotide phosphate, NAADP [140]. It is interesting that acidocalcisomes also contain polyphosphate anions and the enzymes concerned with their synthesis and degradation [141]. As will be discussed later, polyphosphates and pyrophosphates are sources of energy as well as inhibitors of mineral growth and development. Lending support to the possible role of lysosomes in delivering calcium and phosphate ions to extracellular matrices is the observation that lysosome-associated membrane protein (LAMP)1 and LAMP2a are present in matrix vesicles when cells are maintained in an osteogenic medium [142]. This same lysosomal marker protein, LAMP1, is reportedly bound to the enamel protein, amelogenin [143]. Despite the presence of LAMP, there is only sparse experimental evidence to support the notion that matrix vesicles/extracellular vesicles are derived from conventional lysosomes or acidocalcisomes in the vertebrate mineralized tissues. Nevertheless, the intriguing possibility exists that secreted lysosomes, an organelle bridging both the autophagic and endosomal pathways of a cell, maybe a source of calcium and phosphate ions for mineral formation.

The discussion above provides insight into four possible and distinct means by which cells of vertebrate mineralized tissues can take up and transport calcium ions under tight metabolic control. First, many studies show that both the endoplasmic reticulum and the mitochondrion of these cells accumulate elevated levels of calcium ions [15, 16, 44]. Compartmentalization by these intracellular components prevents calcium ions from damaging cytosolic and nuclear regulatory systems. Second, there are other vesicular ion-containing subcellular compartments which are not static dictums but serve as transport vehicles. Whether derived from microvesicles, endosomes, autophagosomes, acidocalcisomes, lysosome-related organelles, or amphisomes, vesicles traffic to the cellular plasma membrane and, together with components of the membrane, ferry their cargo into the extracellular environment. In this context, there is accumulating evidence of crosstalk between the low oxygen status of the tissue and pathways which link endosomes with autophagy [131]. Third, lysosomes may serve as carriers in which calcium will be present in an ionic form, possibly associated with a calcium-binding protein. Finally, organelles containing calcium egress the cell as exosomes or microvesicles and become anchored in the extracellular matrix of the tissue. Additional evidence is presented in the subsequent section of this chapter for the proposal that calcium phosphate deposition in mineralized tissues is initiated as an intracellular process, leading to the presence of mineral in extracellular matrices where it is critical to the formation of bone, dentin, and other vertebrate hard tissues.

6.8 Extracellular Vesicles and Tissue Mineralization

It is clear that healthy cells shed a variety of vesicular organelles, many of which function in critical physiological roles such as cell-cell communication, the delivery of signaling molecules, immune surveillance, and regulation of inflammation. The major types of vesicles released from these viable cells are exosomes and microvesicles, as described above. There is now accumulating evidence that cells of mineralizing tissues produce both exosomes and microvesicles and, indeed, an earlier report has suggested that matrix vesicles represent matrix-anchored forms of both of these examples of extracellular organelles [58]. Which type of vesicle is dominant in complex tissues like developing bone, calcifying cartilage, or another vertebrate mineralized system is unknown, and the distinctions in mechanisms of vesicle formation and sorting in these tissues are likewise incompletely understood.

Despite gaps in knowledge concerning vesicle biogenesis, it is widely recognized that matrix vesicles share a considerable number of molecular characteristics with both exosomes and microvesicles. Indeed, proteomic analyses have identified more than 280–300 different matrix vesicle proteins, many of which are found in extracellular vesicles (Fig. 6.16) [144]. Among those proteins and directly related to their function in chondrocyte-derived extracellular matrix vesicles are SNARES and RABs. The RAB protein family is required for both endosome trafficking and the autophagic process as noted previously [89, 90]. RABS have been documented to modulate mTORC1 activity, which in turn regulates VPS34 and BECLIN-1 [145]. These and several other proteins help in the transport of the ULK1 complex and guide the formation of preautophagosomal membranes, all critical factors in the development of autophagosomes [145]. In enamel, RAB-GTPase activating protein1 (RABGAP1) interacts with the protein, tuftelin (TUFT1), and mediates the mTORC1 signaling pathway, possibly influencing vesicular trafficking in this tissue [146]. A mutation in *RAB33b* has a serious impact on skeletal growth and development. The gene defect is the cause of spondylo-epi-metaphyseal dysplasia, a condition characterized by a disproportionately small stature, abnormalities of the spine and pelvic and long bones, and shortening of the fingers and toes (brachydactyly) [147].

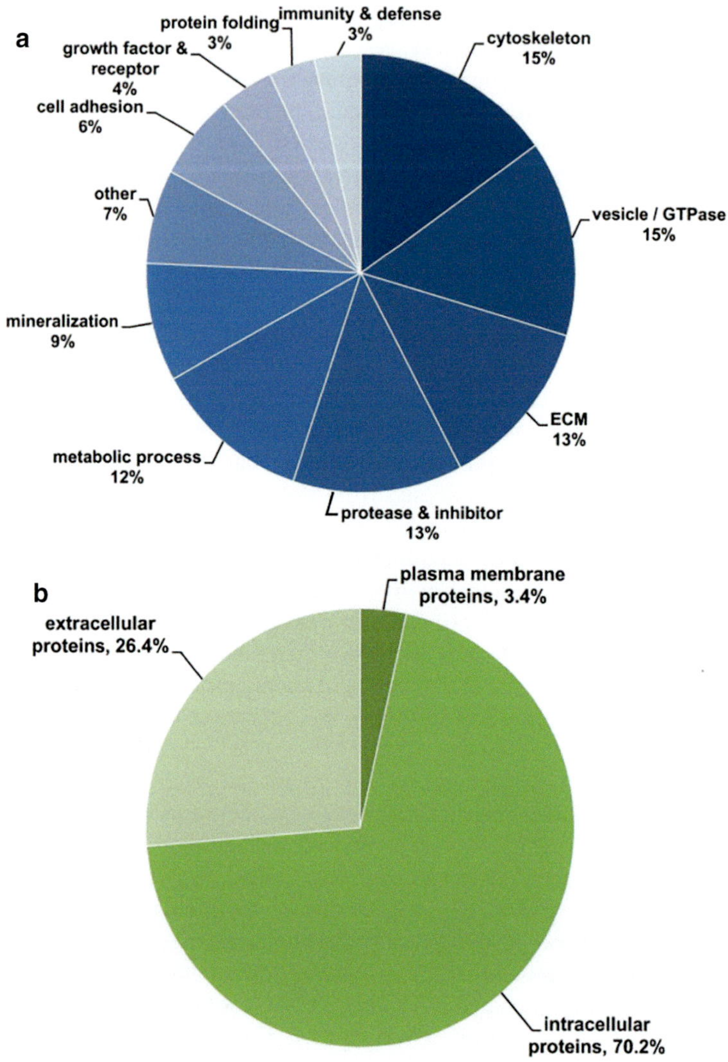

Fig. 6.16 Protein content of mineralizing matrix vesicles. Matrix vesicles were collected by centrifugation from a culture medium of human osteoblast-like cells and vesicle proteins were isolated from tryptic digests and analyzed by mass spectrometry. (**a**) Identification and classification of proteins (~300) into groups representative of their biological or chemical function using gene ontology term assignments. Numerous proteins were concerned functionally with matrix vesicle biogenesis (vesicle/RAB-GTPase) and mineralization. In the latter group were annexins and pyrophosphatases [144]. Those marked as "protease & inhibitor" included matrix metalloproteinases and tissue inhibitors of matrix metalloproteinases and cytoskeletal proteins constituted 15% of the total proteins identified in the vesicle digests [144]. (**b**) Proteins are categorized into groups representative of intracellular, extracellular, and plasma membrane constituents. Over 25% of proteins were characterized as extracellular, a value much greater than those reported for vesicles of nonmineralizing tissues [144]. Not surprisingly, intracellular proteins accounted for the greater part of the composition of matrix vesicles. Since little (3.4%) enrichment of plasma membrane proteins was observed, it is likely that these vesicles were exosomal in origin. (Image adapted from Schmidt et al. [144] with permission)

Other proteins reportedly present in matrix vesicles include cell surface antigens, TNAP, and several annexins (A1, A2, A4–A6, and A11 isoforms) [63, 148]. Annexin A6 may act as a calcium carrier while other annexins have the capacity to bind collagen and matrix vesicles [149], a property which may help align matrix vesicles at certain sites along collagen fibers to promote the onset and propagation of mineral in tissue extracellular matrices [43, 150]. Additional proteins associated with matrix vesicles include PHOSPHO1, NPP1, ENPP, and calcium-binding proteins like calbindins as well as extracellular matrix metalloproteases (MMP-2, -3, -9, and -13) [43, 110, 134]. The role of each of these proteins is thought to be closely linked to vesicle function in mediating mineral deposition and extracellular matrix growth and remodeling.

Additional work has further characterized matrix vesicles. As far as phospholipase in mineralized tissues is concerned, Wuthier and Lipscomb reported significant activity in growth plate cartilage [43], and other groups determined such enzyme action was principally associated with phospholipase A2 isoforms in chondrocytes and matrix vesicles [110, 115]. Investigation of the mechanism of expression of phospholipases in chondrocytes indicates that they respond rapidly to metabolites of vitamin D [151]. Moreover, vitamin D metabolites exert a profound effect on matrix vesicle biogenesis [152]. These effects are both direct, involving a fast response by the vitamin D receptor (1,25-mVDR) in activating phospholipase C, and a slower genomic response [153]. The genomic downstream effect influences protein kinase C (PKC)-dependent phosphorylation of the ERK family of mitogen-activated protein kinases (MAPKs) and promotes the formation of matrix vesicles [153, 154]. Finally, with respect to vesicle biogenesis once again, Hale and Wuthier demonstrated that the release of matrix vesicles from cultured epiphyseal chondrocytes correlated with changes in cellular actin distribution [155]. This finding is particularly relevant to the vesiculation process as the actin cytoskeleton is firmly associated with the cell membrane and hence changes in actin fiber distribution signal alterations in the fluidity of the cell membrane.

Not surprisingly, the lipid profiles of matrix vesicles and exosomes are similar. Differences which exist may relate directly to the function of matrix vesicles in mediating apatite deposition. For example, with regard to matrix vesicles and mineralizing tissues, Wuthier and Lipscomb have suggested the importance of the highly negatively charged phospholipid, phosphatidylserine, in the mineralization of cartilage (Fig. 6.8) [43]. They have shown that the content of this phospholipid in matrix vesicle membranes is greater than that in the plasma membrane [43]. Moreover, the location of phosphatidylserine within the vesicle membrane is unique. Whereas this lipid is found in the outer membrane of most extracellular vesicles, in matrix vesicles prepared from epiphyseal chondrocytes, much of the phosphatidylserine occupies the inner membrane [43]. More than likely, the distribution of phosphatidylserine and other phospholipids is dependent on the activities of many of the translocase enzyme systems in cell membranes as well as interactions with cellular and extracellular proteins, as discussed earlier. Boskey and Dick, Wu et al., and Goldberg et al. have all indicated that the unusual distribution of phosphatidylserine and other phospholipids reflects the functional requirements of the tissue. In this regard, phospholipids and phospholipid-protein complexes, for example, are thought to form part of the apatite mineral crystal nucleating complex in vertebrates [156–158]. Further, these observations lend support to a concept suggesting that, while the formation and membrane lipid composition of extracellular vesicles in mineralized tissues are similar to those of vesicles released from cells of soft tissues, the cargo of matrix vesicles is tissue specific. Thus, osteoblasts, chondrocytes, odontoblasts, and likely ameloblasts, all cells comprising vertebrate mineralized tissues, would be expected to secrete vesicles whose membrane phospholipid distribution is specific to mineralizing cells and a probable determining factor in what cargo is packaged and transported by their composite matrix vesicles. For these cells, calcium and phosphate ions would be the principal matrix vesicle cargo.

In summary of this section of the chapter, numerous studies indicate that cells release multiple types of extracellular vesicles. Biochemical evidence recounted above and morphological evidence noted earlier (see Chap. 5) confirm that both exosomes and microvesicles in mineralizing tissues bud or bleb from the plasma membrane of chondrocytes, osteoblasts, and odontoblasts. On the basis of investigations principally in soft tissues, it may be concluded that exosomes are likely generated from endosomes and their ionic composition is mediated by the release of calcium ions from SOCE and other plasma membrane calcium channels (see Chap. 5). As far as microvesicles are concerned, these structures are formed as a result of changes in the environment around the plasma membrane, causing its destabilization, bending, and blebbing. Although calcium-dependent phospholipases are known to be associated with these processes, whether activation of the enzymes drives the events has yet to be determined. Finally, consistent with earlier discussions, vesicle formation and shedding from the plasma membrane as well as control of SOCE and calcium ion channel activity are dependent on the metabolic state of the cell and responsive to the expression and secretion of the transcription factor, HIF. In the following section of this chapter, consideration is given of cross-talk between the low oxygen status of a tissue and metabolic pathways which link endosome trafficking with regulation of the autophagic pathway and mineralization.

6.9 Functional Implications of Autophagy, Mitochondrial Activity, and Vesicle Biogenesis in Mediating Tissue Mineralization

Many in the research field of vertebrate tissue mineralization are in agreement with the notion that mineral forms in tissues once vesicles leave the cell and become anchored in the extracellular matrix. However, information from studies of both prokaryotic and eukaryotic organisms indicates that the mineralization process likely begins in cells before the appearance of mineral associated with matrix vesicles or collagen fibrils in the tissue extracellular milieu. For example, Rohde and Mayer reported that amorphous calcium phosphate deposits were evident in the cytoplasm and vacuoles of osteoblasts [159]. They wrote that such deposits were "most probably (released) by membrane fusion to the outside of the cells…and secreted via an exocytic process." [159]. This observation [159] and related studies [44–48, 50, 160, 161] raise three questions: First, whether there is a relationship between the activities of the autophagic/endosomal pathway and bone formation and mineralization; second, whether mineral is formed in subcellular organelles of the autophagic/endosomal pathway; and, third, whether material or mineral deposits present in intracellular vesicles exist as free calcium and phosphate ions, amorphous calcium phosphate, or crystalline apatite.

With respect to a relationship between autophagy and bone mineralization, Liu et al. noted that there is a marked increase in autophagy during osteoblast differentiation in mice [161]. Conditional deletion of the gene, focal adhesion kinase family-interacting protein of 200 kDa (*FIP200*), part of a complex (ULK-ATG13-FIP200) which regulates the induction of autophagy, causes a decrease in femoral trabecular bone volume and thickness of the animals [161]. The term, "conditional," indicates that the mutation is silent until activated at a selected time in the life of the mice studied experimentally. In the same report, mutated bone cells isolated from the affected animals demonstrated compromised osteoblast terminal differentiation and mineralization [161]. These findings have provided credence for an earlier investigation showing an association between genes involved with the regulation of the autophagic pathway and a loss of bone mineral density; in the human wrist bone, this relation has led to a measurement which is a potent clinical assessment of osteoporosis [160].

With regard to the question of intracellular mineral deposition and autophagy, Nollet et al. found that osteoblasts at the onset of mineralization exhibited a sixfold increase in autophagy (measured by a change in the lipidated form of LC3-II protein) [47]. Most interestingly, these authors provided experimental evidence that autophagosomes contained apatitic crystallites before the induction of extracellular mineralization in bone (Fig. 6.17) [47]. A more recent study showed that autophagosomes associated with calcium and phosphorus led to the origin of calcified matrix vesicles and extracellular mineralization of osteoarthritic cartilage [162] As might be expected, knockdown of ATG7 and inhibition of BECLIN-1, two gene products required for the early stages of autophagosome formation, resulted in a significant decrease in mineralized nodule formation and, in *ATG5* mutant animals, there was a marked decrease in trabecular bone volume, width, and number [47].

Further support for the involvement of the autophagic system in intracellular mineralization was provided by a study of mineralizing murine osteoblast cultures using scanning electron-assisted dielectric impedance microscopy [50]. This technique images cells in aqueous media with nanoscale resolution, high contrast, and minimal radiation-induced damage. Applying this methodology, Iwayama and colleagues reported that calcium phosphate particles originated from multivesicular bodies [50]. Moreover, by utilizing both intracellular staining techniques and exocytosis inhibitors, they also demonstrated that the mineral particles were transported in lysosomes and secreted from the cells as exosomes (matrix vesicles) [50]. Finally,

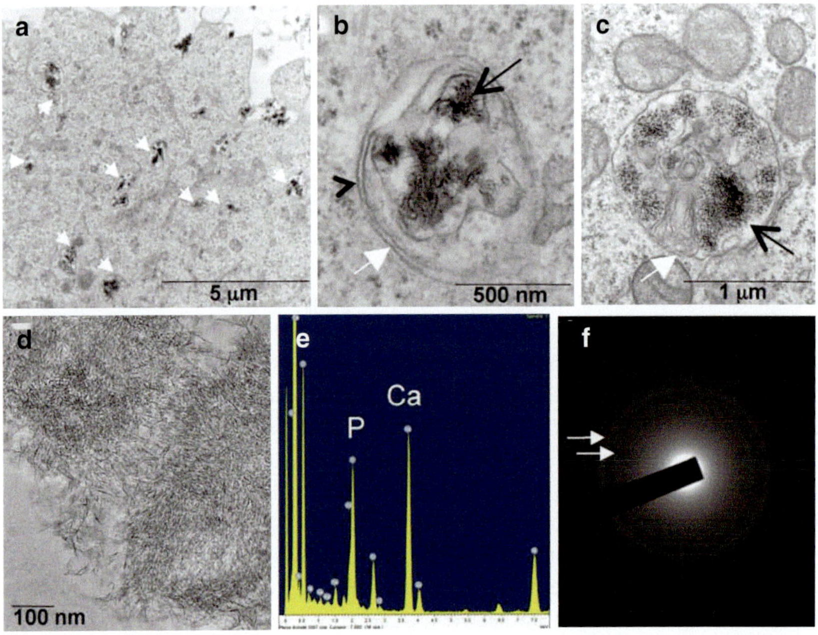

Fig. 6.17 Relationship between autophagy and the initiation of mineralization in cultured bone cells. To ascertain whether mineralization is dependent on the induction of autophagy, bone-like cells were permitted to undergo maturation in vitro. (**a–c**) During the culture period, the cells secreted the autophagic protein, LC3-II, in autophagosomes (white arrows) and the cultures mineralized. (**b**,**c**) Transmission electron microscopy of cultured bone cells, fixed with glutaraldehyde, post-fixed in osmium tetroxide, and sectioned and stained with uranyl acetate and lead citrate, demonstrated dense material (black arrows) within membrane-bound (black arrowhead) autophagic vacuoles. (**d**) A high-resolution image of dense material showed fine stippling. (**e**) Energy dispersive X-ray analysis of intracellular autophagosome-contained densities identified phosphorus (P) and calcium (Ca) as their principal elemental constituents. (**f**) Selected area electron diffraction of the dense material yielded reflections characteristic of hydroxyapatite (arrows). These data suggested that apatitic mineral crystals were associated with autophagic vacuoles [47]. In further work, such vacuoles could move to the plasma membrane as demonstrated by time-lapse confocal imaging (not shown) [89]. (Images adapted from Nollet et al. [47] with permission)

energy-dispersive X-ray microanalysis and Raman spectroscopy indicated that the mineral was amorphous calcium phosphate [50].

Relevant to the question of the structural form of material or mineral deposits found in intracellular vesicles, Mahamid et al. noted that osteogenic cells of the skull of embryonic and newborn mice maintained electron-dense, membrane-bound granules [45]. Based on elemental analysis and electron diffraction of such granules, these workers determined the deposits were a disordered calcium phosphate, a possible mineral phase precursor to hydroxyapatite [45]. A related study conducted by Boonrungsiman et al. showed that vesicles in a mouse osteoblast culture system contained an amorphous calcium phosphate and that intravesicular mineral trafficked to the extracellular culture environment (Fig. 6.18) [46]. Boonrungsiman et al. [46] also reported calcium and phosphorus deposits in the mitochondria of the cells and suggested that these organelles may act as a storage depot for the mineral ions, supporting similar suggestions made far earlier by Lehninger [44]. In addition, Kerschnitzki et al. demonstrated the presence of strikingly large intracellular stores of mineral particles within structures resembling membranous vesicles or mitochondrial networks in embryonic chick bone (Fig. 6.19) [48]. Outside the cells, aggregates of amorphous calcium phosphate were identified but they were not enveloped in vesicle membranes; such aggregates were in intimate contact with collagen fibrils in the tissue (Fig. 6.19) [48].

From the results of experiments described in this section of the chapter, it may be stated that intracellular vesicles generated in the autophagosome/endosome pathway load with calcium ions from reservoirs contained within the Golgi complex or the endoplasmic reticulum of the cell. The vesicles enclose their cargo in the form of disorganized clusters of calcium phosphate ions or an amorphous phase of calcium phosphate. For all vertebrate mineralized tissues except enamel, these intracellular vesicles traffic to the plasma membrane and enter the extracellular matrix carrying intact cargo. As previously noted in Chap. 3 for enamel, vesicle cargo appears in the extracellular matrix following vesicle fusion with the membranes defining the Tomes' process of ameloblasts. Outside the cell, embedded in the extracellular environment maintaining metastable levels of calcium and phosphate ions and putative prenucleation clusters, the contents of the exosomes or microvesicles

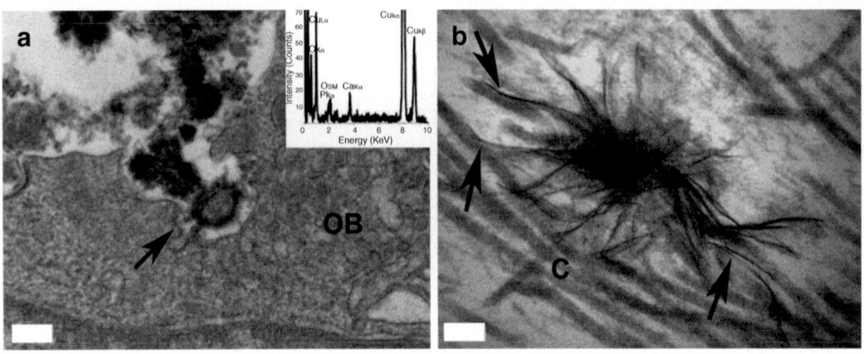

Fig. 6.18 Transmission electron microscopy of a mineralizing mouse calvarial osteoblast culture. Cultures were fixed in glutaraldehyde, post-fixed in osmium tetroxide, embedded, sectioned, and examined in bright field microscopy. (**a**) A section of an osteoblast (OB) shows a plasma membrane invagination (arrow) containing a rounded vesicle outlined by electron-dense material. The vesicle resides in the extracellular tissue space and abuts nodules which appear to be heavily mineralized. Inset. Energy dispersive X-ray analysis of the material within the vesicle demonstrates the presence of calcium and phosphorus, osmium (from fixative), copper (from the microscopic grid holding the section), and ubiquitous carbon. (**b**) Image of a region of extracellular matrix showing banded collagen (C) fibrils and an electron-dense extracellular nodule. Fine dense structures extend from the nodule to several fibrils (arrows). On selected area electron diffraction, such nodules produce a diffraction pattern characteristic of crystalline hydroxyapatite (data not shown). Scale bars = 0.2 μm. (Images adapted from Boonrungsiman et al. [46] with permission)

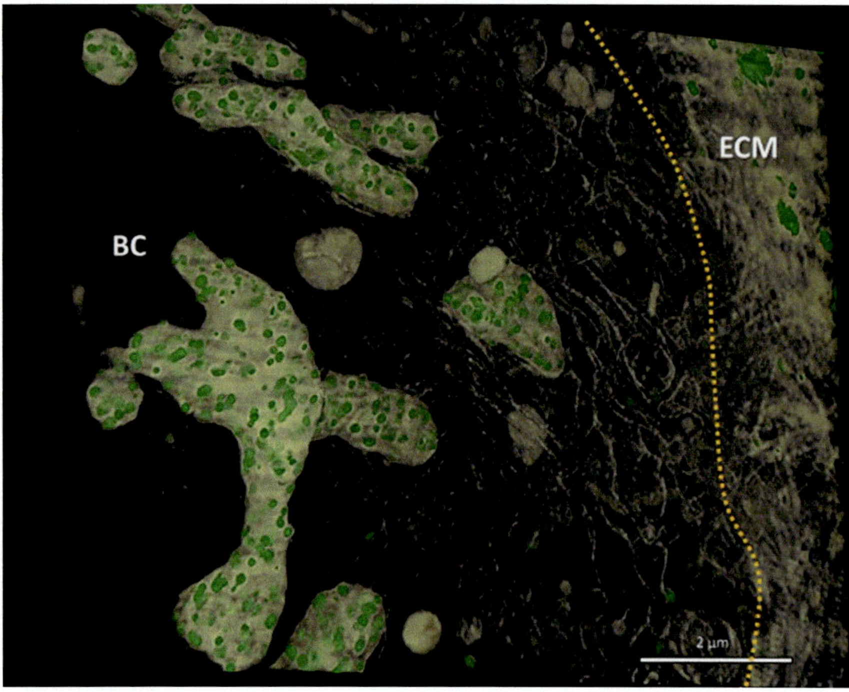

Fig. 6.19 The presence of mineral particles inside rounded or elongated, irregularly shaped organelles respectively resembling vesicles or mitochondria/mitochondrial networks in bone cells (BC) and in the extracellular matrix (ECM) of an embryonic chick femur. In this instance, tissue was prepared for a focused ion beam serial surface view (FIB-SSV) technique. Samples were initially subjected to high-pressure freezing and a freeze-substitution protocol incorporating acetone, glutaraldehyde, osmium, and uranyl acetate, followed by embedding in resin for subsequent ion beam milling. The image is a 3D reconstruction obtained from a volume-rendered data set through a cell which shows intracellular mineral particles (green) located in round, ovoid, and branched membranous structures and extracellular mineral disposed about collagen fibrils. The interface between the cell and the ECM is marked by a yellow broken line. (Image from Kerschnitzki et al. [48] with permission)

(matrix vesicles) undergo further changes as integral membrane proteins, like annexins, may accumulate additional calcium ions within these organelles [163, 164].

Almost all the studies presented earlier in this chapter have indicated that calcium ions contained in secreted endosomes or autophagosomes are fated to serve as extracellular centers of mineralization. While the focus has been on calcium ions, there has been little discussion concerning whether phosphate ions are sequestered within these vesicles and, if so, by what possible mechanism(s). Like that of calcium ions, a possible source of phosphate ions destined for vesicles is the mitochondrion (Fig. 6.11) [44, 80]. Considerable quantities of phosphate ions are utilized in mitochondrial oxidative phosphorylation to form ATP and other nucleotides. A decrease in oxidative phosphorylation caused by events such as limited hypoxia can result in a marked change in the ATP/ADP ratio and a corresponding increase in free phosphate ions. Liberation of these ions from the mitochondria or transport to endoplasmic reticulum membranes through their structurally close-linked mitochondria-associated endoplasmic reticulum membranes and contacts (MAM and MARC, respectively; see Chap. 5) would provide a pathway by which quantities of phosphate ions could be transferred, as required, into forming autophagic or endosomal vesicles. Another reservoir of phosphate ions exists in the plasma membrane microenvironment where the phospholipase-ENPP-TNAP-PHOSPHO1 system releases inorganic phosphate ions from phosphorylated compounds [43, 110, 111]. Ions from this source could be directly incorporated into forming exosomes/matrix vesicles or accumulated by these organelles following phosphate ion release or transport into the extracellular matrix.

At this point, it is plausible to consider that ion levels within the confines of matrix vesicles increase and contribute to intravesicular crystal nucleation and growth, leading to the disruption of vesicle membranes and the release and exposure of their contents to the extracellular milieu. These events may cause local perturbation in the already supersaturated and metastable levels of calcium and phosphate ions in the extracellular tissue fluid as described earlier in Chap. 2. Following such perturbation, these ions may remain free in solution or result in the formation of mineral foci and/or additional stable prenucleation clusters of calcium phosphate in the extracellular milieu. On recruitment of a critical number of solution ions and a consequent shift in local solution thermodynamics, mineral foci and prenucleation clusters may form initial calcium phosphate nuclei. The free ions and prenucleation clusters may also bind to extracellular molecules including collagen, bone sialoprotein, osteopontin, osteocalcin, and proteoglycans. Further crystal nucleation, growth, and development associated principally with collagen may then proceed in the extracellular matrix.

6.10 Exosomes Mediate Cell–Cell Communication

Throughout this chapter, studies have been delineated that provide evidence that cells of mineralizing tissues form and release a wide variety of vesicles. Once these organelles are liberated from the mother cell, the high protein-to-lipid ratio of the vesicles ensures their stability in extracellular tissue fluids [165]. As described above, such organelles are involved in the transit and liberation of calcium and phosphate ions which affect extracellular mineral deposition. Parallel investigations in other fields indicate that these cell-derived vesicles provide a broad spectrum of additional unique physiological functions, including roles in inflammation [166], immune surveillance [167], cancer pathogenesis [168], and transport of mRNA and MIR transcripts and other molecules to neighboring cells where the activity of these target cells is thereby modulated [169].

The address of the target cells is specified by the presence of distinct patterns of their surface receptors and adhesion molecules (like tetraspanins, integrins, proteoglycans, and lectins), which allow docking of vesicles having cognate receptors on their membrane surface. Following vesicle docking and subsequent vesicle-cell membrane fusion at clathrin- or caveolin-rich regions of the target cell plasma membrane, the vesicle cargo of any number of MIRs or proteins can activate one or more signal transduction pathways in promoting functional intercellular communication [49]. In this manner, the complex molecular circuitry required to mediate cell and matrix formation, feed-forward and feed-backward interactions, and maturation and a host of other critical activities may be controlled. In this regard, chondrocyte proliferation and hypertrophy may be exquisitely regulated in the epiphyseal growth plate, for example (see Chap. 3). Here, molecules such as parathyroid hormone-related protein and Indian hedgehog, contained and protected within intracellular vesicles, may be secreted at a singular stage of chondrocyte maturation and influence more distant target chondrocytes [58]. The proposal that vesicles mediate cell–cell communication and influence the function of neighboring and other target cells is supported by the report that chondrocytes express a mineralizing phenotype following their treatment in vitro with matrix vesicles sequestering bioactive molecules such as bone morphogenetic proteins (BMPs), vascular endothelial growth factor (VEGF), and non-collagenous proteins [65].

The concept of exosome regulation of cell–cell communication provides a useful explanation of how secreted proteins like WNT travel to distant cells. Diffusion of WNT is hampered by the presence of lipid molecules, which produce a hydrophobic environment around the protein which slows its transport [170]. Takada and co-workers showed that WNT, initially contained in multivesicular bodies, traffics and attaches to exosomes at their external surface with the involvement of SNARE molecules [170]. The WNT-exosome complex then interacts with cognate receptors on differentiating chondrocytes and mediates downstream changes in the activity of these cells [170].

With regard to the possible role of vesicular transport in affecting the coupling of osteoblast function to bone resorption, a study by Deng et al. indicated that matrix vesicles derived from osteoblasts transport receptor activator of nuclear factor kappa-B ligand, RANKL, the protein which activates bone resorption [171]. Accordingly, when osteoclast-like cells were treated in culture with vesicles collected from osteoblasts, these cells showed a marked increase in RANKL-RANK (receptor activator of nuclear factor kappa-B) signaling and a significant increase in osteoclast development and activity [171].

As briefly noted above, besides nucleic acids, mRNA, and protein molecules, exosomes enclose and shuttle many short (20–24 nucleotides), single-stranded non-coding microRNAs. These MIRs base-pair with the 3′ untranslated region of target genes and regulate about 30% of genes which encode proteins [172]. The mode of action of MIRs is complex and often involves batteries of different MIR species. Individual MIRs recruit proteins (argonaute and GW182) and base-pair with a gene regulatory network or an individual target mRNA. For example, genes concerned with regulating the autophagic gene, *mTOR*, are inactivated (dephosphorylated) by MIR-199 [173].

With respect to vertebrate mineralized tissues, matrix vesicles are enriched with MIR-122-5p, MIR-223-3p, and MIR-451-5p when compared with the mother cell [174, 175]. Each of the MIRs participates in the control of gene expression in differentiating chondrocytes [174]. During osteoblast differentiation, MIR-185-5p negatively regulates the expression of the transcription factor, distal-less homeobox (DLX)2 [176]. Odontogenesis is characterized by expression of MIR-143 and MIR-145, which downregulate *Krüppel-like factor* (*KLF*)*4* and *osterix (OSX)* [177], while amelogenesis is fine-tuned by MIR-21, MIR-31, MIR-135a, MIR-135b, MIR-153, MIR-298, and MIR-488 [178]. Such MIR species direct events involved with cell proliferation and terminal differentiation as well as chondrocyte–osteoblast and odontoblast–ameloblast interactions [174, 176–178]. While these activities are directly related to the development and deposition of mineral in the extracellular matrices of the respective vertebrate tissue, they also contribute to cell–cell communication in complex tissues like bone or dentin and, in doing so, they ensure that mineralization takes place in a specified and orderly manner.

6.11 Perspective: Calcium and Phosphate Ion Efflux from Cells: The Roles of Matrix Vesicles, Extracellular Vesicles, and Other Membrane-invested Transporters in Vertebrate Hard Tissue Mineralization

The earlier discussion of shell formation in invertebrate marine organisms provided an evolutionary basis for reasoning that vesicles are key organelles required for mineral formation of osteoblasts, chondrocytes, odontoblasts, cementoblasts, and tenocytes. In this context, coccolithophores, sea urchins, and other invertebrates form intracellular vesicles, which compartmentalize calcium ions in an amorphous mineral phase, shuttle them to the membranes enveloping the respective cells, and release them into extracellular spaces where they form the mineralized skeleton of the organism. Analogous to coccolithophores and other invertebrates, bone and other vertebrate mineralized tissues accumulate calcium and/or phosphate ions and export the cargo in exosomes or extracellular vesicles from the cell to the extracellular milieu. Once present in the extracellular matrix, these ion-loaded vesicles generate apatite by the continued accrual of calcium and phosphate ions from the metastable tissue fluid. In other words, like that of invertebrates, the process of mineral formation in vertebrate tissues begins in the cell and ends in the extracellular matrix.

It is not possible to comment with certainty on which of the vesicles derived and released by the cells of vertebrate mineralized tissues are equivalent to those described as matrix vesicles. However, based on observational data, it is likely that such cells produce a heterogeneous spectrum of vesicular organelles, and a proportion of them are exosomes which are generated through the intracellular autophagic-endosomal pathway. Another vesicle subset is required for mineralization. For this reason, the general terms, exosomes/microvesicles/extracellular vesicles, are reserved for organelles which are of intracellular origin and subsequently released from the cell, while the term, matrix vesicles, is reserved for those organelles which are present in the extracellular matrix and express characteristics which include high TNAP activity and specific lipids and proteins.

A major objective of this chapter was to review findings supporting the thesis that mineralization normally related to matrix vesicle secretion from cells of mineralized tissues begins within cells and continues to develop once localized in the extracellular matrix. Studies which provided early evidence for this concept were designed to prevent artifact formation resulting from potential calcium ion redistribution in the crowded cytosol, the need for identification of often transient species of mineral, and the requirement that the presence of intracellular mineral should be related to the metabolic state of the cell. More recent investigations have included the use of techniques which were not available at the time when matrix vesicle biogenesis was first reported. Such newer work has shown that, prior to the extracellular mineralization of bone, amorphous calcium phosphate particles originate from multivesicular bodies or are present in osteoblasts as autophagosomes [47]. In cell culture, mineral particles appear to be transported in lysosomes and secreted from the cells through exosomes [45, 46, 50]. Anchored to the extracellular matrix, the secreted, nascent mineral is reported as membrane-free amorphous calcium phosphate aggregates, some of which are in intimate contact with collagen fibrils [48]. Together, the studies noted here provide strong support for the hypothesis that the mineralization process in vertebrate tissues is initiated within cells and completed following interaction with components of the extracellular matrix.

The reports mentioned above indicate that vesicles of different origin may serve as calcium ion shuttles in cells of vertebrate hard and soft tissues. These shuttles include lysosomes, as just cited, and acidocalcisomes, both of which function as components of the internal membranes of the endolysosomal system of cells [45, 46, 50, 179]. Lysosomes receive calcium ions from the endoplasmic reticulum following binding of inositol metabolites to cognate receptors within the endoplasmic reticulum membrane [180]. Acidocalcisomes are similar to lysosomes in that they readily sequester calcium ions, maintain a low pH, express P2X receptors, and contain polyphosphates [181]. These characteristics suggest that acidocalcisomes may also ferry important calcium as well as phosphate and even polyphosphate ions from the cell to the extracellular matrix and as such act as possible partners with other membrane-invested transporters in the vertebrate mineralization process. There is also some evidence that mitochondria form vesicles targeted to the endolysosomal system [182]. Thus, it is plausible that lysosomes, lysosome-related organelles, acidocalcisomes, and possibly mitochondria-related vesicles can accumulate calcium ions released from internal cellular sources. Enclosed in membrane-limited vesicles, calcium ions can be trafficked to the plasma membrane. Following fusion with the plasma membrane, the organellar calcium ions are secreted into the extracellular milieu.

As an addendum to the discussion above, lysosomes may yet have another function in the cells of vertebrate mineralized tissues. In this context, it has been suggested that, following blebbing or vesiculation, the extravasation of lysosomal cargo proteins serves as a membrane-resealing mechanism [183, 184]. Accordingly, during periods of rapid mineralization in which cells such as osteoblasts or hypertrophic chondrocytes generate extracellular vesicles, there would be extensive breaches in their plasma membranes. During these phases of exocytosis, lysosome-related organelles would function to maintain a high level of membrane repair activity [184]. Thus, rather than leaving breaches in the cell membrane following extravasation of cellular contents, which would undoubtedly lead to cell death, the organelles which cause membrane disruption restore membrane integrity and maintain cell viability.

The scope of the endolysosomal system which connects the organelles mentioned above has been extended to include other membrane systems of the cell. This more integrated view is that the internal membranes of the cell are functionally joined with the plasma membrane to form an "endomembrane" system which encompasses not only those vesicles detailed earlier, but also the Golgi apparatus and the enveloping membranes of the nucleus. In this regard, the use of fluorescence microscopy and high-resolution imaging techniques shows that the endomembrane system generates localized domains in the cell [185]. These domains partition activities such as calcium ion accumulation and release from other activities like protein and lipid synthesis, yet interconnecting organelles of intracellular origin with the extracellular environment [185]. Such a more inclusive concept forms the foundation for a unified network which fosters calcium and phosphate ion trafficking from the cell to the extracellular matrix.

Not surprisingly, the highly interconnected endomembrane system includes the investing membranes of both endosomes as well as autophagosomes. This interconnection is of some importance since earlier discussion in the chapter might suggest a disconnect in function between those vesicles originating from the endoplasmic reticulum and those generated de novo in the cytosol by the autophagic pathway. Studies by Baixauli et al. showed that during development there was a coordinated exosome-autophagy response, which regulated both autophagolysosomal activity and release of exosome/extracellular vesicle cargo [186]. Details of this interaction are now becoming clearer in that there are robust interactions between the autophagy genes (*ATG5*, *ATG12,* and *ATG16*) required for autophagosome formation and those genes necessary for lysosome fusion and exosome biogenesis [187]. In addition, ATG5 prevents acidification of the multivesicular body by dissociating the vacuolar ATPase, while ATG16 enhances the fusion of that organelle with the plasma membrane and facilitates exosome release [187, 188]. As a result of the interactions between the endosomal and autophagic systems and the sharing of common proteins, cargo which includes calcium ions can be trafficked between the two types of vesicles and secreted from the cell as both exosomes/extracellular vesicles.

This chapter has described several systems which have evolved for accumulating, storing, transporting, and subsequently secreting calcium with or without phosphate ions from cells. While an autophagic system is involved in this regard, it is not precisely known whether autophagic vacuoles, acidocalcisomes, lysosome-related organelles, amphisomes, or multivesicular bodies are the essential structural elements for these activities. More probable, as mentioned earlier, osteoblasts, chondrocytes, odontoblasts, and other cells of vertebrate mineralized tissues generate populations of heterogeneous organelles, some being necessary for mineralization.

In such context and supporting this concept, the epiphyseal growth plate, for example, produces mineralizing vesicles (matrix vesicles) in its hypertrophic zone as well as non-mineralizing matrix vesicles which have been reported in regions of the tissue which normally do not mineralize [189]. Currently, it is not known if this situation reflects differences in the type of vesicle or if it mirrors functional alterations associated with vesicle structure and contents.

With respect to the likelihood that cells of vertebrate hard tissues generate a heterogeneous family of extracellular vesicles, studies of non-mineralizing systems indicate that there is some control of the type of organelle secreted by cells and passaged into their extracellular matrix or the systemic circulation. As described previously in this chapter, exosomes are regulated by the activity of ARF, phospholipase D2, and phosphatidic acid among several other molecules [86, 113, 115, 116]. Phosphatidic acid leads to syntenin-ALIX activation, which in turn results in intraluminal vesicle budding and exosome release [95]. ARF activation of phospholipase D2 is also required for those organelles generated from the plasma membrane as individual buds or blebs (microvesicles). Additionally, phosphorylation of ERK and activation of MLCK lead to constriction of the plasma membrane of the cell at the base of the microvesicles. As a consequence of this change in membrane dynamics, these organelles are released from the cell [113–117]. The fact that both microvesicles and exosomes share overlapping activating systems suggests that a mechanism may exist for reciprocal control of the type of vesicle generated by the cell [114].

The presence throughout cells of multiple vesicular organelles and vesicular pathways for the transport of calcium ions, calcium phosphate ion clusters, or possibly even nucleated mineral adds compelling support to the concept that the initiation of the mineralization process in vertebrate hard tissues occurs intracellularly and prior to the export of such cargo from the cell [45–48]. Most likely, the trafficking of the large numbers of such ions or ion clusters required for mineralization of the extracellular matrices of these tissues involves the participation of matrix vesicles/extracellular vesicles generated by both the endosomal and the autophagic pathways. Besides producing phagosomes and the membranes required for intact vesicle development and calcium ion transport, autophagy functions to mediate the survival of cells subjected to environmental stress. In this capacity, autophagy is responsive to conditions which are associated with vertebrate mineralization, that is, tissue hypoxia and the expression of the transcription factor, HIF [83, 190]. In this context, a decrease in the ratio of ATP/ADP and a corresponding increase in AMP in cells enhance the activity of AMP kinase [190]. Such changes lead to interference with the activity of mTOR, the major regulator limiting the autophagic pathway, and thereby promote both vesicle formation and cell survival [190]. Since autophagy plays a singular role in vertebrate mineralization, there is no surprise in the observation that undue stress influences matrix mineralization by disturbing the balance between the induction of autophagy and calcium loading by vesicles [17]. This type of amplified vesicular response is not unusual, for example, following the infection of cells with the hepatitis C virus, in which there is elevated autophagic pathway activity and a significant increase in the release of virus-containing exosomes [191].

In bone, dysregulated autophagy is linked to disturbed osteoclast development and function as well as the onset and progression of age-related osteoporosis and possibly Paget's disease [190, 192, 193].

It would be incomplete to leave this discussion of vesicular-calcium efflux without acknowledging that cells may also release calcium and phosphate ions or ion clusters to extracellular matrices as particles not associated with any intracellular lipid–protein (matrix vesicle) membrane system. In this regard, Stains and Gay showed that osteoblast-mediated mineralization required activation primarily of NCX membrane pumps and considered it plausible that "Ca^{2+} enters the cells and is then shuttled to the matrix-facing side of the cell, where it is in turn translocated out of the cell toward the site of mineralization by the Na-Ca exchanger." [17].

Compared with the plasmalemma membrane transporter, PMCA, NCX exhibits a low affinity but high capacity for calcium ion transport. While NCX protein function is viewed as being responsible for calcium ion transport from the cell, it can also mediate the electrogenic transport of ions into the cell depending on membrane potential and ionic concentrations [194]. Based on differences in the K_m of the enzyme for calcium ions across the plasma membrane, it is probable that NCX mediates transport of calcium ions from the cell and into the external environment [195, 196].

Not surprisingly, since the early descriptive studies of Stains, Gay, and colleagues [17, 19], there has been considerable research to advance the understanding of calcium ion movement within cells. As noted previously, studies of soft and hard tissues indicate that the endoplasmic reticulum and possibly mitochondria are the major sources of their cytoplasmic calcium ions [197, 198]. The endoplasmic reticulum is also recognized as serving a principal role in forming endosomes, lysosomes, intraluminal bodies, and additional vesicular cellular components [199]. The endoplasmic reticulum is as well the organelle which initiates phagophore and autosome formation and autophagic flux [200]. As for calcium transiting, the network of tubules and saccules of the endoplasmic reticulum in the filled state maintains a high calcium ion content (up to millimolar quantities), sustained by an influx of calcium ions from the systemic circulation [201]. Following binding of IP3 to its membrane receptors, the endoplasmic reticulum releases calcium ions, which are then taken up by endosomes, lysosomes, or other types of vesicles which subsequently ferry the ions along appropriate endosomal or autophagic pathways as detailed above [180, 202]. Entry of the cytosolic calcium ions into vesicles is enhanced by the presence of annexins, proteins in part comprising the vesicle phospholipid membranes, together with membrane TRPV channel proteins [203]. Other proteins like the calbindins strongly bind and buffer cytosolic calcium ions [204]. Once the level of calcium ions has fallen in the endoplasmic reticulum, SOCE channels open and ORAI proteins in the plasma membrane together with STIM1 proteins present in the lumen of the endoplasmic reticulum allow the refilling of the organelle with calcium ions. As in soft tissues, it is plausible to consider that PMCA is associated with STIM1 and ORAI1. In this case, PMCA would transfer residual cytosolic calcium ions to the extracellular matrix so as to maintain a low resting cytosolic calcium ion concentration.

Compared to what is known about calcium ion transport, much less is understood about phosphate ion homeostasis and transport in cells of vertebrate mineralized tissues. In this regard, mitochondria have been documented to sequester so-called calcium phosphate granules in vivo and in vitro and they have been associated with phosphate ion storage, homeostasis, and mineralization in cells of hard tissues [44, 205–207]. As there are great numbers of mitochondria per cell [208], conceptually the mitochondrial system could allow the transfer of considerable numbers of phosphate ions or calcium phosphate granules from the cell to the extracellular matrix without loss of energy status or the protection required by the cell against damaging high cytosolic calcium flux. Thus, just as the endoplasmic reticulum is concerned with calcium ion metabolism in hard tissue cells, mitochondria may be regarded as the cellular source of phosphate ion metabolism for mineral formation. Supporting such a possibility, Pei et al. reported that a transfer of droplets of "liquid-like" mineral precursors occurred during mitophagy from the mitochondria to autophagosomes or lysosomes in mesenchymal stem cells committed to an osteogenic lineage [209]. In a related manner, transfer of phosphate ions to calcium-loaded cytosolic vesicles of endosomal or autophagic origin would permit these organelles to harbor and transport coalesced, larger intravesicular calcium phosphate particle aggregates. From this perspective, crosstalk between mitochondria and components of the endoplasmic reticulum would provide the ions required for calcium phosphate formation, transit, and secretion within membrane-invested organelles. Immediate evidence that mitochondria participate in phosphate ion homeostasis and metabolism as conjectured above awaits further experimental investigation.

The observations that the cargo of vesicles may contain regulatory elements such as mRNA and MIR in addition to calcium ions [88, 174] have reinforced the concept that these organelles also function in intercellular communication [49, 171, 172]. In this regard, it is intriguing to consider that vesicles generated from one cell type might influence the activities of an adjacent, albeit phenotypically different, cell type. This form of communication could be the basis for the first rudiments of enamel crystals appearing on an existing bed of dentin crystals, a principal consideration supporting the long-standing hypothesis that dentin formation is a requisite for enamel formation in the tooth [210]. One way of explaining such cell and tissue interdependence is that components ferried within exosomes liberated from odontoblasts induce ameloblasts to synthesize proteins linked to enamel formation and mineralization. Such an intercellular association is also suggested by the reports mentioned previously [65, 166–169] and more recently that MIR in exosomes from oncogenic cells boosts metastatic development and tissue angiogenesis and inhibits apoptosis [211]. Although not experimentally documented for developing teeth, such targeting by vesicles from one cell type to another may provide insight into aspects of neuroectoderm–mesoderm interactions which characterize early stages of odontogenesis and amelogenesis. From this perspective and consistent with the present understanding of organelle function, it is likely that extracellular vesicles provide a mechanism for mediating interactions between developing tissues and in this way provide an exquisite level of temporal specificity and control.

References

1. Ray K. From fission to fusion: a perspective on the research that won the Nobel Prize in Physiology or Medicine. J Biosci. 2014;39:3–12.
2. Lowenstam HA, Weiner S. On biomineralization. New York: Oxford University Press; 1989.
3. Marsh ME. Biomineralization in coccolithophores. Gravit Space Biol Bull. 1999;12:5–14.
4. Wilbur KM, Watabe N. Experimental studies on calcification in molluscs and the alga Coccolithus huxleyi. Ann N Y Acad Sci. 1963;109:82–112.
5. Hawkins EK, Lee JJ, Fimiarz DK. Colony formation and sexual morphogenesis in the coccolithophore Pleurochrysis sp. (Haptophyta)(1). J Phycol. 2011;47:1344–9.
6. Mejia R. Will ion channels help coccolithophores adapt to ocean acidification? PLoS Biol. 2011;9:e1001087.
7. Brownlee C, Wheeler GL, Taylor AR. Coccolithophore biomineralization: new questions, new answers. Semin Cell Dev Biol. 2015;46:11–6.
8. Wilt FH, Benson SC. Isolation and culture of micromeres and primary mesenchyme cells. Methods Cell Biol. 2004;74:273–85.
9. Wilt FH. Biomineralization of the spicules of sea urchin embryos. Zool Sci. 2002;19:253–61.
10. Wilt FH, Killian CE, Hamilton P, Croker L. The dynamics of secretion during sea urchin embryonic skeleton formation. Exp Cell Res. 2008;314:1744–52.
11. Vidavsky N, Addadi S, Mahamid J, Shimoni E, Ben-Ezra D, Shpigel M, et al. Initial stages of calcium uptake and mineral deposition in sea urchin embryos. Proc Natl Acad Sci U S A. 2014;111:39–44.
12. Vidavsky N, Masic A, Schertel A, Weiner S, Addadi L. Mineral-bearing vesicle transport in sea urchin embryos. J Struct Biol. 2015;192:358–65.
13. Kornfeld S, Mellman I. The biogenesis of lysosomes. Annu Rev Cell Biol. 1989;5:483–525.
14. Hubbard MJ. Calcium transport across the dental enamel epithelium. Crit Rev Oral Biol Med. 2000;11:437–66.
15. Daverkausen-Fischer L, Pröls F. Regulation of calcium homeostasis and flux between the endoplasmic reticulum and the cytosol. J Biol Chem. 2022;298:102061.
16. Carafoli E. Intracellular calcium homeostasis. Annu Rev Biochem. 1987;56:395–433.
17. Stains JP, Gay CV. Asymmetric distribution of functional sodium-calcium exchanger in primary osteoblasts. J Bone Miner Res. 1998;13:1862–9.
18. Carafoli E, Santella L, Branca D, Brini M. Generation, control, and processing of cellular calcium signals. Crit Rev Biochem Mol Biol. 2001;36:107–260.
19. Stains JP, Weber JA, Gay CV. Expression of Na(+)/Ca(2+) exchanger isoforms (NCX1 and NCX3) and plasma membrane Ca(2+) ATPase during osteoblast differentiation. J Cell Biochem. 2002;84:625–35.
20. Tsumura M, Okumura R, Tatsuyama S, Ichikawa H, Muramatsu T, Matsuda T, et al. Ca2+ extrusion via Na+-Ca2+ exchangers in rat odontoblasts. J Endod. 2010;36:668–74.
21. Sato M, Sobhan U, Tsumura M, Kuroda H, Soya M, Masamura A, et al. Hypotonic-induced stretching of plasma membrane activates transient receptor potential vanilloid channels and sodium-calcium exchangers in mouse odontoblasts. J Endod. 2013;39:779–87.
22. Okumura R, Shibukawa Y, Muramatsu T, Hashimoto S, Nakagawa K, Tazaki M, et al. Sodium-calcium exchangers in rat ameloblasts. J Pharmacol Sci. 2010;112:223–30.
23. Uehara A, Iwamoto T, Nakamura Y, Imanaga I. Forefront of Na+/Ca2+ exchanger studies: physiology and molecular biology of monovalent cation sensitivities in Na+/Ca2+ exchangers. J Pharmacol Sci. 2004;96:19–22.
24. Chan B, Cheng IC, Rozita J, Gorshteyn I, Huang Y, Shaffer I, Chang C, Li W, Lytton J, Den Besten P, Zhang Y. Sodium/(calcium + potassium) exchanger NCKX4 optimizes KLK4 activity in the enamel matrix microenvironment to regulate ECM modeling. Front Physiol. 2023;14:1116091.

25. Robertson SYT, Wen X, Yin K, Chen J, Smith CE, Paine ML. Multiple calcium export exchangers and pumps are a prominent feature of enamel organ cells. Front Physiol. 2017;8:336.
26. Borke JL, Zaki AE, Eisenmann DR, Ashrafi SH, Ashrafi SS, Penniston JT. Expression of plasma membrane Ca++ pump epitopes parallels the progression of enamel and dentin mineralization in rat incisor. J Histochem Cytochem. 1993;41:175–81.
27. El-Yazbi AF, Cho WJ, Schulz R, Daniel EE. Calcium extrusion by plasma membrane calcium pump is impaired in caveolin-1 knockout mouse small intestine. Eur J Pharmacol. 2008;591:80–7.
28. Ong HL, Ambudkar IS. The endoplasmic reticulum-plasma membrane junction: a hub for agonist regulation of Ca(2+) entry. Cold Spring Harb Perspect Biol. 2020;12:a035253.
29. Conrard L, Tyteca D. Regulation of membrane calcium transport proteins by the surrounding lipid environment. Biomolecules. 2019;9:513.
30. James P, Maeda M, Fischer R, Verma AK, Krebs J, Penniston JT, et al. Identification and primary structure of a calmodulin binding domain of the Ca2+ pump of human erythrocytes. J Biol Chem. 1988;263:2905–10.
31. Krapivinsky G, Krapivinsky L, Stotz SC, Manasian Y, Clapham DE. POST, partner of stromal interaction molecule 1 (STIM1), targets STIM1 to multiple transporters. Proc Natl Acad Sci U S A. 2011;108:19234–9.
32. Go CK, Hooper R, Aronson MR, Schultz B, Cangoz T, Nemani N, et al. The Ca(2+) export pump PMCA clears near-membrane Ca(2+) to facilitate store-operated Ca(2+) entry and NFAT activation. Sci Signal. 2019;12:eaaw2627.
33. Micaroni M. The role of calcium in intracellular trafficking. Curr Mol Med. 2010;10:763–73.
34. Kunzelmann K, Nilius B, Owsianik G, Schreiber R, Ousingsawat J, Sirianant L, et al. Molecular functions of anoctamin 6 (TMEM16F): a chloride channel, cation channel, or phospholipid scramblase? Pflugers Arch. 2014;466:407–14.
35. Ehlen HW, Chinenkova M, Moser M, Munter HM, Krause Y, Gross S, et al. Inactivation of anoctamin-6/Tmem16f, a regulator of phosphatidylserine scrambling in osteoblasts, leads to decreased mineral deposition in skeletal tissues. J Bone Miner Res. 2013;28:246–59.
36. Ousingsawat J, Wanitchakool P, Schreiber R, Wuelling M, Vortkamp A, Kunzelmann K. Anoctamin-6 controls bone mineralization by activating the calcium transporter NCX1. J Biol Chem. 2015;290:6270–80.
37. Duong HA, Le KT, Soulema AL, Yueh RH, Scheuner MT, Holick MF, et al. Gnathodiaphyseal dysplasia: report of a family with a novel mutation of the ANO5 gene. Oral Surg Oral Med Oral Pathol Oral Radiol. 2016;121:e123–8.
38. Suzuki J, Umeda M, Sims PJ, Nagata S. Calcium-dependent phospholipid scrambling by TMEM16F. Nature. 2010;468:834–8.
39. Bonucci E. Fine structure and histochemistry of "calcifying globules" in epiphyseal cartilage. Z Zellforsch Mikrosk Anat. 1970;103:192–217.
40. Ali SY, Sajdera SW, Anderson HC. Isolation and characterization of calcifying matrix vesicles from epiphyseal cartilage. Proc Natl Acad Sci U S A. 1970;67:1513–20.
41. Anderson HC, Matsuzawa T, Sajdera SW, Ali SY. Membranous particles in calcifying cartilage matrix. Trans N Y Acad Sci. 1970;32:619–30.
42. Amizuka N, Hasegawa T, Oda K, Luiz de Freitas PH, Hoshi K, Li M, et al. Histology of epiphyseal cartilage calcification and endochondral ossification. Front Biosci (Elite Ed). 2012;4:2085–100.
43. Wuthier RE, Lipscomb GF. Matrix vesicles: structure, composition, formation and function in calcification. Front Biosci (Landmark Ed). 2011;16:2812–902.
44. Lehninger AL. Mitochondria and biological mineralization processes: an exploration. Horiz Biochem Biophys. 1977;4:1–30.
45. Mahamid J, Sharir A, Gur D, Zelzer E, Addadi L, Weiner S. Bone mineralization proceeds through intracellular calcium phosphate loaded vesicles: a cryo-electron microscopy study. J Struct Biol. 2011;174:527–35.

46. Boonrungsiman S, Gentleman E, Carzaniga R, Evans ND, McComb DW, Porter AE, et al. The role of intracellular calcium phosphate in osteoblast-mediated bone apatite formation. Proc Natl Acad Sci U S A. 2012;109:14170–5.
47. Nollet M, Santucci-Darmanin S, Breuil V, Al-Sahlanee R, Cros C, Topi M, et al. Autophagy in osteoblasts is involved in mineralization and bone homeostasis. Autophagy. 2014;10:1965–77.
48. Kerschnitzki M, Akiva A, Ben Shoham A, Asscher Y, Wagermaier W, Fratzl P, et al. Bone mineralization pathways during the rapid growth of embryonic chicken long bones. J Struct Biol. 2016;195:82–92.
49. Asghar S, Litherland GJ, Lockhart JC, Goodyear CS, Crilly A. Exosomes in intercellular communication and implications for osteoarthritis. Rheumatology. 2019;59:57–68.
50. Iwayama T, Okada T, Ueda T, Tomita K, Matsumoto S, Takedachi M, et al. Osteoblastic lysosome plays a central role in mineralization. Sci Adv. 2019;5:eaax0672.
51. Bonucci E. Fine structure of early cartilage calcification. J Ultrastruct Res. 1967;20:33–50.
52. Anderson HC. Molecular biology of matrix vesicles. Clin Orthop Relat Res. 1995;314:266–80.
53. Tanimura A, McGregor DH, Anderson HC. Matrix vesicles in atherosclerotic calcification. Proc Soc Exp Biol Med. 1983;172:173–7.
54. Golub EE, Schattschneider SC, Berthold P, Burke A, Shapiro IM. Induction of chondrocyte vesiculation in vitro. J Biol Chem. 1983;258:616–21.
55. Akisaka T, Gay CV. The plasma membrane and matrix vesicles of mouse growth plate chondrocytes during differentiation as revealed in freeze-fracture replicas. Am J Anat. 1985;173:269–86.
56. Anderson HC. Matrix vesicles and calcification. Curr Rheumatol Rep. 2003;5:222–6.
57. Landis WJ, Silver FH. The structure and function of normally mineralizing avian tendons. Comp Biochem Physiol A Mol Integr Physiol. 2002;133:1135–57.
58. Shapiro IM, Landis WJ, Risbud MV. Matrix vesicles: are they anchored exosomes? Bone. 2015;79:29–36.
59. Garces-Ortiz M, Ledesma-Montes C, Reyes-Gasga J. Presence of matrix vesicles in the body of odontoblasts and in the inner third of dentinal tissue: a scanning electron microscopic study. Med Oral Patol Oral Cir Bucal. 2013;18:e537–41.
60. Takano Y, Sakai H, Baba O, Terashima T. Differential involvement of matrix vesicles during the initial and appositional mineralization processes in bone, dentin, and cementum. Bone. 2000;26:333–9.
61. Sela J, Tamari I, Hirschfeld Z, Bab I. Transmission electron microscopy of reparative dentin in rat molar pulps. Primary mineralization via extracellular matrix vesicles. Acta Anat (Basel). 1981;109:247–51.
62. Lowe J, Bab I, Stein H, Sela J. Primary calcification in remodeling haversian systems following tibial fracture in rats. Clin Orthop Relat Res. 1983;176:291–7.
63. Kirsch T, Harrison G, Golub EE, Nah HD. The roles of annexins and types II and X collagen in matrix vesicle-mediated mineralization of growth plate cartilage. J Biol Chem. 2000;275:35577–83.
64. Cecil RNA, Anderson HC. Freeze-fracture studies of matrix vesicle calcification in epiphyseal growth plate. Metab Bone Dis Rel Res. 1978;1:89–95.
65. Nahar NN, Missana LR, Garimella R, Tague SE, Anderson HC. Matrix vesicles are carriers of bone morphogenetic proteins (BMPs), vascular endothelial growth factor (VEGF), and noncollagenous matrix proteins. J Bone Miner Metab. 2008;26:514–9.
66. Xiao Z, Blonder J, Zhou M, Veenstra TD. Proteomic analysis of extracellular matrix and vesicles. J Proteomics. 2009;72:34–45.
67. Thyberg J, Friberg U. The lysosomal system in endochondral growth. Prog Histochem Cytochem. 1978;10:1–46.
68. Gurunathan S, Kang MH, Jeyaraj M, Qasim M, Kim JH. Review of the isolation, characterization, biological function, and multifarious therapeutic approaches of exosomes. Cells. 2019;8:307.

69. Kalra H, Simpson RJ, Ji H, Aikawa E, Altevogt P, Askenase P, et al. Vesiclepedia: a compendium for extracellular vesicles with continuous community annotation. PLoS Biol. 2012;10:e1001450.
70. Stahl PD, Raposo G. Extracellular vesicles: exosomes and microvesicles, integrators of homeostasis. Physiology (Bethesda). 2019;34:169–77.
71. Barteneva NS, Fasler-Kan E, Bernimoulin M, Stern JN, Ponomarev ED, Duckett L, et al. Circulating microparticles: square the circle. BMC Cell Biol. 2013;14:23.
72. Csordas G, Renken C, Varnai P, Walter L, Weaver D, Buttle KF, et al. Structural and functional features and significance of the physical linkage between ER and mitochondria. J Cell Biol. 2006;174:915–21.
73. Somogyi E, Petersson U, Hultenby K, Wendel M. Calreticulin--an endoplasmic reticulum protein with calcium-binding activity is also found in the extracellular matrix. Matrix Biol. 2003;22:179–91.
74. Petersson U, Somogyi E, Reinholt FP, Karlsson T, Sugars RV, Wendel M. Nucleobindin is produced by bone cells and secreted into the osteoid, with a potential role as a modulator of matrix maturation. Bone. 2004;34:949–60.
75. Lavoie C, Meerloo T, Lin P, Farquhar MG. Calnuc, an EF-hand Ca(2+)-binding protein, is stored and processed in the Golgi and secreted by the constitutive-like pathway in AtT20 cells. Mol Endocrinol. 2002;16:2462–74.
76. Brodeur J, Larkin H, Boucher R, Thériault C, St-Louis SC, Gagnon H, et al. Calnuc binds to LRP9 and affects its endosomal sorting. Traffic. 2009;10:1098–114.
77. Larkin H, Costantino S, Seaman MN, Lavoie C. Calnuc function in endosomal sorting of lysosomal receptors. Traffic. 2016;17:416–32.
78. Naslavsky N, Caplan S. The enigmatic endosome - sorting the ins and outs of endocytic trafficking. J Cell Sci. 2018;131:jcs216499.
79. Raposo G, Stoorvogel W. Extracellular vesicles: exosomes, microvesicles, and friends. J Cell Biol. 2013;200:373–83.
80. Tang C, Wei Y, Gu L, Zhang Q, Li M, Yuan G, et al. Biomineral precursor formation is initiated by transporting calcium and phosphorus clusters from the endoplasmic reticulum to mitochondria. Adv Sci. 2020;7:1902536.
81. Phillips MJ, Voeltz GK. Structure and function of ER membrane contact sites with other organelles. Nat Rev Mol Cell Biol. 2016;17:69–82.
82. Landis WJ, Hauschka BT, Rogerson CA, Glimcher MJ. Electron microscopic observations of bone tissue prepared by ultracryomicrotomy. J Ultrastruct Res. 1977;59:185–206.
83. He C, Klionsky DJ. Regulation mechanisms and signaling pathways of autophagy. Annu Rev Genet. 2009;43:67–93.
84. Murrow L, Debnath J. ATG12-ATG3 connects basal autophagy and late endosome function. Autophagy. 2015;11:961–2.
85. Wesselborg S, Stork B. Autophagy signal transduction by ATG proteins: from hierarchies to networks. Cell Mol Life Sci. 2015;72:4721–57.
86. Liou W, Geuze HJ, Geelen MJ, Slot JW. The autophagic and endocytic pathways converge at the nascent autophagic vacuoles. J Cell Biol. 1997;136:61–70.
87. Ponpuak M, Mandell MA, Kimura T, Chauhan S, Cleyrat C, Deretic V. Secretory autophagy. Curr Opin Cell Biol. 2015;35:106–16.
88. Iavello A, Frech VS, Gai C, Deregibus MC, Quesenberry PJ, Camussi G. Role of Alix in miRNA packaging during extracellular vesicle biogenesis. Int J Mol Med. 2016;37:958–66.
89. Rodriguez-Boulan E, Kreitzer G, Musch A. Organization of vesicular trafficking in epithelia. Nat Rev Mol Cell Biol. 2005;6:233–47.
90. Sheehan P, Waites CL. Coordination of synaptic vesicle trafficking and turnover by the Rab35 signaling network. Small GTPases. 2019;10:54–63.
91. Rothman JE. Intracellular membrane fusion. Adv Second Messenger Phosphoprotein Res. 1994;29:81–96.
92. Astro V, de Curtis I. Plasma membrane-associated platforms: dynamic scaffolds that organize membrane-associated events. Sci Signal. 2015;8:1–11.

93. Hessvik NP, Llorente A. Current knowledge on exosome biogenesis and release. Cell Mol Life Sci. 2018;75:193–208.
94. Casares D, Escribá PV, Rosselló CA. Membrane lipid composition: effect on membrane and organelle structure, function and compartmentalization and therapeutic avenues. Int J Mol Sci. 2019;20:2167.
95. Lambaerts K, Wilcox-Adelman SA, Zimmermann P. The signaling mechanisms of syndecan heparan sulfate proteoglycans. Curr Opin Cell Biol. 2009;21:662–9.
96. Charrin S, le Naour F, Silvie O, Milhiet PE, Boucheix C, Rubinstein E. Lateral organization of membrane proteins: tetraspanins spin their web. Biochem J. 2009;420:133–54.
97. Yanez-Mo M, Barreiro O, Gordon-Alonso M, Sala-Valdes M, Sanchez-Madrid F. Tetraspanin-enriched microdomains: a functional unit in cell plasma membranes. Trends Cell Biol. 2009;19:434–46.
98. Shimaoka M, Kawamoto E, Gaowa A, Okamoto T, Park EJ. Connexins and integrins in exosomes. Cancers (Basel). 2019;11:106.
99. Sims PJ, Wiedmer T. Unraveling the mysteries of phospholipid scrambling. Thromb Haemost. 2001;86:266–75.
100. Sahu SK, Gummadi SN, Manoj N, Aradhyam GK. Phospholipid scramblases: an overview. Arch Biochem Biophys. 2007;462:103–14.
101. Devaux PF, Herrmann A, Ohlwein N, Kozlov MM. How lipid flippases can modulate membrane structure. Biochim Biophys Acta. 2008;1778:1591–600.
102. Ståhl AL, Johansson K, Mossberg M, Kahn R, Karpman D. Exosomes and microvesicles in normal physiology, pathophysiology, and renal diseases. Pediatr Nephrol. 2019;34:11–30.
103. Kalienkova V, Clerico Mosina V, Paulino C. The groovy TMEM16 Family: molecular mechanisms of lipid scrambling and ion conduction. J Mol Biol. 2021;433:166941.
104. Daleke DL. Regulation of transbilayer plasma membrane phospholipid asymmetry. J Lipid Res. 2003;44:233–42.
105. Allhusen JS, Conboy JC. The ins and outs of lipid flip-flop. Acc Chem Res. 2017;50:58–65.
106. Martínez MC, Freyssinet JM. Deciphering the plasma membrane hallmarks of apoptotic cells: phosphatidylserine transverse redistribution and calcium entry. BMC Cell Biol. 2001;2:20.
107. Falzone ME, Rheinberger J, Lee BC, Peyear T, Sasset L, Raczkowski AM, et al. Structural basis of Ca(2+)-dependent activation and lipid transport by a TMEM16 scramblase. Elife. 2019;8:e43229.
108. Aloulou A, Rahier R, Arhab Y, Noiriel A, Abousalham A. Phospholipases: an overview. Methods Mol Biol. 2018;1835:69–105.
109. Shuttleworth TJ. Arachidonic acid, ARC channels, and Orai proteins. Cell Calcium. 2009;45:602–10.
110. Stewart AJ, Leong DTK, Farquharson C. PLA(2) and ENPP6 may act in concert to generate phosphocholine from the matrix vesicle membrane during skeletal mineralization. FASEB J. 2018;32:20–5.
111. Borza R, Salgado-Polo F, Moolenaar WH, Perrakis A. Structure and function of the ecto-nucleotide pyrophosphatase/phosphodiesterase (ENPP) family: tidying up diversity. J Biol Chem. 2022;298:101526.
112. Mendelson K, Evans T, Hla T. Sphingosine 1-phosphate signalling. Development. 2014;141:5–9.
113. Laulagnier K, Grand D, Dujardin A, Hamdi S, Vincent-Schneider H, Lankar D, et al. PLD2 is enriched on exosomes and its activity is correlated to the release of exosomes. FEBS Lett. 2004;572:11–4.
114. Muralidharan-Chari V, Clancy J, Plou C, Romao M, Chavrier P, Raposo G, et al. ARF6-regulated shedding of tumor cell-derived plasma membrane microvesicles. Curr Biol. 2009;19:1875–85.
115. Mebarek S, Abousalham A, Magne D, Do le D, Bandorowicz-Pikula J, Pikula S, et al. Phospholipases of mineralization competent cells and matrix vesicles: roles in physiological and pathological mineralizations. Int J Mol Sci. 2013;14:5036–129.

116. Cocucci E, Meldolesi J. Ectosomes and exosomes: Shedding the confusion between extracellular vesicles. Trends Cell Biol. 2015;25:364–72.
117. Tricarico C, Clancy J, D'Souza-Schorey C. Biology and biogenesis of shed microvesicles. Small GTPases. 2017;8:220–32.
118. Record M, Carayon K, Poirot M, Silvente-Poirot S. Exosomes as new vesicular lipid transporters involved in cell-cell communication and various pathophysiologies. Biochim Biophys Acta. 2014;1841:108–20.
119. Egea-Jimenez AL, Zimmermann P. Phospholipase D and phosphatidic acid in the biogenesis and cargo loading of extracellular vesicles. J Lipid Res. 2018;59:1554–60.
120. Fang Y, Vilella-Bach M, Bachmann R, Flanigan A, Chen J. Phosphatidic acid-mediated mitogenic activation of mTOR signaling. Science. 2001;294:1942–5.
121. Foster DA, Salloum D, Menon D, Frias MA. Phospholipase D and the maintenance of phosphatidic acid levels for regulation of mammalian target of rapamycin (mTOR). J Biol Chem. 2014;289:22583–8.
122. Antonyak MA, Wilson KF, Cerione RA. R(h)oads to microvesicles. Small GTPases. 2012;3:219–24.
123. Lhermusier T, Chap H, Payrastre B. Platelet membrane phospholipid asymmetry: from the characterization of a scramblase activity to the identification of an essential protein mutated in Scott syndrome. J Thromb Haemost. 2011;9:1883–91.
124. Shapiro IM, Adams CS, Freeman T, Srinivas V. Fate of the hypertrophic chondrocyte: microenvironmental perspectives on apoptosis and survival in the epiphyseal growth plate. Birth Defects Res C Embryo Today. 2005;75:330–9.
125. Wang T, Gilkes DM, Takano N, Xiang L, Luo W, Bishop CJ, et al. Hypoxia-inducible factors and RAB22A mediate formation of microvesicles that stimulate breast cancer invasion and metastasis. Proc Natl Acad Sci U S A. 2014;111:E3234–42.
126. Mayorga LS, Cebrian I, Verma M, Hoops S, Bassaganya-Riera J. Reconstruction of endosomal organization and function by a combination of ODE and agent-based modeling strategies. Biol Direct. 2018;13:25.
127. Feng Y, He D, Yao Z, Klionsky DJ. The machinery of macroautophagy. Cell Res. 2014;24:24–41.
128. Galluzzi L, Baehrecke EH, Ballabio A, Boya P, Bravo-San Pedro JM, Cecconi F, et al. Molecular definitions of autophagy and related processes. EMBO J. 2017;36:1811–36.
129. Oku M, Sakai Y. Three distinct types of microautophagy based on membrane dynamics and molecular machineries. Bioessays. 2018;40:e1800008.
130. Mizushima N, Noda T, Yoshimori T, Tanaka Y, Ishii T, George MD, et al. A protein conjugation system essential for autophagy. Nature. 1998;395:395–8.
131. Srinivas V, Shapiro IM. Chondrocytes embedded in the epiphyseal growth plates of long bones undergo autophagy prior to the induction of osteogenesis. Autophagy. 2006;2:215–6.
132. Kim J, Kundu M, Viollet B, Guan KL. AMPK and mTOR regulate autophagy through direct phosphorylation of Ulk1. Nat Cell Biol. 2011;13:132–41.
133. Axe EL, Walker SA, Manifava M, Chandra P, Roderick HL, Habermann A, et al. Autophagosome formation from membrane compartments enriched in phosphatidylinositol 3-phosphate and dynamically connected to the endoplasmic reticulum. J Cell Biol. 2008;182:685–701.
134. Hasegawa T. Ultrastructure and biological function of matrix vesicles in bone mineralization. Histochem Cell Biol. 2018;149:289–304.
135. Srinivas V, Bohensky J, Shapiro IM. Autophagy: a new phase in the maturation of growth plate chondrocytes is regulated by HIF, mTOR and AMP kinase. Cells Tissues Organs. 2009;189:88–92.
136. Claude-Taupin A, Jia J, Mudd M, Deretic V. Autophagy's secret life: secretion instead of degradation. Essays Biochem. 2017;61:637–47.
137. Xing H, Tan J, Miao Y, Lv Y, Zhang Q. Crosstalk between exosomes and autophagy: a review of molecular mechanisms and therapies. J Cell Mol Med. 2021;25:2297–308.

138. Morgan AJ, Platt FM, Lloyd-Evans E, Galione A. Molecular mechanisms of endolysosomal Ca2+ signalling in health and disease. Biochem J. 2011;439:349–74.
139. Rao SK, Huynh C, Proux-Gillardeaux V, Galli T, Andrews NW. Identification of SNAREs involved in synaptotagmin VII-regulated lysosomal exocytosis. J Biol Chem. 2004;279:20471–9.
140. Docampo R, Huang G. Acidocalcisomes of eukaryotes. Curr Opin Cell Biol. 2016;41:66–72.
141. Moreno-Sanchez D, Hernandez-Ruiz L, Ruiz FA, Docampo R. Polyphosphate is a novel pro-inflammatory regulator of mast cells and is located in acidocalcisomes. J Biol Chem. 2012;287:28435–44.
142. Chaudhary SC, Kuzynski M, Bottini M, Beniash E, Dokland T, Mobley CG, et al. Phosphate induces formation of matrix vesicles during odontoblast-initiated mineralization in vitro. Matrix Biol. 2016;52-54:284–300.
143. Tompkins K, George A, Veis A. Characterization of a mouse amelogenin [A-4]/M59 cell surface receptor. Bone. 2006;38:172–80.
144. Schmidt JR, Kliemt S, Preissler C, Moeller S, von Bergen M, Hempel U, et al. Osteoblast-released matrix vesicles, regulation of activity and composition by sulfated and non-sulfated glycosaminoglycans. Mol Cell Proteomics. 2016;15:558–72.
145. Morgan NE, Cutrona MB, Simpson JC. Multitasking rab proteins in autophagy and membrane trafficking: a focus on Rab33b. Int J Mol Sci. 2019;20:3916.
146. Kawasaki N, Isogaya K, Dan S, Yamori T, Takano H, Yao R, et al. TUFT1 interacts with RABGAP1 and regulates mTORC1 signaling. Cell Discov. 2018;4:1.
147. Salian S, Cho TJ, Phadke SR, Gowrishankar K, Bhavani GS, Shukla A, et al. Additional three patients with Smith-McCort dysplasia due to novel RAB33B mutations. Am J Med Genet A. 2017;173:588–95.
148. Thouverey C, Malinowska A, Balcerzak M, Strzelecka-Kiliszek A, Buchet R, Dadlez M, et al. Proteomic characterization of biogenesis and functions of matrix vesicles released from mineralizing human osteoblast-like cells. J Proteomics. 2011;74:1123–34.
149. Minashima T, Small W, Moss SE, Kirsch T. Intracellular modulation of signaling pathways by annexin A6 regulates terminal differentiation of chondrocytes. J Biol Chem. 2012;287:14803–15.
150. Veschi EA, Bolean M, Strzelecka-Kiliszek A, Bandorowicz-Pikula J, Pikula S, Granjon T, et al. Localization of annexin A6 in matrix vesicles during physiological mineralization. Int J Mol Sci. 2020;21:1367.
151. Schwartz Z, Boyan B. The effects of vitamin D metabolites on phospholipase A2 activity of growth zone and resting zone cartilage cells in vitro. Endocrinology. 1988;122:2191–8.
152. Swain LD, Schwartz Z, Caulfield K, Brooks BP, Boyan BD. Nongenomic regulation of chondrocyte membrane fluidity by 1,25-(OH)2D3 and 24,25-(OH)2D3 is dependent on cell maturation. Bone. 1993;14:609–17.
153. Schwartz Z, Ehland H, Sylvia VL, Larsson D, Hardin RR, Bingham V, et al. 1alpha,25-dihydroxyvitamin D(3) and 24R,25-dihydroxyvitamin D(3) modulate growth plate chondrocyte physiology via protein kinase C-dependent phosphorylation of extracellular signal-regulated kinase 1/2 mitogen-activated protein kinase. Endocrinology. 2002;143:2775–86.
154. Sylvia VL, Schwartz Z, Holmes SC, Dean DD, Boyan BD. 24,25-(OH)2D3 regulation of matrix vesicle protein kinase C occurs both during biosynthesis and in the extracellular matrix. Calcif Tissue Int. 1997;61:313–21.
155. Hale JE, Wuthier RE. The mechanism of matrix vesicle formation. Studies on the composition of chondrocyte microvilli and on the effects of microfilament-perturbing agents on cellular vesiculation. J Biol Chem. 1987;262:1916–25.
156. Boskey AL, Dick BL. The effect of phosphatidylserine on in vitro hydroxyapatite growth and proliferation. Calcif Tissue Int. 1991;49:193–6.
157. Wu LN, Genge BR, Sauer GR, Wuthier RE. Characterization and reconstitution of the nucleational complex responsible for mineral formation by growth plate cartilage matrix vesicles. Connect Tissue Res. 1996;35:309–15.

158. Goldberg M, Opsahl S, Aubin I, Septier D, Chaussain-Miller C, Boskey A, et al. Sphingomyelin degradation is a key factor in dentin and bone mineralization: lessons from the fro/fro mouse. The chemistry and histochemistry of dentin lipids. J Dent Res. 2008;87:9–13.
159. Rohde M, Mayer H. Exocytotic process as a novel model for mineralization by osteoblasts in vitro and in vivo determined by electron microscopic analysis. Calcif Tissue Int. 2007;80:323–36.
160. Zhang L, Guo YF, Liu YZ, Liu YJ, Xiong DH, Liu XG, et al. Pathway-based genome-wide association analysis identified the importance of regulation-of-autophagy pathway for ultradistal radius BMD. J Bone Miner Res. 2010;25:1572–80.
161. Liu F, Fang F, Yuan H, Yang D, Chen Y, Williams L, et al. Suppression of autophagy by FIP200 deletion leads to osteopenia in mice through the inhibition of osteoblast terminal differentiation. J Bone Miner Res. 2013;28:2414–30.
162. Yan J, Shen M, Sui B, Lu W, Han X, Wan Q, et al. Autophagic LC3(+) calcified extracellular vesicles initiate cartilage calcification in osteoarthritis. Sci Adv. 2022;8:eabn1556.
163. Genge BR, Wu LN, Wuthier RE. Differential fractionation of matrix vesicle proteins. Further characterization of the acidic phospholipid-dependent Ca2(+)-binding proteins. J Biol Chem. 1990;265:4703–10.
164. Gerke V, Moss SE. Annexins: from structure to function. Physiol Rev. 2002;82:331–71.
165. Hade MD, Suire CN, Suo Z. Mesenchymal stem cell-derived exosomes: applications in regenerative medicine. Cells. 2021;10:1959.
166. Console L, Scalise M, Indiveri C. Exosomes in inflammation and role as biomarkers. Clin Chim Acta. 2019;488:165–71.
167. Zitvogel L, Regnault A, Lozier A, Wolfers J, Flament C, Tenza D, et al. Eradication of established murine tumors using a novel cell-free vaccine: dendritic cell-derived exosomes. Nat Med. 1998;4:594–600.
168. Kalluri R. The biology and function of exosomes in cancer. J Clin Invest. 2016;126:1208–15.
169. Nawaz M, Fatima F, Nazarenko I, Ekström K, Murtaza I, Anees M, et al. Extracellular vesicles in ovarian cancer: applications to tumor biology, immunotherapy and biomarker discovery. Expert Rev Proteomics. 2016;13:395–409.
170. Takada R, Satomi Y, Kurata T, Ueno N, Norioka S, Kondoh H, et al. Monounsaturated fatty acid modification of Wnt protein: its role in Wnt secretion. Dev Cell. 2006;11:791–801.
171. Deng L, Wang Y, Peng Y, Wu Y, Ding Y, Jiang Y, et al. Osteoblast-derived microvesicles: a novel mechanism for communication between osteoblasts and osteoclasts. Bone. 2015;79:37–42.
172. Filipowicz W, Bhattacharyya SN, Sonenberg N. Mechanisms of post-transcriptional regulation by microRNAs: are the answers in sight? Nat Rev Genet. 2008;9:102–14.
173. Elkayam E, Faehnle CR, Morales M, Sun J, Li H, Joshua-Tor L. Multivalent recruitment of human argonaute by GW182. Mol Cell. 2017;67:646–58.
174. Lin Z, McClure MJ, Zhao J, Ramey AN, Asmussen N, Hyzy SL, et al. MicroRNA contents in matrix vesicles produced by growth plate chondrocytes are cell maturation dependent. Sci Rep. 2018;8:3609.
175. Asmussen NC, Cohen DJ, Lin Z, McClure MJ, Boyan BD, Schwartz Z. Specific microRNAs found in extracellular matrix vesicles regulate proliferation and differentiation in growth plate chondrocytes. Calcif Tissue Int. 2021;109:455–68.
176. Chang H, Wang Y, Liu H, Nan X, Wong S, Peng S, et al. Mutant Runx2 regulates amelogenesis and osteogenesis through a miR-185-5p-Dlx2 axis. Cell Death Dis. 2017;8:3221.
177. Liu H, Lin H, Zhang L, Sun Q, Yuan G, Zhang L, et al. miR-145 and miR-143 regulate odontoblast differentiation through targeting Klf4 and Osx genes in a feedback loop. J Biol Chem. 2013;288:9261–71.
178. Yin K, Hacia JG, Zhong Z, Paine ML. Genome-wide analysis of miRNA and mRNA transcriptomes during amelogenesis. BMC Genomics. 2014;15:998.
179. Klumperman J, Raposo G. The complex ultrastructure of the endolysosomal system. Cold Spring Harb Perspect Biol. 2014;6:a016857.

180. Morgan AJ, Davis LC, Wagner SK, Lewis AM, Parrington J, Churchill GC, et al. Bidirectional Ca(2)(+) signaling occurs between the endoplasmic reticulum and acidic organelles. J Cell Biol. 2013;200:789–805.
181. Docampo R, Moreno SN. Acidocalcisomes. Cell Calcium. 2011;50:113–9.
182. Soubannier V, McLelland GL, Zunino R, Braschi E, Rippstein P, Fon EA, et al. A vesicular transport pathway shuttles cargo from mitochondria to lysosomes. Curr Biol. 2012;22:135–41.
183. Castro-Gomes T, Corrotte M, Tam C, Andrews NW. Plasma membrane repair is regulated extracellularly by proteases released from lysosomes. PLoS One. 2016;11:e0152583.
184. Atanassoff AP, Wolfmeier H, Schoenauer R, Hostettler A, Ring A, Draeger A, et al. Microvesicle shedding and lysosomal repair fulfill divergent cellular needs during the repair of streptolysin O-induced plasmalemmal damage. PLoS One. 2014;9:e89743.
185. Csordas G, Weaver D, Hajnoczky G. Endoplasmic reticulum-mitochondrial contactology: Structure and signaling functions. Trends Cell Biol. 2018;28:523–40.
186. Baixauli F, Lopez-Otin C, Mittelbrunn M. Exosomes and autophagy: Coordinated mechanisms for the maintenance of cellular fitness. Front Immunol. 2014;5:403.
187. Guo H, Chitiprolu M, Roncevic L, Javalet C, Hemming FJ, Trung MT, et al. Atg5 disassociates the V1V0-ATPase to promote exosome production and tumor metastasis independent of canonical macroautophagy. Dev Cell. 2017;43:716–30.
188. Xu J, Camfield R, Gorski SM. The interplay between exosomes and autophagy - partners in crime. J Cell Sci. 2018;131:jcs215210.
189. Kirsch T, Nah HD, Shapiro IM, Pacifici M. Regulated production of mineralization-competent matrix vesicles in hypertrophic chondrocytes. J Cell Biol. 1997;137:1149–60.
190. Shapiro IM, Layfield R, Lotz M, Settembre C, Whitehouse C. Boning up on autophagy: the role of autophagy in skeletal biology. Autophagy. 2014;10:7–19.
191. Mori H, Fukuhara T, Ono C, Tamura T, Sato A, Fauzyah Y, et al. Induction of selective autophagy in cells replicating hepatitis C virus genome. J Gen Virol. 2018;99:1643–57.
192. Xiao L, Xiao Y. The autophagy in osteoimmonology: self-eating, maintenance, and beyond. Front Endocrinol. 2019;10:490.
193. Yin X, Zhou C, Li J, Liu R, Shi B, Yuan Q, et al. Autophagy in bone homeostasis and the onset of osteoporosis. Bone Res. 2019;7:28.
194. Khananshvili D. Sodium-calcium exchangers (NCX): molecular hallmarks underlying the tissue-specific and systemic functions. Pflugers Arch. 2014;466:43–60.
195. Carafoli E. Membrane transport of calcium: an overview. Methods Enzymol. 1988;157:3–11.
196. Khananshvili D. Basic and editing mechanisms underlying ion transport and regulation in NCX variants. Cell Calcium. 2020;85:102131.
197. Iannotti JP, Brighton CT, Stambough JL, Storey BT. Calcium flux and endogenous calcium content in isolated mammalian growth-plate chondrocytes, hyaline-cartilage chondrocytes, and hepatocytes. J Bone Joint Surg Am. 1985;67:113–20.
198. Somlyo AP, Bond M, Somlyo AV. Calcium content of mitochondria and endoplasmic reticulum in liver frozen rapidly in vivo. Nature. 1985;314:622–5.
199. Friedman JR, Dibenedetto JR, West M, Rowland AA, Voeltz GK. Endoplasmic reticulum-endosome contact increases as endosomes traffic and mature. Mol Biol Cell. 2013;24:1030–40.
200. Schwarz DS, Blower MD. The endoplasmic reticulum: structure, function and response to cellular signaling. Cell Mol Life Sci. 2016;73:79–94.
201. Stathopulos PB, Ikura M. Structural aspects of calcium-release activated calcium channel function. Channels (Austin). 2013;7:344–53.
202. Pihán P, Hetz C. Getting intimate: lysosomes and ER rendezvous to control autophagy. Cell Calcium. 2020;91:102249.
203. Mirsaeidi M, Gidfar S, Vu A, Schraufnagel D. Annexins family: insights into their functions and potential role in pathogenesis of sarcoidosis. J Transl Med. 2016;14:89.
204. Schmidt H. Three functional facets of calbindin D-28k. Front Mol Neurosci. 2012;5:25.
205. Greenawalt JW, Rossi CS, Lehninger AL. Effect of active accumulation of calcium and phosphate ions on the structure of rat liver mitochondria. J Cell Biol. 1964;23:21–38.

206. Becker GL, Chen CH, Greenawalt JW, Lehninger AL. Calcium phosphate granules in the hepatopancreas of the blue crab Callinectes sapidus. J Cell Biol. 1974;61:316–26.
207. Lehninger AL. The possible role of mitochondria and phosphocitrate in biological calcification. In: Westbroek P, de Jong EW, editors. Biomineralization and biological metal accumulation. Dordrecht: D. Reidel Publishing Co.; 1983. p. 107–21.
208. Cole LW. The evolution of per-cell organelle number. Front Cell Dev Biol. 2016;4:85.
209. Pei DD, Sun JL, Zhu CH, Tian FC, Jiao K, Anderson MR, et al. Contribution of mitophagy to cell-mediated mineralization: revisiting a 50-year-old conundrum. Adv Sci. 2018;5:1800873.
210. Smith CE, Hu Y, Hu JC, Simmer JP. Ultrastructure of early amelogenesis in wild-type, Amelx(-/-), and Enam(-/-) mice: enamel ribbon initiation on dentin mineral and ribbon orientation by ameloblasts. Mol Genet Genomic Med. 2016;4:662–83.
211. Padda J, Khalid K, Khedr A, Patel V, Al-Ewaidat OA, Tasnim F, et al. Exosome-derived microRNA: efficacy in cancer. Cureus. 2021;13:e17441.

Collagen-based Mineralization of Bones, Teeth, and Other Vertebrate Skeletal Tissues

7

In physics we have dealt hitherto only with periodic crystals. To a humble physicist's mind these are very interesting and complicated objects; they constitute one of the most fascinating and complex material structures by which inanimate nature puzzles his wits. Yet, compared with the aperiodic crystal, they are rather plain and dull. The difference in structure is of the same kind as that between an ordinary wallpaper in which the same pattern is repeated again and again in regular periodicity and a masterpiece of embroidery, say a Raphael tapestry, which shows no dull repetition, but an elaborate, coherent, meaningful design traced by the great master. [1]

Glossary

5″ untranslated region of mRNA The mRNA region upstream from the initiation codon.

Chondrichthyes A taxonomic class of fish characterized by the acquisition of a cartilaginous skeleton.

Discoidin domain receptor A non-integrin receptor tyrosine kinase which recognizes collagens as its ligands.

Ehlers-Danlos syndromes A group of inherited collagen diseases which affect the structure of connective tissues, including bones, skin, and blood vessels, and lead to laxity of joints and thin and loose skin, among other characteristics.

GLUT transporter A protein which transports glucose across the plasma membrane.

Glutamine A serum amino acid, similar to glutamic acid, whose carboxyl group is replaced by an amide group.

Glycolysis A catabolic pathway which converts glucose into pyruvate, thereby releasing energy in the form of ATP and NADH. When the final component is lactic acid, the process is termed anaerobic glycolysis.

Hydroxyproline An amino acid derived from proline which undergoes hydroxylation as a post-translational event.

Octacalcium phosphate and amorphous calcium phosphate Possible precursor mineral phases of hydroxyapatite.

I. M. Shapiro, W. J. Landis, *Mechanisms of Mineralization of Vertebrate Skeletal and Dental Tissues*, https://doi.org/10.1007/978-3-031-34304-9_7

Osteichthyes A taxonomic class of fish characterized by the presence of a bony endoskeleton.

Osteogenesis imperfecta A disorder of bone and other connective tissues commonly caused by mutations in the COL1A1 and COL1A2 chains of type I collagen (*COLI*) and referred to as brittle bone disease.

Oxidative phosphorylation The process of generating energy (ATP and NADH) in the mitochondrion.

Pyrrolidine A cyclic tetrahydropyrrole resulting from the interaction of proline and hydroxyproline residues and whose formation during collagen synthesis stabilizes collagen secondary structure.

Quarter-stagger An arrangement of individual collagen molecules in fibrils, fibers, and fiber bundles, structured such that they overlap by approximately one-fourth of their length.

Reactive oxygen species Unstable and highly reactive oxygen-containing molecules or free radicals which interact with other cellular molecules and whose accumulation in cells may damage nucleic acids and proteins and lead to cell death.

Receptor tyrosine kinases A large family of cell membrane receptors, important in both normal and pathological processes, which have a high affinity for hormones, growth factors, and chemokines.

Schmid metaphyseal chondrodysplasia A genetic disorder of growth resulting from mutations in type X collagen (*COLX*) and leading to dwarfism.

Telopeptides Short non-helical peptide sequences at both the C- and N-termini of the collagen molecule.

Transphosphorylation In receptor biology, the process by which a phosphoryl group is transferred from one protein component of a receptor to a second protein component.

Abbreviations and Proper Names for Genes/Proteins and Other Molecules

BAG-75	Bone acidic glycoprotein-75
BSP	Bone sialoprotein
COL1A1	Type I collagen alpha 1 chain
COL2A1	Type II collagen alpha 1 chain
COLI/COL1	Type I collagen
COLII/COL2	Type II collagen
COLX/COL10	Type X collagen
COLXA1	Type X collagen alpha 1 chain
DDR	Discoidin domain receptor
GLUT1	Glucose transporter 1
GM-CSF	Granulocyte macrophage-colony stimulating factor
HIF	Hypoxia inducible factor
IL-4	Interleukin-4
JNK	c-Jun N-terminal kinase
MAPK	Mitogen-activated protein kinase
MMP	Matrix metalloproteinase

NADH	Nicotinamide adenine dinucleotide hydride; also coenzyme-1 or reduced NAD^+
NF-kB	Nuclear factor kappa-light-chain-enhancer of activated B cells
OCN	Osteocalcin
OPN	Osteopontin
RUNX	Runt-related transcription factor
SIBLING	Small integrin-binding ligand N-linked glycoprotein
SLC	Solute carrier family
SLRP	Small leucine-rich repeat *protein*
SOX9	Sex determining region Y (SRY)-box 9
TCA cycle	Tricarboxylic acid cycle
TNAP	Tissue non-specific alkaline phosphatase
WNT	Wingless-related integration site

7.1 Bone, Soup, and Glue

In previous chapters of this book, the importance of mineralized tissues has been considered in the evolution of the vertebrate kingdom. In this regard, the appearance of bones and teeth was noted to provide animals with a marked biological advantage regarding predation, protection, and survival. Humans killed animals for their food content of protein and fat and for their skin, serving as a means of warmth and safeguard against the ravages of weather. Animal bones in the Paleolithic era and more advanced times were used for many other purposes, including ornaments, religious symbols, and musical instruments. Larger bones from mammoths, for example, were utilized as shelters. Resourceful hominids developed tools and weapons from animal bones to extend their domains well beyond small locales. The Bible describes the exploits of Samson, who "found a new jawbone of an ass," so he reached out and took it and killed a thousand men with it (Fig. 7.1).

While man the hunter went about animal predation and episodes of fighting his fellow man, *Homo sapiens* was also examining more closely the nature of bone itself. It was soon recognized that the marrow contained within bone was a ready source of nutrients which could be obtained by boiling bone in water. By the sixteenth century, extraction of bone marrow led to its decomposition and the formation of a sticky residue, glue, which was of immediate practical value. The subsequent nutritional and growing industrial uses of bone prompted the famous nineteenth century physical chemist, Berzelius, to perform the first gravimetric analysis of this tissue. He reported that 100 parts of human bone contained 53.04 parts phosphate of lime with a little fluoride of calcium, 11.30 parts carbonate of lime, 1.20 parts soda and chloride of sodium, 1.16 parts false phosphate of magnesia, and 33.30 parts animal matter (cartilage and vessels) [2]. For further informative reading of the history of the chemical analysis of bone, Dorozhkin has written an outstanding review of the topic [3]. If Berzelius had had access to modern technology, he would have noted that a high percentage of organic molecules within the bone soup and even the bone glue were amino acids of which glycine, proline, and hydroxyproline predominated.

Fig. 7.1 Samson slaying the Philistines. Artist: Julius Schnorr von Carolsfeld; engraver: Z. Scheckel; source: *Die Bibel in Bildern,* Plate 80. Judges 15, King James Bible, "And he found a new jawbone of an ass, and put forth his hand, and took it, and slew a thousand men therewith. And Samson said, With the jawbone of an ass, heaps upon heaps, with the jaw of an ass have I slain a thousand men." (Credit: The Hathi Digital Trust Library and the Columbia University Library)

While it is now well recognized that bone is a two-phase composite system, the Berzelius study was the first to show that at least a third of bone matter is organic. Such a component may be visualized by a common home or laboratory study of treating chicken bones with vinegar. This process dissolves away the mineral, leaving behind a rubbery but tough demineralized structure with the same shape and dimensions of the original bone. This type of experiment demonstrates that bone has an acid-soluble mineral phase and a vinegar-insoluble organic matrix phase which permeates the whole of the bone. Indeed, two centuries from the time of Berzelius the composition of 35% of the dry weight of bone and 31% of the dry weight of dentin is known to be protein. In both tissues, over 90% of the protein is collagen.

Chen et al. [4] put the strong interaction between the two phases of bone on a scientific footing. These investigators carefully removed the mineral phase of bone and reported that, despite the loss of apatite, the demineralized bone retained the integrity of its form, a result suggesting that it was the organic protein phase which held the bone together. However, removal of the protein (leaving the mineral behind) yielded retention at the macroscopic level of many of the morphologic features of the tissue. A closer examination of the protein components of the organic matrix indicated that they knit together to create a tight continuous structure [4]. Likewise, the individual crystals of the mineral phase interconnected to form a mesh of apatite (Fig. 7.2). In other words, bone is a two-phase composite material, in which the integration of both organic and mineral components together maintains the structure and integrity of the tissue. Such a composite, coherent interwoven structure is presumed to exist for dentin, cementum, and the other normal as well as abnormal vertebrate mineralized tissues (see Chap. 10).

The very intimate relationship between the two phases comprising bone and other vertebrate mineralized tissues is consistent with the concept that collagen, either passively or actively, directs the arrangement and disposition of mineral within extracellular matrices. If that be the case, then this consideration raises the possibility that collagen molecules may serve as heterogeneous nucleators for apatite formation (see Chap. 2). Furthermore, the idea introduces the question as to why some collagenous tissues mineralize and others do not.

To address such basic concerns, it is first necessary to understand the structure, assembly, and organization of collagen and then ascertain how the molecule participates in the process of biological mineralization. Parenthetically, while these

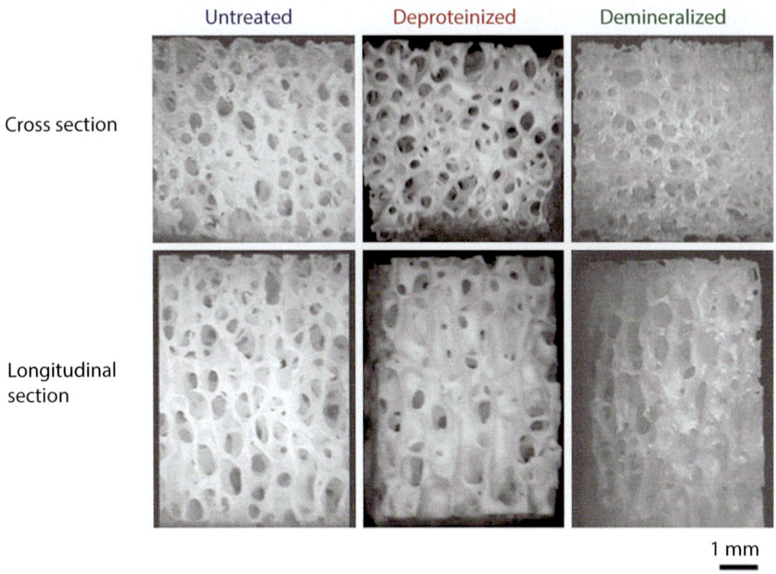

Fig. 7.2 Stereomicrographs illustrating the structure of cancellous bone from a bovine femur following deproteinization and demineralization. Vertebrate bone consists of a composite organic–inorganic mixture principally of COLI and a carbonated apatite, respectively, along with water and small amount of lipid. These components form a complex structure, ordered hierarchically from sub-micron to millimeter length scales. To gain insight into possible organic–inorganic interaction, cancellous bone was treated with 2.6 wt% sodium hypochlorite for 14 days to remove collagen and other proteins ("deproteinized") or 0.6 N hydrochloric acid for 6 days ("demineralized") to remove mineral. Results of imaging show little structural difference between untreated, deproteinized, and demineralized samples. Even after complete deproteinization, the native trabecular bone structure was maintained through mineral crystals fusing together to form a continuous network. Likewise, complete sample demineralization demonstrated that collagen sheets exhibited structural integrity. (Image from Chen et al. [4] with permission)

characteristics relate directly to the collagen species which are present in bone, dentin, cementum, and the mineralizing tendons of certain avians (principally type I collagen, COLI) as well as those which are found in the hypertrophic region of the cartilaginous growth plate (type X collagen, COLX), they also pertain to the major collagen species of mineralizing cartilage (type II collagen, COLII). Collagen is not present in significant quantities in enamel or enameloid.

7.2 General Comments about Collagen and Collagen Structure

Collagen (Greek, "kolla" or "glue") is the name of a family of proteins which appear in both vertebrates as well as invertebrates. It is the most prominent of all proteins found in vertebrates, and it is present in a multitude of tissues, including the skin, lung, eye, intervertebral disc, blood vessel, cartilage, dentin, cementum, tendon, ligament, and bone (see Table 7.1). The major collagen species are those which form fibrils (COLI,

Table 7.1 Summary of common collagen genes and proteins present in soft and mineralized tissues and their relationship to disease

Protein type	Gene	Protein composition	Tissue	Disease
Type I collagen	*COL1A1* *COL1A2*	$\alpha1(I)_2\ \alpha2(I)$	Skin, tendon, bone, dentin, cementum, vasculature, ligaments, annulus fibrosus, cornea	Osteogenesis imperfecta, Ehlers-Danlos syndrome, osteoporosis, degenerative disc disease
Type II collagen	*COL2A1*	$\alpha1(II)_3$	Cartilage, vitreous humor, nucleus pulposus of the intervertebral disc	Spondyloepimetaphyseal dysplasia, achondrogenesis, hypochondrogenesis, Ehlers-Danlos syndrome, Kniest dysplasia, Stickler syndrome, degenerative disc disease
Type III collagen	*COL3A1*	$\alpha1(III)_3$	Skin, tissue reticular fibers, vasculature	Ehlers-Danlos syndrome
Type IX collagen	*COL9A1* *COL9A2* *COL9A3*	$\alpha1(IX)\ \alpha2(IX)\ \alpha3(IX)$	Cartilage, vitreous humor, cornea	Osteoarthrosis, multiple epiphyseal dysplasia
Type X collagen	*COL10A*	$\alpha1(X)_3$	Epiphyseal cartilage	Schmid metaphyseal chondrodysplasia
Type XI collagen	*COL11A1* *COL11A2* *COL11A3*	$\alpha1(XI)\ \alpha2(XI)\ \alpha3(XI)$	Cartilage, intervertebral disc	Chondrodysplasia, osteoarthrosis, degenerative disc disease
Type XII collagen	*COL12A1*	$\alpha1(XII)_3$	Periosteum, perichondrium, tendon, ligaments, skin	Mixed myopathy/Ehlers-Danlos syndrome

II, III, V, and XI), and these fibrils function to accommodate applied tensile forces on or within their respective tissues. There is also some support for the view that COLX, a unique and transient species found only in the growth plates of bones, together with COLII and other extracellular matrix constituents, is organized to support those forces applied to centers of longitudinal growth [5]. The term, "transient," is used here to indicate that, when growth plate lengthening and development are terminated, the mineral and organic matrix, including COLII and COLX, are removed by osteo(chondro) clasts and replaced by bone and the tissue components of the marrow cavity.

About 45 separate genes in either the human or mouse genome encode a family of 28 proteins, all of which are described as "collagens." In common with other gene families, over 1000 mutations of collagen genes have been described which lead to disease states affecting both soft and hard tissues [6]. For instance, osteogenesis imperfecta or brittle bone disease is the result of mutations in the *COLI* genes (*COL1A1* and *COL1A2)* [7] or, more rarely, in other proteins such as cartilage-associated protein (CRTAP) or prolyl 3-hydroxylase 1 (P3H1/LEPRE1) [8]. Mutations in *COLI*, *III*, *V*, and *XII* cause Ehlers-Danlos syndrome [9]. Certain of these conditions are discussed further in Chap. 10. Other examples include a mutation in *COL2A1* which causes dwarfism (Kniest dysplasia) [10] and a frameshift mutation in *COLXA1* which prevents trimer formation and induces Schmid metaphyseal chondrodysplasia, a disease state common in Mormon populations which impairs childhood growth (Table 7.1) [11].

A distinguishing feature of the collagen family of proteins is their formation by the association of three similar, but not necessarily identical, polypeptide chains. The type of protein species is identified by a Roman number (I–XXVIII), which simply reflects the order of discovery. Individual collagen chains are designated with the lower-case symbol, α (alpha). To illustrate the nomenclature, COLI, a trimer encoded by two genes, *COL1A1* and *COL1A2,* is comprised of two α1(I) chains and one α2(I) chain; COLII is a homotrimer of three identical α1(II) chains encoded by *COL2A1;* and COLX is a homotrimer of three α1(X) chains transcribed from the *COLXA1* gene (Table 7.1) [12].

An unusual characteristic of the family of collagens is that they maintain a common amino acid structural motif, the repeated appearance of a triplet of the amino acid sequence, glycine-X-Y, where proline and hydroxyproline regularly occupy X and Y positions, respectively (Fig. 7.3) [13]. Glycine constitutes about one-third of the amino acid residues in each of the three individual α-chains of the protein. As noted earlier (Table 7.1), two of the α-chains of COLI are identical [designated as 2α1(I)] and the third [α2(I)] differs slightly from the others in amino acid composition and the number of bound carbohydrate residues [12].

Complementary to glycine comprising collagen is proline, which has just been mentioned above. The chemistry of proline is notable because, unlike most other amino acids, it forms two connections in a protein, one through the common peptide bond which links amino acids to one another, and the other through its pyrrolidine side chain which has an almost flattened shape. As a result of these characteristics, proline induces a slight kink in each collagen chain, and the many proline residues lead to a helical twist to the molecule [14–16]. Because of the restriction in movement, especially rotation, and the bulkiness of their side chains, both the proline and

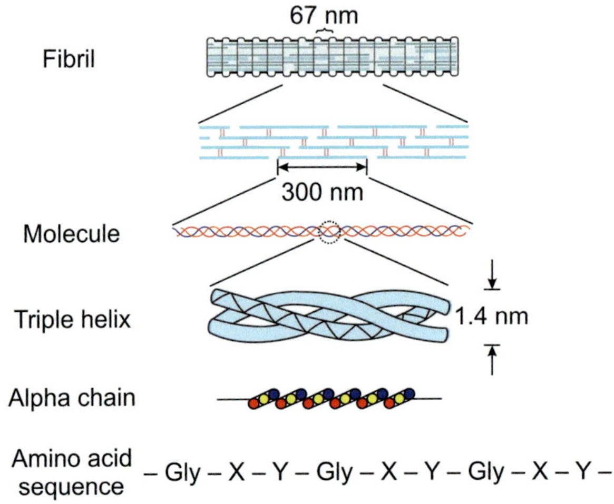

Fig. 7.3 The structural organization of the collagen family of proteins. Collagens are comprised of three individual protein chains (α-chains) consisting of a series of amino acid residues. The chains associate and adopt a triple helical configuration. Each chain winds around the other two to form a superhelix of about 1.4 nm in diameter, which may become extended or fibrous in shape. A very unusual feature of the collagens in general is that they share a common structural motif at the amino acid level: The amino acid sequence describing an α-chain is characterized by the repeat triplet of residues, glycine-X-Y, where proline and hydroxyproline frequently appear in the X and Y positions, respectively. The collagen molecule, comprised of its complement of three α-chains, assembles into groups which then constitute collagen fibrils. For COL1 and II, molecules approximately 300 nm in length assemble in a specific manner which results in a banding pattern with a periodicity of about 67 nm (64–70 nm). (Image adapted from Eyre [13] with permission)

hydroxyproline residues impart rigidity to the collagen helix and are mainly responsible for the fiber-like shape of the molecule (Fig. 7.4) [17]. Of note, hydroxyproline is generated from proline residues, as needed by the cell. Thus, hydroxyproline synthesis is a post-translational event, and as such the levels of hydroxyproline are very low in cells. Since few other proteins incorporate this hydroxylated amino acid, it is a valuable indicator of the collagen content of a tissue.

The sequence of amino acids which characterizes a collagen peptide chain causes the formation of a left-handed helix, which then associates with other newly synthesized collagen chains to form a right-handed triple helical molecule. The pyrrolidine side chains of both proline and hydroxyproline as well as the side chains of many of the amino acids present in the X- and Y-positions of the glycine-X-Y motif are directed principally outward from the collagen helix [18]. These groups also extend from the surfaces of aggregates of individual collagen molecules. Like other proteins, large numbers of hydrogen bonds are formed principally through peptide (NH and C=O) groups (Fig. 7.5) [19]. Although these bonds are weak, they serve to promote and stabilize interactions between the individual collagen α-chains as well as within each chain.

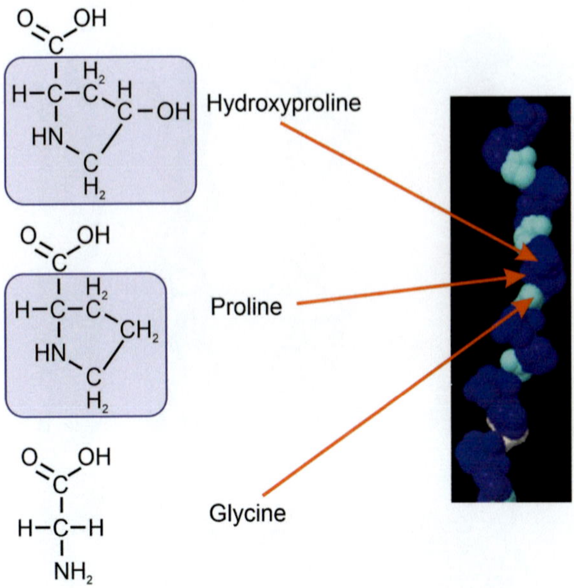

Fig. 7.4 The chemical formulae of the three common amino acids of collagen, glycine–proline–hydroxyproline, showing the pyrrolidine side chains of proline and hydroxyproline (pale blue box). These three amino acids are responsible for forming a rigid helix resembling the shape of a fiber characterizing a peptide chain of collagen. Three such peptide chains associate to create the collagen molecule. (Image adapted from Rau and Kajzar [17] with permission)

Of the family of 28 collagens, COLI is the most abundant in Nature. Indeed, COLI is critical to mineral formation in all of those skeletal tissues noted above except for mineralizing cartilage—in this tissue, the major protein is COLII. It would not be unrealistic to state that COLI is present in one form or another in all the tissues of the body, even those which do not undergo mineralization. As noted earlier, collagen is the principal protein of the mineralized skeleton and teeth of vertebrates: endochondral and membrane bone, calcified cartilage and tendon, Sharpey fibers of the periodontal ligament, dentin, and cementum. In each of these tissues, the individual α-chains spontaneously assume a helical shape, which then self-assembles with other α-chains to form a superhelix, a rod-like, fibrous structure (Fig. 7.5). In describing these associations, Nestler et al. [20] emphasized that it would be "much more appropriate to characterize collagen as viscoelastic or a semi-flexible rod" rather than a "rigid particle," and Orgel and Madhurapantula [21] reported that "some regions of the molecule are more flexible and prone to variations in helix character (or not) than others."

Although the individual chains of these collagens form helical structures and show some similarity to each other, no instances of collagens comprised of mixed α-chains from different collagen types forming a triple helix together have been documented. In other words, α-chains of COLII do not substitute for or replace

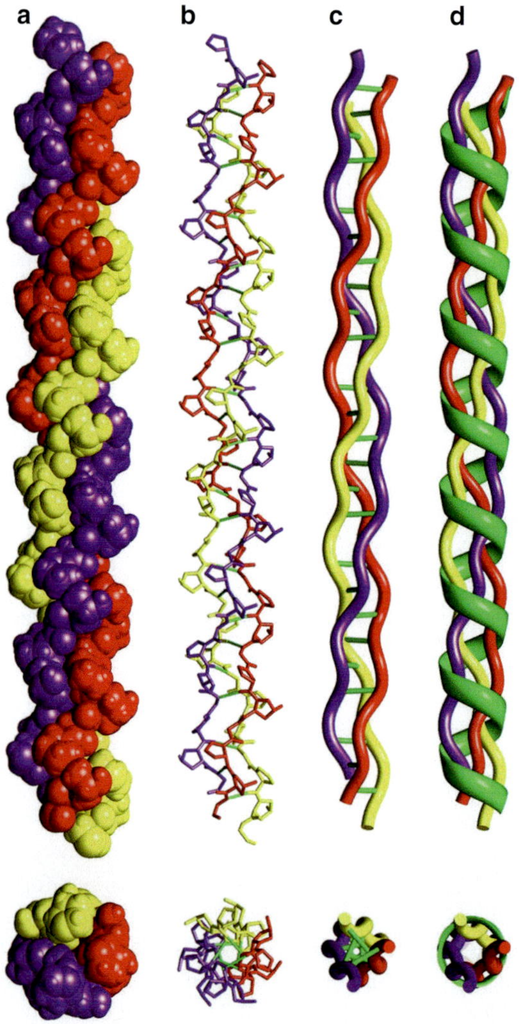

Fig. 7.5 Different representations of a collagen triple helix with the amino acid sequence, (proline-hydroxyproline-glycine)$_{10}$. The three polypeptide chains are shown as a trailing strand (purple), a middle strand (red), and a leading strand (yellow), and each panel shows a side and top view of the same representation. The strands are designated by their vertical position in the direction from the N- to the C-terminus (top to bottom of the helix, respectively) to the C-terminus (bottom of the helix). (**a**) Space-filling diagram of the right-handed supercoil of each individual strand. (**b**) Stick diagram where hydrogen atoms have been removed for clarity and hydrogen bonds are illustrated as green sticks connecting the strands. Hydrogen bonding topology may be followed along a helical path between Gly-N-H groups and Pro (X)-C=O groups. Such a path progresses from trailing strand to middle strand, from middle strand to leading strand, and from leading strand to one step (triplet) ahead on the trailing strand. (**c**) Ribbon diagram of the ladder of hydrogen bonds perpendicular to the helix axis, following their own helical path and connecting alternate pairs of strands at each step. (**d**) Ribbon diagram including the left-handed superhelix (green), which describes mathematically the collagen triple helix as a continuous helix of repeating units across the three individual strands. The structural model shown is an idealized 7_5 triple helix. (Image and details adapted from Bella [19] with permission)

α-chains of COLI within their respective triple helices. The chains differ, for example, in amino acid composition, sequence, charge, and attachment of carbohydrate moieties. On the other hand, whether in bone or cartilage, for instance, different collagen species are always present as components together in a tissue. Even in a tissue-like tendon where COLI predominates, COLII, III, V, XI, and XII have also been described [22, 23]. Heterotypic collagen fibrils, in which, for example, COLI copolymerizes with COLIII and V, among others, are common in a variety of tissues as well [24, 25]. Aside from the presence of the minor collagen species, collagen fibrils are additionally associated with other proteins comprising a particular tissue. In this case, interactions occur between collagen and several non-collagenous proteins, and these events may profoundly affect the regulation of mineralization and many other aspects of cell structure and function. Such molecular interplay is described in some detail in Chap. 8 of this book.

Remarkably, 11 peptide sequences of COLI have been isolated from demineralized fossil bones of 2 dinosaur skeletons. Some were recovered from a specimen of *Brachylophosaurus canadensis*, a daunting Hadrosaur which inhabited a region of what is now Montana about 80 million years ago [26], and others were found in fossils of an even more formidable dinosaur, *Tyrannosaurus rex* [27]. The yielded sequences of COLI peptides mapped to closely packed fibrillar regions of the protein which were assumed to be protected from enzymatic degradation during fossilization. Interestingly, four of the peptide sequences are now known to be involved with cell–collagen interactions and tissue remodeling [28], topics discussed in greater detail below. While these results are indeed quite notable, it must be stated that they are controversial and have been questioned in more recent reports [29, 30].

While considerable emphasis is placed on the fact that the three α-chains of collagen together comprise a helical complex, in fact there are non-helical regions of the protein. These segments are described as being non-collagenous. Commonly, such non-collagenous amino acid stretches are present in the nascent non-processed collagen molecule at both its N- and C-termini. These regions are required for organizing and assembling the collagen α-chains and the formation of crosslinks. As described below, such non-helical aspects of the molecule are subsequently removed by the action of proteolytic enzymes during collagen secretion and localization in the tissue extracellular matrix [31].

Since collagen molecules comprise 90% or more of the extracellular organic matrix of bone, dentin, cementum, and other related vertebrate mineralized tissues, it is not surprising that collagen interacts and complexes with cells and other extracellular and matricellular non-collagenous molecules, including small integrin-binding ligand N-linked glycoproteins (SIBLINGs), proteoglycans, small leucine-rich repeat proteins (SLRPs), and even plasma membrane protein receptors (see Chap. 8). Orgel et al. have noted that, while as many as 50 ligands bind the COLI molecule, only about half of the possible ligands can be mapped with a high degree of confidence to the collagen triple-helical sequences [32]. Of the more important charted sites are those which designate interactions with the matrix proteins, decorin [33] and fibronectin [34], hydroxylysine residues involved in intermolecular crosslinking [35], and a matrix metalloproteinase (MMP-1) cleavage site [36].

Besides binding other extracellular proteins, many collagenous proteins communicate with cells through integrin receptors [37]. In this regard, collagen molecules contain a number of RGD (arginine–glycine–aspartic acid) motifs which interact with RGD-dependent cell membrane integrins such as α5β1 and αv-heterodimers [38, 39]. Of the over 20 different types of integrins, at least 4 of them (α1β1, α2β1, α10β1, and α11β1) associate with selective RGD-independent domains of the native collagen molecule [40]. Among these, there is a triple-helical amino acid motif, glycine–phenylalanine–hydroxyproline–glycine–glutamic acid–arginine, which serves as a high-affinity integrin-binding domain within COLI and several other collagen molecules [41]. Sipilä et al. [42] list numerous additional collagen motifs which bind integrins. These include, for instance, glycine–arginine–hydroxyproline–glycine–glutamic acid–arginine, glycine–leucine–hydroxyproline–glycine–glutamic acid–arginine, and many others. Each of these RGD-dependent and RGD-independent integrin-binding motifs serves a critical function in cell–extracellular matrix interaction through collagens as major extracellular components of vertebrate tissues, as is described below.

In the context of collagen motifs, certain specific regions of the collagen molecule, the discoidin domain receptor family (DDR1 and DDR2) binding sites, ligate and activate the DDR receptors (Fig. 7.6) [21, 43–47]. These triple helical domains are present on a number of different collagen isoforms. According to Leitinger [48], DDR1 isoforms "bind and signal in response to both fibrillar and non-fibrillar collagens, whereas DDR2 is mostly activated by fibrillar collagens." Each of the two DDR isoforms participates in the regulation of the activity of bone and tooth cells, while DDR2 is specific for COLX [49]. That collagen could serve as a ligand for receptor tyrosine kinases is surprising as most commonly these receptors bind ligands which are secreted and soluble, such as growth factors, cytokines, and hormones, but not ligands like collagen which are very large insoluble extracellular proteins [50]. Activated by such ligands, the tyrosine kinase receptors regulate critical aspects of cell function, including proliferation, differentiation, cell cycle control, and cell survival.

Once collagen is bound to the extracellular domain of a DDR molecule, there is transphosphorylation of multiple tyrosine residues in its cytoplasmic domain [45, 51], followed by downstream signaling involving the MAPK, phosphoinositide 3-kinase, JNK, and NF-kB pathways [40, 52]. These pathways regulate events which include cell adhesion, migration, proliferation, and matrix remodeling (Fig. 7.6). Inhibition of DDR activation prevents or slows remodeling of the extracellular matrix. Moreover, abnormal DDR signaling can raise MMP13 activity and promote a loss of articular cartilage, a major feature of osteoarthritis [53, 54].

An additional factor in the mineralization of vertebrate tissues concerns the synthesis of collagen by cells of hard tissues. Osteoblasts utilize the anaerobic glycolytic pathway to generate energy for collagen biosynthesis. This observation has been confirmed a number of times using isolated osteoblasts and organ cultures of bone [55]. Anerobic energy metabolism by osteoblasts depends on the reduction of glucose to form lactate and NADH. Utilization of the anaerobic pathway to generate ATP instead of the more efficient TCA cycle is surprising since bone is permeated

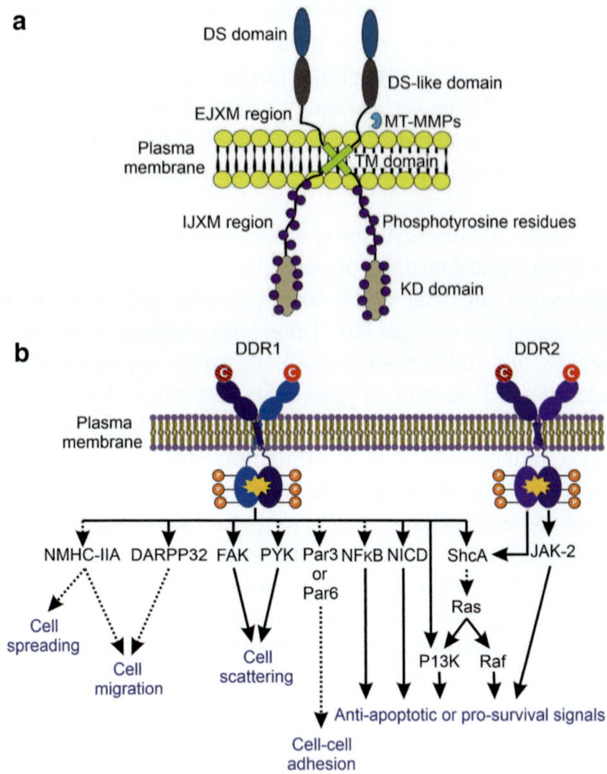

Fig. 7.6 (**a**) Schematic representation of the discoidin domain receptor (DDR). The DDR molecule spans the extracellular membrane of osteoblasts and other cell types [46, 47]. It is a dimer consisting of an extracellular discoidin (DS) domain and an extracellular DS-like domain which mediates ligand (collagen) binding. A transmembrane (TM) domain links the receptor to the interior of the cell. A region identified as the EJXM (extracellular juxtamembrane) is a site for cleavage by membrane-type matrix metalloproteinases (MT-MMPs), which remove the entire extracellular region of the receptor and thereby regulate receptor activation [47]. An intracellular juxtamembrane (IJXM) and a catalytic kinase domain (KD) comprise the cytoplasmic aspect of the receptor molecule and consist of several tyrosine residues which are phosphorylated during receptor activation [46]. (**b**) Schematic of downstream events linked to activation of the DDR1 and DDR2 receptors by collagen. Binding of collagen (C) to the receptors promotes a number of signal transduction events mediated by kinases which catalyze the transfer of phosphoryl groups (P). Collagen-DS domain binding results in the transphosphorylation of the cytoplasmic aspects of the receptors and leads to downstream signaling [45]. Such signals affect cell spreading and migration, cell–cell adhesion, and antiapoptotic and survival activities. Cell spreading is inhibited by DDR signaling, cell migration is promoted by DDR signaling and the other processes shown are either inhibited or promoted. Indirect activity is identified by dashed lines in the schematic and direct effects are indicated by solid lines. Schematic key is as follows: NMHC-IIA (non-muscle myosin IIA heavy chain), DARPP32 (dopamine- and cAMP-regulated phosphoprotein), FAK (focal adhesion kinase), PYK (protein tyrosine kinase), Par3/Par6 (cell polarity regulators), NFκB (nuclear factor-kappa B), NICD (Notch1 intracellular domain), P13K (phosphatidylinositol 3-kinase), ShcA (SH2-containing transforming protein A), Raf (Raf protein), Ras (Ras protein), JAK-2 (Janus kinase-2). (Schematics adapted from Orgel and Madhurapantula [21] with permission)

by blood vessels which deliver oxygen to the cells synthesizing collagen. Possibly the reliance on glycolysis may serve to minimize the production of reactive oxygen species and spare amino acids for cellular anabolic synthesis rather than use them for energy production. At the stage when mineral is being deposited in the collagenous extracellular matrix, the glycolytic cells exhibit an increase in mitochondrial oxidative phosphorylation, thus transitioning to a more efficient form of energy generation [56].

The activity of the glycolytic pathway is responsive to hypoxia inducible factor-1 (HIF-1). This transcription factor promotes glucose entry into the cell by means of the GLUT transporter, an event which downregulates mitochondrial activity. From this perspective, signaling through HIF-1 regulates collagen synthesis and bone formation in vivo [57, 58]. Aside from glucose, glutamine appears to play a major role in energy generation. Glutamine is principally imported into the cell through the transporter protein, solute carrier family (SLC) member 1A5 [59]. In addition to its role as an important metabolic fuel, glutamine may be a critical substrate for collagen synthesis during mineralization, rather than at an earlier stage when cells secrete their premineralized extracellular matrix [60]. Glutamine metabolism is regulated by both HIF-1 and Wingless-related integration site (WNT) in bone [61] and by sex determining region Y (SRY)-box 9 (SOX9) in cartilage [62].

In the context related to the discussion of glutamine above, the amino acid, proline, comprises nearly one-third of the residues of collagen as previously noted. Its presence and abundance in collagen raise a question as to the source of this amino acid. If proline were derived exclusively from serum, its levels would be low compared with the levels of glutamine. On the other hand, proline can be synthesized from glutamine although proline synthesis is energetically very costly: Aside from NADPH, proline requires the oxidation of eight molecules of ATP/mol of product [63]. Despite the energetic cost, there is evidence that serum proline is directly incorporated into nascent protein and, when the demand is high, its availability determines the rate of collagen biosynthesis [64]. More than likely, when demand cannot be met, then glutamine supports and fulfills this role.

7.3 Structure, Organization, and Assembly of Collagen

Based on COLI molecules, collagen is derived from a precursor molecule, procollagen, which is secreted from cells and then processed by enzymatic cleavage of its N- and C-terminal peptides, as briefly mentioned previously. The cleaved N- and C-terminal fragments (telopeptides) are useful clinical indicators of collagen synthesis. Canty et al. [65] have shown that the final processing steps of procollagen occur close to, or within, the cell plasma membrane. At these membrane locales, protein molecules are secreted into the extracellular matrix. Most likely, cleavage of the peptides from procollagen occurs at the external aspect of the cell membrane of fibroblasts, vascular endothelial cells, osteoblasts, tenocytes, odontoblasts, cementoblasts, and chondrocytes. As a result, there is secretion of considerable numbers of individual triple helical collagen molecules, approximately 300 nm in length (comprised of 1038 amino acids) and 1.2–1.5 nm in diameter [66]. In the extracellular environment, collagen molecules associate with one another and spontaneously aggregate by lateral bonding and crosslinking to form arrays which are "quarter-staggered" according to the well accepted concept originally proposed in 1963 by Hodge and Petruska [67]. A summary diagram with additional details of collagen synthesis and secretion [68, 69] and its assembly in extracellular tissue spaces [70, 71] is presented in Fig. 7.7.

In the Hodge and Petruska model [67], the ends (234 amino acids, approximately 27 nm in length) of each collagen molecule overlap and crosslink with the ends of other molecules (Fig. 7.8) [72]. Dimers, trimers, and larger groupings associate laterally and crosslink further so that gaps or hole zones of about 40 nm are formed between them (Fig. 7.8). The combined overlap and gap length is 67 nm, which varies from about 64 to 70 nm depending on the degree of hydration of collagen and the vertebrate species examined [73]. In general, as this region constitutes approximately one-fourth of the 300 nm length of the collagen molecule, it is conventionally regarded as the "quarter-staggered" collagen model aggregated in two dimensions. In three dimensions, the arrays of molecules associate such that their overlap and hole zones align and crosslink in strict registration with each other [74]. The assemblage, microfibrils [75], creates a series of continuous void spaces or channels appearing periodically every 64–70 nm and extending through the array of molecules (Fig. 7.9) [67, 74, 76, 77]. The ends of the channels are suggested to open at the surfaces of the aggregated arrays [78, 79], which may take the form of collagen microfibrils, fibrils, fibers, and higher ordered, more complex hierarchical structures (Fig. 7.9). These openings or porosities form possible pathways between the outside and the inside of collagen fibrils for the diffusion of ions or possible ion clusters of small atomic radius and the subsequent deposition of mineral within collagen. The 64–70 nm periodicity of the molecular or higher ordered arrays gives rise in the electron microscope to a banded appearance which is the ultrastructural hallmark of collagen fibrils. Periodicity may be enhanced by staining collagen with dilute solutions such as osmium tetroxide, lead citrate, phosphotungstic acid, or uranyl acetate, whose heavy metals bind to the charged amino acid side chains of the protein (Fig. 7.10) [80].

A principal goal of this chapter is to provide information about collagen–mineral interaction. In that regard, an understanding of the nature of the mineral is necessary along with a description of how mineral may be accommodated by collagen given its structure, organization, and assembly as presented above. As was indicated earlier, pure geologic hydroxyapatite has a stoichiometry of $Ca_{10}(PO_4)_6(OH)_2$ [81]. However, the stoichiometry and composition of biological apatite is quite variable [81]. For example, in human cortical bone, there is reportedly only about 20% of the hydroxide content which would comprise stoichiometric apatite [82], and the mineral character is now known to be a poorly crystalline apatite, principally comprised of calcium, phosphate, and carbonate ions [81]. The mineral may also contain a host of other cations and anions which can substitute for any of these three major ions and/or fill voids, or vacancies, in the apatite crystal lattice [81]. Such vacancies contribute to an often structurally imperfect lattice along with point defects of

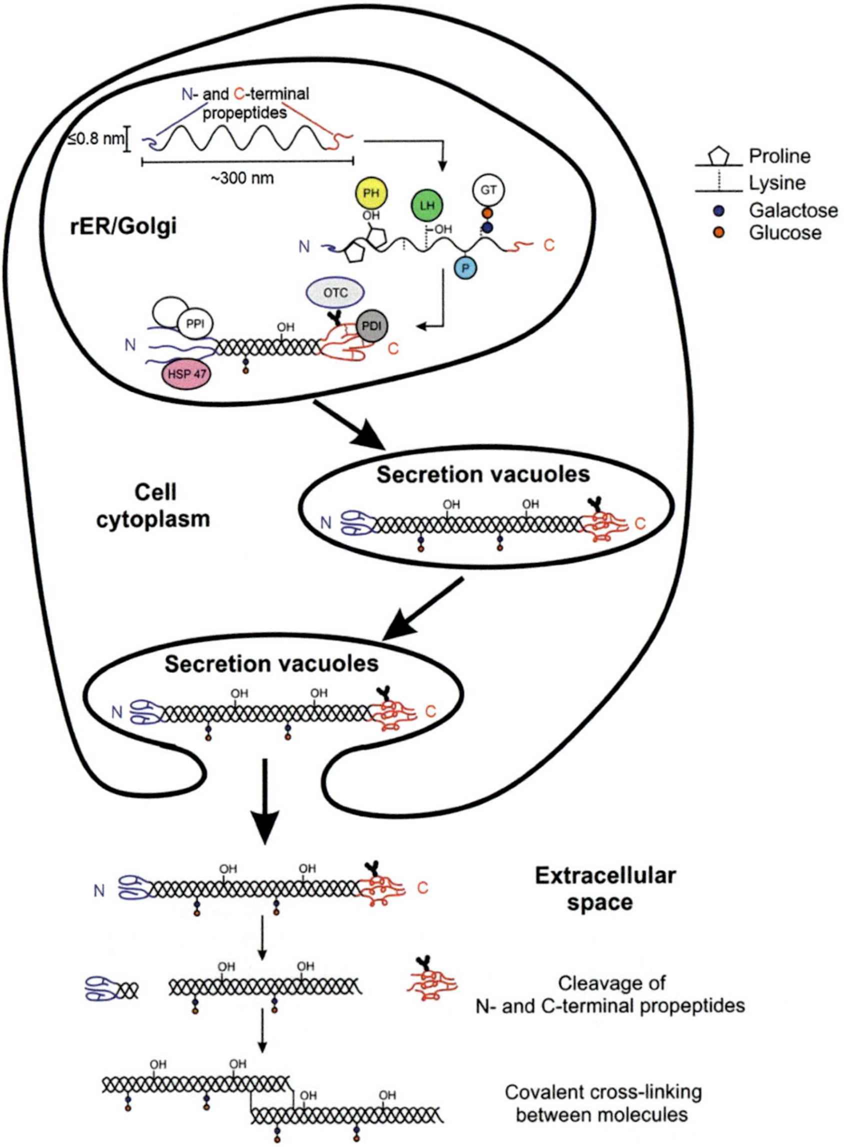

Fig. 7.7 A schematic illustration of major events in the process of collagen synthesis and secretion, typified here by the fibril-forming COLI. Following collagen gene transcription in the cell nucleus, collagen mRNA is translated in the rough endoplasmic reticulum (rER) to α-chains of propeptides having N- and C-termini. The chains are post-translationally modified by hydroxylation of lysine and proline residues (LH and PH sites, respectively), glycosylation (GT sites) and phosphorylation (P sites) in the rER and they interact through intramolecular hydrogen bonding to form helical triple-stranded monomeric procollagen molecules [68]. Prior to monomer formation, C-terminal propeptides of three α-chains are sulfated and stabilized as globules by means of intrachain disulfide linkages. An oligosaccharyl transferase complex (OTC) adds a carbohydrate group to the globular C-terminus, which then initiates and directs the alignment and formation of the

(continued)

Fig. 7.7 (continued) α-chains into the procollagen triple helix. Proper folding and progressive formation of the chains from C- to N-termini of the nascent procollagen molecules are modulated by protein disulfide isomerase (PDI), peptidyl-prolyl cis-trans-isomerase (PPI), and a collagen-specific chaperone molecule, heat shock protein 47 (HSP 47). Procollagen molecules are next packaged in the Golgi and enclosed in secretory vacuoles which bleb from Golgi membranes and are transported through the cell cytoplasm to the cell membrane. Vacuoles fuse with the cell membrane and ultimately secrete their procollagen contents by exocytosis to the extracellular milieu. N- and C-terminal propeptides of procollagen are cleaved by specific peptidases (ADAMTS-2, 3, and 14 at the N-terminus and BMP-1/mTolloid family members at the C-terminus) to yield collagen and N- and C-terminal propeptide molecules [70, 71]. Collagen molecules then self-assemble through covalent crosslinking at the non-helical N-telopeptide head of one molecule and the non-helical C-telopeptide tail of another to form a short overlap region of both. Additional crosslinking of dimers and trimers leads to the quarter-staggered collagen model described by Hodge and Petruska [67]. Schematic key is as follows: PH (prolyl hydroxylase), LH (lysyl hydroxylase), GT (hydroxylysyl galactosyltransferase and galactosylhydroxylysyl glucosyltransferase). (Schematic and principal details adapted from Gelse et al. [68] and Shoulders and Raines [69] with permission)

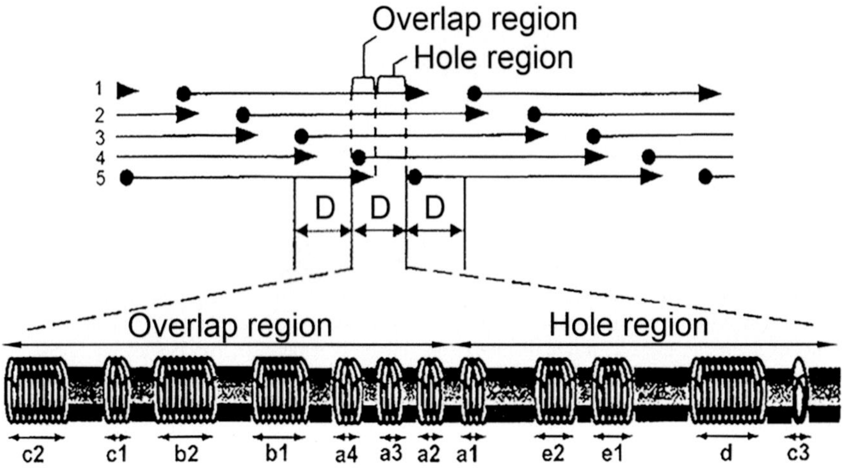

Fig. 7.8 Two-dimensional schematic of the microfibrillar packing pattern of COLI molecules. The molecules are unidirectional and depicted as arrows with their N- and C-termini shown as a small sphere and an arrowhead, respectively. All collagen molecules, 4.46 D in length and identified by segments 1–5, are packed into the so-called quarter-stagger pattern in their long-axis direction [67]. In this model, the five collagen molecules of a microfibril are shifted with respect to each other by a distance D equal to 67 nm. The distance D is comprised of a hole region of approximately 0.54 D and an overlap region of approximately 0.46 D [67, 75]. In a cross section of a microfibril, the hole zone includes collagen molecular segments 1–4 and the overlap zone consists of segments 1–5. When collagen molecules are stained with metal ions and then viewed by electron microscopy, a series of 12 dark bands interrupted by light bands may be observed across the axis of the fibril. The dark bands are designated sequentially c2 through c3. The pattern of banding in the D period has been enlarged at the bottom of the figure to show mechanically flexible charged regions denoted by springs and rigid regions depicted by cylinders [72]. The flexible regions are devoid of proline and hydroxyproline and they alternate with the rigid regions which contain these imino acids. (Schematic adapted from Silver and Landis [72] with permission)

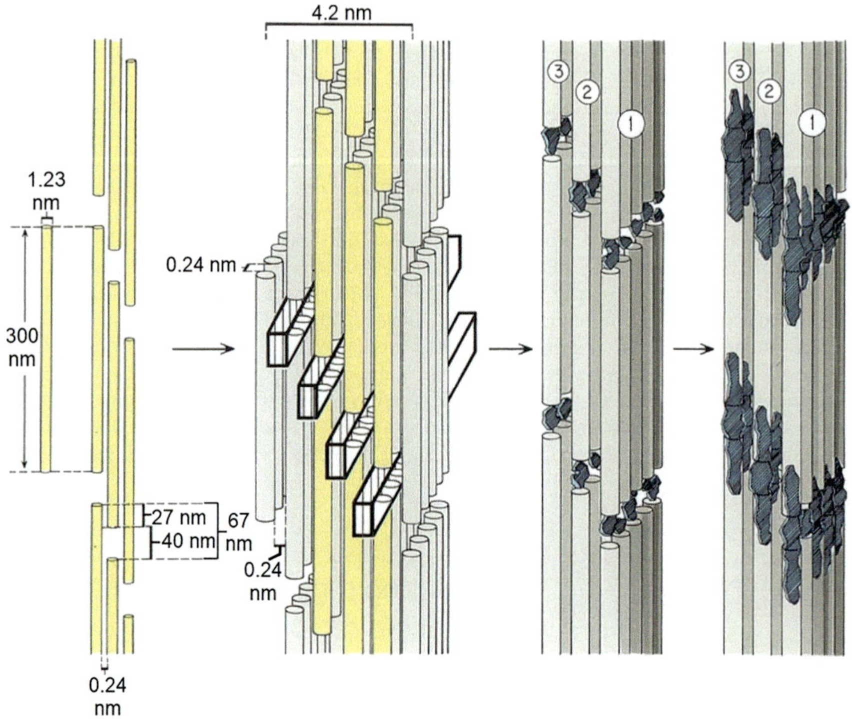

Fig. 7.9 Schematic of intrafibrillar collagen mineralization showing the 2D and 3D organization, assembly, and initial mineralization of COLI molecules. The diagram follows the Hodge and Petruska [67] and Landis et al. [74] models which suggest the formation of channels or gaps within assemblages of collagen. Such channels are considered the principal sites in which nucleation of mineral crystals occurs within collagen. The pore spaces (0.24 nm) [76] between adjacent collagen molecules constitute narrow regions whose capacity to accommodate mineral is unknown. A single collagen molecule, 1.23 nm in diameter and 300 nm in length [67, 76], is shown at the left of the diagram. Several such molecules (yellow) assemble into a 2D aggregate, crosslinked into a quarter-staggered array with characteristic hole (40 nm) and overlap (27 nm) zones [67]. Packing of consecutive 2D arrays into 3D assemblages occurs with strict registration of all hole and overlap zones so that channels are created throughout the model [74]. Once nucleated principally in the collagen hole zones, the crystals (blue) grow preferentially in their *c*-crystal axis direction. The (100) planes describing the largest developing face of the crystals lie generally parallel (within ~15° to 20°) to each other and to the collagen molecules associated with the crystals. Other features of the assembly and mineralization of collagen have been described in detail previously [74]. (Schematic adapted from Landis and Jacquet [77] with permission)

atoms, dislocations, and discontinuities. Some of these variations in composition and structure reflect the nature of the local environment including, for instance, the ion availability in the solution surrounding and bathing the crystals and the degree of substitution of other cations and anions for calcium, phosphate, and carbonate in the apatite lattice. The final chemical composition of the mineral in biological

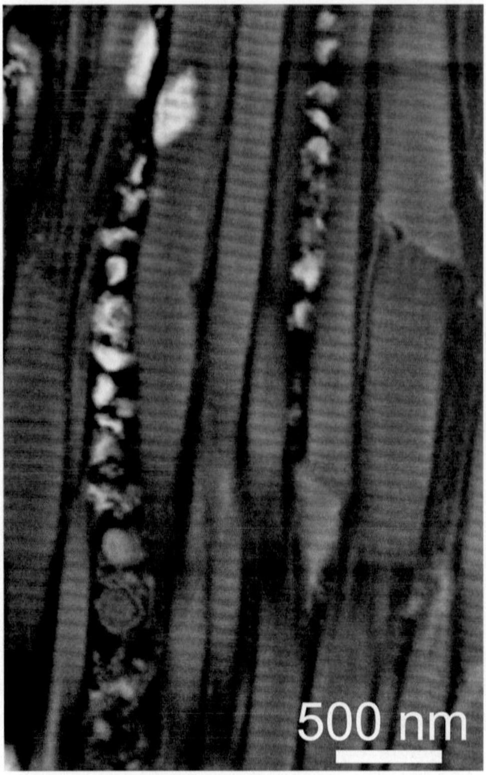

Fig. 7.10 COLI fibrils viewed in longitudinal profile by focused ion beam-scanning electron microscopy and illustrating the 64–70 nm periodicity characteristic of the rod-like protein. The tissue section shows fibrils from a normal avian leg tendon undergoing early stages of mineralization. Several small irregular-, oblate-, or globular-shaped structures, suggested as putative matrix vesicles neighboring the fibrils, and a few fibrils appear associated with mineral (white image regions). The sample was prepared by high pressure freezing and automated freeze substitution and was stained with osmium tetroxide-uranyl acetate to provide increased contrast to the collagen period. (Image from Zou et al. [80] with permission)

apatite, then, will be a combination of ions which have come from solution to replace the three major ions comprising stoichiometric apatite and/or filled available apatite lattice vacancies.

The individual crystallites of apatite comprising normal bone mineral are among the smallest known in Nature. Their dimensions are ~20 to 50 nm in length, 12–20 nm in width and 2–4 nm in thickness [74, 83]. Based on careful electron microscopy and electron and X-ray diffraction studies, the crystallites exhibit the shape of long, thin, flat plates, or platelets in vivo [74, 84]. Apatite crystallites comprising dentin, cementum, and avian calcifying tendon have similar dimensions and shape, although such dimensions will change depending on the developmental stage of the tissue and whether the tissue is newly formed or reforming after partial dissolution [85]. Crystallites in enamel are also apatitic but they are ribbon-shaped and orders of magnitude longer, wider, and thicker than their counterparts in the collagenous-based vertebrate mineralized tissues, bone, dentin, and so on [86].

Given their small size, platelet shape, and the presence of lattice vacancies, defects, dislocations, discontinuities, and ion substitutions which occur in the otherwise generally well-ordered apatite from collagenous vertebrate mineralized tissues, the crystals are enveloped in a relatively disordered hydration layer enriched in HPO_4^{2-} ions and possibly hydrated PO_4^{3-} ions [84, 87]. In addition to such a disordered phase, the interface between apatite crystals may be even more complex because they are not independent particles but exist as complex aggregates [4], and their interface may be occupied by negatively charged citrate ions, which bind to the apatitic platelets and hold them together [88].

With regard to citrate, it has been known for many years that considerable quantities of this tricarboxylic acid are present in bone [89]. Citrate is adsorbed to the surface of apatitic crystallites, forming salt bridges between two of its three carboxyl groups and calcium ions of the mineral as well as hydrogen bonds between the citrate hydroxyl group and the mineral surface [88, 90]. These multiple interactions help stabilize the crystal structure [88, 90]. Davies et al. proposed the idea that the mineral appearing at the interface between apatitic crystals in the presence of citrate has the characteristics of octacalcium phosphate [91]. Here, citrate would be expected to replace HPO_4^{2-} in octacalcium phosphate and, in this case, the structure of the mineral becomes yet additionally complicated with octacalcium phosphate, citrate, and the hydration layer forming a discontinuous zone (Fig. 7.11) [87]. Since complexing with octacalcium phosphate limits crystal growth, citrate may be responsible for the platelet-shaped morphology of the crystals of bone as well as the stacking of individual apatite crystallites that is a characteristic of mineralizing bone [91]. The report that citrate can help to stabilize amorphous calcium phosphate provides further insight into the overall complexity of vertebrate mineral [92]. These and other investigations suggest that citrate is important, if not critical, in controlling apatite crystal shape, chemistry, and aggregation; modulating mineral phase stability; possibly guiding mineral phase transitions; and providing an understanding of mineralized tissue pathophysiology [88, 91–95]. From results of additional work in vitro, citrate may also modify the charge of collagen molecules and enhance their mineralization [96]. Citrate presence in vivo also implies the participation of respiring citrate-generating cells in defining the architecture of the mineral phase of bone and other normal vertebrate hard tissues.

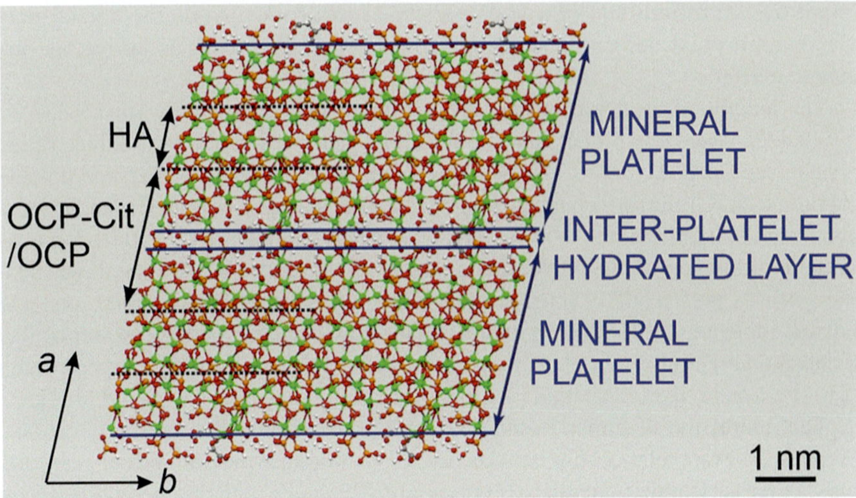

Fig. 7.11 A model proposed for bone mineral consisting of unit cells of octacalcium phosphate ($Ca_8H_2(PO_4)_6{\bullet}5H_2O$; OCP) and citrate (Cit) split along the (200) plane and sandwiching unit cells of hydroxyapatite (HA), resulting in a parallel arrangement of mineral platelets alternating with hydrated layers (the hydrated layers of the OCP-citrate unit cells) (dotted blue lines). Only one in three of the OCP-citrate unit cells contains citrate; remaining ones are modeled with OCP with a unit cell expanded to match that of OCP-citrate. Solid blue lines indicate the surfaces of the mineral platelets (defined so that all calcium ions are contained within the mineral platelets), which are bridged by citrate anions. In practice, the position and orientation of the citrate anions in the hydrated layers will be disordered. Orthogonal crystal planes are designated a and b. Atoms are calcium (green), phosphorus (orange), oxygen (red), and hydrogen (white). (Diagram and details adapted from Davies et al. [91] with permission)

7.4 Organic–Inorganic Interaction and the Role of Collagen in Nucleating Mineral

Studies of the intimate and complex interactions between the inorganic mineral and organic proteinaceous phases of bone and other hard tissues have led to consideration of whether collagen, the principal organic constituent of these tissues, or a collagen motif is primarily responsible for apatite nucleation. To address the point, knowledge is needed of aspects of the spatial relationship between the unit cell of apatite and the physical and chemical characteristics of the collagen molecule. To begin, the atomic structure of both inorganic and organic phases is described below.

Crystallographers use the term, "unit cell," to describe the spatial lattice of atoms which produces an ordered symmetrical structure in three dimensions. The apatite unit cell is complex, comprised of at least 18 atoms which exhibit a hexagonal symmetry (see Chap. 2). Such symmetry indicates that each of the sides of the crystal rotated 60° will resemble one another. However, in all mineralized tissues, stoichiometric apatite, $Ca_{10}(PO_4)_6(OH)_2$, is not "pure" and often contains impurities (substitutions) of a wide variety of other ions, while the crystal may also exhibit point defects, dislocations, and discontinuities in its lattice, as discussed earlier.

The lengths of the three crystal axes of the apatite unit cell are $a_0 = 9.432$ Å, $b_0 = 9.432$ Å, and $c_0 = 6.881$ Å, while the (100) plane defines the largest face of the crystal (see Chap. 2) [81, 91, 97, 98]. The *ac* and *bc* crystal planes of apatite possess calcium-rich (C) sites which are positively charged while the *ab* crystal plane is more abundant in phosphate-rich (P) sites which are negatively charged [99, 100]. Because of these differences in the predominant charge on the planes of the crystallites, apatite surfaces preferentially interact with positively or negatively charged ions and organic molecules [101]. In this regard, Shaw et al. demonstrated, for example, that a charged carboxyl group on a fragment of the major enamel protein, amelogenin (LRAP), bound hydroxyapatite, possibly through the C sites of the crystal [102].

The interaction of mineral with COLI molecules raises two central questions: What is the exact relationship between these two principal constituents of bone and most other vertebrate mineralized tissues and what does such an interaction reveal about the mechanism of mineralization? With respect, first, to the spatial location of apatite, initial reports that the mineral resided within the structure of the collagen molecule were met with considerable skepticism from the scientific community. However, as numerous ultrastructural investigations later documented, mineral association in COLI occurs within fibrils, fibers, and their higher ordered aggregates (intrafibrillar mineralization) and at their surfaces and into the spaces between collagen structures (interfibrillar mineralization) [77, 103, 104].

Details of the relationship between COLI and mineral were established in part using electron diffraction techniques, a procedure in which electrons impinging on a microscopic sample interact with the atoms of mineral crystals and are diffracted. Resulting diffraction patterns provide information on crystal identity, structure, orientation, and other parameters. These studies revealed that the character of the mineral associated with collagen differed according to its location. If the mineral were intrafibrillar, the crystals of apatite are highly ordered, generally coherent and parallel to one another (within approximately 15–20°), and aligned with their crystallographic *c*-axis in the same direction as the long axes of collagen with which the crystals associate [74]. A more recent model of intrafibrillar mineral formation in COLI suggests greater variation in crystal coherence [105]. In contrast, interfibrillar mineralization generates diffraction patterns which are consistent with apatite crystals randomly oriented and disorganized and having no preferential alignment with collagen [106]. Formation of intrafibrillar and interfibrillar mineralization was reported to be independent of each other, that is, deposition of mineral could occur within and outside collagen molecules separately in time and location [103, 106, 107]. Very recent work indicates that mineral in vertebrate tissues appears to form first outside fibrils and then propagates to nearby adjacent fibril surfaces and within fibrils [80]. This result will be briefly elaborated below.

From conventional and high-voltage transmission electron microscopy and three-dimensional (tomographic) reconstruction of high voltage micrographs, the presence of mineral crystals appears initially and principally associated with the COLI hole zone regions but also to some extent with the overlap regions of the channels in COLI array assemblages (larger quarter-staggered associations of collagen molecules) (Fig. 7.9) [74, 79, 108, 109]. As mineralization progresses, apatite crystals grow preferentially in length along their crystallographic *c*-axes and in the same direction as the long axes of the collagen structures with which they associate. The crystals putatively may also grow in width by the fusion of neighboring particles within the same channel [74]. Crystal thickness is likely governed by the collagen molecules which border or neighbor the collagen channels (Fig. 7.9) [74]. Once nucleated, individual crystals increase in size through the recruitment and addition of more and more calcium and phosphate ions from the surrounding metastable tissue fluid [110]. With mineral formation and crystal growth, it is thought that there may be cleavage of intramolecular and intermolecular collagen bonds and disruption of the polypeptide chains and molecules or that the mineral may envelop or occlude collagen chains, thereby maintaining existing protein crosslinking within the mineral [111].

7.5 The Chemistry of the COLI Hole Zone and Nascent Mineral Formation

As was discussed earlier, when a bone or vertebrate mineralized tissue (other than enamel) is treated to remove its mineral, the remaining, principally collagenous structure is a replica of the untreated tissue. The structural integrity of the replica results from the presence of collagen molecules which are linked to each other through a variety of bonds to form a mechanically robust facsimile of the original tissue. The intact, untreated tissue is comprised of bonds between the mineral and the molecules principally of collagen in the tissue extracellular matrices. To understand more clearly the interaction between mineral and collagen, high resolution microscopic and advanced computer techniques have yielded fascinating conceptual insights into inorganic–organic bonding and, very importantly, the possible role which the collagen molecule plays in mediating mineral deposition in vertebrate hard tissues.

With respect again to the aggregation and assembly of COLI molecules, it has been found that six molecules associate with one another to form a so-called quasi-hexagonal grouping which is considered the collagen microfibril noted previously [75]. Within a microfibril, each triple helical collagen molecule, approximately 300 nm in length and 1.23 nm in diameter, as also mentioned above, overlaps and binds to its neighbors. With reference to Fig. 7.12, collagen molecules 2, 3, and 4 designate molecules staggered 234 amino acids, 2 × 234 amino acids, and 3 × 234 amino acids, respectively, from molecule 1. The order of neighboring molecules is 1-4-1-3-2-3 to form the quasi-hexagonal structure, and this configuration of the 6 molecules surrounds each collagen hole zone (Fig. 7.12) [72, 75, 77, 112]. Associations of microfibrils lead to the formation of larger units called fibrils, which in turn group together to form fibers, and aggregates of fibers form fiber bundles in the development of increasing collagen structural hierarchy. The 64–70 nm periodicity of collagen, the fundamental *D*-stagger described by Hodge and Petruska [67] and Hodge [113], is maintained throughout the assemblages and is evident microscopically in fibrils, fibers, and fiber bundles.

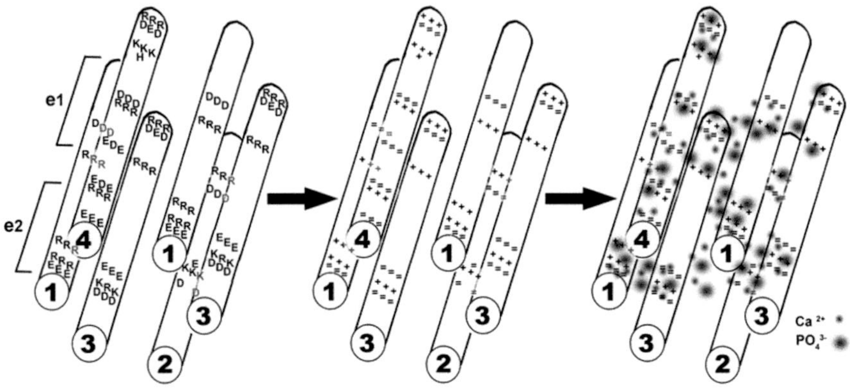

Fig. 7.12 Depiction of the arrangement of the e1 and e2 bands of high charge density in the neighborhood of COLI molecules which comprise a portion of the hole zone. The six collagen molecular segments bordering the hole zone are shown in pseudohexagonal packing [75, 112] with the specific location and distribution of the charged amino acids comprising the e1 and e2 bands of the individual molecular segments determined by collagen sequence data [72]. Collagen molecular segments labeled 1–4 designate the relative stagger of 234 amino acid residues compared to segment 1. Thus, segments 2, 3, and 4 are, respectively, staggered 234, 2 × 234, and 3 × 234 residues from the beginning of the triple helix defining segment 1. The left grouping of molecular segments above shows the location of each charged amino acid residue comprising the six segments surrounding the hole zone. The middle grouping of segments illustrates the charge at each amino acid site. The amino acids are identical or have the same charge in almost all instances at the adjacent sites in α1(I) and α2(I) chains comprising the collagen molecules. In the right grouping of segments, the charged amino acids are presented as attracting and binding individual calcium and phosphate ions to respective negative and positive side chains of the corresponding residues. Sufficient numbers of ions lead to nucleation in the vicinity of the hole zones, as proposed and described by Silver and Landis [72]. (Schematic from Landis and Jacquet [77] with permission)

In the microfibrillar grouping of COLI molecules through higher orders of collagen association, the amino acid sequence of collagen neighbors surrounding each hole zone reveals a unique character of residues with critical implications for the nucleation of mineral (Fig. 7.12). The basic repeating collagen tripeptide of amino acids (glycine-X-Y), mentioned earlier in this chapter and where proline and hydroxyproline constitute the respective X- and Y-positions for much of the primary sequence of collagen, is effectively uncharged. However, X- and Y-positions may be occupied by other amino acids, including the various known charged residues, which then impart charged collagen sequences in the tripeptide motif.

The staggered association between collagen molecules ordered as 1-4-1-3-2-3 leads to a series of domains or regions which have high charge density interrupted by regions with low charge density [72]. Such regions correspond to and describe the 12 bands as shown in Fig. 7.8 comprising the hole and overlap zones detailed first by Hodge and Petruska [67] and subsequently by Silver et al. [114] in terms of elastic energy storage. As noted previously, these regions are made prominent and may be visualized when heavy metals stains are used to improve their electron density and they yield the resulting characteristic periodicity of collagen under the electron microscope [115, 116]. Since the sequence of amino acids is known for each of the collagen molecules and its neighbors comprising the microfibril [72], it has been possible to define several notable features about its structure (Fig. 7.12) [72]. These include (1) numerous sites in which the same charged amino acid (glutamic acid, aspartic acid, arginine, lysine and hydroxylysine) is adjacent to one another in all three helical peptide chains [2α1(I)] and α2(I)] constituting a collagen molecule; (2) sites in which the adjacent amino acids of the 2α1(I) and the α2(I) chains are different but are of the same charge; (3) sites in which two or three amino acids of the same charge are close to the same two or three amino acids in the three peptide chains; (4) sites in which two or three glutamic or aspartic acid residues are close to their respective counterpart two or three aspartic or glutamic acid residues in the three peptide chains; (5) sites in which the same two or three amino acids of one charge are close to the same two or three amino acids of the opposite charge in the three peptide chains; and (6) sites in which lysine or arginine residues are replaced by a hydroxylysine residue to contribute identically charged groups in the three peptide chains [72].

The features defined above for the charged amino acids, together with their spatial locations surrounding and lining the hole zones and channel regions of collagen assemblages, are important and provide insight into apatite nucleation. As noted previously, because the collagen molecules are staggered in their neighboring 1-4-1-3-2-3 configuration, the stereochemistry of the charged amino acid residues concentrates charge density in the specific 12 domains or regions of the collagen hole and overlap zones described by Hodge and Petruska (Fig. 7.12) [67]. Thus, the charged amino acids provide potential binding sites for calcium and phosphate ions present in the tissue fluid circulating throughout the hole zone channels. When they are bound to these amino acids, these ions are thought to be brought into stereochemical positions which make possible critical nucleation events leading to apatite crystal formation [72, 111].

Insights into the relationship between the ordered 1-4-1-3-2-3 molecular domains about the collagen hole zone channels, ion transport and mineralization have been the subject of molecular dynamics simulation (in silico) experiments [111, 117]. This approach provides a means for investigating atoms and molecules as they move physically in their environment. By computer modeling, the ensemble of atoms and molecules interacts dynamically for a known time interval and the final state of atom and molecule interplay may be observed and recorded. In the case of the collagen neighborhood above, molecular dynamics simulation showed that the side chains of each of the charged amino acid residues of the collagen molecules are principally directed toward and into the spaces of the channel (Fig. 7.13) [111]. This is an important result as it would indicate that any calcium and phosphate ions bound to the side chains of the charged residues would be located within the volume of the channel and positioned so they could potentially interact if the calcium and phosphate ions were close enough to each other [111].

The simulation method also demonstrated that the side chains are different in length and direction for different charged amino acids and extend to different distances into the channel (Fig. 7.13) [111]. The consequence of this result is that the channel then varies in its average diameter, becoming effectively winding or

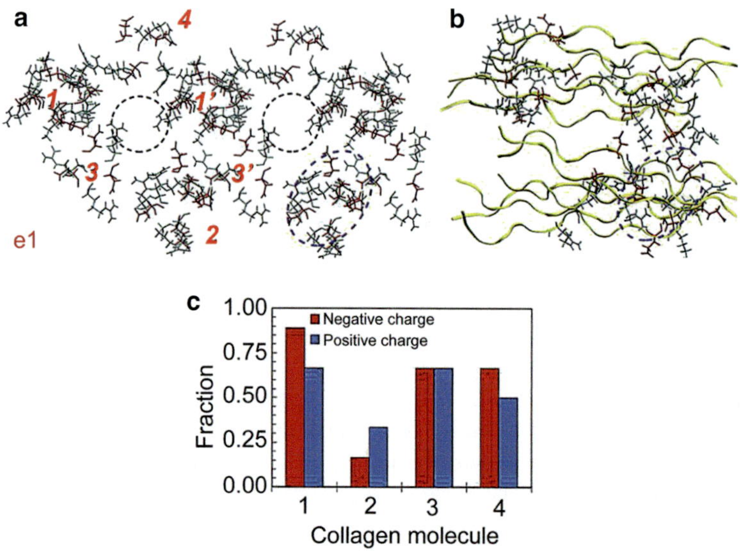

Fig. 7.13 (**a**) Molecular dynamics simulation of the distribution of side chains of charged residues of COLI molecules comprising the e1 band about the hole zones in a fully hydrated system also containing sodium and chloride ions. COLI molecular segments are designated 1-4-1-3-2-3 following Orgel et al. [75] Side chains are shown as positively (blue) and negatively (red) charged groups as viewed along the crystallographic *c*-axis of collagen. (**b**) View parallel to the collagen *c*-axis and where the backbone of the various molecular segments is shown by yellow ribbons. Positions of the collagen hole zones are depicted by dashed black circles and the blue ellipse indicates an area of closely interacting charged side chains. (**c**) Plot of the fraction of charged side chains comprising the e1 band directed toward the hole zone for each of the collagen molecules surrounding the hole region. Most positively and negatively charged side chains point toward the hole zone. (Image and details adapted from Xu et al. [111] with permission)

meandering and wider and narrower along and through the collagen arrays it accesses [117]. As the channels in collagen arrays are the principal sites in which apatite mineral crystals are nucleated [74, 79, 108, 109], the location of such crystals within the winding nature and variable diameter of the channels would likely not follow a linear arrangement within the channels but would instead be twisting, turning, and convoluted. This effect may influence the ability of nuclei or small crystals to interact with each other in the same channel and fuse to form larger mineral crystals.

To restate these findings which relate collagen structure and mineral formation conceptually, the locations of charged amino acids surrounding and lining the hole zone regions of collagen molecular assemblages may provide sites for the nucleation of crystals of apatite [72, 111]. Indeed, the side chains of charged amino acids could serve to bind calcium and phosphate ions from solution, and molecular dynamics simulation indicates that these bound ions are close enough to interact with each other, forming the nidus of a predicted amorphous phase of calcium phosphate, ~1.3–1.6 nm in size, in the collagen hole zones and channels (Fig. 7.13) [111]. Following nucleation, the additional recruitment of calcium and phosphate ions from solution leads ultimately to transformation of an amorphous to a poorly crystalline calcium phosphate phase. Such a transition may be accompanied by possible metastable precursors such as octacalcium phosphate or another calcium phosphate phase. Amorphous calcium phosphate and perhaps other mineral phases may be mediated and stabilized by the presence of non-collagenous proteins such as osteocalcin (OCN), which has been localized to the collagen hole zones [118], so that phase changes of nascent crystal alignment and orientation may be carefully controlled [118]. Further growth and development of apatite crystals extending beyond the hole zones into overlap zones would lead to the presence of mineral throughout collagen channels.

Not all side chains of the charged amino acids of collagen point toward the void spaces created by collagen hole zones [111]. In this instance, these chains could also bind calcium and phosphate ions from the supersaturated solution and mediate nucleation in collagen regions other than those defining the holes and overlaps of the channels. Such nucleation away from the channels would potentially result in further mineral formation between and about intrafibrillar collagen chains. Here, the mineral might break collagen bonds or occlude, surround and encase collagen chains, as noted previously, to complete intrafibrillar collagen-mineral interaction.

It should be noted parenthetically that the discussion above of conceptual intrafibrillar collagen–mineral ion interactions has not addressed any possibility of collagen binding of putative prenucleation clusters of calcium phosphate [104, 119–121], exosomes or matrix vesicles which may appear or reside in the extracellular milieu of vertebrate mineralized tissues. Exosomes and matrix vesicles are considered too large in size to enter the channels which open to the intrafibrillar spaces of collagen [122]. Prenucleation clusters, on the other hand, may be sufficiently small (nanometer-sized [119]) to diffuse into and through hole and overlap zones of collagen channels to interact with the side chains of charged amino acids and induce nucleation in a manner related to that described previously for individual calcium and phosphate ions.

Compared with intrafibrillar events of collagen mineralization, details of interfibrillar mineralization are less well known. For example, the surface structure and stereochemistry of collagen fibrils, fibers, or other higher ordered aggregates are poorly understood. On the other hand, hole zone channels are thought to open at numerous surface sites to provide pathways and access for tissue fluid supersaturated with calcium and phosphate ions to flow between interfibrillar and intrafibrillar collagen spaces [106, 110]. Further, side chains of amino acids such as proline and hydroxyproline occupying X- and Y-positions of the glycine-X-Y motif are outwardly directed from each composite collagen helix and thus from the collagen surfaces of fibrils, fibers, and fiber bundles. The surfaces of collagen structures, then, would likely be decorated and characterized in part by several sites of positive and negative charge attributable to the presence of charged amino acid residues residing in X- and Y-sites. The localized accumulation of surface charge in this environment would favor calcium and phosphate binding from ions or prenucleation clusters in solution and thereby promote the propagation of mineral over collagen surfaces and throughout the extracellular matrix (Fig. 7.14) [67, 72, 77, 106, 109, 113, 118, 119, 123, 124].

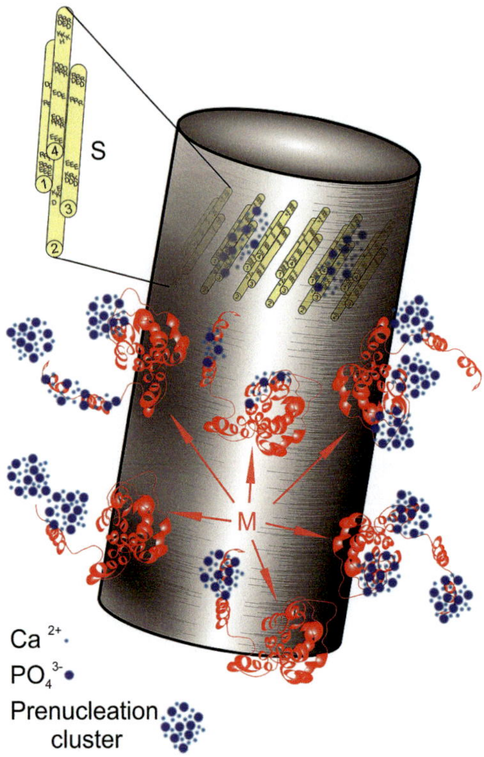

Fig. 7.14 Schematic diagram of both intrafibrillar and interfibrillar mineralization of a COLI fibril (grey cylinder). Numbers of collagen molecules structurally organized in parallel arrangement with ordered hole and overlap zones comprise the collagen fibril. Intrafibrillar mineral formation is represented by four segments (S, yellow) of the e1 and e2 bands of collagen molecules, slightly shaded and hidden to indicate their location within the larger cylindrical collagen fibril. As described previously, these two segments and their 10 additional segments of high charge density comprising the hole and overlap zones of collagen are suggested to be mineralized by means of calcium and phosphate ion binding to the template of their composite charged residues [67, 72, 113, 123, 124]. Events of apatite crystal nucleation, growth and development occur to form a series of platelets whose (100) faces are approximately parallel to each other (within ~15° to 20°) and to the long axes of collagen molecules with which they associate. Propagation of the platelets and their increases in size with time ultimately lead to a poorly crystalline apatite network throughout the internal collagen structure. Interfibrillar mineralization is proposed to occur through two possibilities: a) binding of calcium and phosphate ions or their putative prenucleation clusters [109, 119] to side chains of charged amino acid residues of collagen exposed at the collagen surfaces and/or b) binding of calcium and phosphate ions or prenucleation clusters to certain non-collagenous molecules (M, red), themselves bound and attached to various collagen surface sites [106, 118]. Six such molecules, for example, bone sialoprotein, are shown conceptually in the diagram. These molecules may initially bind calcium phosphate ions and prenucleation clusters from solution and transport them to collagen surfaces or the molecules themselves may become bound to collagen surfaces and subsequently attract and bind calcium and phosphate ions and clusters. As with intrafibrillar mineralization, propagation of apatite crystals and their increasing size with time provide a means of developing interfibrillar mineralization of a poorly crystalline network over the extracellular spaces between collagen fibrils and their higher ordered structures. (Schematic and details adapted from Landis and Jacquet [77] with permission)

With reference to extracellular matrices, these regions between collagen structures are also uncertain regarding their exact composition. Located and bathed in the solution supersaturated in calcium and phosphate ions, there exists a host of non-collagenous molecules such as bone sialoprotein (BSP), osteopontin (OPN), OCN, albumin, citrate, fetuin-A (α2-HS glycoprotein), bone acidic glycoprotein-75 (BAG-75) and many others, whose presence depends on the tissue and species in question and its respective age and state of maturation. Certain of these non-collagenous proteins, for instance BSP, OPN, and OCN, may bind calcium and phosphate ions in solution and transport them to collagen surfaces (Fig. 7.14) [77]. These three proteins have been shown by immunocytochemical studies to be localized to tissue spaces between collagen or on collagen surfaces [118]. Moreover, they have also been suggested to influence not only formation (BSP, OCN) [118, 125–129] but also inhibition (OPN) [130–132] of mineralization of vertebrate tissues. Details of these and additional roles of non-collagenous proteins are presented in Chap. 8. In this structural, biochemical and biological context, mineralization between neighboring collagen fibrils occurs to the extent that, like the events of intrafibrillar mineral formation, deposition of apatite ultimately involves all available space.

As noted earlier, formation of intrafibrillar and interfibrillar mineralization has been suggested to be independent of each other. Electron microscopic images of collagen in longitudinal profile have shown mineral both within and outside the fibrils at different and distinctly separate extracellular tissue locations. Accordingly, it was presumed that mineral nucleation occurred at multiple extracellular sites unrelated to one another in both time and space [103, 106, 107].

However, this assumption may not be valid. Very recently, focused ion beam-scanning electron microscopy (FIB-SEM) and tomographic reconstruction methods have been used to examine mineral deposits in vertebrate tissues [80]. Images reconstructed in three dimensions showed that mineralization of collagen appeared to begin outside the protein (interfibrillar mineralization) and propagated to collagen surfaces and then into or within the fibrils (intrafibrillar mineralization) at the point of collagen surface mineralization. The mineral continued to propagate at intrafibrillar sites located at a distance from the surface site(s) of propagation. Thus, intrafibrillar and interfibrillar mineralization sites may be dependent on each other and interrelated through a common collagen surface site [80]. Once initiated within a fibril, intrafibrillar mineralization propagates along and within fibrils, while interfibrillar mineralization may separately propagate further outside fibrils. Subsequently, the vicinity of the extracellular tissue spaces containing the sites of interfibrillar mineral, intrafibrillar mineral, and their common collagen surface sites fills with mineral, and these events occurring at multiple locations and different times throughout the matrix ultimately result in mineralization of the entire extracellular tissue space, as mentioned previously.

7.6 COLX and Mineralization of Epiphyseal Cartilage

While much of the preceding discussion relates to COLI, the most common of collagen species, another collagen species, COLX, a non-fibrillar, network-forming collagen, is found in the depths of the cartilaginous growth plate of vertebrates. In this instance, as shown by Tryfonidou et al. [133] for example, COLX is prominent in growth plate regions which undergo mineralization prior to their removal and replacement by bone (Fig. 7.15). The anatomical presence, location, and uniqueness of this collagen species imply that it has a specific role in the mineralization process, and a great effort has been made to understand its relationship with mineral deposition in this tissue.

Like COLII, three identical polypeptide α-chains comprise COLX and they adopt a helical structure in which the α-chains wind around each other to form a superhelix. COLX consists of 682 amino acids and hence it is smaller than COLI (1038 residues). The gene structure also differs from multi-exon COLI in that only three exons are present for COLX. One exon encodes the 5″ untranslated region of

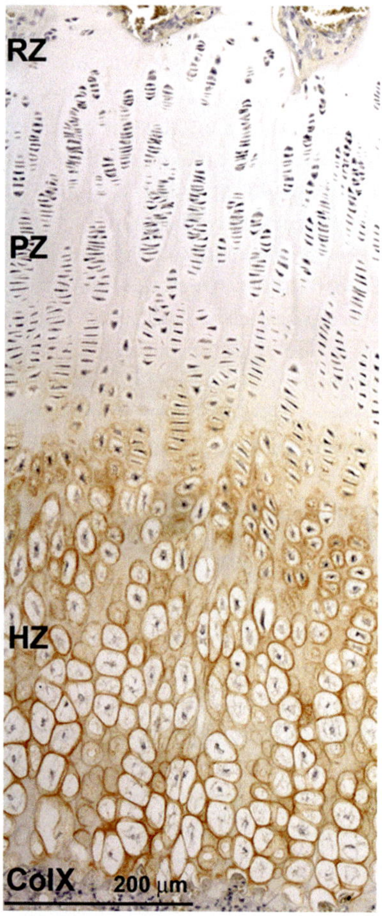

Fig. 7.15 A longitudinal section through the growth plate of a rib from a dog stained with an antibody against COLX. Antibody staining (brown) indicates the presence of COLX in the extracellular matrix of hypertrophic chondrocytes. Growth plate zones are identified as the resting zone (RZ), proliferative zone (PZ), and hypertrophic zone (HZ). (Image from Tryfonidou et al. [133] with permission)

COLX mRNA, and a second exon codes for the signal peptide and the amino terminus non-helical region of the protein (38 residues). A third exon dictates the amino acids which comprise the remainder of the helical region of the protein (463 residues) together with amino acids (161 residues) which form a large non-triple-helical region at the carboxy end of the protein [134].

Once synthesized and secreted from hypertrophic chondrocytes of the growth plate, the COLX chains form hexagon-shaped supramolecular aggregates [135, 136]. Kwan et al. [137] have suggested that, "Type X collagen may act as a pericellular network retaining or trapping the necessary types and amounts of matrix components, including the MV [matrix vesicle] and proteoglycans, into a correct location of the hypertrophic zone of the cartilage, promoting initiation of normal mineralization." Whether COLX has greater tensile strength than that of COLII has not been determined, nor is there information on how COLX is distributed in the growth plate in relationship to applied forces of the tissue. Thus, it is difficult to conclude that one of the roles for COLX is structural.

The observation that *COLX* is expressed only by hypertrophic chondrocytes and not by osteoblasts, odontoblasts, cementocytes, or calcifying tendon cells is intriguing. As will be discussed in the following sections of this book, most vertebrate mineralized tissues express a multitude of genes (*SOX9, COLI, TNAP [Tissue non-specific alkaline phosphatase], OCN,* and *RUNX [Runt-related transcription factor]2,* for instance). While hypertrophic chondrocytes also express these genes, they additionally express *COLX*. Reginato et al. [138] have shown that encoded COLX comprises about 6% of the total extractable collagen in the avian epiphyseal growth plate.

The specificity and abundance of the protein in hypertrophic cartilage raises the question as to whether *COLX* expression is a singular evolutionary event which confines its role to the mineralizing epiphysis or whether a more widespread expression existed but was subsequently lost as tissue specialization occurred over time. To address this question, the phylogenetic expression of the gene was examined in a number of ancestral vertebrate tissues. Based on the classification briefly discussed in Chap. 1 that divides evolution of the vertebrate skeleton into chondrichthyans (cartilaginous fishes) and osteichthyans (bony fishes), there is evidence that *COLX* is expressed in both chondrocytes and osteoblasts of bony animals like the frog, medaka and zebrafish (osteichthyans) [139–142]. The gene is also expressed in the developing teeth and scales of the chondrichthyan, the catshark (*Scyliorhinus canicula*) (Fig. 7.16), and in the teeth of two osteichthyans, the zebrafish (*Danio rerio)* and the clawed frog (*Xenopus tropicalis*) [143]. Debiais-Thibaud et al. [143] have written that, "These findings support the origin of *Col10a1* before the divergence of osteichthyans and chondrichthyans" and "demonstrate its ancestral association with mineralization of both the odontode skeleton and the endoskeleton." [143]. Debiais-Thibaud et al. [143] have also proposed that *COLX* expression is a hallmark of an ancestral biomineralization event, specific to early vertebrates. Since the COLX (protein) is not secreted in the teeth and bone of higher organisms, it is conceivable that *COLX* has been silenced or lost during evolution. Alternatively, the function of this gene in osteoblasts and cells of the developing dentition of higher organisms may have been replaced through redundant function provided by other genes and their proteins.

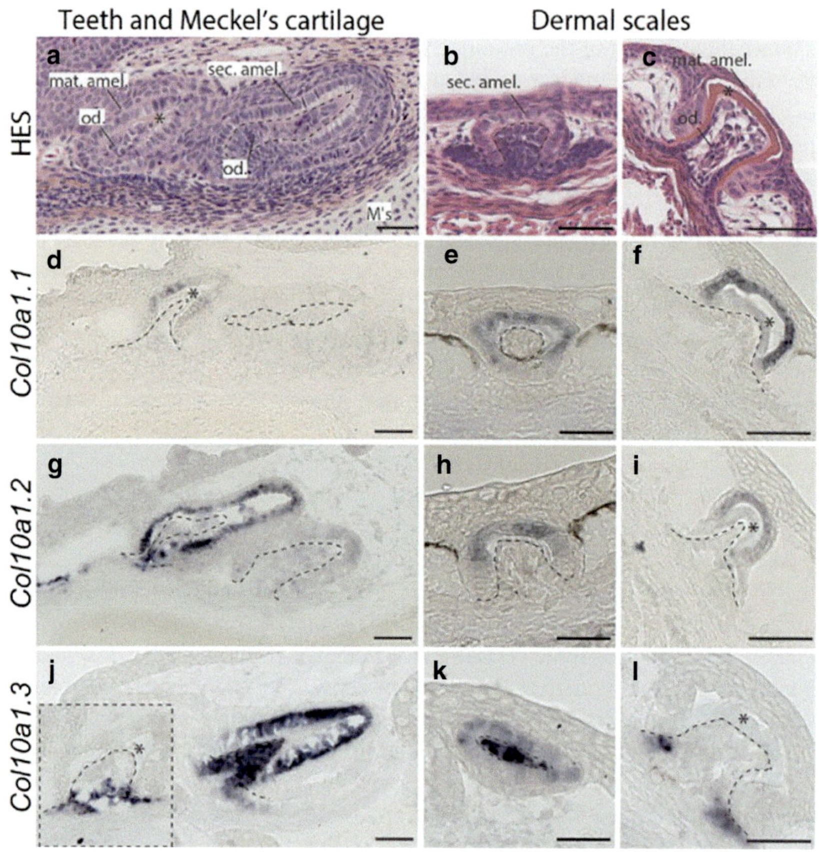

Fig. 7.16 Detection by in situ hybridization of COLX RNA in sections of developing teeth and scales of the catshark, *Scyliorhinus canicula*. Frozen sections of the teeth and scales of embryos at stage 33 were collected and stained with hematoxylin, eosin and saffron (HES) or treated with COLX riboprobes which contained 1–3 tandem repeats mimicking duplication events which occurred in the evolution of the chondrichthyan stem lineage (Col10a1.1–Col10a1.3). (**a–c**). HES histological staining of developing teeth and dermal scales. Tooth development stages included the presence of columnar secretory ameloblasts having translucent secretory vesicles at their secretory pole (**a**, **b**) and the maturation stage with cubic-shaped ameloblasts without cytoplasmic secretory vesicles and adjacent to mineralizing extracellular matrix (**a**, **c**). At these stages, the mesenchymal compartment is comprised of differentiating and secretory odontoblasts (**a**). Transverse sections showed dermal scale development with secretory ameloblasts adjacent to undifferentiated mesenchymal cells (**b**) and maturation ameloblasts adjacent to secretory odontoblasts at a later stage of scale development (**c**). (**d–l**) Hybridization of the riboprobe to antisense COLI RNA in tissue sections. *Col10a1.1* and *Col10a1.2* were transcribed in the epithelium of developing teeth at two successive stages of ameloblast differentiation, the secretory and maturation stages (**d**, **g**, respectively). *Col10a1.3* showed expression in both the epithelium (ameloblasts) and mesenchyme (odontoblasts) at the secretory stage of developing tooth buds (**j**), but its expression was later restricted to the odontoblasts of older tooth buds (**j**, dashed square). These expression patterns in teeth were comparable to those in developing scales in counterpart sections with restricted expression of *Col10a1.1* and *Col10a1.2* in the secretory and maturation ameloblasts (**e**, **f**, **h**, **i**). Asterisks mark mineralizing matrix, observed as empty space caused by decalcification during specimen processing. Dotted lines indicate limits of mesenchymal compartments. Abbreviations in the panels: mat. Amel., maturation stage ameloblasts; M's, Meckel's cartilage; notoch, notochord; od., odontoblasts; sec. Amel., secretory ameloblasts. Scale bars = 50 μm. (Images and details adapted from Debiais-Thibaud et al. [143] with permission)

The observations just described concerning evolutionary changes in the mineralization toolkit prompt the question as to the importance of COLX in endochondral mineral formation of higher vertebrates. Clearly, as noted above, COLX is secreted in epiphyseal tissue in vivo, as well as in chondrocyte cell cultures, and it is present in isolated matrix vesicle preparations. These findings reinforce the assumption that the protein plays a critical role in mechanisms linked to apatite deposition. Of course, an optimal test of a possible causal relationship between *COLX* and apatite deposition is to mutate or silence the gene and assess its effects on bone formation. In this context, as mentioned previously, a COLX frameshift mutation leads to Schmid metaphyseal chondrodysplasia, characterized by short limbs [11]. Kwan et al. suppressed *COLX* in transgenic mice and demonstrated that the mutant animals exhibited craniofacial abnormalities affecting the development of their chondrocranium and mandible [137]. At the ultrastructural level, the pericellular network around the hypertrophic chondrocytes in the tissues was disrupted, a result leading to the conclusion that *COLX* was required for "the compartmentalization of matrix components to the hypertrophic zone of growth cartilage, providing the proper environment for mineralization." [137]. On the other hand, Rosati et al. reported that the loss of *COLX* exerted minimal effects on the development of growth cartilage and its mineralization [144].

The studies by Warman et al. [11], Kwan et al. [137], Rosati et al. [144], Lee et al. [145], and others just mentioned represent some of the research literature which has still left the role of *COLX* and cartilage mineralization uncertain and unresolved. Under these circumstances, the involvement of *COLX* has been examined during tissue development, in particular its influence on stem cell differentiation and cartilage extracellular matrix synthesis. For example, a knock-down of stem cells which differentiate into hypertrophic chondrocytes showed that the absence of *COLX* reduced both COLII and glycosaminoglycan levels in vitro [146]. This result pointed to a role of *COLX* in regulating the expression of these two major components of the epiphyseal cartilaginous extracellular matrix. Another investigation exploring the potential of *COLX* in directing the fate of uncommitted stem cells documented that the expression of this gene was required for cartilage formation [147]. However, the work revealed little about the mechanism(s) underlying these developmental changes [147]. Thus, from all these various perspectives, *COLX* appears to participate in a number of aspects concerning the development and maturation of cells of the epiphyseal plate, but its precise functions remain largely undefined, including any putative control or regulation of the mineralization of growth cartilage.

7.7 Perspective: Collagen-based Mineralization of Bones, Teeth, and Other Vertebrate Skeletal Tissues

The overall goal of this chapter was to consider the function of major collagen species within the extracellular matrix of vertebrate tissues in the context of the mineralization process. Much of the discussion concerned COLI since this molecule comprises approximately 90% of the proteins and additional molecules of the organic matrix of bone, dentin, cementum, and certain avian tendons. The approach to understanding the role of COL1 was that, rather than being a passive recipient of mineral, its molecular structure provides critical sites for binding calcium and phosphate ions or their putative prenucleation clusters as well as promoting the transition of an amorphous mineral precursor phase to a final, poorly crystalline apatite. Since it is known that COLI has both integrin- and DDR-binding sequences which bridge collagen and its parent cells [21, 38], collagen can also directly influence the cells involved in the formation and resorption of the developing or remodeling extracellular matrix of vertebrate mineralized tissues. Further, during such processes, the availability of these binding sequences would be influenced by the growth of mineral within the collagen molecules or their ordered assemblages and on their surfaces and between them. Accordingly, the presence of accruing extracellular mineral would feedback and modify the impact of COLI on cell function. In this way, as mineral encloses critical amino acid motifs comprising individual collagen molecules and their higher ordered structures, interactions between collagen and its neighboring cells would change and provide a mechanism to fine tune the cellular response in generating a mineralized tissue.

While much of this chapter focuses on COLI, there is a brief discussion of a second collagen species, COLX, a protein which transiently occupies the extracellular matrix of developing endochondral bone. Compared with COLI, much less is known about this collagen isoform, and its function in mineralization is far from clear. However, since loss of function mutations in *COLX* is associated with cartilage and mineral abnormalities and disease [11, 137] and the normal absence of the gene is correlated with a suppression of mineral formation in certain cartilaginous tissues [141, 142], it is clear that *COLX* expression is required for tissue mineralization. Further detailed understanding into the relationship between COLI, COLX, and vertebrate tissue mineralization is presented below.

Considering first the major species, COLI, examples of studies using high resolution and novel microscopy [80, 105, 109], gene and protein investigations [118, 129], X-ray scattering and X-ray and electron diffraction analysis [105, 148, 149], molecular dynamics simulation [111, 117], and other approaches [148–150] have provided refined information and insight concerning the chemistry and organization of ions which constitute the apatite crystal lattice, the structure of the collagen molecule and the interaction between mineral and collagen at atomic, molecular, and macromolecular levels. These investigations have yielded (a) three-dimensional structural analyses of the protein; (b) information about the protein environment around its gap and overlap zones; (c) the charge distributions surrounding the gap and overlap zones; and (d) support for the concept that a relationship exists between amino acid charge and the nucleation and growth of apatite crystals. In addition, the investigations have suggested that channels of varying diameter exist within collagen aggregates which form fibrils, fibers, and fiber bundles and that the channels permit ion or small mineral cluster entry into such collagen groupings. All of these new findings have profoundly impacted the understanding of the function of collagen in relation to the initiation of mineralization leading to the formation of bone, dentin, or other vertebrate mineralized tissues. Moreover, knowledge of the nature of this biphasic organic–inorganic system has inspired the development of unique composite biomimetic or bioinspired materials for use in medicine, dentistry, and industry.

The possibility has been discussed previously that collagen may serve as a nucleating protein which could seed a mineral phase by lowering the free energy ($-\Delta G$) of the system it comprises. However, numerous studies have presented evidence that triple helical collagen was not itself an apatite-nucleating molecule and the interactions related to a number of other negatively charged non-collagenous moieties or proteins fulfilled this role [151–153]. For example, Deshpande and Beniash [151] found that a charged model peptide (polyaspartic acid) promoted mineral formation in collagen. This result appeared not to be artifactual because the crystallographic *c*-axes of the apatite and the longitudinal axis of the collagen molecules were aligned and maintained features similar to those found in vivo [151]. Further, the earliest mineral detected in the model was interpreted as an amorphous phase which rapidly transformed to a more conventional apatite [151]. Additional investigations have reported concepts of vertebrate mineral nucleation differing from the principal role of collagen [81, 107, 109, 114, 148] and its specific stereochemistry [72, 74, 77, 106, 111, 154, 155] as a fundamental means and basic template for apatite nucleation, growth and development. In this regard, the transport pathways of mineral particles mediated by membrane-invested vesicles [156], a polymer-induced liquid-precursor process [157], interactions between water and mineral [150, 158], and the confinement by intermolecular channels or spaces within or between collagen [105, 149, 159–161] in regulating and controlling mineral ion association are other cogent considerations in the scheme of apatite nucleation and the mineralization of collagen.

Regardless of the exact mechanism, these results and reports each emphasize that collagen is a critical component of the vertebrate mineralization process. Supporting this conclusion, mutations in the collagen molecule such as those found in osteogenesis imperfecta (brittle bone disease) impact both the alignment of the mineral in the collagen gap zone and the biomechanical properties of the tissue, often leading to multiple bone fractures [7, 162, 163]. Osteogenesis imperfecta is discussed in greater detail in Chap. 10 of this book.

As noted in this and the previous book chapter, understanding of events which lead to mineral formation has been developed in large part through observations generated by physical–chemical and other means of experimentation in vitro. Compared with studies of cells, tissues or whole animals conducted in vivo, such investigations in vitro are relatively easily carried out, very reproducible, inexpensive, and precisely controlled. Of course, the approaches in vitro do not necessarily mimic or represent events in the human body and hence must be regarded as "useful" only. Nevertheless, one very important outcome of such work is that the findings provide a firm foundation on which hypotheses concerning living tissues may be proposed and tested through further research.

The literature reports described in this chapter have generated insights into how and why mineral is associated with COLI, nucleated in the collagen gap zone and then propagated throughout the protein and the extracellular milieu of most vertebrate hard tissues. Also, the studies indicate the potential importance of non-collagenous proteins in promoting the mineralization process. Many of these non-collagenous proteins are synthesized and secreted by cells contained within the mineralizing tissue itself. These molecules serve to regulate (both promote and inhibit) mineralization and, like citrate or OCN, may stabilize a particular mineral phase. The absence of one or more of the molecules may explain why skin and a host of other vertebrate tissues which contain COLI do not normally mineralize. Finally, modeling studies have focused on mineral formation mediated by ions present in the metastable extracellular tissue fluid of vertebrate systems. One mode of delivery of these ions to the hole zone channels of collagen is as small calcium phosphate aggregates or putative prenucleation clusters, generated by the cells occupying the environmental niche of the mineralizing tissue.

At the beginning of this perspective, it was noted that hypertrophic chondrocytes in the developing epiphyses of vertebrate long bones secrete COLX. Because of its uniqueness of expression and abundance in this example and others, most investigators have assumed that this protein plays a dominant role in the formation and mineralization of the extracellular matrix of epiphyseal cartilage. However, from an unequivocal point of view and despite the fact that studies of this collagen isoform have continued since it was discovered more than 35 years ago [164], it is still difficult to be certain of the precise role or roles of *COLX* in the mineralization process. One possible clue to its specific function is that *COLX* is expressed almost simultaneously with *COLII* during chondrogenic differentiation [165]. As a result of these apparently linked developmental events, both collagen isoforms generate a matrix which accepts calcium ions, calcium phosphate clusters, and matrix vesicles and fosters the maturation of nascent mineral into mature apatite platelets. Moreover, since cell function is dependent on feedback from matrix proteins, as mentioned earlier in this chapter, changes in the composition of the extracellular organic matrix would be expected to influence most anabolic functions, including those related to mineral deposition. From this perspective, *COLX* impacts the function of hypertrophic chondrocytes in terms of the secretion of a unique matrix which favors mineral formation.

The relative importance of *COLX* compared with *COLII* in promoting the matrix activities described above is unknown. However, as previously noted in this chapter, mutational studies show that, in the absence of *COLX*, disturbances occur in cartilage matrix composition, bone, and bone marrow development and function [138], and possibly even changes in immune cell function [166–168]. As far as the effects of this gene on matrix vesicle-mediated mineralization, Kirsch and Wuthier [169] reported that *COLX* and *COLII* promoted calcium ion influx into matrix vesicles. However, not all investigators are convinced that this is the case. Grskovic et al. [170], for instance, showed that knock-down of *annexin 5* and *6* genes together with *COLX* failed to influence the structure of the growth plate or the distribution of its matrix vesicles. Indeed, a study by Rosati et al. [144] indicated that deletion of the gene did not adversely affect long bone growth or development.

Accordingly, as concerns persist regarding a specific role of *COLX* in mediating the mineralization response in epiphyseal cartilage, additional concepts have been developed and experimentation conducted to determine more convincingly that this protein may influence epiphyseal growth and mineralization. One possibility is that *COLX* may exert systemic effects, rather than immediate cellular effects, which relate to hypertrophic chondrocyte function and it may modulate the deposition of mineral in this indirect manner. Grskovic et al. reported that the protein affected cytokine production: In the absence of *COLX*, splenic cells were found to increase secretion of the cytokines, interleukin-4 (IL-4), and granulocyte macrophage-colony stimulating factor (GM-CSF) [170]. These cytokines influence chondrocyte, osteoblast, and osteoclast function. Studies by Frost et al. showed that IL4 promoted osteoblast differentiation and alkaline phosphatase (TNAP) expression, while inhibiting proliferation [171]. GM-CSF serves as a potent inhibitor of osteoclast function and hence acts as a brake on tissue remodeling [172]. Thus, the effects of these two cytokines appear to demonstrate a putative systemic route for *COLX* activity. In a related context, *COLX* is expressed by stromal stem cells undergoing chondrogenic differentiation [146]. If their expression of *COLX* can be prevented, then these cells may remain in a prechondrogenic state. Possibly, maintenance of this state may be of value in the replacement of articular cartilage in patients suffering from osteoarthritis [147, 173]. In the growth plate, a similar type of action might help regulate the orderly maturation of chondrocytes as they transition to their terminally differentiated state.

As a summary of these differing views of the role of *COLX*, since mutation in this gene is a cause of Schmid metaphyseal chondrodysplasia, a condition in which there are both short limbs and stature [11], it is safe to state that *COLX* is required for normal endochondral bone development and growth. From this [11] and among other deletion, mutation or dietary-deficient studies [137, 138], it is also clear that the absence or alterations of the protein lead to a change in the organization of the epiphyseal extracellular matrix, an event which would undoubtedly influence the deposition and organization of apatite crystallites. Since extracellular matrix components feed back on cell function, any changes in extracellular matrix structure would be expected to alter cell-related activities required for ion accumulation and exosome formation. There is also the possibility that *COLX* has a systemic effect influencing not only differentiated chondrocytes, osteoblasts, and osteoclasts but also stem cells and circulating splenic cells [168, 174]. Of course, if viewed from an evolutionary perspective with the progressively increasing complexity of the extracellular matrix, what may have been earlier functions of COLX in bone, dentin, and other vertebrate mineralized tissues could have become subsumed by other matrix components such as the SIBLINGs, SLRPs, and additional noncollagenous proteins and molecules. The role of these minor proteins and molecules in skeletal and tooth mineralization is discussed in considerable detail in the next book chapter.

As a final point of discussion in the present chapter concerning the mineralization of bones, teeth, and other vertebrate skeletal tissues, the deposition of apatite has been described above in terms of extracellular matrix interactions between ions of calcium and phosphate and different collagen species (principally COLI, COLII, and COLX) in the presence of a considerable number of non-collagenous proteins and other molecules. Such ions and their putative pre-nucleation clusters as well as small non-collagenous proteins and additional molecules have been regarded as comprising the extracellular fluid spaces of the variety of vertebrate tissues where diffusion of these matrix constituents plays a critical role leading to mineral formation. A new study utilizing FIB-SEM and tomographic reconstruction mentioned above has revealed an extracellular matrix architecture in calcifying cartilage and bone which provides novel insight and perspective into the transport and delivery of ions and small molecules throughout the extracellular milieu [175]. In this instance, an extensive and complex network of small (~40 nm diameter) nanochannels has been found to permeate the principally collagenous cartilage and bone matrices of the mouse femoral head (Fig. 7.17) [175]. The nanochannels surround mineralization domains, which appear to undergo a maturation suggested to resemble the sintering process of traditional materials [175]. Sintering induces particle growth at the expense of the porous spaces between them. The nanochannels are proposed to be conduits functioning to allow passage of ions and small molecules deep into extracellular matrices and supporting mineralization even at the most remote sites within the extracellular spaces of vertebrate hard tissues [175]. Such a nanoscale network may be fundamental in understanding the transport and delivery of ions and molecules to collagen as an early step in matrix mineralization of vertebrates. How the movement and delivery of ions and molecules to the extracellular milieu by means of secondary channels possibly compare or relate to like processes mediated by exosomes and matrix vesicles described earlier is a subject for further research investigation.

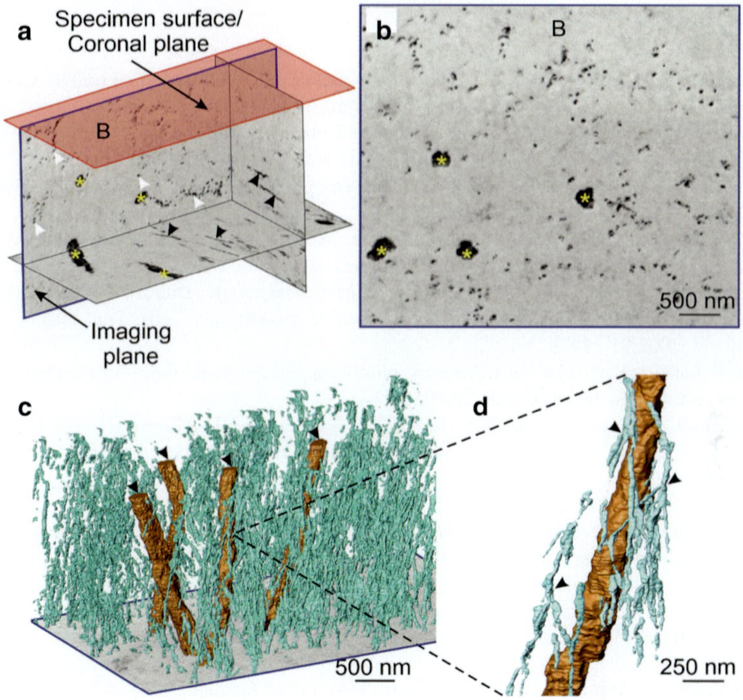

Fig. 7.17 The extensive extracellular matrix nanochannel network comprising the growth plate cartilage and metaphyseal bone interface of a mouse femoral head. The tissue was fixed in 4% paraformaldehyde, embedded at low temperature, stained with iodine vapor, and examined by FIB-SEM. Images of specimen slices were reconstructed in 3D. (**a**) A rendered image showing three orthogonal planes and a reference plane (red/transparent, specimen surface) in the metaphyseal bone region (B) at the cartilage–bone interface. Small, dark pores were detected in the imaging plane of the bone tissue (white arrowheads). In the other two orthogonal planes, such pores corresponded to dark nanochannels (black arrowheads) along with four bone canaliculi (yellow asterisks) also observed in the tissue. These ultrastructural features were similar to those found in the cartilage aspects of the femoral head as well. (**b**) A representative electron backscattered image of a region of the bone domain (B) close to that shown in (**a**) and presenting the four bone canaliculi (yellow asterisks) and several associated fine, dark pore-like features in this volume of tissue. (**c**) 3D reconstruction corresponding to (**a**) and (**b**) demonstrates the four canaliculi (light brown, denoted by black arrowheads) and a dense network of nanochannels (cyan) oriented principally in the same direction as that of the canaliculi. The fine, dark pore-like features found in sections like that in (**b**) are transverse profiles of the nanochannels as they intersect the section. (**d**) Enlargement of a small region of the network of nanochannels in (**c**) illustrates one of the canaliculi (light brown) and multiple nanochannels (cyan) interconnected with and surrounding it. Nanochannels possess numerous branch points or nodes (black arrowheads), a characteristic similar to the branching of the nanochannels in the calcified cartilage of the mouse femoral head. (Images adapted from Tang et al. [175] with permission)

References

1. Schrödinger E. What is life? The physical aspect of the living cell. Cambridge: Cambridge University Press; 1944. A work based on lectures (1944) delivered under the auspices of the Dublin Institute for Advanced Studies, Trinity College, Dublin, Ireland, February, 1943.
2. Berzelius JJ, Marchand RF. Analysis of human bones. Philos Mag Ser. 1843;1(22):236–7.
3. Dorozhkin SV. A detailed history of calcium orthophosphates from 1770s till 1950. Mater Sci Eng C Mater Biol Appl. 2013;33:3085–110.
4. Chen PY, Toroian D, Price PA, McKittrick J. Minerals form a continuum phase in mature cancellous bone. Calcif Tissue Int. 2011;88:351–61.
5. Villemure I, Chung MA, Seck CS, Kimm MH, Matyas JR, Duncan NA. The effects of mechanical loading on the mRNA expression of growth-plate cells. Stud Health Technol Inform. 2002;91:114–8.
6. Myllyharju J, Kivirikko KI. Collagens, modifying enzymes and their mutations in humans, flies and worms. Trends Genet. 2004;20:33–43.
7. Marini JC, Forlino A, Cabral WA, Barnes AM, San Antonio JD, Milgrom S, et al. Consortium for osteogenesis imperfecta mutations in the helical domain of type I collagen: regions rich in lethal mutations align with collagen binding sites for integrins and proteoglycans. Hum Mutat. 2007;28:209–21.
8. Marini JC, Cabral WA, Barnes AM. Null mutations in LEPRE1 and CRTAP cause severe recessive osteogenesis imperfecta. Cell Tissue Res. 2010;339:59–70.
9. Dalgleish R. The human type I collagen mutation database. Nucleic Acids Res. 1997;25:181–7.
10. Winterpacht A, Hilbert M, Schwarze U, Mundlos S, Spranger J, Zabel BU. Kniest and stickler dysplasia phenotypes caused by collagen type II gene (COL2A1) defect. Nat Genet. 1993;3:323–6.
11. Warman ML, Abbott M, Apte SS, Hefferon T, McIntosh I, Cohn DH, et al. A type X collagen mutation causes Schmid metaphyseal chondrodysplasia. Nat Genet. 1993;5:79–82.
12. Ricard-Blum S. The collagen family. Cold Spring Harb Perspect Biol. 2011;3:a004978.
13. Eyre DR. Collagen: molecular diversity in the body's protein scaffold. Science. 1980;207:1315–22.
14. Bella J, Liu J, Kramer R, Brodsky B, Berman HM. Conformational effects of Gly-X-Gly interruptions in the collagen triple helix. J Mol Biol. 2006;362:298–311.
15. Okuyama K, Wu G, Jiravanichanun N, Hongo C, Noguchi K. Helical twists of collagen model peptides. Biopolymers. 2006;84:421–32.
16. Chow WY, Bihan D, Forman CJ, Slatter DA, Reid DG, Wales DJ, et al. Hydroxyproline ring pucker causes frustration of helix parameters in the collagen triple helix. Sci Rep. 2015;5:12556.
17. Rau I, Kajzar F. Biopolymers for application in photonics. Vestnik Volgogradskogo gosudarstvennogo universiteta Serija 10 Innovatcionnaia deiatel'nost'; 2014.
18. Kubyshkin V. Stabilization of the triple helix in collagen mimicking peptides. Org Biomol Chem. 2019;17:8031–47.
19. Bella J. Collagen structure: new tricks from a very old dog. Biochem J. 2016;473:1001–25.
20. Nestler FH, Hvidt S, Ferry JD, Veis A. Flexibility of collagen determined from dilute solution viscoelastic measurements. Biopolymers. 1983;22:1747–58.
21. Orgel JPRO, Madhurapantula RS. A structural prospective for collagen receptors such as DDR and their binding of the collagen fibril. Biochim Biophys Acta Mol Cell Res. 2019;1866:118478.
22. Nimni ME, Harkness RD. Molecular structure and function of collagen. In: Nimni ME, editor. Collagen. Boca Raton, FL: CRC Press; 1988.
23. Silver FH, Kato YP, Ohno M, Wasserman AJ. Analysis of mammalian connective tissue: relationship between hierarchical structures and mechanical properties. J Long-Term Eff Med Implants. 1992;2:165–98.

24. Cameron GJ, Alberts IL, Laing JH, Wess TJ. Structure of type I and type III heterotypic collagen fibrils: an X-ray diffraction study. J Struct Biol. 2002;137:15–22.
25. Spiess K, Zorn TM. Collagen types I, III, and V constitute the thick collagen fibrils of the mouse decidua. Microsc Res Tech. 2007;70:18–25.
26. Schweitzer MH, Zheng W, Organ CL, Avci R, Suo Z, Freimark LM, et al. Biomolecular characterization and protein sequences of the Campanian hadrosaur B. canadensis. Science. 2009;324:626–31.
27. Asara JM, Schweitzer MH, Freimark LM, Phillips M, Cantley LC. Protein sequences from mastodon and Tyrannosaurus rex revealed by mass spectrometry. Science. 2007;316:280–5.
28. San Antonio JD, Schweitzer MH, Jensen ST, Kalluri R, Buckley M, Orgel JPRO. Dinosaur peptides suggest mechanisms of protein survival. PLoS One. 2011;6:e20381.
29. Buckley M, Walker A, Ho SY, Yang Y, Smith C, Ashton P, et al. Comment on "Protein sequences from mastodon and Tyrannosaurus rex revealed by mass spectrometry". Science. 2008;319:33.
30. Buckley M, Warwood S, van Dongen B, Kitchener AC, Manning PL. A fossil protein chimera; difficulties in discriminating dinosaur peptide sequences from modern cross-contamination. Proc Biol Sci. 2017;284:20170544.
31. Kadler KE, Baldock C, Bella J, Boot-Handford RP. Collagens at a glance. J Cell Sci. 2007;120:1955–8.
32. Orgel JP, San Antonio JD, Antipova O. Molecular and structural mapping of collagen fibril interactions. Connect Tissue Res. 2011;52:2–17.
33. Orgel JP, Eid A, Antipova O, Bella J, Scott JE. Decorin core protein (decoron) shape complements collagen fibril surface structure and mediates its binding. PLoS One. 2009;4:e7028.
34. Erat MC, Slatter DA, Lowe ED, Millard CJ, Farndale RW, Campbell ID, et al. Identification and structural analysis of type I collagen sites in complex with fibronectin fragments. Proc Natl Acad Sci U S A. 2009;106:4195–200.
35. Yamauchi M, Taga Y, Hattori S, Shiiba M, Terajima M. Analysis of collagen and elastin cross-links. Methods Cell Biol. 2018;143:115–32.
36. Chung L, Dinakarpandian D, Yoshida N, Lauer-Fields JL, Fields GB, Visse R, et al. Collagenase unwinds triple-helical collagen prior to peptide bond hydrolysis. EMBO J. 2004;23:3020–30.
37. Xu Y, Gurusiddappa S, Rich RL, Owens RT, Keene DR, Mayne R, et al. Multiple binding sites in collagen type I for the integrins alpha1beta1 and alpha2beta1. J Biol Chem. 2000;275:38981–9.
38. Ruoslahti E. RGD and other recognition sequences for integrins. Annu Rev Cell Dev Biol. 1996;12:697–715.
39. Taubenberger AV, Woodruff MA, Bai H, Muller DJ, Hutmacher DW. The effect of unlocking RGD-motifs in collagen I on pre-osteoblast adhesion and differentiation. Biomaterials. 2010;31:2827–35.
40. Leitinger B. Transmembrane collagen receptors. Annu Rev Cell Dev Biol. 2011;27:265–90.
41. Knight CG, Morton LF, Peachey AR, Tuckwell DS, Farndale RW, Barnes MJ. The collagen-binding A-domains of integrins alpha(1)beta(1) and alpha(2)beta(1) recognize the same specific amino acid sequence, GFOGER, in native (triple-helical) collagens. J Biol Chem. 2000;275:35–40.
42. Sipilä KH, Drushinin K, Rappu P, Jokinen J, Salminen TA, Salo AM, et al. Proline hydroxylation in collagen supports integrin binding by two distinct mechanisms. J Biol Chem. 2018;293:7645–58.
43. Shrivastava A, Radziejewski C, Campbell E, Kovac L, McGlynn M, Ryan TE, et al. An orphan receptor tyrosine kinase family whose members serve as nonintegrin collagen receptors. Mol Cell. 1997;1:25–34.
44. Vogel W, Gish GD, Alves F, Pawson T. The discoidin domain receptor tyrosine kinases are activated by collagen. Mol Cell. 1997;1:13–23.
45. Lemeer S, Bluwstein A, Wu Z, Leberfinger J, Müller K, Kramer K, et al. Phosphotyrosine mediated protein interactions of the discoidin domain receptor 1. J Proteome. 2012;75:3465–77.

46. Fu HL, Valiathan RR, Arkwright R, Sohail A, Mihai C, Kumarasiri M, et al. Discoidin domain receptors: unique receptor tyrosine kinases in collagen-mediated signaling. J Biol Chem. 2013;288:7430–7.
47. Fu HL, Sohail A, Valiathan RR, Wasinski BD, Kumarasiri M, Mahasenan KV, et al. Shedding of discoidin domain receptor 1 by membrane-type matrix metalloproteinases. J Biol Chem. 2013;288:12114–29.
48. Leitinger B. Discoidin domain receptor functions in physiological and pathological conditions. Int Rev Cell Mol Biol. 2014;310:39–87.
49. Leitinger B, Kwan AP. The discoidin domain receptor DDR2 is a receptor for type X collagen. Matrix Biol. 2006;25:355–64.
50. Lemmon MA, Schlessinger J. Cell signaling by receptor tyrosine kinases. Cell. 2010;141:1117–34.
51. Borza CM, Pozzi A. Discoidin domain receptors in disease. Matrix Biol. 2014;34:185–92.
52. Valiathan RR, Marco M, Leitinger B, Kleer CG, Fridman R. Discoidin domain receptor tyrosine kinases: new players in cancer progression. Cancer Metastasis Rev. 2012;31:295–321.
53. Sunk I-G, Bobacz K, Hofstaetter JG, Amoyo L, Soleiman A, Smolen J, et al. Increased expression of discoidin domain receptor 2 is linked to the degree of cartilage damage in human knee joints: a potential role in osteoarthritis pathogenesis. Arthritis Rheum. 2007;56:3685–92.
54. Manning LB, Li Y, Chickmagalur NS, Li X, Xu L. Discoidin domain receptor 2 as a potential therapeutic target for development of disease-modifying osteoarthritis drugs. Am J Pathol. 2016;186:3000–10.
55. Felix R, Neuman WF, Fleisch H. Aerobic glycolysis in bone: lactic acid production by rat calvaria cells in culture. Am J Phys. 1978;234:C51–5.
56. Komarova SV, Ataullakhanov FI, Globus RK. Bioenergetics and mitochondrial transmembrane potential during differentiation of cultured osteoblasts. Am J Physiol Cell Physiol. 2000;279:C1220–9.
57. Regan JN, Lim J, Shi Y, Joeng KS, Arbeit JM, Shohet RV, et al. Up-regulation of glycolytic metabolism is required for HIF1α-driven bone formation. Proc Natl Acad Sci U S A. 2014;111(23):8673–8.
58. Stegen S, Laperre K, Eelen G, Rinaldi G, Fraisl P, Torrekens S, et al. HIF-1α metabolically controls collagen synthesis and modification in chondrocytes. Nature. 2019;565:511–5.
59. Shen L, Sharma D, Yu Y, Long F, Karner CM. Biphasic regulation of glutamine consumption by WNT during osteoblast differentiation. J Cell Sci. 2021;134:jcs251645.
60. Brown PM, Hutchison JD, Crockett JC. Absence of glutamine supplementation prevents differentiation of murine calvarial osteoblasts to a mineralizing phenotype. Calcif Tissue Int. 2011;89:472–82.
61. Karner CM, Long F. Wnt signaling and cellular metabolism in osteoblasts. Cell Mol Life Sci. 2017;74:1649–57.
62. Stegen S, Rinaldi G, Loopmans S, Stockmans I, Moermans K, Thienpont B, et al. Glutamine metabolism controls chondrocyte identity and function. Dev Cell. 2020;53:530–544.e8.
63. Li P, Wu G. Roles of dietary glycine, proline, and hydroxyproline in collagen synthesis and animal growth. Amino Acids. 2018;50:29–38.
64. Karna E, Szoka L, Huynh TYL, Palka JA. Proline-dependent regulation of collagen metabolism. Cell Mol Life Sci. 2020;77:1911–8.
65. Canty EG, Lu Y, Meadows RS, Shaw MK, Holmes DF, Kadler KE. Coalignment of plasma membrane channels and protrusions (fibripositors) specifies the parallelism of tendon. J Cell Biol. 2004;165:553–63.
66. Lodish H, Berk A, Zipursky SL, Matsudaira P, Baltimore D, Darnell J. Collagen: the fibrous proteins of the matrix. In: Darnell JE, editor. Molecular cell biology. 4th ed. New York: WH Freeman & Co Ltd; 2000.
67. Hodge AJ, Petruska J. Recent studies with the electron microscope on ordered aggregates of the tropocollagen macromolecule. In: Ramachandran GN, editor. Aspects of protein structure. New York: Academic Press; 1963. p. 289–300.

68. Gelse K, Pöschl E, Aigner T. Collagens-structure, function, and biosynthesis. Adv Drug Deliv Rev. 2003;55:1531–46.
69. Shoulders MD, Raines RT. Collagen structure and stability. Annu Rev Biochem. 2009;78:929–58.
70. Prockop DJ, Sieron AL, Li SW. Procollagen N-proteinase and procollagen C-proteinase. Two unusual metalloproteinases that are essential for procollagen processing probably have important roles in development and cell signaling. Matrix Biol. 1998;16:399–408.
71. Holmes DF, Lu Y, Starborg T, Kadler KE. Collagen fibril assembly and function. Curr Top Dev Biol. 2018;130:107–42.
72. Silver FH, Landis WJ. Deposition of apatite in mineralizing vertebrate extracellular matrices: a model of possible nucleation sites on type I collagen. Connect Tissue Res. 2011;52:242–54.
73. Lees S. Mineralization of type I collagen. Biophys J. 2003;85:204–7.
74. Landis WJ, Song MJ, Leith A, McEwen L, McEwen BF. Mineral and organic matrix interaction in normally calcifying tendon visualized in three dimensions by high-voltage electron microscopic tomography and graphic image reconstruction. J Struct Biol. 1993;110:39–54.
75. Orgel JP, Irving TC, Miller A, Wess TJ. Microfibrillar structure of type I collagen in situ. Proc Natl Acad Sci U S A. 2006;103:9001–5.
76. Lees S. Considerations regarding the structure of the mammalian mineralized osteoid from viewpoint of the generalized packing model. Connect Tissue Res. 1987;16:281–303.
77. Landis WJ, Jacquet R. Association of calcium and phosphate ions with collagen in the mineralization of vertebrate tissues. Calcif Tissue Int. 2013;93:329–37.
78. Katz EP, Li ST. Structure and function of bone collagen fibrils. J Mol Biol. 1973;80:1–15.
79. Weiner S, Traub W. Organization of hydroxyapatite crystals within collagen fibrils. FEBS Lett. 1986;206:262–6.
80. Zou Z, Tang T, Macías-Sánchez E, Sviben S, Landis WJ, Bertinetti L, et al. Three-dimensional structural interrelations between cells, extracellular matrix, and mineral in normally mineralizing avian leg tendon. Proc Natl Acad Sci U S A. 2020;117:14102–9.
81. Glimcher MJ, Krane SM. The organization and structure of bone, and the mechanism of calcification. In: Remachandran GN, Gould BS, editors. Treatise on collagen. New York: Academic Press; 1968. p. 68–251.
82. Cho G, Wu Y, Ackerman JL. Detection of hydroxyl ions in bone mineral by solid-state NMR spectroscopy. Science. 2003;300:1123–7.
83. Weiner S, Price PA. Disaggregation of bone into crystals. Calcif Tissue Int. 1986;39:365–75.
84. Moradian-Oldak J, Weiner S, Addadi L, Landis WJ, Traub W. Electron imaging and diffraction study of individual crystals of bone, mineralized tendon and synthetic carbonate apatite. Connect Tissue Res. 1991;25:219–28.
85. Boskey AL. Bone mineral crystal size. Osteoporos Int. 2003;14(Suppl 5):S16–21.
86. Kirkham J, Brookes SJ, Shore RC, Bonass WA, Smith DA, Wallwork ML, et al. Atomic force microscopy studies of crystal surface topology during enamel development. Connect Tissue Res. 1998;38:91–100; discussion 139–145.
87. Rey C, Combes C, Drouet C, Glimcher MJ. Bone mineral: update on chemical composition and structure. Osteoporos Int. 2009;20:1013–21.
88. Hu Y-Y, Rawal A, Schmidt-Rohr K. Strongly bound citrate stabilizes the apatite nanocrystals in bone. Proc Natl Acad Sci U S A. 2010;107:22425–9.
89. Hartles RL. Citrate in mineralized tissues. Adv Oral Biol. 1964;1:225–53.
90. Wang Z, Xu Z, Zhao W, Chen W, Miyoshi T, Sahai N. Isoexergonic conformations of surface-bound citrate regulated bioinspired apatite nanocrystal growth. ACS Appl Mater Interfaces. 2016;8:28116–23.
91. Davies E, Müller KH, Wong WC, Pickard CJ, Reid DG, Skepper JN, et al. Citrate bridges between mineral platelets in bone. Proc Natl Acad Sci U S A. 2014;111:E1354–63.
92. Delgado-López JM, Bertolotti F, Lyngsø J, Pedersen JS, Cervellino A, Masciocchi N, et al. The synergic role of collagen and citrate in stabilizing amorphous calcium phosphate precursors with platy morphology. Acta Biomater. 2017;49:555–62.

93. Irizarry AR, Yan G, Zeng Q, Lucchesi J, Hamang MJ, Ma YL, et al. Defective enamel and bone development in sodium-dependent citrate transporter (NaCT) Slc13a5 deficient mice. PLoS One. 2017;12:e0175465.
94. Chen H, Wang Y, Dai H, Tian X, Cui ZK, Chen Z, et al. Bone and plasma citrate is reduced in osteoporosis. Bone. 2018;114:189–97.
95. Granchi D, Baldini N, Ulivieri FM, Caudarella R. Role of citrate in pathophysiology and medical management of bone diseases. Nutrients. 2019;11:2576.
96. Jiang W, Griffanti G, Tamimi F, McKee MD, Nazhat SN. Multiscale structural evolution of citrate-triggered intrafibrillar and interfibrillar mineralization in dense collagen gels. J Struct Biol. 2020;212:107592.
97. Kay MI, Young RA, Posner AS. Crystal structure of hydroxyapatite. Nature. 1964;204:1050–2.
98. Mathew M, Takagi S. Structures of biological minerals in dental research. J Res Natl Inst Stand Technol. 2001;106:1035–44.
99. Kawasaki T, Takahashi S, Ikeda K. Hydroxyapatite high-performance liquid chromatography: column performance for proteins. Eur J Biochem. 1985;152:361–71.
100. Kandori K, Miyagawa K, Ishikawa T. Adsorption of immunogamma globulin onto various synthetic calcium hydroxyapatite particles. J Colloid Interface Sci. 2004;273:406–13.
101. Kandori K, Fudo A, Ishikawa T. Study on the particle texture dependence of protein adsorption by using synthetic micrometer-sized calcium hydroxyapatite particles. Colloid Surf B. 2002;24:145–53.
102. Shaw WJ, Campbell AA, Paine ML, Snead ML. The COOH terminus of the amelogenin, LRAP, is oriented next to the hydroxyapatite surface. J Biol Chem. 2004;279:40263–6.
103. Landis WJ. Mineral characterization in calcifying tissues: atomic, molecular and macromolecular perspectives. Connect Tissue Res. 1996;34:239–46.
104. Ma YX, Hoff SE, Huang XQ, Liu J, Wan QQ, Song Q, et al. Involvement of prenucleation clusters in calcium phosphate mineralization of collagen. Acta Biomater. 2020;120:213.
105. Xu Y-F, Nudelman F, Eren ED, Wirix MJM, Cantaert B, Nijhuis WH, et al. Intermolecular channels direct crystal orientation in mineralized collagen. Nat Commun. 2020;11:5068.
106. Landis WJ, Hodgens KJ, Song MJ, Arena J, Kiyonaga S, Marko M, et al. Mineralization of collagen may occur on fibril surfaces: evidence from conventional and high-voltage electron microscopy and three-dimensional imaging. J Struct Biol. 1996;117:24–35.
107. Landis WJ. An overview of vertebrate mineralization with emphasis on collagen-mineral interaction. Gravit Space Biol Bull. 1999;12:15–26.
108. Arsenault AL. Image analysis of collagen-associated mineral distribution in cryogenically prepared Turkey leg tendons. Calcif Tissue Int. 1991;48:56–62.
109. Nudelman F, Pieterse K, George A, Bomans PH, Friedrich H, Brylka LJ, et al. The role of collagen in bone apatite formation in the presence of hydroxyapatite nucleation inhibitors. Nat Mater. 2010;9:1004–9.
110. Christoffersen J, Landis WJ. A contribution with review to the description of mineralization of bone and other calcified tissues in vivo. Anat Rec. 1991;230:435–50.
111. Xu Z, Yang Y, Zhao W, Wang Z, Landis WJ, Cui Q, et al. Molecular mechanisms for intrafibrillar collagen mineralization in skeletal tissues. Biomaterials. 2015;39:59–66.
112. Orgel JP, Miller A, Irving TC, Fischetti RF, Hammersley AP, Wess TJ. The in situ supermolecular structure of type I collagen. Structure. 2001;9:1061–9.
113. Hodge AJ. Molecular models illustrating the possible distributions of 'holes' in simple systematically staggered arrays of type I collagen molecules in native-type fibrils. Connect Tissue Res. 1989;21:137–47.
114. Silver FH, Freeman JW, Horvath I, Landis WJ. Molecular basis for elastic energy storage in mineralized tendon. Biomacromolecules. 2001;2:750–6.
115. Chapman JA. The staining pattern of collagen fibrils. I. An analysis of electron micrographs. Connect Tissue Res. 1974;2:137–50.
116. Chapman JA, Hardcastle RA. The staining pattern of collagen fibrils. II. A comparison with patterns computer-generated from the amino acid sequence. Connect Tissue Res. 1974;2:151–9.

117. Xu Z, Zhao W, Wang Z, Yang Y, Sahai N. Structure analysis of collagen fibril at atomic-level resolution and its implications for intra-fibrillar transport in bone biomineralization. Phys Chem Chem Phys. 2018;20:1513–23.
118. Chen L, Jacquet R, Lowder E, Landis WJ. Refinement of collagen-mineral interaction: a possible role for osteocalcin in apatite crystal nucleation, growth and development. Bone. 2015;71:7–16.
119. Dey A, Bomans PH, Müller FA, Will J, Frederik PM, de With G, et al. The role of prenucleation clusters in surface-induced calcium phosphate crystallization. Nat Mater. 2010;9:1010–4.
120. Gebauer D, Cölfen H. Prenucleation clusters and non-classical nucleation. Nano Today. 2011;6:564–84.
121. Habraken WJEM, Tao J, Brylka LJ, Friedrich H, Bertinetti L, Schenk AS, et al. Ion-association complexes unite classical and non-classical theories for the biomimetic nucleation of calcium phosphate. Nat Commun. 2013;4:1507.
122. Toroian D, Lim JE, Price PA. The size exclusion characteristics of type I collagen: implications for the role of noncollagenous bone constituents in mineralization. J Biol Chem. 2007;282:22437–47.
123. Meek KM, Chapman JA, Hardcastle RA. The staining pattern of collagen fibrils. Improved correlation with sequence data. J Biol Chem. 1979;254:10710–4.
124. Landis WJ, Silver FH. The structure and function of normally mineralizing avian tendons. Comp Biochem Physiol A Mol Integr Physiol. 2002;133:1135–57.
125. Tye CE, Hunter GK, Goldberg HA. Identification of the type I collagen-binding domain of bone sialoprotein and characterization of the mechanism of interaction. J Biol Chem. 2005;280:13487–92.
126. Gordon JA, Tye CE, Sampaio AV, Underhill TM, Hunter GK, Goldberg HA. Bone sialoprotein expression enhances osteoblast differentiation and matrix mineralization in vitro. Bone. 2007;41:462–73.
127. Poundarik AA, Boskey A, Gundberg C, Vashishth D. Biomolecular regulation, composition and nanoarchitecture of bone mineral. Sci Rep. 2018;8:1191.
128. Simon P, Grüner D, Worch H, Pompe W, Lichte H, El Khassawna T, et al. First evidence of octacalcium phosphate@osteocalcin nanocomplex as skeletal bone component directing collagen triple-helix nanofibril mineralization. Sci Rep. 2018;8:13696.
129. Chen L, Childs RD, Landis WJ. Correlations between gene expression and mineralization in the avian leg tendon. Bone. 2019;121:42–59.
130. Hunter GK, Kyle CL, Goldberg HA. Modulation of crystal formation by bone phosphoproteins: structural specificity of the osteopontin-mediated inhibition of hydroxyapatite formation. Biochem J. 1994;300:723–8.
131. Boskey AL, Spevak L, Paschalis E, Doty SB, McKee MD. Osteopontin deficiency increases mineral content and mineral crystallinity in mouse bone. Calcif Tissue Int. 2002;71:145–54.
132. Addison WN, Masica DL, Gray JJ, McKee MD. Phosphorylation-dependent inhibition of mineralization by osteopontin ASARM peptides is regulated by PHEX cleavage. J Bone Miner Res. 2010;25:695–705.
133. Tryfonidou MA, Hazewinkel HA, Riemers FM, Brinkhof B, Penning LC, Karperien M. Intraspecies disparity in growth rate is associated with differences in expression of local growth plate regulators. Am J Physiol Endocrinol Metab. 2010;299:E1044–52.
134. LuValle P, Ninomiya Y, Rosenblum ND, Olsen BR. The type X collagen gene. Intron sequences split the 5′-untranslated region and separate the coding regions for the non-collagenous amino-terminal and triple-helical domains. J Biol Chem. 1988;263:18378–85.
135. Schmid TM, Linsenmayer TF. Immunoelectron microscopy of type X collagen: supramolecular forms within embryonic chick cartilage. Dev Biol. 1990;138:53–62.
136. Kwan AP, Cummings CE, Chapman JA, Grant ME. Macromolecular organization of chicken type X collagen in vitro. J Cell Biol. 1991;114:597–604.

137. Kwan KM, Pang MK, Zhou S, Cowan SK, Kong RY, Pfordte T, et al. Abnormal compartmentalization of cartilage matrix components in mice lacking collagen X: implications for function. J Cell Biol. 1997;136:459–71.
138. Reginato AM, Shapiro IM, Lash JW, Jimenez SA. Type X collagen alterations in rachitic chick epiphyseal growth cartilage. J Biol Chem. 1988;263:9938–45.
139. Albertson RC, Yan Y-L, Titus TA, Pisano E, Vacchi M, Yelick PC, et al. Molecular pedomorphism underlies craniofacial skeletal evolution in Antarctic notothenioid fishes. BMC Evol Biol. 2010;10:4.
140. Renn J, Winkler C. Characterization of collagen type 10a1 and osteocalcin in early and mature osteoblasts during skeleton formation in medaka. J Appl Ichthyol. 2010;26:196–201.
141. Eames BF, Amores A, Yan Y-L, Postlethwait JH. Evolution of the osteoblast: skeletogenesis in gar and zebrafish. BMC Evol Biol. 2012;12:27.
142. Aldea D, Hanna P, Munoz D, Espinoza J, Torrejon M, Sachs L, et al. Evolution of the vertebrate bone matrix: an expression analysis of the network forming collagen paralogues in amphibian osteoblasts. J Exp Zool B Mol Dev Evol. 2013;320:375–84.
143. Debiais-Thibaud M, Simion P, Ventéo S, Muñoz D, Marcellini S, Mazan S, et al. Skeletal mineralization in association with type X collagen expression is an ancestral feature for jawed vertebrates. Mol Biol Evol. 2019;36:2265–76.
144. Rosati R, Horan GS, Pinero GJ, Garofalo S, Keene DR, Horton WA, et al. Normal long bone growth and development in type X collagen-null mice. Nat Genet. 1994;8:129–35.
145. Lee JK, Gegg CA, Hu JC, Reddi AH, Athanasiou KA. Thyroid hormones enhance the biomechanical functionality of scaffold-free neocartilage. Arthritis Res Ther. 2015;17:28.
146. Knuth CA, Andres Sastre E, Fahy NB, Witte-Bouma J, Ridwan Y, Strabbing EM, et al. Collagen type X is essential for successful mesenchymal stem cell-mediated cartilage formation and subsequent endochondral ossification. Eur Cell Mater. 2019;38:106–22.
147. Petit A, Wang HT, Girard-Lauriault P-L, Wertheimer MR, Antoniou J, Mwale F. Novel insights into the mechanism of decreased expression of type X collagen in human mesenchymal stem cells from patients with osteoarthritis cultured on nitrogen-rich plasma polymers: implication of cyclooxygenase-1. J Biomed Mater Res A. 2010;94:744–50.
148. Wang Y, Azaïs T, Robin M, Vallée A, Catania C, Legriel P, et al. The predominant role of collagen in the nucleation, growth, structure and orientation of bone apatite. Nat Mater. 2012;11:724–33.
149. Kim D, Lee B, Thomopoulos S, Jun Y-S. The role of confined collagen geometry in decreasing nucleation energy barriers to intrafibrillar mineralization. Nat Commun. 2018;9:962.
150. Veis A, Dorvee JR. Biomineralization mechanisms: a new paradigm for crystal nucleation in organic matrices. Calcif Tissue Int. 2013;93:307–15.
151. Deshpande AS, Beniash E. Bioinspired synthesis of mineralized collagen fibrils. Cryst Growth Des. 2008;8:3084–90.
152. Boskey AL, Villarreal-Ramirez E. Intrinsically disordered proteins and biomineralization. Matrix Biol. 2016;52–54:43–59.
153. Bouleftour W, Juignet L, Bouet G, Granito RN, Vanden-Bossche A, Laroche N, et al. The role of the SIBLING, bone sialoprotein in skeletal biology - contribution of mouse experimental genetics. Matrix Biol. 2016;52-54:60–77.
154. Cohen-Solal L, Maroteaux P, Glimcher MJ. Presence of gamma-glutamyl phosphate in the collagens of bone and other calcified tissues and its absence in the collagens of unmineralized tissues. In: Veis A, editor. The chemistry and biology of mineralized connective tissues. New York: Elsevier; 1981. p. 7–11.
155. Yalak G, Olsen BR. Proteomic database mining opens up avenues utilizing extracellular protein phosphorylation for novel therapeutic applications. J Transl Med. 2015;13:125.
156. Kerschnitzki M, Akiva A, Ben Shoham A, Asscher Y, Wagermaier W, Fratzl P, et al. Bone mineralization pathways during the rapid growth of embryonic chicken long bones. J Struct Biol. 2016;195:82–92.
157. Jee SS, Culver L, Li Y, Douglas EP, Gower LB. Biomimetic mineralization of collagen via an enzyme-aided PILP process. J Cryst Growth. 2010;312:1249–56.

158. Von Euw S, Chan-Chang T-H-C, Paquis C, Haye B, Pehau-Arnaudet G, Babonneau F, et al. Organization of bone mineral: the role of mineral–water interactions. Geosciences. 2018;8:466.
159. Olszta MJ, Cheng X, Jee SS, Kumar R, Kim Y-Y, Kaufman MJ, et al. Bone structure and formation: a new perspective. Mater Sci Eng R Rep. 2007;58:77–116.
160. Li Y, Thula TT, Jee S, Perkins SL, Aparicio C, Douglas EP, et al. Biomimetic mineralization of woven bone-like nanocomposites: role of collagen cross-links. Biomacromolecules. 2012;13:49–59.
161. Reznikov N, Bilton M, Lari L, Stevens MM, Kröger R. Fractal-like hierarchical organization of bone begins at the nanoscale. Science. 2018;360:6388.
162. Landis WJ. Tomographic imaging of collagen-mineral interaction: implications for osteogenesis imperfecta. Connect Tissue Res. 1995;31:287–90.
163. Fratzl P, Paris O, Klaushofer K, Landis WJ. Bone mineralization in an osteogenesis imperfecta mouse model studied by small-angle x-ray scattering. J Clin Invest. 1996;97:396–402.
164. Schmid TM, Linsenmayer TF. Immunohistochemical localization of short chain cartilage collagen (type X) in avian tissues. J Cell Biol. 1985;100:598–605.
165. Knuth CA, Witte-Bouma J, Ridwan Y, Wolvius EB, Farrell E. Mesenchymal stem cell-mediated endochondral ossification utilising micropellets and brief chondrogenic priming. Eur Cell Mater. 2017;34:142–61.
166. Jacenko O, Chan D, Franklin A, Ito S, Underhill CB, Bateman JF, et al. A dominant interference collagen X mutation disrupts hypertrophic chondrocyte pericellular matrix and glycosaminoglycan and proteoglycan distribution in transgenic mice. Am J Pathol. 2001;159:2257–69.
167. Jacenko O, Roberts DW, Campbell MR, McManus PM, Gress CJ, Tao Z. Linking hematopoiesis to endochondral skeletogenesis through analysis of mice transgenic for collagen X. Am J Pathol. 2002;160:2019–34.
168. Sweeney E, Campbell M, Watkins K, Hunter CA, Jacenko O. Altered endochondral ossification in collagen X mouse models leads to impaired immune responses. Dev Dyn. 2008;237:2693–704.
169. Kirsch T, Wuthier RE. Stimulation of calcification of growth plate cartilage matrix vesicles by binding to type II and X collagens. J Biol Chem. 1994;269:11462–9.
170. Grskovic I, Kutsch A, Frie C, Groma G, Stermann J, Schlötzer-Schrehardt U, et al. Depletion of annexin A5, annexin A6, and collagen X causes no gross changes in matrix vesicle-mediated mineralization, but lack of collagen X affects hematopoiesis and the Th1/Th2 response. J Bone Miner Res. 2012;27:2399–412.
171. Frost A, Jonsson KB, Brändström H, Ljunghall S, Nilsson O, Ljunggren O. Interleukin (IL)-13 and IL-4 inhibit proliferation and stimulate IL-6 formation in human osteoblasts: evidence for involvement of receptor subunits IL-13R, IL-13Ralpha, and IL-4Ralpha. Bone. 2001;28:268–74.
172. Balani D, Aeberli D, Hofstetter W, Seitz M. Interleukin-17A stimulates granulocyte–macrophage colony-stimulating factor release by murine osteoblasts in the presence of 1,25-dihydroxyvitamin D3 and inhibits murine osteoclast development in vitro. Arthritis Rheum. 2013;65:436–46.
173. Mwale F, Rampersad S, Richard H, Guoying Y, Al Rowas S, Madiraju P, et al. The constitutive expression of type x collagen in mesenchymal stem cells from osteoarthritis patients is reproduced in a rabbit model of osteoarthritis. J Tissue Eng. 2011;2011:587547.
174. Sweeney E, Roberts D, Corbo T, Jacenko O. Congenic mice confirm that collagen X is required for proper hematopoietic development. PLoS One. 2010;5:e9518.
175. Tang T, Landis WJ, Raguin E, Werner P, Bertinetti L, Dean M, et al. A 3D network of nanochannels for possible ion and molecule transit in mineralizing bone and cartilage. Adv Nanobiomed Res. 2022;2:2100162.

The Role of Non-collagenous Proteins and Other Matrix Molecules in Vertebrate Mineralization

8

Knowing what we know now from X-ray and related studies of the fibrous proteins, how they are built from long polypeptide chains with linear patterns drawn to a grand scale, how these chains can contract and take up different configurations by intramolecular folding, how the chain-groups are penetrated by, and their sidechains react with, smaller co-operating molecules, and finally how they can combine so readily with nucleic acid molecules and still maintain the fibrous configuration, it is but natural to assume, as a first working hypothesis at least, that they form the long scroll on which is written the pattern of life. [1]

Glossary

Adipokine A cytokine secreted by fat (adipose) cells.

Fam20C/casein kinase II A constitutively active serine/threonine protein kinase, concerned with the regulation of numerous cellular activities such as cell proliferation and survival.

Glycosaminoglycans or glycans Negatively charged linear polysaccharides comprised of repeating disaccharide units.

Hypophosphatemia Low phosphate ion concentration in the blood.

Matricellular proteins Non-structural proteins which bind to matrix proteins and cell membrane receptors.

Paralogous genes Genes or clusters of genes at different chromosomal locations which are the result of gene duplication.

Protein secondary structure Domains of a protein which are stabilized by hydrogen bonding between atoms comprising the protein polypeptide backbone and excluding the amino acid side chains. Common protein secondary structure includes polypeptide chain coiling or folding into configurations such as an α-helix or a β-pleated sheet.

Redundancy The occurrence of a similar, overlapping or identical function produced by proteins encoded by different genes.

Tamm-Horsfall glycoprotein The most abundant urinary protein secreted by cells of the kidney.

I. M. Shapiro, W. J. Landis, *Mechanisms of Mineralization of Vertebrate Skeletal and Dental Tissues*, https://doi.org/10.1007/978-3-031-34304-9_8

Abbreviations and Proper Names for Genes/Proteins and Other Molecules

ACAN	Aggrecan
ASARM	Acidic serine-aspartate-rich motif/MEPE-associated peptide
BGN	Biglycan
BMP	Bone morphogenetic protein
BSP	Bone sialoprotein
CCN	Abbreviation from three matricellular protein family members, CYR61 (cysteine-rich angiogenic protein 61; CCN1), CTGF (connective tissue growth factor; CCN2), and NOV (nephroblastoma overexpressed; CCN3). The complete family is comprised of six matricellular proteins (CCN1-6), each having a specific conserved domain of prominent cysteine residues
COLI	Type I collagen
COLII	Type II collagen
COMP	Cartilage oligomeric matrix protein
DCN	Decorin
DGP	Dentin phosphoglycoprotein
DLX5	Distal-less homeobox 5
DMP	Dentin matrix protein
DPP	Dentin phosphoprotein (phosphophoryn)
DSP	Dentin sialoprotein
DSPP	Dentin sialophosphoprotein
FGF	Fibroblast growth factor
FN	Fibronectin
LCN	Lipocalin
MEPE	Matrix extracellular phosphoglycoprotein
MGP	Matrix gla protein
MMP	Matrix metalloproteinase
OCN	Osteocalcin
OPN	Osteopontin
PHEX	Phosphate-regulating neutral endopeptidase, X-linked
RGD	Arginine-glycine-aspartic acid, the three amino acid motif of matrix proteins which binds to transmembrane proteins (integrins)
RSPO	R-spondin (Roof plate-specific spondin)
RUNX	Runt-related transcription factor
SIBLING	Small integrin-binding ligand, N-linked glycoprotein
SLRP	Small leucine-rich repeat proteoglycan
SOST	Sclerostin
SPARC	Secreted protein acidic and rich in cysteine (osteonectin)
TGF	Transforming growth factor
TNAP	Tissue non-specific alkaline phosphatase
TSP	Thrombospondin
WNT	Wingless-related integration site

8.1 Evolution and Function of Non-collagenous Proteins and Other Molecules of Vertebrate Mineralized Tissues

Because of their complexity, diversity, and scarcity, non-collagenous proteins of the vertebrate mineralized tissues have not been studied quite as frequently as the proteins of the collagen family of proteins. Nevertheless, investigations in vitro and in vivo have provided numerous critical insights into their function in the skeleton and dentition. The experiments in vitro were not facile undertakings: Difficulties encountered in measurement and analyses included, for example, addressing the high calcium phosphate background from specimen mineral content and the elevated protein signal from the collagenous constituents of these tissues. Historically, much of this type of work in vitro occurred before the molecular biology revolution which included mouse genomics which could be used to define the identity and probe the role of a particular protein (or gene) in an experimental system in vivo. At the present time, the use of gene sequencing systems, "gain and loss of function" studies, ultra-sensitive proteomic techniques, and several other approaches has revealed a great deal more about these molecules, from their evolution from common ancestral genes to the details of protein structure, location, organization, regulation, and function.

The predominant protein of bone, dentin, cementum, and tendon is type I collagen (COLI), one of the most abundant molecules in Nature. COLI comprises approximately 90% of the extracellular matrix of all normally mineralizing vertebrate tissues, other than that of calcifying cartilage and enamel. The residual extracellular composition of bone, dentin, cementum, and tendon is therefore less than about 10% of the total protein content and 1% of the dry tissue weight. Not surprisingly, these molecules are collectively known as "non-collagenous." The prefix "non-" is somewhat unfortunate as it suggests that these molecules are of lesser importance than collagen when in fact their functions appear to be quite the opposite. Many of the non-collagenous proteins are related structurally and thereby grouped together as small integrin-binding ligand, N-linked glycoproteins (SIBLINGs) [2]. Other related non-collagenous molecules in vertebrate tissues include the family of small, leucine-rich proteoglycans (SLRPs), as well as additional proteoglycans and proteins.

Since the discoveries of these various non-collagenous proteins and molecules, their roles in the mineralization process remain conjectural. Studies in vitro suggest that purified collagen, which is normally phosphorylated, fail to induce physiologic mineral formation; for instance, while there is some evidence that phosphorylated residues in other extracellular matrix molecules of vertebrate mineralized tissues promote apatite deposition [3–5]. Some investigators have proposed that collagen is required for mineralization, serving as the template for apatite nucleation without the presence of additional molecules [6–10]. Others believe that specific non-collagenous molecules are necessary, if not essential, for mineralization to occur [11].

In reference again to phosphorylated non-collagenous molecules in extracellular matrices, George and Veis commented that, "What is important is the ability of the cells involved to deliver proteins bearing exposed highly negatively charged sequences to a structural matrix in the extracellular environment, so that even relatively short but concentrated anionic sequences on an otherwise hydrophobic molecule may suffice. Such molecules are, in fact, abundant in invertebrate systems as well as vertebrates" [12]. Thus, the interaction between phosphorylated collagenous and non-collagenous proteins may be the basis for the formation of a specialized extracellular matrix supporting an early calcium phosphate mineral phase which subsequently grows and develops into crystalline apatite in all collagenous vertebrate mineralized tissues.

The above generalized overview of the possible interaction between collagenous and non-collagenous proteins raises the question of what types of non-collagenous proteins and other molecules are present in the organic matrices of the vertebrate mineralized tissues. In terms of types and numbers of matrix proteins, Schreiweis et al. used proteomics to classify the major components of the extracellular matrix of the rat femur, a model of mammalian bone [13]. A partial listing of these proteins and additional molecules in metaphyseal and diaphyseal bone is presented in Table 8.1. Of the more than 100 peptide fragments identified in the metaphysis and 25 in the diaphysis, the most common non-collagenous proteins were the precursors to the SIBLINGs, osteopontin (OPN), dentin matrix protein (DMP1), bone sialoprotein (BSP), and matrix extracellular phosphoglycoprotein (MEPE), as well as osteocalcin (OCN), matrix gla protein (MGP), and thrombospondin (TSP) [13]. In a subsequent extensive proteomic analysis of the 208 proteins identified in mouse dentin, Jágr et al. reported the particular abundance of non-collagenous proteins and molecules which included dentin sialoprotein (DSP), dentin sialophosphoprotein (DSPP), DMP1, OCN, biglycan (BGN), and decorin (DCN) [14]. Of those many proteins and molecules noted here, however, they constitute only a small portion of the total number of non-collagenous species contributing to the organic matrix composition, structure and function of vertebrate mineralized tissues.

8.2 Some General Comments about the SIBLINGs and Enamel Proteins

In humans, all the members of the SIBLING family of proteins are encoded by genes present on chromosome 4 in its 4q21–25 domain, more specifically, 4q21 (Fig. 8.1) [15–17]. Kawasaki et al. considered that many of the non-collagenous proteins evolved from a common gene ancestor, *SPARC* (secretory protein acidic and rich in cysteine)/osteonectin (Fig. 8.1) [19, 20]. Phylogenetic analysis showed that, early in vertebrate evolution when a mineralized bony skeleton developed in jawless fish during Paleozoic times, there was segmental duplication of *SPARC* on chromosome 4 and the formation of a new gene, *SPARC-like-1* (or *SPARCL1*) [19,

Table 8.1 Partial listing of unique and common acid-soluble proteins in metaphyseal and diaphyseal bone

Protein annotation	Number of distinct peptides in metaphyseal bone	Number of distinct peptides in diaphyseal bone
α2-HS-glycoprotein (fetuin) precursor	31	23
Collagen α1(I)	29	17
Osteopontin (OPN) precursor	25	13
Collagen α2(I) chain precursor	24	16
Collagen α1(II)	10	2
Dentin matrix phosphoprotein 1 (DMP1) precursor	9	7
Matrix extracellular phosphoglycoprotein, MEPE (Osteoregulin)	9	3
Procollagen type XI α2	8	5
Superoxide dismutase 1	8	2
Osteocalcin (OCN) precursor	5	6
β-2-glycoprotein I precursor	3	1
Spp (secreted phosphoprotein)-24-precursor	2	3
Bone sialoprotein (BSP)	2	2
Neutrophil antibiotic peptide NP-4 precursor	2	2
Insulin-like growth factor binding protein 5 precursor	1	7
Matrix Gla protein (MGP) precursor	1	5
Splice isoform 1 of insulin-like growth factor I precursor	1	3
S100 calcium binding protein A9 (calgranulin B)	1	2

Metaphyseal and diaphyseal femurs of 6-month-old female rats were extracted with 10% formic acid, fractionated by reversed phase chromatography and analyzed by mass spectrometry [13]. Protein identification was accomplished using database searching programs. False positive identifications were controlled by comparing searches against reversed versions of the protein databases and estimating p-values [13]. Not shown in the table are small numbers of peptides exclusively identified in the femoral diaphyseal region consisting largely of cortical bone: These include thrombospondin-1 (TSP-1), pleiotrophin, sclerostin (SOST), and α-2 antiplasmin. In addition to the proteins listed in the table, metaphyseal bone contained skeletal muscle α-actin, SPARC (osteonectin) precursor, tissue non-specific alkaline phosphatase (TNAP) isozyme precursor, stathmin 1 and 2, calmodulin 3, BSP II precursor, enzymes associated with bone turnover (tartrate resistant acid phosphatase 5 precursor, cathepsin D and K precursors, and MMP 13), plasma transport proteins (hemopexin precursor and vitamin D-binding protein precursor), and certain ribosomal proteins. Table and protein data adapted from Schreiweis et al. [13] with permission

21]. *SPARCL1* subsequently underwent further duplication to form the secretory calcium-binding phosphoprotein (*SCPP*) gene family (Fig. 8.1) [20]. Notably, *SPARC* is expressed at high levels in osteoblasts and odontoblasts of modern vertebrates, especially during the mineralization of fetal bone [22, 23]. The high negative charge of this acidic protein enhances its ability to bind calcium ions and it also promotes collagen fibril formation and secretion [24].

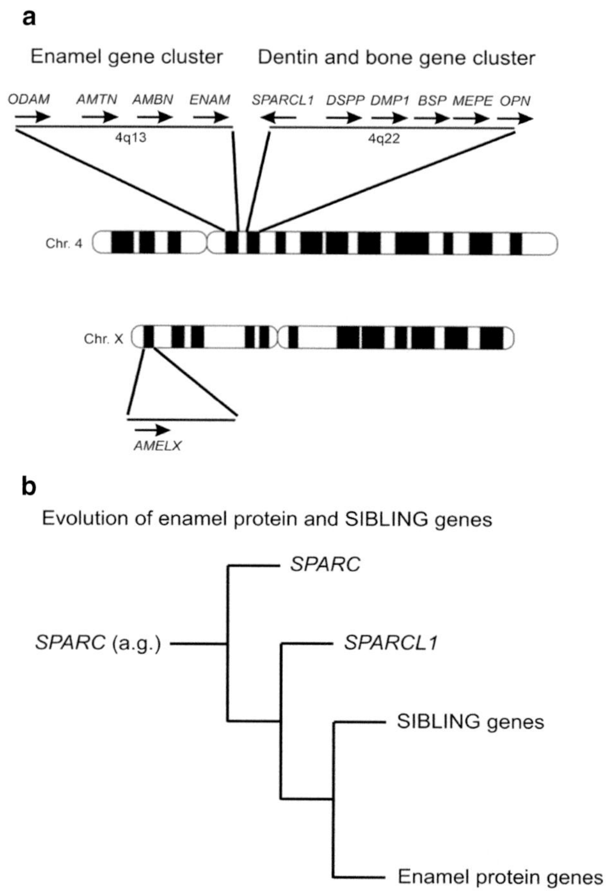

Fig. 8.1 Evolution of genes encoding proteins of bones and teeth. (**a**) Schematic showing the clustering of genes on chromosome 4 at 4q22 that encode the proteins, SPARCL1 (SPARC-like-1 protein or SC1, a relative of SPARC/osteonectin), DSPP (dentin sialophosphoprotein), DMP1 (dentin matrix acidic phosphoprotein 1), BSP (bone sialoprotein), MEPE (matrix extracellular phosphoglycoprotein), and OPN (osteopontin/SPP1, secreted phosphoprotein 1). The gene domain for these proteins, which in part include the SIBLING protein family (DSPP, DMP1, BSP, MEPE, and OPN), is also reported in human as 4q21 [15] and 4q21–25 [16]. On the same chromosome at position 4q13 are the enamel genes: *ODAM* (odontogenic, ameloblast-associated), *AMTN* (amelotin), *AMBN* (ameloblastin), and *ENAM* (enamelin). Arrows beneath each gene indicate its direction of transcription. (Schematic from Yoshizaki and Yamada [17] with permission.) Sire et al. suggested that *AMEL* (amelogenin) was formed during genomic duplication of *AMBN* or *ENAM* and translocated to chromosome X (*AMELX*) [18]. (**b**) Schematic of the evolution of *SPARC/osteonectin* to form SIBLING and enamel protein genes. *SPARCL1* has been proposed to evolve from an ancestral *SPARC* gene (*SPARC a.g.*). Duplication of *SPARCL1* on chromosome 4 then resulted in two clusters of genes, the SIBLINGs and the enamel protein genes (*AMBN*, *ENAM*, and *AMEL*) [18, 19]. (Schematic adapted from Sire et al. [18] with permission)

Human chromosome 4 shows a cluster of two gene families, the putatively more ancient cluster of *SIBLING* genes and a more recent cluster which encodes the enamel matrix proteins (ameloblastin, amelotin, and odontogenic- and ameloblast-associated proteins) (Fig. 8.1) [18, 19]. Genes comprising chromosome 4 also encode milk casein (casein, s1-casein) and salivary proteins (statherin, histatin 1, histatin 3, hypothetical protein LOC401137, follicular dendritic cell secreted peptide, proline-rich 5, 3 and 1, and mucin 7) [18, 19]. The gene encoding the major protein constituent of enamel, amelogenin, is located on the human X- and Y-chromosomes. What is surprising is that, in one way or another, almost all these proteins are concerned with calcium stabilization and/or mineralization. In the following pages, the nature of several of these proteins and other important non-collagenous molecules of the extracellular matrix of vertebrate mineralized tissues will be reviewed.

To continue the discussion of non-collagenous proteins in vertebrate mineralized tissues, it is notable that they comprise a highly acidic group of molecules as a result of their abundance of aspartic and glutamic acid residues [25]. Also, since many of their serine and threonine residues are phosphorylated, these negatively charged groups contribute to the overall anionic nature of the proteins and they facilitate binding to calcium and other cations as well as to apatite crystals themselves. In some non-collagenous proteins, the presence of negatively charged γ-carboxyglutamic acid residues, along with charged carboxyl groups on carbohydrate chains, permits interaction with calcium ions and promotes attachment to calcium-rich surfaces of apatite crystals. These structural and chemical characteristics endow the non-collagenous constituents with regulatory roles in vertebrate mineral formation, on the one hand supporting nucleation of mineral crystals through calcium ion binding and on the other hand interacting with calcium at crystal surfaces to modulate mineral growth and development. Based on such considerations, the concept has emerged that these anionic non-collagenous molecules are principal regulators of mineralization, and they do so by initially promoting calcium binding and then limiting mineral growth and phase transition [26].

In terms of their function, the SIBLING proteins may be divided into a number of different groups. One group, OPN and BSP, is abundantly secreted by osteoblasts, chondrocytes, and odontoblasts. Most likely, the major role of both encoded proteins is to participate in the mineralization of bone, cartilage, dentin, and other vertebrate mineralized tissues. OPN has a wider sphere of activity than BSP, including involvement in lactation and the proliferation of cancer cells [27]. Two additional SIBLINGs, MEPE and DMP, may be separately grouped among this family of proteins. They are most concerned with mineralization and phosphate ion homeostasis. Interestingly, these and other SIBLING proteins contain an Acidic Serine-Aspartate-Rich MEPE (ASARM) peptide motif [28]. This motif interacts with the proteolytic enzyme PHEX (phosphate-regulating gene with homologies to endopeptidases on the X-chromosome) and protects SIBLINGs from enzymatic degradation [28]. If *PHEX* is defective or low in activity, then MEPE is cleaved by enzymes (cathepsins), releasing the ASARM peptide into the circulation (Fig. 8.2). This negatively charged peptide binds apatite and inhibits mineral formation and also induces fibroblast growth factor *(FGF)23* expression [28]. An ASARM-dependent rise in FGF23 decreases the tubular reabsorption of phosphate by the kidney and, accompanied by changes in levels of vitamin D, lowers the systemic concentration of phosphate ions to cause a form of rickets (Fig. 8.2) [29]. Children suffering from this form of hereditary rickets (X-chromosome-linked hypophosphatemic rickets) are often short in stature as a consequence of abnormalities in chondrocyte terminal differentiation, and their bones are fragile and of low mineral density [30].

Another unique characteristic of the SIBLINGs is their three-dimensional architecture. Since the SIBLINGs do not exhibit a globular, fibrous, or transmembrane

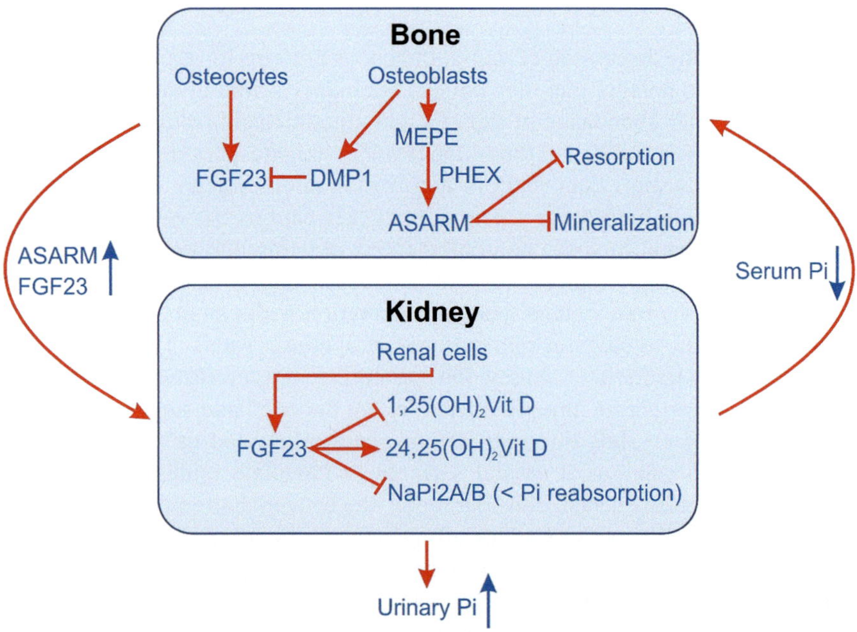

Fig. 8.2 Role of matrix extracellular phosphoglycoprotein (MEPE) and FGF23 in bone mineral homeostasis. MEPE and other SIBLINGs such as OPN and DMP1 contain the peptide sequence, ASARM. When released from these non-collagenous proteins by the enzyme, PHEX, in bone, the ASARM peptide interacts with apatite and prevents crystal growth and mineralization; and it also down-regulates the activity of osteoclasts to slow bone resorption and remodeling. If the systemic levels of the ASARM peptide rise, there is a profound change in kidney function as the ASARM peptide promotes the expression of *FGF23* by renal cells. FGF23 is additionally released into the circulation by osteocytes embedded in bone. However, this source of FGF23 is inhibited by DMP1 so as to provide a regulatory brake on systemic levels of the protein. The effects of kidney- and bone-derived FGF23 on kidney function are two-fold. First, the protein inhibits the generation of the active metabolite of vitamin D (1,25 dihydroxycholecalciferol) while promoting the formation of the inactive metabolite (24,25 dihydroxycholecalciferol). Second, FGF23 increases the activity of NaPiT2A/B, the kidney phosphate ion transporter, thereby enhancing the secretion of phosphate ions into the urine. Simultaneously, phosphate is no longer reabsorbed into the serum and serum phosphate levels fall. The resulting decrease in the serum phosphate ion concentration feeds back onto bone cells to prevent new bone formation and mineralization and to decrease remodeling activity

conformation, they do not fit nicely into any of these well-recognized categories of protein structure. The absence of a defined structure is attributable to electrostatic interactions between regions or domains of similarly charged amino acid residues within the same SIBLING protein [31]. Thus, SIBLING structure is classified as intrinsically disordered, a feature shared by a considerable number of other proteins (Fig. 8.3).

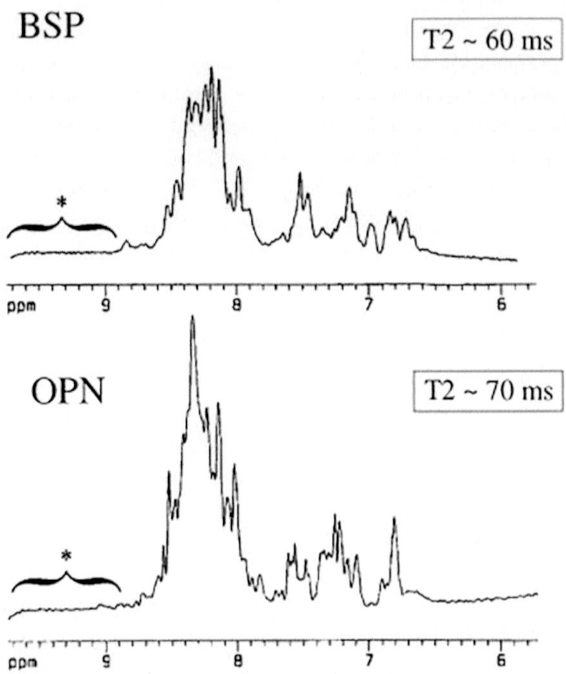

Fig. 8.3 Evidence that bone sialoprotein (BSP) and osteopontin (OPN) exhibit intrinsically disordered structure. Purified recombinant human BSP (top panel) and OPN (bottom panel) were examined by one-dimensional proton nuclear magnetic resonance spectroscopy (NMR), which provides information of the atomic structure of a molecule caused by the flow of electrons around a magnetic nucleus [2]. Spectra shown were collected by applying a magnetic pulse to the protein in solution and causing nuclei in different environments to resonate at slightly different field strengths or frequencies. As a result of the pulse, the dynamics of specific chemical environments around vibrating molecular groups can be visualized while the time taken for excited protons to realign in the fluctuating magnetic field, the relaxation time or T2, can be measured. The horizontal axis of the spectra shows the chemical shift in parts per million (ppm), while the spectral spread is indicative of the proton transverse relaxation time (T2) of the peptide chain. In comparison with structured proteins, which would exhibit numerous significant peaks in the ppm range greater than ~8.8 (brackets with *), both BSP and OPN yielded little structure over the experimental NMR time frame. Each peptide also showed only a narrow spread of its backbone amide peaks in the spectral range of ~7.8–8.6 ppm and of its side chain amide peaks in the range of ~6.6–7.4 ppm, results suggestive of limited structural order for the two proteins. Further, the proton T2 relaxation times were determined as ~60 ms for BSP and 70 ms for OPN. These values are consistent with an extended and flexible structure in solution rather than a compact structure (T2 typically ~5–20 ms) for the two proteins. The summary data [2] demonstrate that each protein in solution is flexible and unstructured along its full length, having no significant peptide regions which maintain a consistent structural form beyond a few milliseconds. (Image from Fisher et al. [2] with permission)

While the lack of order may signify disorganization of a protein, in fact many biologically important proteins fail to assume a defined three-dimensional structure. Such a conformation, despite its disorder, reflects the primary sequence of the protein, one in which there are regions of high positive or negative charge and marked hydrophobicity leading to an intrinsically unstructured state [32]. However, adoption of a flexible and partially or completely unfolded tertiary structure can be an advantage as it favors the formation of low affinity complexes with multiple molecular species and, in the case of proteins of mineralized tissues, it allows associations with ions and crystals [33]. Disordered structure also adds to other protein functions which include serving as a transcription factor, a translation factor or a regulator of a transcription factor, for example [34]. Further, the unstructured state need not be permanent as a protein may become ordered and change its functionality when it is associated with a binding protein, ion or mineral [35]. In this regard, Uversky notes, "With their exceptional spatio-temporal heterogeneity and high conformational flexibility, intrinsically disordered protein regions/intrinsically disordered proteins represent complex systems that act at the edge of chaos and are specifically tunable by various means" [36]. Thus, the absence of a higher ordered structure of a SIBLING protein or other molecule is not a requisite for effective function.

An excellent example of intrinsic structural disorder and multiple functionality for a protein is provided by the SIBLING, OPN [2]. When it is fully phosphorylated, OPN is a random coil; when bound to mineral, OPN becomes somewhat organized with the appearance of β-sheet regions [37]. Boskey et al. [38] explored this change in structure by fragmenting milk OPN and showed that the intact protein and its terminal fragments promoted apatite formation while its central fragment inhibited mineralization. Likewise, the intrinsically disordered DMP1 promotes mineral formation when bound to collagen but inhibits crystal formation when free in solution [39]. Kalmar et al. [40] commented that "almost all proteins which are closely related to hydroxyapatite formation, regulation and orientation during growth are mostly or fully disordered." Finally, as will be discussed later, posttranslational modification of disordered proteins, particularly those undergoing phosphorylation reactions, plays major roles in vertebrate mineralization (and elsewhere) in both generating the disordered structure of a protein and influencing the interaction of the protein with other proteins and molecules, calcium ions, and mineral crystals. In the following pages, additional characteristics of some of the more important non-collagenous molecules involved in mineralization of extracellular matrices of vertebrate tissues will be detailed. Since the number of these molecules is very large, discussion will be confined to the major SIBLINGs, SLRPs, and matricellular proteins represented by members of the TSP family and the liver protein, albumin. Four additional important non-collagenous proteins which also affect vertebrate mineralization, OCN, MGP, sclerostin (SOST) and fetuin, are addressed later in this chapter and in Chap. 9 of this book.

8.3 Bone Sialoprotein (BSP)

While *SPARC* may represent the common ancestral gene of the SIBLINGs, BSP is the most abundant of the non-collagenous constituents of bone (about 8–12% of total SIBLING proteins) and is also present in cementum, dentin, and calcifying cartilage and tendon [41–45]. The molecular weight of BSP is approximately 34,000 Daltons (about half the weight of albumin and twice that of the protein subunits of hemoglobin).

Like other SIBLING proteins, BSP is intrinsically disordered and contains glycosylated, phosphorylated, and sulfated groups [46]. The disordered structure of the protein is such that amino acid residues located at the N-terminus of the protein are predominantly basic, domains at the C-terminus are principally acidic, and the middle portion of the molecule is unusual in its many repeating sequences of polyglutamic acid along with multiple serine residues [47, 48]. BSP has been described as "unstructured" and having a "flexible conformation in solution, with no α-helical and β-sheet structure present." (Fig. 8.4) [2]. In another description of the protein, Wuttke *et al.* indicated that the molecule existed as a "globule linked to a thread-like structure, with 5% α-helix, 32% β-sheet, 17% β-turn, and 46% random coil" [50].

BSP influences the mineralization of bone, cartilage, dentin, and enamel, and it promotes acellular cementum formation and tooth attachment to bone [51]. The effects of BSP on bone are complex. When *BSP* gene expression is ablated, long bone growth is reported to be defective [52]. However, there is some confusion concerning details of the effects of deletion of the *BSP* gene on bone. In their studies of the absence of *BSP* in mouse bones, Bouleftour et al. [52] and Malaval et al. [53] documented an increased trabecular bone mass and lowered rates of bone turnover. In a parallel study, Holm et al. found that there was delayed bone development and mineralization [54]. To explain their findings, these investigators concluded that BSP promotes cell-cycle progression of chondrocytes, the timing of mineral onset, and mineral density [54]. It is also possible that these differences in the effects of BMP could involve the different genetic backgrounds of the animals used experimentally and the potential overlapping functions (called "redundancy" in molecular biology) of other SIBLING proteins.

The primary (amino acid) structure of BSP provides further insights into its function. Inspection of the protein shows that it contains a triplet of amino acids (arginine-glycine aspartic acid, RGD) which binds membrane receptors (integrin receptors of the type $\alpha v\beta 3$ or $\alpha v\beta 5$) on osteoprogenitor cells [43, 48, 49] and regulates their differentiation into osteoblasts [55, 56]. In terms of mechanism of action, the BSP molecule exhibits an overall negative charge attributable to its extended stretches of glutamic acid and phosphorylated serine and threonine residues as well as the presence of sulfated tyrosine. The nucleation activity is believed to reside primarily in two polyglutamate domains found in the amino-terminal half of the molecule [57]. A recent study also implicates sialic acid groups of BSP in mineral formation. Sialic acid is a nine-carbon, negatively charged carbohydrate linked to the glycan components of the protein molecule, its presence being the principal reason for naming BSP a sialoprotein. Xu et al. have shown that the removal of the sialic acid groups from BSP prevents or slows bone mineralization [58]. A short distance from the sialic acid, a stretch of hydrophobic amino acids (residues 19–46) binds collagen fibrils and fiber surfaces; the protein is not found within the collagen molecule [45, 49]. Tethered to collagen fibrils and fibers in this manner, BSP avidly binds calcium ions and nucleates hydroxyapatite [49, 59]. The RGD sequence of amino acids permits cell binding, so the molecule can thus serve as a bridge between the cell, collagen, and the mineral (Fig. 8.4) [49].

Because of its molecular size, it is likely that BSP is excluded from entering arrays of collagen molecules [60]. This possibility provides an interesting insight into the mechanism(s) of intrafibrillar collagen mineralization and any probable role for BSP in such an event. Indeed, the collagen-binding domain of BSP localizes to sites along the surfaces of COLI fibrils or outside the fibrils [61]. In this association, the protein is free to bind calcium ions or clusters [62] and thereby promote interfibrillar collagen mineral formation [10]. There is evidence that other non-collagenous proteins like OCN (molecular weight ~ 5800 Daltons) permeate collagen fibrils and

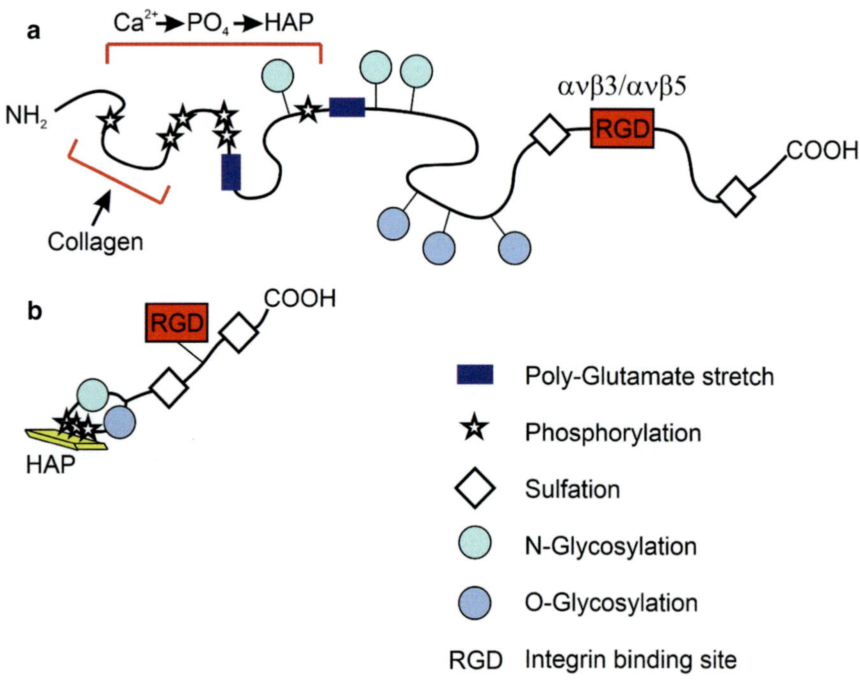

Fig. 8.4 Schematic of the structure of bone sialoprotein (BSP) and its possible interaction with mineral. (**a**) Major features of the protein in an unfolded disordered state. At the NH_2 end of the protein, there is a motif for binding collagen located close to a highly phosphorylated sequence of amino acids which binds calcium (Ca^{2+}) ions. When complexed with phosphate ions (PO_4), the calcium-phosphate ion pairs can nucleate hydroxyapatite (HAP). Toward the COOH end of the protein, an RGD-enriched motif (arginine-glycine-aspartic acid) enhances binding to integrin receptors. (**b**) Conceptual drawing of the interaction of BSP with hydroxyapatite, shown as a thin platelet. (Image adapted from Bouleftour et al. [49] with permission)

their aggregates [60], and OCN molecules have been directly localized inside fibrils as well as outside as noted above [61]. The presence of OCN within collagen fibrils and their higher ordered aggregates may mediate aspects of both intrafibrillar and interfibrillar collagen mineralization as will be described below in this chapter.

8.4 Osteopontin (OPN)

OPN is an intrinsically disordered, highly charged, small (molecular weight 34,000 Daltons) phosphorylated glycoprotein, comprising about 10% of the non-collagenous proteins of the extracellular matrix of vertebrate mineralized tissues other than enamel. Its abundant number of acidic and phosphorylated amino acid residues (25%) endows the protein with a strong negative charge. Its function has intrigued basic scientists and clinicians alike as OPN is prominently localized to the extracellular matrices of both mineralizing and non-mineralizing tissues [63, 64] and associated with a variety of pathologies, including osteoporosis, kidney stone formation, cancer, inflammation, and atherosclerosis, as well as bone fracture repair. With age, the phosphorylated state of OPN changes, decreasing 30% by the ninth decade of life [65]. The decline in OPN phosphorylation with age has been suggested as the cause for the increase in the propensity of bone to fracture, especially in relationship to osteoporosis [65].

OPN is a secreted protein, present in urine, milk, bile, and other fluids of the body where it binds to calcium, inhibiting mineral precipitation [66, 67]. It is also found in the extracellular matrices of bone (deposited principally by osteoblasts), the growth plate (hypertrophic chondrocytes) and articular cartilage (chondrocytes), cementum (cementocytes), dentin (odontoblasts), and the normally mineralizing tendons (tenocytes) of some avian species [44–46, 68, 69]. In bone and mineralizing avian leg tendon, OPN, like BSP, is localized on the surfaces of COLI fibrils and fibers in extracellular matrix sites [61, 70, 71]. OPN is not present within the intrafibrillar spaces of collagen [60, 61].

A structural map of the intrinsically unstable OPN molecule shows that the protein is characterized by the presence of multiple amino acid residues which can undergo phosphorylation, thereby enhancing the anionic character of the protein (Fig. 8.5) [72]. The acquisition of charged phosphate residues increases the electronegativity as well as the hydrophilicity of the protein, allowing it to bind calcium ions and attach to apatite crystals and mineral [72]. While 29 possible phosphorylation sites are present on rat bone OPN, only 10–11 residues are phosphorylated and thus many of its potential phosphorylation sites remain naive [77]. Residues which can be phosphorylated are in groups separated by regions of non-phosphorylated amino acids. Phosphorylation is mediated by the kinase enzyme Fam20C and possibly casein kinase II [78]. These kinases are located in the Golgi apparatus of osteoblasts, odontoblasts, and other cells, and they target the OPN amino acid motif, serine-X-glutamic, acid/phosphoserine [79]. If *Fam20C* is mutated, the loss of kinase activity and the associated loss of OPN phosphorylation leads to a bone dysplasia (Raines syndrome) with dental anomalies, ectopic calcification, and hypophosphatemia [80].

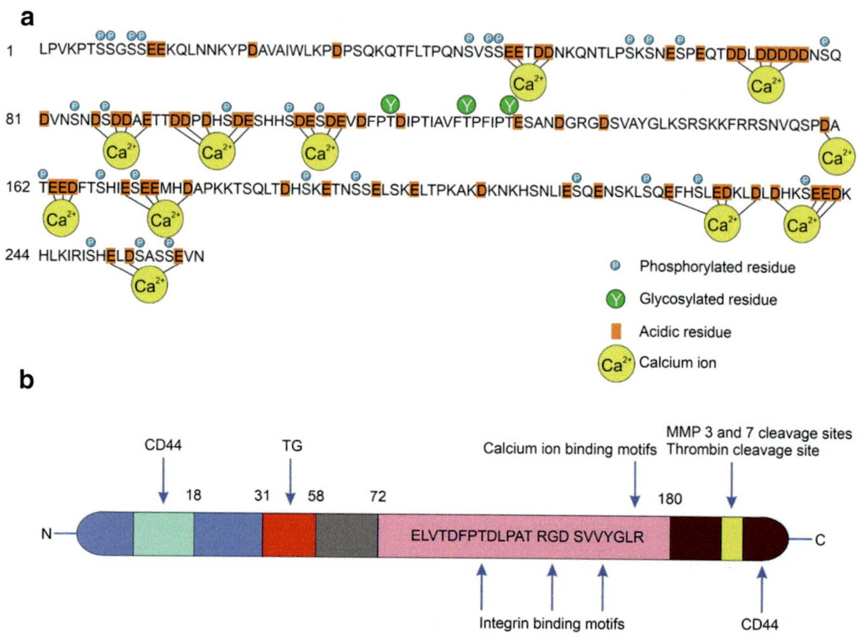

Fig. 8.5 Schematic of the primary amino acid sequence and domain structure of osteopontin. (**a**) Bovine milk osteopontin has 28 possible phosphorylation sites (blue circles), numerous acidic residues (orange), proposed calcium-binding motifs (black lines and red circles), and glycosylation sites (green circles). The several charged phosphate groups cause condensation of the molecule because of minimized internal electrostatic repulsion. In this circumstance, osteopontin "would be shielded from premature and/or undesired adsorption to ECM components as well as from binding to other proteins because of unspecific ionic interactions" [65]. (Schematic adapted from Kläning et al. [72] with permission.) (**b**) Schematic of the domain structure of human osteopontin (canonical isoform 1A) [73]. The protein contains an arginine-glycine-aspartate (RGD) and serine-valine-valine-tyrosine-glutamate-leucine-arginine (SVVYGLR) integrin-binding sequence as well as a sequence which binds transglutaminase (TG) [74]. Once activated, TG promotes the formation of isopeptide bonds between OPN molecules and enhances type I collagen binding to the protein. A sequence ELVTDFPTDLPAT also binds integrin ($\alpha 4\beta 1$) [75]. Heparan sulfate receptor sequences (CD44) are located toward both the amino (N)- and carboxyl (C)-ends of the protein. These sequences additionally bind hyaluronan, collagen, MMPs, and stem cell receptors [76]. A calcium ion binding domain appears near the C-terminus of the protein. Thrombin and MMP cleavage sites on the protein are found toward the C-terminus as well. Amino acid sequence numbers are given in (**a**) and (**b**)

Kidney stone formation represents another common mineralization defect which is linked to the presence of OPN. Together with Tamm-Horsfall glycoprotein, OPN is a major component of the mineral deposits which can accumulate in the kidney and cause extreme pain and loss of organ function (Fig. 8.6) [81]. The mineral deposits are commonly comprised of calcium phosphate and calcium oxalate crystals [81]. Predictably, OPN serves as an inhibitor of oxalate nucleation, affecting the aggregation behavior, lattice characteristics, and growth of the forming crystals of calcium oxalate (Fig. 8.7) [82, 83]. Most likely, the long stretches of

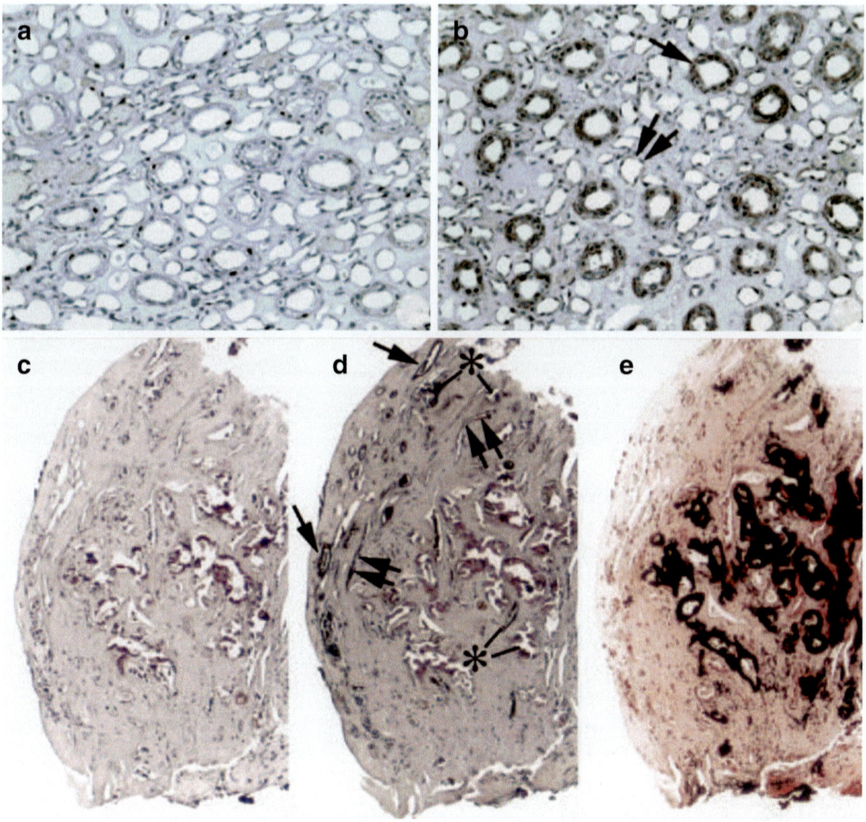

Fig. 8.6 OPN antibody distribution in human kidney tissue. (**a**, **b**) Light microscopy of biopsy sections of normal kidney papillary tissue from human patients who do not form kidney stones. Tissue in (**a**) has been processed in the absence of OPN antibodies and illustrates general morphology. Tissue in (**b**) has been incubated with OPN antibody, which is localized to the inner medullary collecting ducts (arrow) and thin loops of Henle (double arrow). (**c–e**) A set of serial sections of kidney papillary tissue biopsied from calcium oxalate stone-forming individuals. General morphology of tissue in the absence of OPN antibody is presented in (**c**), while tissue incubated in OPN antibody in (**b**) illustrates the presence of OPN over inner medullary collecting ducts (arrows) and thin loops of Henle (double arrows) as well as sites of crystal deposition (*). (**e**) The localization of calcium oxalate as detected by the Yasue method, which stains oxalate a brown-black color. Sites in (**d**) which are OPN-positive may be correlated with the oxalate stain in (**e**). Results of this study indicate that the "Intensity of cellular staining for osteopontin did not differ in stone formers vs. controls, or in areas with plaque vs. areas without plaque" [81]. Thus, while OPN was associated with kidney function and oxalate stone formation, other factors play a more prominent role in regulating mineral deposition. Original magnifications = ×600 (**a** and **b**); ×60 (**c–e**). (Images adapted from Evan et al. [81] with permission)

charged sites on OPN interact with calcium atoms in the forming mineral and prevent crystal growth or modulate mineral crystallinity [84]. However, the pathogenesis of stone formation does not rest on the notion that OPN inhibits mineral nucleation and crystal growth. Other factors directly related to aberrant mineral

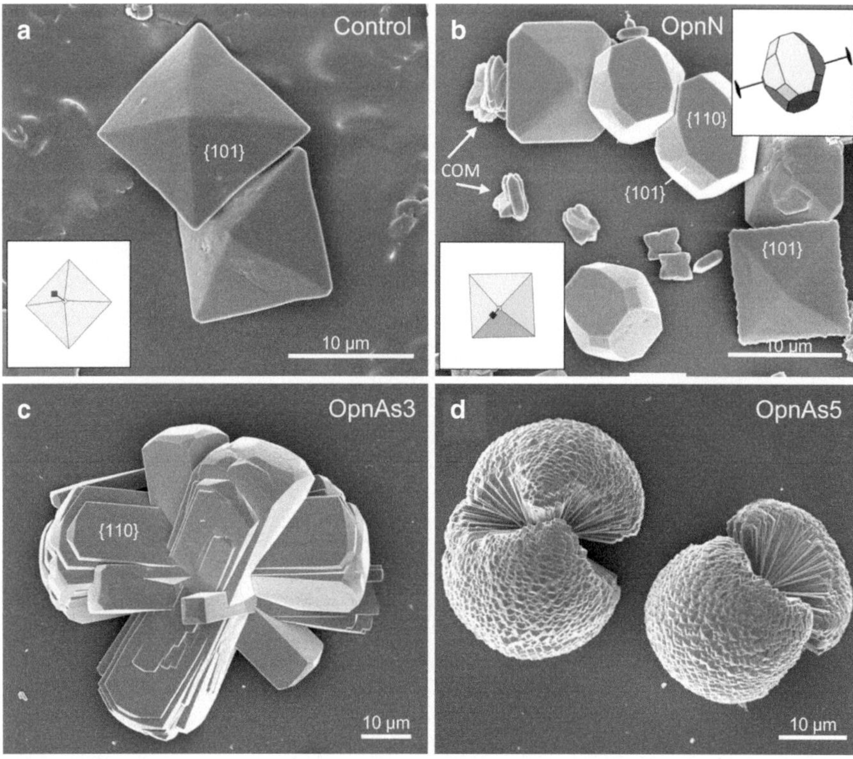

Fig. 8.7 Effect of OPN on the crystal morphology of calcium oxalate (dihydrate) crystals. (**a**, **b**) and (**c**, **d**) illustrate scanning electron microscopic (SEM) images of crystals formed in a calcium chloride-sodium oxalate solution in the presence of human OPN-derived peptides. The peptides inhibited oxalate crystal growth and influenced the morphology of those crystals developed in the medium. The degree of inhibition was dependent on the extent of phosphorylation of the peptide. (**a**) The di-pyramidal morphology and {101} face of calcium oxalate crystals formed in the absence of any OPN peptide. The inset presents the computed crystallographic model of calcium oxalate dihydrate. (**b**) Crystals grown for 14 days in the presence of an OPN ASARM peptide (12.5 μM) with neutral charge (OpnN). Here, crystals display the di-pyramidal shape characteristic of normal calcium oxalate dihydrate or develop a new face {110} suggestive of calcium oxalate monohydrate (COM). The insets show the predicted topography of the two different crystals. (**c** and **d**) The crystal morphology formed in the presence of phosphorylated peptides of OPN, OpnAs3 (an ASARM peptide with three phosphoserine residues) and OpnAs5 (an ASARM peptide with five phosphoserine residues), respectively. Large crystal aggregates appear as radiating spherulites, composed of calcium oxalate crystals with {110} faces following incubation with OpnAs3 in solution (**c**). In contrast, incubation with OpnAs5 generates "mushroom- and dumbbell-shaped" structures comprised of thin, elongated individual oxalate crystals of similar size. (Summary results and images adapted from Chien et al. [82] with permission)

deposition in the kidney include trauma, infection, the accumulation of products of oxalate and citrate metabolism, and the influence of MGP and parathyroid hormone [85]. Thus, while OPN is of some consequence in kidney stone formation, deposition of calcium phosphate or oxalate is governed by many other local and systemic factors.

The fact that OPN is involved in multiple biological systems and impacts numerous biological events raises the question as to its importance in bone mineralization. To address this point, Rittling et al. deleted the *OPN* gene in mice [86]. Surprisingly, no overt phenotypic abnormalities were observed in either the bones or the teeth of the experimental animals [86]. However, a subsequent and closer examination of the tissues in other mice revealed increases in crystal size, crystal perfection, and mineral content [87]. Further studies in mice reported loss of proteoglycans and structural alterations in articular cartilage, as well as an increase in bone mineral content [88]. Together, these and several additional investigations emphasize the concept that a principal role of OPN is to inhibit mineral formation and apatite crystal growth and development [68, 70, 87, 89]. As was discussed earlier, the inhibitory character of OPN is dependent on and modulated through phosphorylation of its ASARM motif, which binds directly to mineral and can affect crystal structure (Fig. 8.5) [38, 82, 90]. Cleavage by PHEX of phosphorylated ASARM restores normal mineralization [90].

In addition to its affinity for calcium ions and apatite, OPN contains RDG and other sequences which bind receptor proteins ($\alpha v\beta 3$ integrins, for example) of the plasma membrane of osteoclasts and other cell types (Fig. 8.5) [70, 73, 75, 91–94]. Through these interactions, OPN activates signaling pathways which regulate osteoclast and osteoblast proliferation, differentiation and function [95, 96]. OPN also contains binding motifs for transglutaminase and CD44 (Fig. 8.5) [73, 74, 76, 94]. Once bound to OPN, the enzyme, transglutaminase-2, catalyzes the formation of intermolecular and intramolecular (isopeptide) crosslinks between glutamyl and lysyl residues of OPN molecules [97]. The resulting high molecular weight OPN aggregates in bone cell culture inhibit apatite crystal growth and mineralization and demonstrate increased collagen binding [98] and they promote cell adhesion, spreading, and migration mediated through integrins [99]. In these several ways, OPN integrates interaction with mineralized tissue cells, the principal organic component, collagen, and the calcium ions and mineral comprising the extracellular matrices of vertebrates [97, 100].

In summary, OPN is an intrinsically disordered, non-collagenous SIBLING family member comprising extracellular matrices and synthesized by cells of vertebrate mineralized tissues. The protein contains numerous acidic and phosphorylated amino acid residues which endow it with a strong negative charge. The anionic character of OPN modulates calcium binding and leads to inhibition of mineralization as a major OPN function. Further, certain collagen- and cell-binding sequence motifs of OPN influence collagen interaction with the extracellular matrix and mediate activity of mineralized tissue cells. In addition to the involvement by OPN in these roles, the actions of the protein have been linked to numerous disease states which include kidney stone formation, cardiovascular disease, diabetes, and cancer.

8.5 Matrix Extracellular Phosphoglycoprotein (MEPE)

The *MEPE* gene is a member of the SIBLING family and encodes a matrix extracellular phosphoglycoprotein found in osteoblasts, osteocytes, and odontoblasts [101–103]. There is also a low expression level of *MEPE* in non-mineralizing tissues such

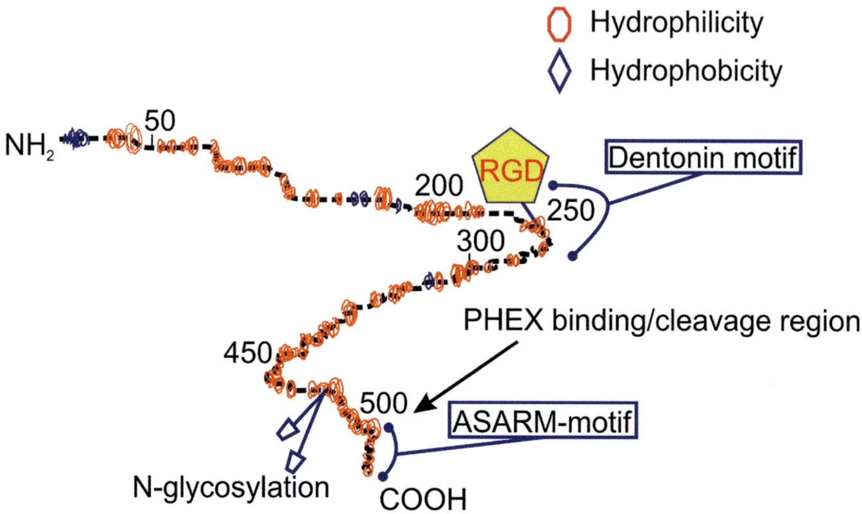

Fig. 8.8 Schematic of the secondary structure of matrix extracellular phosphoglycoprotein, MEPE. By means of computation techniques, the secondary structure of the protein was determined from its primary amino acid sequence [28]. This methodology predicted turns in the primary structure and indicated regions of hydrophilicity (ellipsoids, red) and hydrophobicity (diamonds, blue). The structure contains a signal peptide in the vicinity of its amino terminus as well as a region of cell binding through an RGD sequence. The molecule is also comprised of N-glycosylation sites, a PHEX binding site, and an ASARM motif. MEPE additionally maintains a region defined as a dentonin motif, around which peptides have been synthesized to promote dentin formation. Amino acid residue numbers are given along the protein backbone. (Schematic adapted from Rowe [28] with permission)

as the lung, brain, and kidney [104]. MEPE is a secreted phosphoprotein of molecular weight 45,000 Daltons and, like other SIBLING family members, it is enriched in phosphorylated serine residues (Fig. 8.8). The protein contains an RGD motif for integrin cell receptor binding [104, 105] and residues for attachment of carbohydrates (N-glycosylation sites). The carboxy-terminal end of the protein consists in part of a phosphorylated ASARM domain. When released from MEPE following its cleavage by cathepsin B, the free ASARM peptide mediates apatite binding and inhibits mineralization [104]. If *MEPE* expression is ablated in animal models, the resulting phenotype exhibits an increase in mineral apposition rates and greater bone mass, trabecular number, and thickness [106]. Overexpression of *MEPE* leads to decreased bone mass and bone remodeling as well as increased growth plate width [31, 107]. The relationship between ASARM, FGF23, and renal defects (X-linked hypophosphatemic rickets) has been mentioned earlier in this chapter.

8.6 Dentin Matrix Protein (DMP1)

The name of this SIBLING protein suggests that it is specific for dentin. It is not, and in fact it is quite promiscuous. *DMP1* is expressed by osteocytes and osteoblasts (bone), hypertrophic chondrocytes (cartilage) and, of course, odontoblasts (dentin) as well as some non-mineralizing tissues [108–116]. Expression of *DMP1* in dentin varies as a function of the development and maturation state of this tissue. In rodent teeth, the gene is expressed in early differentiating odontoblasts. At the onset of dentin matrix mineralization, *DMP1* expression is down-regulated and it is absent when dentin is fully mineralized [117]. Its absence is a cause of periodontal tissue breakdown and may lead to dental deformities. The expression of *DMP1* in bone parallels that of dentin.

Structurally, DMP1 consists of 473 amino acid residues which are unusually hydrophilic [12]. The molecule is acidic in character because of its numerous glutamic and aspartic acid residues. These two amino acids frequently neighbor serine residues, more than half of which are phosphorylated [12]. The molecule also contains several sites for the attachment of carbohydrate groups (glycans) mediated by an oxygen linkage (O-glycosylation) and one site for attachment through a nitrogen (N-glycosylation) bond. Aside from these domains, RGD sequences in DMP1 bind to integrin receptors ($\alpha v\beta 3$) in cell membranes, an event which up-regulates phosphorylation of intracellular kinase enzyme systems (p38 MAPK) leading to gene transcription [118]. Taken up by osteoblasts through endocytosis, DMP1 increases intracellular free calcium ion concentration; this loading process in turn induces transcription of *RUNX2* and other transcription factors to promote further differentiation of osteoblasts [119].

Like MEPE and other SIBLINGs, DMP1 has an RGD motif for integrin binding and its C-terminus region contains the ASARM domain [120] which binds both apatite and collagen [12, 121–123]. The ASARM peptide carries a high negative charge because of its numerous aspartic acid as well as glutamic acid residues [124]. Released from DMP1 by cysteine proteases, the free ASARM peptide is a potent mineralization inhibitor and is important in suppressing renal calcification as noted previously.

The high affinity for calcium binding mediated by multiple phosphorylation reactions shows that DMP1 assumes an extended configuration in the absence of calcium ions. In the presence of calcium, the protein changes first to an extended conformation with a globular domain at one end of the molecule and then to a dimer with globular domains at each end of the structure [124]. In this latter form, two polyglutamic acid regions of DMP1 appear to fold together, creating a specific locale for calcium binding and possible apatite nucleation [39, 124].

Full-length DMP1 is reported to nucleate apatite in the presence of COLI in vitro and the N-terminus of the protein inhibited formation of apatite and stabilized amorphous calcium phosphate nuclei developed in the experimental system [125]. Binding to amorphous calcium phosphate is likely influenced by the high negative charge density at the N-terminal region of DMP1 [125]. Related to these observations was earlier work examining certain conceptual interactions between DMP1 and calcium phosphate clusters through use of small-angle X-ray scattering and atomic force microscopy [124]. This study demonstrated that DMP1 inhibited calcium phosphate precipitation in vitro by stabilizing mineral nanoparticles in solution (Fig. 8.9) [124]. Further, mineral clusters stabilized through binding to self-assembled DMP1 molecules could be targeted to collagen by means of the specific collagen-binding domains of DMP1 [124]. In this manner, in tissues such as dentin and bone, DMP1-associated clusters may lead to collagen mineralization. In support of this model, *DMP1* deletion in vivo was found to increase apatite crystal size and decrease mineral content and overall bone mineralization status [126].

While an inhibitory relationship between DMP1 and mineralization has been well documented, there remains some uncertainty to the mechanism of this action. Investigations of DMP1 gene knockouts in animals models [127, 128], human gene studies [115], and in vitro reports [129] indicate that DMP1 participates in both mineral formation and inhibition of bone and dentin deposition. This duality of possible DMP1 activity is not fully understood, but an explanation may be that only small amounts of the full-length protein are present in the two tissues. Instead, there is a predominance of DMP1 fragments which appear as a result of proteolysis, and the full-length and fragmented molecules then perform different functions [130, 131].

The functionality of biologically active DMP1 fragments just noted is interesting in several respects. A fragment, for example, derived from the NH_2-terminus of DMP1 contains a chain of sulfated sugar (chondroitin sulfate) groups similar to that found in proteoglycans (PG) comprising the extracellular matrices of many vertebrate mineralized tissues [130]. The distribution of this fragment (DMP1-PG) in the rat femoral head suggests that this molecular species engages in chondrogenesis [69, 131]. Like the full-length protein, DMP1-PG is a mineralization inhibitor and is localized in predentin, the non-mineralized region of forming dentin [129, 132]. This observation provides insight concerning the negative effect of DMP1 on mineralization. A positive effect of the protein is linked to the biological activities of two other fragments of the protein, a small 37,000 Dalton molecular weight fragment from the NH_2 terminus of the molecule and a 57,000 Dalton fragment from the COOH end. The NH_2-terminal product is found in predentin and in the non-mineralized regions of the growth plate, and the COOH fragment is present in mineralized dentin and in the calcifying regions of the epiphysis [132, 133]. Both of these peptides exhibit negatively charged domains which support mineral formation in assays in vitro [129]. Accordingly, as one of a number of non-collagenous proteins, DMP1 serves as a powerful regulator of mineral formation.

Probably the most critical role of DMP1 is its connection to phosphate ion homeostasis. The *DMP1* null mouse is hypophosphatemic. The mutation is the cause of autosomal-recessive hypophosphatemia rickets [115, 128, 134]. Inactivation of the gene increases expression and secretion of FGF23 by osteocytes (Fig. 8.2) [115, 116]. The marked increase in blood levels of FGF23, a hormone which regulates serum phosphate ion levels together with parathyroid hormone and vitamin D, prevents phosphate reabsorption by the kidney (Fig. 8.2). The resulting fall in the serum phosphate ion concentration can be so extreme as to prevent mineral deposition in the extracellular matrix of bone and interfere with mineral deposition in dentin, cementum and even enamel [127].

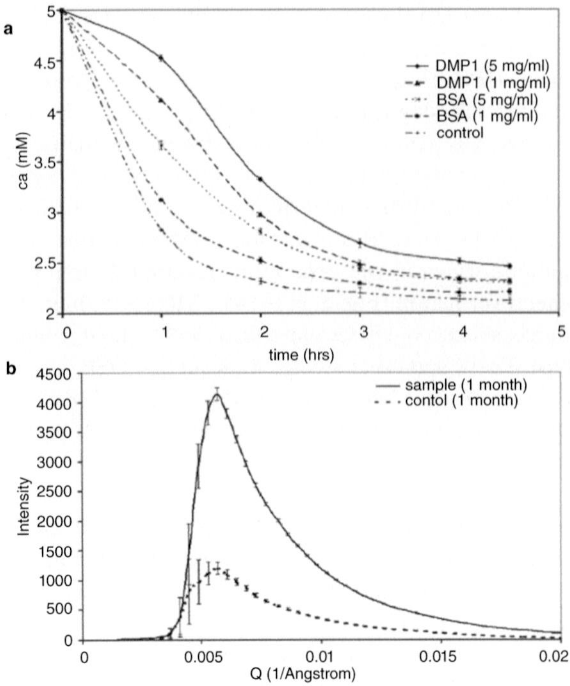

Fig. 8.9 DMP1 inhibition of calcium phosphate precipitation in vitro. (**a**) The result of a mineral precipitation assay conducted with various concentrations of DMP1 added to a supersaturated calcium phosphate solution. In the absence of DMP1, there was precipitation of mineral which depleted the solution of calcium ions over about a 3-h period. DMP1 retarded the spontaneous precipitation of calcium phosphate in the same time period and retardation was concentration-dependent. A control protein (bovine serum albumin, BSA) also bound calcium ions and, while slowing mineral precipitation, was not as efficient as DMP1. (**b**) A plot of a small-angle X-ray scattering (SAXS) study which provides details of the conformational nature and development of calcium phosphate nanoparticles or clusters in solution. SAXS analysis showed that, over a 1-month period, greater numbers of nanoparticles were formed in the presence of DMP1 (0.5 μg/mL) in a supersaturated calcium phosphate solution than in the same solution without DMP1. Thus, in the presence of DMP1, calcium phosphate nanoparticle precipitation was slowed and the nanoclusters were stabilized by the protein and remained suspended in the aqueous solution. The SAXS study confirmed the finding that DMP1 could sequester and stabilize calcium phosphate clusters formed de novo. (Images adapted from He et al. [124] with permission)

8.7 Dentin Sialophosphoprotein (DSPP), Dentin Sialoprotein (DSP), Dentin Phosphoprotein (DPP, DMP2), and Dentin Glycoprotein (DGP)

The title of this section of the chapter suggests that there are as many as three different dentin SIBLINGs, but this is not the case. With regard first to DSPP, this SIBLING protein is found in dentin at relatively high levels and in bone and cementum at relatively low levels [109, 135–137]. Cleavage of DSPP by BMP-1 and other enzymes (zinc metallopeptidases) [138] results in two fragments, one containing the DSPP amino terminus (dentin sialoprotein, DSP) and the other the carboxy terminus (dentin phosphoprotein, DPP) (Fig. 8.10) [135, 141, 142]. As a result of the cleavage reactions in vivo, the full-length form of DSPP is sparse in abundance while both DSP and DPP are readily detected in dentin extracellular matrices [69, 139, 143, 144]. Cleavage in the pig also occurs at the C-terminal end of DSP by matrix metalloproteinases (MMP-2 and MMP-20), resulting in a third fragment, dentin glycoprotein (DGP) (Fig. 8.10) [144]. DPP contains the ASARM peptide of DSPP [138].

DSPP is required for dentin and bone mineralization. Deletion of *DSPP* results in defective mineral formation (Fig. 8.11) [145]. Moreover, mutation of the *DSPP* human gene produces dentinogenesis imperfecta, a disease in which dentin is hypomineralized [146]. These and other studies indicate that DSP is required for dentin mineralization [147], whereas DMP1 is more closely linked to ossification.

DSP, the N-terminal domain of DSPP, is the most abundant proteoglycan in dentin (Fig. 8.12) [144, 149]. It is not highly phosphorylated with only three such predicted sites in its amino acid sequence, and it is not highly conserved [144]. The single cysteine residue covalently joins two chains of DSP into a dimer [144]. Studies in vitro indicate that DSP is relatively ineffective in regulating mineralization [150]. However, DSP is highly glycosylated [144] with an unusually abundant sialic acid content (~9%) [151] thought to link the protein with others in extracellular matrices [152]. In this regard, it has been suggested that DSP may interact with collagen though its glycosaminoglycan chains, as do other proteoglycans [153], and thereby possibly regulate mineral formation [144, 154].

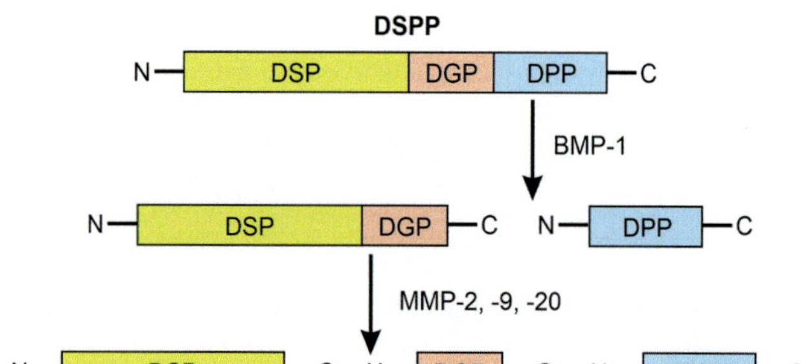

Fig. 8.10 Schematic diagram showing the relationship among dentin sialophosphoprotein (DSPP), dentin sialoprotein (DSP), dentin phosphoprotein (DPP), and dentin glycoprotein (DGP). Odontoblasts and osteocytes secrete high levels of the full-length DSPP; lower levels of the protein are synthesized by osteoblasts, while transient levels are secreted by ameloblasts. The DSPP transcript is cleaved by an enzyme, pro-protein convertase, and BMP-1 to generate DPP, a highly phosphorylated protein and a major constituent of the non-collagenous proteins of dentin [138]. The DSP–DGP protein complex undergoes further processing by MMPs to yield DSP and DGP separately [139, 140]. DGP contains an ASARM peptide located at the C-terminus of the protein [140]. (Schematic adapted from Yamakoshi and Simmer [141] with permission)

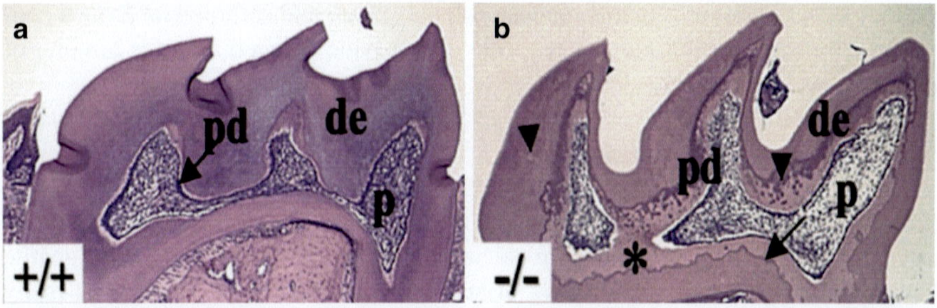

Fig. 8.11 Histological appearance of demineralized molars from a wild-type (**a**, +/+) and dentin sialophosphoprotein (DSPP)-knockout (**b**, –/–) young mouse. A section through a tooth shows (**a**) the normal pulp chamber (p) enclosed by dentin (de) with a narrow line of predentin (pd, arrow) separating the other two tissues. A section of a molar from the DSPP–/– mouse maintains an enlarged pulp chamber and predentin which is more extensive (*) and irregular (arrow) than that of its counterpart control. The widened predentin also contains unusual globular dentin (arrowheads). The dentin itself was abnormal with large areas of interglobular mineralization (not shown). Scale bars unavailable. (Images from Sreenath et al. [145] with permission)

The DGP fragment, 81 amino acids in length, lies between the DSP and DPP segments of DSPP in porcine dentin (Fig. 8.10) [140]. It contains four phosphorylated serine residues and one glycosylated asparagine [140]. Disialidated glycan species are the most abundant, attached to the asparagine along with mono- and un-sialidated forms. The importance of this peptide is unknown as DGP has not been found in any species other than the pig.

In contrast to DGP, DPP (also known as DMP2 and dentin phosphophoryn), first reported by Veis and Perry [155], is the most abundant (up to 50%) non-collagenous matrix protein of dentin [156, 157]. DPP is localized to the mineralization front of the dentin extracellular matrix [109, 158, 159], and it is also present in undifferentiated ameloblasts of rat molars [160] and mouse embryonic kidney and lung [161] as well as in postnatal alveolar bone [12]. The key structural feature of DPP is that it is an unusually polyanionic molecule with an extremely high content of aspartic acid and serine residues, accounting for at least 75% of its total amino acids (Fig. 8.12) [148, 162–164]. Characteristic repeat sequences of aspartate-serine-serine and serine-aspartate are evident particularly at its C-terminus (Fig. 8.12) [148, 165]. Equally unique, up to 90% of the serine residues are phosphorylated [166, 167]. The protein is transcribed from the 3′ end of the *DSPP* gene [135, 164].

In the rat, DPP is not uniformly phosphorylated but has three apparent regions of high, moderate, and low phosphorylation; the highly phosphorylated form contains ~200 attached phosphate groups [166]. The multiple, highly negatively charged residues repel one another and conformationally create an extended backbone of ridges where the numerous phosphate and carboxylate groups are ordered along the peptide backbone [165]. In this manner, the phosphate and carboxylate groups contributed by serine and aspartate, respectively, provide charged sites for binding of counterions, particularly calcium.

Functionally, DPP has been proposed as a sink for calcium, modulating apatite nucleation, growth and inhibition (Fig. 8.13) [12, 39]. Domains of DPP comprised of the many serine and aspartate residues just mentioned (Fig. 8.12) are reported with different phosphate and carboxylate interaction edges, a result suggesting they have different binding properties for calcium ions and the various surfaces of apatite crystals [165]. Molecular modeling of structural motifs of DPP, rich in its many aspartate and phosphoserine residues, shows that such motifs interact with, and favorably adapt to, the known crystal structure of hydroxyapatite or octacalcium phosphate [163]. The latter is a possible intermediate calcium phosphate phase in the maturation of apatite from its precursors. Moreover, other modeling studies suggest that these motifs might provide a scaffold for orientation and growth of mineral crystals [163].

```
   1 MKIITYFCIW AVAWAIPVPQ SKPLERHVEK SMNLHLLARS NVSVQDELNA SGTIKESGVL
  61 VHEGDRGRQE NTQDGHKGEG NGSKWAEVGG KSFSTYSTLA NEEGNIEGWN GDTGKAETYG
 121 HDGIHGKEEN ITANGIQGQV SIIDNAGATN RSNTNGNTDK NTQNGDVGDA GHNEDVAVVQ
 181 EDGPQVAGSN NSTDNEDEII ENSCRNEGNT SEITPQINSK RNGTKEAEVT PGTGEDAGLD
 241 NSDGSPSGNG ADEDEDEGSG DDEDEEAGNG KDSSNNSKGQ EGQDHGKEDD HDSSIGQNSD  DSP
 301 SKEYYDPEGK EDPHNEVDGD KTSKSEENSA GIPEDNGSQR IEDTQKLNHR ESKRVENRIT
 361 KESETHAVGK SQDKGIEIKG PSSGNRNITK EVGKGNEGKE DKGQHGMILG KGNVKTQGEV
 421 VNIEGPGQKS EPGNKVGHSN TGSDSNSDGY DSYDFDDKSM QGDDPNSSDE SNGNDDANSE
 481 SDNNSSSRGD ASYNSDESKD NGNGSDSKGA EDDDSDSTSD TNNSDSNGNG NNGNDDNDKS
 541 DSGKGKSDSS DSDSSDSSNS SDSSDSSDSD SSDSNSSSDS DSSDSDSSDS SDSDSSDSSN
 601 SSDSSDSSDS SDSSDSSDSS DSKSDSSKSE SDSSDSDSKS DSSDSNSSDS SDNSDSSDSS
 661 NSSNSSDSSD SSDSSDSSSS SDSSSSSDSS NSSDSSDSSD SSNSSESSDS SDSSDSDSSD
 721 SSDSSNSNSS DSDSSNSSDS SDSSDSSDSS NSSDSSDSSD SSNSSDSSDS SDSSDSSDSS
 781 NSSDSNDSSN SSDSSDSSNS SDSSNSSDSS DSSDSSDSDS SNSSDSSNSS DSSDSSNSSD
 841 SSDSSDSSDS SDSDSSNRSD SSNSSDSSDS SDSSNSSDSS DSSDSSDSNE SSNSSDSSDS  DPP
 901 SNSSDSDSSD SSNSSDSSDS SNSSDSSESS NSSDNSNSSD SSNSSDSSDS SDSSNSSDSS
 961 NSGDSSNSSD SSDSNSSDSS DSSNSSDSSD SSDSSDSSDS SDSSNSSDSS DSSDSSDSSN
1021 SSDSSNSSDS SDSSDSSDSS DSSDSSNSSD SSDSSDSSDS SDSSGSSDSS DSSDSSDSSD
1081 SSDSSDSSDS SDSSESSDSS DSSDSSDSSD SSDSSDSSDS SDSSDSSDSS NSSDSSDSSD
1141 SSDSSDSSDS SDSSDSSDSS DSSDSSDSSD SSDSSDSSDS SDSNESSDSS DSSDSSDSSN
1201 SSDSSDSSDS SDSTSDSNDE SDSQSKSGNG NNNGSDSDSD SEGSDSNHST SDD
```

Fig. 8.12 The amino acid sequence of dentin sialophosphoprotein (DSPP) isolated from human dentin. Each of the letters corresponds to a single amino acid: For example, serine = S, aspartic acid = D, asparagine = N. The amino acid sequence was deduced from the open reading frame of the *DSPP* gene. Green letters show the signal sequence of the protein and black and blue letters represent the DSP and the DPP portions of the protein, respectively. An endopeptidase binding site (positions 452–454–461–462) is marked in red letters: At this site, DSPP is cleaved into its two separate DSP and DPP moieties. Within DPP, an extensive serine-aspartate (DSS) repeat sequence is denoted in brown. Casein kinase I and II phosphorylate about 90% of the DPP seryl groups, endowing the protein with a high negative charge and an extended and largely unstructured conformation. DPP is a major non-collagenous component of dentin, and the protein is secreted by ameloblasts, cementocytes, and osteocytes in addition to odontoblasts [148]. (Image sequence from George and Veis [12] as adapted from Gu et al. [149] with permission)

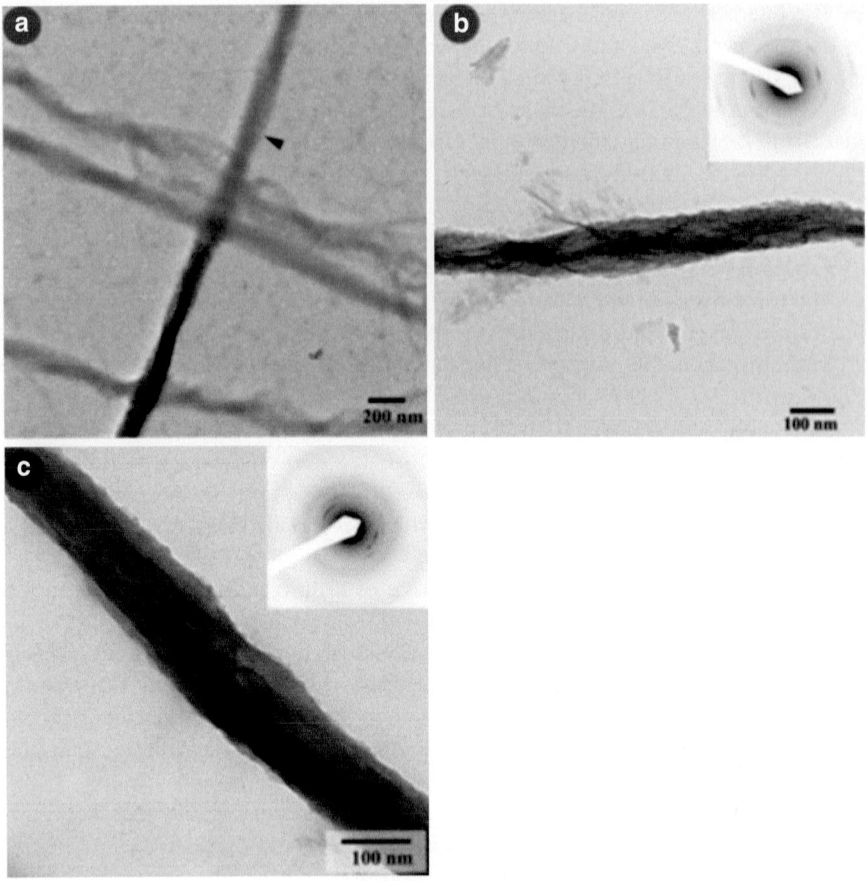

Fig. 8.13 Effect of DPP on the mineralization of collagen in vitro. A gel of purified rat tail collagen was incubated with 1.67 mM $CaCl_2$, 9.5 mM phosphate, and 127 mM NaCl at pH 7.7 for 3 hr. in the presence of 125 μg/ml phosphorylated DPP. (**a–c**) Illustrate transmission electron microscopy images of the collagen fibrils and mineral. (**a**) A single collagen fibril which is both mineralized (dark, heavy density) and non-mineralized (grey density, arrow) has a minimal difference in its diameter, a result suggesting that the mineral is present in the intrafibrillar spaces within the fibril. (**b**, **c**) Enlarged images of mineralized collagen, again showing mineral in collagen intrafibrillar spaces. Image insets present electron diffraction patterns produced by the mineral and demonstrating that the mineral crystal *c*-axes are oriented parallel to the long axis of the collagen fibril associated with the mineral. Such images suggest phosphorylated DPP may lower the activity of calcium ions in a solution with collagen and lead to mineral inhibition in interfibrillar spaces (between collagen fibrils) while facilitating mineral presence in intrafibrillar spaces of collagen. (Images adapted from Deshpande et al. [39] with permission)

The interaction of DPP with collagen appears to be of considerable importance to its role in mineralization. Relevant to its presence in mineralized dentin, DPP crosslinks and binds to COLI in vivo [168–170]. In solution, the interaction is with the e-band in the COLI hole zone [168]. Other studies in vitro support the concept that DPP binds both collagen and calcium ions (Fig. 8.13) [124, 171–173]. Phosphorylated DPP-induced collagen aggregation and formation of platelet-shaped apatite crystals [124, 172]. Experimentally, non-phosphorylated DPP cross-links collagen monomers electrostatically, increasing their rate of formation and growth in length without any effect on apatite nucleation [124]. Together, these several investigations demonstrate that phosphorylation is required for apatite nucleation and growth associated with collagen [124, 172].

With respect to its role as a nucleation factor, DPP at high concentration (0.45 mg/mL) binds one calcium ion for each aspartic and phosphorylated serine residue, while at low concentration (0.05 mg/mL) calcium ion binding is reduced by two-thirds [5, 174]. This result was attributed to a conformational change in DPP which influences its ability to bind to surfaces of apatite where higher DPP concentrations lead to extended β-sheet formation and a greater adsorption on the crystals [5, 12].

DPP preferentially binds to the (100) face of apatite, permitting growth in the crystal *c*-axis direction [175]. In the absence of serine phosphorylation in DPP, apatite nucleation is reduced [176] as well as DPP adsorption to apatite crystal surfaces [12]. As phosphate levels vary in a tissue, phosphorylation of DPP in dentin would likewise vary [177] and thereby alter calcium binding, apatite nucleation, and growth. It has therefore been suggested that DPP may both inhibit and enhance mineralization depending on its phosphorylation state [12].

Finally, in addition to its many aspartate and serine residues and repeat sequences noted previously, DPP contains an RGD motif, common to other SIBLING family members, near the N-terminus of the protein (Fig. 8.12). This motif is available for DPP-cell binding and, as such, would lead to the regulation of events as encompassing as cell proliferation, migration, and differentiation [178–180]. In this context, in certain model cell lines and human mesenchymal stem cells, DPP binds cell membrane integrin receptors and mediates gene expression and differentiation through MAPK signaling pathways [181] in a manner like that of DMP1 [118]. DPP also is reported to promote SMAD1 phosphorylation, the result of which leads to up-regulation of transcription factors, DLX5 and RUNX2 [182]. It may be recalled that DLX drives *RUNX2* expression during osteogenesis and odontogenesis (see Chap. 4).

Thus, DPP is a unique structural protein which is proposed to be critically involved in numerous aspects of mineralization, including its binding to both collagen and calcium ions. DPP also binds to apatite crystals, and it affects their nucleation and growth, depending on the state of phosphorylation of the protein. Additionally, gene expression, cell proliferation, migration, and differentiation are all suggested to be mediated by DPP binding to cell membrane integrins, resulting in the enhancement of osteogenesis and odontogenesis and promotion of a mineralized phenotype in bone, dentin, and other vertebrate hard tissues.

8.8 Matricellular Proteins: Thrombospondins (TSP), Cartilage Oligomeric Matrix Protein (COMP), and R-Spondin (RSPO)

Matricellular proteins are molecules localized to the surface of cells and/or extracellular matrix where they mediate cell function by interacting with cell-surface receptors, cytokines, proteases, or hormones [183, 184]. Among such molecules, TSPs comprise a group of five matricellular proteins concerned with tissue growth and remodeling. Hankenson et al. reported that all five TSPs are present in cartilage and bone [185]. Other proteins with matricellular activity and nature include SPARC (osteonectin), periostin, and the CCN (connective tissue growth factor, cysteine-rich protein, and nephroblastoma overexpressed gene) family of regulatory proteins [186].

The TSP proteins are large, structurally complex and modular (Fig. 8.14) [187]. TSP-1, for example, consists of three similar peptide chains containing motifs (type 1–3 repeat regions) which bind matrix proteins (collagen and fibronectin (FN)), cell receptors (β3 integrin, CD47 and CD36a), growth factors (TGF-β, bFGF, and platelet derived growth factor), proteases, heparan sulfate proteoglycan (syndecans), the lipoprotein-related receptor protein, and yet additional molecules [188, 189]. The functional roles for the five different TSP proteins vary considerably. TSP-1 is in part associated with inhibition of mineralization, for instance: The treatment of cells in vitro with oligonucleotides which are inhibitory to the TSP-1 sequences leads to increased mineralization and deposition of extracellular proteins [190]. Moreover, work in vivo shows that, while the phenotype of *TSP-1* null mice is affected by the loss of the *TSP-1* gene, the skeletal changes of the animals are mild [185]. In contrast, a transient reduction in *TSP-2* expression produces marked increases in *COLI, RUNX2*, *osterix*, *TNAP*, *BSP*, and *OCN* expression, results leading to the suggestion that TSP-2 is involved with matrix assembly and mineralization [191].

TSP-5 is also known as cartilage oligomeric matrix protein (COMP), a protein abundant in cartilage and many other soft and hard connective tissues [192]. The molecule is reported to be 524,000 Daltons in molecular weight and comprised of five identical chains which interact to form a coiled coil, four EGF-like regions and eight TSP repeating domains [193]. COMP binds to COLI, COLII, COLIX and FN, modulating COLI and COLII fibril formation [194]. COMP also interacts with aggrecan (ACAN), a major extracellular matrix component of cartilage and, as such, COMP promotes the organization and assembly of the matrix to maintain cartilage mechanical strength [195]. COMP is regarded as a biological indicator of the turnover, breakdown and loss of cartilage and is used as a diagnostic marker of arthritis [193]. Two dysplasias are associated with *COMP* mutations: A serious growth impairment, pseudoachondroplasia, related to abnormal functioning of the endoplasmic reticulum and a less serious condition influencing the function of the growth plate. This latter abnormality, multiple epiphyseal dysplasia, is linked to matrix synthesis and assembly [196].

One additional matricellular family of four secreted proteins may be briefly mentioned, the R-spondins (RSPO1–4), which exert effects in bone and other tissues

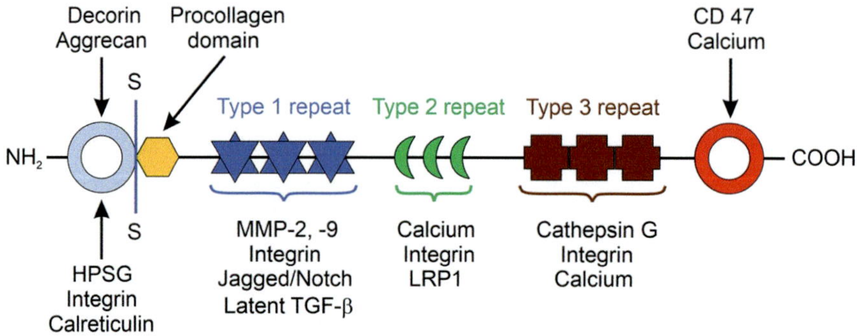

Fig. 8.14 Schematic of the structure of one of the three peptide strands which comprise the structure of the multidomain protein, thrombospondin-1 (TSP-1). Each strand consists of several amino acid domains, including NH_2 (amino)- and COOH (carboxyl)-termini, an interchain disulfide knot (S–S), a procollagen region, and repeat regions, 1–3. The interchain disulfide knot and additional cysteinyl groups interact with the other two peptide strands to form the homotrimeric TSP-1 protein. The motifs at the N-terminal region of TSP-1 bind calreticulin and integrins, as well as heparan sulfate proteoglycan (HSPG), aggrecan, and decorin. These interactions permit TSP-1 to interact with other proteoglycans such as syndecan, perlecan, and versican. Located between the N-terminus and a type I repeat region is a procollagen domain which serves to facilitate collagen subunit assembly. The procollagen region is followed by the type I repeat region comprised of three domains, each 57 amino acids in length and framed by six cysteinyl groups. The type I repeat region interacts with cell integrins, matrix metalloproteinase (MMP)-2 and -9, and latent transforming growth factor-β (TGF-β). The MMPs promote matrix turnover, while TSP-1 binding to the TGF-β complex promotes growth factor activation. The type I domains are involved in the regulation of various signaling pathways through the transcription factors, Jagged and Notch. These two proteins are components of an evolutionarily conserved signal transduction pathway concerned with cell fate. A type II repeat region follows the type I repeat and consists of three motifs, each 50–60 amino acids in length. These motifs bind calcium ions and interact with low-density lipoprotein receptor related protein (LRP)1 and integrins. LRP1 is involved in activation of the WNT/β-catenin signaling pathway, while integrins provide a mechanism for cell binding and activation. The third domain (type III repeat) consists of seven groups, each 38 amino acids in length, which serve as important calcium ion binding sites. The arg-gly-asp-ala sequence in the type III repeat region interacts with αvβ3 integrins and is the site for cell attachment and spreading. Other structural aspects of the type III repeat interact with fibroblast growth factor (FGF)-2 and calcium ions, contain sequences which bind additional integrins, and activate proteolytic enzymes such as cathepsins. The C-terminal domain of TSP-1 interacts with the cell surface antigen CD47 and calcium. Together, these various motifs and domains of TSP-1 regulate a multiplicity of functions through cellular membrane receptors, other extracellular molecules, including collagen and glycosaminoglycans, matrix-regulating enzymes, and growth factors. Accordingly, TSP-1 can respond to positional and environmental cues which mediate bone and cartilage matrix composition and remodeling, cell/matrix interactions, and cell adhesion, motility, proliferation, differentiation, maturation, and apoptosis. Further TSP-1 ion binding and possible association with nascent mineral and collagen influence tissue mineralization. (Summary results and schematic adapted from Resovi et al. [187] with permission)

and contain components of the TSP modular structure (type 1 repeat domain) [197, 198]. These molecules range between ~230 and 270 amino acid residues in extent and consist of two neighboring cysteine-rich furan-like domains adjacent to the repeat domain of TSP. The C-termini of RSPOs maintain a domain which is comprised of numerous basic peptides [198]. The respective domains of each of the RSPO members in various mammalian species have a high degree of sequence similarity, an observation underlying suggested protein functions which are evolutionarily conserved [199]. The RSPOs are bound to heparan-sulfated proteoglycans and modulate attachment to osteoblasts and other cells as well as extracellular matrices. A principal function of these proteins is promotion of morphogenesis, including osteogenesis in bone, and stem cell growth. If *RSPOs* are mutated or absent, they cause a wide range of developmental disorders in affected tissues. The RSPO family activates and regulates canonical wingless-related integration site (WNT) signaling through interactions with leucine-rich repeat-containing G-protein-coupled receptors (LGR)4/5/6 and frizzled receptors [200]. RSPO members also regulate non-canonical WNT pathway events [201].

8.9 Proteoglycans and Small Leucine-Rich Repeat Proteoglycans (SLRPs), Biglycan (BGN), and Decorin (DCN)

Another important group of non-collagenous molecules present in vertebrate mineralized tissues are those of proteoglycans and related species. The proteoglycans are a family of protein-carbohydrate complexes which are found as transmembrane and extracellular matrix molecules in cells. In cartilage, extracellular proteoglycans form densely packed, gel-like, hydrophilic, space-filling structures which mechanically and electrostatically anchor collagen, SIBLING and other molecules in place in the extracellular landscape. Much of the volume of proteoglycans is attributable to chains of modified sugar molecules (or more precisely glycosaminoglycans or glycans) of varying length attached to a core protein [202]. Multiple core proteins are themselves attached to a long chain sugar polymer, hyaluronic acid (hyaluronan), through a link protein [203]. These interactions form huge, exceedingly high molecular weight macromolecular proteoglycan aggregates (Fig. 8.15) [204, 205].

The most common of these hyaluronic acid-binding proteoglycans in cartilage and other tissues is ACAN (Fig. 8.16) [206]. ACAN may be described structurally as consisting of a complex core protein characterized by the presence of stable globular regions (G1–3) formed by sulfur-hydrogen bonds of the amino acid, cysteine. G1 at the NH_2-terminus of the protein core associates with the link protein and binds hyaluronan. This region of the molecule also binds to a receptor on chondrocytes (CD44), thus providing a bridge and allowing crosstalk between cartilage extracellular matrices and the cells [207]. G3, present at the COOH end of the core protein,

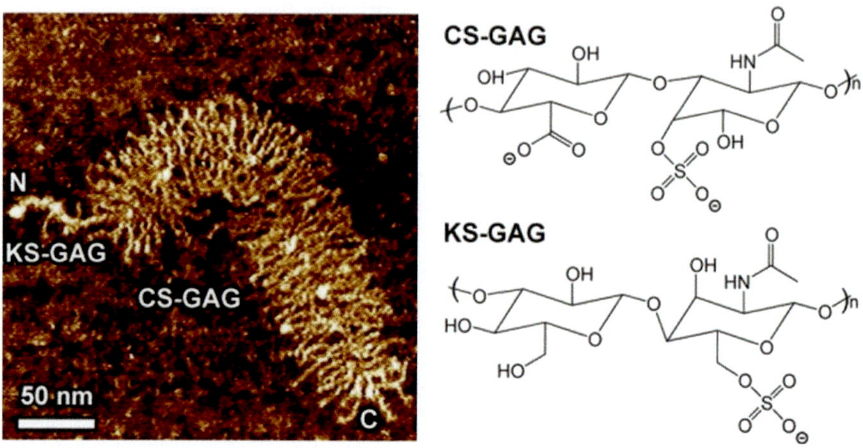

Fig. 8.15 Atomic force microscopic image (tapping mode) of an aggrecan monomer from fetal epiphyseal cartilage. Aggrecan is a complex macromolecule consisting of numerous glycosaminoglycan molecules attached to a polysaccharide core molecule of hyaluronic acid. The left panel presents one such monomer, showing the N (amino)- and C (carboxyl)-termini of aggrecan and its longer chondroitin sulfate glycosaminoglycan (CS-GAG) and shorter keratan sulfate glycosaminoglycan (KS-GAG) side chains. The chemical structure of chondroitin sulfate is shown as a repeating disaccharide of glucuronic acid linked by a β-1, 4-glycosidic bond to *N*-acetyl galactosamine. The GAG is sulfated at position C4 or 6. Like chondroitin sulfate, the keratan sulfate GAG is a linear polymer formed by repeating disaccharide units of galactose linked to *N*-acetylglucosamine with charged sulfate groups at C6. (Images from Ng et al. [204] and adapted by Rojas et al. [205]. A link to the image by Rojas et al. [205] is provided (https://pubs.acs.org/doi/10.1021/bm401611b) with a notice to readers that further permission related to the material excerpted should be directed to the American Chemical Society)

associates with other matrix components such as the protein, fibulin. Between G2 and a G3 are the attachment sites of the glycosaminoglycan chains of the ACAN complex [207].

Glycosaminoglycans are commonly repeating disaccharide units or chains comprised of one amino sugar (hexosamine) and one acid sugar (uronic acid), and the acid residue is often sulfated (Fig. 8.16). In cartilage and other vertebrate mineralized tissues, the actual composition of the chains varies with the most abundant glycan being chondroitin sulfate. Other glycans include a small number of dermatan sulfate and keratan sulfate molecules [207, 208]. For chondroitin sulfate, as many as a hundred or more of these disaccharides can form a single glycan chain. The electrical repulsion of the negatively charged carboxyl groups of the sugar-based uronic acids and sulfate groups produces a rigid, highly extended conformation of the molecular aggregate, and these sites tightly bind hydrated cations such as calcium [209]. The extended nature of each of these aggregates and their bound hydrated ions leads to the large volume structure which characterizes proteoglycans (molecular weight range 10^5–10^7 Daltons). In addition, the glycan chains interact with growth factors like TGF-β as well as with cell surface proteins such as CD44 mentioned above, thereby influencing gene transcription and post-translational events [11].

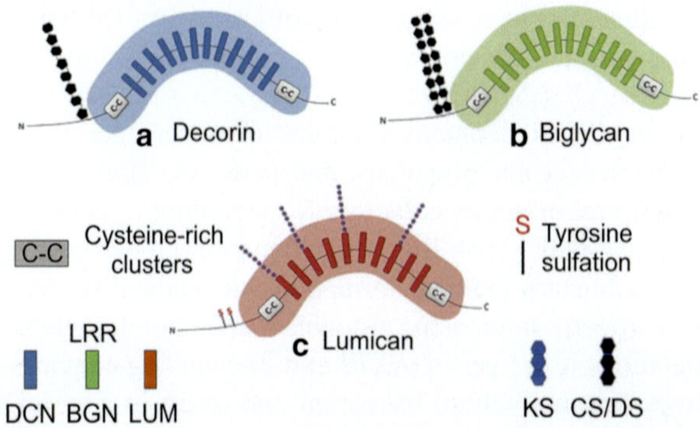

Fig. 8.16 Schematic of the molecular configuration of three small leucine-rich proteoglycans (SLRPs): Decorin, biglycan, and lumican. The common features of the three molecules include their curve-shaped core protein to which variable numbers of carbohydrate units are bound and flanked by cysteinyl-rich motifs (C–C). The core proteins are unusual in that they contain 10–12 domains of leucine-rich repeats (LRR). Decorin (**a**; DCN) consists of a single chondroitin sulfate (CS)/dermatan sulfate (DS) glycosaminoglycan (GAG) chain. Decorin has a large array of binding partners which include signaling molecules, other extracellular molecules, and growth factors. Biglycan (**b**; BGN) possesses two CS/DS GAG chains and is important in bone formation and mineralization. Biglycan is required for collagen fibril formation. Lumican (**c**; LUM) contains a core protein which binds four keratan sulfate (KS) and tyrosine sulfate residues. Lumican is necessary for bone collagen fibrillogenesis. The amino (N)- and carboxyl (C)-termini of the molecules are marked. (Schematic adapted from Appunni et al. [206] with permission)

When calcium ions are bound to the highly negatively charged proteoglycan aggregates (and phosphate ions repelled), mineralization is inhibited. On the other hand, bound calcium ions may be released on the degradation of proteoglycan aggregates. In this case, these ions would become available to interact with phosphate ions to form early phases of mineral as possible precursors to apatite crystals in mineralizing cartilage. As mineral is deposited into the cartilage extracellular matrices, enzymes are activated which break down the glycan hydrophilic domains and provide space for mineral deposition and growth. Further, a recent study in vitro of demineralized murine dental (periodontal) tissues which were then remineralized in the presence or absence of glycosaminoglycans demonstrated that loss of these molecules decreased the rate of remineralization [210]. This result suggested that glycosaminoglycans enhanced mineral deposition in the dental tissues and supported their role as facilitating mineralization in vertebrate tissues rather than inhibiting it [210].

Gene defect analyses and gene deletion studies in animal models have provided direct information about the importance of ACAN in mineralization. For example, a disease in chickens (nanomelia) caused by a point mutation results in a shortened ACAN core protein which fails to be correctly processed and secreted by chondrocytes; thus, the amount of ACAN is abnormally reduced in cartilage extracellular matrices [211]. If the gene defect is homozygous, the disease is lethal. In mice, a mutation in the gene encoding the G1 region of the ACAN core protein leads to impairment of cartilage formation and bone growth (cartilage matrix deficiency) [212]. Loss of *ACAN* and its link protein results in growth failure and aberrant bone and cartilage in the skull and facial bones of mice [213]. These studies demonstrate that *ACAN* and link protein are necessary for proper maintenance of chondrocyte structure, the collagen extracellular matrix network in cartilage, and normal development of endochondral-directed growth plate mineralization.

As noted earlier, all cells contain a number of transmembrane heparan sulfate proteoglycans like syndecans, glypicans, and perlecan. The extracellular heparan sulfate chains can bind or act as cofactors for signaling molecules such as BMP, WNT, Hedgehog, and FGF [214]. Binding of heparan sulfate sugar residues to each of these proteins influences gradient formation, bioavailability, accumulation, and turnover, thereby directly influencing osteoblast and chondrocyte metabolic activity. In bone, mutations in the genes (*EXT1* and *2*) encoding enzymes like glycosyltransferases block heparan sulfate formation and result in abnormal chondrocyte proliferation with the formation of osteophytes (hereditary multiple exostoses) [215]. Moreover, transmembrane proteins like syndecans are required for exosome (matrix vesicle) formation [216]. The importance of this event in mineral formation is discussed elsewhere in this book.

In addition to ACAN, other proteoglycans found in vertebrate mineralized tissues include the family of SLRPs (36,000–42,000 Dalton molecular weight). Eighteen different SLRPs have now been documented in vertebrate extracellular matrices, many of which are found in mineralized tissues [217]. Some of the family members are fibromodulin, keratocan, osteoglycin, osteoadherin, epiphycan, BGN, DCN, and lumican. In mineralizing tissues most attention has been directed toward the function of BGN and DCN. Their presence has been localized to epiphyseal cartilage, subchondral bone, and the periosteum among other tissue examples (Fig. 8.16) [218].

The structure of all SLRPs is characterized by regions of limited length (about 20–40 amino acids) in which the hydrophobic amino acid, leucine, predominates in a repeating motif. The leucine repeats exemplified in three SLRPs, BGN, DCN, and lumican occupy the central concave protein core region of these small molecules and form hydrophobic β-sheets, while the outer convex surface of the molecules contains multiple α-helices (Fig. 8.16) [206, 219]. The protein core (~38,000 Daltons in molecular weight) is flanked by two cysteine-rich regions which are conserved in SLRP family members (Fig. 8.16) [220, 221]. Of considerable importance, the leucine-rich repeating domains are utilized as protein-binding motifs for molecular recognition, including signal transduction, cell adhesion and several additional processes [219]. With respect to each of these SLRPs, much of the biologic activity of DCN is related to its structure as a monomer, but it, like BGN, has been reported as a dimer in solution, in which its hydrophobic surfaces associate with each other and sequester its leucine repeating units [220]. This structural feature has led to the concept that the monomer is the biologically active molecule, whereas the dimer with its buried leucine repeats serves as the stabilized, less-active form of the molecule [222]. An additional defining structural hallmark of the SLRPs is the number and composition of their carbohydrate side chains. DCN and BGN have one and two glycosaminoglycan chains, respectively, each chain being ~40,000 Daltons in molecular weight (Fig. 8.16) [221]. Their bound carbohydrates are chondroitin or dermatan sulfate. Lumican, a related SLRP, has four keratan sulfate chains attached to its core protein; osteoglycin, another SLRP, contains an attached keratan sulfate chain; and epiphycan maintains one dermatan sulfate glycan.

SLRPs can be found deep within tissue extracellular matrices or more commonly in pericellular matrices where they bind cytokines and growth factors. Ligands bound include TGF-β [223], BMP-4 [224], WISP-1 (WNT-inducible secreted protein-1) [225], TNF-α (tumor necrosis factor-α) [226], and IGF-1 (insulin-like growth factor-I) [227]. Because of their intimate interactions with cells, SLRPs mediate receptor (outside-in) signaling and thereby influence both matrix synthesis and mineralization. DCN may combine with BSP to induce mineral formation in collagenous matrices [228].

Among their multiple functions, SLRPs impact collagen cross-linking and promote collagen fibrillogenesis and extracellular matrix assembly [229]. With respect to DCN as an example, this small proteoglycan binds to and "decorates" COLI fibrils and the fibrils of numerous other collagen species (COLII, III, IV, V, VI, XII, and XIV) [209]. Binding involves complex association between the leucine repeats of DCN and the d and e1 gap zone bands of COLI [230]. DCN in this manner enhances the lateral association of collagen molecules to form fibrils, fibers and other higher ordered aggregates. Binding also requires DCN glycosaminoglycan chains, which help regulate interfibrillar distances between these collagen macromolecules [153]. Further, DCN sterically hinders the attachment of degradative enzymes (MMP-9) to collagen substrates, so that the proteoglycan protects collagen from proteolysis [231]. Finally, *DCN* gene deletion results in collagen fibrils which have variable diameter and are often abnormally fused, characteristics leading to matrix structure and organization of limited strength and stiffness [232].

Besides DCN, BGN is the other major SLRP found in vertebrate mineralized tissues. BGN has been reported in bone, cartilage and dentin matrices [233–235]. As noted above, its protein core (~38,000 Daltons in molecular weight) and its general structure are similar to that of DCN, but BGN contains two chondroitin sulfate chains (hence, the name, "bi-glycan") compared to one glycan chain of DCN (Fig. 8.16) [221, 233]. BGN binds COLI and COLII [236, 237], BMP-4 [224] and TGF-β [238]. The precise role of BGN in vertebrate mineralized tissues is unclear. It appears to affect collagen structure and organization and has a high affinity of binding to apatite [221]. At low concentrations in solution studies, the molecule is reported to enhance formation of mineral while, at higher concentrations, BGN inhibits crystal growth and development [221]. It has been suggested that BGN also serves a role of transducing forces in association with osteocytes in bone matrices [11].

BGN gene deletion in mice results in delayed osteogenesis, reduced collagen production and bone mass, a slow loss of bone density (osteopenia, a condition resembling osteoporosis), and a marked effect on odontogenesis leading to decreased dentin mineralization [236, 239]. These data support *BGN* as functionally mediating cellular events which regulate bone and dentin formation and mineralization [240]. A reduction in expression of *BGN* in humans is frequently accompanied by early-onset osteoporosis and limited bone growth (Turner Syndrome) [241]. Deletion of *DCN*, on the other hand, has a minimum effect on bone phenotype [240] but, as noted above, yields altered collagen fibril sizes and aggregation accompanied by reduced biomechanical strength and stiffness of extracellular matrices [232]. When expression of both *BGN* and *DCN* is ablated, the effects on bone and collagen occur early and are more severe when compared to the knockout of either gene alone [242].

The SLRP family members are important, albeit minor, components with certain overlapping activities in vertebrate mineralized tissues. They play roles in collagen fibrillogenesis and extracellular matrix structure and architecture, and their presence is therefore required for matrix formation and collagen-related mineralization. Moreover, they bind critical growth factors and thereby help regulate molecules like BMP or WNT in mediating mineralization of bone and dentin. In addition, there is a growing understanding that SLRPs like DCN promote autophagy even when there is nutrient sufficiency [#352, 243]. Hence, the possibility exists that these small proteoglycans also serve to regulate events leading to exosome formation, egress from cells, and subsequent participation in the initiation of physiological mineralization as described elsewhere in this book.

8.10 Osteocalcin (OCN)

OCN or bone gla (γ-carboxyglutamic acid) protein and related MGP belong to a group of vitamin K-dependent calcium-binding proteins which bear the same gla-structural nature as that of blood clotting proteins such as prothrombin and Factors VII, IX, and X as well as of growth arrest-specific protein 6 (GAS6) and the matricellular protein, periostin. Using phylogenetic analysis, Laizé et al. [244] provided evidence that OCN and MGP are more closely associated with each other than to the other proteins just mentioned and have a common origin. These workers suggested that the γ-carboxyglutamic acid-containing proteins appeared very early in vertebrate evolution through gene duplication and then sequence divergence, and they also showed that MGP was associated with the time of emergence of cartilage and predated OCN, related to the time of development of bone [244]. The action of OCN in this regard is interesting: Unlike the highly phosphorylated extracellular matrix proteins and several EF-hand cellular proteins which bind calcium ions in vertebrate mineralized tissues, OCN is unusual in that its negatively charged γ-carboxyglutamic acid residues perform the same function.

OCN is synthesized by osteoblasts with smaller amounts produced by odontoblasts and hypertrophic chondrocytes, cells of the mineralizing avian tendon, as well as the nucleus pulposus of the intervertebral disc, dental pulp, and adipose tissue [45, 245–247]. The protein, a major component of bone (1% of its total protein [248]), is secreted into the systemic circulation such that its presence in serum correlates well with bone formation. Moreover, as will be discussed below, there is accumulating evidence that OCN may maintain an endocrine function following its processing.

Considering its molecular structure, OCN is a relatively small protein (less than 50 amino acid residues and molecular weight ~ 5800 Daltons). The central region of the protein is conserved over several species and consists of two antiparallel α-helices connected by a β-turn and stabilized by a disulfide bond [249–251]. In the presence of vitamin K following transcription, its three glutamate residues (in human) are carboxylated in the C^{γ} position [251]. Prior to release from bone and secretion into the bloodstream, OCN undergoes non-enzymatic processing in the acidic environment mediated by osteoclasts, an event causing loss of one or more of the negatively charged gla groups of the molecule. The circulating, undercarboxylated form of OCN may be monitored to provide a valuable clinical biomarker of skeletal health.

The expression of *OCN* coincides with differentiation of osteoblasts and tenocytes prior to and during mineral formation in bone and avian tendon extracellular matrices [45, 252]. OCN has also been reported in pathological mineralization such as urolithiasis [253]. Studies of *OCN* knockout models [254, 255] yielded no overt skeletal phenotype but suggested that OCN was a negative regulator of bone formation. Work in vitro [256] gave results in which mineralization was inhibited, possibly by delaying apatite nucleation.

In mineralizing avian tendons, immunocytochemical analysis of OCN in its carboxylated form demonstrated localization of the protein at the surfaces of, outside, and within COLI fibrils of this tissue [11]. More specifically in these fibrils, OCN was associated with the collagen hole zone regions where it was proposed to play a possible role in stabilizing putative early forming amorphous mineral ion aggregates and the subsequent nucleation, growth, and development of apatite crystals during intrafibrillar collagen mineralization [61]. OCN immunolocalized at the surfaces of collagen or outside the fibrils was suggested to influence mineral deposition in these aspects of the avian tendon but by means yet to be explored [61]. Additional recent studies have proposed close cooperativity between OCN and OPN and between OCN and octacalcium phosphate in mediating mineralization of collagen [257, 258].

Mineralized tissues and apatite specifically are recognized for their capacity to store or bind small molecules as well as numerous anions and cations [7, 259]. In this context, the presence of OCN at sites within and along collagen fibrils may reflect sequestration of the molecule in an inactive protein form until such time that remodeling of the surrounding mineralized tissue matrices occurs. At that point, OCN may be released to the circulation to function hormonally at sites other than those which are mineralizing [45, 61].

As mentioned earlier, while OCN is expressed and secreted by osteoblasts and odontoblasts, it also circulates in the serum in undercarboxylated and uncarboxylated forms [260], and the variable content of circulating OCN has been associated with several pathologies [261, 262]. The functions of OCN differ with its structural forms as carboxylated OCN reportedly relates to mineralization [61, 249, 250, 263, 264] and undercarboxylated OCN serves a hormonal role. The concept that OCN could serve as a hormone is based on the observation that animals in which *OCN* is deleted experimentally become obese and exhibit decreased β-cell proliferation, glucose intolerance, and insulin resistance [265]. A decarboxylated species of OCN, possibly formed by the acidic conditions developed during osteoclastic bone resorption, may regulate insulin secretion by cells of the pancreas and modify the insulin sensitivity of adipocytes [266, 267]. Indeed, if the OCN receptor (a type of G-protein-coupled receptor) in the β-cells of the pancreas is mutated, there is evidence of insulin resistance [268]. Moreover, OCN is suggested to promote *insulin* gene expression and enhance insulin protein secretion [269]. The expression of *OCN* is down-regulated by the transcription factor, FOXO, which is itself a target of insulin [270]. Thus, a complex metabolic feedback loop is formed with other hormones which regulate glucose metabolism [270].

Related to these observations concerning the regulation of glucose homeostasis, two other proteins resident in bone also serve as hormones which affect the metabolism of distant tissues. One of them, the protein, lipocalin (LCN)-2, influences glucose homeostasis and energy metabolism by enhancing appetite-suppressing signaling in the brain [271]. The second protein, SOST, is secreted principally by osteocytes. If the *SOST* gene is deleted, there is an increase in insulin sensitivity and alterations in the ratio of anabolic to catabolic metabolism in adipocytes [272]. In their review paper summarizing the role of osteoblasts in energy metabolism, Dirckx et al. [273] commented, "Osteoblasts communicate with other organ systems to coordinate bone formation with the overall energy balance" (Fig. 8.17). From an

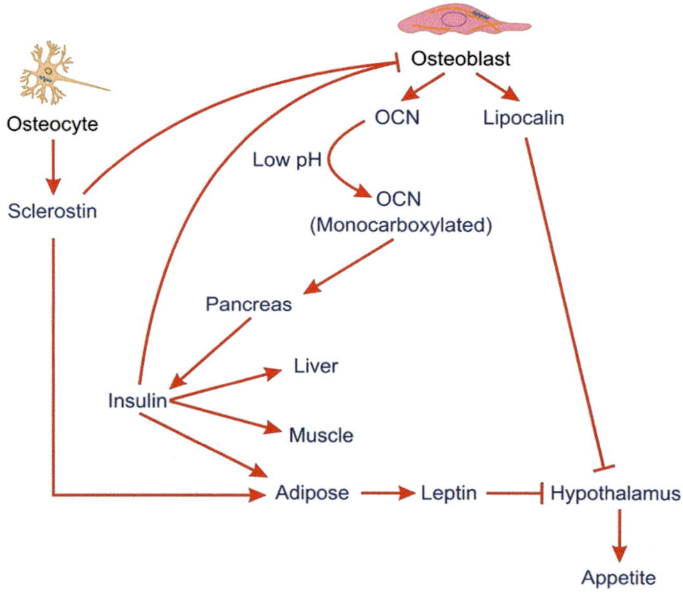

Fig. 8.17 Schematic of metabolic circuits linking bone cell function with insulin release and nutrient intake. Bone cells generate three proteins which influence the metabolism of distant tissues: OCN, lipocalin, and sclerostin. Osteoblasts secrete OCN, which is decarboxylated in the acidic environment of remodeling bone and transported to the β-cells of the pancreas. There it binds a G-protein-coupled receptor to increase insulin secretion. The insulin response triggers four major effects: it promotes glucose uptake by muscle, liver, and adipose tissue; it sensitizes each of the soft tissues to insulin; it promotes leptin secretion by adipocytes; and it feeds back on osteoblasts to inhibit new bone formation and additional OCN release. Accompanying these changes, sclerostin, a protein mediated by osteocytes, enhances adipocyte function in increasing leptin secretion from adipose cells while also inhibiting osteoblast function. Unlike OCN, lipocalin from osteoblasts acts directly on the hunger centers of the brain (hypothalamus) to block nutrient intake. (Schematic modeled from Dirckx et al. [273] with permission)

evolutionary perspective, it is reasonable to speculate that the endocrine actions of bone-derived OCN, LCN, and SOST probably evolved in response to the energetic demands of a constantly remodeling skeleton. Finally, while far removed from bone, there is also some evidence to indicate that OCN may also influence testosterone synthesis, increase neurotransmitter release in the brain, and modulate cognate activity at the level of the hippocampus [274, 275].

The complexity of OCN action bears additional discussion. One approach to understanding the multiple proposed effects of OCN is to divide or categorize matrix proteins (named the matrisome by Hynes and Naba [276]) into different functional groups: Proteins required for collagen stability; proteins associated with mineral formation and growth; enzymes required for tissue maturation, turnover and cell adhesion; and matricellular proteins which convey chemical messages (growth factors and hormones) to cells. In this context, OCN would fall into two groups, proteins involved with mineral formation and growth and with conveying chemical signals to cells. As noted above, aside from serving with other non-collagenous proteins as a regulator of mineralization and as a clinical indicator of osteogenesis and bone turnover, additional evidence suggests that OCN acts as a hormone regulating the activities of distant tissues, specifically the secretory pancreas, adipose, muscle and brain tissue, and possibly the reproductive organs [265, 274, 275, 277–280]. These putative roles of OCN and functions of other non-collagenous proteins have, however, been deduced in many instances from mouse models where genes can be manipulated with some ease and results are often freely extrapolated to assessments of human health and disease. While the mouse and human genomes are similar, the homeostatic responses of the two species to environmental and systemic challenges vary substantially. Thus, the effects and importance of OCN as a possible critical endocrinal factor would be expected to hold differing significance between these species. As an example, mouse studies indicate a direct relationship between OCN and insulin release from the pancreas [279]. However, when this hypothesis was examined in a human population, the normal diurnal changes in insulin levels were not mirrored by systemic changes in OCN concentrations [281]. Considering the intricacies of insulin regulation of glucose metabolism, this result was not unanticipated, but it points to the dangers of oversimplification of the effects of multiple interactive molecules which regulate multifaceted biological systems and of the interpretation of certain experimental findings involving OCN as well as other non-collagenous proteins.

The role of OCN in regulating aspects of skeletal mineralization and its more general impact on endocrine function has been explored by a considerable number of investigators using gene deletion techniques. In the Karsenty laboratory, Ducy et al. deleted exon 4 of the *OCN* gene in mice and reported that the animals had increased bone formation, were glucose intolerant, and maintained higher blood glucose levels and lower insulin serum levels compared to control mice [254]. In other words, OCN exhibited both endocrinal and skeletal effects. In a later study from the same laboratory, it was noted that the suggested endocrine nature of *OCN* in mice could be extended further as the gene promoted the synthesis of testosterone, and thus male fertility [274]. However, these findings were not universally accepted. A subsequent report, for example, in which *OCN* in rats was completely disrupted utilizing CRISPR/Cas9 technology showed no increase in cortical bone density and no evidence of glucose intolerance or insulin resistance [282]. Likewise, in a somewhat similar study in mice and also applying CRISPR/Cas9 methodology, Diegel et al. concluded that bone mass or strength was not influenced by *OCN* [283]. It was also reported that there was no change in testosterone levels or impaired male fertility in the animals [283]. Noting that *OCN* exists in mice as a gene cluster, Moriishi et al. used gene targeting to inactivate major components of the cluster globally in genomic DNA [284]. It was determined in this work that the gene product served to align forming apatitic crystals within collagen fibers in the mouse bones, but no changes were found in endocrinal activities impacting glucose metabolism, testosterone synthesis, and the maintenance of muscle mass [284]. In a very recent summary review of these conflicting reports, Komori commented that, rather than decreasing bone quality, OCN promoted bone strength by enhancing the interplay between collagen fibrils and the mineral phase [285]. Komori also remarked that OCN failed to function as a hormone in the pancreas, testis and muscle [285]. Such results are supportive of data and conclusions from other sources suggesting that OCN enhanced bone mineralization but was not a hormone [283, 286]. The discrepancies in the comparative analyses and the reproducibility of respective results from these many investigations may possibly be explained by the phenotypic variation among the strains of the animals, and clearly further studies are needed to resolve the present uncertainty regarding the several proposed roles of OCN [284, 285].

8.11 Sclerostin (SOST)

SOST is a glycoprotein expressed and secreted predominantly by osteocytes but also by hypertrophic chondrocytes and cementocytes [287, 288]. SOST is additionally detected in kidney, lung, heart, and aorta, but at low levels in these tissues [289]. SOST consists of a 213-amino acid sequence which has a distant similarity to the BMP antagonist, noggin, and forms a C-terminal cysteine knot of three intertwined disulfide bridges [288, 290]. On reviewing the history of SOST, Tanaka and Matsumoto link its discovery to the etiology of two pathologies characterized by high bone mass, sclerosteosis (an osteopetrosis with syndactyly) and van Buchem disease (a hyperostosis involving abnormalities of the skull and mandible) [291]. Both disorders implicated a loss-of-function mutation in the *SOST* gene [291]. Deletion of *SOST* in a murine animal model leads to a phenotype of high appendicular and axial bone density and increased bone strength involving both cortical and trabecular bone [292]. These and other findings lend strong support to the concept that SOST negatively regulates bone formation and mineralization.

Insight into a mechanism by which SOST controls bone cell activity and mineral deposition has been gained through studies of osteoprogenitor cells, osteoblasts, osteoclasts and osteocytes. In osteoprogenitor cells and osteoblasts, SOST suppresses WNT signaling by binding co-receptor low-density lipoprotein receptor-related protein 5 and 6 (LRP5/6) at the plasma membrane of these cells and preventing LRP5/6 interaction with the WNT receptor, frizzled (see Chap. 4) [293]. Since canonical WNT signaling is critical for osteogenesis, its suppression profoundly limits bone cell function and mineral deposition (see Chap. 4). Thus, high secretion levels of SOST inhibit bone formation and mineralization and, conversely, low levels of SOST enhance bone formation and mineralization (see Chap. 4) [293]. Another result of SOST suppression of WNT signaling is that osteoprogenitor cells and osteoblasts reduce secretion of osteoprotegerin. This soluble decoy receptor prevents receptor activator of nuclear factor kappa-B ligand (RANKL) binding to osteoclasts, slowing osteoclastogenesis and bone resorption (see Chap. 4) [294]. In this instance and consistent with its additional activities described above, SOST functions as a potent inhibitor of skeletal mineralization and enhances bone resorption. Clinical implications of such effects are discussed below.

Not unexpectedly, the regulation of *SOST* is complex and involves several transcription factors mentioned earlier in this book. Among the most important positive transcription regulators of *SOST* are osterix (OSX) and runt-related transcription factor (RUNX)2, while negative regulators include parathyroid hormone, estrogen and histone deacetylases (HDAC). Sirtuin 1, a HDAC which is sensitive to the hypoxic state of bone cells, suppresses *SOST* expression [295, 296].

SOST secretion by osteocytes may explain the observation that biomechanical force or stimulation enhances bone anabolism. In this context, Robling et al. showed that mechanical loading of bone tissue in vivo promoted osteocyte expression of *SOST* while unloading had the opposite outcome [297]. In other words, SOST secretion by osteocytes provides the chemical signal (likely mediated through WNT/LRP5/6 signaling) which transduces the consequence of applied forces on the architecture of skeletal tissues.

Finally, the therapeutic importance of SOST as a treatment for bone fragility associated with post-menopausal osteoporosis and cancers such as multiple myeloma has not been overlooked. Such conditions often result in painful fractures with high morbidity and mortality. SOST is secreted into the systemic circulation as it directly influences bone cells, and numerous pre-clinical and clinical studies have demonstrated strong correlations between serum levels of SOST and factors related to fracture risk, bone mineral density, age and sex [298]. Therapeutically, the anti-sclerostin monoclonal antibody, romosozumab, now in phase III clinical trials, has been shown quite effective in lowering the incidence of vertebral and nonvertebral fractures while increasing bone density and suppressing resorptive activity [299]. This SOST antibody is now viewed as a promising new agent for treatment of a range of common conditions which involve diseases affecting both the axial and appendicular skeletal systems.

8.12 Albumin

Albumin belongs to a family of non-glycosylated, water-soluble globular proteins present in the serum and plasma of blood. It is synthesized in the liver and is the most abundant of the circulating plasma proteins [300]. Human serum albumin (HSA) has marked affinity to many ion and molecular species, including calcium and other metal ions, fatty acids, amino acids, metabolites and certain drugs [301]. It is found in both bone tissue fluid and mineralized bone matrices [302]. The small size of HSA permits its penetration into hydrated collagen fibrils [60], where it can interact with calcium ions and newly forming mineral. Several papers demonstrate that HSA inhibits apatite crystal growth [303, 304]. These studies, however, do not provide a precise mechanism of crystal interaction and inhibition. Among the possibilities are albumin-mediated aggregation of crystals, effects of albumin surface charge or conformational state, or recognition and binding between specific HSA functional groups and molecular motifs of apatite crystal faces [303, 305], While HSA and fetuin-A (see Chap. 9) are somewhat complementary in their character of inhibiting mineral ions or small mineral crystals in the circulation, HSA is distinct from fetuin-A in their relative abundance during vertebrate development, mode of action and sensitivity to micromolar ion concentrations [306, 307].

8.13 Perspective: Role of Non-collagenous Proteins and Other Matrix Molecules in Vertebrate Mineralization

The several non-collagenous proteins and other molecules whose details of structure, function, and other properties or characteristics are presented here by no means constitute an exhaustive list of the family members of mineralized tissues. Those discussed above are considered the more abundant, the more critical for promoting, inhibiting or maintaining the state of mineralization in tissues, or the more well documented. There are many, many other less prominent or less well understood examples of individual molecules or family members whose effects on mineralization are yet incompletely described. For instance, this list might include additions such as lipids (in particular, phospholipids) since many of these molecules can be released only following degradation of the inorganic phase of vertebrate mineralized tissues, and it must be presumed that these lipids are components of the extracellular hard tissue matrix rather than cell-related molecules [308]. Other matrix molecules related to mineralization but are not considered here include, for instance, several collagen family members, statherins, bone acidic glycoprotein-75, vimentin, and numerous extracellular enzymes and growth factors (see Table 8.1).

The selectivity of the non-collagenous molecular characters chosen to be included more prominently in this chapter reflects the fundamental and critical concept that mineralization must be an extremely carefully regulated and controlled process in both the vertebrates and invertebrates as well as in other life forms, including plants and bacteria. The excess of normally circulating calcium and phosphate ions in vertebrate fluids, for instance, and of related ions in the interstitial milieu of other species, must be monitored constantly and accommodated with appropriate inhibitors, stabilizers, and promotors so as to avoid the unfortunate fate of Lot's wife, Edith, forever cast as a lifeless statue of mineral salt.

In this regard, perhaps two additional non-collagenous proteins may be mentioned here as important and related to the function of these molecules in the mineralization events of vertebrate tissues. Among the matrisome constituents, the cross-linking enzyme, transglutaminase (TG), catalyzes covalent bond formation and stabilizes individual SIBLINGS and leads to the formation of their higher molecular weight complexes. For example, on treatment with TG-2, both BSP and OPN form oligomers which enhance osteoblast recruitment and adhesion [309, 310]. SIBLINGS in turn regulate enzymes associated with matrix breakdown and turnover. Thus, BSP activates MMP-2, which has a wide range of substrates, including collagen, elastin, TGF-β, and other MMPs; OPN activates MMP-3, causing enzymatic digestion of collagens and cartilage proteoglycans as well as activation of procollagenase; and DMP1 up-regulates yet another metalloproteinase, MMP-9, which alters its own specific matrix protein substrates while also promoting cytokine and chemokine function [311]. Together these enzymatic activities are requirements for tissue remodeling and replacement as aspects of the normal events of vertebrate mineralization.

Aside from regulating activities of various enzymes, another important non-collagenous protein function is to interact with and promote fibrillogenesis of COLI, the major extracellular protein of bone and other vertebrate mineralized tissues. Aggregation, processing and assembly of collagen molecules, the stabilization of their higher ordered structures, and the formation of charged domains of collagen fibrils permit calcium and phosphate ion binding and apatite nucleation (for details, see Chap. 7). In this regard, SPARC/osteonectin is documented to interact with both collagen fibrils and apatite crystals [309]. If *SPARC* is deleted, then osteopenia occurs with a decline in bone mineral density and a loss of bone mechanical properties. Along with the marked changes in bone structure, there is a decrease in COLI content of mineralizing tissues, fibril diameter is smaller than normal and possible alterations occur in matrix-cell signaling mediated by discoidin domain receptors of collagen [184, 312, 313]. Likewise, if the level of expression of *SPARC* is increased, then collagen content levels are elevated [314]. Such studies point to a complex relationship between *SPARC* and collagen synthesis and homeostasis.

As mentioned early in this chapter, a primary concern, if not controversy, in the mineralization field for some time has been whether collagen, notably COLI or COLII, alone and in the absence of non-collagenous proteins and other molecules is sufficient to induce mineralization in vertebrate tissues. There have been papers supporting the concept [6–10] and those arguing against it [11]. Presently, there is limited resolution to the question. However, the relatively recent immunocytochemical work of Chen et al. [61] provides evidence that the non-collagenous proteins, BSP, OPN, and OCN, are localized to specific sites with respect to COLI and where each of these three particular molecules may mediate aspects of mineralization. In this instance, all three of these proteins are localized between collagen fibrils or on the fibril outer surfaces (interfibrillar localization), whereas OCN is also localized specifically within collagen fibrils in their hole zones (intrafibrillar localization) [61]. The roles of the proteins are not completely understood as noted elsewhere in this chapter: BSP is considered a possible critical factor in apatite crystal nucleation, OPN a possible factor in apatite crystal inhibition, and OCN a factor in nucleation, growth, and development of crystals. All three molecules have the potential to bind calcium ions and transport them in the extracellular fluid. In this respect, they may move and bind to the surfaces of collagen fibrils and thereby bring calcium to collagen surface sites for nucleation events in these interfibrillar locales.

The observation that OCN is located precisely in collagen hole zones [61], where other studies have demonstrated these intrafibrillar regions to be sites of apatite crystal nucleation [315, 316], is potentially of great significance. OCN, by virtue of its unique structure and presence of γ-carboxyglutamic acid residues, may serve to bind calcium ions in the hole zone milieu or to bind incipient apatite phases to stabilize them and prevent their further growth [61]. Additionally, OCN may direct the growth and development of initial apatite crystals as platelet-shaped constituents whose crystallographic *c*-axes are directed parallel to the collagen long axes with which they associate [61]. In any case, it is clear that collagen associates spatially and temporally with OCN as well as BSP and OPN, thereby enhancing aspects of apatite crystal nucleation, stabilization, growth, and development. Based on these results, collagen may be regarded as the principal template for mineral formation in vertebrate mineralized tissues.

Another uncertainty in vertebrate mineralization is the means by which the spaces between collagen fibrils (interfibrillar spaces) become fully mineralized. It may be conceived that apatite nucleation events occur within collagen fibrils in association with both OCN and collagen, and that similar events occur on collagen surfaces as a result of the binding of calcium and phosphate ions to collagen directly or as a consequence of calcium ion transport to the collagen surfaces by BSP, OPN, and OCN [10, 61]. From both these intrafibrillar and surface sites of collagen, apatite crystals may nucleate, and their growth and development may propagate throughout the intrafibrillar collagen regions as well as to some extent in the interfibrillar spaces. However, interfibrillar volume may be variable and large at different times of tissue maturation. In this case, there remains a question as to how such spaces become completely mineralized even though they are presumably complemented with an abundance (supersaturated levels) of calcium and phosphate ions and are bound and transported by various molecules, including BSP, OPN, and OCN. It may be that, as vertebrate tissues mature, their interfibrillar volumes ultimately become packed with collagen fibrils of increasing volume, leaving relatively little space between them. In this instance, apatite nucleation mediated at collagen surfaces would propagate crystal growth to be sufficiently and completely space-filling.

With respect again to mineralization, there is support for the view that other charged proteins like BSP, OPN, and OCN may serve to mediate apatite nucleation associated with collagen. DMP1, secreted by osteoblasts, odontoblasts, cementoblasts, ameloblasts, and osteocytes [111] and being a SIBLING protein family member like BSP and OPN, exhibits related functions to stabilize nascent calcium phosphate and affect apatite deposition. Even in a non-phosphorylated form, DMP1 provokes apatite formation [317]. In these examples, it is certainly quite likely that several non-collagenous proteins and other matrix molecules need to be present for marrying the inorganic phase with the organic matrix and thereby optimizing mineral formation in vertebrate hard tissues.

The presence and action of multiple non-collagenous molecules exerting their effects on vertebrate mineral may be understood in the context of proteins like OCN and OPN which coaggregate in the collagenous mineralized matrices to form "dilatational bands," binding neighboring mineral aggregates through calcium ion-mediated complexes [318, 319]. If these multiprotein complexes are not formed or are disturbed, the interactions between the organic matrix and the mineral phase may be decreased, a result which in turn decreases biomechanical properties of the tissue such as the overall strength of bone [320]. The requirement for multiple non-collagenous molecules to enhance mineralization may also be deduced from studies of null or targeted mutations, which suggest a high degree of redundancy among them. In other words, different genes generate varying individual non-collagenous proteins and other molecules with close or identical function necessary for enhancing and promoting certain of the events leading to proper mineral deposition in a tissue. Thus, deletion of an individual gene would cause a minimal change in skeletal phenotype because other encoded proteins fulfil the functional role of the missing protein. In this regard, studies in vitro indicate that the redundancy of function in vertebrate mineralized tissues is concerned with the regulation of collagen fibrillogenesis, the aggregation and stabilization of early mineral ions, and/or the nucleation, growth, and development of apatite from its precursor phases.

The possibility of redundant non-collagenous molecules, notably in the SIBLING or SLRP families, might explain particular results of gene knockout experiments with animal models in which the loss of a single such matrix constituent often leads to a very mild phenotype, whereas a double knockout causes a marked phenotype change. As an example mentioned previously in a related context, mutational studies show that the deletion of the SLRP, *BGN*, yields a weak phenotype [240], whereas double knockouts of *BGN* and *DCN* produce a phenotype in which there are serious skeletal deficits reflecting changes in collagen fibril structure and organization as well as in the character of mineral [242]. Most likely, the presence in vertebrate hard tissues of several non-collagenous molecules with overlapping functions provides a mechanism for oversight, control and tuning a highly complex system so that its final mineralization product is successfully and faithfully elaborated.

The matricellular proteins are also important in aspects of vertebrate mineralization. Bornstein described these non-collagenous proteins as "modular, extracellular proteins whose functions are achieved by binding to matrix proteins as well as to cell surface receptors, or to other molecules such as cytokines and proteases which interact, in turn, with the cell surface" [321]. The matricellular protein family includes TSPs, CCNs, periostin, RSPOs, SLRPs, and certain SIBLINGs. In mineralized tissues, these matricellular proteins bind and, in some cases, activate growth factors such as TGF-β, FGF, and VEGF [322, 323]. The role and significance of these factors in the regulation of bone, dentin, and cementum formation have been discussed earlier. Moreover, since they are often present in the matrices of such tissues in an inactive form, their activation is required so that they can respond to local or systemic signals and influence cell function. The numerous matricellular proteins are also responsive to microRNA and other epigenetic mechanisms (DNA methylation, histone acetylation/deacetylation) [324–326]. Aside from cytokine signaling, the matricellular proteins contain multiple RGD sequences which interact with cell integrin receptors; for example, OPN binds to and interacts with multiple cell surface receptors ($\alpha v\beta 1$, $\alpha v\beta 3$, $\alpha v\beta 5$, $\alpha 9\beta 1$, and CD44), signaling cell adhesion, survival, and migration [327]. Related directly to intracellular events which trigger mineralization, TSP changes endoplasmic reticulum function and responds to loss of calcium through STIM1 [328–330]. Accordingly, TSP and other matricellular proteins have both matrix and cell-modulating effects associated with mineral deposition in bone and other vertebrate hard tissues [328–330].

Non-collagenous proteins interact with both calcium and phosphate ions in numerous events of vertebrate mineral deposition, and their regulation of calcium has been detailed previously in this chapter. The control of phosphate ion levels during mineralization may be achieved through the mechanism of enzymatic cleavage of pyrophosphate or phosphate esters at sites of mineralization as mentioned (see Chap. 9). A second mechanism is designed to regulate systemic phosphate levels, which in turn influence the tissue fluid concentration of the anion. In bone, this systemic effect is mediated by the enzyme, PHEX, which cleaves ASARM from MEPE as well as other SIBLINGs and even the salivary protein, statherin. ASARM subsequently leads to kidney excretion of phosphate ions by promoting FGF23 secretion [28]. This charged peptide can also bind hydroxyapatite crystals to inhibit mineral growth and development [331]. In summary, the role of non-collagenous proteins and other matrix molecules in vertebrate tissue mineralization is multifaceted and comprehensive, ultimately directed toward exquisite and critical regulation, control, and coordination of proper extracellular matrix formation and its integration with mineral deposition.

References

1. Astbury WT, Bell FO. Some recent developments in the X-ray study of proteins and related structures. Cold Spring Harb Symp Quant Biol. 1938;6:109–21.
2. Fisher LW, Torchia DA, Fohr B, Young MF, Fedarko NS. Flexible structures of SIBLING proteins, bone sialoprotein, and osteopontin. Biochem Biophys Res Commun. 2001;280:460–5.
3. Volpin D, Veis A. Differences between CNBr peptides of soluble and insoluble bovine collagens. Biochem Biophys Res Commun. 1971;44:804–12.
4. Volpin D, Veis A. Cyanogen bromide peptides from insoluble skin and dentin bovine collagens. Biochemist. 1973;12:1452–64.
5. Stetler-Stevenson WG, Veis A. Bovine dentin phosphophoryn: calcium ion binding properties of a high molecular weight preparation. Calcif Tissue Int. 1987;40:97–102.
6. Glimcher MJ, Krane SM. A treatise on collagen. In: Ramachandran GN, Gould BS, editors. Biology of collagen. New York: Academic Press; 1968. p. 68–251.

7. Glimcher MJ. Bone: nature of the calcium phosphate crystals and cellular, structural, and physical chemical mechanisms in their formation. Rev Mineral Geochem. 2006;64:223–82.
8. Silver FH, Landis WJ. Deposition of apatite in mineralizing vertebrate extracellular matrices: a model of possible nucleation sites on type I collagen. Connect Tissue Res. 2011;52:242–54.
9. Wang Y, Azaïs T, Robin M, Vallée A, Catania C, Legriel P, et al. The predominant role of collagen in the nucleation, growth, structure and orientation of bone apatite. Nat Mater. 2012;11:724–33.
10. Landis WJ, Jacquet R. Association of calcium and phosphate ions with collagen in the mineralization of vertebrate tissues. Calcif Tissue Int. 2013;93:329–37.
11. Robey PG. Noncollagenous bone matrix proteins. In: Bilezikian J, Raisz L, Martin TJ, editors. Principles of bone biology. 3rd ed. London: Elsevier; 2008. p. 335–49.
12. George A, Veis A. Phosphorylated proteins and control over apatite nucleation, crystal growth, and inhibition. Chem Rev. 2008;108:4670–93.
13. Schreiweis MA, Butler JP, Kulkarni NH, Knierman MD, Higgs RE, Halladay DL, et al. A proteomic analysis of adult rat bone reveals the presence of cartilage/chondrocyte markers. J Cell Biochem. 2007;101:466–76.
14. Jágr M, Ergang P, Pataridis S, Kolrosová M, Bartoš M, Mikšík I. Proteomic analysis of dentin-enamel junction and adjacent protein-containing enamel matrix layer of healthy human molar teeth. Eur J Oral Sci. 2019;127:112–21.
15. Alam I, Padgett LR, Ichikawa S, Alkhouli M, Koller DL, Lai D, et al. SIBLING family genes and bone mineral density: association and allele-specific expression in humans. Bone. 2014;64:166–72.
16. MacDougall M, Dong J, Acevedo AC. Molecular basis of human dentin diseases. Am J Med Genet A. 2006;140:2536–46.
17. Yoshizaki K, Yamada Y. Gene evolution and functions of extracellular matrix proteins in teeth. Orthod Waves. 2013;72:1–10.
18. Sire JY, Delgado S, Fromentin D, Girondot M. Amelogenin: lessons from evolution. Arch Oral Biol. 2005;50:205–12.
19. Kawasaki K, Buchanan AV, Weiss KM. Gene duplication and the evolution of vertebrate skeletal mineralization. Cells Tissues Organs. 2007;186:7–24.
20. Kawasaki K, Weiss KM. SCPP gene evolution and the dental mineralization continuum. J Dent Res. 2008;87:520–31.
21. Venkatesh B, Lee AP, Ravi V, Maurya AK, Lian MM, Swann JB, et al. Elephant shark genome provides unique insights into gnathostome evolution. Nature. 2014;505:174–9.
22. Salonen J, Domenicucci C, Goldberg HA, Sodek J. Immunohistochemical localization of SPARC (osteonectin) and denatured collagen and their relationship to remodelling in rat dental tissues. Arch Oral Biol. 1990;35:337–46.
23. Martinek N, Shahab J, Sodek J, Ringuette M. Is SPARC an evolutionarily conserved collagen chaperone? J Dent Res. 2007;86:296–305.
24. Pottgiesser J, Maurer P, Mayer U, Nischt R, Mann K, Timpl R, et al. Changes in calcium and collagen IV binding caused by mutations in the EF hand and other domains of extracellular matrix protein BM-40 (SPARC, osteonectin). J Mol Biol. 1994;238:563–74.
25. Boskey AL. Biomineralization: an overview. Connect Tissue Res. 2003;44:5–9.
26. Huang B, Sun Y, Maciejewska I, Qin D, Peng T, McIntyre B, et al. Distribution of SIBLING proteins in the organic and inorganic phases of rat dentin and bone. Eur J Oral Sci. 2008;116:104–12.
27. Hao C, Cui Y, Owen S, Li W, Cheng S, Jiang WG. Human osteopontin: potential clinical applications in cancer (review). Int J Mol Med. 2017;39:1327–37.
28. Rowe PS. Regulation of bone-renal mineral and energy metabolism: the PHEX, FGF23, DMP1, MEPE ASARM pathway. Crit Rev Eukaryot Gene Expr. 2012;22:61–86.
29. Bacchetta J, Bardet C, Prié D. Physiology of FGF23 and overview of genetic diseases associated with renal phosphate wasting. Metabolism. 2020;103:153865.
30. Bitzan M, Goodyer PR. Hypophosphatemic rickets. Pediatr Clin N Am. 2019;66:179–207.

31. Staines KA, MacRae VE, Farquharson C. The importance of the SIBLING family of proteins on skeletal mineralisation and bone remodelling. J Endocrinol. 2012;214:241–55.
32. Wright PE, Dyson HJ. Intrinsically unstructured proteins: re-assessing the protein structure-function paradigm. J Mol Biol. 1999;293:321–31.
33. Kriwacki RW, Hengst L, Tennant L, Reed SI, Wright PE. Structural studies of p21Waf1/Cip1/Sdi1 in the free and Cdk2-bound state: conformational disorder mediates binding diversity. Proc Natl Acad Sci U S A. 1996;93:11504–9.
34. Jeffery CJ. Moonlighting proteins. Trends Biochem Sci. 1999;24:8–11.
35. Dunker AK, Brown CJ, Lawson JD, Iakoucheva LM, Obradović Z. Intrinsic disorder and protein function. Biochemistry. 2002;41:6573–82.
36. Uversky VN. Intrinsically disordered proteins and their "mysterious" (meta)physics. Front Phys. 2019;7:10.
37. Gericke A, Qin C, Spevak L, Fujimoto Y, Butler WT, Sørensen ES, et al. Importance of phosphorylation for osteopontin regulation of biomineralization. Calcif Tissue Int. 2005;77:45–54.
38. Boskey AL, Christensen B, Taleb H, Sørensen ES. Post-translational modification of osteopontin: effects on in vitro hydroxyapatite formation and growth. Biochem Biophys Res Commun. 2012;419:333–8.
39. Deshpande AS, Fang PA, Zhang X, Jayaraman T, Sfeir C, Beniash E. Primary structure and phosphorylation of dentin matrix protein 1 (DMP1) and dentin phosphophoryn (DPP) uniquely determine their role in biomineralization. Biomacromolecules. 2011;12:2933–45.
40. Kalmar L, Homola D, Varga G, Tompa P. Structural disorder in proteins brings order to crystal growth in biomineralization. Bone. 2012;51:528–34.
41. Bianco P, Fisher LW, Young MF, Termine JD, Robey PG. Expression of bone sialoprotein (BSP) in developing human tissues. Calcif Tissue Int. 1991;49:421–6.
42. Chen J, Shapiro HS, Wrana JL, Reimers S, Heersche JNM, Sodek J. Localization of bone sialoprotein (BSP) expression to sites of mineralized tissue formation in fetal rat tissues by in situ hybridization. Matrix. 1991;11:133–43.
43. Ganss B, Kim RH, Sodek J. Bone sialoprotein. Crit Rev Oral Biol Med. 1999;10:79–98.
44. Moses KD, Butler WT, Qin C. Immunohistochemical study of small integrin-binding ligand, N-linked glycoproteins in reactionary dentin of rat molars at different ages. Eur J Oral Sci. 2006;114:216–22.
45. Chen L, Childs RD, Landis WJ. Correlations between gene expression and mineralization in the avian leg tendon. Bone. 2019;121:42–59.
46. Qin C, Baba O, Butler WT. Post-translational modifications of SIBLING proteins and their roles in osteogenesis and dentinogenesis. Crit Rev Oral Biol Med. 2004;15:126–36.
47. Ohnishi T, Nakamura O, Ozawa M, Arakaki N, Muramatsu T, Daikuhara Y. Molecular cloning and sequence analysis of cDNA for a 59 kD bone sialoprotein of the rat: demonstration that it is a counterpart of human alpha 2-HS glycoprotein and bovine fetuin. J Bone Miner Res. 1993;8:367–77.
48. Shapiro HS, Chen J, Wrana JL, Zhang Q, Blum M, Sodek J. Characterization of porcine bone sialoprotein: primary structure and cellular expression. Matrix. 1993;13:431–40.
49. Bouleftour W, Juignet L, Bouet G, Granito RN, Vanden-Bossche A, Laroche N, et al. The role of the SIBLING, bone sialoprotein in skeletal biology - contribution of mouse experimental genetics. Matrix Biol. 2016;52-54:60–77.
50. Wuttke M, Müller S, Nitsche DP, Paulsson M, Hanisch FG, Maurer P. Structural characterization of human recombinant and bone-derived bone sialoprotein. Functional implications for cell attachment and hydroxyapatite binding. J Biol Chem. 2001;276:36839–48.
51. Foster BL, Soenjaya Y, Nociti FH Jr, Holm E, Zerfas PM, Wimer HF, et al. Deficiency in acellular cementum and periodontal attachment in bsp null mice. J Dent Res. 2013;92:166–72.
52. Bouleftour W, Boudiffa M, Wade-Gueye NM, Bouët G, Cardelli M, Laroche N, et al. Skeletal development of mice lacking bone sialoprotein (BSP)--impairment of long bone growth and progressive establishment of high trabecular bone mass. PLoS One. 2014;9:e95144.
53. Malaval L, Monfoulet L, Fabre T, Pothuaud L, Bareille R, Miraux S, et al. Absence of bone sialoprotein (BSP) impairs cortical defect repair in mouse long bone. Bone. 2009;45:853–61.

54. Holm E, Aubin JE, Hunter GK, Beier F, Goldberg HA. Loss of bone sialoprotein leads to impaired endochondral bone development and mineralization. Bone. 2015;71:145–54.
55. Cooper LF, Yliheikkilä PK, Felton DA, Whitson SW. Spatiotemporal assessment of fetal bovine osteoblast culture differentiation indicates a role for BSP in promoting differentiation. J Bone Miner Res. 1998;13:620–32.
56. Gordon JA, Tye CE, Sampaio AV, Underhill TM, Hunter GK, Goldberg HA. Bone sialoprotein expression enhances osteoblast differentiation and matrix mineralization in vitro. Bone. 2007;41:462–73.
57. Goldberg HA, Warner KJ, Li MC, Hunter GK. Binding of bone sialoprotein, osteopontin and synthetic polypeptides to hydroxyapatite. Connect Tissue Res. 2001;42:25–37.
58. Xu L, Zhang Z, Sun X, Wang J, Xu W, Shi L, et al. Glycosylation status of bone sialoprotein and its role in mineralization. Exp Cell Res. 2017;360:413–20.
59. Tye CE, Hunter GK, Goldberg HA. Identification of the type I collagen-binding domain of bone sialoprotein and characterization of the mechanism of interaction. J Biol Chem. 2005;280:13487–92.
60. Toroian D, Lim JE, Price PA. The size exclusion characteristics of type I collagen: implications for the role of noncollagenous bone constituents in mineralization. J Biol Chem. 2007;282:22437–47.
61. Chen L, Jacquet R, Lowder E, Landis WJ. Refinement of collagen-mineral interaction: a possible role for osteocalcin in apatite crystal nucleation, growth and development. Bone. 2015;71:7–16.
62. Dey A, Bomans PH, Müller FA, Will J, Frederik PM, de With G, et al. The role of prenucleation clusters in surface-induced calcium phosphate crystallization. Nat Mater. 2010;9:1010–4.
63. Nomura S, Wills AJ, Edwards DR, Heath JK, Hogan BL. Developmental expression of 2ar (osteopontin) and SPARC (osteonectin) RNA as revealed by in situ hybridization. J Cell Biol. 1988;106:441–50.
64. Butler WT. The nature and significance of osteopontin. Connect Tissue Res. 1989;23:123–36.
65. Sroga GE, Vashishth D. Phosphorylation of extracellular bone matrix proteins and its contribution to bone fragility. J Bone Miner Res. 2018;33:2214–29.
66. Aubin JE, Liu F, Malaval L, Gupta AK. Osteoblast and chondroblast differentiation. Bone. 1995;17:S77–83.
67. Zaka R, Williams CJ. Role of the progressive ankylosis gene in cartilage mineralization. Curr Opin Rheumatol. 2006;18:181–6.
68. Sodek J, Ganss B, McKee MD. Osteopontin. Crit Rev Oral Biol Med. 2000;11:279–303.
69. Sun Y, Ma S, Zhou J, Yamoah AK, Feng JQ, Hinton RJ, et al. Distribution of small integrin-binding ligand, N-linked glycoproteins (SIBLING) in the articular cartilage of the rat femoral head. J Histochem Cytochem. 2010;58:1033–43.
70. McKee MD, Nanci A. Osteopontin and the bone remodeling sequence. Colloidal-gold immunocytochemistry of an interfacial extracellular matrix protein. Ann N Y Acad Sci. 1995;760:177–89.
71. McKee MD, Nanci A. Osteopontin: an interfacial extracellular matrix protein in mineralized tissues. Connect Tissue Res. 1996;35:197–205.
72. Kläning E, Christensen B, Sørensen ES, Vorup-Jensen T, Jensen JK. Osteopontin binds multiple calcium ions with high affinity and independently of phosphorylation status. Bone. 2014;66:90–5.
73. Briones-Orta MA, Avendaño-Vázquez SE, Aparicio-Bautista DI, Coombes JD, Weber GF, Syn W-K. Osteopontin splice variants and polymorphisms in cancer progression and prognosis. Biochim Biophys Acta Rev Cancer. 2017;1868:93–108.
74. Yokosaki Y, Matsuura N, Sasaki T, Murakami I, Schneider H, Higashiyama S, et al. The integrin alpha(9)beta(1) binds to a novel recognition sequence (SVVYGLR) in the thrombin-cleaved amino-terminal fragment of osteopontin. J Biol Chem. 1999;274:36328–34.

75. Bayless KJ, Davis GE. Identification of dual alpha 4beta1 integrin binding sites within a 38 amino acid domain in the N-terminal thrombin fragment of human osteopontin. J Biol Chem. 2001;276:13483–9.
76. Senbanjo LT, Chellaiah MA. CD44: a multifunctional cell surface adhesion receptor is a regulator of progression and metastasis of cancer cells. Front Cell Dev Biol. 2017;5:18.
77. Keykhosravani M, Doherty-Kirby A, Zhang C, Brewer D, Goldberg HA, Hunter GK, et al. Comprehensive identification of post-translational modifications of rat bone osteopontin by mass spectrometry. Biochemist. 2005;44:6990–7003.
78. Tagliabracci VS, Engel JL, Wen J, Wiley SE, Worby CA, Kinch LN, et al. Secreted kinase phosphorylates extracellular proteins that regulate biomineralization. Science. 2012;336:1150–3.
79. Tagliabracci VS, Wiley SE, Guo X, Kinch LN, Durrant E, Wen J, et al. A single kinase generates the majority of the secreted phosphoproteome. Cell. 2015;161:1619–32.
80. Fradin M, Stoetzel C, Muller J, Koob M, Christmann D, Debry C, et al. Osteosclerotic bone dysplasia in siblings with a Fam20C mutation. Clin Genet. 2011;80:177–83.
81. Evan AP, Coe FL, Rittling SR, Bledsoe SM, Shao Y, Lingeman JE, et al. Apatite plaque particles in inner medulla of kidneys of calcium oxalate stone formers: osteopontin localization. Kidney Int. 2005;68:145–54.
82. Chien YC, Mansouri A, Jiang W, Khan SR, Gray JJ, McKee MD. Modulation of calcium oxalate dihydrate growth by phosphorylated osteopontin peptides. J Struct Biol. 2018;204:131–44.
83. De Yoreo JJ, Qiu SR, Hoyer JR. Molecular modulation of calcium oxalate crystallization. Am J Physiol Renal Physiol. 2006;291:F1123–31.
84. O'Young J, Chirico S, Al Tarhuni N, Grohe B, Karttunen M, Goldberg HA, et al. Phosphorylation of osteopontin peptides mediates adsorption to and incorporation into calcium oxalate crystals. Cells Tissues Organs. 2009;189:51–5.
85. Aggarwal KP, Narula S, Kakkar M, Tandon C. Nephrolithiasis: molecular mechanism of renal stone formation and the critical role played by modulators. Biomed Res Int. 2013;2013:292953.
86. Rittling SR, Matsumoto HN, McKee MD, Nanci A, An XR, Novick KE, et al. Mice lacking osteopontin show normal development and bone structure but display altered osteoclast formation in vitro. J Bone Miner Res. 1998;13:1101–11.
87. Boskey AL, Spevak L, Paschalis E, Doty SB, McKee MD. Osteopontin deficiency increases mineral content and mineral crystallinity in mouse bone. Calcif Tissue Int. 2002;71:145–54.
88. Matsui Y, Iwasaki N, Kon S, Takahashi D, Morimoto J, Matsui Y, et al. Accelerated development of aging-associated and instability-induced osteoarthritis in osteopontin-deficient mice. Arthritis Rheum. 2009;60:2362–71.
89. Hunter GK, Kyle CL, Goldberg HA. Modulation of crystal formation by bone phosphoproteins: structural specificity of the osteopontin-mediated inhibition of hydroxyapatite formation. Biochem J. 1994;300:723–8.
90. Addison WN, Masica DL, Gray JJ, McKee MD. Phosphorylation-dependent inhibition of mineralization by osteopontin ASARM peptides is regulated by PHEX cleavage. J Bone Miner Res. 2010;25:695–705.
91. McKee MD, Farach-Carson MC, Butler WT, Hauschka PV, Nanci A. Ultrastructural immunolocalization of noncollagenous (osteopontin and osteocalcin) and plasma (albumin and alpha 2HS-glycoprotein) proteins in rat bone. J Bone Miner Res. 1993;8:485–96.
92. Lee JY, Choo JE, Choi YS, Park JB, Min DS, Lee SJ, et al. Assembly of collagen-binding peptide with collagen as a bioactive scaffold for osteogenesis in vitro and in vivo. Biomaterials. 2007;28:4257–67.
93. Rodriguez DE, Thula-Mata T, Toro EJ, Yeh YW, Holt C, Holliday LS, et al. Multifunctional role of osteopontin in directing intrafibrillar mineralization of collagen and activation of osteoclasts. Acta Biomater. 2014;10:494–507.
94. Butti R, Ghosh P, Totakura KVS, Venkata RNN, Nimma R, Kundu GC. Role of osteopontin in tumor microenvironment: a new paradigm in cancer therapy. In: Gandhi V, Mehta K,

Grover R, Pathak S, Aggarwal B, editors. Multi-targeted approach to treatment of cancer. Adis: Cham; 2015. p. 113–25.

95. Reinholt FP, Hultenby K, Oldberg A, Heinegård D. Osteopontin--a possible anchor of osteoclasts to bone. Proc Natl Acad Sci U S A. 1990;87:4473–5.
96. Schwab EH, Halbig M, Glenske K, Wagner A-S, Wenisch S, Cavalcanti-Adam EA. Distinct effects of RGD-glycoproteins on integrin-mediated adhesion and osteogenic differentiation of human mesenchymal stem cells. Int J Med Sci. 2013;10:1846–59.
97. Christensen B, Zachariae ED, Scavenius C, Kløverpris S, Oxvig C, Petersen SV, et al. Transglutaminase 2-catalyzed intramolecular cross-linking of osteopontin. Biochemist. 2016;55:294–303.
98. Hoac B, Nelea V, Jiang W, Kaartinen MT, McKee MD. Mineralization-inhibiting effects of transglutaminase-crosslinked polymeric osteopontin. Bone. 2017;101:37–48.
99. Higashikawa F, Eboshida A, Yokosaki Y. Enhanced biological activity of polymeric osteopontin. FEBS Lett. 2007;581:2697–701.
100. Chau DY, Collighan RJ, Verderio EA, Addy VL, Griffin M. The cellular response to transglutaminase-cross-linked collagen. Biomaterials. 2005;26:6518–29.
101. Nampei A, Hashimoto J, Hayashida K, Tsuboi H, Shi K, Tsuji I, et al. Matrix extracellular phosphoglycoprotein (MEPE) is highly expressed in osteocytes in human bone. J Bone Miner Metab. 2004;22:176–84.
102. Gluhak-Heinrich J, Pavlin D, Yang W, MacDougall M, Harris SE. MEPE expression in osteocytes during orthodontic tooth movement. Arch Oral Biol. 2007;52:684–90.
103. Wei X, Ling J, Wu L, Liu L, Xiao Y. Expression of mineralization markers in dental pulp cells. J Endod. 2007;33:703–8.
104. Rowe PS, de Zoysa PA, Dong R, Wang HR, White KE, Econs MJ, et al. MEPE, a new gene expressed in bone marrow and tumors causing osteomalacia. Genomics. 2000;67:54–68.
105. Petersen DN, Tkalcevic GT, Mansolf AL, Rivera-Gonzalez R, Brown TA. Identification of osteoblast/osteocyte factor 45 (OF45), a bone-specific cDNA encoding an RGD-containing protein that is highly expressed in osteoblasts and osteocytes. J Biol Chem. 2000;275:36172–80.
106. Gowen LC, Petersen DN, Mansolf AL, Qi H, Stock JL, Tkalcevic GT, et al. Targeted disruption of the osteoblast/osteocyte factor 45 gene (OF45) results in increased bone formation and bone mass. J Biol Chem. 2003;278:1998–2007.
107. David V, Martin A, Hedge AM, Rowe PS. Matrix extracellular phosphoglycoprotein (MEPE) is a new bone renal hormone and vascularization modulator. Endocrinology. 2009;150:4012–23.
108. George A, Sabsay B, Simonian PA, Veis A. Characterization of a novel dentin matrix acidic phosphoprotein. Implications for induction of biomineralization. J Biol Chem. 1993;268:12624–30.
109. D'Souza RN, Cavender A, Sunavala G, Alvarez J, Ohshima T, Kulkarni AB, et al. Gene expression patterns of murine dentin matrix protein 1 (Dmp1) and dentin sialophosphoprotein (DSPP) suggest distinct developmental functions in vivo. J Bone Miner Res. 1997;12:2040–9.
110. MacDougall M, Gu TT, Luan X, Simmons D, Chen J. Identification of a novel isoform of mouse dentin matrix protein 1: spatial expression in mineralized tissues. J Bone Miner Res. 1998;13:422–31.
111. Toyosawa S, Shintani S, Fujiwara T, Ooshima T, Sato A, Ijuhin N, et al. Dentin matrix protein 1 is predominantly expressed in chicken and rat osteocytes but not in osteoblasts. J Bone Miner Res. 2001;16:2017–26.
112. Feng JQ, Zhang J, Dallas SL, Lu Y, Chen S, Tan X, et al. Dentin matrix protein 1, a target molecule for Cbfa1 in bone, is a unique bone marker gene. J Bone Miner Res. 2002;17:1822–31.
113. Feng JQ, Huang H, Lu Y, Ye L, Xie Y, Tsutsui TW, et al. The dentin matrix protein 1 (Dmp1) is specifically expressed in mineralized, but not soft, tissues during development. J Dent Res. 2003;82:776–80.

114. Terasawa M, Shimokawa R, Terashima T, Ohya K, Takagi Y, Shimokawa H. Expression of dentin matrix protein 1 (DMP1) in nonmineralized tissues. J Bone Miner Metab. 2004;22:430–8.
115. Feng JQ, Ward LM, Liu S, Lu Y, Xie Y, Yuan B, et al. Loss of DMP1 causes rickets and osteomalacia and identifies a role for osteocytes in mineral metabolism. Nat Genet. 2006;38:1310–5.
116. Qin C, D'Souza R, Feng JQ. Dentin matrix protein 1 (DMP1): new and important roles for biomineralization and phosphate homeostasis. J Dent Res. 2007;86:1134–41.
117. Hao J, Zou B, Narayanan K, George A. Differential expression patterns of the dentin matrix proteins during mineralized tissue formation. Bone. 2004;34:921–32.
118. Wu H, Teng PN, Jayaraman T, Onishi S, Li J, Bannon L, et al. Dentin matrix protein 1 (DMP1) signals via cell surface integrin. J Biol Chem. 2011;286:29462–9.
119. Eapen A, Sundivakkam P, Song Y, Ravindran S, Ramachandran A, Tiruppathi C, et al. Calcium-mediated stress kinase activation by DMP1 promotes osteoblast differentiation. J Biol Chem. 2010;285:36339–51.
120. Martin A, David V, Laurence JS, Schwarz PM, Lafer EM, Hedge AM, et al. Degradation of MEPE, DMP1, and release of SIBLING ASARM-peptides (minhibins): ASARM-peptide(s) are directly responsible for defective mineralization in HYP. Endocrinology. 2008;149:1757–72.
121. He G, Dahl T, Veis A, George A. Nucleation of apatite crystals in vitro by self-assembled dentin matrix protein 1. Nat Mater. 2003;2:552–8.
122. He G, Dahl T, Veis A, George A. Dentin matrix protein 1 initiates hydroxyapatite formation in vitro. Connect Tissue Res. 2003;44:240–5.
123. He G, George A. Dentin matrix protein 1 immobilized on type I collagen fibrils facilitates apatite deposition in vitro. J Biol Chem. 2004;279:11649–56.
124. He G, Gajjeraman S, Schultz D, Cookson D, Qin C, Butler WT, et al. Spatially and temporally controlled biomineralization is facilitated by interaction between self-assembled dentin matrix protein 1 and calcium phosphate nuclei in solution. Biochemist. 2005;44:16140–8.
125. Gajjeraman S, Narayanan K, Hao J, Qin C, George A. Matrix macromolecules in hard tissues control the nucleation and hierarchical assembly of hydroxyapatite. J Biol Chem. 2007;282:1193–204.
126. Ling Y, Rios HF, Myers ER, Lu Y, Feng JQ, Boskey AL. DMP1 depletion decreases bone mineralization in vivo: an FTIR imaging analysis. J Bone Miner Res. 2005;20:2169–77.
127. Ye L, MacDougall M, Zhang S, Xie Y, Zhang J, Li Z, et al. Deletion of dentin matrix protein-1 leads to a partial failure of maturation of predentin into dentin, hypomineralization, and expanded cavities of pulp and root canal during postnatal tooth development. J Biol Chem. 2004;279:19141–8.
128. Ye L, Mishina Y, Chen D, Huang H, Dallas SL, Dallas MR, et al. Dmp1-deficient mice display severe defects in cartilage formation responsible for a chondrodysplasia-like phenotype. J Biol Chem. 2005;280:6197–203.
129. Gericke A, Qin C, Sun Y, Redfern R, Redfern D, Fujimoto Y, et al. Different forms of DMP1 play distinct roles in mineralization. J Dent Res. 2010;89:355–9.
130. Qin C, Brunn JC, Cook RG, Orkiszewski RS, Malone JP, Veis A, et al. Evidence for the proteolytic processing of dentin matrix protein 1. Identification and characterization of processed fragments and cleavage sites. J Biol Chem. 2003;278:34700–8.
131. Huang B, Maciejewska I, Sun Y, Peng T, Qin D, Lu Y, et al. Identification of full-length dentin matrix protein 1 in dentin and bone. Calcif Tissue Int. 2008;82:401–10.
132. Maciejewska I, Qin D, Huang B, Sun Y, Mues G, Svoboda K, et al. Distinct compartmentalization of dentin matrix protein 1 fragments in mineralized tissues and cells. Cells Tissues Organs. 2009;189:186–91.
133. Maciejewska I, Cowan C, Svoboda K, Butler WT, D'Souza R, Qin C. The NH2-terminal and COOH-terminal fragments of dentin matrix protein 1 (DMP1) localize differently in the compartments of dentin and growth plate of bone. J Histochem Cytochem. 2008;57:155–66.
134. Lorenz-Depiereux B, Benet-Pages A, Eckstein G, Tenenbaum-Rakover Y, Wagenstaller J, Tiosano D, et al. Hereditary hypophosphatemic rickets with hypercalciuria is caused

by mutations in the sodium-phosphate cotransporter gene SLC34A3. Am J Hum Genet. 2006;78:193–201.
135. MacDougall M, Simmons D, Luan X, Nydegger J, Feng J, Gu TT. Dentin phosphoprotein and dentin sialoprotein are cleavage products expressed from a single transcript coded by a gene on human chromosome 4. Dentin phosphoprotein DNA sequence determination. J Biol Chem. 1997;272:835–42.
136. Qin C, Brunn JC, Cadena E, Ridall A, Tsujigiwa H, Nagatsuka H, et al. The expression of dentin sialophosphoprotein gene in bone. J Dent Res. 2002;81:392–4.
137. Fisher LW, Jain A, Tayback M, Fedarko NS. Small integrin binding ligand N-linked glycoprotein gene family expression in different cancers. Clin Cancer Res. 2004;10:8501–11.
138. Tsuchiya S, Simmer JP, Hu JC, Richardson AS, Yamakoshi F, Yamakoshi Y. Astacin proteases cleave dentin sialophosphoprotein (Dspp) to generate dentin phosphoprotein (Dpp). J Bone Miner Res. 2011;26:220–8.
139. Yuan G, Chen L, Feng J, Yang G, Ni Q, Xu X, et al. Dentin sialoprotein is a novel substrate of matrix metalloproteinase 9 in vitro and in vivo. Sci Rep. 2017;7:42449.
140. Yamakoshi Y, Hu JC, Fukae M, Zhang H, Simmer JP. Dentin glycoprotein: the protein in the middle of the dentin sialophosphoprotein chimera. J Biol Chem. 2005;280:17472–9.
141. Yamakoshi Y, Simmer JP. Structural features, processing mechanism and gene splice variants of dentin sialophosphoprotein. Jpn Dent Sci Rev. 2018;54:183–96.
142. Prasad M, Butler WT, Qin C. Dentin sialophosphoprotein in biomineralization. Connect Tissue Res. 2010;51:404–17.
143. Qin C, Brunn JC, Baba O, Wygant JN, McIntyre BW, Butler WT. Dentin sialoprotein isoforms: detection and characterization of a high molecular weight dentin sialoprotein. Eur J Oral Sci. 2003;111:235–42.
144. Yamakoshi Y, Hu JC, Fukae M, Iwata T, Kim JW, Zhang H, et al. Porcine dentin sialoprotein is a proteoglycan with glycosaminoglycan chains containing chondroitin 6-sulfate. J Biol Chem. 2005;280:1552–60.
145. Sreenath T, Thyagarajan T, Hall B, Longenecker G, D'Souza R, Hong S, et al. Dentin sialophosphoprotein knockout mouse teeth display widened predentin zone and develop defective dentin mineralization similar to human dentinogenesis imperfecta type III. J Biol Chem. 2003;278:24874–80.
146. Kim JW, Hu JC, Lee JI, Moon SK, Kim YJ, Jang KT, et al. Mutational hot spot in the DSPP gene causing dentinogenesis imperfecta type II. Hum Genet. 2005;116:186–91.
147. Suzuki S, Sreenath T, Haruyama N, Honeycutt C, Terse A, Cho A, et al. Dentin sialoprotein and dentin phosphoprotein have distinct roles in dentin mineralization. Matrix Biol. 2009;28:221–9.
148. Ritchie H, Wang L. Sequence determination of an extremely acidic rat dentin phosphoprotein. J Biol Chem. 1996;271:21695–8.
149. Gu K, Chang S, Ritchie HH, Clarkson BH, Rutherford RB. Molecular cloning of a human dentin sialophosphoprotein gene. Eur J Oral Sci. 2000;108:35–42.
150. Boskey A, Spevak L, Tan M, Doty S, Butler W. Dentin sialoprotein (DSP) has limited effects on in vitro apatite formation and growth. Calcif Tissue Int. 2000;67:472–8.
151. Ritchie HH, Hou H, Veis A, Butler WT. Cloning and sequence determination of rat dentin sialoprotein, a novel dentin protein. J Biol Chem. 1994;269:3698–702.
152. Schauer R. Sialic acids as regulators of molecular and cellular interactions. Curr Opin Struct Biol. 2009;19:507–14.
153. Iozzo RV. Matrix proteoglycans: from molecular design to cellular function. Annu Rev Biochem. 1998;67:609–52.
154. Dechichi P, Biffi JCG, Moura CCG, de Ameida AW. A model of the early mineralization process of mantle dentin. Micron. 2007;38:486–91.
155. Veis A, Perry A. The phosphoprotein of the dentin matrix. Biochemist. 1967;6:2409–16.
156. Dickson IR, Dimuzio MT, Volpin D, Ananthanarayanan S, Veis A. The extraction of phosphoproteins from bovine dentin. Calcif Tissue Int. 1975;19:51–61.

157. MacDougall M, Zeichner-David M, Slavkin HC. Production and characterization of antibodies against murine dentine phosphoprotein. Biochem J. 1985;232:493–500.
158. Takagi Y, Fujisawa R, Sasaki S. Identification of dentin phosphophoryn localization by histochemical stainings. Connect Tissue Res. 1986;14:279–92.
159. Rabie AM, Veis A. An immunocytochemical study of the routes of secretion of collagen and phosphophoryn from odontoblasts into dentin. Connect Tissue Res. 1995;31:197–209.
160. Hao J, He G, Narayanan K, Zou B, Lin L, Muni T, et al. Identification of differentially expressed cDNA transcripts from a rat odontoblast cell line. Bone. 2005;37:578–88.
161. Alvares K, Kanwar YS, Veis A. Expression and potential role of dentin phosphophoryn (DPP) in mouse embryonic tissues involved in epithelial-mesenchymal interactions and branching morphogenesis. Dev Dyn. 2006;235:2980–90.
162. Chang SR, Chiego D, Clarkson BH. Characterization and identification of a human dentin phosphophoryn. Calcif Tissue Int. 1996;59:149–53.
163. Dahlin S, Ångström J, Linde A. Dentin phosphoprotein sequence motifs and molecular modeling: conformational adaptations to mineral crystals. Eur J Oral Sci. 1998;106:239–48.
164. Ritchie HH, Wang L-H, Knudtson K. A novel rat 523 amino acid phosphophoryn: nucleotide sequence and genomic organization. BBA-Gen Subjects. 2001;1520:212–22.
165. George A, Bannon L, Sabsay B, Dillon JW, Malone J, Veis A, et al. The carboxyl-terminal domain of phosphophoryn contains unique extended triplet amino acid repeat sequences forming ordered carboxyl-phosphate interaction ridges that may be essential in the biomineralization process. J Biol Chem. 1996;271:32869–73.
166. Butler WT, Bhown M, DiMuzio MT, Cothran WC, Linde A. Multiple forms of rat dentin phosphoproteins. Arch Biochem Biophys. 1983;225:178–86.
167. Lee SL, Veis A, Glonek T. Dentin phosphoprotein: An extracellular calcium-binding protein. Biochemist. 1977;16:2971–9.
168. Traub W, Jodaikin A, Arad T, Veis A, Sabsay B. Dentin phosphophoryn binding to collagen fibrils. Matrix. 1992;12:197–201.
169. Dahl T, Sabsay B, Veis A. Type I collagen-phosphophoryn interactions: specificity of the monomer-monomer binding. J Struct Biol. 1998;123:162–8.
170. Huq NL, Loganathan A, Cross KJ, Chen Y-Y, Johnson NI, Willetts M, et al. Association of bovine dentine phosphophoryn with collagen fragments. Arch Oral Biol. 2005;50:807–19.
171. Stetler-Stevenson WG, Veis A. Type I collagen shows a specific binding affinity for bovine dentin phosphophoryn. Calcif Tissue Int. 1986;38:135–41.
172. George A, Hao J. Role of phosphophoryn in dentin mineralization. Cells Tissues Organs. 2005;181:232–40.
173. Wang S, Yang Y, Wang R, Kong X, Wang X. Mineralization of calcium phosphate controlled by biomimetic self-assembled peptide monolayers via surface electrostatic potentials. Bioact Mater. 2020;5:387–97.
174. Lee SL, Veis A. Cooperativity in calcium ion binding to repetitive, carboxylate-serylphosphate polypeptides and the relationship of this property to dentin mineralization. Int J Pept Protein Res. 1980;16:231–40.
175. Furedi-Milhofer H, Moradian-Oldak J, Weiner S, Veis A, Mintz KP, Addadf L. Interactions of matrix proteins from mineralized tissues with octacalcium phosphate. Connect Tissue Res. 1994;30:251–64.
176. He G, Ramachandran A, Dahl T, George S, Schultz D, Cookson D, et al. Phosphorylation of phosphophoryn is crucial for its function as a mediator of biomineralization. J Biol Chem. 2005;280:33109–13314.
177. Verdelis K, Lukashova L, Yamauchi M, Atsawasuwan P, Wright JT, Peterson MGE, et al. Changes in matrix phosphorylation during bovine dentin development. Eur J Oral Sci. 2007;115:296–302.
178. Bègue-Kirn C, Ruch JV, Ridall AL, Butler WT. Comparative analysis of mouse DSP and DPP expression in odontoblasts, preameloblasts, and experimentally induced odontoblast-like cells. Eur J Oral Sci. 1998;106:254–9.

179. Feng JQ, Luan X, Wallace J, Jing D, Ohshima T, Kulkarni AB, et al. Genomic organization, chromosomal mapping, and promoter analysis of the mouse dentin sialophosphoprotein (Dspp) gene, which codes for both dentin sialoprotein and dentin phosphoprotein. J Biol Chem. 1998;273:9457–64.
180. Gu K, Chang SR, Slaven MS, Clarkson BH, Rutherford RB, Ritchie HH. Human dentin phosphophoryn nucleotide and amino acid sequence. Eur J Oral Sci. 1998;106:1043–7.
181. Jadlowiec J, Koch H, Zhang X, Campbell PG, Seyedain M, Sfeir C. Phosphophoryn regulates the gene expression and differentiation of NIH3T3, MC3T3-E1, and human mesenchymal stem cells via the integrin/MAPK signaling pathway. J Biol Chem. 2004;279:53323–30.
182. Jadlowiec JA, Zhang X, Li J, Campbell PG, Sfeir C. Extracellular matrix-mediated signaling by dentin phosphophoryn involves activation of the Smad pathway independent of bone morphogenetic protein. J Biol Chem. 2006;281:5341–7.
183. Bornstein P. Matricellular proteins: An overview. J Cell Commun Signal. 2009;3:163–5.
184. Murphy-Ullrich JE, Sage EH. Revisiting the matricellular concept. Matrix Biol. 2014;37:1–14.
185. Hankenson KD, Sweetwyne MT, Shitaye H, Posey KL. Thrombospondins and novel TSR-containing proteins, r-spondins, regulate bone formation and remodeling. Curr Osteoporos Rep. 2010;8:68–76.
186. Yeger H, Perbal B. The CCN family of genes: a perspective on CCN biology and therapeutic potential. J Cell Commun Signal. 2007;1:159–64.
187. Resovi A, Pinessi D, Chiorino G, Taraboletti G. Current understanding of the thrombospondin-1 interactome. Matrix Biol. 2014;37:83–91.
188. Chen H, Sottile J, O'Rourke KM, Dixit VM, Mosher DF. Properties of recombinant mouse thrombospondin 2 expressed in Spodoptera cells. J Biol Chem. 1994;269:32226–32.
189. Chen H, Sottile J, Strickland DK, Mosher DF. Binding and degradation of thrombospondin-1 mediated through heparan sulphate proteoglycans and low-density-lipoprotein receptor-related protein: localization of the functional activity to the trimeric N-terminal heparin-binding region of thrombospondin-1. Biochem J. 1996;318:959–63.
190. Ueno A, Miwa Y, Miyoshi K, Horiguchi T, Inoue H, Ruspita I, et al. Constitutive expression of thrombospondin 1 in MC3T3-E1 osteoblastic cells inhibits mineralization. J Cell Physiol. 2006;209:322–32.
191. Alford AI, Terkhorn SP, Reddy AB, Hankenson KD. Thrombospondin-2 regulates matrix mineralization in MC3T3-E1 pre-osteoblasts. Bone. 2010;46:464–71.
192. Di Cesare PE, Fang C, Leslie MP, Tulli H, Perris R, Carlson CS. Expression of cartilage oligomeric matrix protein (COMP) by embryonic and adult osteoblasts. J Orthop Res. 2000;18:713–20.
193. Tseng S, Reddi AH, Di Cesare PE. Cartilage oligomeric matrix protein (COMP): a biomarker of arthritis. Biomark Insights. 2009;4:33–44.
194. Halász K, Kassner A, Mörgelin M, Heinegård D. COMP acts as a catalyst in collagen fibrillogenesis. J Biol Chem. 2007;282:31166–73.
195. Haleem-Smith H, Calderon R, Song Y, Tuan RS, Chen FH. Cartilage oligomeric matrix protein enhances matrix assembly during chondrogenesis of human mesenchymal stem cells. J Cell Biochem. 2012;113:1245–52.
196. Posey KL, Coustry F, Hecht JT. Cartilage oligomeric matrix protein: COMPopathies and beyond. Matrix Biol. 2018;71-72:161–73.
197. Bell SM, Schreiner CM, Wert SE, Mucenski ML, Scott WJ, Whitsett JA. R-spondin 2 is required for normal laryngeal-tracheal, lung and limb morphogenesis. Development. 2008;135:1049–58.
198. Jin YR, Yoon JK. The R-spondin family of proteins: emerging regulators of WNT signaling. Int J Biochem Cell Biol. 2012;44:2278–87.
199. Kim KA, Wagle M, Tran K, Zhan X, Dixon MA, Liu S, et al. R-Spondin family members regulate the Wnt pathway by a common mechanism. Mol Biol Cell. 2008;19:2588–96.
200. Wang D, Huang B, Zhang S, Yu X, Wu W, Wang X. Structural basis for R-spondin recognition by LGR4/5/6 receptors. Genes Dev. 2013;27:1339–44.

201. Ohkawara B, Glinka A, Niehrs C. Rspo3 binds syndecan 4 and induces Wnt/PCP signaling via clathrin-mediated endocytosis to promote morphogenesis. Dev Cell. 2011;20:303–14.
202. Chandran PL, Horkay F. Aggrecan, an unusual polyelectrolyte: review of solution behavior and physiological implications. Acta Biomater. 2012;8:3–12.
203. Zu W, Robey PG, Boskey AL. The biochemistry of bone. In: Marcus R, Feldman D, Nelson DA, Rosen CJ, editors. Osteoporosis. Burlington: Elsevier; 2007. p. 191–204.
204. Ng L, Grodzinsky AJ, Patwari P, Sandy J, Plaas A, Ortiz C. Individual cartilage aggrecan macromolecules and their constituent glycosaminoglycans visualized via atomic force microscopy. J Struct Biol. 2003;143:242–57.
205. Rojas FP, Batista MA, Lindburg CA, Dean D, Grodzinsky AJ, Ortiz C, et al. Molecular adhesion between cartilage extracellular matrix macromolecules. Biomacromolecules. 2014;15:772–80.
206. Appunni S, Anand V, Khandelwal M, Gupta N, Rubens M, Sharma A. Small leucine rich proteoglycans (decorin, biglycan and lumican) in cancer. Clin Chim Acta. 2019;491:1–7.
207. Kiani C, Chen L, Wu YJ, Yee AJ, Yang BB. Structure and function of aggrecan. Cell Res. 2002;12:19–32.
208. Nap RJ, Szleifer I. Structure and interactions of aggrecans: statistical thermodynamic approach. Biophys J. 2008;95:4570–83.
209. Lamoureux F, Baud'huin M, Duplomb L, Heymann D, Rédini F. Proteoglycans: key partners in bone cell biology. BioEssays. 2007;29:758–71.
210. Wojtas M, Lausch AJ, Sone ED. Glycosaminoglycans accelerate biomimetic collagen mineralization in a tissue-based in vitro model. Proc Natl Acad Sci U S A. 2020;117:12636.
211. Vertel BM. The ins and outs of aggrecan. Trends Cell Biol. 1995;5:458–64.
212. Watanabe H, Kimata K, Line S, Strong D, Gao LY, Kozak CA, et al. Mouse cartilage matrix deficiency (cmd) caused by a 7 bp deletion in the aggrecan gene. Nat Genet. 1994;7:154–7.
213. Watanabe H, Yamada Y. Chondrodysplasia of gene knockout mice for aggrecan and link protein. Glycoconj J. 2002;19:269–73.
214. Esko JD, Selleck SB. Order out of chaos: assembly of ligand binding sites in heparan sulfate. Annu Rev Biochem. 2002;71:435–71.
215. Pacifici M. Hereditary multiple exostoses: new insights into pathogenesis, clinical complications, and potential treatments. Curr Osteoporos Rep. 2017;15:142–52.
216. Friand V, David G, Zimmermann P. Syntenin and syndecan in the biogenesis of exosomes. Biol Cell. 2015;107:331–41.
217. Schaefer L, Iozzo RV. Biological functions of the small leucine-rich proteoglycans: from genetics to signal transduction. J Biol Chem. 2008;283:21305–9.
218. Neill T, Schaefer L, Iozzo RV. Decoding the matrix: instructive roles of proteoglycan receptors. Biochemistry. 2015;54:4583–98.
219. Kobe B, Deisenhofer J. A structural basis of the interactions between leucine-rich repeats and protein ligands. Nature. 1995;374:183–6.
220. McEwan PA, Scott PG, Bishop PN, Bella J. Structural correlations in the family of small leucine-rich repeat proteins and proteoglycans. J Struct Biol. 2006;155:294–305.
221. Boskey AL, Robey PG. The regulatory role of matrix proteins in mineralization of bone. In: Marcus R, Dempster D, Cauley J, Feldman D, editors. Osteoporosis. 4th ed. London: Elsevier; 2013. p. 235–55.
222. Gubbiotti MA, Vallet SD, Ricard-Blum S, Iozzo RV. Decorin interacting network: a comprehensive analysis of decorin-binding partners and their versatile functions. Matrix Biol. 2016;55:7–21.
223. Hildebrand A, Romarís M, Rasmussen LM, Heinegård D, Twardzik DR, Border WA, et al. Interaction of the small interstitial proteoglycans biglycan, decorin and fibromodulin with transforming growth factor beta. Biochem J. 1994;302:527–34.
224. Chen XD, Fisher LW, Robey PG, Young MF. The small leucine-rich proteoglycan biglycan modulates BMP-4-induced osteoblast differentiation. FASEB J. 2004;18:948–58.

225. Desnoyers L, Arnott D, Pennica D. WISP-1 binds to decorin and biglycan. J Biol Chem. 2001;276:47599–607.
226. Tufvesson E, Westergren-Thorsson G. Tumour necrosis factor-α interacts with biglycan and decorin. FEBS Lett. 2002;530:124–8.
227. Schönherr E, Sunderkötter C, Iozzo RV, Schaefer L. Decorin, a novel player in the insulin-like growth factor system. J Biol Chem. 2005;280:15767–72.
228. Hunter GK, Poitras MS, Underhill TM, Grynpas MD, Goldberg HA. Induction of collagen mineralization by a bone sialoprotein–decorin chimeric protein. J Biomed Mater Res. 2001;55:496–502.
229. Kalamajski S, Oldberg A. The role of small leucine-rich proteoglycans in collagen fibrillogenesis. Matrix Biol. 2010;29:248–53.
230. Orgel JP, Eid A, Antipova O, Bella J, Scott JE. Decorin core protein (decoron) shape complements collagen fibril surface structure and mediates its binding. PLoS One. 2009;4:e7028.
231. Ueda K, Yoshimura K, Yamashita O, Harada T, Morikage N, Hamano K. Possible dual role of decorin in abdominal aortic aneurysm. PLoS One. 2015;10:e0120689.
232. Hoshi K, Kemmotsu S, Takeuchi Y, Amizuka N, Ozawa H. The primary calcification in bones follows removal of decorin and fusion of collagen fibrils. J Bone Miner Res. 1999;14:273–80.
233. Fisher LW, Termine JD, Young MF. Deduced protein sequence of bone small proteoglycan I (biglycan) shows homology with proteoglycan II (decorin) and several nonconnective tissue proteins in a variety of species. J Biol Chem. 1989;264:4571–6.
234. Bianco P, Fisher LW, Young MF, Termine JD, Robey PG. Expression and localization of the two small proteoglycans biglycan and decorin in developing human skeletal and non-skeletal tissues. J Histochem Cytochem. 1990;38:1549–63.
235. Orsini G, Ruggeri A, Mazzoni A, Papa V, Mazzotti G, Di Lenarda R, et al. Immunohistochemical identification of decorin and biglycan in human dentin: a correlative field emission scanning electron microscopy/transmission electron microscopy study. Calcif Tissue Int. 2007;81:39–45.
236. Xu T, Bianco P, Fisher LW, Longenecker G, Smith E, Goldstein S, et al. Targeted disruption of the biglycan gene leads to an osteoporosis-like phenotype in mice. Nat Genet. 1998;20:78–82.
237. Vynios DH, Papageorgakopoulou N, Sazakli H, Tsiganos CP. The interactions of cartilage proteoglycans with collagens are determined by their structures. Biochimie. 2001;83:899–906.
238. Schönherr E, Hausser HJ. Extracellular matrix and cytokines: a functional unit. Dev Immunol. 2000;7:89–101.
239. Goldberg M, Septier D, Rapoport O, Iozzo RV, Young MF, Ameye LG. Targeted disruption of two small leucine-rich proteoglycans, biglycan and decorin, excerpts divergent effects on enamel and dentin formation. Calcif Tissue Int. 2005;77:297–310.
240. Young MF, Bi Y, Ameye L, Chen XD. Biglycan knockout mice: new models for musculoskeletal diseases. Glycoconj J. 2002;19:257–62.
241. Geerkens C, Vetter U, Just W, Fedarko NS, Fisher LW, Young MF, et al. The X-chromosomal human biglycan gene BGN is subject to X inactivation but is transcribed like an X-Y homologous gene. Hum Genet. 1995;96:44–52.
242. Corsi A, Xu T, Chen XD, Boyde A, Liang J, Mankani M, et al. Phenotypic effects of biglycan deficiency are linked to collagen fibril abnormalities, are synergized by decorin deficiency, and mimic Ehlers-Danlos-like changes in bone and other connective tissues. J Bone Miner Res. 2002;17:1180–9.
243. Buraschi S, Neill T, Iozzo RV. Decorin is a devouring proteoglycan: remodeling of intracellular catabolism via autophagy and mitophagy. Matrix Biol. 2019;75–76:260–70.
244. Laizé V, Martel P, Viegas CS, Price PA, Cancela ML. Evolution of matrix and bone gamma-carboxyglutamic acid proteins in vertebrates. J Biol Chem. 2005;280:26659–68.
245. Bronckers AL, Gay S, Finkelman RD, Butler WT. Immunolocalization of Gla proteins (osteocalcin) in rat tooth germs: comparison between indirect immunofluorescence, peroxidase-antiperoxidase, avidin-biotin-peroxidase complex, and avidin-biotin-gold complex with silver enhancement. J Histochem Cytochem. 1987;35:825–30.

246. McKee MD, Glimcher MJ, Nanci A. High-resolution immunolocalization of osteopontin and osteocalcin in bone and cartilage during endochondral ossification in the chicken tibia. Anat Rec. 1992;234:479–92.
247. Saraiya M, Nasser R, Zeng Y, Addya S, Ponnappan RK, Fortina P, et al. Reversine enhances generation of progenitor-like cells by dedifferentiation of annulus fibrosus cells. Tissue Eng Part A. 2010;16:1443–55.
248. Hauschka PV, Lian JB, Gallop PM. Direct identification of the calcium-binding amino acid, gamma-carboxyglutamate, in mineralized tissue. Proc Natl Acad Sci U S A. 1975;72:3925.
249. Hauschka PV, Carr SA. Calcium-dependent alpha-helical structure in osteocalcin. Biochemist. 1982;21:2538–47.
250. Hauschka PV. Osteocalcin: the vitamin K-dependent Ca2+-binding protein of bone matrix. Haemostasis. 1986;16:258–72.
251. Hauschka PV, Lian JB, Cole DE, Gundberg CM. Osteocalcin and matrix Gla protein: vitamin K-dependent proteins in bone. Physiol Rev. 1989;69:990–1047.
252. Aronow MA, Gerstenfeld LC, Owen TA, Tassinari MS, Stein GS, Lian JB. Factors that promote progressive development of the osteoblast phenotype in cultured fetal rat calvaria cells. J Cell Physiol. 1990;143:213–21.
253. McKee MD, Nancl A, Khan SR. Ultrastructural immunodetection of osteopontin and osteocalcin as major matrix components of renal calculi. J Bone Miner Res. 1995;10:1913–29.
254. Ducy P, Desbois C, Boyce B, Pinero G, Story B, Dunstan C, et al. Increased bone formation in osteocalcin-deficient mice. Nature. 1996;382:448–52.
255. Boskey AL, Gadaleta S, Gundberg C, Doty SB, Ducy P, Karsenty G. Fourier transform infrared microspectroscopic analysis of bones of osteocalcin-deficient mice provides insight into the function of osteocalcin. Bone. 1998;23:187–96.
256. Hunter GK, Hauschka PV, Poole AR, Rosenberg LC, Goldberg HA. Nucleation and inhibition of hydroxyapatite formation by mineralized tissue proteins. Biochem J. 1996;317:59–64.
257. Poundarik AA, Boskey A, Gundberg C, Vashishth D. Biomolecular regulation, composition and nanoarchitecture of bone mineral. Sci Rep. 2018;8:1191.
258. Simon P, Grüner D, Worch H, Pompe W, Lichte H, El Khassawna T, et al. First evidence of octacalcium phosphate@osteocalcin nanocomplex as skeletal bone component directing collagen triple–helix nanofibril mineralization. Sci Rep. 2018;8:13696.
259. Burger C, Zhou H-W, Wang H, Sics I, Hsiao BS, Chu B, et al. Lateral packing of mineral crystals in bone collagen fibrils. Biophys J. 2008;95:1985–92.
260. Cairns JR, Price PA. Direct demonstration that the vitamin K-dependent bone Gla protein is incompletely gamma-carboxylated in humans. J Bone Miner Res. 1994;9:1989–97.
261. Lee AJ, Hodges S, Eastell R. Measurement of osteocalcin. Ann Clin Biochem. 2000;37:432–46.
262. Razny U, Fedak D, Kiec-Wilk B, Goralska J, Gruca A, Zdzienicka A, et al. Carboxylated and undercarboxylated osteocalcin in metabolic complications of human obesity and prediabetes. Diabetes Metab Res Rev. 2017;33:e2862.
263. Camarda AJ, Butler WT, Finkelman RD, Nanci A. Immunocytochemical localization of γ-carboxyglutamic acid-containing proteins (osteocalcin) in rat bone and dentin. Calcif Tissue Int. 1987;40:349–55.
264. Hoang QQ, Sicheri F, Howard AJ, Yang DS. Bone recognition mechanism of porcine osteocalcin from crystal structure. Nature. 2003;425:977–80.
265. Lee NK, Sowa H, Hinoi E, Ferron M, Ahn JD, Confavreux C, et al. Endocrine regulation of energy metabolism by the skeleton. Cell. 2007;130:456–69.
266. Ferron M, Wei J, Yoshizawa T, Del Fattore A, DePinho RA, Teti A, et al. Insulin signaling in osteoblasts integrates bone remodeling and energy metabolism. Cell. 2010;142:296–308.
267. Sabek OM, Nishimoto SK, Fraga D, Tejpal N, Ricordi C, Gaber AO. Osteocalcin effect on human β-cells mass and function. Endocrinology. 2015;156:3137–46.
268. De Toni L, Guidolin D, De Filippis V, Tescari S, Strapazzon G, Santa Rocca M, et al. Osteocalcin and sex hormone binding globulin compete on a specific binding site of GPRC6A. Endocrinology. 2016;157:4473–86.

269. Liu D-M, Mosialou I, Liu J-M. Bone: another potential target to treat, prevent and predict diabetes. Diabetes Obes Metab. 2018;20:1817–28.
270. Rached M-T, Kode A, Silva BC, Jung DY, Gray S, Ong H, et al. FoxO1 expression in osteoblasts regulates glucose homeostasis through regulation of osteocalcin in mice. J Clin Investig. 2010;120:932.
271. Mera P, Ferron M, Mosialou I. Regulation of energy metabolism by bone-derived hormones. Cold Spring Harb Perspect Med. 2018;8:a031666.
272. Kim SP, Frey JL, Li Z, Kushwaha P, Zoch ML, Tomlinson RE, et al. Sclerostin influences body composition by regulating catabolic and anabolic metabolism in adipocytes. Proc Natl Acad Sci U S A. 2017;114:E11238.
273. Dirckx N, Moorer MC, Clemens TL, Riddle RC. The role of osteoblasts in energy homeostasis. Nat Rev Endocrinol. 2019;15:651–65.
274. Oury F, Sumara G, Sumara O, Ferron M, Chang H, Smith CE, et al. Endocrine regulation of male fertility by the skeleton. Cell. 2011;144:796–809.
275. Khrimian L, Obri A, Ramos-Brossier M, Rousseaud A, Moriceau S, Nicot A-S, et al. Gpr158 mediates osteocalcin's regulation of cognition. J Exp Med. 2017;214:2859–73.
276. Hynes RO, Naba A. Overview of the matrisome--an inventory of extracellular matrix constituents and functions. Cold Spring Harb Perspect Biol. 2012;4:a004903.
277. Oury F, Khrimian L, Denny CA, Gardin A, Chamouni A, Goeden N, et al. Maternal and offspring pools of osteocalcin influence brain development and functions. Cell. 2013;155:228–41.
278. Mera P, Laue K, Ferron M, Confavreux C, Wei J, Galán-Díez M, et al. Osteocalcin signaling in myofibers is necessary and sufficient for optimum adaptation to exercise. Cell Metab. 2016;23:1078–92.
279. Karsenty G, Mera P. Molecular bases of the crosstalk between bone and muscle. Bone. 2018;115:43–9.
280. Berger JM, Singh P, Khrimian L, Morgan DA, Chowdhury S, Arteaga-Solis E, et al. Mediation of the acute stress response by the skeleton. Cell Metab. 2019;30:890–902.e8.
281. Basu R, Peterson J, Rizza R, Khosla S. Effects of physiological variations in circulating insulin levels on bone turnover in humans. J Clin Endocrinol Metab. 2011;96:1450–5.
282. Lambert LJ, Challa AK, Niu A, Zhou L, Tucholski J, Johnson MS, et al. Increased trabecular bone and improved biomechanics in an osteocalcin-null rat model created by CRISPR/Cas9 technology. Dis Model Mech. 2016;9:1169–79.
283. Diegel CR, Hann S, Ayturk UM, Hu JCW, Lim K-E, Droscha CJ, et al. An osteocalcin-deficient mouse strain without endocrine abnormalities. PLoS Genet. 2020;16:e1008361.
284. Moriishi T, Ozasa R, Ishimoto T, Nakano T, Hasegawa T, Miyazaki T, et al. Osteocalcin is necessary for the alignment of apatite crystallites, but not glucose metabolism, testosterone synthesis, or muscle mass. PLoS Genet. 2020;16:e1008586.
285. Komori T. Functions of osteocalcin in bone, pancreas, testis, and muscle. Int J Mol Sci. 2020;21:7513.
286. Manolagas SC. Osteocalcin promotes bone mineralization but is not a hormone. PLoS Genet. 2020;16:e1008714.
287. Winkler DG, Sutherland MK, Geoghegan JC, Yu C, Hayes T, Skonier JE, et al. Osteocyte control of bone formation via sclerostin, a novel BMP antagonist. EMBO J. 2003;22:6267–76.
288. van Bezooijen RL, Roelen BAJ, Visser A, van der Wee-Pals L, de Wilt E, Karperien M, et al. Sclerostin is an osteocyte-expressed negative regulator of bone formation, but not a classical BMP antagonist. J Exp Med. 2004;199:805–14.
289. Collette NM, Yee CS, Murugesh D, Sebastian A, Taher L, Gale NW, et al. Sost and its paralog Sostdc1 coordinate digit number in a Gli3-dependent manner. Dev Biol. 2013;383:90–105.
290. Brunkow ME, Gardner JC, Van Ness J, Paeper BW, Kovacevich BR, Proll S, et al. Bone dysplasia sclerosteosis results from loss of the SOST gene product, a novel cystine knot-containing protein. Am J Hum Genet. 2001;68:577–89.
291. Tanaka S, Matsumoto T. Sclerostin: from bench to bedside. J Bone Miner Metab. 2021;39:332–40.

292. Li X, Ominsky MS, Niu QT, Sun N, Daugherty B, D'Agostin D, et al. Targeted deletion of the sclerostin gene in mice results in increased bone formation and bone strength. J Bone Miner Res. 2008;23:860–9.
293. Baron R, Kneissel M. WNT signaling in bone homeostasis and disease: from human mutations to treatments. Nat Med. 2013;19:179–92.
294. Nakashima T, Hayashi M, Fukunaga T, Kurata K, Oh-Hora M, Feng JQ, et al. Evidence for osteocyte regulation of bone homeostasis through RANKL expression. Nat Med. 2011;17:1231–4.
295. Cohen-Kfir E, Artsi H, Levin A, Abramowitz E, Bajayo A, Gurt I, et al. Sirt1 is a regulator of bone mass and a repressor of Sost encoding for sclerostin, a bone formation inhibitor. Endocrinology. 2011;152:4514–24.
296. Plotkin LI, Bellido T. Osteocytic signalling pathways as therapeutic targets for bone fragility. Nat Rev Endocrinol. 2016;12:593–605.
297. Robling AG, Niziolek PJ, Baldridge LA, Condon KW, Allen MR, Alam I, et al. Mechanical stimulation of bone in vivo reduces osteocyte expression of Sost/sclerostin. J Biol Chem. 2008;283:5866–75.
298. Clarke BL, Drake MT. Clinical utility of serum sclerostin measurements. Bonekey Rep. 2013;2:361.
299. Vasiliadis ES, Evangelopoulos D-S, Kaspiris A, Benetos IS, Vlachos C, Pneumaticos SG. The role of sclerostin in bone diseases. J Clin Med. 2022;11:806.
300. Sugio S, Kashima A, Mochizuki S, Noda M, Kobayashi K. Crystal structure of human serum albumin at 2.5 Å resolution. Protein Eng Des Sel. 1999;12:439–46.
301. Peters T Jr. Serum albumin. Adv Protein Chem. 1985;37:161–245.
302. Triffitt JT, Owen M. Preliminary studies on the binding of plasma albumin to bone tissue. Calcif Tissue Int. 1977;23:303–5.
303. Garnett J, Dieppe P. The effects of serum and human albumin on calcium hydroxyapatite crystal growth. Biochem J. 1990;266:863–8.
304. Gilman H, Hukins DWL. Seeded growth of hydroxyapatite in the presence of dissolved albumin. J Inorg Biochem. 1994;55:21–30.
305. Hlady V, Füredi-Milhofer H. Adsorption of human serum albumin on precipitated hydroxyapatite. J Colloid Interface Sci. 1979;69:460–8.
306. Schafer C, Heiss A, Schwarz A, Westenfeld R, Ketteler M, Floege J, et al. The serum protein alpha 2-Heremans-Schmid glycoprotein/fetuin-A is a systemically acting inhibitor of ectopic calcification. J Clin Invest. 2003;112:357–66.
307. Jahnen-Dechent W, Heiss A, Schäfer C, Ketteler M. Fetuin-A regulation of calcified matrix metabolism. Circ Res. 2011;108:1494–509.
308. Boyan BD, Schwartz Z, Swain LD, Khare A. Role of lipids in calcification of cartilage. Anat Rec. 1989;224:211–9.
309. Delany AM, Hankenson KD. Thrombospondin-2 and SPARC/osteonectin are critical regulators of bone remodeling. J Cell Commun Signal. 2009;3:227–38.
310. Forsprecher J, Wang Z, Goldberg HA, Kaartinen MT. Transglutaminase-mediated oligomerization promotes osteoblast adhesive properties of osteopontin and bone sialoprotein. Cell Adhes Migr. 2011;5:65–72.
311. Fedarko NS, Jain A, Karadag A, Fisher LW. Three small integrin binding ligand N-linked glycoproteins (SIBLINGs) bind and activate specific matrix metalloproteinases. FASEB J. 2004;18:734–6.
312. Bradshaw AD. The role of SPARC in extracellular matrix assembly. J Cell Commun Signal. 2009;3:239–46.
313. Carafoli F, Bihan D, Stathopoulos S, Konitsiotis AD, Kvansakul M, Farndale RW, et al. Crystallographic insight into collagen recognition by discoidin domain receptor 2. Structure. 2009;17:1573–81.
314. Trombetta-Esilva J, Bradshaw AD. The function of SPARC as a mediator of fibrosis. Open Rheumatol J. 2012;6:146–55.

315. McEwen BF, Song MJ, Landis WJ. Quantitative determination of the mineral distribution in different collagen zones of calcifying tendon using high voltage electron microscopic tomography. J Comput Assist Microsc. 1991;3:201–10.
316. Landis WJ, Song MJ, Leith A, McEwen L, McEwen BF. Mineral and organic matrix interaction in normally calcifying tendon visualized in three dimensions by high-voltage electron microscopic tomography and graphic image reconstruction. J Struct Biol. 1993;110:39–54.
317. Tartaix PH, Doulaverakis M, George A, Fisher LW, Butler WT, Qin C, et al. In vitro effects of dentin matrix protein-1 on hydroxyapatite formation provide insights into in vivo functions. J Biol Chem. 2004;279:18115–20.
318. Ritter NM, Farach-Carson MC, Butler WT. Evidence for the formation of a complex between osteopontin and osteocalcin. J Bone Miner Res. 1992;7:877–85.
319. Nikel O, Laurencin D, McCallum SA, Gundberg CM, Vashishth D. NMR investigation of the role of osteocalcin and osteopontin at the organic-inorganic interface in bone. Langmuir. 2013;29:13873–82.
320. Poundarik AA, Diab T, Sroga GE, Ural A, Boskey AL, Gundberg CM, et al. Dilatational band formation in bone. Proc Natl Acad Sci U S A. 2012;109:19178.
321. Bornstein P. Diversity of function is inherent in matricellular proteins: an appraisal of thrombospondin 1. J Cell Biol. 1995;130:503–6.
322. Schultz-Cherry S, Murphy-Ullrich JE. Thrombospondin causes activation of latent transforming growth factor-beta secreted by endothelial cells by a novel mechanism. J Cell Biol. 1993;122:923–32.
323. Alcaraz LB, Exposito J-Y, Chuvin N, Pommier RM, Cluzel C, Martel S, et al. Tenascin-X promotes epithelial-to-mesenchymal transition by activating latent TGF-β. J Cell Biol. 2014;205:409–28.
324. Wang X, Liu J, Wang Z, Huang Y, Liu W, Zhu X, et al. Periostin contributes to the acquisition of multipotent stem cell-like properties in human mammary epithelial cells and breast cancer cells. PLoS One. 2013;8:e72962.
325. Dogar AM, Semplicio G, Guennewig B, Hall J. Multiple microRNAs derived from chemically synthesized precursors regulate thrombospondin 1 expression. Nucleic Acid Ther. 2014;24:149–59.
326. Dobson JR, Taipaleenmäki H, Hu Y-J, Hong D, van Wijnen AJ, Stein JL, et al. hsa-mir-30c promotes the invasive phenotype of metastatic breast cancer cells by targeting NOV/CCN3. Cancer Cell Int. 2014;14:73.
327. Wang KX, Denhardt DT. Osteopontin: role in immune regulation and stress responses. Cytokine Growth Factor Rev. 2008;19:333–45.
328. Posey KL, Veerisetty AC, Liu P, Wang HR, Poindexter BJ, Bick R, et al. An inducible cartilage oligomeric matrix protein mouse model recapitulates human pseudoachondroplasia phenotype. Am J Pathol. 2009;175:1555–63.
329. Coustry F, Posey KL, Liu P, Alcorn JL, Hecht JT. D469del-COMP retention in chondrocytes stimulates caspase-independent necroptosis. Am J Pathol. 2012;180:738–48.
330. Lynch JM, Maillet M, Vanhoutte D, Schloemer A, Sargent MA, Blair NS, et al. A thrombospondin-dependent pathway for a protective ER stress response. Cell. 2012;149:1257–68.
331. Addison WN, Nakano Y, Loisel T, Crine P, McKee MD. MEPE-ASARM peptides control extracellular matrix mineralization by binding to hydroxyapatite: an inhibition regulated by PHEX cleavage of ASARM. J Bone Miner Res. 2008;23:1638–49.

Local and Systemic Regulation of Mineralization: Role of Coupling Factors, Pyrophosphate, Polyphosphates, Vitamin D, Fetuin, Matrix Gla Protein, and Osteopontin

9

Of all the constituents of the human body, bone is the hardest, the driest, the earthiest and the coldest; and, excepting only the teeth, it is devoid of sensation. God, the great Creator of all things, formed its substance to this specification with good reason, intending it to be like a foundation for the whole body; for in the fabric of the human body bones perform the same function as do walls and beams in houses, poles in tents and keels and ribs in boats. [1]

Glossary

Acidocalcisome A lysosome-like acidic organelle rich in calcium ions and polyphosphates.

Ankylosis Abnormal stiffness or immobility of a joint resulting from adhesion, rigidity or fusion of the bones of the joint.

Basic multicellular unit A term used to describe regions of bone undergoing replacement and renewal.

Calciprotein particles A fetuin complex containing calcium and phosphate ions, present in the serum.

Chemokines Proteins which induce directed chemotaxis.

Chromatin remodeling Changes in chromatin structure permitting RNA transcription.

CYP24A1 An enzyme which converts raised or excessive levels of $1\alpha,25(OH)_2D_3$ to the less active $24,25(OH)_2D_3$ metabolite.

CYP27B1 An enzyme which generates the active metabolite, $1\alpha,25(OH)_2D_3$.

Cystatin domain in fetuin Regions of fetuin close to the N-terminus of the protein which form a sheet of negatively charged acidic amino acids with four conserved cysteines. The cystatin domains bind calcium ions and calcium phosphate ion clusters.

Enamel knot A cytokine signaling center active during tooth development.

EPHB4-EPHRINB2 Coupling factors, ephrins, on osteoblasts (EPHB4) and osteoclasts (EPHRINB2) promoting bi-directional signaling between these cells.

I. M. Shapiro, W. J. Landis, *Mechanisms of Mineralization of Vertebrate Skeletal and Dental Tissues*, https://doi.org/10.1007/978-3-031-34304-9_9

Exosomes Small membrane-bound organelles generated in the endosomal/autophagic/lysosomal pathway and released from cells. Exosomes contain several characteristic membrane proteins such as CD 9 and CD 63 as well as LAMP-1 and LAMP-2, proteins linked to the autophagic pathway. When trapped in the extracellular matrix of calcifying tissues, these organelles may function as matrix vesicles.

Integrins Transmembrane receptors which bind proteins of the extracellular matrix and soluble ligands in order to facilitate cell-cell function and cell-extracellular matrix adhesion.

Keutel syndrome A MGP mutation causing aberrant mineralization of soft tissues, together with midfacial hypoplasia and hearing loss.

Kinase An enzyme which phosphorylates substrates.

MGP γ-carboxylation Post-translational modification of the γ-carboxylate group of glutamate by vitamin K in matrix gla protein (MGP).

Multivesicular body An intracellular organelle formed from the aggregation of late-stage endosomes and which fuses with the plasma membrane to liberate intraluminal vesicles/exosomes to the extracellular matrix.

Osteocalcin (bone gla protein) A vitamin K-dependent non-collagenous γ-carboxyglutamic acid-containing protein secreted by osteoblasts, odontoblasts, cementocytes, and tenocytes (in mineralizing tendons of certain avian species).

Osteoprotegerin A protein secreted by osteoblasts which binds RANKL and inhibits osteoclast formation and resorption. Osteoprotegerin is also known as a "decoy receptor."

Osterix (SP7) A transcription factor involved with osteoblast differentiation.

Phosphoanhydride bond A high energy bond formed between two phosphate groups by loss of water.

Polyphosphate An anionic polymer comprised of phosphate monomers, linked together by high-energy phosphoanhydride bonds.

Post-menopausal osteoporosis A bone disease associated with a decrease in estrogen levels and marked by elevated osteoclastic resorption.

PTHR (type I) A parathyroid hormone receptor present on osteoblasts and a member of the G protein-coupled superfamily which induces two signal transduction pathways, cAMP/protein kinase A and phospholipase C.

Pyrophosphate Two phosphate groups held together by an acid anhydride bond.

Sclerostin A bone protein encoded by the gene, *SOST*.

Semaphorin A secreted and membrane-associated protein which regulates both osteoblast and osteoclast function within the BMU.

Vitamin K2 An organic molecule which promotes the carboxylation of the γ-glutamyl residues of proteins such as MGP.

Warfarin An antagonist of vitamin K.

Abbreviations and Proper Names for Genes/Proteins and Other Molecules

$1\alpha,25(OH)_2D$	1α,25-Dihydroxyvitamin D_3
$24,25(OH)_2D_3$	24,25-Dihydroxyvitamin D_3
$25(OH)D_3$	25-Hydroxyvitamin D_3
AHSG	Fetuin, α2HS-glycoprotein

ADP	Adenosine diphosphate
AMP	Adenosine monophosphate
ANK	Progressive ankylosis protein
ATP	Adenosine triphosphate
ASARM	Acidic serine-aspartate-rich motif/MEPE-associated peptide
BGN	Biglycan
BMP	Bone morphogenetic protein
BMU	Basic multicellular unit
cAMP	Cyclic AMP
C/EBP	CCAAT/enhancer-binding protein
COLI	Type I collagen
COLII	Type II collagen
CSF-1	Colony stimulating factor-1
DMP1	Dentin matrix protein-1
ENPP	Ectonucleotide pyrophosphatase/phosphodiesterase
EPH	Ephrin
FGF23	Fibroblast growth factor 23
HDAC	Histone deacetylase
LAMP	Lysosome-associated membrane glycoprotein
MCP-1	Monocyte chemoattractant protein-1
MGP	Matrix gla protein
NFAT	Nuclear factor activator of T-cells
NF-κB	Nuclear factor-kappa beta
OCN	Osteocalcin
OPN	Osteopontin
PHEX	Phosphate regulating endopeptidase homologue, X-linked
PHOSPHO1	Phosphoethanolamine/phosphocholine phosphatase
PTH	Parathyroid hormone
RANK	Receptor activator of nuclear factor κB
RANKL	Receptor activator of nuclear factor kB ligand
RUNX	Runt-related transcription factor
SOST	Sclerostin
TGF-β	Transforming growth factor-β
TNAP	Tissue non-specific alkaline phosphatase
TNF	Tumor necrosis factor
TRAF6	TNF receptor-associated factor 6
VDR	Vitamin D receptor
VDRE	Vitamin D response element

9.1 Walls, Beams, Poles, Keels, and Ribs

Most students or practitioners of the biomedical sciences have at one time or another seen reproductions of Vesalius' extraordinary hand-drawn illustrations of the human body. His insights into human anatomy revolutionized the practice of medicine and surgery and helped to brush away the teachings of Hippocrates and Galen where treatment involved astrological signs and the balancing of the bodily humors. However, as the quotation above states ("...bones perform the same function as do walls and beams in houses, poles in tents and keels and ribs in boats."), the skeleton was regarded as the foundation of the body rather than a vital tissue. Of course, as is now known, nothing could be further from the truth. Bone, cartilage, dentin, and cementum are living tissues, and even enamel, which has no living cells in its mature state, is the product of cellular activity. Moreover, far from being "the hardest, the driest, the earthiest and the coldest" of the tissues of the body, each of these mineralized tissues responds to environmental, metabolic, and genetic factors. Many of these same influences control cellular events which lead to mineralization (osteoblasts, chondrocytes, odontoblasts, cementocytes, ameloblasts) and demineralization (osteoclasts and chondroclasts).

The overarching goal of all these groups of cells is to work together to produce and maintain a mineralized tissue which is responsive to the functional needs of the organism. However, because of the high calcium phosphate ion activity product of serum and extracellular fluid, there is a systemic as well as local need for agents which prevent aberrant precipitation of mineral in both hard and soft tissues. This requirement is met by the release of inhibitory molecules which block mineral precipitation within the metastable tissue fluid. Some of this inhibitory activity is caused by the presence of inorganic ions like magnesium and small organic molecules like citrate, an anion generated as a reaction product in mitochondria (see Chap. 7). Other agents like serum albumin and prothrombin bind available calcium ions. Although their binding activity is weak, their concentrations are sufficient to lower the calcium phosphate ion activity product and thereby decrease the risk of aberrant mineral formation.

While each of these agents listed above is of importance, four potent systemic inhibitors of vertebrate mineralization exist, the proteins, fetuin (gene *AHSG*), matrix Gla protein (*MGP*) and osteopontin (*OPN*), and the pyrophosphate anion. In the oral cavity, statherin and other proline-rich proteins of saliva also serve to block mineral deposition. Together, all these agents prevent dystrophic mineral deposition and additionally play a role in regulating physiological mineralization.

In the following sections of this chapter, the regulation of mineralization is considered in terms of coordinated cellular control and the systemic and local effects of several molecules found in the vertebrate hard tissues. Interactions will be reviewed first between multiple cell types which prevent aberrant mineralization during the replacement of old bone with new bone, this turnover process being common to all tissues of the body and to the specialized needs of the skeleton [2]. Second, the mechanisms will be discussed by which the pyrophosphate anion and the proteins, fetuin and MGP, inhibit abnormal mineral deposition while at the same time regulate new bone formation. The role of vitamin D will also be described in the context of its contribution to the regulation of hard tissue growth and development. In the final chapter of this book, details will be presented as they relate to consequences in vertebrate tissues resulting from deficiencies or disturbances of these molecules, including OPN.

9.2 Regulated Bone Formation and Mineralization within the Basic Multicellular Unit (BMU): Coupling Factors

Thus far in this book, there has been discussion about the mechanism by which calcium and phosphate ions accumulate in cells and subsequently are delivered to the extracellular matrix where they form the mineral phase of bone and other calcified vertebrate tissues. The first section of this chapter addresses a more general question, the mechanism by which there is coordinated and regulated replacement of old bone by new bone. In other words, processes are explored by which cells (osteoblasts, osteoclasts, osteocytes, immune cells, and bone lining cells) occupying localized tissue niches work cooperatively to foster controlled new bone formation. The goal of these replacement activities is to maintain the structural, biological, and biomechanical properties of the whole tissue without changing the dimensions, architecture, or density of the bone.

Frost used the terminology, "basic multicellular units (BMU)," to characterize regions of bone undergoing replacement and renewal [3]. Within the BMU, localized bone remodeling is characterized by an initial phase of rapid resorption (days, possibly weeks), followed by a slower phase of new bone formation and mineralization (weeks–months). The BMU contains all the cell types discussed above, with osteoclasts (having a lifespan of about 14 days) degrading and removing existing and often damaged tissue and osteoblasts (having a lifespan of approximately 3 months) secreting a replacement organic matrix (osteoid) which subsequently mineralizes [4]. It is noteworthy that osteocytes, the long-lived and terminally differentiated form of an osteoblast, are the major cell type in bone. As presented in Chaps. 3, 4, and 8, these cells regulate osteoblasts by secreting sclerostin (SOST), which suppresses wingless-related integration site (WNT) signaling by binding low density lipoprotein receptor (LRP)5/6 [5]. In addition, as discussed in some detail below, osteocytes secrete osteoprotegerin (OPG), a molecule which competes with receptor activator of nuclear factor-kappa B ligand (RANKL) for receptor activator of nuclear factor-kappa B (RANK) found on osteoclasts [6, 7]. In trabecular bone, this renewal process takes place in a tunnel-like structure or as a surface excavation, for example, the osteoclastic resorption pits on the inner surface of the growing skull. Such a process endows the organism with powerful evolutionary advantages, among them the ability to accommodate variations in applied mechanical forces, a mechanism to rid the tissue of regions of defective or aged bone, and a system to adapt to changes in the local environment.

In the early 1980s at a Gordon Research Conference in New Hampshire, Rodan and Martin proposed that bone formation was coupled molecularly and functionally to that of bone resorption [8]. This idea, based on discussions with several other leaders in the bone biology field which included Drs. Boyde, Feldman, Heershe, Kahn, Mechanic, Mundy, Price, Raisz, Rowe, and Teitelbaum, has become a basic precept of skeletal biology: That bone removal and formation occur in concert with close molecular "coupling" between the bone-forming osteoblasts and the bone-resorbing osteoclasts. Sims and Martin likened the coupling concept to a locomotive by which a "train carriage follows the engine" and "Osteoclasts are the engine driving remodeling and, through coupling, osteoblasts are the cellular carriage that follows." [9] Osteoclasts were chosen to be the "engine" as bone resorption activity triggers osteoblast-mediated bone formation. Also, since the processes of bone formation and resorption occur at very different rates, the impact of coupling is the "summation of the contributions from BMUs throughout the body." [10] Fig. 9.1 shows the intimate relationship between an osteoclast on a bone surface and a neighboring osteoblast [12].

An expansive constellation of agents including parathyroid hormone (PTH), RANKL, macrophage colony-stimulating factor (MCSF), platelet-derived growth factor (PDGF), insulin-like growth factors (IGFs), phosphoethanolamine/phosphocholine phosphatase (PHOSPHO1), vitamin D metabolites, and other growth factors (transforming growth factor-beta [TGF-β], WNT/β-catenin, fibroblast growth factors [FGFs], bone morphogenetic proteins [BMPs]), together with microRNAs and male and female sex hormones, serve as coupling agents or influence one or both arms of the bone remodeling process [9, 11–13]. In this context, osteoblasts, bone lining cells, and monocytes/macrophages within the BMU release chemokines such as CCL2, CXCL7, CCL8, and CCL9 [14]. These

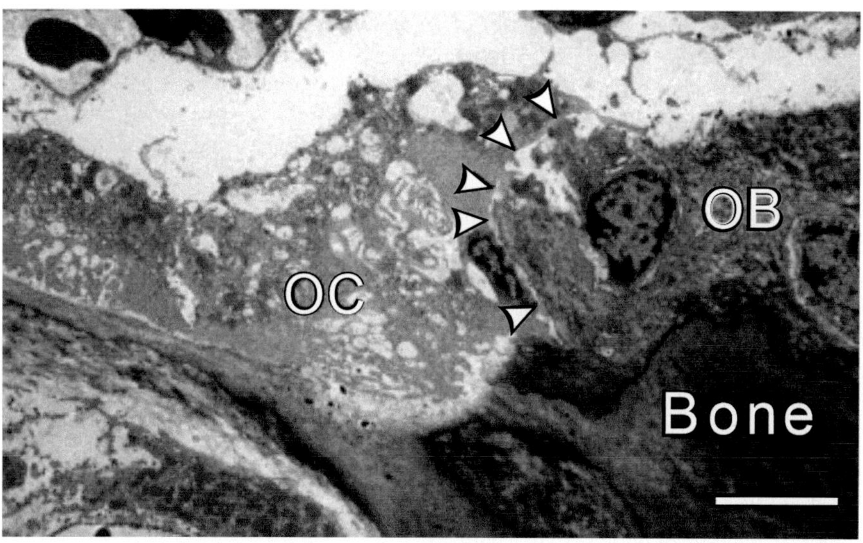

Fig. 9.1 Transmission electron microscopic image of osteoclast-osteoblast interaction within a basic multicellular unit (BMU) located in the metaphyseal region of a tibia from a 14-week-old mouse. The photomicrograph shows direct contact (arrowheads) between an osteoblast (OB) and an osteoclast (OC) with a well-developed ruffled border along the surface of bone. This type of cell-cell contact would permit membrane-bound ligands and receptors to interact and initiate intracellular signaling [11]. Other communication means exist between the two cell types, including passage of small water-soluble molecules through gap junctions and release of diffusible paracrine factors. During resorption beneath their ruffled border, osteoclasts liberate sclerostin, RANKL, TGF-β, bone morphogenetic proteins (BMPs) and insulin-like growth factor II (IGF-II) from the demineralizing bone matrix [11, 12]. Scale bar = 5 μm. (Image modified from Matsuo and Irie [12] with permission)

molecules are present at sites of bone turnover and hence are regarded as integral components of the bone resorption response. Indeed, chemokines CXCL1 and CCL12 participate in the mechanism which drives osteoclast migration and fusion [15]. One particular chemokine, CCL2 (MCP-1), deserves special mention as it promotes osteoclast generation [16]. MCP-1 is present at sites of tooth eruption [17] and its expression increases following treatment with PTH [18]. Aside from proteins, lipidic molecules, including leukotrienes and prostaglandins, also activate resorption. Prostaglandins (in particular prostaglandin E_2, prostacyclin and prostaglandin F2α) stimulate osteoclast formation and differentiation, in some cases through RANKL [19].

In addition to these agents, resorption causes the release of certain proteins from bone itself and these molecules further influence the remodeling process. For example, SOST, a protein secreted by osteocytes, antagonizes the mineralizing effect of the transcription factor, WNT/β-catenin, delaying or inhibiting the proliferation, differentiation, and survival of osteoblasts [20]. Also, the release of TGF-β from resorbing bone enhances the recruitment of osteoblast precursors to the BMU while stimulating osteoclasts to express WNT1, WNT 10b, and BMP [21, 22]. How each of these factors integrates into the overall activities of cells in the BMU is a mystery, although it is clear that a disturbance in the balance between bone formation and resorption leads to skeletal disease.

Mechanisms employed to link these two activities include direct contact between ligands present on one cell type and receptors on the second cell type (Fig. 9.1), secretion of diffusible ligands across gap junctions, generation of powerful osteoclast-activating factors, and, as mentioned above, release of bound bioactive proteins from the bone matrix which influences cell function [23]. Without doubt, the most critical coupling factors expressed by cells which occupy the BMU (osteoblast progenitor cells, marrow stromal cells, osteocytes, monocytes, and T and B cells) are colony stimulating factor (CSF-1) and RANKL, a transmembrane protein and a member of the tumor necrosis factor (TNF) superfamily. RANKL is synthesized principally by bone-forming cells. Once bound, the RANK receptor recruits an adaptor molecule, TNF receptor associated factor (TRAF)6, which facilitates activation of the transcription factors, nuclear factor-kappa beta (NF-κB), nuclear factor of activated T-cells c1 (NFATc1), and activator protein-1 (AP-1) [24]. These proteins drive cell proliferation and the survival of mature osteoclasts, while blocking apoptosis in progenitor populations [24]. Not surprisingly, their mutagenesis suppresses osteoclast formation and results in over-mineralization of bone, leading to a condition described as osteopetrosis [25].

The activity of RANKL and its effect on bone resorption is dependent on yet another bone protein, osteoprotegerin (Latin, "protogere," to protect), noted earlier. This protein, expressed by bone as well as a number of soft tissues, [26] protects against untoward bone resorption by binding RANKL and decreasing its availability. In this way, OPG serves as a "decoy receptor" for the powerful signaling molecule, RANKL. The observation that RANKL is also present on the plasma membrane of activated T-cell lymphocytes is intriguing. There is gathering evidence that within the BMU, these cells may function to prevent bone infection by up-regulating an adaptive immune response [27].

In concert with RANKL, CSF-1 is synthesized by osteoblasts as well as bone marrow progenitor cells. When bound to its receptor, colony stimulating factor receptor (c-FMS), on osteoclast progenitors, a protein complex (growth factor receptor-bound protein 2, phosphatidylinositol 3 kinase/AKT, and phospholipase γ [PLCγ]) is formed which generates signals inducing expression of RANK [24]. Indeed, deletion or mutation of the gene encoding CSF-1 causes a condition in which there is a marked reduction in osteoclast number and activity [28]. As a consequence, bone becomes dense with a loss of marrow spaces; in the jaw, such osteopetrotic bone results in failure of the teeth to erupt [29]. From these and other studies, it is clear that CSF-1 plays an important role in the BMU by increasing the number and differentiation of osteoclasts and enhancing osteoclast resorptive activity [30]. Together with RANKL and co-stimulatory molecules, these ligands promote the genesis and viability of mature osteoclasts and their bone resorptive function [28].

Of the ligands listed above that promote bone resorption, the most important is PTH. While a high and continual release of PTH is a cause of extensive bone resorption (as in hyperparathyroidism), lower and intermittent PTH levels maintain bone homeostasis. Under these conditions, a hormone pulse is anabolic in that it enhances new bone formation while at the same time accelerating removal of old bone [18]. This dual action is achieved through the binding of the hormone to its receptor (type I PTHR, a member of the G protein-coupled receptor superfamily) on osteoblasts. Hormone binding induces two signal transduction pathways, cAMP/protein kinase A (PKA) and phospholipase C (PLC). Signaling through both the pathways promotes osteoblast-mediated bone formation and osteoclast-dependent bone resorption [31]. The anabolic effect of parathyroid hormone receptor signaling is influenced by β-catenin, which switches signaling from cAMP to the PLC pathway [32].

The activities of a second coupling system, ephrinB4-ephrinB2 (EPHB4-EPHB2), are also sensitive to PTH [33]. In this case, EPHB4 is present in the plasma membrane of osteoblasts while EPHB2 is in the enveloping membrane of osteoclasts. Cell-cell signaling is bidirectional: Forward-signaling EPHB4 promotes osteoblast differentiation and reverse-signaling through EPHB2 inhibits osteoclast differentiation and activity. The net effect of EPHB4-EPHB2 is resorption of bone [34]. Other osteoblast-osteoclast coupling systems include the semaphorins, a family of proteins which influence both bone formation and resorption among other functions (Fig. 9.2) [38].

In addition to the coupling mechanisms described above, there is epigenetic regulation of osteoblast-osteoclast function. As mentioned in Chap. 4, by remodeling chromatin structure, histone acetylation/deacetylation reactions influence the accessibility of transcription factors to target genes. Moreover, transcription factors themselves are targets for acetylation/deacetylation. Thus, acetylation of common osteoblast transcription factors such as p53, SMADs, and NF-κB influences their interactions with not only their target genes but also other transcription factors [39, 40]. In bone, histone deacetylase (HDAC)5 negatively influences the expression and secretion of SOST [7]. When released from bone, SOST inhibits bone formation and stimulates RANKL-dependent resorptive activity. In fact, the effect on SOST is indirect as HDAC5

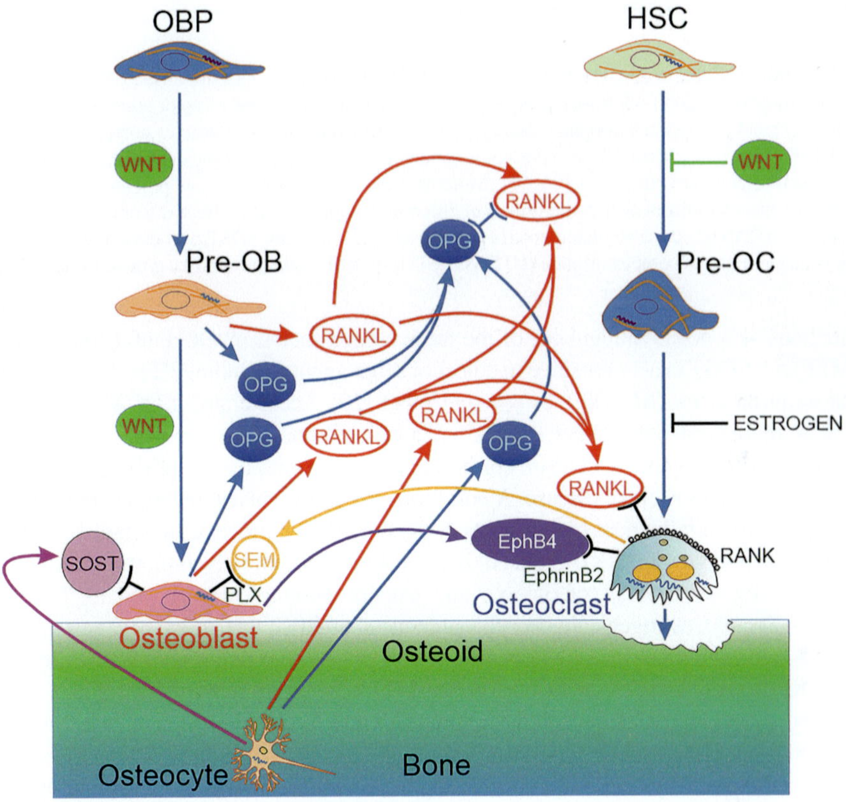

Fig. 9.2 Schematic of aspects of coupling of bone formation to bone resorption. In the presence of agents such as CSF-1 and RANKL, hematogenous stem cells (HSC) commit to the osteoclast lineage. The pathway from these cells to osteoclasts includes osteoclast precursor cells (Pre-OC) whose development is in part mediated (inhibited) by WNT (See Chap. 4). Active osteoclasts attach to the bone surface, where they form ruffled borders, become multinucleated and resorb bone. In a different pathway, osteoblast progenitor cells (OBP) arise from stromal stem cells driven to differentiate by parathyroid hormone, bone morphogenetic protein and other proteins (not shown). OBPs further develop into preosteoblasts (Pre-OB) and then active osteoblasts in part under control (enhanced) also by WNT. Adherent to the bone surface, these cells express the osteoblast phenotype and synthesize osteoid, which is then mineralized. As osteoid becomes mineralized, osteoblasts are incorporated into the bone matrix and express the phenotype of an osteocyte. The process of bone remodeling includes bone resorption by osteoclasts followed by bone formation by osteoblasts. Strict regulation and control of remodeling are directed by cell-cell communication and coupling between osteoblasts and osteoclasts. Such coupling involves the synthesis and secretion of the protein, RANKL, by Pre-OBs, osteoblasts and osteocytes. RANKL binds to its cognate receptor, RANK, located along the plasma membrane of osteoclasts, and the binding promotes osteoclast resorption on the bone surface. The level of RANKL is regulated by osteoprotegerin (OPG), an additional protein secreted by both osteoblasts and their progenitors as well as osteocytes. OPG binds RANKL and hence acts as a "decoy receptor," preventing RANK/RANKL binding which would otherwise stimulate osteoclast resorption [25]. Thus, the RANK/RANKL/OPG signaling pathway serves as a major regulatory mechanism which couples osteoblasts, osteoclasts and osteocytes in the bone remodeling process. Other molecular interactions control remodeling. In addition to OPG/RANKL binding, estrogen serves as a "brake" on osteoclast resorptive activity. A decrease in estrogen levels in the BMU promotes bone resorption. Control is also exerted by

Fig. 9.2 (continued)

EPHB4 from osteoblasts and its interaction with EPHRINB2, a cognate receptor on osteoclasts, as well as by semaphorin 4D (SEM) from osteoclasts and its osteoblast membrane receptor protein, plexin-B1 (PLX) [33, 34]. Additional coupling factors (not shown) include other molecules comprising EPH-EPHRIN pairs; semaphorin 3A and its receptor, sphingosine-1-phosphate; complement factor 3a; WNT 10b; BMP6; and cardiotrophin-1 [35–37]. Moreover, the bone matrix contains proteins which also influence the recruitment and commitment of osteoprogenitor cells and affect osteoblastic activity. Aside from RANKL, these regulatory matrix proteins include sclerostin (SOST, a product of osteocytes), the collagen triple helix repeat peptide (CTHRC1), TGF-β, BMP2 and insulin-like growth factor [9].

functions as a negative regulator of the transcription factor, myeloid elf-1-like factor (MEF)2C; SOST is also sensitive to a host of other agents including PTH, runt-related transcription factor (RUNX2), osterix and WNT [20]. Accordingly, if HDAC5 activity is slow, there is decreased bone density and a lack of bone resorption [41].

Aside from epigenetic control, chronic interference with the coupling activities of bone cells causes skeletal disease. A common example involves the female hormone, estrogen, which influences both aspects of the remodeling process: In doing so, estrogen first suppresses RANKL-induced osteoclast differentiation by inhibiting RANKL secretion by osteocytes, osteoblasts, and osteoprogenitor cells as well as T- and B-cells. Second, estrogen enhances the generation of OPG; this effect increases OPG/RANKL binding, and thereby decreases the actual concentration and availability of RANKL for activating osteoclast resorption [42]. Fig. 9.2 presents a schematic of factors regulating osteoclast-osteoblast interactions and the relationships between estrogen, OPG, RANKL, and osteoclastic activity [9, 25, 33–37]. In the post-menopausal state, the lowered systemic levels of estrogen release the RANKL "brake" on bone resorption. This result leads to an increase in osteoclast activity and a corresponding decrease in bone mass [43, 44]. It is worth commenting that such post-menopausal osteoporosis is far more complicated than simply a RANKL constraint as summarized above. The condition also involves regulation of bone cell apoptosis, expansion of a T-cell response, and an elevated expression of TNFα and IL (interleukin)-7 [43, 45]. How these processes and molecules together affect and control the mineralization step in the BMU is a topic of considerable therapeutic interest.

The carefully regulated and controlled replacement of old bone by new bone in vertebrates is not exclusively restricted to the coupled, coordinated activity of osteoblasts and osteoclasts as described above. A third principal type of bone cell, the osteocyte, exerts its own effects on the process and is structurally coupled itself to other osteocytes, osteoblasts, and osteoclasts to orchestrate the full complement of cellular actions on bone formation, resorption, and maintenance (Fig. 9.2). Indeed, the osteocyte has been regarded in many respects as the master cell among the three and in addition serves highly sophisticated mechanosensory and endocrine functions [7].

Osteocytes comprise greater than 90% of the total number of cells of bone (osteoblasts account for about 5% while osteoclasts constitute 2–3%). Buenzli and Sims estimated that human bone contained 42×10^9 osteocytes with 23×10^{12} cell-cell connections covering a surface area of 215 m^2 [46]. Osteocytes are terminally differentiated osteoblasts which ultimately become incorporated or encased into mineralized bone as secretion of the organic osteoid matrix and mineral deposition occur. Factors which promote osteocyte formation include calcium signaling, extracellular fluid flow, hydrostatic pressure, and oxygen tension in the tissue [47]. As was discussed earlier, the hypoxic environment of bone causes the expression and activation of hypoxia-inducible factor (HIF-1α) and the appearance of the osteocytic phenotype, possibly through WNT and BMP signaling pathways [48].

During the transition of osteoblasts to osteocytes, there is an increase in the expression and secretion of matrix metalloproteinases, enzymes which cleave collagen and other proteins present in the bone extracellular matrix. How the maturing osteoblast becomes subsumed into bone is effectively unknown, but most likely the process involves mineralization of osteoid deposited around the cell accompanied by activation of these metalloproteinases. Proteolysis by these matrix molecules would putatively provide pathways and space for the transitioning cell to establish its lacunar residence with additional space created for the extension of its long, sprouting dendritic processes. These processes communicate with other osteocytes in the mineralized matrix, osteoblasts at the bone periphery, and osteoclasts in their resorption cavities (Fig. 9.3) [7, 49].

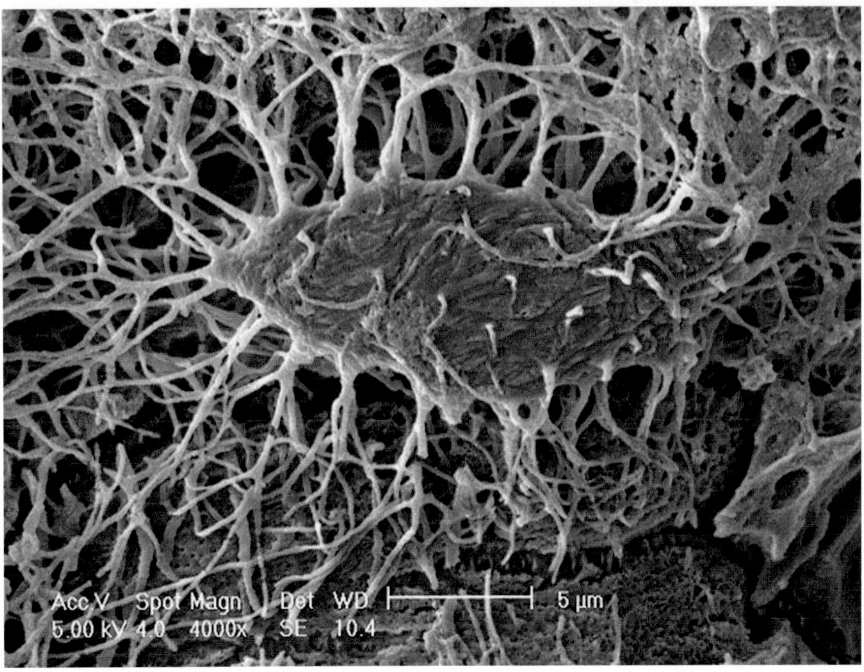

Fig. 9.3 A scanning electron micrograph of a single osteocyte in its lacunar space within a murine bone extracellular matrix and showing its extensive lacunocanalicular network after acid-etching of the specimen embedding material. The network provides the means for cell-cell communication throughout the bone matrix. (Image from L. Bonewald [49] with permission)

The phenotype of an osteocyte differs from that of its osteoblast parental cell. With osteoblast terminal differentiation, osteocytes down-regulate osteoblast genes such as *type I collagen* (*COLI*), *tissue non-specific alkaline phosphatase* (*TNAP*), and *osteocalcin* (*OCN*) [7]. Once they have assumed a dendritic morphology, osteocytes then express a new spectrum of genes, including *connexin43*, required for gap junction function; *podoplanin*, needed for cytoplasmic process formation; and mineralization-related genes, *dentin matrix protein* (*DMP*)*1* and *phosphate regulating endopeptidase homologue, X-linked* (*PHEX*) [7]. Other genes expressed by the differentiated osteocyte include *SOST,* acting to inhibit bone formation; *FGF23,* involved with phosphate homeostasis; *RANKL;* and *OPG*, which modulates normal RANK/RANKL signaling between osteoblasts and osteoclasts as noted previously [7]. When released from osteocytes, SOST binds to membrane receptors on osteoblasts to diminish WNT signaling, while FGF23 regulates phosphate metabolism to inhibit new bone formation. Described above, OPG binding to RANKL secreted by osteoblasts and osteocytes reduces RANKL availability for osteoclast activation, thereby hindering bone resorption. These activities are coupled locally with decreased osteoblast gene expression and protein secretion of COLI, TNAP, and OCN, as mentioned earlier. The latter osteoblast-associated events further interrupt bone deposition and mineralization and the regulation and control of old-to-new bone replacement and renewal within the BMU.

9.3 Regulation of Mineralization: Pyrophosphate and Polyphosphates

The above discussion centers on events which occur in the BMU, whereby osteoblasts, osteoclasts, osteocytes, bone-lining cells, immune cells, and systemic factors regulate bone turnover, a process involving both bone formation and resorption. While the "crosstalk" between the different cell types which occupy the bone niche is extraordinarily effective in controlling and coordinating the two processes, the high calcium and phosphate ion activity product of the tissue fluid and the propensity of the ions to precipitate mineral mandate a second level of regulatory control. This level of local control is mediated by inhibitory molecules contained within the tissue itself, as well as molecules released into the BMU environment. These inhibitory factors maintain bone homeostasis while preventing aberrant mineralization of soft tissues. Of course, disruption of the regulatory systems causes both functional and mineralization deficits.

Several key molecules which regulate vertebrate mineralization are discussed in this chapter, pyrophosphate and polyphosphates, vitamin D, fetuin (α2HS-glycoprotein, AHSG), matrix gla protein (MGP), and osteopontin (OPN). Other than vitamin D, these molecules are inhibitory to mineral deposition and have overlapping and thus redundancies of function. The most powerful of such inhibitors is the anion, pyrophosphate. Chemically, the molecule is very simple, two molecules of orthophosphate linked together by a phosphoanhydride bond (Fig. 9.4). The

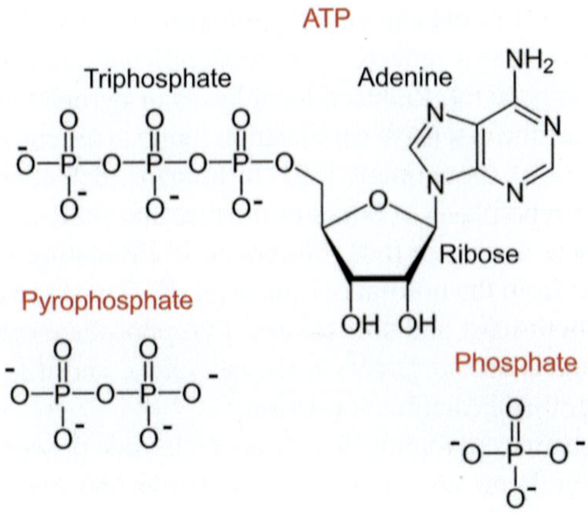

Fig. 9.4 Chemical structure of adenosine triphosphate (ATP), pyrophosphate and the orthophosphate anion

electrostatic repulsion between the phosphoric acid residues and the negatively charged oxygen endows the molecule with a high energy status, thereby accounting for the prefix, *pyro* (Greek, "fire"). Pyrophosphate prevents apatite formation and growth and hence serves as a potent systemic regulator (inhibitor) of tissue mineralization. Ibsen and Birkedal noted that this anion slowed crystal nucleation and the rate of transformation from an amorphous to a crystalline apatite material [50]. Pyrophosphate also increased the temporal stability of the amorphous phase by binding to the 002 crystal plane of apatite [50].

Much of the circulating pyrophosphate anion derives from ATP, the primary energy-rich molecule in human tissues. Not surprisingly, the turnover of ATP is very high and kilograms of ADP, AMP, and pyrophosphate are generated in the liver daily [51]. ATP is transported into the systemic circulation from the liver as well as the kidney and other tissues by the membrane transporter protein, ATP binding cassette subfamily C member 6 (ABCC6) [52]. ATP is then rapidly broken down to pyrophosphate (and AMP) by the enzyme, ectonucleotide pyrophosphatase/phosphodiesterase (ENPP/NPP)1 (Fig. 9.5) [52].

While ENPP1 is the major regulator of ATP catabolism, two other extracellular proteins, ectonucleoside triphosphate diphosphohydrolase-1 (CD39) and ecto-5′-nucleotidase (CD73) mediate nucleotide breakdown. CD39 catalyzes the conversion of ATP and ADP to AMP, while CD73 converts AMP to adenosine, a nucleoside which inhibits TNAP and reduces pyrophosphate degradation [53, 54]. Of these nucleotide degradative enzymes, ENPP1 is present in osteoblasts, chondrocytes and matrix vesicles [55].

Recently, it has been documented that, like ABCC6, progressive ankylosis protein (ANK) mediates cellular release of ATP (Fig. 9.5) [56]. It is estimated that ANK and ABCC6 generate about 90% of the total pyrophosphate of the systemic circulation [57]. The absence of functional *ANK* in experimental mutant mice leads to pyrophosphate levels in plasma which are approximately 25% lower than those in wild-type controls, [58] a reflection of the significant contribution to pyrophosphate produced by this factor. Reduced local levels of pyrophosphate in such *ANK*-deficient mice explain the extensive calcification found in the cartilage, intervertebral disk, and joint space of the animals [59]. In humans, defective ANK transporter expression leads to hyperplasia of bones of the face and skull, a condition described as craniometaphyseal dysplasia [60]. Moreover, if circulating levels of pyrophosphate are decreased from the normal plasma level of ~2.6 μM [61], there is aberrant mineralization of both hard and soft tissues. Pyrophosphate exerts a similar anti-mineralization property in the urinary tract and saliva, and it is included in dentifrices to slow mineralized calculus formation.

While ENPP1 liberates pyrophosphate from nucleoside phosphates, TNAP, a phenotypic indicator of bone and tooth mineralization and ectoenzyme catalyzing dephosphorylation and transphosphorylation reactions, releases phosphate ions from pyrophosphate, ATP and AMP (Fig. 9.5) [59, 62, 63]. The critical importance of the phosphate anion in the mineralization scheme has been discussed earlier in this book. Besides its central role in apatite formation, an elevation in the activity of phosphate ions influences the expression of genes linked to mineralization. Thus, a localized rise in phosphate ion concentration promotes the differentiation of odontoblasts, [64] cementocytes [65] and osteoblasts [66]. Phosphate ions enhance expression of the phosphate transporter, PIT-1, and the mineralization inhibitors, MGP and OPN [67, 68]. In this way, an increase in cellular phosphate ion levels triggers a molecular feedback loop which, like an accelerator and braking system in an automobile, defines a finely tuned and smooth mechanism for deposition of mineral in the extracellular matrix. From this perspective, intracellular phosphate ion accumulation can be viewed as a key regulatory step in the initiation of mineral deposition [69].

In the context of nucleoside tri- and di-phosphates as possible sources of phosphate ions for bone formation and turnover, it is important to note that polymers of phosphate ions, polyphosphates, exist within the cells of vertebrate mineralized tissues. Unlike pyrophosphates, which are two phosphate ions linked together by a high-energy bond, polyphosphates (phosphagens) form linear polymers of a hundred or so phosphate units joined by a similar phosphoanhydride bond (these types of molecules, briefly mentioned in Chap. 1, were considered as a possible prebiotic energy source). Kinases, and phosphagen kinases in particular, can be used to transfer the free energy ($-\Delta G_0$) of the bonds to generate other energy-rich molecules, such as ATP. One such kinase, creatine kinase, is present in calcifying cartilage [70].

Aside from the intrinsic energy value of polyphosphates, the multiple negatively charged phosphate groups promote ionic interactions with calcium and other anions and lead to the formation of granules (acidocalcisomes). Omelon et al. advanced the idea that these granules serve as a reservoir of phosphate ions for the mineralization process [71]. This hypothesis was supported by the finding that phosphate polymers exist in osteoblasts, and their levels are higher than the levels in cells of non-mineralizing tissues [72–75]. Moreover, other cells within the BMU, such as platelets, could possibly release polyphosphates into the bone microenvironmental niche [76]. Whether polyphosphates act as the brake or the accelerator in the mineralization process is not clear. As was noted earlier, pyrophosphate, which is the smallest

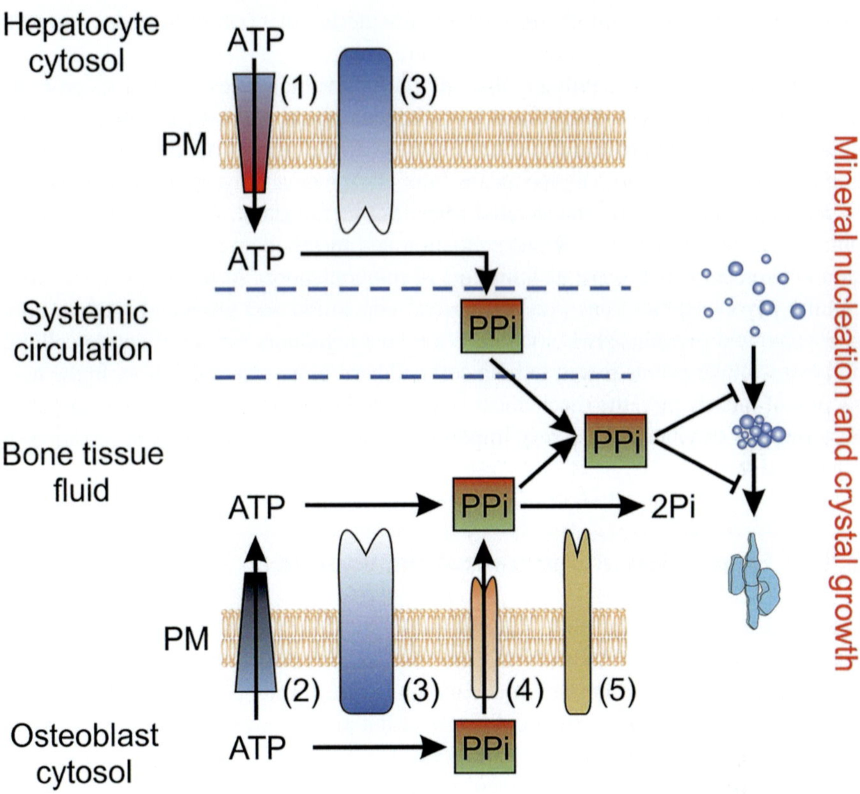

Fig. 9.5 Schematic of inorganic pyrophosphate (PPi) metabolism by liver cells (hepatocytes) and osteoblasts. Hepatocytes utilize an ABCC6 transporter (1) in their plasma membrane (PM) to secrete considerable quantities of ATP into the hepatic circulation. A plasma membrane-bound ectonucleotide pyrophosphatase-phosphodiesterase (ENPP1) (3) converts the ATP into PPi and AMP (AMP not shown). Within mineralizing bone, osteoblasts synthesize and then secrete ATP into bone tissue fluid by means of a P2 purinergic transporter (2) in the osteoblast plasma membrane (PM). Like that in hepatocytes, a plasma membrane-bound ENPP1 (3) in osteoblasts also catalyzes the conversion of ATP into PPi and AMP (AMP not shown). Some intracellular ATP is additionally broken down to PPi, which is then released from the osteoblast by a multi-pass transmembrane protein transporter, progressive ankylosing protein, ANK (4). The bone tissue fluid thus receives PPi from both the systemic circulation and from bone cells. In almost all mineralizing tissues, the turnover of PPi is rapid because of the catalytic activity of membrane-anchored tissue non-specific alkaline phosphatase (TNAP) (5). This enzyme severs the phosphodiester bond of PPi into two molecules of orthophosphate (2Pi). The accumulated PPi serves as a powerful inhibitor of both nucleation of mineral and crystal growth.

member of the polyphosphate family, is a potent inhibitor of apatite formation. Likewise, even in the presence of increased levels of phosphate, low concentrations of polyphosphate (50 μmol) inhibit osteoblast-mediated mineral formation in vitro [77]. On the other hand, Muller et al. reported that cultures of osteoblast-like cells incubated with polyphosphate exhibit a ten-fold increase in the intracellular ATP level accompanied by an elevation in the secretion of ATP into the extracellular milieu [76]. While the metabolic fate of this nucleoside triphosphate is unknown, there are some, albeit limited, grounds for considering that polyphosphates and ATP may contribute to hydroxyapatite formation.

In summary, while it is unlikely that polyphosphate molecules themselves promote apatite formation, these long-chain phosphagens could serve to enhance calcium uptake into cells and provide a source of energy (ATP) for cells mediating tissue mineralization. In contrast to polyphosphate, both phosphate and pyrophosphate ions play increasingly complex and interrelated roles in mineralization. A rise in intracellular phosphate ions influences cell differentiation and metabolism as well as the expression of proteins which serve as inhibitors of mineralization. At the same time, extracellular pyrophosphate ions prevent mineral deposition and growth. Together, both phosphate and pyrophosphate ions form a robust regulatory circuit affecting cellular and extracellular events linked to both soft and hard tissue mineralization. In the next section of this chapter, the mechanism is described by which phosphate ions regulate the synthesis of vitamin D, a very important agent required for the mineralization of bones and teeth.

9.4 Regulation of Mineralization: Vitamin D

Undoubtedly, each of the nine water-soluble and four fat-soluble vitamins in the body influence bone and tooth formation. Two such vitamins, fat-soluble vitamin D (cholecalciferol) and water-soluble vitamin K, are nutrients which influence not only the synthesis of the extracellular matrix but also the transport and metabolism of the major ions of bone and tooth mineral.

Vitamin D is a sterol which influences calcium and phosphate homeostasis, mineral deposition, and several other physiological activities, including muscle contraction, neuromuscular coordination, and hormone release. While vitamin D is a common dietary constituent, deficiency of the nutrient resulting from inadequate intake or insufficient exposure to sunlight is common. In children, the vitamin deficiency causes a growth defect, rickets, with poorly mineralized bones and an increase in bone fracture rate. Low vitamin D levels may also lead to rickets (osteomalacia) in adults. During war time in Britain, the effects of the vitamin deficiency were known, and children were fed cod-liver oil as a prophylaxis against developing the weakened bones and slower growth characteristic of the disease state. In those cases where the defect is attributable to a nutritional deficiency, treatment with vitamin D decreases fracture incidence while increasing bone mineral density [78, 79].

From work by Holick et al., the active form of vitamin D has been identified as 1,25-dihydroxyvitamin D_3 (designated as 1,25$(OH)_2D_3$, 1α,25$(OH)_2D_3$ or 1,25-dihydroxycholecalciferol) [80]. This metabolite is synthesized by the stepwise hydroxylation of 7-dehydrocholecalciferol: The first hydroxylation occurs in the skin, under the influence of sunlight. Bound to a vitamin D binding protein, the molecule is transported to the liver where it is hydroxylated in its C25 position by CYP2R1, an isozyme of the cytochrome P450 system, to form 25-hydroxyvitamin

D_3 (25(OH)D_3), the major circulating form of vitamin D. A further hydroxylation at C1 by the kidney enzyme CYP27B1 generates the active metabolite, 1,25$(OH)_2D_3$. This final hydroxylation is regulated by PTH, calcium and phosphate ions, and fibroblast growth factor FGF23. A third kidney enzyme, CYP24A1, converts high or excess levels of 1,25$(OH)_2D_3$ to the less active 24,25$(OH)_2D_3$ or 1,24,25$(OH)_2D_3$ metabolites (Fig. 9.6) [81].

Like the kidney, osteoblasts can also synthesize the active vitamin D metabolite. Human osteoblasts express the kidney proteins, cubulin and megalin, required for

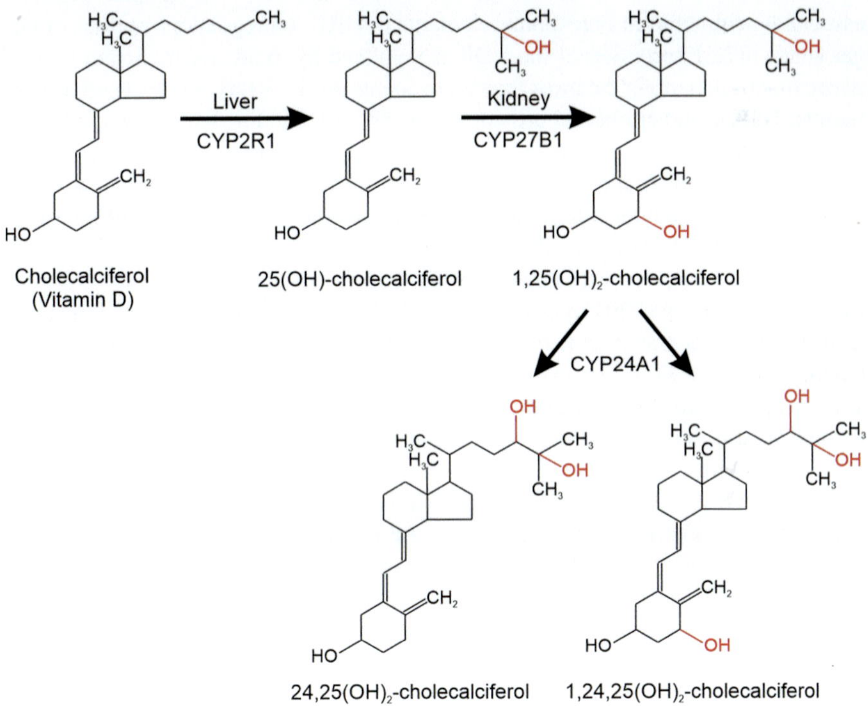

Fig. 9.6 Schematic of the metabolism of cholecalciferol (vitamin D_3). Cholecalciferol is generated photochemically in the skin from the cholesterol derivative, 7-dehydrocholesterol, or it is ingested as a nutrient. In the liver, cholecalciferol undergoes hydroxylation at its C25 position by the enzyme, CYP2R1, and is converted to 25(OH)-cholecalciferol, the major circulating metabolite of vitamin D in humans. A second hydroxylation takes place in the kidney where 25(OH)-cholecalciferol is catalyzed by the enzyme, CYP27B1, to form the polar metabolite, 1,25-dihydroxycholecalciferol, the most active form of the vitamin. CYP27B1 activity is also exhibited by bone cells and is dependent on parathyroid hormone and insulin-like growth factor 1. A third enzyme, CYP24A1, converts 1,25-dihydroxycholecalciferol in the kidney to the less active metabolites, 24,25- and 1,24,25-dihydroxycholecalciferol. CYP24A1 activity is regulated by 1,25-dihydroxycholecalciferol, calcium ions and FGF23. Vitamin D critically affects homeostasis of calcium and phosphate ions and mediates mineral deposition and other physiological activities in vertebrates.

the uptake of $25(OH)D_3$ and the secretion of 1α-hydroxylase, CYP27B1. Thus, bone tissue can both accumulate $25(OH)D_3$ and synthesize $1,25(OH)_2D_3$ [82, 83]. Moreover, osteoblasts also express CYP24A1, the enzyme which degrades $1,25(OH)_2D_3$ [84] to the 24-hydroxylated products, $24,25(OH)_2D_3$ and $1,24,25(OH)_2D_3$. These latter two metabolites are biologically active and enhance the mineralization-related activities of mesenchymal stem cells and osteoblasts [85, 86]. Transcription of vitamin D responsive genes involves the engagement of the active metabolites with the nuclear vitamin D receptor (VDR) and the formation of a heterodimer with the retinoid receptor, X (RXR). Transcription is initiated when the VDR heterodimer complex, together with a number of co-receptor proteins, associates with vitamin D response elements (VDRE) on regulatory regions of target genes [87]. Expression of the VDR is regulated by many agents linked to hard tissue function: Among the most important agents are $1,25(OH)_2D_3$, PTH, glucocorticoids, TGF-β, and epidermal growth factor [88, 89]. Binding sites for these agents can be thousands of base pairs away from the start site of the gene and even within the first two introns in the mouse genome [90, 91].

Vitamin D influences osteoblast activity in many ways. In differentiating osteoblasts, the vitamin promotes the interaction of the transcription factor, CCAAT/enhancer-binding protein (C/EBP)β, with RUNX2 to promote the expression of the *OCN* gene [92]. C/EBPβ also enhances VDR-dependent 24(OH)ase transcription [92, 93]. Together with PTH, vitamin D promotes *FGF23* transcription in osteocytes. Since *FGF23* regulates phosphate metabolism and vitamin D hydroxylation at the level of the kidney, these activities form a circuit controlling bone formation and mineralization [94]. Adding even more complexity to this regulatory loop, vitamin D inhibits PHEX, the protein which suppresses *FGF23* expression [95]. Finally, vitamin D-dependent binding of the VDR to the VDRE causes chromatin remodeling to facilitate the recruitment of histone acetyl transferases (HAT) and demethylases (DME), enzymes which further influence transcriptional activity [96, 97]. To summarize all these interactive relationships is difficult, but it is clear that vitamin D plays a major role in transcriptional regulation of both bone matrix gene expression and, most importantly, systemic phosphate ion homeostasis.

Besides the conventional nuclear response mediated by the VDR complex as mentioned above, there is convincing evidence that vitamin D can activate osteoblasts by binding to membrane receptors [98, 99]. Unlike the slow response of sterols, this effect is rapid and mediated by proteins which include a $1,25(OH)_2D_3$ membrane-associated protein disulfide isomerase, leading to phospholipase A2 and protein kinase C activation and rapid calcium signaling [100, 101].

Along with its iconic role as a required metabolite for bone and cartilage mineralization, in the context of a BMU, vitamin D exerts an antithetical damping effect on mineral deposition. In this regard, cells treated with $1,25(OH)_2D_3$ suppress mineralization by increasing the expression of OPN and pyrophosphate. The increase in pyrophosphate anions in osteoblasts is attributable to the expression of the enzyme, ENPP1, which generates pyrophosphate from ATP [102]. In addition, the rate of osteoclastic bone resorption increases because of RANKL secretion by osteoprogenitor cells [103, 104]. At first glance, the observation that vitamin D both inhibits mineral deposition and promotes bone removal may seem at odds with the stated role of this nutritional factor in promoting bone formation. However, if considered against the broad function of the molecule and likely its evolutionary importance, it is clear that the overriding importance of vitamin D is first to regulate systemic phosphate ion metabolism, second to integrate signals from a host of agents intimately involved with the control of mineral formation, and third to maintain the blood calcium level at the expense of bone calcium and skeletal integrity.

9.5 Regulation of Mineralization: Fetuin-A (α2HS-Glycoprotein, AHSG)

The above discussion emphasizes that systemic factors such as vitamin D, PTH and FGF23 form coupled molecular circuits which control the systemic activities of calcium and phosphate ions and the mineralization-related functions of cells in the BMU. These molecules, together with localized products of metabolism like pyrophosphate, maintain serum ion homeostasis while preventing the untoward precipitation of aberrant foci of mineral in extracellular matrices. With respect to such non-physiological abnormal mineral deposits in systems designed for mineral formation, there is a need for multiple agents to block the precipitation of mineral plaques in soft tissues. Accordingly, the following sections of this chapter are devoted to understanding mechanisms by which the molecules, fetuin and MGP, form an interactive system inhibiting mineral deposition in vertebrate tissues.

The notion of systemic control of mineral deposition is not new. Over 70 years ago, Pedersen discovered a protein in calf serum which blocked calcium phosphate precipitation [105]. Synthesized in the liver, this inhibitory molecule was found to be secreted into the bloodstream and accumulated in bone and dental tissues. Pedersen called the molecule, fetuin; its human homologue is fetuin-A, also known as α2-Heremans-Schmid (HS) glycoprotein (gene nomenclature, *AHSG*) [105]. The protein is present in high levels in fetal blood, and lower levels of the protein are present in human adult serum.

Fetuin is a member of a class of proteins called cystatin domain-containing proteins. The presence of cystatin domains in fetuin prevents the precipitation of mineral from solutions containing elevated activities of calcium and phosphate ions (Fig. 9.7) [106]. The protein also inhibits the deposition of mineral in tendons [107]. The molecular mechanism by which fetuin influences calcium phosphate precipitation is fascinating. Two cystatin-like domains close to one end (N-terminus) of the protein define a region or a sheet of extended negatively charged acidic amino acids (enriched with aspartic acid, glutamic acid, and phosphoserine residues) with four evolutionarily conserved cysteine residues forming two disulfide bonds [108]. The cystatin domains bind calcium ions and calcium phosphate ion clusters. Distances between negatively charged amino acids in the

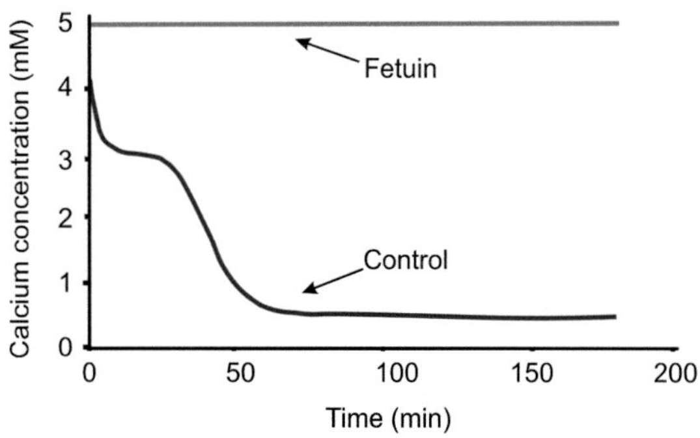

Fig. 9.7 Kinetics of mineral formation in the presence and absence of fetuin. The plot shows the result of an experiment to examine the impact of fetuin on mineral formation in a solution containing 5 mM each of calcium and phosphate ions at pH 7.4 and room temperature. In the absence of fetuin, the calcium ion concentration in a control solution decreased after 1 min and then, following a lag period of about 30 min, dropped dramatically and formed a mineral precipitate. In the presence of fetuin (5 mg/ml), no changes in calcium ion concentration were observed as the solution calcium ion concentration remained at 5 mM. Thus, fetuin completely inhibited the biphasic precipitation of calcium phosphate mineral in the solution. (Image from Price and Lim [106] with permission)

cystatin domains of fetuin are such that they can bind to positively charged calcium atoms on the surface of apatite crystals, related apatite isoforms or amorphous calcium phosphate. By interacting in this manner with calcium ions and these calcium phosphate species, fetuin limits subsequent growth of ion clusters and nucleation centers, thereby preventing aberrant mineral formation. The association between calcium ions and ion clusters and the cystatin domains of fetuin causes the formation of "transiently soluble, colloidal complexes with calcium and phosphate" that Jahnen-Dechent and his colleagues call "calciprotein particles" (Fig. 9.8) [109, 110].

The calciprotein particle complex is thought to consist of a central mineral core enveloped by fetuin. On a volume basis, the particles are comprised of 75% mineral and 25% fetuin [111]. Studies in vitro show that the first formed calciprotein particles (50–100 nm in diameter) gradually transform or mature into particles about 2–3 times larger. The slow maturation of the particles is sufficient for phagocytic cells to remove the calciprotein particles from the systemic circulation [110]. The physiological importance of these stabilized particles is that fetuin ultimately blocks nucleation of mineral as noted above and serves as a powerful inhibitor of spurious mineral formation. In the absence of the circulating protein, deposition of ectopic mineral occurs in tissues as diverse as testis, pancreas and lung (Fig. 9.9) [112]. Fetuin also provides some protection from possible cytotoxic effects of aberrant mineral as the protein interacts with the plasma membrane of vascular smooth muscle cells to minimize membrane disruption [113].

Possibly just as important as its mineral binding sites, fetuin contains a cytokine-binding region close to the disulfide sequence in the N-terminal cystatin domain.

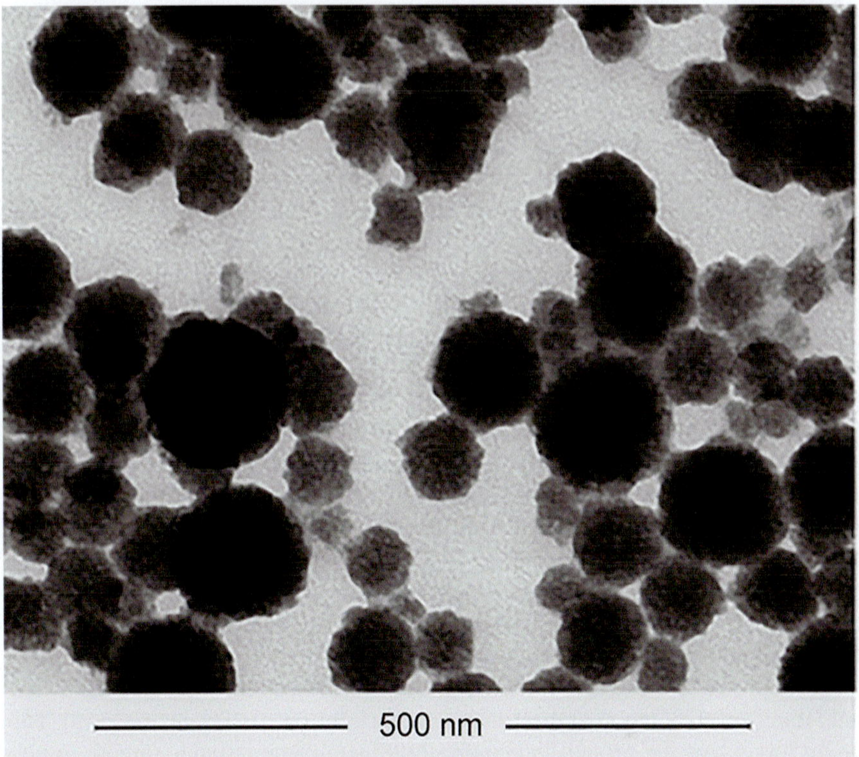

Fig. 9.8 Electron microscopic image of primary calciprotein particles containing fetuin-A and amorphous calcium phosphate. The particles were isolated from a calcium phosphate solution treated with human serum and vary in size (50–100 nm diameter). With time in vitro, primary calciprotein particles are found to convert to larger secondary particles (100–300 nm in size). Such secondary particles are comprised of a core of mineral enclosed by a monolayer of fetuin-A. Fetuin-A in serum and in solution has a very strong affinity for calcium phosphate and binds the mineral phase in a stable colloidal structure to prevent crystal nucleation and growth, thereby inhibiting mineralization. Fetuin-A is one of the most abundant non-collagenous proteins present in bone and one of the most effective inhibitors of mineral formation in the tissue. (Image from Brylka and Jahnen-Dechent [109] with permission)

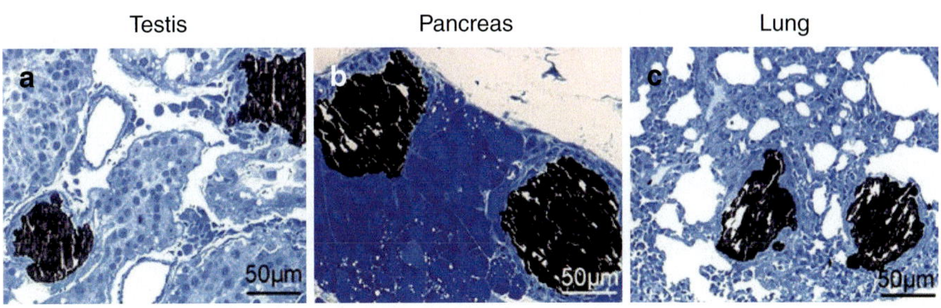

Fig. 9.9 Light microscopic appearance of soft tissues in fetuin-A knockout mice. In animals absent of fetuin, soft tissue samples of testes, pancreas and lung are found with generally spherical-shaped ectopic mineral deposits. Sections shown are undecalcified, plastic-embedded and stained by the von Kossa method for the presence of phosphate as an indicator of mineral (black). (Images adapted from Jahnen-Dechent et al. [109] with permission)

Demetriou and colleagues reported that fetuin bound BMP-2, BMP-4, TGF-β1 and TGF-β2 and it also blocked TGF-β1 activity [114]. Since all these cytokines profoundly influence bone and tooth cell function, fetuin binding would be expected to affect osteogenesis, odontogenesis and mineral deposition (See below) [115].

While the types of studies in vitro described above have provided very useful insights into fetuin structure and function, gene overexpression and mutation studies have led to insights into the physiologic function of the protein. Indeed, deletion of the *AHSG* gene results in mineral deposition in multiple soft tissues (prostate, pancreas, and lung) [116]. Large mineral particles (~50 μm in diameter) are present in such tissues (Fig. 9.9), and specific histological and immunohistochemical staining of the particles lends support for the notion that they grow by slow accretion of mineral on an underlying organic matrix [112]. A more detailed discussion of mineral formation in soft tissues is presented in Chap. 10.

Gene deletion studies show that the loss of the *AHSG* gene product increases cortical bone thickness with little effect on bone density or mineral quality [117, 118]. In addition, there are growth plate deficits with delayed and limited epiphyseal bone development and excessive mineralization of calcified cartilage [117, 118]. These fetuin-dependent defects likely reflect the inability of the low levels of protein to stabilize circulating mineral ions and their clusters effectively, the consequence of which leads to abnormal terminal differentiation and biogenesis of matrix vesicles of growth plate chondrocytes [118].

In addition to the interactions between fetuin, cytokines and mineral ions or clusters, Kapustin et al. reported that fetuin directly influences cell function [119]. This group found that the protein in ectopically mineralizing vascular smooth muscle cells was taken up by endosomes and trafficked to both lysosomes and multivesicular bodies (see Chap. 6). Vesicles released from these cells were similar to exosomes in size and contained several characteristic exosomal membrane proteins (CD 9 and CD 63) as well as proteins linked to the autophagic pathway (lysosome-associated membrane glycoprotein [LAMP]-1 and -2). Thus, it is likely that fetuin affects both endosomal and autophagic pathway activity. If pathway activity is compromised, then functional deficits in exosome filling and release would be expected, an event which would impact physiological mineralization and precipitate non-skeletal calcific deposits. In Chap. 10, examples and the clinical consequences of fetuin dystrophic mineralization are discussed.

It is clear from numerous studies in vitro and in vivo that the serum protein, fetuin, is an important inhibitor of soft tissue calcification, and it plays a significant role in modulating epiphyseal plate maturation and long bone mineralization in vertebrates. Fetuin prevents mineral formation in different ways: Through its cystatin domains, fetuin interacts and binds with calcium ions and ion clusters as well as nascent calcium phosphate particles in the circulation. Such interaction forms circulating colloidal complexes of calciprotein particles, removing the mineral species from further reactivity and hindering mineral accumulation or deposition at extracellular sites. Through a cytokine-binding region near the cystatin motif of the protein, fetuin also reduces the activity of cytokines required for exosome release and subsequent bone formation and growth. Ultimately, macrophages and osteoclasts phagocytose the calciprotein particles. Fetuin, then, supports a novel sequence of molecular and cellular events which maintain mineral ion levels sufficient for normal mineralization of vertebrate hard tissues but not excessive to induce ectopic mineralization of soft tissues.

9.6 Regulation of Mineralization: Matrix Gla Protein (MGP) and Leafy Green Vegetables

As indicated previously, the evolution of the skeleton was accompanied by the development of systems to both regulate physiologic bone and tooth formation and prevent aberrant deposition of mineral. Systemic agents like fetuin and certain local factors, especially proteins synthesized by cells of bone and the dental tissues, inhibit non-physiological mineralization. Of the local agents, MGP, a molecule found in the extracellular matrix of the skeleton and tissues of the tooth, serves as a potent inhibitor of the mineralization process.

MGP is but one of a family of 16 gla (γ-carboxyglutamic acid)-containing proteins in the human body (Fig. 9.10) [120]. Compared with many other blood proteins, MGP is of low molecular weight, about 15 kDa (albumin is three times larger). The protein is present in serum, and, as its name suggests, it is also a component of the extracellular matrix of bone, dentin, cartilage and other vertebrate mineralized tissues. Investigators have shown that vascular smooth muscle cells, chondrocytes, odontoblasts and osteoblasts, cementocytes, and even deer antler cells synthesize the protein [121]. There are no reports of its activity or presence in forming enamel. MGP should not be confused with another gla-containing protein, osteocalcin or bone gla protein (OCN; see Chap. 8) [122]. It is likely that both proteins evolved from a common ancestral gene: MGP appeared during the evolution of cartilaginous fishes and preceded OCN, which was probably present at the time of the later emergence of bony fishes (Osteichthyes) [123]. While the two proteins have similar molecular features, they exhibit separate and non-overlapping biological roles [124].

To recapitulate an earlier discussion concerning the synthesis of MGP, post-translational reactions serve to carboxylate individual glutamate residues of a protein (Fig. 9.10). Critical requirements for the carboxylation reaction include the enzyme, γ-carboxylase, and substrates, carbon dioxide and oxygen. Vitamin K (K = "koagulation," German), in the form of hydroquinone, serves as a cofactor for the reaction; it is present in leafy vegetables as phylloquinone (vitamin K1), which is converted by gut bacteria to menaquinone (vitamin K2). In humans, vitamin K2 is a co-factor in the carboxylation of the γ-glutamyl residues of MGP [125]. The high negative charges mediated by the γ-carboxylated groups of MGP facilitate binding of the protein to free calcium ions, charged phospholipids, and calcium atoms present on the surfaces of apatite crystals. Of note, vitamin K-dependent carboxylation activates many of the blood coagulation proteins (factors VII, IX, X, and prothrombin), which in their carboxylated form also bind calcium ions.

Vitamin K is required for both bone remodeling within the BMU: The vitamin supports osteogenesis while inhibiting bone resorption [126]. In humans, high

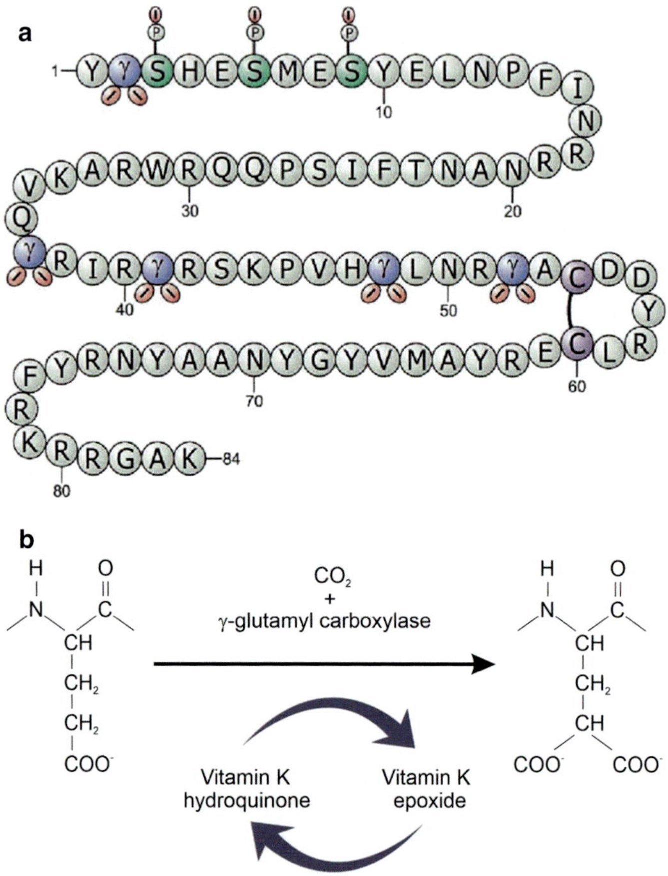

Fig. 9.10 (**a**) Diagram of the amino acid sequence of human matrix gla protein (MGP) and showing (red) the residues which undergo post-translational vitamin K dependent γ-carboxylation. Carboxylation of glutamate (E) occurs at positions 2, 37, 41, 48 and 52; serine residues at 3, 6 and 9 are phosphorylated by a casein kinase. (**b**) Schematic of vitamin K-dependent carboxylation of glutamyl residues of MGP. The carboxylation reaction is catalyzed by γ-glutamyl carboxylase with vitamin K acting as a co-enzyme. The fully reduced vitamin, as the hydroquinone, donates an electron pair to the carboxylation reaction. The resulting reduced compound (vitamin K epoxide) is then regenerated using electrons donated by NADH. (Diagram (**a**) from Schurgers et al. [120] with permission)

levels of vitamin K appear to mitigate aberrant mineral deposition [127]. Rats fed vitamin K-enriched diets show a reversal of arterial calcification, [128] while inhibition of vitamin K by the drug, warfarin (coumadin), increases dystrophic mineral deposition in blood vessels [129]. Of particular interest, it is also reported that, in addition to protecting against cardiovascular disease, vitamin K may extend the human lifespan [130]. Hence, if longevity is a concern, adults, as well as children, should be encouraged to eat leafy vegetables as a good source of phylloquinone or foods high in fermented soybeans, a rich source of menaquinone.

Luo et al. conducted the first definitive study of the role of MGP in tissue mineralization [131]. Deletion of the gene encoding MGP in mice led to overt mineral deposition in the arterial tree. The vessel walls were weak and of low elasticity, and arteries ruptured in animals 2-months-old. At the microscopic level, vessel walls contained cells with hypertrophic phenotypes embedded in a cartilage-like extracellular matrix characterized by type II collagen and high levels of proteoglycans. Matrix vesicle-like structures were apparent at early sites of mineral formation in the vessels. Defects in the growth plate of the animals included disorganization of chondrocyte columns and the presence of mineralized deposits which extended into proliferating zones of the tissue. These changes resulted in mice with short stature, osteopenia, and multiple fractures, features clearly demonstrating that MGP was a key inhibitor of mineralization [131].

Price and colleagues carried out a study related to that of Luo et al., [131] but they fed mice the vitamin K antagonist, warfarin, instead of deleting the *MGP* gene [132]. In this way, they were able to confirm that decarboxylated MGP led to profound increases in the mineralization status of blood vessels, heart valves, and cartilage [130]. Moreover, Schurgers and colleagues showed that, in the presence of warfarin, vascular smooth muscle cells exhibited an increase in calcium deposition and the release of matrix vesicle-like particles enriched in calcium ions [133]. In a more recent study, Marulanda et al. demonstrated that MGP prevented mineral deposition in soft tissues which included permanent cartilage and major blood vessels (Fig. 9.11) [134]. In the absence of the protein, there is deposition of amorphous calcium phosphate in these tissues [134]. To summarize, the loss of *MGP* expression or even the presence of the protein deficient in normal γ-carboxylation causes aberrant mineral precipitation in non-skeletal sites in animal models and the human body.

While rare, mutations in *MGP* (Keutel syndrome) present as anomalous mineralization of soft tissues, together with midfacial hypoplasia and hearing loss [135]. MGP protein deficiency is often the result of loss of function of enzymes required for carboxylation reactions mediated by vitamin K [136] or the effect of prenatal warfarin [137]. In both cases, the phenotype is similar to that described above for defects in the *MGP* gene.

Yagami and colleagues explored the mechanism by which MGP influences the mineralization process [138]. They found that overexpression of *MGP* decreased mineral deposition associated with epiphyseal chondrocytes. In bones of the developing chick limb, virally driven *MGP* overexpression blocked cartilage mineralization and delayed chondrocyte hypertrophy and the formation of diaphyseal intramembranous bone (Fig. 9.12). In other words, while the protein is a local inhibitor of mineralization, it has a more widespread role in regulating many aspects of endochondral and intramembranous ossification. Whether MGP triggers changes at the level of the endosome, autophagy and exosome formation is not known.

Like fetuin, MGP inhibits the activity of cytokines, in particular, BMPs, which actively enhance the mineralization of bone and promote dystrophic mineralization of soft tissues and atherosclerotic plaque [139]. Using a site-directed mutagenesis

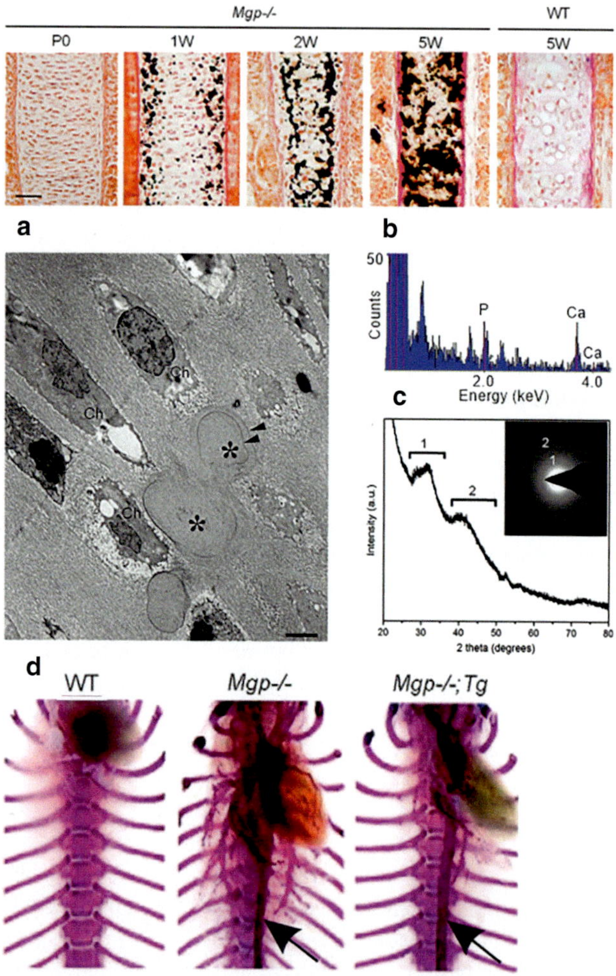

Fig. 9.11 Effects of deletion of *MGP* on the mineralization of nasal cartilage and vascular tissue in mice. Since MGP inhibits mineralization, deletion of the gene would be expected to promote the deposition of mineral in both permanent (non-mineralizing) cartilages as well as soft tissues. To test this hypothesis, the nasal septum, ribs and blood vessels of an MGP knockout ($Mgp^{-/-}$) mouse model were examined microscopically. The series of five images at the top of the figure compares cartilage mineralization in nasal septum of knockout mice over an initial (P0) to 5-week (W) period and of a normal (wild type [WT]) mouse at 5 W. There is a progressive increase in abnormal cartilage mineralization in knockout mice as detected in sections of von Kossa- and van Gieson-stained plastic embedded specimens. The WT animal does not mineralize over 5 W. Scale bar = 40 μm. (**a**) Transmission electron micrograph of the nasal cartilage of an *MGP* knockout mouse. Chondrocytes (Ch) are observed together with rounded mineral deposits (*) enclosed by incremental lines (arrowheads), suggestive of episodic mineral deposition. (**b**) Energy-dispersive X-ray microanalysis of the deposits shown in (**a**) yields peak signals of calcium and phosphorus. (**c**) X-ray diffraction analysis of deposits in (**a**) suggests the presence principally of an amorphous calcium phosphate phase with two peaks (1, 2) indicative of early crystallization. Inset, an electron diffraction pattern of deposits in (**a**) showing two diffuse rings of reflections (1, 2) of an amorphous calcium phosphate phase. (**d**) Histological view of rib cages from mice 5-weeks-old and stained with alizarin red and Alcian blue. Ectopic mineral is present in the descending aorta of the $Mgp^{-/-}$ mice (arrowheads) but not in the WT animal. Weak expression of the *MGP* transgene in knockout mice ($Mgp^{-/-};Tg$) failed to prevent vascular calcification (arrowheads). Scale bar in (**a**) = 100 μm. (Images adapted from Marulanda et al. [134] with permission)

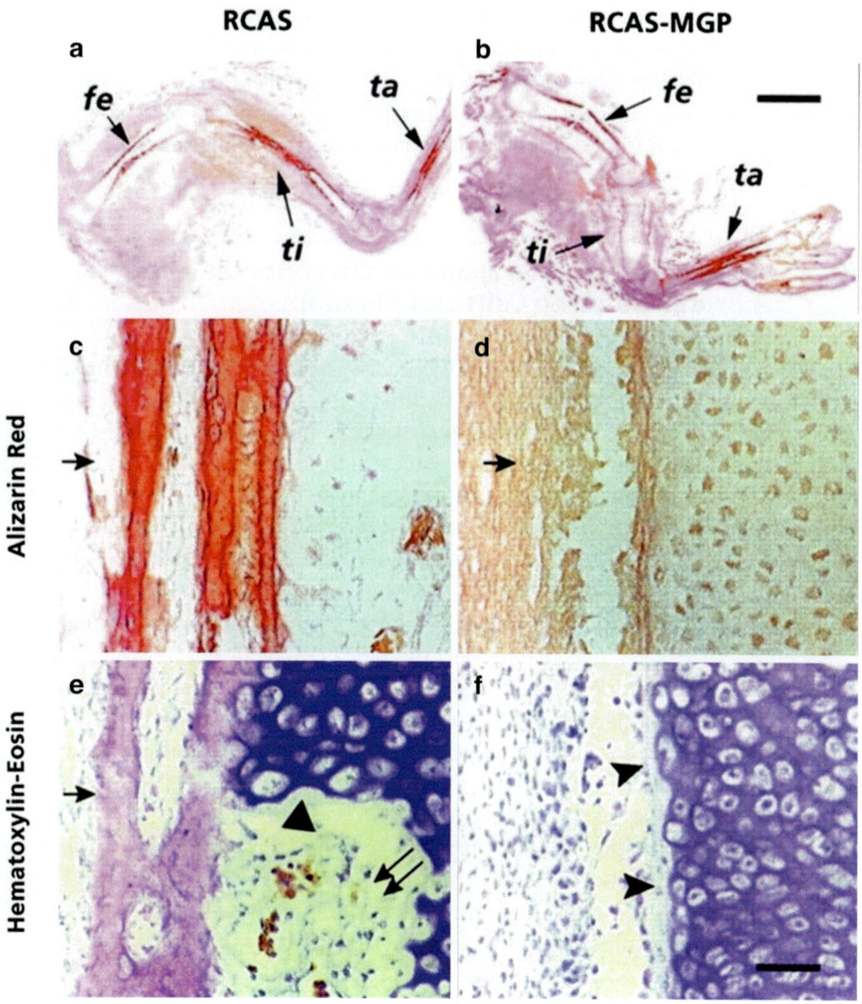

Fig. 9.12 *MGP* overexpression during long bone development in embryonic chicks. (**a**) In a control limb (RCAS, a retroviral vector), the femur (fe), tibia (ti), and tarsal (ta) elements exhibit normal morphology and their diaphyses stain positively with alizarin red. (**c**, **e**) The elements consist in part of an intramembranous bone collar (**c** and **e**, arrow), fully hypertrophic chondrocytes (**e**, arrowhead) and invading marrow cells (**e**, double arrows). (**b**) In a limb overexpressing *MGP* (RCAS-MGP), the tibia (ti) is completely cartilaginous and does not stain with alizarin red. (**d**, **f**) The tibial diaphysis has neither a bone collar (**d**, arrow) nor mature hypertrophic chondrocytes (**f**, arrowheads). Scale bars (**a** and **b**) = 2.5 mm; (**c–f**) = 80 μm. (Images from Yagami et al. [138] with permission)

technique, which permitted analysis of the role of different regions of the *MGP* gene, Yao et al. showed that a conserved proline residue (proline 64) was bound to and "functionally intertwined" with BMP4 in bovine endothelial cells [139]. The results of this study suggest that obstructing BMP signaling by MGP might be an innovative approach to prevent aberrant mineralization, a common consequence of vascular disease [140].

9.7 Regulation of Mineralization: Osteopontin (OPN), a Bridge between Cells and Mineral

The SIBLING protein, osteopontin (OPN), is an additional powerful inhibitor of vertebrate mineralization. The name of this acidic phosphorylated protein was meant to imply that it formed a bridge (Latin, "pontem") between cells, mineral, and the extracellular matrix in the tissues in which it has been detected, including bone, articular and growth plate cartilage, cementum, dentin, and the normally mineralizing tendons of certain avian species [141–145]. As noted in Chap. 8, OPN is a highly regulated, multifunctional protein with activities involved with immune function, inflammation, chemotaxis, and apoptosis. OPN exists as matrix-bound and also as a circulating and secreted soluble protein. In bone, it associates with extracellular matrix proteins like fibronectin, type I collagen, and osteocalcin [146]. Enzymes (transglutaminases) link multiple OPN molecules together to form high molecular weight complexes, [147] similar to those present in the calcified arteries of OPN-deficient mice [148]. The protein is linked to the pathogenesis of numerous diseases, among them cancer metastases, autoimmune disease, deposition of kidney stones, cardiovascular disease, diabetes, and osteoarthritis [149].

Besides these multiple linked forms of the protein, OPN undergoes considerable post-translational modification and may be both glycosylated and phosphorylated to variable degrees [150]. Phosphorylation of the acidic serine-aspartate-rich motif/MEPE-associated peptide (ASARM) motif of OPN is critical, leading to binding of the protein with mineral and inducing subsequent changes in apatite structure [151–153]. The inhibition of mineralization is regulated by the cleavage of ASARM peptides by PHEX (see Chap. 8) [151]. On phosphorylation, the highly charged protein, bound to calcium ions and mineral, may be detected at tissue sites where there is active mineral deposition [154]. Interaction by OPN with apatite is through weak ionic bonds which form between negatively charged phosphorylated amino acids of the protein and positively charged calcium atoms on the surface of the mineral [155]. In addition to its binding mineral, physicochemical studies demonstrate that OPN also blocks crystal growth [150, 152, 156].

The phosphorylation status of OPN appears to be related to opposing functions of the protein. OPN inhibits mineralization when the molecule is phosphorylated, but its inhibitory activity diminishes on dephosphorylation [157, 158]. To examine these different mineralization effects, Pampena et al. investigated various peptide fragments of OPN and found that two, each containing phosphoserine, were particularly inhibitory to hydroxyapatite formation [155]. Building on these findings, Boskey et al. developed a model system to mimic the physical-chemical aspects of mineralization in vitro and confirmed that some OPN protein fragments promoted mineralization while others inhibited mineral deposition and dephosphorylation caused a loss of mineralization activity [152]. These observations add strength to the notion that phosphorylated OPN is an inhibitor of mineralization but specific domains of the protein may support apatite formation.

Besides its post-translational state, an additional factor which may be important for the opposing mineralization-related functions of OPN is the three-dimensional intrinsically disordered structure of the protein (see Chap. 8). Such disorder allows OPN to assume many different conformations and, in its most flexible configuration, the molecule may interact with other proteins and charged components of cell membranes. In this regard, the protein may bind αvβ1 and αvβ5 integrin receptors as well as CD44, the hyaluronic acid receptor, on the osteoblast membrane. Since the phosphorylated protein also engages αvβ3 integrin receptors on osteoclasts, influencing the motility of these cells, there is the likelihood that OPN interaction with both osteoblasts and osteoclasts inhibits mineralization while it also promotes bone remodeling [159].

As has been discussed previously, the most definitive way to examine the importance of a particular protein is to mutate or delete the encoding gene. Ritting et al. used this approach to assess the role of OPN in skeletal formation [160]. Following targeted disruption of the *OPN* locus in mice, the mutant animals appeared to be unaffected developmentally and showed no overt structural changes in their bones and teeth [160]. The study, on the other hand, reported modified osteoclast formation and suggested that OPN affected osteoclast differentiation but was not necessary for normal development and bone deposition in the animals [160]. A correlative analysis of the mutant mice indicated increased mineral size and crystallinity and over-mineralized bones [161]. These findings strongly support the physical-chemical studies noted above concluding that OPN is predominantly an inhibitor of mineralization.

In summary, OPN is a multifunctional, highly phosphorylated protein with a wide range of functions. In dentin and bone, OPN is synthesized by odontoblasts and osteoblasts, respectively; osteoclasts and other cells of additional vertebrate mineralized tissues also express this protein. OPN binds integrin receptors on the osteoblast membrane and RGD motifs in the protein bind αvβ3 integrin receptors on osteoclasts. Thus, as a secreted or bound protein, OPN functions in the BMU, enhancing the interactions of osteoblasts and osteoclasts. Like other intrinsically disordered SIBLING proteins, OPN exerts two effects on mineralization: it stabilizes apatite and inhibits new mineral formation and growth, both events dependent on the phosphorylation status of the protein. Finally, while mutational studies indicate a requirement of OPN for mineral development, there is no overt phenotype exhibited by the affected animals. In these instances, other SIBLING proteins likely compensate for OPN in regulating mineralization-related events.

9.8 Perspective: Local and Systemic Regulation of Mineralization and the Role of Coupling Factors, Pyrophosphate, Polyphosphates, Vitamin D, Fetuin, Matrix Gla Protein, and Osteopontin

As described and emphasized earlier in this book, the systemic circulation is normally supersaturated with respect to calcium and phosphate ion concentrations and should readily precipitate a calcium phosphate mineral phase in the absence of constraints, not unlike the misfortune of Lot's wife, Edith. The formation of mineral, therefore, is subject to exquisite regulation and control at several levels of structure and solution chemistry. Such rigorous oversight preventing aberrant mineralization occurs through two principal and critical modes as this chapter details: The coordinated regulation of bone homeostasis in the BMU, that is, new bone formation, mineralization, and degradation, and the control of mineral deposition by both systemic and local factors. This former event at the level of the BMU involves groups of osteoblasts which line the bone surface and maintain osteoid synthesis and secretion to support subsequent apatite deposition. The bone building activities of these cells spatially and temporally are correlated and coupled with bone resorptive functions of osteoclasts, which communicate with and recruit their counterpart osteoblasts to complete the process of bone remodeling.

In the event that the equilibrium between bone formation and bone resorption is disturbed, clinical disease develops. Indeed, the most common human skeletal disorder, post-menopausal osteoporosis, is attributable to a decrease in bone apposition and an increase in resorption activity. As was pointed out earlier, many of the molecules or agents involved in regulating these complex coupled events are known and their effects occur systemically or locally. Systemic regulation involves estrogen, PTH, fetuin, and the vitamin D metabolite, 1,25-dihydroxycholecalciferol. At the local level of the BMU, control is mediated by RANKL, OPG, semaphorins, SOST, FGF23 and EPHB4-EPHB2. At the local bone surface, integrin-binding proteins such as fibronectin and OPN promote cell adhesion and attachment to extracellular tissue matrices while the charged anion, pyrophosphate, and calcium-binding proteins like MGP serve to prevent aberrant mineralization.

In dental tissues, coupled cell-cell interactions govern the activities of odontoblasts, ameloblasts and cementocytes. There is overwhelming evidence that ameloblast activity is dependent on molecular cues from odontoblasts as well as from the surrounding mesoderm. Likewise, niche conditions, as well as molecular signals from the underlying mesoderm, regulate predentin and cementum synthesis, secretion and mineral deposition. While coupling factors for enamel and dentin formation and mineralization have yet to be defined, studies of the enamel knot (see Chap. 4) suggest that such factors are similar to those in bone.

In the context of cell-cell interactions and molecular coupling, osteoblasts and osteoclasts coordinate bone formation and resorption through various means: Close contacts which permit the interplay of membrane-localized ligands and receptors between the cells; secretion of diffusible signaling molecules such as RANKL from osteoblasts to receptors on osteoclast surfaces as well as release of SOST, TGF-β,

BMPs and other molecules to the intercellular milieu; and passage of water-soluble elements across gap junctions, for example. Osteocytes also participate in coupling despite their being buried in the mineralized bone matrix. These cells, often overlooked in the coupling process, [46] comprise and maintain dominant numbers in trabecular and cortical bone, and they utilize an extensive lacuno-canalicular network for cell-cell communication.

It is now evident that osteocytes are sensitive to the oxemic status of the tissue in which they reside, and they regulate bone formation through WNT and BMP signaling pathways and through the synthesis and secretion of SOST and FGF23 [48]. SOST binds to membrane receptors on osteoblasts to inhibit WNT signaling and thereby to decrease mineral formation. In the forming tooth, SOST has a similar function in inhibiting dentin formation [162]. FGF23 slows mineralization in several different ways: By depleting the amount of phosphate ions available for apatite formation, suppressing vitamin D metabolite synthesis, and acting directly on bone cell anabolic function independently of systemic effects [163]. As noted above, osteocytes also secrete OPG, which complexes with RANKL and makes this ligand less available to activate osteoclasts and bone resorption. Thus, osteocytes, through the secretion of SOST and FGF23, together with other coupling molecules such as OPG and DMP1, play a central role in the regulation of osteoblast-osteoclast interactions which significantly influence critical events leading to new bone formation and degradation.

In addition to the coordinated coupling effects exerted principally by osteoblasts, osteoclasts and osteocytes in the BMU, a second level of mineralization control involves the activities of a group of proteins which inhibit mineral deposition in the extracellular matrix. Since the calcium phosphate ion product is high in this domain, rendering tissue fluid metastable, mineral deposition can be considered a critical default state in vertebrate tissues. Without these inhibitory proteins, then, even a slight increase in the metastable state would induce heterotopic mineral deposition in both skeletal and extraskeletal sites. Thus, there is an absolute need for molecular inhibitors which both regulate mineralized tissue formation, resorption and turnover and prevent aberrant mineral precipitation in non-skeletal tissues. Of several such inhibitory or regulatory molecules, pyrophosphate, vitamin D, fetuin, and MGP are now known to be particularly important at the systemic or local levels of vertebrate tissue structure and function.

Unlike vitamin D, fetuin or MGP, pyrophosphate is a very small molecule generated from circulating (mainly hepatic) ATP as well as ATP secreted by bone cells. As a result of its size, pyrophosphate is freely accessible to cells of the BMU. The anion is the end product of reactions mediated by ENPP1, which hydrolyses ATP to release pyrophosphate and AMP [62]. The actual levels of pyrophosphate decrease at mineralizing sites because of the activity of TNAP, which catalyzes the dissociation of all ester phosphates into orthophosphate ions [164]. Steady state levels of pyrophosphate therefore depend on differences in the rates of activities of the two enzymes, ENPP1 and TNAP [59]. A study utilizing vascular smooth muscle cells showed that the rate of pyrophosphate hydrolysis to phosphate is ten times slower than the rate of pyrophosphate synthesis [165]. Thus, even when TNAP expression levels are high, pyrophosphate accumulates at sites of mineralization and exerts its considerable inhibitory effects on mineral deposition.

Although the pyrophosphate anion also carries a strong negative charge, it can diffuse into fibrous proteins like collagen, and, like fetuin and MGP, limit mineral deposition at micromolar concentrations. Adsorbed onto specific sites of the apatite crystal, pyrophosphate inhibits crystal growth and development as well as subsequent mineral phase transitions [166]. Within bone itself, the localized generation of pyrophosphate by osteocytes may prevent mineralization of their lacunar space. In that regard, Hajjawi et al. noted that deletion of *ENPP1* caused a significant reduction in the number and size of osteocyte lacunae [167].

Vitamin D is a multi-functional sterol which mediates many physiological actions, including the regulation and control of mineral homeostasis. Its nutritional deficiency can lead to bones which are characteristically weak, poorly mineralized, and easily fractured. Synthesis and processing by the kidney and osteoblasts of the active metabolite of vitamin D, 1,25-dihydroxyvitamin D_3 (1,25$(OH)_2D_3$), ultimately leads to the two hydroxylated products, 24,25$(OH)_2D_3$ and 1,24,25$(OH)_2D_3$, which can both affect osteoblast activity in systemic and local control of mineralization events. In this regard, vitamin D mediates expression of OCN in differentiating osteoblasts by enhancing C/EBPβ and RUNX2 interaction, and it also binds to osteoblast membranes to activate the cells and support bone formation. On the other hand, vitamin D increases pyrophosphate and the expression of OPN in osteoblasts in vitro. The latter molecules are both inhibitory to mineral deposition. Further, vitamin D promotes FGF23 transcription in osteocytes, and in turn FGF23 inhibits mineralization as noted previously. Additionally, vitamin D suppresses PHEX, which negatively regulates FGF23 expression. These complex and seemingly opposing functions for vitamin D demonstrate that this sterol is central to several critical activities in vertebrate tissues, principal of which is proper maintenance of systemic phosphate metabolism accompanied as well by precise and strict control and regulation of mineral formation and circulating blood calcium levels.

Fetuin functions as a powerful inhibitor of mineralization. Indeed, fetuin is ten times more powerful than albumin, another prominent circulating calcium-binding protein in vertebrates. In contrast to MGP, fetuin binds calcium and phosphate ions in colloidal complexes (calciprotein particles) and thereby prevents the ions from cluster formation and development of nucleation centers [111]. Fetuin is excluded from a densely packed tissue like compact bone or dentin; instead, the protein interacts with cells lining the superficial surfaces of tissues or organs. Toroian et al. reported that the small molecular dimensions of hydrated collagen fibrils restricted entry of larger fetuin molecules into intrafibrillar spaces of the fibrils [168]. Nonetheless, fetuin exerts a major effect on cells on the surface of the osteoid in the BMU [169].

Formation of calciprotein particles is somewhat at variance with conventional thinking, which views mineral ions in the circulation as either freely diffusible or attached to albumin. Clearly, however, there is ample evidence documenting the enormous capacity of these particles to complex mineral ions and additionally suggesting that such particles influence both formative and resorptive activities of cells in the BMU. For example, the calciprotein particles may serve to transport calcium-phosphate ions to sites of mineralization. Further, the dimensions of these particles are such that they can be taken up by macrophages and other cells and cleared from the circulation.

The phenotype of the *AHSG* gene knockout in animal models is less severe than that of *MGP* and it presents as a small decrease in bone length, indicating compromised epiphyseal growth [117]. The actual phenotype of these animals is probably confounded by the fact that fetuin associates with growth factors like TGF-β and BMP, [114] and hence it is difficult to parse changes attributable to fetuin from those resulting from changes in associated proteins. Even so, the reported phenotypic changes would be expected to influence the strength of bones over a lifetime or manifest functional failure following a major challenge to this tissue.

MGP is synthesized by osteoblasts and chondrocytes, secreted systemically and localized within the matrices of bone and cartilage [131]. This protein is comprised of a number of γ-carboxylated glutamate residues in its amino acid sequence and these moieties in other gla-containing proteins such as OCN avidly bind calcium from solution [124] and may be immunolocalized in vivo to mineral or collagen fibrils in sections of vertebrate tissues [122, 170]. Unlike the colloidal complexes formed on binding of calcium and phosphate ions by fetuin, [109] a molecular mechanism describing MGP inhibition of mineral deposition in vivo remains unclear, whether by MGP interaction with mineral ions or their clusters, poisoning of apatite crystal surfaces, stabilization of early calcium phosphate mineral phases, or some other set of events. However, despite this uncertainty in understanding the precise basis for MGP action, the inhibitory effects of the protein on mineralization of vertebrate tissues are well recognized through numerous investigations in vitro and in vivo and as noted in this chapter. In *MGP* knockout studies of mice, there is expression of a distinct phenotype: Abnormal calcification in cartilage and arteries of the animals, short bone length, osteopenia, fractures, and death caused by rupture of the calcified aorta [119]. In humans, the recessive disorder of *MGP* expression is characterized by unusual cartilage mineralization, hearing loss, midfacial hypoplasia, and ectopic mineral deposition. Determining how these phenotypic manifestations occur will require more complete knowledge of structural, molecular biological, and biochemical analyses of MGP and its association with specific ions, cells, tissues, and matrices.

Finally, studies of fetuin, MGP, and other inhibitory proteins emphasize the widespread cellular utilization of endomembranous vesicles to generate mineralizing matrix vesicle-like particles through either the exosomal pathway or membrane blebbing. Such reports document accumulation of intracellular calcium and phosphate ions and then their release to extracellular matrices either as free ions or more likely ions compartmentalized within vesicles. The absence of a protein like fetuin or MGP significantly impairs intracellular calcium transport, an action which commonly leads to disease. Details of these and other forms of aberrant mineralization are discussed in the following chapter of this book.

References

1. Vesalius A. On the fabric of the human body. Basel; 1543.
2. Courpron P, Meunier P, Vignon G. Dynamics of bone remodeling explained by Harold Frost. Theory of the B.M.U. (basic multicellular unit). Nouv Press Med. 1975;4:421–4.
3. Frost HM. Perspectives: a proposed general model of the "mechanostat" (suggestions from a new skeletal-biologic paradigm). Anat Rec. 1996;244:139–47.
4. Manolagas SC. Birth and death of bone cells: basic regulatory mechanisms and implications for the pathogenesis and treatment of osteoporosis. Endocr Rev. 2000;21:115–37.
5. Poole KE, van Bezooijen RL, Loveridge N, Hamersma H, Papapoulos SE, Löwik CW, et al. Sclerostin is a delayed secreted product of osteocytes that inhibits bone formation. FASEB J. 2005;19:1842–4.
6. Lacey DL, Timms E, Tan HL, Kelley MJ, Dunstan CR, Burgess T, et al. Osteoprotegerin ligand is a cytokine that regulates osteoclast differentiation and activation. Cell. 1998;93:165–76.
7. Bonewald LF. The amazing osteocyte. J Bone Miner Res. 2011;26:229–38.
8. Rodan GA, Martin TJ. Role of osteoblasts in hormonal control of bone resorption--a hypothesis. Calcif Tissue Int. 1981;33:349–51.
9. Sims NA, Martin TJ. Osteoclasts provide coupling signals to osteoblast lineage cells through multiple mechanisms. Annu Rev Physiol. 2019;82:507–29.
10. Sims NA, Martin TJ. Coupling the activities of bone formation and resorption: a multitude of signals within the basic multicellular unit. Bonekey Rep. 2014;3:481.
11. Chen X, Wang Z, Duan N, Zhu G, Schwarz EM, Xie C. Osteoblast-osteoclast interactions. Connect Tissue Res. 2018;59:99–107.
12. Matsuo K, Irie N. Osteoclast-osteoblast communication. Arch Biochem Biophys. 2008;473:201–9.
13. Siddiqui JA, Partridge NC. Physiological bone remodeling: systemic regulation and growth factor involvement. Physiology (Bethesda). 2016;31:233–45.
14. Xiao W, Wang Y, Pacios S, Li S, Graves DT. Cellular and molecular aspects of bone remodeling. Front Oral Biol. 2016;18:9–16.
15. Onan D, Allan EH, Quinn JM, Gooi JH, Pompolo S, Sims NA, et al. The chemokine Cxcl1 is a novel target gene of parathyroid hormone (PTH)/PTH-related protein in committed osteoblasts. Endocrinology. 2009;150:2244–53.
16. Wu AC, Morrison NA, Kelly WL, Forwood MR. MCP-1 expression is specifically regulated during activation of skeletal repair and remodeling. Calcif Tissue Int. 2013;92:566–75.
17. Wise GE, Frazier-Bowers S, D'Souza RN. Cellular, molecular, and genetic determinants of tooth eruption. Crit Rev Oral Biol Med. 2002;13:323–34.
18. Li X, Liu H, Qin L, Tamasi J, Bergenstock M, Shapses S, et al. Determination of dual effects of parathyroid hormone on skeletal gene expression in vivo by microarray and network analysis. J Biol Chem. 2007;282:33086–97.
19. Lee HH, Moon Y, Shin JS, Lee JH, Kim TW, Jang C, et al. A novel mPGES-1 inhibitor alleviates inflammatory responses by downregulating PGE2 in experimental models. Prostaglandins Other Lipid Mediat. 2019;144:106347.
20. Delgado-Calle J, Sato AY, Bellido T. Role and mechanism of action of sclerostin in bone. Bone. 2017;96:29–37.
21. Pederson L, Ruan M, Westendorf JJ, Khosla S, Oursler MJ. Regulation of bone formation by osteoclasts involves Wnt/BMP signaling and the chemokine sphingosine-1-phosphate. Proc Natl Acad Sci U S A. 2008;105:20764–9.
22. Weivoda MM, Ruan M, Hachfeld CM, Pederson L, Howe A, Davey RA, et al. Wnt signaling inhibits osteoclast differentiation by activating canonical and noncanonical cAMP/PKA pathways. J Bone Miner Res. 2016;31:65–75.
23. Everts V, Delaisse JM, Korper W, Jansen DC, Tigchelaar-Gutter W, Saftig P, et al. The bone lining cell: its role in cleaning Howship's lacunae and initiating bone formation. J Bone Miner Res. 2002;17:77–90.

24. Takayanagi H, Kim S, Koga T, Nishina H, Isshiki M, Yoshida H, et al. Induction and activation of the transcription factor NFATc1 (NFAT2) integrate RANKL signaling in terminal differentiation of osteoclasts. Dev Cell. 2002;3:889–901.
25. Xing L, Bushnell TP, Carlson L, Tai Z, Tondravi M, Siebenlist U, et al. NF-kappaB p50 and p52 expression is not required for RANK-expressing osteoclast progenitor formation but is essential for RANK- and cytokine-mediated osteoclastogenesis. J Bone Miner Res. 2002;17:1200–10.
26. Wada T, Nakashima T, Hiroshi N, Penninger JM. RANKL-RANK signaling in osteoclastogenesis and bone disease. Trends Mol Med. 2006;12:17–25.
27. Walsh MC, Kim N, Kadono Y, Rho J, Lee SY, Lorenzo J, et al. Osteoimmunology: interplay between the immune system and bone metabolism. Annu Rev Immunol. 2006;24: 33–63.
28. Mun SH, Park PSU, Park-Min KH. The M-CSF receptor in osteoclasts and beyond. Exp Mol Med. 2020;52:1239–54.
29. Marks SC Jr, Wojtowicz A, Szperl M, Urbanowska E, MacKay CA, Wiktor-Jedrzejczak W, et al. Administration of colony stimulating factor-1 corrects some macrophage, dental, and skeletal defects in an osteopetrotic mutation (toothless, tl) in the rat. Bone. 1992;13:89–93.
30. Sakai H, Chen Y, Itokawa T, Yu KP, Zhu ML, Insogna K. Activated c-Fms recruits Vav and Rac during CSF-1-induced cytoskeletal remodeling and spreading in osteoclasts. Bone. 2006;39:1290–301.
31. Qin L, Raggatt LJ, Partridge NC. Parathyroid hormone: a double-edged sword for bone metabolism. Trends Endocrinol Metab. 2004;15:60–5.
32. Yang Y, Wang B. Disruption of beta-catenin binding to parathyroid hormone (PTH) receptor inhibits PTH-stimulated ERK1/2 activation. Biochem Biophys Res Commun. 2015;464:27–32.
33. Allan EH, Hausler KD, Wei T, Gooi JH, Quinn JM, Crimeen-Irwin B, et al. EphrinB2 regulation by PTH and PTHrP revealed by molecular profiling in differentiating osteoblasts. J Bone Miner Res. 2008;23:1170–81.
34. Zhao C, Irie N, Takada Y, Shimoda K, Miyamoto T, Nishiwaki T, et al. Bidirectional ephrinB2-EphB4 signaling controls bone homeostasis. Cell Metab. 2006;4:111–21.
35. Hayashi M, Nakashima T, Taniguchi M, Kodama T, Kumanogoh A, Takayanagi H. Osteoprotection by semaphorin 3A. Nature. 2012;485:69–74.
36. Durdan MM, Azaria RD, Weivoda MM. Novel insights into the coupling of osteoclasts and resorption to bone formation. Semin Cell Dev Biol. 2022;123:4–13.
37. Grewe JM, Knapstein PR, Donat A, Jiang S, Smit DJ, Xie W, et al. The role of sphingosine-1-phosphate in bone remodeling and osteoporosis. Bone Res. 2022;10:34.
38. Negishi-Koga T, Shinohara M, Komatsu N, Bito H, Kodama T, Friedel RH, et al. Suppression of bone formation by osteoclastic expression of semaphorin 4D. Nat Med. 2011;17:1473–80.
39. Tu AW, Luo K. Acetylation of Smad2 by the co-activator p300 regulates activin and transforming growth factor beta response. J Biol Chem. 2007;282:21187–96.
40. Pejanovic N, Hochrainer K, Liu T, Aerne BL, Soares MP, Anrather J. Regulation of nuclear factor kappaB (NF-kappaB) transcriptional activity via p65 acetylation by the chaperonin containing TCP1 (CCT). PLoS One. 2012;7:e42020.
41. Wein MN, Spatz J, Nishimori S, Doench J, Root D, Babij P, et al. HDAC5 controls MEF2C-driven sclerostin expression in osteocytes. J Bone Miner Res. 2015;30:400–11.
42. Khosla S, Oursler MJ, Monroe DG. Estrogen and the skeleton. Trends Endocrinol Metab. 2012;23:576–81.
43. Verborgt O, Tatton NA, Majeska RJ, Schaffler MB. Spatial distribution of Bax and Bcl-2 in osteocytes after bone fatigue: complementary roles in bone remodeling regulation? J Bone Miner Res. 2002;17:907–14.
44. Manolagas SC, Parfitt AM. What old means to bone. Trends Endocrinol Metab. 2010;21:369–74.

45. Weitzmann MN. T-cells and B-cells in osteoporosis. Curr Opin Endocrinol Diabetes Obes. 2014;21:461–7.
46. Buenzli PR, Sims NA. Quantifying the osteocyte network in the human skeleton. Bone. 2015;75:144–50.
47. Dallas SL, Prideaux M, Bonewald LF. The osteocyte: an endocrine cell … and more. Endocr Rev. 2013;34:658–90.
48. Zhu J, Tang Y, Wu Q, Ji YC, Feng ZF, Kang FW. HIF-1alpha facilitates osteocyte-mediated osteoclastogenesis by activating JAK2/STAT3 pathway in vitro. J Cell Physiol. 2019;234:21182–92.
49. Bonewald LF. Scientists make no bones about first study of osteocyte cultures on space station. 2011. https://phys.org/news/2015-04-scientists-bones-osteocyte-cultures-space.html.
50. Ibsen CJS, Birkedal H. Pyrophosphate-inhibition of apatite formation studied by in situ X-ray diffraction. Fortschr Mineral. 2018;8:65–73.
51. Russell RG, Wadstrom LB, Lindstedt S, Care AD, Bisaz S, Fleisch H. The origin of inorganic pyrophosphate in urine. Clin Sci. 1969;37:419–29.
52. Russell RG. Metabolism of inorganic pyrophosphate (PPi). Arthritis Rheum. 1976;19:465–78.
53. Plesner L. Ecto-ATPases: identities and functions. Int Rev Cytol. 1995;158:141–214.
54. Deaglio S, Dwyer KM, Gao W, Friedman D, Usheva A, Erat A, et al. Adenosine generation catalyzed by CD39 and CD73 expressed on regulatory T cells mediates immune suppression. J Exp Med. 2007;204:1257–65.
55. Terkeltaub RA. Inorganic pyrophosphate generation and disposition in pathophysiology. Am J Physiol Cell Physiol. 2001;281:C1–C11.
56. Ralph D, van de Wetering K, Uitto J, Li Q. Inorganic pyrophosphate deficiency syndromes and potential treatments for pathologic tissue calcification. Am J Pathol. 2022;192:762–70.
57. Jansen RS, Duijst S, Mahakena S, Sommer D, Szeri F, Váradi A, et al. ABCC6-mediated ATP secretion by the liver is the main source of the mineralization inhibitor inorganic pyrophosphate in the systemic circulation-brief report. Arterioscler Thromb Vasc Biol. 2014;34:1985–9.
58. Szeri F, Niaziorimi F, Donnelly S, Fariha N, Tertyshnaia M, Patel D, et al. The mineralization regulator ANKH mediates cellular efflux of ATP, not pyrophosphate. J Bone Miner Res. 2022;37:1024–31.
59. Harmey D, Hessle L, Narisawa S, Johnson KA, Terkeltaub R, Millán JL. Concerted regulation of inorganic pyrophosphate and osteopontin by akp2, enpp1, and ank: an integrated model of the pathogenesis of mineralization disorders. Am J Pathol. 2004;164:1199–209.
60. Reichenberger E, Tiziani V, Watanabe S, Park L, Ueki Y, Santanna C, et al. Autosomal dominant craniometaphyseal dysplasia is caused by mutations in the transmembrane protein ANK. Am J Hum Genet. 2001;68:1321–6.
61. O'Neill WC, Sigrist MK, McIntyre CW. Plasma pyrophosphate and vascular calcification in chronic kidney disease. Nephrol Dial Transplant. 2010;25:187–91.
62. Hessle L, Johnson KA, Anderson HC, Narisawa S, Sali A, Goding JW, et al. Tissue-nonspecific alkaline phosphatase and plasma cell membrane glycoprotein-1 are central antagonistic regulators of bone mineralization. Proc Natl Acad Sci U S A. 2002;99:9445–9.
63. Fakhry M, Roszkowska M, Briolay A, Bougault C, Guignandon A, Diaz-Hernandez JI, et al. TNAP stimulates vascular smooth muscle cell trans-differentiation into chondrocytes through calcium deposition and BMP-2 activation: possible implication in atherosclerotic plaque stability. Biochim Biophys Acta Mol basis Dis. 2017;1863:643–53.
64. Bourgine A, Beck L, Khoshniat S, Wauquier F, Oliver L, Hue E, et al. Inorganic phosphate stimulates apoptosis in murine MO6-G3 odontoblast-like cells. Arch Oral Biol. 2011;56:977–83.
65. Foster BL, Nociti FH Jr, Swanson EC, Matsa-Dunn D, Berry JE, Cupp CJ, et al. Regulation of cementoblast gene expression by inorganic phosphate in vitro. Calcif Tissue Int. 2006;78:103–12.
66. Yoshiko Y, Candeliere GA, Maeda N, Aubin JE. Osteoblast autonomous Pi regulation via Pit1 plays a role in bone mineralization. Mol Cell Biol. 2007;27:4465–74.

67. Khoshniat S, Bourgine A, Julien M, Petit M, Pilet P, Rouillon T, et al. Phosphate-dependent stimulation of MGP and OPN expression in osteoblasts via the ERK1/2 pathway is modulated by calcium. Bone. 2011;48:894–902.
68. Beck GR Jr, Moran E, Knecht N. Inorganic phosphate regulates multiple genes during osteoblast differentiation, including Nrf2. Exp Cell Res. 2003;288:288–300.
69. Shapiro IM, Boyde A. Mineralization of normal and rachitic chick growth cartilage: vascular canals, cartilage calcification and osteogenesis. Scanning Microsc. 1987;1:599–606.
70. Funanage VL, Carango P, Shapiro IM, Tokuoka T, Tuan RS. Creatine kinase activity is required for mineral deposition and matrix synthesis in endochondral growth cartilage. Bone Miner. 1992;17:228–36.
71. Omelon S, Georgiou J, Henneman ZJ, Wise LM, Sukhu B, Hunt T, et al. Control of vertebrate skeletal mineralization by polyphosphates. PLoS One. 2009;4:e5634.
72. Lust G, Seegmiller JE. A rapid, enzymatic assay for measurement of inorganic pyrophosphate in biological samples. Clin Chim Acta. 1976;66:241–9.
73. McGuire MB, Colman CH, Baghat N, Russell RG. Radiometric measurement of pyrophosphate in cell cultures. Biochem Soc Trans. 1980;8:529–30.
74. Lust G, Faure G, Netter P, Seegmiller JE. Increased pyrophosphate in fibroblasts and lymphoblasts from patients with hereditary diffuse articular chondrocalcinosis. Science. 1981;214:809–10.
75. Leyhausen G, Lorenz B, Zhu H, Geurtsen W, Bohnensack R, Muller WE, et al. Inorganic polyphosphate in human osteoblast-like cells. J Bone Miner Res. 1998;13:803–12.
76. Muller WE, Tolba E, Feng Q, Schroder HC, Markl JS, Kokkinopoulou M, et al. Amorphous Ca(2)(+) polyphosphate nanoparticles regulate the ATP level in bone-like SaOS-2 cells. J Cell Sci. 2015;128:2202–7.
77. Ariganello MB, Omelon S, Variola F, Wazen RM, Moffatt P, Nanci A. Osteogenic cell cultures cannot utilize exogenous sources of synthetic polyphosphate for mineralization. J Cell Biochem. 2014;115:2089–102.
78. Lips P, van Schoor NM. The effect of vitamin D on bone and osteoporosis. Best Pract Res Clin Endocrinol Metab. 2011;25:585–91.
79. Lips P, Gielen E, van Schoor NM. Vitamin D supplements with or without calcium to prevent fractures. Bonekey Rep. 2014;3:512.
80. Holick MF, Schnoes HK, DeLuca HF. Identification of 1,25-dihydroxycholecalciferol, a form of vitamin D3 metabolically active in the intestine. Proc Natl Acad Sci U S A. 1971;68:803–4.
81. Jones G, Strugnell SA, DeLuca HF. Current understanding of the molecular actions of vitamin D. Physiol Rev. 1998;78:1193–231.
82. Atkins GJ, Anderson PH, Findlay DM, Welldon KJ, Vincent C, Zannettino AC, et al. Metabolism of vitamin D3 in human osteoblasts: evidence for autocrine and paracrine activities of 1 alpha,25-dihydroxyvitamin D3. Bone. 2007;40:1517–28.
83. Geng S, Zhou S, Glowacki J. Effects of 25-hydroxyvitamin D(3) on proliferation and osteoblast differentiation of human marrow stromal cells require CYP27B1/1alpha-hydroxylase. J Bone Miner Res. 2011;26:1145–53.
84. Anderson PH. Vitamin D activity and metabolism in bone. Curr Osteoporos Rep. 2017;15:443–9.
85. van der Meijden K, Lips P, van Driel M, Heijboer AC, Schulten EA, den Heijer M, et al. Primary human osteoblasts in response to 25-hydroxyvitamin D3, 1,25-dihydroxyvitamin D3 and 24R,25-dihydroxyvitamin D3. PLoS One. 2014;9:e110283.
86. Curtis KM, Aenlle KK, Roos BA, Howard GA. 24R,25-dihydroxyvitamin D3 promotes the osteoblastic differentiation of human mesenchymal stem cells. Mol Endocrinol. 2014;28:644–58.
87. Bouillon R, Carmeliet G, Verlinden L, van Etten E, Verstuyf A, Luderer HF, et al. Vitamin D and human health: lessons from vitamin D receptor null mice. Endocr Rev. 2008;29:726–76.
88. Reinhardt TA, Horst RL. Parathyroid hormone down-regulates 1,25-dihydroxyvitamin D receptors (VDR) and VDR messenger ribonucleic acid in vitro and blocks homologous up-regulation of VDR in vivo. Endocrinology. 1990;127:942–8.

89. van Leeuwen JP, Birkenhager JC, Vink-van Wijngaarden T, van den Bemd GJ, Pols HA. Regulation of 1,25-dihydroxyvitamin D3 receptor gene expression by parathyroid hormone and cAMP-agonists. Biochem Biophys Res Commun. 1992;185:881–6.
90. Carlberg C, Polly P. Gene regulation by vitamin D3. Crit Rev Eukaryot Gene Expr. 1998;8:19–42.
91. Jurutka PW, Bartik L, Whitfield GK, Mathern DR, Barthel TK, Gurevich M, et al. Vitamin D receptor: key roles in bone mineral pathophysiology, molecular mechanism of action, and novel nutritional ligands. J Bone Miner Res. 2007;22:V2–10.
92. Gutierrez S, Javed A, Tennant DK, van Rees M, Montecino M, Stein GS, et al. CCAAT/enhancer-binding proteins (C/EBP) beta and delta activate osteocalcin gene transcription and synergize with Runx2 at the C/EBP element to regulate bone-specific expression. J Biol Chem. 2002;277:1316–23.
93. Dhawan P, Peng X, Sutton AL, MacDonald PN, Croniger CM, Trautwein C, et al. Functional cooperation between CCAAT/enhancer-binding proteins and the vitamin D receptor in regulation of 25-hydroxyvitamin D3 24-hydroxylase. Mol Cell Biol. 2005;25:472–87.
94. Quarles LD. Skeletal secretion of FGF-23 regulates phosphate and vitamin D metabolism. Nat Rev Endocrinol. 2012;8:276–86.
95. Hines ER, Kolek OI, Jones MD, Serey SH, Sirjani NB, Kiela PR, et al. 1,25-dihydroxyvitamin D3 down-regulation of PHEX gene expression is mediated by apparent repression of a 110 kDa transfactor that binds to a polyadenine element in the promoter. J Biol Chem. 2004;279:46406–14.
96. Quack M, Carlberg C. Ligand-triggered stabilization of vitamin D receptor/retinoid X receptor heterodimer conformations on DR4-type response elements. J Mol Biol. 2000;296:743–56.
97. Carlberg C, Molnar F. Vitamin D receptor signaling and its therapeutic implications: genome-wide and structural view. Can J Physiol Pharmacol. 2015;93:311–8.
98. Boyan BD, Bonewald LF, Sylvia VL, Nemere I, Larsson D, Norman AW, et al. Evidence for distinct membrane receptors for 1 alpha,25-(OH)(2)D(3) and 24R,25-(OH)(2)D(3) in osteoblasts. Steroids. 2002;67:235–46.
99. Nemere I, Safford SE, Rohe B, DeSouza MM, Farach-Carson MC. Identification and characterization of 1,25D3-membrane-associated rapid response, steroid (1,25D3-MARRS) binding protein. J Steroid Biochem Mol Biol. 2004;89–90:281–5.
100. Farach-Carson MC, Xu Y. Microarray detection of gene expression changes induced by 1,25(OH)(2)D(3) and a Ca(2+) influx-activating analog in osteoblastic ROS 17/2.8 cells. Steroids. 2002;67:467–70.
101. Doroudi M, Schwartz Z, Boyan BD. Membrane-mediated actions of 1,25-dihydroxy vitamin D3: a review of the roles of phospholipase A2 activating protein and Ca(2+)/calmodulin-dependent protein kinase II. J Steroid Biochem Mol Biol. 2015;147:81–4.
102. Yang D, Turner AG, Wijenayaka AR, Anderson PH, Morris HA, Atkins GJ. 1,25-Dihydroxyvitamin D3 and extracellular calcium promote mineral deposition via NPP1 activity in a mature osteoblast cell line MLO-A5. Mol Cell Endocrinol. 2015;412:140–7.
103. Raisz LG, Trummel CL, Holick MF, DeLuca HF. 1,25-dihydroxycholecalciferol: a potent stimulator of bone resorption in tissue culture. Science. 1972;175:768–89.
104. Yang D, Anderson PH, Wijenayaka AR, Barratt KR, Triliana R, Stapledon CJM, et al. Both ligand and VDR expression levels critically determine the effect of 1alpha,25-dihydroxyvitamin-D3 on osteoblast differentiation. J Steroid Biochem Mol Biol. 2018;177:83–90.
105. Pedersen KO. Ultracentrifugal and electrophoretic studies on fetuin. J Phys Colloid Chem. 1947;51:164–71.
106. Price PA, Lim JE. The inhibition of calcium phosphate precipitation by fetuin is accompanied by the formation of a fetuin-mineral complex. J Biol Chem. 2003;278:22144–52.
107. Price PA, Toroian D, Lim JE. Mineralization by inhibitor exclusion: the calcification of collagen with fetuin. J Biol Chem. 2009;284:17092–101.

108. Araki T, Yoshioka Y, Schmid K. The position of the disulfide bonds in human plasma alpha 2 HS-glycoprotein and the repeating double disulfide bonds in the domain structure. Biochim Biophys Acta. 1989;994:195–9.
109. Brylka L, Jahnen-Dechent W. The role of fetuin-A in physiological and pathological mineralization. Calcif Tissue Int. 2013;93:355–64.
110. Herrmann M, Schafer C, Heiss A, Graber S, Kinkeldey A, Buscher A, et al. Clearance of fetuin-A--containing calciprotein particles is mediated by scavenger receptor-A. Circ Res. 2012;111:575–84.
111. Heiss A, DuChesne A, Denecke B, Grotzinger J, Yamamoto K, Renne T, et al. Structural basis of calcification inhibition by alpha 2-HS glycoprotein/fetuin-A. Formation of colloidal calciprotein particles. J Biol Chem. 2003;278:13333–41.
112. Jahnen-Dechent W, Schafer C, Ketteler M, McKee MD. Mineral chaperones: a role for fetuin-A and osteopontin in the inhibition and regression of pathologic calcification. J Mol Med (Berl). 2008;86:379–89.
113. Dautova Y, Kozlova D, Skepper JN, Epple M, Bootman MD, Proudfoot D. Fetuin-A and albumin alter cytotoxic effects of calcium phosphate nanoparticles on human vascular smooth muscle cells. PLoS One. 2014;9:e97565.
114. Demetriou M, Binkert C, Sukhu B, Tenenbaum HC, Dennis JW. Fetuin/alpha2-HS glycoprotein is a transforming growth factor-beta type II receptor mimic and cytokine antagonist. J Biol Chem. 1996;271:12755–61.
115. Binkert C, Demetriou M, Sukhu B, Szweras M, Tenenbaum HC, Dennis JW. Regulation of osteogenesis by fetuin. J Biol Chem. 1999;274:28514–20.
116. Schafer C, Heiss A, Schwarz A, Westenfeld R, Ketteler M, Floege J, et al. The serum protein alpha 2-Heremans-Schmid glycoprotein/fetuin-A is a systemically acting inhibitor of ectopic calcification. J Clin Invest. 2003;112:357–66.
117. Szweras M, Liu D, Partridge EA, Pawling J, Sukhu B, Clokie C, et al. Alpha 2-HS glycoprotein/fetuin, a transforming growth factor-beta/bone morphogenetic protein antagonist, regulates postnatal bone growth and remodeling. J Biol Chem. 2002;277:19991–7.
118. Seto J, Busse B, Gupta HS, Schafer C, Krauss S, Dunlop JW, et al. Accelerated growth plate mineralization and foreshortened proximal limb bones in fetuin-A knockout mice. PLoS One. 2012;7:e47338.
119. Kapustin AN, Chatrou ML, Drozdov I, Zheng Y, Davidson SM, Soong D, et al. Vascular smooth muscle cell calcification is mediated by regulated exosome secretion. Circ Res. 2015;116:1312–23.
120. Schurgers LJ, Cranenburg EC, Vermeer C. Matrix Gla-protein: the calcification inhibitor in need of vitamin K. Thromb Haemost. 2008;100:593–603.
121. Cancela ML, Williamson MK, Price PA. The putative RGD-dependent cell adhesion activity of matrix Gla protein is due to higher molecular weight contaminants. J Biol Chem. 1994;269:12185–9.
122. Bronckers AL, Farach-Carson MC, Van Waveren E, Butler WT. Immunolocalization of osteopontin, osteocalcin, and dentin sialoprotein during dental root formation and early cementogenesis in the rat. J Bone Miner Res. 1994;9:833–41.
123. Cancela ML, Laize V, Conceicao N. Matrix Gla protein and osteocalcin: from gene duplication to neofunctionalization. Arch Biochem Biophys. 2014;561:56–63.
124. Murshed M, Schinke T, McKee MD, Karsenty G. Extracellular matrix mineralization is regulated locally; different roles of two gla-containing proteins. J Cell Biol. 2004;165:625–30.
125. Stenflo J, Suttie JW. Vitamin K-dependent formation of gamma-carboxyglutamic acid. Annu Rev Biochem. 1977;46:157–72.
126. Yamaguchi M, Weitzmann MN. Vitamin K2 stimulates osteoblastogenesis and suppresses osteoclastogenesis by suppressing NF-kappaB activation. Int J Mol Med. 2011;27:3–14.
127. Brandenburg VM, Reinartz S, Kaesler N, Kruger T, Dirrichs T, Kramann R, et al. Slower progress of aortic valve calcification with vitamin K supplementation: results from a prospective interventional proof-of-concept study. Circulation. 2017;135:2081–3.

128. Schurgers LJ, Spronk HM, Soute BA, Schiffers PM, DeMey JG, Vermeer C. Regression of warfarin-induced medial elastocalcinosis by high intake of vitamin K in rats. Blood. 2007;109:2823–31.
129. McCabe KM, Booth SL, Fu X, Shobeiri N, Pang JJ, Adams MA, et al. Dietary vitamin K and therapeutic warfarin alter the susceptibility to vascular calcification in experimental chronic kidney disease. Kidney Int. 2013;83:835–44.
130. McCann JC, Ames BN. Vitamin K, an example of triage theory: is micronutrient inadequacy linked to diseases of aging? Am J Clin Nutr. 2009;90:889–907.
131. Luo G, Ducy P, McKee MD, Pinero GJ, Loyer E, Behringer RR, et al. Spontaneous calcification of arteries and cartilage in mice lacking matrix GLA protein. Nature. 1997;386:78–81.
132. Price PA, Faus SA, Williamson MK. Warfarin causes rapid calcification of the elastic lamellae in rat arteries and heart valves. Arterioscler Thromb Vasc Biol. 1998;18:1400–7.
133. Schurgers LJ, Spronk HM, Skepper JN, Hackeng TM, Shanahan CM, Vermeer C, et al. Post-translational modifications regulate matrix Gla protein function: importance for inhibition of vascular smooth muscle cell calcification. J Thromb Haemost. 2007;5:2503–11.
134. Marulanda J, Eimar H, McKee MD, Berkvens M, Nelea V, Roman H, et al. Matrix Gla protein deficiency impairs nasal septum growth, causing midface hypoplasia. J Biol Chem. 2017;292:11400–12.
135. Weaver KN, El Hallek M, Hopkin RJ, Sund KL, Henrickson M, Del Gaudio D, et al. Keutel syndrome: report of two novel MGP mutations and discussion of clinical overlap with arylsulfatase E deficiency and relapsing polychondritis. Am J Med Genet A. 2014;164A:1062–8.
136. Pauli RM, Lian JB, Mosher DF, Suttie JW. Association of congenital deficiency of multiple vitamin K-dependent coagulation factors and the phenotype of the warfarin embryopathy: clues to the mechanism of teratogenicity of coumarin derivatives. Am J Hum Genet. 1987;41:566–83.
137. Pettifor JM, Benson R. Congenital malformations associated with the administration of oral anticoagulants during pregnancy. J Pediatr. 1975;86:459–62.
138. Yagami K, Suh JY, Enomoto-Iwamoto M, Koyama E, Abrams WR, Shapiro IM, et al. Matrix GLA protein is a developmental regulator of chondrocyte mineralization and, when constitutively expressed, blocks endochondral and intramembranous ossification in the limb. J Cell Biol. 1999;147:1097–108.
139. Yao Y, Shahbazian A, Bostrom KI. Proline and gamma-carboxylated glutamate residues in matrix Gla protein are critical for binding of bone morphogenetic protein-4. Circ Res. 2008;102:1065–74.
140. Malhotra R, Burke MF, Martyn T, Shakartzi HR, Thayer TE, O'Rourke C, et al. Inhibition of bone morphogenetic protein signal transduction prevents the medial vascular calcification associated with matrix Gla protein deficiency. PLoS One. 2015;10:e0117098.
141. Sodek J, Ganss B, McKee MD. Osteopontin. Crit Rev Oral Biol Med. 2000;11:279–303.
142. Qin C, Baba O, Butler WT. Post-translational modifications of sibling proteins and their roles in osteogenesis and dentinogenesis. Crit Rev Oral Biol Med. 2004;15:126–36.
143. Moses KD, Butler WT, Qin C. Immunohistochemical study of small integrin-binding ligand, N-linked glycoproteins in reactionary dentin of rat molars at different ages. Eur J Oral Sci. 2006;114:216–22.
144. Sun Y, Ma S, Zhou J, Yamoah AK, Feng JQ, Hinton RJ, et al. Distribution of small integrin-binding ligand, N-linked glycoproteins (SIBLING) in the articular cartilage of the rat femoral head. J Histochem Cytochem. 2010;58:1033–43.
145. Chen L, Childs RD, Landis WJ. Correlations between gene expression and mineralization in the avian leg tendon. Bone. 2019;121:42–59.
146. Giachelli CM, Steitz S. Osteopontin: a versatile regulator of inflammation and biomineralization. Matrix Biol. 2000;19:615–22.
147. Hoac B, Nelea V, Jiang W, Kaartinen MT, McKee MD. Mineralization-inhibiting effects of transglutaminase-crosslinked polymeric osteopontin. Bone. 2017;101:37–48.

148. Kaartinen MT, Murshed M, Karsenty G, McKee MD. Osteopontin upregulation and polymerization by transglutaminase 2 in calcified arteries of matrix Gla protein-deficient mice. J Histochem Cytochem. 2007;55:375–86.
149. Cheng C, Gao S, Lei G. Association of osteopontin with osteoarthritis. Rheumatol Int. 2014;34:1627–31.
150. Hunter GK. Role of osteopontin in modulation of hydroxyapatite formation. Calcif Tissue Int. 2013;93:348–54.
151. Addison WN, Masica DL, Gray JJ, McKee MD. Phosphorylation-dependent inhibition of mineralization by osteopontin ASARM peptides is regulated by PHEX cleavage. J Bone Miner Res. 2010;25:695–705.
152. Boskey AL, Christensen B, Taleb H, Sørensen ES. Post-translational modification of osteopontin: effects on in vitro hydroxyapatite formation and growth. Biochem Biophys Res Commun. 2012;419:333–8.
153. Chien YC, Mansouri A, Jiang W, Khan SR, Gray JJ, McKee MD. Modulation of calcium oxalate dihydrate growth by phosphorylated osteopontin peptides. J Struct Biol. 2018;204:131–44.
154. McKee MD, Nanci A. Osteopontin and the bone remodeling sequence. Colloidal-gold immunocytochemistry of an interfacial extracellular matrix protein. Ann N Y Acad Sci. 1995;760:177–89.
155. Pampena DA, Robertson KA, Litvinova O, Lajoie G, Goldberg HA, Hunter GK. Inhibition of hydroxyapatite formation by osteopontin phosphopeptides. Biochem J. 2004;378:1083–7.
156. Ibsen CJS, Gebauer D, Birkedal H. Osteopontin stabilizes metastable states prior to nucleation during apatite formation. Chem Mater. 2016;28:8550–5.
157. Hunter GK, Kyle CL, Goldberg HA. Modulation of crystal formation by bone phosphoproteins: structural specificity of the osteopontin-mediated inhibition of hydroxyapatite formation. Biochem J. 1994;300:723–8.
158. Hunter GK, Hauschka PV, Poole AR, Rosenberg LC, Goldberg HA. Nucleation and inhibition of hydroxyapatite formation by mineralized tissue proteins. Biochem J. 1996;317:59–64.
159. Chellaiah MA, Hruska KA. The integrin alpha(v)beta(3) and CD44 regulate the actions of osteopontin on osteoclast motility. Calcif Tissue Int. 2003;72:197–205.
160. Rittling SR, Matsumoto HN, McKee MD, Nanci A, An XR, Novick KE, et al. Mice lacking osteopontin show normal development and bone structure but display altered osteoclast formation in vitro. J Bone Miner Res. 1998;13:1101–11.
161. Boskey AL, Spevak L, Paschalis E, Doty SB, McKee MD. Osteopontin deficiency increases mineral content and mineral crystallinity in mouse bone. Calcif Tissue Int. 2002;71:145–54.
162. Liao C, Ou Y, Wu Y, Zhou Y, Liang S, Wang Y. Sclerostin inhibits odontogenic differentiation of human pulp-derived odontoblast-like cells under mechanical stress. J Cell Physiol. 2019;234:20779–89.
163. Sitara D, Kim S, Razzaque MS, Bergwitz C, Taguchi T, Schuler C, et al. Genetic evidence of serum phosphate-independent functions of FGF-23 on bone. PLoS Genet. 2008;4:e1000154.
164. Vimalraj S. Alkaline phosphatase: structure, expression and its function in bone mineralization. Gene. 2020;754:144855.
165. Villa-Bellosta R. Synthesis of extracellular pyrophosphate increases in vascular smooth muscle cells during phosphate-induced calcification. Arterioscler Thromb Vasc Biol. 2018;38:2137–47.
166. Addison WN, Azari F, Sorensen ES, Kaartinen MT, McKee MD. Pyrophosphate inhibits mineralization of osteoblast cultures by binding to mineral, up-regulating osteopontin, and inhibiting alkaline phosphatase activity. J Biol Chem. 2007;282:15872–83.
167. Hajjawi MO, MacRae VE, Huesa C, Boyde A, Millan JL, Arnett TR, et al. Mineralisation of collagen rich soft tissues and osteocyte lacunae in Enpp1 (−/−) mice. Bone. 2014;69:139–47.

168. Toroian D, Lim JE, Price PA. The size exclusion characteristics of type I collagen: implications for the role of noncollagenous bone constituents in mineralization. J Biol Chem. 2007;282:22437–47.
169. Jahnen-Dechent W, Schinke T, Trindl A, Muller-Esterl W, Sablitzky F, Kaiser S, et al. Cloning and targeted deletion of the mouse fetuin gene. J Biol Chem. 1997;272:31496–503.
170. Chen L, Jacquet R, Lowder E, Landis WJ. Refinement of collagen-mineral interaction: a possible role for osteocalcin in apatite crystal nucleation, growth and development. Bone. 2015;71:7–16.

Aberrant Mineral Deposition in Soft and Hard Tissues

10

I definitely deny that any pathological process, i.e., any life-process taking place under unfavourable circumstances, is able to call forth qualitatively new formations lying beyond the customary range of forms characteristic of the species. All pathological formations are either degenerations, transformations, or repetitions of typical physiological structures [1].

Glossary

Advanced glycation end-products (AGEs) Compounds formed non-enzymatically when proteins or lipids are treated with sugars.

Amphisome A vesicle formed by fusion of an endosome with an autophagosome.

Autophagy The intracellular system for ensuring cell survival through removal of degenerate organelles and non-functional proteins.

Autosomal mutation A mutation of a gene present on non-sex chromosomes.

CRAC channel Calcium release-activated calcium ion channel in the plasma membrane of cells which opens in response to a decrease in calcium ions in the endoplasmic reticulum.

Dentin dysplasia A disorder of dentin caused by mutations in genes which include *SSUH2*, *SMOC2*, and *VPS4B*.

Dystrophic mineralization Localized non-skeletal mineralization often associated with soft tissue which has been damaged or impaired and in the presence of normal blood calcium levels.

Ectonucleotide pyrophosphatase-phosphodiesterase Enzymes which hydrolyze ATP and other nucleoside triphosphates to generate the pyrophosphate anion and nucleoside monophosphate.

Ehlers-Danlos syndrome A genetic mutation affecting type I collagen crosslinking of connective tissues, resulting in fragile skin and lax tendons and ligaments.

Fibrodysplasia ossificans progressiva A rare genetic disease in which mineral is deposited in muscle and other soft tissues.

I. M. Shapiro, W. J. Landis, *Mechanisms of Mineralization of Vertebrate Skeletal and Dental Tissues*, https://doi.org/10.1007/978-3-031-34304-9_10

Genome-wide association study Whole genome analysis usually associated with single nucleotide polymorphism to detect genetic variants linked to a specific trait.

Heparan sulfate proteoglycan A proteoglycan containing a small number of polysaccharide chains (heparan) attached to a core protein located in the plasma membrane of the cell. The charged heparan chains interact with a number of matrix ligands. Examples include syndecan, glypican, and perlecan.

Heterotopic mineralization Abnormal formation of organized bone within soft, normally non-osseous, sound extraskeletal tissues.

Hypercalcemia Elevation in blood calcium ion level.

Hyperphosphatemia Elevation in blood phosphate ion level.

Lipoproteins Lipid–protein complexes present in serum.

Metastatic mineralization Widespread deposition of mineral in normal soft tissues and associated with elevated serum levels of calcium.

Mönckeberg's media sclerosis Mineralization of the medial layer of blood vessels.

Myofilaments A string of myosin or actin filaments in muscle cells.

NOD-, LRR-, and pyrin domain-containing protein (NLRP)3 inflammasome An intracellular multiprotein complex which triggers the production of pro-inflammatory cytokines IL-1β and IL-18.

ORAI A family of plasma membrane pore-forming proteins of the calcium release-activated calcium (CRAC) channel which associate with STIM and mediate calcium ion influx into a cell.

Osteogenesis imperfecta A family of genetic diseases of collagen characterized by skeletal fragility (so-called brittle bone) and multiple fractures of the long bones.

Osteomimicry A term used to describe the appearance of mineral-forming cells in breast tissue.

Pericytes Cells which are closely associated with endothelial cells lining blood vessels.

Peyronie's disease A relatively rare disorder in which fibrous plaques develop in the soft connective tissue of the penis, causing pain, curvature, and dysfunction of the organ.

Prolyl 3-hydroxylase An enzyme which catalyzes post-translational modification of proline residues of collagen.

Proto-oncogene A normal gene regulating the growth and proliferation of cells or the inhibition of apoptosis; following a gain-of-function mutation, a proto-oncogene may be up-regulated and lead to enhanced cell growth and predisposition to cancer.

PTX A tumor-suppressor gene concerned with regulating inflammation and encoding the protein, pentraxin 3.

Rapamycin An immune-suppressor drug which is a specific inhibitor of mTOR (mammalian or mechanistic target of rapamycin).

Retinoids A class of chemical agents derived from vitamin A.

RNA oligonucleotides Short RNA sequences which can be used to block RNA expression.

SIBLING A member of the family of small integrin-binding ligand, N-linked glycoproteins present in mineralized tissues.

SMADs A family of proteins of similar structure which are the principal transducers of signals for receptors of the TGF-β superfamily.

STIM1/2 Sensor proteins resident in the endoplasmic reticulum which regulate calcium ion homeostasis in a cell by controlling store-operated calcium channels in the plasma membrane.

TRP A transient receptor potential ion channel protein in the plasma membrane of a cell.

Abbreviations and Proper Names for Genes/Proteins and Other Molecules

ACAN	Aggrecan
ACVR1	Activin receptor type 1, a family member of the bone morphogenetic protein (BMP) type 1 receptors; also known as activin receptor-like kinase-2 (*ALK2*)
BMP	Bone morphogenetic protein
BMU	Basic multicellular unit
BRIL	Bone-restricted interferon-induced transmembrane protein-like
CD	Cluster of differentiation
CRTAP	Cartilage-associated protein
COLI	Type I collagen
COLII	Type II collagen
COLV	Type V collagen
COLX	Type X collagen
COLXVII	Type XVII collagen
COLIA1 and *COLIA2*	Genes encoding the α1(I) and α2(I) chains of type I collagen
COLIIA1	Gene encoding the α1(II) chains of type II collagen
FGF	Fibroblast growth factor
FKBP12	12-kDa binding protein for FK506, an immunosuppressant drug, and which binds to intracellular calcium release channels and to the TGF-β1 receptor
FOP	Fibrodysplasia ossificans progressiva
GWAS	Genome-wide association study
HIF	Hypoxia inducible factor
IFITM5	Interferon-induced transmembrane protein 5
IL	Interleukin
MGP	Matrix gla protein
MMP	Matrix metalloproteinase
MSX	Muscle segment homeobox
mTORC1	Mammalian or mechanistic target of rapamycin complex 1
NLRP3	NOD-, LRR-, and pyrin domain-containing protein 3
OCN	Osteocalcin
OI	Osteogenesis imperfecta
OPN	Osteopontin

OSX	Osterix
P3H1	Prolyl 3-hydroxylase
PIT-1 (*SLC20A1*)	Sodium-dependent phosphate cotransporter
PTH	Parathyroid hormone
PTX	Pentraxin
RAB	*Ras*-related proteins in brain; protein members of the RAS superfamily of small GTPases
RAR	Retinoic acid receptor
RUNX	Runt-related transcription factor
SIBLING	Small integrin-binding ligand, N-linked glycoprotein
SLC	Solute carrier family
SMAD	Mothers against decapentaplegic, homologue of *Drosophila* gene
STIM1	Stromal interaction molecule 1
SORT1	Sortilin 1
SOX9	Sex determining region Y (SRY)-box 9
TGF-β	Transforming growth factor-β
TMEM38B	Transmembrane protein 38B
TNF	Tumor necrosis factor
TNAP	Tissue non-specific alkaline phosphatase
TRIC-B	Trimeric intracellular cation channel type B protein
TRP	Transient receptor potential ion family channel protein
VPS4B	Vacuolar protein sorting 4 homologue B
WNT	Wingless-related integration site

10.1 Prevalence and Impact of Aberrant Mineralization on Tissue Function

This book began with a discussion of mechanisms of mineralization by considering the fate of Lot's wife, a metaphor for aberrant mineralization. In the presence of kin and accompanying angels, deity-catalyzed deposition of minerals into soft tissues converted a pliant Edith into a hard and fragile human monument to disobedience. According to Levine, "One might ask whether Lot's wife, in turning back to watch fire and brimstone consume everything she valued, experienced an acute imbalance between phosphate and pyrophosphate that led to her sudden and tragic transformation into a pillar of salt" [2]. Whether Edith was mineralized with sodium chloride or the ions mentioned above, the deposition of an aberrant mineral into soft tissues would interfere with the mechanical properties of the tissue, disrupt ionic balance, confound the needs of cells for physiologic levels of calcium and phosphate ions, and promote cell, tissue, and organismal death.

While Lot's wife represents the extreme, numerous types of aberrant mineralization have been described; the literature is replete with descriptions of apatite, magnesium apatite, brushite, dicalcium phosphate, octacalcium phosphate, and calcium oxalates, for example, in soft tissues. Aside from the Biblical event, anthropologists describe blood vessel mineralization in a frozen and long-dead human who exhibited robust radiographic evidence of mineral deposits in the remnants of his aorta and carotid artery [3]. Over a century ago, it was reported that the muscular layer of some arteriosclerotic arteries contained mineral deposits (this condition is now known as Mönckeberg's media sclerosis) [4]. Other common sites of this form of mineralization include blood vessels, heart valves, breast, and kidney. Undoubtedly, the most common form of aberrant mineralization is the deposition of calculus on the surfaces of teeth. This type of mineralization coated the teeth of hominins living 1.2 million years ago [5]. Calculus comprises a mixture of calcified mineral phases (apatite, octacalcium phosphate, and dicalcium phosphate) in an organic matrix consisting of salivary proteins like mucins, food debris, and bacterial products. These mineral deposits are without useful function: Mandel and Gaffar regarded calculus as a "slow-release device discharging toxic and pathogenic products into the soft tissues" [6]. Together, calculus and the other forms of soft tissue mineralization serve no purpose other than to reduce and destroy tissue function.

In this context, why is there a propensity for mineral to form on teeth and in the soft tissues of the body? As has been discussed earlier, in the presence of nucleating agents such as proteins and lipids, the calcium and phosphate ion activity product in the serum, saliva, and tissue fluid is sufficient to cause the spontaneous formation of mineral. Unless checked, mineral will be deposited in the soft tissues of Edith, the blood vessels of a 5000-year-old human frozen and preserved in ice, and even around the teeth of modern-day men and women.

It is most surprising that pathological forms of mineralization are rare and, while many soft tissues accumulate small quantities of aberrant mineral, in general they elicit minimal tissue deficits. However, there are several well-described conditions in which pathological mineral deposition interferes with cellular metabolism, a situation which can lead to loss of function, pain, and death. The pathogenesis of these disorders, in the light of cellular, genetic, and protein regulation of mineralization, is discussed further in the following pages of this chapter. Consideration is given to four aspects of these disorders: dystrophic mineralization, heterotopic mineralization, microcalcifications associated with breast cancer, and defective mineralization related to mutations in extracellular matrix proteins. These topics were chosen knowing that numerous abnormalities afflict the vertebrate mineralized tissues and many of them have already been mentioned in earlier chapters of this book. The exemplars highlighted here for discussion emphasize that mineralization is highly regulated and complex and not simply a default event dependent on changes in the activities of serum calcium and phosphate ions.

10.2 Disorders of Mineralization: Dystrophic Mineralization

Previous chapters of this book have emphasized that mineral deposition in extracellular tissue matrices is exquisitely sensitive to systemic factors and that even minor changes in the activities of regulatory agents may promote a disease phenotype which can affect both skeletal and extraskeletal tissues. Virchow, one of the founders of modern pathology, recognized that mineral could accumulate in tissues like atherosclerotic plaque without any observable change in factors governing calcium and phosphate homeostasis. He regarded this form of mineralization, dystrophic mineralization, as a physiological response to local inflammatory agents and tissue necrosis [7]. Dystrophic mineralization is now regarded as occurring in tissues already predisposed to functional impairment and damage, including scarred, degenerated, or necrotic tissues, although the degree of injury or abnormality may be slight and not necessarily overt. This type of mineral deposition is localized and not widespread in the body and blood calcium levels are normal. Since such mineralization differs from physiological mineralization involving sound and unimpaired tissues, it is worth examining the nature of the dystrophic process and possibly utilizing this information to gain insight and understanding of the mechanism by which apatite is deposited in bone, cartilage, dentin, and other normally mineralizing vertebrate tissues.

With respect to the disposition of dystrophic mineral, while this type of mineral may form in any tissue of the body, it is most frequently found in blood vessels, muscle, kidney, skin, heart, heart valves [8], and, as Virchow noted [7], atherosclerotic plaque (Fig. 10.1). Moreover, pre-existing conditions such as diabetes, hypertension, and kidney disease can promote and exacerbate such aberrant mineral formation (Fig. 10.2) [9]. Not surprisingly, the presence of dystrophic mineral can dramatically change the mechanical properties of the affected tissues and obstruct blood flow [10]. If deposited in vascular endothelium or heart valves, such abnormal mineral alters the elastic moduli of these tissues and leads to an inability to accommodate applied biomechanical forces (Fig. 10.3). As a result, there is general loss of tissue function and, in blood vessels, development of tissue tears accompanied by an unfortunate and critical tendency toward vessel wall rupture. Even if vessel walls remain intact, the mineral can aggravate a wide range of medical conditions, including congestive heart failure, hypertension, and cardiac hypertrophy. When mineral appears in the kidney, there may be concomitant changes in bone and bone density with a loss of mechanical strength and increased fracture risk. In this and other instances, the presence of dystrophic mineral is often a manifestation of an underlying disease, and the mineral can initiate pathogenic changes leading to further deficits in tissue function.

A number of investigators have explored the nature of the dystrophic mineral formed by cells of vascular and valvular tissues. The mineral appearance and character vary from diffuse apatitic deposits, which are totally unlike bone, to ossified structures which closely mimic bone. Hruska et al. and Vattikuti and Towler, for instance, showed that, when the medial layer of a blood vessel mineralizes, a tissue is formed which resembles membrane bone [12, 13]. In contrast, intimal mineralization of atherosclerotic plaque causes the initial formation of a cartilaginous tissue phase followed by the development of an endochondral-like tissue approximating trabecular bone (Fig. 10.3) [11]. In many such instances, the dystrophic mineral is associated with matrix vesicle-like structures [14]. Commenting on the presence of mineral in vascular material, Abedin et al. noted, "The concept that vascular calcification is related to osteogenesis is not unexpected, given the interaction of vascular and bone cells in normal embryonic skeletogenesis" [15]. In this regard, one possible role of blood vessels is to provide a vehicle for the delivery of stem cells, growth factors, metabolites, and oxygen to the tissue region undergoing mineralization. These observations concerning the pathogenesis of aberrant mineral elicit many

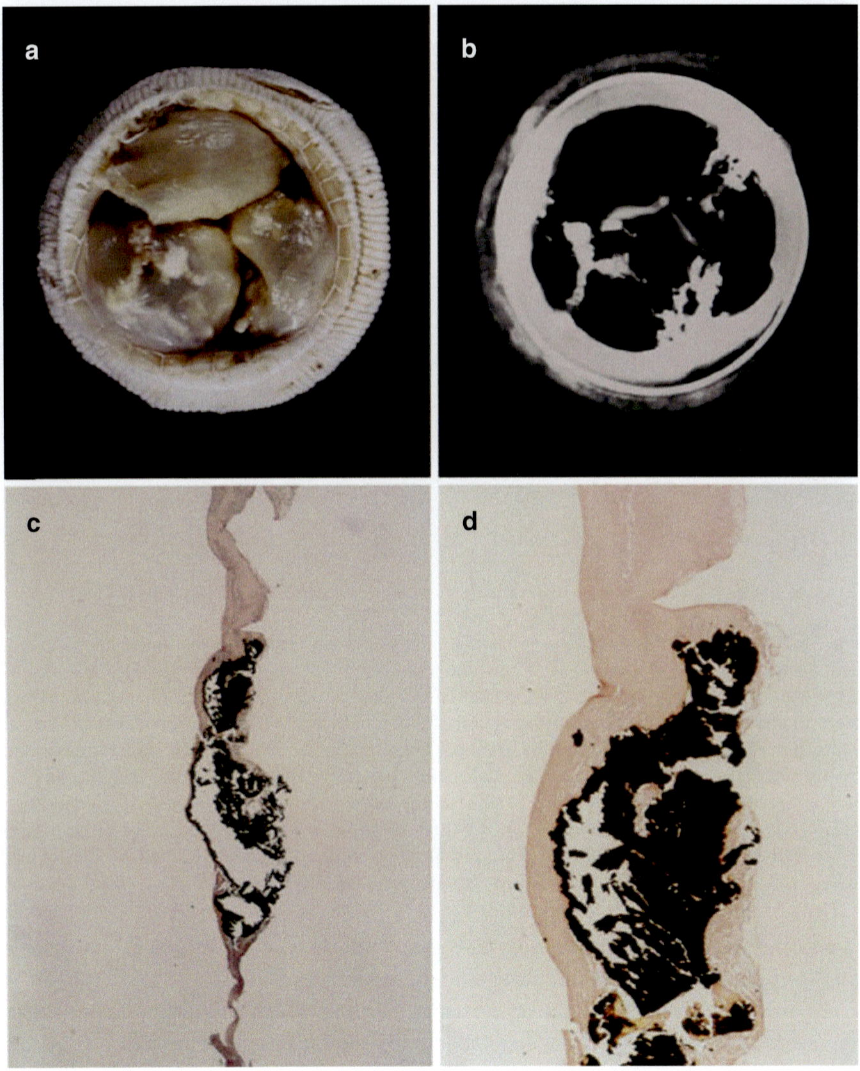

Fig. 10.1 Evidence of aberrant mineral deposition within and around a porcine cardiac valve xenograft seven years after implantation. (**a**) Surface view of the failed mitral valve xenograft with mineral deposits on the valve surface and elsewhere. (**b**) X-ray radiogram of the xenograft showing opaque mineral (white). (**c**) Histological section through the xenograft following von Kossa staining for mineral (black) (original mag ×12). (**d**) Enlargement of a portion of the area shown in (**c**) and illustrating prominent mineralization of the valve (original magnification, ×36). (Images adapted from Pettenazzo et al. [8] with permission)

questions: what initiating event provokes mineral formation; is the type of mineral hypoxia-dependent; is there a population of latent stem cells within the affected tissue; is activation of progenitor cells dependent on growth factors; is there evidence of intracellular ion accumulation; what transcription factors may be involved; are

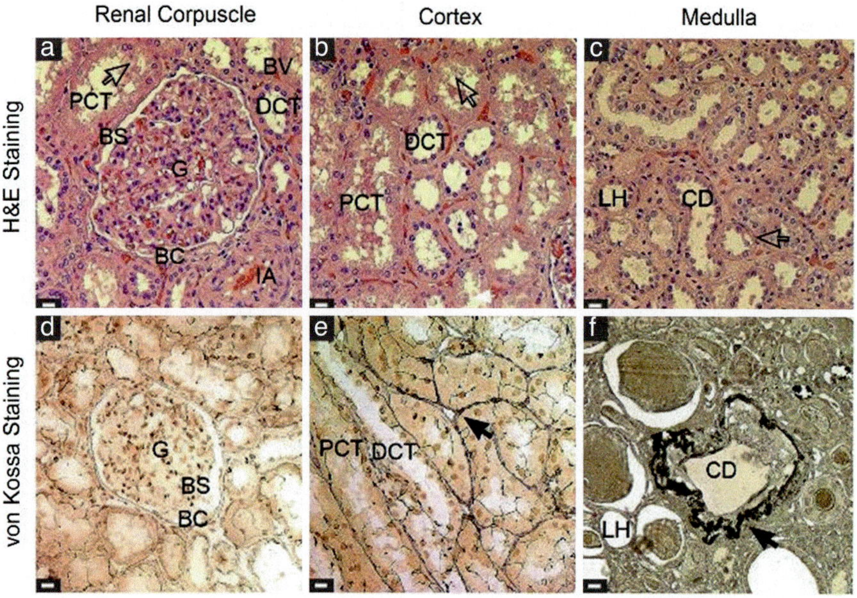

Fig. 10.2 Histology of human kidney tissue showing aberrant mineral deposition. Mineral is evident throughout the entire tissue except for the renal corpuscle. (**a–c**) Histological appearance of the renal corpuscle, cortex, and medulla stained with hematoxylin and eosin. The hollow arrows point to areas of tissue damage. (**d–f**) Sections stained with von Kossa solution showing mineral deposition (black). While mineral (black arrows) is present in the cortex (**e**), the stain is heaviest in the medulla (**f**), especially around the basal membrane and interstitium surrounding the loop of Henle (LH) and the collecting duct (CD). Mineral was not evident in the glomerulus (G), Bowman's space (BS), or Bowman's capsule (BC) of the renal corpuscle (D). Examples of a blood vessel (BV), a distal convoluted tubule (DCT), a proximal collecting tubule (PCT), and an interlobular artery (IA) are shown. Scale bars = 10 μm. (Images adapted from Wong et al. [9] with permission)

matrix vesicles required; and, responding to Virchow's observations [7], is there a role for inflammatory cells in dystrophic mineralization?

Of the questions raised above, possibly the one which initially attracted most attention was the source of cells responsible for mediating a mineralization response in mature, complex, and functioning soft tissues. Indeed, clones of "calcifying vascular cells" could be isolated from the medial vascular wall [16] and even the endothelial lining of the blood vessel [17]. In line with studies by Brighton et al. [18] and Caplan [19], pericytes were thought to represent the stem cell population which could commit to both osteogenic and chondrogenic lineages. Subsequently, more thought has been given to osteogenic activities of endogenous vascular smooth muscle cells as well as valvular interstitial cells which express a myofibrillar phenotype, referring to the presence of intracellular chains of actin and myosin filaments [20]. It was postulated that such cells in these tissues commit to or transdifferentiate into the mineral-forming cells which lead to bone or cartilage formation [21]. In this regard, it should also be mentioned that cells with a fibroblast-like morphology derived from circulating monocytes, exhibiting a cluster of differentiation (CD)14, CD45, and CD34 marker proteins, can commit to several different mesenchymal

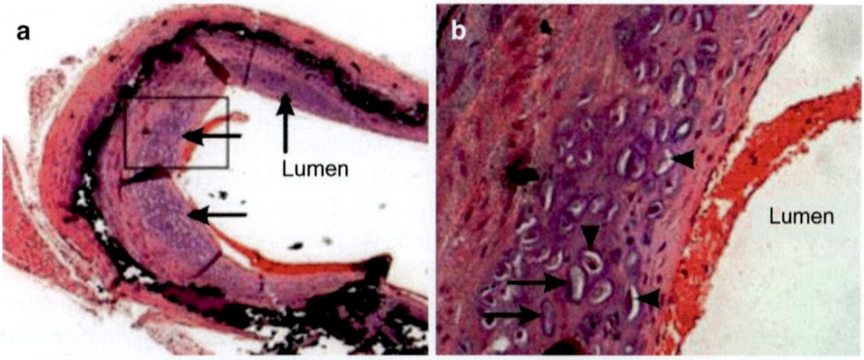

Fig. 10.3 Chondrocyte-mediated mineralization of the rat aorta following chronic kidney failure. (**a**) A transverse section through the aorta, stained with von Kossa solution and showing a band of circumferential mineral (black) in the medial layer. Arrows point to chondrocyte-like cells (original magnification ×50). (**b**) Enlargement of the boxed region in (**a**). Large swollen cells are apparent which resemble chondrocytes morphologically. Arrowheads identify capsular regions around the cells and chondrocyte-like lacunae. In some sections (not shown), mineral was deposited about the cells while immunohistochemical staining indicated the expression by the cells of SOX9 and COLII, specific markers for chondrocytes. Original magnification ×320. (Images adapted from Neven et al. [11] with permission)

lineages [22]. Since induction of such cells to differentiate led to secretion of osteoblast and chondrocyte-specific proteins [22], there is the possibility that these cells can provide a population of osteo- and chondroprogenitors.

A major characteristic in the initiation of dystrophic mineralization by cells derived from vascular and valvular tissues is their loss of myofilaments, a decrease in α-smooth muscle actin and expression of *type II collagen (COLII)* and *sex determining region Y (SRY)-box 9* (*SOX9*), providing evidence of transition to a chondrocyte phenotype [17]. Observational studies showed these transitioning cells resided in a cartilaginous matrix [14]. Confirming this result, Neven and co-workers reported that vascular smooth muscle cells in uremic rats exhibited an increase in expression of *COLII*, *SOX9*, and *aggrecan* (*ACAN*), while noting that osterix (OSX), a marker for osteoblasts, and molecules comprising the wingless-related integration site (WNT)/β-catenin pathway were suppressed [23]. This finding suggested that chondrogenesis preceded osteogenesis during dystrophic mineral formation [23]. However, such a sequence of events may not represent the only pathway to osteogenesis: Chen determined that populations of cells from the aorta can commit directly to an osteogenic lineage [24]. In other words, depending on their environment, stem cells exhibit plasticity in their progression toward osteogenesis.

A second question in defining dystrophic mineralization relates to understanding what factor or factors initiate the cellular changes delineated above. It is known that bone morphogenetic protein (BMP) isoforms are robustly expressed in cells in atherosclerotic plaque and other vascular lesions. Of these, BMP2 promotes the commitment of myofibrillar muscle cells to a mineralizing phenotype [25]. Using a hyperlipidemic and diabetic mouse model which exhibited vascular mineralization, Towler et al. reported that BMP2 enhanced the expression of *muscle segment homeobox* (*MSX*)*2* in a subset of myofibroblasts [26]. Expression of this transcription factor is a requirement for somatic cell reprogramming into other lineages [26]. In other words, by promoting the expression of *MSX2*, BMP2 serves to increase the presence and number of cells, which can induce a mineralizing phenotype in those tissues investigated. Another BMP isozyme, BMP7, is expressed in smooth muscle cells [27]. Unlike BMP2, however, BMP7 maintains the phenotype of the cells and thus prevents their differentiation to an osteoblastic lineage [27]. Together, these observations indicate that BMP isoforms govern and finely tune the phenotype of cells which are activated by environmental changes in affected valves and blood vessels.

The osteogenic activities of BMPs on vascular cell function have also been extensively documented. BMP2, for example, promotes expression of *Runt-related transcription factor 2* (*RUNX2*), which targets the *OSX* promoter and fosters the expression of many genes required for endochondral bone formation [28]. These genes include *ACAN*, *type I collagen* (*COLI*), and *COLII*, *osteopontin* (*OPN*), *osteocalcin* (*OCN*), *tissue non-specific alkaline phosphatase* (*TNAP*), and *sodium-dependent phosphate cotransporter* (*PIT*)*-1/solute carrier family 20* (*SLC20*)*A1* [29]. The BMP2-mediated increase in PIT expression and secretion provides the phosphate ions required for the synthesis of bone mineral [30].

As a general summary, induction of BMP family members mediates the transdifferentiation of somatic cells into an osteogenic lineage and exerts a decisive effect on the expression of other genes required for the synthesis and secretion of extracellular matrix molecules necessary for vertebrate tissue mineralization. These events raise the related question as to what factor or factors induce or mediate BMP expression. In this context, hypoxia- and turbulence-induced biomechanical changes in the wall of blood vessels are known to promote BMP release and stem cell commitment [31]. Likewise, within an atheromatous plaque (the fat-laden deposits which accumulate in vessels), the presence of "bad cholesterol," a component of oxidized low-density lipoproteins (LDLs), can also trigger BMP release [32]. Other factors promoting BMP release include the interaction of advanced glycation end-products (AGE), compounds generated during chronic diabetes, atherosclerosis, and kidney disease, with receptors on cell membranes [33, 34]. Mineralization associated with AGE is thought to take place around apoptotic cell remnants, with possible crosstalk between BMP and SMAD 1/5/8 signaling and protein kinase B (PKB/AKT)/mammalian or mechanistic target of rapamycin (mTOR) pathways [35, 36]. Such a response leads to calcium accumulation, exosome formation, and secretion of matrix vesicles (Fig. 10.4), details of which are discussed in Chap. 6. At this late stage, the rate and extent of mineralization is governed by pyrophosphate and expression and secretion of TNAP, progressive ankylosis protein homologue (ANK), ectonucleotide triphosphate pyrophosphatase/phosphodiesterase (ENPP)1, OPN, matrix gla protein (MGP), and fetuin (or AHSG, α2HS glycoprotein), among other molecules. How each of these factors influences mineral development is described in Chap. 9 and later sections of this chapter.

An additional aspect of the pathogenesis of vascular mineral deposition concerns the question of whether there is a genetic predisposition to dystrophic mineralization. This possibility has been addressed using a genetic approach called a genome-wide association study (GWAS). Such a study can involve thousands of subjects and analyses of numerous chromosomal regions to identify genetic variants which may be

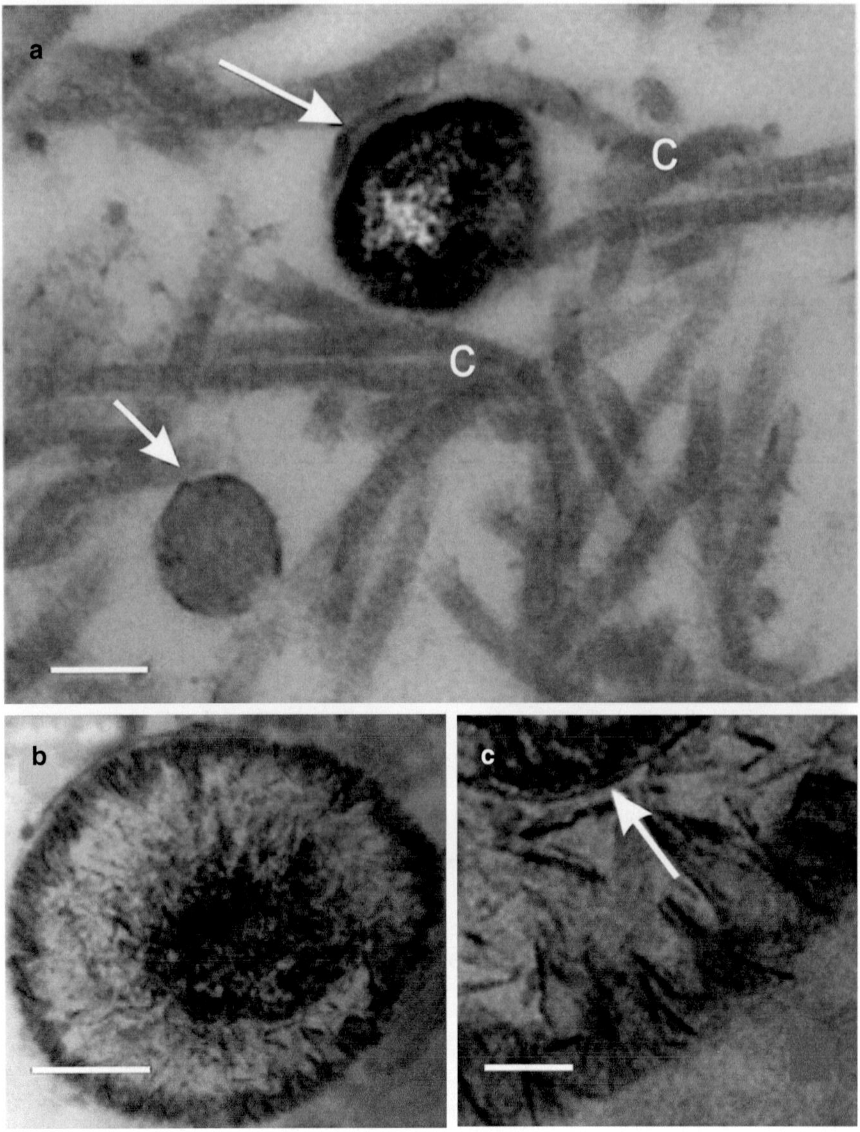

Fig. 10.4 Transmission electron microscopy and energy dispersive X-ray elemental (EDX) microanalysis of mineral present in atherosclerotic plaques from the human carotid artery following endarterectomy. (**a**) Microscopy of the plaque matrix shows the presence of matrix vesicles and numerous collagen fibrils (C). Samples were fixed with glutaraldehyde and osmium tetroxide and sections were stained with uranyl acetate and lead citrate. The longer arrow identifies a vesicle with a relatively high electron-dense membrane and content. The shorter arrow points to another matrix vesicle with a less electron-dense membrane/content. (**b**, **c**) Enlargements of a matrix vesicle comprised of dense spicules in both the vesicle membrane and the center of the vesicle (arrow). EDX analysis of unstained vesicles demonstrates the presence of both calcium and phosphorus [14]. Scale bars = 100 nm (**a**, **b**), 25 nm (**c**). (Images adapted from Bobryshev et al. [14] with permission)

inherited and coupled with a specific trait or disease risk [37, 38]. In the instance of vascular mineral formation, GWAS found single nucleotide polymorphisms around a specific locus of the gene *SORT1*, encoding the protein, sortilin (SORT1), on human chromosome 1 (1p13.3) [39]. SORT1 is a member of the vacuolar protein sorting 10 protein (VPS10P) domain receptor family, located within the membranes of the trans-Golgi network of vascular cells [40]. As its name suggests, SORT1 functions to bind and sort ligands for transport into lysosomes, endosomes, exosomes, and the plasma membrane [40–43]. Binding is mediated through a molecular ten-bladed (β)-propeller (Fig. 10.5) [44]. *SORT1* polymorphism is linked to many of the arterial and tissue deficits associated with vascular disease (high plasma low-density lipoprotein [LDL] and cholesterol levels, and clinical evidence of myocardial infarction and aortic aneurysm) and extraskeletal mineralization [37, 39]. More definitively, *SORT1* expression has been found to be elevated in mineralized arteries and atherosclerotic plaque, and deletion of *SORT1* decreases vascular tissue mineralization [42]. SORT1 also promotes fibrosis and mineralization in aortic valve disease (Fig. 10.6) [45].

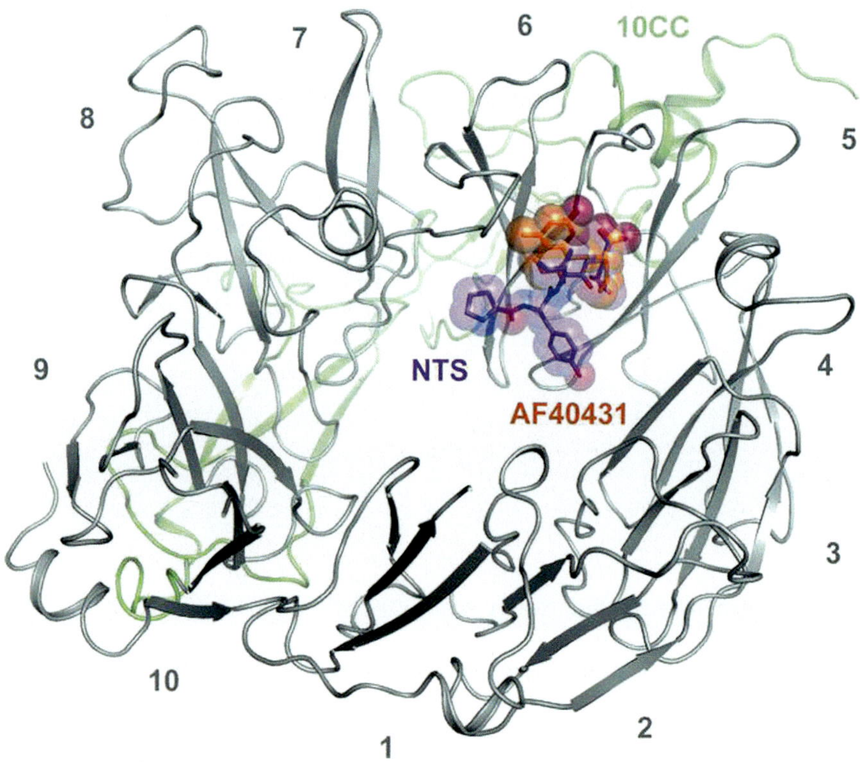

Fig. 10.5 Molecular architecture determined by X-ray crystallography of sortilin, a transmembrane glycoprotein member of the vacuolar protein sorting 10 protein (VPS10P) family of receptors, viewed from an upper open side of its tunnel (2.7 Å resolution). Sortilin is comprised of a β-propeller structure of ten blades (1–10) and ten cysteine-rich domains (10CC), which probably serve to stabilize the β-propeller. The blades of the β-propeller and the 10CC domain are represented in gray and green, respectively. The neurotensin (NTS) binding site of sortilin, in the center of the tunnel of the β-propeller, is depicted as a purple ball-and-stick model linked to AF40431 (orange sticks and spheres), the first small molecule ligand identified to be bound to sortilin. (Image from Andersen et al. [44] with permission)

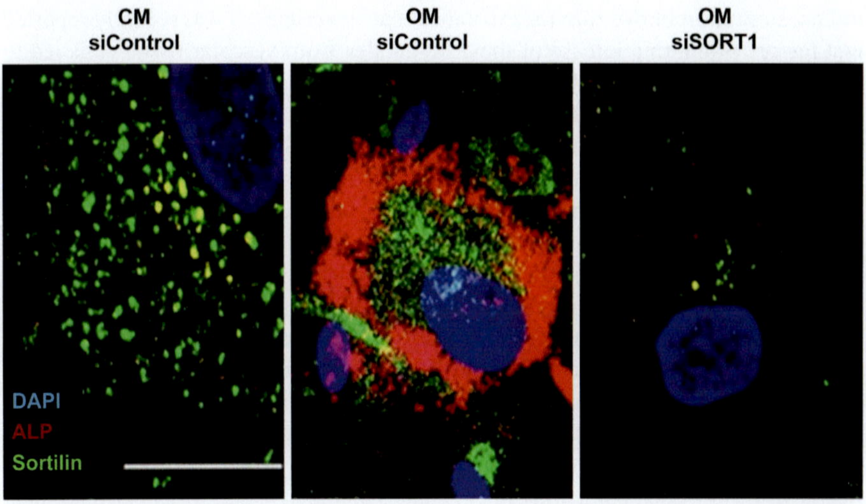

Fig. 10.6 Immunofluorescence analysis of sortilin in human primary interstitial cells obtained from aortic valves collected from autopsy donors. Isolated interstitial cells were maintained in culture in a calcifying osteogenic medium (OM) or a control (non-osteogenic) medium (CM) for periods of time up to three weeks. For a brief time during culture, cells were transfected with small interfering (si) RNA, either 25 nm control siRNA (siControl) or 25 nm SORT1 siRNA (siSORT1), the latter to silence *SORT1*, the gene expressing sortilin. Cells were subsequently stained with antibodies to alkaline phosphatase, a marker of mineralization (ALP, red), and sortilin (yellow) as well as DAPI (4,6-diaminidino-2-phenylindole), a nuclear/DNA indicator (blue), and examined. Compared to CM (left panel), the OM resulted in co-secretion of ALP and sortilin under siControl (middle panel). On sortilin silencing in OM (right panel), fluorescent signals for both sortilin and ALP decreased notably. Scale bar = 25 μm. (Experimental details and images from Iqbal et al. [45] with permission)

In addition to SORT1 as a critical cell constituent necessary for protein trafficking, RAS-related proteins in brain (RAB) function in a related manner and are required for autophagy, especially autophagosome formation and maturation [46]. Moreover, RAB proteins enhance fusion of autophagosomes with endosomes to form the hybrid organelle, the amphisome [46, 47], and they have been reported to facilitate the convergence of the endosomal and autophagic systems to form cellular components which share characteristics of both pathways [46]. This latter observation explains how calcium ions located within the endoplasmic reticulum can be partitioned to vesicles originating in the autophagic system (see Chap. 6). Consistent with these findings, the Golgi-endoplasmic reticulum of cells would package calcium and phosphate ion clusters and TNAP, which then could become intermixed with vesicles generated by the autophagic pathway. Regardless of their source, intracellular vesicles which are associated with the appropriate proteins (RABs, soluble *N*-ethylmaleimide-sensitive fusion protein-attachment protein receptors [SNAREs], SORT1) are transported and delivered to specific sites on the cell plasma membrane as described previously. Vesicles containing ions and proteins necessary for mineralization are next expelled from the cell as exosomes/matrix vesicles, subsequently generating mineral foci which grow and develop in the extracellular milieu. In support of this role for exosomes/matrix vesicles, it was recently reported that factors promoting release of these organelles from vascular tissue cells led to aberrant mineralization [48].

From the results presented above, it is clear that a propensity for mineral deposition occurs in soft tissues undergoing pathological changes. In instances when aberrant soft tissue mineralization occurs, its structure and organization vary from amorphous granular material to that resembling membrane and endochondral bone. As noted earlier, circulating proteins like fetuin, tissue-localized proteins like MGP, and the pyrophosphate anion inhibit such calamitous forms of mineral. In many soft tissues, locally generated environmental factors trigger the expression of transcription factors which target genes closely associated with bone formation. Thus, there is increased expression of cytokines like BMP-2, which in turn upregulate genes required for somatic cell reprogramming prior to osteogenesis. In terms of a possible pathway for mineral deposition, SORT1 partitions cargo, particularly TNAP, for transport in endosomes [49]. SORT1 also promotes phagosome maturation, prevents fusion with lysosomal compartments, regulates autophagy, and fosters development and secretion of exosomes containing TNAP [49]. These exosomes promote mineral deposition in soft tissue extracellular matrices. Additionally, SORT1 influences the inflammatory milieu by decreasing secretion of pro-inflammatory cytokines in these tissues [49]. In this way, cellular interactions with the extracellular matrix, together with the local environment, conspire to cause dystrophic mineral deposition in blood vessels and other soft tissues. As a further note, it is important to state that it is not known how SORT1 and the RAB-GTPases participate in the normal physiologic mineralization of bone, cartilage, tooth, and other vertebrate tissues. However, it is likely that TNAP is transported from these cells of hard tissues through the same (exosomal) pathways as those described in mineralizing vascular tissues. If that is the case, then an important step in the mineralization cascade would be delineated.

10.3 Disorders of Mineralization: Heterotopic Ossification

Another form of abnormal vertebrate mineralization has been described as heterotopic ossification. The term "heterotopic" (Greek, "hetero" or "other" and "topos" or "place") refers to the formation of ectopic bone at a non-osseous site of the body and, unlike dystrophic mineralization, histology clearly indicates that the new tissue is well organized and in most cases appears as lamellar bone. A condition of heterotopic ossification was first described in a patient in 1692 by G. Patin [50], a physician in Paris, and was much later named myositis ossificans progressiva. Eisenstein et al. [51], quoting an earlier statement of Brackett [52], noted that, "The most common form (of heterotopic ossification) was an irregular mushrooming with a tendency to spurs on the inner aspect of the femur. Occasionally sharp exostoses were seen. These often were sharp enough and long enough to cause sufficient pain to warrant their removal" [51]. More recently because of conflicts in the Middle East and Afghanistan, combat blast injuries have increased and with them a marked rise in the incidence of heterotopic ossification in both soldiers and civilian populations (Fig. 10.7).

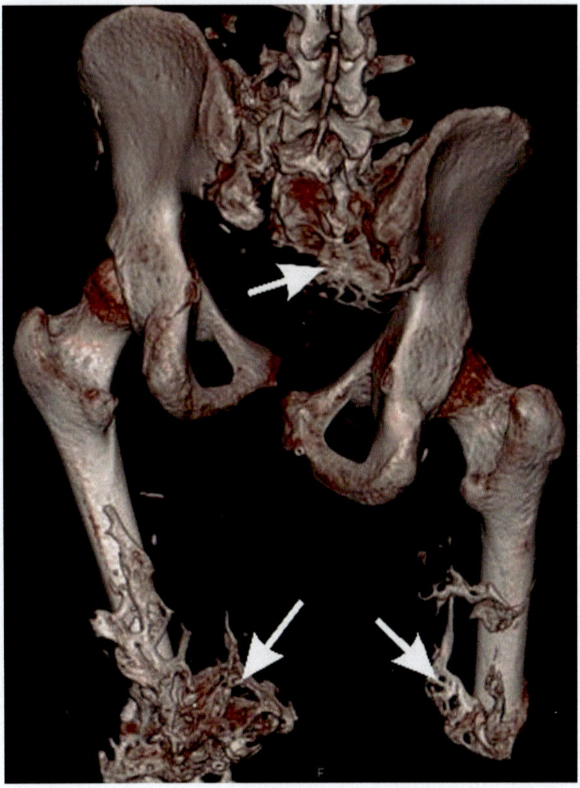

Fig. 10.7 Heterotopic ossification of the skeleton of a combat-injured patient. Three-dimensional (3D) computed tomographic (CT) reconstruction of the bones of the pelvis and femora. White arrows point to regions of heterotopic ossification in the mid-diaphysis of the femurs and sacrum. (Image from Eisenstein et al. [51] with permission)

There are two distinct forms of this type of ectopic ossification. The most common is progressive osseous heteroplasia, acquired usually as a result of some manner of tissue insult and typically exhibiting the structural characteristics of mature intramembranous bone [53]. Although the condition was long ago reported in muscle [50], its scope is not limited to this tissue. Such extraosseous bone is also found in ligaments, skin, and indeed in any connective tissue [54, 55]. In some instances, a bone-like tissue may form in the spinal cord, a condition identified clinically as neurogenic heterotopic ossification [56]. Factors initiating progressive osseous heteroplasia include trauma (burns, fractures, and dislocation), amputation, inflammation, and surgery, especially after total hip replacement surgery and brain and spinal cord injury [57]. The second type of heterotopic ossification is very uncommon: this condition, fibrodysplasia ossificans progressiva (FOP), renamed from myositis ossificans progressiva noted above, is marked by deposition of endochondral bone at extraskeletal sites and is caused by a gain-of-function mutation of the BMP receptor [58].

The pathogenesis of the common, acquired heterotopic disorder is directly associated with non-osseous tissue injury although such injury rarely leads directly to tissue mineralization. How the injurious event results in the deposition of an aberrant mineralized tissue with the hallmark of cortical or even endochondral bone has intrigued many investigators, and uncertainty persists about the cause and origin of the osteogenic response. With respect to the pathogenesis of heterotopic ossification, it is known that injury initiates a chronic inflammatory reaction characterized by the invasion of white blood cells (mainly monocytes, macrophages, and some lymphocytes) into the traumatized soft tissue [59]. Macrophages, especially those expressing the M2-activated phenotype, appear to promote aberrant mineralization in soft connective tissues [60]. Unlike the M1 state characterized by the expression of pro-inflammatory cytokines, the M2 macrophage secretes cytokines such as BMP4, transforming growth factor-β (TGF-β), and others which are closely associated with mineral deposition [61]. Additionally, macrophages release oncostatin, a cytokine promoting the commitment of stromal stem cells to osteoblasts [62].

Besides those invasive cells just mentioned above and in line with metastatic mineralization, attention in this context has focused on other major cell types populating soft tissue niches: pericytes, vascular endothelial cells, multipotent mesenchymal stromal cells, skeletal muscle cells, and hematopoietic cells. Studies with stem cell marker proteins such as CD133 antigen, tyrosine-protein kinase receptor 2 (TIE2), glial high-affinity glutamate transporter (GLAST), and glioma-associated oncogene (GLI)1 have been used to try to pinpoint presumptive osteogenic stem cells [63, 64]. These investigations by genetic lineage tracing indicate that subpopulations of cells expressing the marker genes, *TIE2*, *GLAST*-creERT, and *GLI1*, behave like niche-dwelling skeletal/progenitor cells [63, 64].

Another approach to addressing the stem cell question in extraskeletal mineralization has been to evaluate the phenotype of soft tissue cells cultured in medium treated with BMPs and other cytokines. Probably the most interesting findings of such work with skeletal muscle suggest that satellite cells, a cell type present on the surface of muscle fibers concerned with muscle repair and regeneration, can express the properties of a stem cell [65, 66]. Hashimoto and colleagues showed such progenitor cells secreted both the muscle lineage-specific proteins, PAX7 and MYOD, and osteogenic proteins, TNAP and RUNX2 [65]. Even in the absence of BMP, these post-natal myogenic precursor cells can undergo terminal osteogenic differentiation and their osteogenic characteristics are downregulated when muscle repair is stimulated [65, 67]. The switch between these two pathways appears to be responsive to RHO-GTPase [65].

There remains uncertainty as to whether soft tissue cells such as those noted here can become osteogenic without first expressing the phenotypic characteristics of a pluripotent or even a multipotent stem cell. When compared with pluripotent cells, multipotent post-embryonic stem cells express a limited spectrum of differentiation genes and retain the ability to commit to a restricted number of cell lineages. In this regard, Furmanik et al. considered that stress signals caused a partial change in the phenotype of vascular smooth muscle cells (multipotency) and influenced expression of muscle contraction-specific genes [68]. These changes were sufficient to cause extracellular vascular mineralization mediated by vesicles [68].

Besides the migration of the various types of white blood cells noted above into non-osseous tissue and the nature of the resident multipotent stromal cells, there are other considerations regarding the origin of heterotopic ossification. Changes may occur, for example, in the nature of the tissue environment resulting from excessive mechanical force or skin or muscle burns. These factors promote a hypoxic state and induction of hypoxia inducible factor, HIF-1α [69]. The expression of this critical transcription factor in combination with the low oxygen tension, accompanied by inflammatory cell invasion and the actual presence of mineral, provides a persistent imbalance in local ionic homeostasis [70]. Such a set of events would be expected to influence resident soft tissue cells to adopt an osteogenic phenotype marked by the secretion of matrix vesicles serving to enhance heterotopic mineral deposition in the extracellular matrix [70]. Agarwal et al. also indicated that HIF-1α is critical for both trauma-induced and genetic forms of heterotopic ossification [69].

Finally, BMP has also been studied in possibly mediating systemic effects leading to a heterotopic response in soft tissues following injury. In this context, for example, levels of BMP2 are elevated in serum of patients after traumatic brain injury and bone fracture [71]. Similarly, in an animal model of spinal cord injury, expression of *BMP2*, *4*, *7*, and *9* is increased [72, 73]. Cells contributing to elevated *BMP2* expression include activated (T-cell) lymphocytes, which secrete cytokines (tumor necrosis factor [TNF]-α, TGF-β, interferon [IFN]-γ, and interleukin [IL]-17) whose effects in turn lead to BMP2 liberation [62]. Thus, the presence of these cytokines and growth factors together with ionic changes in the local environment of an injured tissue primes surviving multipotent soft tissue cells to express a mineralizing phenotype. Once osteogenesis has commenced, the cells express a wide spectrum of osteoblastic genes which include *BMP2*, *TNAP*, *SMADs*, *BSP*, and *OCN* [74]. In aortic valve cells, there is also production of exosomes/matrix vesicles characterized by annexins, CD9, CD63, LAMP-1, and LAMP-2 [75, 76]. These exosomes/matrix vesicles generate a mineralized extracellular matrix in the otherwise soft aortic tissue. Summarizing the available evidence, Lees-Shepard and Goldhamer concluded that acquired heterotopic ossification may be modeled as "initiated locally by tissue-resident progenitors and … circulating progenitors participate in subsequent bone growth and remodeling" [77]. Much more research needs to be conducted to establish clearly which cell or cells or other factors might lead to the development of the osteogenic phenotype in progressive osseous heteroplasia.

As briefly introduced above, a much less common form of heterotopic ossification is represented by the unusual and medically serious disorder, fibrodysplasia ossificans progressiva (FOP). Like other forms of heterotopic ossification, this genetic disease is characterized by mineral deposition in muscle and other soft tissues, leading to the formation of a "second skeleton" in the body (Fig. 10.8) [78]. The cause of this condition, revealed by the exacting work of Shore, Xu, Feldman, Kaplan, and colleagues, is a missense mutation by a single nucleotide substitution in the gene which encodes a member of the BMP family of receptors, activin receptor type 1 (ACVR1; also known as ALK2 or activin receptor-like kinase-2) [58].

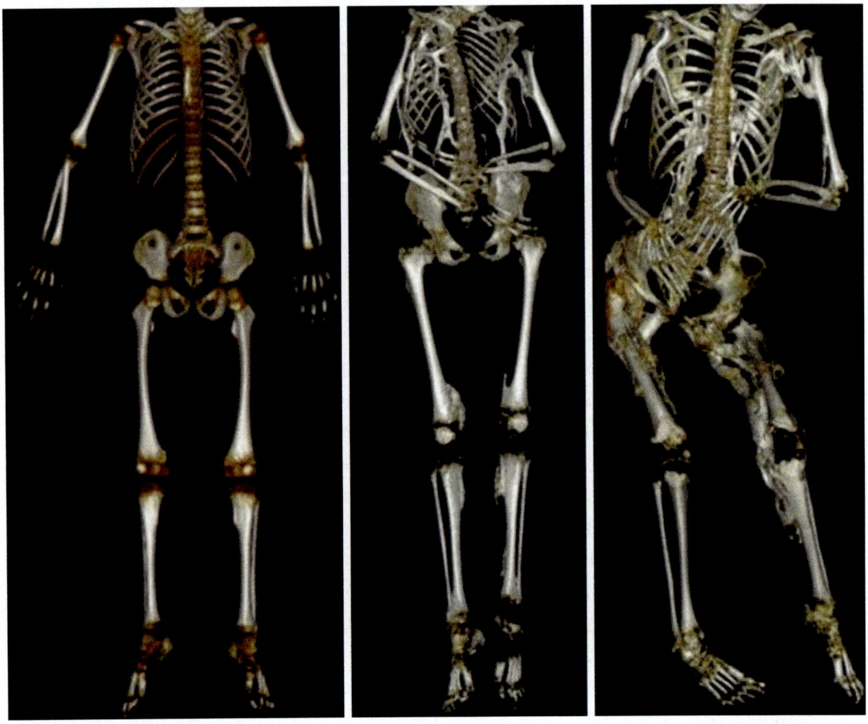

Fig. 10.8 A visual representation of the increase in heterotopic mineralization with time (4, 10, and 31 years) in three individuals suffering from fibrodysplasia ossificans progressiva. Whole-body computed tomography excluding the head was used to assess the quantity of heterotopic bone. The scans shown were acquired in the cranio-to-caudal axis from the head to the feet. With time, both the number of regions exhibiting abnormal bone deposits (heterotopic ossification, HO) and the volume of new bone increase (new bone volume is presented in mm^3). Associated with the increase in heterotopic bone is profound functional impairment of the individual. (Images adapted from Pignolo et al. [78] with permission)

Under normal circumstances, ACVR1 interacts with type II receptors of the BMP protein family to form a complex consisting of receptors which bind activins and BMPs, respectively [58, 79]. The missense mutation in FOP affects the kinase activation domain of ACVR1, causing a gain-of-function such that, following ligand–receptor binding, the receptor becomes locked in an "on" position. The consequence of this condition results in over-activation of the SMAD1/5/8 rather than the SMAD 2/3 signaling pathway (Fig. 10.9). An aberrant complex thus translocates to the nucleus, further enhancing activin signaling [85]. FKBP12 (12-kDa binding protein for FK506) is a negative regulator of the type I receptor and, if overexpressed in model systems, FKBP12 can rescue the effects of many of the ACVR mutations (Fig. 10.9) [86]. This example of FKBP12-BMP-receptor/SMAD interaction is but one of numerous instances of the system of signal transduction involving the TGF-β superfamily of growth factors, including the BMPs, which are critical to mineralization, embryogenesis, organogenesis, and many other fundamental biological processes (Fig. 10.9) [79–86].

Several therapeutic techniques are now being explored for the treatment of the mineralization-associated defect in FOP. These include application of a kinase inhibitor of the BMP type I receptor, which would suppress SMAD1/5/8 signaling [87]; an RNA interference (RNAi) polynucleotide that targets the mutated receptor [88]; and an HIF-1α inhibitor that prevents endochondral ossification [88, 89]. In addition to these approaches, pharmacological intervention of the mTOR complex by rapamycin both inhibits autophagy and enhances stromal stem cell commitment to the osteogenic pathway [90–93]. Likewise, mutation of the mTOR complex protein impairs osteogenesis and causes severe trabecular bone loss in mouse stromal stem cells [94]. Conversely, bone anabolic proteins like WNT ligands activate mTORC1 through phosphoinositide 3-kinase (PI3K)-AKT signaling and stimulate osteogenesis [91]. From this perspective, manipulation of mTOR and its associated proteins provides a rational means to regulating heterotopic mineralization [92, 93].

The clinical utility of each of the treatment methods described above has not been established. However, retinoid agonists, agents which block chondrogenesis by binding nuclear hormone receptors (retinoic acid receptors, RARs), key regulators of gene transcription, are undergoing clinical trials. In the presence of endogenous ligand, the RAR-retinoic X receptor heterodimer acts as a transcriptional activator to enhance chondrogenesis [95, 96]. However, in its unbound form, this receptor complex acts as a transcriptional repressor [95, 96]. A synthetic retinoid agonist specific for RAR (the α-isoform) prevents BMP2-induced heterotopic ossification [95]. The drug reduced BMP signaling, lowering SMAD1/5/8 phosphorylation in mouse fibroblasts [96], and inhibited the recruitment of immune and inflammatory cells in a subcutaneous mouse model of heterotopic ossification [97]. Using a mouse knock-in model expressing the mutant ALK2 gene, Pacifici has developed a retinoid-γ agonist, Palovarotene, which significantly reduces heterotopic bone formation [98]. In clinical trials, this drug indicated effectiveness in treating FOP patients and is now approved for use in adults and children by Health Canada [98]. There is also a good chance that, by blocking chondrogenesis, Palovarotene could be used to inhibit or prevent aberrant osteogenesis and heterotopic ossification resulting from gunshot wounds or other forms of trauma [98]. Whether a similar approach could be used to prevent metastatic mineralization has not yet been explored.

The findings related to heterotopic ossification described above impact present understanding of normal biological mineralization in several ways. These include, first, the central role of BMP as an agent which induces progenitor cells to express an osteogenic/chondrogenic phenotype. Second, the mineralization process may involve more than one cell type. This observation is consistent with the metabolic activities in the basic multicellular unit (BMU) where considerable numbers of different cells are required to both mediate and control the mineralization response. Third, injury and other examples of insult or disease to the body which often lead to heterotopic mineral deposition point to the importance of local environmental factors (hypoxia, pH) and inflammation in regulating mineral induction. Finally, there is a commonality in terms of gene expression which is directly associated with heterotopic mineral formation. Regardless of its initiating cause, heterotopic ossification requires the expression of TNAP, BMPs, SMADs, matrix metalloproteinases (MMPs), COLI, COLII, type X collagen (COLX), and several others of the same genes which are directly involved with physiologic mineralization of bone, cartilage, dentin, and additional vertebrate hard tissues. Thus, heterotrophic ossification, as well as dystrophic mineralization, shares many of the molecular biological, biochemical, and structural features which also characterize the processes defining normal vertebrate mineral deposition.

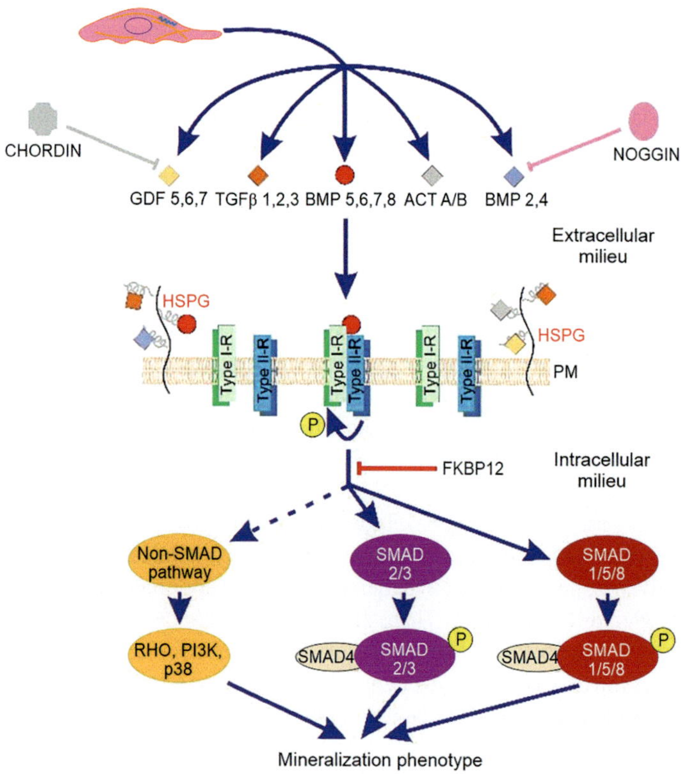

Fig. 10.9 Schematic representation of the signal transduction system utilized by members of the transforming growth factor-β (TGF-β) superfamily of growth factors and leading to a mineralization phenotype. This family includes the bone morphogenetic protein (BMP) family members (BMP 5, 6, 7, 8, and BMP 2, 4), TGF-β members (TGF-β 1, 2, 3), activins (ACT A/B), and growth and differentiation factors (GDF 5, 6, 7). Cell-secreted ligands (TGF-βs, BMPs, ACTs, and GDFs) released to the extracellular milieu bind to plasma membrane (PM) receptors to form heterotetrameric protein receptor complexes [79–81]. BMP and GDF proteins are the ligands which exert the most powerful effect on the mineralization of bones and teeth. Protein receptors consist most commonly

(continued)

Fig. 10.9 (continued)

of dimer (homomeric) pairs of two types of serine/threonine kinases, I-R and II-R, located close to each other in the PM [82]. Occasionally, preformed I-R and II-R receptor pairs cluster in heteromeric form in the PM [83]. The various ligands bind to type I-R or type II-R receptors to form a heterotetrameric signaling platform leading to three downstream SMAD or non-SMAD pathways and mineralization. Activation of a particular downstream pathway is complicated and determined by a specific ligand and its manner of receptor binding [82, 83]. In the case of BMP 5–8 binding as illustrated, these family members most frequently associate with type II-R receptors and, in so doing, cause recruitment of nearby type I-R receptors [84]. In this structural assemblage, autophosphorylation (P) of the type I-R receptor occurs by the type II-R receptor. Autophosphorylation within the protein receptor complex induces phosphorylation of an intracellular SMAD1/5/8 complex, which then binds SMAD 4. This cluster of molecules translocates to the cell nucleus and upregulates transcription of genes linked to mineralization (including *OCN*, *ON*, and *TNAP*) [79, 80, 82, 84]. In the case of TGF-β, the SMAD 1/5/8 complex may be activated as described for BMP 5–8, but TGF-β binding to PM receptors may also activate an SMAD 2/3 pathway. SMAD 2/3 becomes phosphorylated, binds in turn to SMAD 4, moves to the cell nucleus, and promotes mineralization [80, 82]. The SMAD 2/3 pathway may additionally be activated when certain ligands bind to the preformed I-R and II-R receptor pairs in the PM. Alternatively, TGF-β and other specific ligands binding PM receptors may activate a non-SMAD pathway whose targets include RHO, p38 MAPK, and the phosphoinositide 3-kinase (PI3K)/AKT/mTOR pathway, also leading to mineralization [79, 83, 85]. The availability of the TGF-β family of ligands is tightly controlled by inhibitory matrix proteins such as chordin and noggin as well as by matricellular, plasma membrane-incorporated proteoglycans, including heparan sulfate proteoglycan (HSPG), which also bind the various ligands [79]. In addition, FKBP12, acting at the intracellular level, is a negative regulator (inhibitor) of the type I receptor and overexpressed FKBP12 can modulate the consequences of many possible ligand and receptor mutations such as those in ACVR [84, 86]. The activities of the signal transduction pathways and their relationships to the SMAD pathways are context (tissue)-driven, and they provide critical functionality in numerous roles besides mineralization, including embryogenesis, cell proliferation and differentiation, inflammation, immune response, and organogenesis, for example. (Schematic based partly on Yadin et al. [79] with permission)

10.4 Disorders of Mineralization: Microcalcifications Associated with Breast Cancer

Aberrant mineralization occurs in breast tissue, often associated with blood vessels (Fig. 10.10) [99]. Unlike many other extraskeletal deposits of mineral elsewhere in the body, such abnormal mineral accretions do not cause biomechanical problems. These deposits are unusual as they are commonly found as thin streaks during mammography (Fig. 10.10) and their presence raises the important question as to whether the micron-sized areas of mineralization ("microcalcifications"), evident in 15% of women undergoing routine screening for breast cancer, predict tumor malignancy. Indeed, Scott et al. underscore microcalcifications as one of the earliest critical diagnostic markers of cancer of the breast and relate the general belief that such deposits may promote cancer cell proliferation and neighboring tissue invasion [100]. Furthermore, the chemical composition of microcalcifications may be a major factor involved with the pathogenesis of breast carcinoma [100].

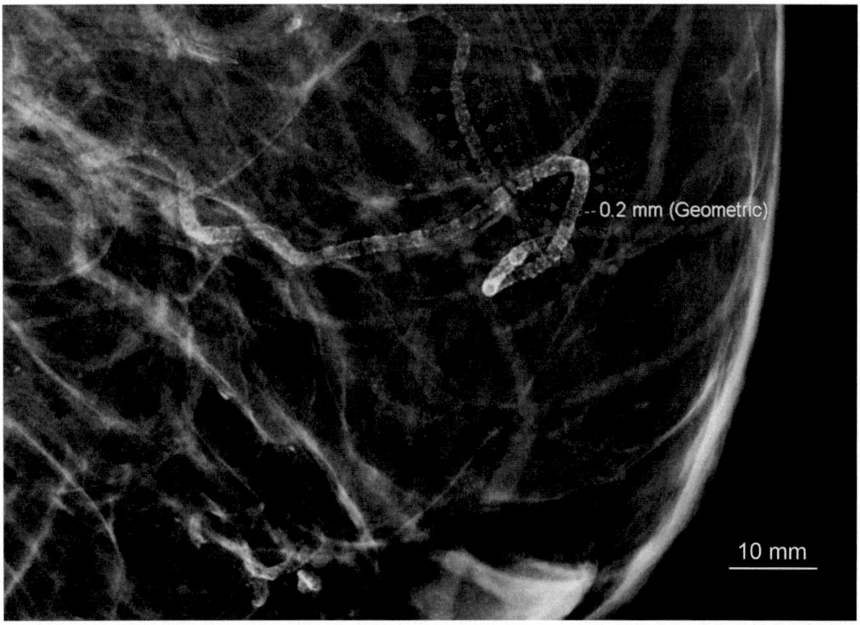

Fig. 10.10 Mammographic X-ray image of human breast tissue showing vascular mineralization. This mediolateral oblique view demonstrates the presence of branching blood vessels with linear and serpentine mineral deposits. Red arrows trace the path of a single convoluted vessel having a diameter of ~0.2 mm. Mineral deposits in this 60-year-old patient were benign. (Image kindly provided by Jason Shames, M.D., M.B.S., Division of Breast Imaging, Thomas Jefferson University, Philadelphia, PA)

In this context, in a relatively recent report by Nyante et al. involving invasive cancer, such mineral deposits were "positively associated with higher tumor grade and the presence of an in situ tumor component," and inversely associated with factors which included hormone receptor negativity and tumor size [101]. Henrot et al. have used mammographically identified calcification to separate ductal carcinoma in situ from benign lesions of the breast [102]. Thus, both the deposition and composition of mineral salts abnormally appearing in the breast are of considerable diagnostic and pathoetiologic interest and value.

Regarding compositional analysis, two different types of minerals have been identified in breast microcalcifications. In benign and malignant lesions in situ, the mineral has been found to be a combination of carbonated hydroxyapatite and magnesium-rich hydroxyapatite [103]; in other benign lesions, the mineral phase is principally calcium oxalate [103, 104]. The presence and importance of calcium oxalate in benign lesions are currently unclear, however, since a very recent survey of over 200 mineral deposits in breast tissue failed to confirm this mineral [105]. With respect to apatitic mineral, the carbonate content is reportedly higher in the microcalcifications from benign compared to malignant lesions [106].

Another observed change in mineral composition is related to the presence of magnesium in breast calcifications. Magnesium content of the mineral appeared to increase with respect to the pathological state of the lesion, from benign to malignant and invasive [100]. This finding prompted a detailed investigation of the apatite comprising breast lesions, and analysis led to the suggestion that the mineral may contain whitlockite, a magnesium-substituted calcium phosphate [100]. In a subsequent study conducted by the same laboratory, the initial finding was revised to note that there was "a statistically significant inverse correlation between Mg:Ca ratio and percentage of magnesium whitlockite" [107]. Thus, the importance of magnesium ions in apatite is currently unknown, although it is doubtful that magnesium-enriched mineral is more toxic than carbonated apatite. While discussion continues concerning elemental composition in apatite, Gosling et al. showed a change in apatite lattice structure with malignancy and that malignancy was accompanied by an increase in apatite crystallinity (coherence length, crystallite size, and non-uniform strain) [105].

The documented changes in chemical and physical characteristics of apatitic mineral in breast microcalcifications would be expected to influence the activities of cells in the vicinity of such deposits. This suggestion is supported by the observation that chondrocytes, fibroblasts, and synovial cells in contact with the mineral undergo what can best be described as a "stress" response [108]. The "stress" is mediated by either endocytosis of the mineral into the lysosomal compartment of the cell or direct contact of nanocrystals with the plasma membrane of adjacent cells [109]. In the latter case, apatitic crystals may induce a multitude of changes, many of which are associated with skeletal disease. Chondrocytes, macrophages, or fibroblasts activated by endogenous mineral in vitro are reported to produce tumor necrosis factor (TNF)-α and IL-6 and IL-1B/1β, for example [110, 111]. Such factors induce the expression of MMPs and ADAMTS (a disintegrin and metalloproteinase with thrombospondin motifs), enzymes involved with degradation of cartilage extracellular matrix and the development of osteoarthritis [112].

Apart from the effects on cells of microcalcifications which contact the plasma membrane directly, these calcium phosphate nanocrystals appear to stimulate vesicular activity in cancer cells [113]. In this instance and following phagocytosis, the mineral is solubilized within the acidic milieu of lysosomes, an action which would result in the intracellular release of calcium ions. Such a rapid or extensive increase in calcium ion content could likely overcome the buffering capacity of the cell and negatively influence functions as diverse as cell proliferation, apoptosis, production of inflammatory cytokines, and MMP activation. Moreover, the increase in cell calcium ions within the hypoxic environment encompassing many tumors would also influence energy generation. Studies by Xue et al. show that nanocrystalline apatite interferes with electron transport along the respiratory chain [114]. With the loss of this activity, ATP levels would decrease and calcium ions would either accumulate in the cell or cause cell death. The greater levels of cell calcium would enhance the release of TNF-α, nitric oxide (NO), and reactive oxygen species (ROS). If death ensued, cell and molecular remnants would act as nucleating surfaces for further accumulation of extracellular mineral. A more controlled increase in intracellular calcium ion levels would prompt the endoplasmic reticulum and mitochondria to act as calcium buffers and subsequently provoke the release from the cell of calcium ions sequestered within vesicles (exosomes). Discharge of membrane vesicles has been well documented for a number of mammary cell lines and appears to be correlated with tumor invasiveness [115–117].

While the above discussion concerns the effects of apatite on breast cancer cells, the presence of extracellular mineral can also lead to a local inflammatory response. In cartilage, for example, apatite activates primary macrophages through the NOD-, LRR-, and pyrin domain-containing protein (NLRP)3 inflammasome and causes the secretion of IL-1B and IL-18 [118]. Additionally, since breast cancer cells secrete IL-6, IL-8, and cyclooxygenase-2 together with prostaglandin E2, mineral would exacerbate the release of these inflammatory mediators and contribute in a substantial way to the expression of a malignant phenotype.

As with studies noted above concerning osteogenic progenitors, little is known of the cell type within a cancerous lesion which provokes aberrant microcalcifications. Since mineralization often follows vascular elements, the suggestion has been made that uncommitted mesenchymal cells, possibly pericytes, commit to the "osteoblastic" lineage and initiate mineral deposition [119]. A modification of this notion is that cancer cells undergo an epithelial–mesenchymal transition and secrete BMPs (BMP2, BMP4) which then stimulate stem cell commitment to osteogenesis [120]. The process has been described by Scimeca et al. as breast cancer "osteomimicry" [120]. This research group has noted that the morphological characteristics of breast cells at the mineralization front resembled those of authentic osteoblasts with a "large cytoplasm, a huge rough reticulum and cytoplasm rich in vesicles containing electron-dense granules composed of hydroxyapatite" [120]. Experimentally, cancer cells treated with BMP express many of the genes associated with bone mineralization [121, 122].

Several mechanisms have been proposed by which agents like BMP promote microcalcifications. Among these, BMP increases the level of expression of the transient receptor potential (TRP) superfamily of ion channel proteins [123], thus providing an immediate route for calcium ion entry into breast cancer cells [124] and the potential subsequent formation of nanocrystals. Indeed, breast cancer progression has been correlated with alterations in cytosolic calcium ion homeostasis mediated by calcium ORAI channels and/or the endoplasmic reticulum calcium ion sensor, stromal interaction molecule 1 (STIM1) [125, 126]. Sustained calcium ion transients and elevated intracellular calcium ion activity promote both cell proliferation and apoptosis, cardinal events in neoplasia. Further, these transients influence endoplasmic reticulum filling activities, overcome normal cellular calcium–protein buffering, and alter endosomal calcium trafficking, thereby conspiring to enhance an abnormal deposition of extracellular mineral [127]. In a related context, Cross and colleagues introduced a new insight into microcalcification formation in breast tissue by suggesting that the process is linked to lactation: thus, "a developmentally determined program of gene expression controls the massive transcellular mobilization of Ca^{2+} from the blood into milk by the coordinated action of calcium transporters, including pumps, channels, sensors, and buffers" [127].

Aside from BMPs, Bonavita et al. have reported that breast cancer cells express the gene *pentraxin* (*PTX*) which encodes the protein PTX3, a member of the TNF family [128]. *PTX* is a tumor-suppressor gene concerned with regulating inflammation [128]. More relevant to osteomimicry, *PTX* is involved with extracellular matrix remodeling, promotes differentiation of vascular endothelial cells and smooth muscle cells, enhances expression of *RUNX2*, and leads to mineral deposition [129, 130]. Since these cell types may undergo mineralization, the identification of *PTX* in breast tissue in relation to extracellular microcalcifications is not unexpected, especially as breast tumors are often accompanied by the expression of inflammatory mediators. From these several perspectives of inflammation, tumor suppression, and mineralization, PTX3 likely plays a general role in both aberrant mineralization and tumor cell microcalcifications.

An interesting additional clinical feature of *PTX* expression and PTX3 secretion is related to Peyronie's disease, a penile disorder causing painful erections, abnormal curvature of the organ, and erectile dysfunction among other characteristics. Although an os penis (penile bone) is absent in humans, the etiology of Peyronie's disease is related to the formation of obstructive mineral-containing plaques or scars in the tunica albuginea, along the principally collagenous soft tissue shaft of the penis. The inflammatory cascade resulting in this condition includes both the local and systemic secretion of PTX3. When the mineralization is treated with pentoxifylline, the drug stabilizes or reduces plaque mineral levels and promotes normal penile function [131].

Breast cancer cells also express a number of *small integrin-binding ligand, N-linked glycoprotein* (*SIBLING*) genes and their corresponding secreted proteins. Of these proteins, OPN directly relates to the appearance of microcalcifications [132]. As the majority of microcalcifications found in OPN-positive tumors are malignant with poor prognosis, some studies have utilized aspects of OPN biology to hinder tumor growth. In this regard, mice harboring cancerous tissue reverse tumor growth when they are treated with an RNA oligonucleotide which blocks *OPN* expression [133], and silencing the *OPN* gene inhibits breast carcinoma invasion and growth [134]. In addition, a humanized anti-OPN antibody has been reported to inhibit breast cancer cell adhesion, migration, invasion, and colony formation and to suppress spontaneous metastasis in vivo [135]. Results of these studies point to the close involvement of this particular SIBLING protein with both microcalcifications and malignancy.

This section of the chapter concerned the regulation of abnormal mineral deposits (microcalcifications) and their possible role in mediating the malignant state of cells in breast tissue. It is clear that there is considerable crosstalk between physical and chemical compositional characteristics of the extracellular nanocrystals and the fate of breast cancer cells. Numerous molecular biological, biochemical, and structural changes occur in breast tissue cells which lead to microcalcification formation and provide insight into possible mechanisms of such mineral deposition. Among these changes are the plasticity of cells surrounding breast tissue, the altered intracellular handling of calcium ions by tumor cells, upregulation of a local inflammatory response, and the induction of a mineralizing phenotype by molecules such as BMP and PTX3. In addition, extracellular proteins like OPN appear to modulate microcalcification deposition. These summary data reveal many similarities between breast cancer microcalcifications, dystrophic mineralization, heterotopic ossification, and physiological bone mineral homeostasis. While there is still much to be learned about microcalcifications, the molecular interactions and the processes involved in forming them, as well as their relationship with the onset and progression of breast tissue tumorigenesis, add substance to concepts and ideas expressed in other chapters of this book. More directly related to the mineralization process, it is likely that deposition of microcalcifications represents a modification of pathways leading to normal bone mineral formation and turnover.

10.5 Disorders of Mineralization: Defective Mineralization Caused by Mutations in Extracellular Matrix Proteins

As noted above, aberrant mineral formation in vascular or breast tissues maintains certain features in common with physiological mineralization of vertebrate skeletal tissues. The latter deposition of mineral is distinct, of course, from the former in involving the association of apatite with specific proteins comprising the matrix of the appendicular and axial skeleton and the dentition of the body. Interactions principally between calcium and phosphate ions and such proteins, as discussed previously in this book, lead to nucleation, growth, and development of apatite while generating the unique features and properties which characterize bone, cartilage, dentin, and other tissues which normally mineralize in the vertebrates. Not surprisingly, defects in connective tissue proteins, especially those which alter the structure and post-translational modifications of the most prominent extracellular matrix constituent, COLI, affect the critical processes of apatite deposition. In this part of the chapter, protein structural changes manifested by well defined diseases of the extracellular matrices of bones and teeth will be described as they influence apatite mineral deposition. In this case, a shift in emphasis will be made from elucidating the responses of the cells which populate hard tissues to assessing the importance of the extracellular matrix as the recipient of mineral ions and crystals.

As was detailed in Chap. 7 of this book, COLI is the major extracellular protein comprising the normally mineralizing vertebrate connective tissues of bone, dentin, cementum, and the tendons in certain avian species. While COLI is a very stable protein with a long evolutionary history, over 800 sporadic mutations have been documented in the molecule, affecting its structure, macromolecular formation, and its subsequent mineralization [136–138]. Mutations in COLI coding genes, for example, are sufficient to cause the brittle bone family of diseases or osteogenesis imperfecta (OI) (Fig. 10.11) [136, 139]. Overt symptoms of OI include marked skeletal fragility with multiple fractures, especially of the long bones, and spinal complications which include vertebral compression and bending (kyphosis and scoliosis) (Fig. 10.12) [140, 141]. As collagen is also a primary constituent of the soft tissues of the body, its defects additionally impact lung function, hearing acuity, and even the color of the sclera of the eyes.

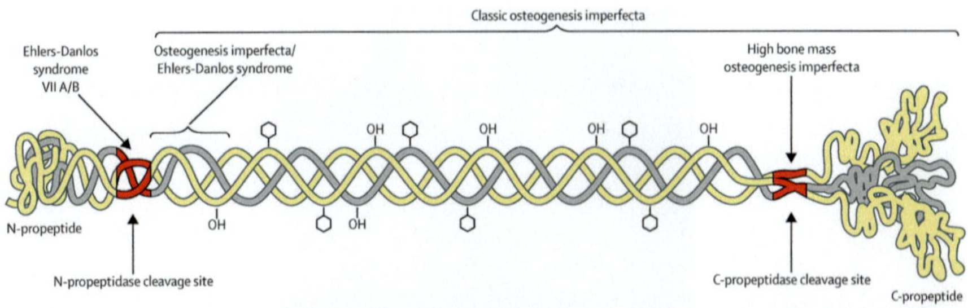

Fig. 10.11 Schematic of the COLI molecule showing sites of mutations causing clinical disease. Collagen mutations in the triple helical and the terminal C-propeptide regions of the molecule result in classical osteogenesis imperfecta. Those mutations occurring close to the junction of the helical region with the N-propeptide lead to an osteogenesis imperfecta/Ehlers-Danlos variant. Ehlers-Danlos syndrome VII A/B is attributable to mutations influencing the structure of the protein at its N-propeptide cleavage site. Those mutations present at the C-propeptide cleavage site produce a clinical abnormality of high bone mass osteogenesis imperfecta. Glycosylation sites are indicated by hexagons along the collagen triple helix and OH designates a hydroxyl group linked to proline or lysine residues. (Schematic from Forlino and Marini [136] with permission)

The molecular basis of the most common of OI conditions is an autosomal mutation of genes encoding the α1(I) and α2(I) chains of COLI (*COLIA1* and *COLIA2*) [138]. A variant of this condition, Ehlers-Danlos syndrome, affects connective tissues and is seen clinically as skin fragility and lax ligaments and tendons [142]. Trying to understand the relation between the location of a mutation and the severity of the OI pathology, Marini et al. have suggested that, if the mutations occur within the first 200 residues of COLI, they are non-lethal; if the mutations occur in helix positions 691–823 and 910–964, they can be lethal [139]. In the latter case, the increased seriousness of *COLI* mutations in OI was proposed to be caused by alterations in ligand-binding sites for integrins, MMPs, fibronectin, and cartilage oligomeric matrix protein [139].

Most frequently, the mutations in OI result in the replacement of glycine by a bulky amino acid in the collagen polypeptide chains, a change thereby interfering with the formation of a stable three-dimensional structure of the collagen triple helix [143]. Although the individual chains of the mutant collagen molecules in OI self-assemble in nearly normal configuration, the fibrils in fact have abnormal crosslinks resulting from alterations in lysine hydroxylation [144]. Another factor affecting the stability of bone in OI is the absence of the α2(I) chains from some heterotrimeric collagen molecules [145]. In a mouse model of the disease, it was shown that these homotrimers formed kinks along the collagen polypeptide chains, which reduced intermolecular crosslinking and decreased the mechanical strength of the protein and the bone of the animals [145]. Since unstable collagen molecules are inefficiently secreted into extracellular tissue spaces, they accumulate intracellularly where they may lead to endoplasmic reticulum stress and apoptosis [146].

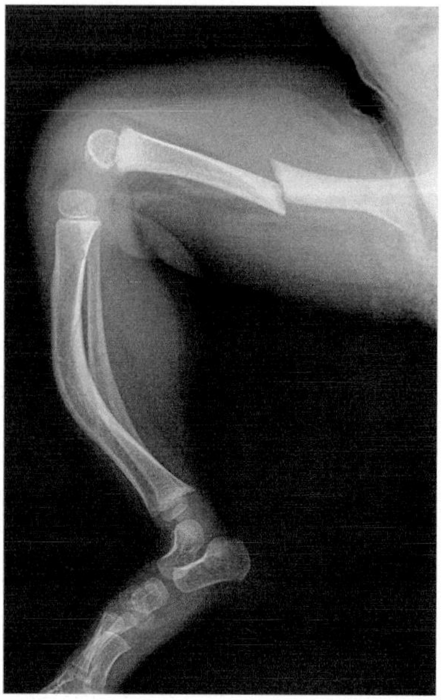

Fig. 10.12 Lateral X-ray image of the leg of a child with osteogenesis imperfecta. The radiograph shows a fracture in the diaphyseal region of the femur. The tibia and fibula show extensive bowing. (Image from Tauer et al. [140] with permission)

Some OI mutations, on the other hand, cause defective intracellular assembly of individual collagen chains into proper triple helical conformation, and they, too, may then affect the extracellular process of collagen fibril formation [147]. In addition, because collagen interacts with many additional proteins comprising the extracellular matrix and it serves as an anchorage site for cells, numerous abnormalities may also occur in the architecture of the affected tissues. Since the protein abnormalities influence many other tissue components, the effects of mutations in collagen molecules are broad and the impact of OI on mineralization becomes more difficult to interpret and treat.

From a mechanistic viewpoint, the structural changes resulting from substitutions of glycine and other COLI amino acid residues dramatically modify OI bone architecture, resulting in, for example, thinning of cancellous bone trabeculae, a decrease in osteocyte number, and an expansion of the marrow and vascular elements of the tissue (Fig. 10.13) [148]. At the ultrastructural level, there are notable changes in the gap and overlap zone environments of collagen fibrils, regions which are critical for mineral nucleation, growth, and deposition as discussed in Chap. 7. As a result, development of apatite associated with collagen is altered and the mineral platelets are smaller and thinner than normal [149–151]. These changes contribute significantly to the OI pathogenesis of this mineralization-related disease.

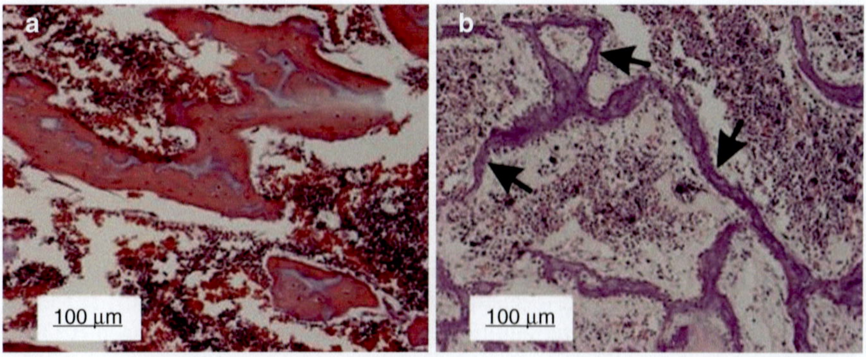

Fig. 10.13 Microstructure of osteogenesis imperfecta bone. (**a**) Normal cancellous bone showing islands of trabeculae (red) with some retained cartilage (pink) surrounded by diffusely stained marrow and blood cells. (**b**) Osteogenesis imperfecta bone characterized by thinned trabeculae (arrows) with few osteocytes and an expanded marrow cavity. Tissue samples were demineralized and stained with hematoxylin and eosin. (Images adapted from Nijhuis et al. [148] with permission)

In a detailed study of bone mineralization in OI, Fratzl et al. investigated a mouse model of the disease and showed that the apatite crystals were smaller, more tightly packed and not as well aligned as the crystals analyzed from counterpart normal animals [152]. Indeed, the reported levels of mineralization were higher in the model than the levels in controls probably because of the close packing of the small apatite crystals [152]. In a later study, Vanleene et al. confirmed that, despite the increased level of mineralization in the disease, the bones in OI were mechanically stiffer (lower modulus of elasticity) than normal bones, a result again presumably attributable to the smaller, highly packed and disoriented apatite crystals [153]. These observations concerning the brittle nature of their bones in part explain why OI patients suffer from multiple fractures. The data also provide evidence that an intact, unaltered extracellular matrix is a requisite for normal bone mineralization.

In addition to the mutations in *COLI* which influence bone structure and lead to the common forms of OI noted above, there are rarer forms of OI (types V and VI) caused by dominant mutations in genes required for osteoblast differentiation and subsequent mineralization. For example, a single nucleotide mutation at the 5′ untranslated start site of the gene, *interferon-induced transmembrane protein (IFITM)5*, creates a new, five-residue amino acid elongation at the protein N-terminus and leads to OI type V [154, 155]. The expression of IFITM5 is osteoblast-specific, having a pattern like that of osterix (OSX) and affecting early-stage bone growth in experimental mouse models [156, 157]. For these reasons, IFITM5 is also known as bone-restricted IFITM-like protein (BRIL) [156, 158]. The phenotype of OI type V resulting from *IFITM5/BRIL* mutation is highly variable, but patients present with mineralization abnormalities, bone fragility, low bone mass, and hyperplastic callus formation during bone fracture healing [155,

158, 159]. The precise mechanism by which genetic changes in the expression of *IFITM5/BRIL* influence collagen structure and OI type V is still unknown although the importance of the protein as a positive regulator of bone mineralization is well documented [160]. In this regard, overexpression of *IFITM5/BRIL* promotes TNAP secretion and mineral deposition, while knockdown of the gene reduces mineralization [156].

In the past 15 years, more and more information has been forthcoming to help understand recessive forms of OI resulting from mutations in genes required for post-translational modifications of the collagen molecule. If such modifications are incomplete, they impair helix formation and stability of COLI and COLII as well as the stability of other collagen family members such as type V collagen (COLV) [161]. The gene mutations affect the activities of enzymes such as prolyl 3-hydroxylase (P3H1) and the proper function of cartilage-associated protein (CRTAP) [138, 162, 163]. P3H1 together with CRTAP and the protein, cyclophilin, form an enzyme complex which hydroxylates prolyl residues of COLI and functions as a molecular chaperone [138, 162, 163]. Furthermore, mutations in this complex and in heat shock protein (HSP)47, a collagen-specific protein chaperone encoded by the *serine proteinase inhibitor family H (SERPINH)1* gene, may also lead to severe forms of OI [138, 164].

Because of the defective collagen, the mineral content and density of the bone in OI are elevated by a "loss of control over crystal nucleation and growth" [163]. Mutations in *transmembrane protein* (*TMEM*)*38B* also cause a recessive form of OI [165]. The gene in this instance encodes the protein, trimeric intracellular cation channel type B (TRIC-B), required for the proper maintenance of intracellular calcium in the endoplasmic reticulum, calcium trafficking, collagen post-translational reactions, and mineralization [166]. A TRIC-B deficiency leads to dysregulated calcium ion accumulation and release, and the mutation is found clinically, especially among Bedouins, with a bone phenotype similar to that of OI [165].

Like bone, enamel can be affected by mutations which cause mechanical brittleness and hypocalcification. These changes, identified as a group of disorders known collectively as amelogenesis imperfecta, result in a structurally and functionally defective tissue which can break and chip away from its underlying dentin core [167]. While these mutations are rare, they most often occur in those genes encoding and elaborating the enamel matrix proteins amelogenin (AMELX), enamelin (ENAM), and ameloblastin (AMBN). Other mutations include genes encoding proteases required for processing of enamel matrix constituents (Kallikrein-related peptidase 4 [KLK4] and MMP20); genes encoding adhesion proteins (integrin b6, laminins [LAMA3 and LAMAB3], type XVII collagen [COLXVII]); genes encoding proteins required for the transport of calcium and other ions into extracellular matrices (WD repeat-containing protein 72 [WDR72] and SLC24A4); and a gene linked to apatite nucleation (*chromosome 4 open reading frame 26* or *C4ORF26*) [168]. Indeed, 18 mutations are associated with amelogenesis imperfecta disorders [168]. The matrix of fully mineralized enamel contains very little protein, and it is distinct in that manner from the protein-rich matrices of bone, dentin, cementum, and calcifying tendon present in certain avians. However, a failure in even one of its matrix components is likely to lead to substantial deficits in the structure of enamel, like any flaw in matrix proteins of other normally mineralizing tissues of vertebrates.

Dentinogenesis imperfecta refers to another family of diseases caused by mutations in genes encoding proteins of the dentin extracellular matrix [169]. These disorders are classified into those with mutations in *COLIA1* and *COLIA2* and those resulting from abnormalities in *dentin sialophosphoprotein* (*DSPP*), encoding a major non-collagenous protein comprising the dentin extracellular milieu [170]. Clinically, dentinogenesis imperfecta can affect the primary and secondary dentition; dentin structure is weak, and teeth are mobile with abnormalities in root structure and can undergo premature exfoliation. Microscopic analysis of the condition also shows that the characteristic dentin tubular structure is lost in the deeper layers of the tooth [171].

In addition to dentinogenesis imperfecta, Chen et al. have described a further mineralization disorder in dentin resulting from mutations in proteins related to those of extracellular matrices [172]. In this case, dentin dysplasia is a condition caused by mutations in several other dentin genes (*SSU-2 homologue* [*SSUH2*], *SPARC-related modular calcium-binding protein 2* [*SMOC2*], and *vacuolar protein sorting 4 homologue B* [*VPS4B*]) [172]. One of these, *VPS4B*, encodes a protein component of the endosomal sorting complexes required for transport (ESCRT) machinery of the cell and for multivesicular body formation and exosome biogenesis (see Chap. 6 for discussion of the importance of these organelles in the initiation of mineralization) [173].

In summary for bone, dentin, enamel, or other vertebrate mineralized tissues, it is clear that a small change in the structure of a single extracellular matrix protein may cause extraordinary alterations in the biochemistry, architecture, mechanical properties, and critical function of that tissue. In a tissue like enamel, where its extracellular matrix also contains enzymes which process a variety of proteins in the presence of intact structural constituents, delayed or failed expression of such enzymes can be catastrophic. In other words, for hard tissue formation, there exists exquisite coordination between those intracellular events which generate ions or ion clusters and particles of mineral and the molecular configuration and organization of the extracellular proteins associated with such mineral. While yet to be completely delineated, feedback loops from the extracellular matrix to the osteoblast, chondrocyte, odontoblast, ameloblast, or other cells comprising a vertebrate mineralized tissue ensure that mineral deposition and matrix synthesis and turnover are carefully integrated and highly synchronized. Even without definitive feedback information, the studies which have been reported currently provide unequivocal evidence that the proteins of the extracellular matrix exist as critical components of the mineralization process and that minimal alteration in protein structure following gene mutation, inadequate post-translational processing, or other related abnormal events may result in severe deficits in tissue structure and function.

10.6 Perspective: Aberrant Mineral Deposition in Soft and Hard Tissues

The final chapter of this book has been designed to highlight conditions, diseases, and pathologies which affect mineral formation, particularly aberrant mineral deposition in soft tissues. While these various disorders provide further details of an encompassing normal and abnormal mineralization "landscape" and are of obvious interest to the clinician, they fulfill two other requirements. First, they exemplify the complexities of the process of biological mineralization in terms of the roles of key genes and their transcription factors, hormones, and inflammatory and other proteins in regulating the cellular and extracellular availability of calcium and phosphate ions. Second, while the type of mineralization is aberrant in the several examples described here, such deposition nevertheless builds on the underlying hypothesis that mineralization is a cell-mediated process, dependent on calcium accumulation in vesicles, responsive to selected growth factors, and initiated by the release of calcium-loaded exosomes and calcium phosphate ion complexes from the cell into an extracellular matrix constructed in a specific molecular architecture. From this perspective, individual disease states can be readily understood as representing deviations from underlying evolutionarily conserved physiological pathways.

In terms of complexity of regulation, as has been discussed previously, within the confines of a BMU of forming or remodeling bone, a battery of hormones (parathyroid hormone [PTH], vitamin D, estrogen), cytokines (fibroblast growth factor [FGF], BMP), coupling factors (receptor activator of nuclear factor κB ligand [RANKL]/receptor activator of nuclear factor κB [RANK], semaphorins, ephrins [EPHs]), and systemic and local inhibitory molecules (pyrophosphate, fetuin, MGP, OPN, other SIBLING proteins), together with sclerostin (SOST), osteoprotegerin (OPG), and microRNAs (miRNAs), play their own and overlapping roles in controlling and fine-tuning physiological mineral deposition. In almost all cases and despite redundancy, a loss of function in any one of these molecules leads to a defect or failure in the events to either remove bone or form new bone. Ultimately, the disequilibrium between bone formation and resorption results in disease.

Aberrant mineral deposition in soft tissues of the body has been distinguished above as dystrophic mineralization, heterotopic mineralization, microcalcifications associated with breast cancer, and defective mineralization related to mutations in extracellular matrix proteins. Each disorder presents its own specific features, yet all possess certain common characteristics, including those defining aspects of physiological mineral formation. Dystrophic mineral deposition, especially in atheromatous plaque, is linked to the low oxygen status of a tissue, the generation of oxygen radicals and advanced glycation end-products, as well as the presence of oxidized lipids of low-density lipoprotein. These factors incite an inflammatory response in affected soft tissues mediated by cytokines such as TNF-α and IL-1B which facilitate endothelial-to-mesenchymal cell transitions, and they promote BMP-dependent osteogenic differentiation [174] and mineralization [175]. With these changes in phenotype, there is a marked increase in intracellular ion trafficking, induction of endoplasmic reticulum stress, and capacitance-loading of calcium by means of CRAC channels [176]. As a result of trafficking, the exosomes released from calcifying vascular cells contain calcium ions as well as low levels of inhibitory proteins such as fetuin and MGP. Once anchored in the extracellular matrix, these vesicles become primed for rapid mineral formation. At the same time, inflammatory proteins cause the migration of macrophages into the tissue walls. These cells release a spectrum of enzymes, including MMPs and cysteine endoproteases, which cause tissue remodeling, a process which prepares the extracellular matrix for subsequent events linked to further mineral growth and development [177, 178].

Diseases like heterotopic ossification provide supporting evidence that BMPs, members of the TNF-β superfamily of proteins, play major roles in directing physiological bone formation. The expression of BMPs is related to both trauma-induced and genetic forms of heterotopic mineralization [77, 179]. Besides regulation through interactions with their receptors, the activity of these proteins is carefully controlled by a host of secreted molecules such as noggin and chordin, mentioned previously, as well as gremlin, cerberus, and twisted gastrulation-1, which bind to and compete for BMP membrane receptors [180]. In dentinogenesis, microRNAs control (inhibit) noggin expression [181], while BMP regulates microRNA expression [182]. Other intracellular feedback loops exist which control BMP-mediated signal transduction. These loops involve the SMAD proteins with their transcriptional co-regulators, RUNX2, and several downstream regulatory elements such as mTOR, the protein which mediates autophagy, regulates osteogenesis, and indirectly controls cell death as described in Chap. 6 [183].

A somewhat different aspect of the induction of mineralization in soft tissues is that many of the factors noted above, together with lipid metabolites, cause an increase in reactive oxygen species, which in turn activates an autophagic survival response [184, 185]. Indeed, when autophagic genes like *autophagy-related gene (ATG)5* are deleted or blocked, there is a rapid increase in cell death [186].

If autophagy is inhibited, there is a decrease in mineral deposition [187]. Likewise, an inhibitor of autophagosome formation impedes phosphate-dependent cell death [188]. From a mechanistic viewpoint, as has been discussed in some detail earlier, autophagy and endosome recycling are processes which allow intracellular vesicles to fill with calcium ions. Vesicles fated to deliver ions or proteins to the mineralization front are sorted by SORT1. In this regard, Maeda et al. reported that overexpression of *SORT1* caused the acceleration of mineralization during osteogenic differentiation without affecting alkaline phosphatase activity [189]. Such processes traffic ions within exosomes or matrix vesicles to the extracellular matrix where they undergo further processing to form the mineral phase of the tissue.

While an autophagic response represents a key step in the cell-mediated aberrant mineralization process affecting vascular and other soft tissues, some debate yet persists concerning the cell type which commits to an osteogenic or chondrogenic phenotype in such instances. Although cells have been documented to migrate to sites of mineralization [190], Tintut and colleagues suggested that, within blood vessels, approximately 10–30% of smooth muscle cells had the capacity to become "calcifying vascular cells," phenotypically similar to pericytes [191]. A putative relationship between pericytes and adult stem cells has yet to be fully resolved, but these vascular cells may commit to several different lineages, possibly after dedifferentiation [191, 192]. Once committed, such cells respond to osteogenic stimuli and express many of the phenotypic indicators of bone-forming cells, including critical transcription factors, elevated TNAP activity, and the secretion of proteins like OCN and OPN [191].

As alluded to earlier in this chapter and to extrapolate the discussion of mineral deposition in non-osseous tissues, a fundamental question may be asked as to how a cell, exquisitely adapted to and performing specific functions in its soft tissue environment, can undergo a dramatic change to exhibit a new phenotype. Such a transformation is readily observed in tissues like the cartilaginous growth plate where, as discussed earlier, terminally differentiated chondrocytes may assume the phenotype of an osteoblast. The transdifferentiation process may simply involve changes in the activity of one or more critical cellular transcription factors as, for example, in the transdifferentiation mediated by MYOD1 of retinal pigmented epithelial cells to myoblasts [193]. On the other hand, this type of phenotypic switch may require substantial reorganization of the genome which may necessitate passage into a less differentiated stage before commitment by the cell to a new phenotype. Behavior of this nature requires that genes representative of the existing differentiated state of a cell be first downregulated or repressed to those characteristic of a more primitive or basic stem cell. After reaching this state, the upregulation of expression of subsequently appropriate genes leads to the appearance of a new phenotype. In bone fracture repair, such a transformation between chondrocytes and osteoblasts is known to require expression of stem cell pluripotency genes (*SOX2*, *OCT4*, and *NANOG*) as well as genes directly linked to the osteogenic phenotype (*RUNX2*, *OSX*, *OPN*, *OCN*, and *COLI*) [194].

It is thought that the transformational changes described above are mediated by factors which enhance nuclear chromatin remodeling such that the chromatin presents a "configuration which is more accessible to transcription factors and chromatin-modifying enzymes, permitting global changes in transcription" [195]. Factors which are critical for genome reprogramming include the presence of exogenous cytokines such as BMP and mechanical forces affecting the cell [196, 197]. Vining and Mooney have established that changes in the mechanical properties of a matrix direct stem cell differentiation [197]. These mechanical changes sensed by integrins, G-coupled protein receptors, and growth factor receptors mediate transcriptional changes resulting in the expression of pluripotency genes just noted [198]. This type of epigenetic regulation causes changes in chromatin structure involving histones and non-coding RNAs as well as key transcription factor activity [199].

The pathogenesis of metastatic mineralization differs considerably from mineralization associated with transdifferentiation and involves the dysregulation of serum calcium and phosphate ions principally by vitamin D, FGF23, and PTH [200]. Thus, there is an elevation in the serum phosphate ion concentration accompanied by a fall in the serum calcium level. In response to the decrease in calcium ion levels, PTH secretion increases and triggers osteoclastic bone resorption, liberating calcium and phosphate ions into the circulation and restoring or possibly increasing the ion activity product of calcium and phosphate ions in the blood. As a result of these changes, the elevated ion activity product encourages soft tissue mineralization while at the same time hard tissues lose mineral and become osteopenic or potentially osteoporotic [201].

Metastatic mineralization and tumorous breast tissue, however, need not involve fluctuations in hormonal levels or even development of an osteoporotic state. Both age and systemic disease can predispose soft tissues to form mineralized deposits, which then exacerbate the disease state [202]. In blood vessels already weakened by the presence of atheromatous plaques, for example, mineral deposition leads to loss of elasticity and other biomechanical parameters with consequences which include elevated blood pressure and development of possible aneurysms [203, 204]. In cardiac valves, mineral deposition causes a change in tissue stiffness which can result in heart failure [205]. In breast tissue, the presence of microcalcifications, while still relatively poorly understood, probably has minimum effects on tissue function but may influence the pathogenesis as well as the prognosis of the disease state [206].

Based on the preceding information and consistent with events which govern physiologic mineralization, the possibility exists that similar pathways are activated during the development of the several forms of aberrant soft tissue mineralization described here. Thus, vascular cells, for example, may express osteoblast-like characteristics and lysosomal/endosomal/autophagic vacuoles load with calcium, to be transported subsequently to the mineralization front of the tissue. There is also the possibility that aberrant mineralization results from blockage of the cell survival pathway. In this case, following cell death, charged sites on membrane remnants and residual DNA and lipids could serve as nucleation sites for mineral deposition.

Current knowledge of mechanisms of aberrant mineralization indicates that numerous types of molecular events regulate the interactions between cells and matrices of vertebrate tissues. For bone, cartilage, tooth, and other examples of mineralizing tissues, the nature of these feedforward and feedback circuits reflects the complexity of underlying control systems. Unlike the accelerator and the brake pedals of a car, which function in a digital go-or-stop mode, the biological signals refined through millions of years of evolutionary experiments are analog, subtle, overlapping, and redundant. However, despite meticulous elucidation of many of these complicated and involved molecular circuits, there is still much to be learned and important questions remain to be clarified about them. These include, for instance, determining the relationship between autophagy and endosomal cycling, exploring the provocative role of sortilin-like proteins in physiological mineralization, and a host of others. Applied to the mysteries of the mineralization process, it is worth recalling a comment of Sir Winston Churchill: "This is not the end, this is not even the beginning of the end, this is just perhaps the end of the beginning" [207]. Indeed, as the "beginning" traffics toward the "end," the physician and scientist will have access to an armamentarium of molecular approaches for treating mineralization defects, thereby decreasing and eliminating human pain and suffering. Hopefully, this book will catalyze the reader to delve further into mechanisms driving this fascinating aspect of human skeletal and dental biology.

References

1. Virchow R. Cellular-Pathologie. In: Virchow R, Reinhardt BEH, editors. Archiv für pathologische Anatomie und Physiologie und fur klinische Medizin. 1855;13–14, translated by Rather LJ. Redwood City: Stanford University Press; 1938.
2. Levine MA. Pathological calcification and the mystery of Lot's wife. Cell Cycle. 2015;14:3354–5.
3. Murphy WA Jr, Nedden Dz D, Gostner P, Knapp R, Recheis W, Seidler H. The iceman: discovery and imaging. Radiology. 2003;226:614–29.
4. Mönckeberg JG. Über die reine Mediaverkalkung der Extremitätenarterien und ihr Verhalten zur Arteriosklerose. Virchows Arch Pathol Anat Physiol Klin Med. 1903;171:141–67.
5. Hardy K, Radini A, Buckley S, Blasco R, Copeland L, Burjachs F, et al. Diet and environment 1.2 million years ago revealed through analysis of dental calculus from Europe's oldest hominin at Sima del Elefante, Spain. Naturwissenschaften. 2017;104:2.
6. Mandel ID, Gaffar A. Calculus revisited. A review. J Clin Periodontol. 1986;13:249–57.
7. Virchow R. Cellular pathology. As based upon physiological and pathological histology. Lecture XVI—atheromatous affection of arteries. Nutr Rev. 1889;47:23–5.
8. Pettenazzo E, Deiwick M, Thiene G, Molin G, Glasmacher B, Martignago F, et al. Dynamic in vitro calcification of bioprosthetic porcine valves: evidence of apatite crystallization. J Thorac Cardiovasc Surg. 2001;121:500–9.
9. Wong TY, Wu CY, Martel J, Lin CW, Hsu FY, Ojcius DM, et al. Detection and characterization of mineralo-organic nanoparticles in human kidneys. Sci Rep. 2015;5:15272.
10. Cardoso L, Kelly-Arnold A, Maldonado N, Laudier D, Weinbaum S. Effect of tissue properties, shape and orientation of microcalcifications on vulnerable cap stability using different hyperelastic constitutive models. J Biomech. 2014;47:870–7.
11. Neven E, Dauwe S, De Broe ME, D'Haese PC, Persy V. Endochondral bone formation is involved in media calcification in rats and in men. Kidney Int. 2007;72:574–81.
12. Hruska KA, Mathew S, Saab G. Bone morphogenetic proteins in vascular calcification. Circ Res. 2005;97:105–14.
13. Vattikuti R, Towler DA. Osteogenic regulation of vascular calcification: an early perspective. Am J Physiol Endocrinol Metab. 2004;286:E686–96.
14. Bobryshev YV, Killingsworth MC, Lord RS, Grabs AJ. Matrix vesicles in the fibrous cap of atherosclerotic plaque: possible contribution to plaque rupture. J Cell Mol Med. 2008;12:2073–82.
15. Abedin M, Tintut Y, Demer LL. Vascular calcification: mechanisms and clinical ramifications. Arterioscler Thromb Vasc Biol. 2004;24:1161–70.
16. Steitz SA, Speer MY, Curinga G, Yang HY, Haynes P, Aebersold R, et al. Smooth muscle cell phenotypic transition associated with calcification: upregulation of Cbfa1 and downregulation of smooth muscle lineage markers. Circ Res. 2001;89:1147–54.
17. Demer LL, Tintut Y. Vascular calcification: pathobiology of a multifaceted disease. Circulation. 2008;117:2938–48.
18. Brighton CT, Lorich DG, Kupcha R, Reilly TM, Jones AR, Woodbury RA II. The pericyte as a possible osteoblast progenitor cell. Clin Orthop Relat Res. 1992:287–99.
19. Caplan AI. All MSCs are pericytes? Cell Stem Cell. 2008;3:229–30.
20. Liu AC, Gotlieb AI. Characterization of cell motility in single heart valve interstitial cells in vitro. Histol Histopathol. 2007;22:873–82.
21. Farrington-Rock C, Crofts NJ, Doherty MJ, Ashton BA, Griffin-Jones C, Canfield AE. Chondrogenic and adipogenic potential of microvascular pericytes. Circulation. 2004;110:2226–32.
22. Kuwana M, Okazaki Y, Kodama H, Izumi K, Yasuoka H, Ogawa Y, et al. Human circulating CD14+ monocytes as a source of progenitors that exhibit mesenchymal cell differentiation. J Leukoc Biol. 2003;74:833–45.

23. Neven E, Persy V, Dauwe S, De Schutter T, De Broe ME, D'Haese PC. Chondrocyte rather than osteoblast conversion of vascular cells underlies medial calcification in uremic rats. Arterioscler Thromb Vasc Biol. 2010;30:1741–50.
24. Chen NX, Duan D, O'Neill KD, Wolisi GO, Koczman JJ, Laclair R, et al. The mechanisms of uremic serum-induced expression of bone matrix proteins in bovine vascular smooth muscle cells. Kidney Int. 2006;70:1046–53.
25. Ahrens M, Ankenbauer T, Schroder D, Hollnagel A, Mayer H, Gross G. Expression of human bone morphogenetic proteins-2 or -4 in murine mesenchymal progenitor C3H10T1/2 cells induces differentiation into distinct mesenchymal cell lineages. DNA Cell Biol. 1993;12:871–80.
26. Towler DA, Bidder M, Latifi T, Coleman T, Semenkovich CF. Diet-induced diabetes activates an osteogenic gene regulatory program in the aortas of low density lipoprotein receptor-deficient mice. J Biol Chem. 1998;273:30427–34.
27. Kang YH, Jin JS, Yi DW, Son SM. Bone morphogenetic protein-7 inhibits vascular calcification induced by high vitamin D in mice. Tohoku J Exp Med. 2010;221:299–307.
28. Nakashima K, Zhou X, Kunkel G, Zhang Z, Deng JM, Behringer RR, et al. The novel zinc finger-containing transcription factor osterix is required for osteoblast differentiation and bone formation. Cell. 2002;108:17–29.
29. Wallin R, Wajih N, Greenwood GT, Sane DC. Arterial calcification: a review of mechanisms, animal models, and the prospects for therapy. Med Res Rev. 2001;21:274–301.
30. Suzuki A, Ghayor C, Guicheux J, Magne D, Quillard S, Kakita A, et al. Enhanced expression of the inorganic phosphate transporter Pit-1 is involved in BMP-2-induced matrix mineralization in osteoblast-like cells. J Bone Miner Res. 2006;21:674–83.
31. Frank DB, Abtahi A, Yamaguchi DJ, Manning S, Shyr Y, Pozzi A, et al. Bone morphogenetic protein 4 promotes pulmonary vascular remodeling in hypoxic pulmonary hypertension. Circ Res. 2005;97:496–504.
32. Scimeca M, Anemona L, Granaglia A, Bonfiglio R, Urbano N, Toschi N, et al. Plaque calcification is driven by different mechanisms of mineralization associated with specific cardiovascular risk factors. Nutr Metab Cardiovasc Dis. 2019;29:1330–6.
33. Wang Z, Jiang Y, Liu N, Ren L, Zhu Y, An Y, et al. Advanced glycation end-product Nε-carboxymethyl-lysine accelerates progression of atherosclerotic calcification in diabetes. Atherosclerosis. 2012;221:387–96.
34. Ren X, Shao H, Wei Q, Sun Z, Liu N. Advanced glycation end-products enhance calcification in vascular smooth muscle cells. J Int Med Res. 2009;37:847–54.
35. Verma N, Manna SK. Advanced glycation end products (AGE) potently induce autophagy through activation of RAF protein kinase and nuclear factor kappaB (NF-kappaB). J Biol Chem. 2016;291:1481–91.
36. Sartori R, Gregorevic P, Sandri M. TGFβ and BMP signaling in skeletal muscle: potential significance for muscle-related disease. Trends Endocrinol Metab. 2014;25:464–71.
37. O'Donnell CJ, Kavousi M, Smith AV, Kardia SL, Feitosa MF, Hwang SJ, et al. Genome-wide association study for coronary artery calcification with follow-up in myocardial infarction. Circulation. 2011;124:2855–64.
38. Jones GT, Tromp G, Kuivaniemi H, Gretarsdottir S, Baas AF, Giusti B, et al. Meta-analysis of genome-wide association studies for abdominal aortic aneurysm identifies four new disease-specific risk loci. Circ Res. 2017;120:341–53.
39. Jones GT, Bown MJ, Gretarsdottir S, Romaine SP, Helgadottir A, Yu G, et al. A sequence variant associated with sortilin-1 (SORT1) on 1p13.3 is independently associated with abdominal aortic aneurysm. Hum Mol Genet. 2013;22:2941–7.
40. Gu F, Crump CM, Thomas G. Trans-Golgi network sorting. Cell Mol Life Sci. 2001;58:1067–84.
41. Carlo AS, Nykjaer A, Willnow TE. Sorting receptor sortilin-a culprit in cardiovascular and neurological diseases. J Mol Med (Berl). 2014;92:905–11.

42. Goettsch C, Hutcheson JD, Aikawa M, Iwata H, Pham T, Nykjaer A, et al. Sortilin mediates vascular calcification via its recruitment into extracellular vesicles. J Clin Invest. 2016;126:1323–36.
43. Talbot H, Saada S, Naves T, Gallet PF, Fauchais AL, Jauberteau MO. Regulatory roles of Sortilin and SorLA in immune-related processes. Front Pharmacol. 2018;9:1507.
44. Andersen JL, Schroder TJ, Christensen S, Strandbygard D, Pallesen LT, Garcia-Alai MM, et al. Identification of the first small-molecule ligand of the neuronal receptor sortilin and structure determination of the receptor-ligand complex. Acta Crystallogr D Biol Crystallogr. 2014;70:451–60.
45. Iqbal F, Schlotter F, Becker-Greene D, Lupieri A, Goettsch C, Hutcheson JD, et al. Sortilin enhances fibrosis and calcification in aortic valve disease by inducing interstitial cell heterogeneity. Eur Heart J. 2023;44:885.
46. Szatmari Z, Kis V, Lippai M, Hegedus K, Farago T, Lorincz P, et al. Rab11 facilitates crosstalk between autophagy and endosomal pathway through regulation of Hook localization. Mol Biol Cell. 2014;25:522–31.
47. Filimonenko M, Stuffers S, Raiborg C, Yamamoto A, Malerod L, Fisher EM, et al. Functional multivesicular bodies are required for autophagic clearance of protein aggregates associated with neurodegenerative disease. J Cell Biol. 2007;179:485–500.
48. Qian Y, Li L, Sun Z, Liu J, Yuan W, Wang Z. A multi-omics view of the complex mechanism of vascular calcification. Biomed Pharmacother. 2021;135:111192.
49. Mortensen MB, Kjolby M, Gunnersen S, Larsen JV, Palmfeldt J, Falk E, et al. Targeting sortilin in immune cells reduces proinflammatory cytokines and atherosclerosis. J Clin Invest. 2014;124:5317–22.
50. Patin G, Lettres choisies de Feu de M, Guy P. docteur en medecine Kessinger Legacy Reprints; 1692.
51. Eisenstein N, Stapley S, Grover L. Post-traumatic heterotopic ossification: an old problem in need of new solutions. J Orthop Res. 2018;36:1061–8.
52. Brackett EG. Care of the amputated in the United States. The Medical Department of the United States Army in the World War. 1927;XI:713–48.
53. Kaplan FS, Craver R, MacEwen GD, Gannon FH, Finkel G, Hahn G, et al. Progressive osseous heteroplasia: a distinct developmental disorder of heterotopic ossification. Two new case reports and follow-up of three previously reported cases. J Bone Joint Surg Am. 1994;76:425–36.
54. Kaplan FS, Shore EM. Progressive osseous heteroplasia. J Bone Miner Res. 2000;15:2084–94.
55. Shore EM, Kaplan FS. Inherited human diseases of heterotopic bone formation. Nat Rev Rheumatol. 2010;6:518–27.
56. Jensen LL, Halar E, Little JW, Brooke MM. Neurogenic heterotopic ossification. Am J Phys Med. 1987;66:351–63.
57. Vanden Bossche L, Vanderstraeten G. Heterotopic ossification: a review. J Rehabil Med. 2005;37:129–36.
58. Shore EM, Xu M, Feldman GJ, Fenstermacher DA, Cho TJ, Choi IH, et al. A recurrent mutation in the BMP type I receptor ACVR1 causes inherited and sporadic fibrodysplasia ossificans progressiva. Nat Genet. 2006;38:525–7.
59. Chen L, Deng H, Cui H, Fang J, Zuo Z, Deng J, et al. Inflammatory responses and inflammation-associated diseases in organs. Oncotarget. 2018;9:7204–18.
60. Muñoz J, Akhavan NS, Mullins AP, Arjmandi BH. Macrophage polarization and osteoporosis: a review. Nutrients. 2020;12:2999.
61. Davies OG, Liu Y, Player DJ, Martin NRW, Grover LM, Lewis MP. Defining the balance between regeneration and pathological ossification in skeletal muscle following traumatic injury. Front Physiol. 2017;8:194.
62. Rifas L. T-cell cytokine induction of BMP-2 regulates human mesenchymal stromal cell differentiation and mineralization. J Cell Biochem. 2006;98:706–14.

63. Kan C, Chen L, Hu Y, Ding N, Li Y, McGuire TL, et al. Gli1-labeled adult mesenchymal stem/progenitor cells and hedgehog signaling contribute to endochondral heterotopic ossification. Bone. 2018;109:71–9.
64. Kan C, Ding N, Yang J, Tan Z, McGuire TL, Lu H, et al. BMP-dependent, injury-induced stem cell niche as a mechanism of heterotopic ossification. Stem Cell Res Ther. 2019;10:14.
65. Hashimoto N, Kiyono T, Wada MR, Umeda R, Goto Y, Nonaka I, et al. Osteogenic properties of human myogenic progenitor cells. Mech Dev. 2008;125:257–69.
66. Julien A, Perrin S, Martínez-Sarrà E, Kanagalingam A, Carvalho C, Luka M, et al. Skeletal stem/progenitor cells in periosteum and skeletal muscle share a common molecular response to bone injury. J Bone Miner Res. 2022;37:1545–61.
67. Wada MR, Inagawa-Ogashiwa M, Shimizu S, Yasumoto S, Hashimoto N. Generation of different fates from multipotent muscle stem cells. Development. 2002;129:2987–95.
68. Furmanik M, Chatrou M, van Gorp R, Akbulut A, Willems B, Schmidt H, et al. Reactive oxygen-forming Nox5 links vascular smooth muscle cell phenotypic switching and extracellular vesicle-mediated vascular calcification. Circ Res. 2020;127:911–27.
69. Agarwal S, Loder S, Brownley C, Cholok D, Mangiavini L, Li J, et al. Inhibition of Hif1alpha prevents both trauma-induced and genetic heterotopic ossification. Proc Natl Acad Sci U S A. 2016;113:E338–47.
70. Bohner M, Maazouz Y, Ginebra MP, Habibovic P, Schoenecker JG, Seeherman H, et al. Sustained local ionic homeostatic imbalance caused by calcification modulates inflammation to trigger heterotopic ossification. Acta Biomater. 2022;145:1–24.
71. Wang YK, Sun WF, Liu XG, Deng J, Yan BE, Jiang WY, et al. Comparative study of serum levels of BMP-2 and heterotopic ossification in traumatic brain injury and fractures patients. Zhongguo Gu Shang. 2011;24:399–403.
72. Setoguchi T, Yone K, Matsuoka E, Takenouchi H, Nakashima K, Sakou T, et al. Traumatic injury-induced BMP7 expression in the adult rat spinal cord. Brain Res. 2001;921:219–25.
73. Cui ZS, Zhao P, Jia CX, Liu HJ, Qi R, Cui JW, et al. Local expression and role of BMP-2/4 in injured spinal cord. Genet Mol Res. 2015;14:9109–17.
74. Ryoo HM, Lee MH, Kim YJ. Critical molecular switches involved in BMP-2-induced osteogenic differentiation of mesenchymal cells. Gene. 2006;366:51–7.
75. Bakhshian Nik A, Hutcheson JD, Aikawa E. Extracellular vesicles as mediators of cardiovascular calcification. Front Cardiovasc Med. 2017;4:78.
76. Cui L, Rashdan NA, Zhu D, Milne EM, Ajuh P, Milne G, et al. End stage renal disease-induced hypercalcemia may promote aortic valve calcification via Annexin VI enrichment of valve interstitial cell derived-matrix vesicles. J Cell Physiol. 2017;232:2985–95.
77. Lees-Shepard JB, Goldhamer DJ. Stem cells and heterotopic ossification: lessons from animal models. Bone. 2018;109:178–86.
78. Pignolo RJ, Baujat G, Brown MA, De Cunto C, Di Rocco M, Hsiao EC, et al. Natural history of fibrodysplasia ossificans progressiva: cross-sectional analysis of annotated baseline phenotypes. Orphanet J Rare Dis. 2019;14:98.
79. Yadin D, Knaus P, Mueller TD. Structural insights into BMP receptors: specificity, activation and inhibition. Cytokine Growth Factor Rev. 2016;27:13–34.
80. Massagué J. TGFbeta signaling: receptors, transducers, and Mad proteins. Cell. 1996;85:947–50.
81. Zou H, Wieser R, Massagué J, Niswander L. Distinct roles of type I bone morphogenetic protein receptors in the formation and differentiation of cartilage. Genes Dev. 1997;11:2191–203.
82. Knaus P, Sebald W. Cooperativity of binding epitopes and receptor chains in the BMP/TGFbeta superfamily. Biol Chem. 2001;382:1189–95.
83. Nohe A, Hassel S, Ehrlich M, Neubauer F, Sebald W, Henis YI, et al. The mode of bone morphogenetic protein (BMP) receptor oligomerization determines different BMP-2 signaling pathways. J Biol Chem. 2002;277:5330–8.

84. Sangadala S, Devereaux EJ, Presciutti SM, Boden SD, Willet NJ. FK506 induces ligand-independent activation of the bone morphogenetic protein pathway and osteogenesis. Int J Mol Sci. 2019;20:1900.
85. Hino K, Horigome K, Nishio M, Komura S, Nagata S, Zhao C, et al. Activin-A enhances mTOR signaling to promote aberrant chondrogenesis in fibrodysplasia ossificans progressiva. J Clin Invest. 2017;127:3339–52.
86. Machiya A, Tsukamoto S, Ohte S, Kuratani M, Fujimoto M, Kumagai K, et al. Effects of FKBP12 and type II BMP receptors on signal transduction by ALK2 activating mutations associated with genetic disorders. Bone. 2018;111:101–8.
87. Shimono K, Tung WE, Macolino C, Chi AH, Didizian JH, Mundy C, et al. Potent inhibition of heterotopic ossification by nuclear retinoic acid receptor-γ agonists. Nat Med. 2011;17:454–60.
88. Kaplan J, Kaplan FS, Shore EM. Restoration of normal BMP signaling levels and osteogenic differentiation in FOP mesenchymal progenitor cells by mutant allele-specific targeting. Gene Ther. 2012;19:786–90.
89. Lv B, Li F, Han J, Fang J, Xu L, Sun C, et al. Hif-1α overexpression improves transplanted bone mesenchymal stem cells survival in rat MCAO stroke model. Front Mol Neurosci. 2017;10:80.
90. Cosin-Roger J, Simmen S, Melhem H, Atrott K, Frey-Wagner I, Hausmann M, et al. Hypoxia ameliorates intestinal inflammation through NLRP3/mTOR downregulation and autophagy activation. Nat Commun. 2017;8:98.
91. Chen J, Long F. mTOR signaling in skeletal development and disease. Bone Res. 2018;6:1.
92. Kaplan FS, Zeitlin L, Dunn SP, Benor S, Hagin D, Al Mukaddam M, et al. Acute and chronic rapamycin use in patients with Fibrodysplasia Ossificans Progressiva: a report of two cases. Bone. 2018;109:281–4.
93. Wentworth KL, Masharani U, Hsiao EC. Therapeutic advances for blocking heterotopic ossification in fibrodysplasia ossificans progressiva. Br J Clin Pharmacol. 2019;85:1180–7.
94. Singha UK, Jiang Y, Yu S, Luo M, Lu Y, Zhang J, et al. Rapamycin inhibits osteoblast proliferation and differentiation in MC3T3-E1 cells and primary mouse bone marrow stromal cells. J Cell Biochem. 2008;103:434–46.
95. Shimono K, Morrison TN, Tung WE, Chandraratna RA, Williams JA, Iwamoto M, et al. Inhibition of ectopic bone formation by a selective retinoic acid receptor alpha-agonist: a new therapy for heterotopic ossification? J Orthop Res. 2010;28:271–7.
96. Uchibe K, Son J, Larmour C, Pacifici M, Enomoto-Iwamoto M, Iwamoto M. Genetic and pharmacological inhibition of retinoic acid receptor γ function promotes endochondral bone formation. J Orthop Res. 2017;35:1096–105.
97. Sinha S, Uchibe K, Usami Y, Pacifici M, Iwamoto M. Effectiveness and mode of action of a combination therapy for heterotopic ossification with a retinoid agonist and an anti-inflammatory agent. Bone. 2016;90:59–68.
98. Pacifici M. Hereditary multiple exostoses: are there new plausible treatment strategies? Expert Opin Orphan Drugs. 2018;6:385–91.
99. Lai KC, Slanetz PJ, Eisenberg RL. Linear breast calcifications. AJR Am J Roentgenol. 2012;199:W151–7.
100. Scott R, Stone N, Kendall C, Geraki K, Rogers K. Relationships between pathology and crystal structure in breast calcifications: an in situ X-ray diffraction study in histological sections. NPJ Breast Cancer. 2016;2:16029.
101. Nyante SJ, Lee SS, Benefield TS, Hoots TN, Henderson LM. The association between mammographic calcifications and breast cancer prognostic factors in a population-based registry cohort. Cancer. 2017;123:219–27.
102. Henrot P, Leroux A, Barlier C, Genin P. Breast microcalcifications: the lesions in anatomical pathology. Diagn Interv Imaging. 2014;95:141–52.
103. Frappart L, Boudeulle M, Boumendil J, Lin HC, Martinon I, Palayer C, et al. Structure and composition of microcalcifications in benign and malignant lesions of the breast: study by

light microscopy, transmission and scanning electron microscopy, microprobe analysis, and X-ray diffraction. Hum Pathol. 1984;15:880–9.

104. Scimeca M, Giannini E, Antonacci C, Pistolese CA, Spagnoli LG, Bonanno E. Microcalcifications in breast cancer: an active phenomenon mediated by epithelial cells with mesenchymal characteristics. BMC Cancer. 2014;14:286.
105. Gosling S, Scott R, Greenwood C, Bouzy P, Nallala J, Lyburn ID, et al. Calcification microstructure reflects breast tissue microenvironment. J Mammary Gland Biol Neoplasia. 2019;24:333–42.
106. Haka AS, Shafer-Peltier KE, Fitzmaurice M, Crowe J, Dasari RR, Feld MS. Diagnosing breast cancer by using Raman spectroscopy. Proc Natl Acad Sci U S A. 2005;102:12371–6.
107. Scott R, Kendall C, Stone N, Rogers K. Elemental vs. phase composition of breast calcifications. Sci Rep. 2017;7:136.
108. Wang J, Wang L, Fan Y. Adverse biological effect of TiO(2) and hydroxyapatite nanoparticles used in bone repair and replacement. Int J Mol Sci. 2016;17:798.
109. Sun Y, Zeng XR, Wenger L, Cheung HS. Basic calcium phosphate crystals stimulate the endocytotic activity of cells--inhibition by anti-calcification agents. Biochem Biophys Res Commun. 2003;312:1053–9.
110. McCarthy GM, Augustine JA, Baldwin AS, Christopherson PA, Cheung HS, Westfall PR, et al. Molecular mechanism of basic calcium phosphate crystal-induced activation of human fibroblasts. Role of nuclear factor kappab, activator protein 1, and protein kinase c. J Biol Chem. 1998;273:35161–9.
111. Reuben PM, Brogley MA, Sun Y, Cheung HS. Molecular mechanism of the induction of metalloproteinases 1 and 3 in human fibroblasts by basic calcium phosphate crystals. Role of calcium-dependent protein kinase C alpha. J Biol Chem. 2002;277:15190–8.
112. Nguyen C, Lieberherr M, Bordat C, Velard F, Come D, Liote F, et al. Intracellular calcium oscillations in articular chondrocytes induced by basic calcium phosphate crystals lead to cartilage degradation. Osteoarthr Cartil. 2012;20:1399–408.
113. Morgan MP, Cooke MM, Christopherson PA, Westfall PR, McCarthy GM. Calcium hydroxyapatite promotes mitogenesis and matrix metalloproteinase expression in human breast cancer cell lines. Mol Carcinog. 2001;32:111–7.
114. Xue Y, Chen Q, Sun J. Hydroxyapatite nanoparticle-induced mitochondrial energy metabolism impairment in liver cells: in vitro and in vivo studies. J Appl Toxicol. 2017;37:1004–16.
115. Ginestra A, La Placa MD, Saladino F, Cassara D, Nagase H, Vittorelli ML. The amount and proteolytic content of vesicles shed by human cancer cell lines correlates with their in vitro invasiveness. Anticancer Res. 1998;18:3433–7.
116. Dolo V, Ginestra A, Cassara D, Ghersi G, Nagase H, Vittorelli ML. Shed membrane vesicles and selective localization of gelatinases and MMP-9/TIMP-1 complexes. Ann N Y Acad Sci. 1999;878:497–9.
117. Cox RF, Morgan MP. Microcalcifications in breast cancer: lessons from physiological mineralization. Bone. 2013;53:437–50.
118. Jin C, Frayssinet P, Pelker R, Cwirka D, Hu B, Vignery A, et al. NLRP3 inflammasome plays a critical role in the pathogenesis of hydroxyapatite-associated arthropathy. Proc Natl Acad Sci U S A. 2011;108:14867–72.
119. Liu J, Liao S, Diop-Frimpong B, Chen W, Goel S, Naxerova K, et al. TGF-beta blockade improves the distribution and efficacy of therapeutics in breast carcinoma by normalizing the tumor stroma. Proc Natl Acad Sci U S A. 2012;109:16618–23.
120. Scimeca M, Bonfiglio R, Menichini E, Albonici L, Urbano N, De Caro MT, et al. Microcalcifications drive breast cancer occurrence and development by macrophage-mediated epithelial to mesenchymal transition. Int J Mol Sci. 2019;20:E5633.
121. Liu F, Bloch N, Bhushan KR, De Grand AM, Tanaka E, Solazzo S, et al. Humoral bone morphogenetic protein 2 is sufficient for inducing breast cancer microcalcification. Mol Imaging. 2008;7:175–86.

122. Tan CC, Li GX, Tan LD, Du X, Li XQ, He R, et al. Breast cancer cells obtain an osteomimetic feature via epithelial-mesenchymal transition that have undergone BMP2/RUNX2 signaling pathway induction. Oncotarget. 2016;7:79688–705.
123. Li X, Lu W, Fu X, Zhang Y, Yang K, Zhong N, et al. BMP4 increases canonical transient receptor potential protein expression by activating p38 MAPK and ERK1/2 signaling pathways in pulmonary arterial smooth muscle cells. Am J Respir Cell Mol Biol. 2013;49:212–20.
124. Mandavilli S, Singh BB, Sahmoun AE. Serum calcium levels, TRPM7, TRPC1, microcalcifications, and breast cancer using breast imaging reporting and data system scores. Breast Cancer (Dove Med Press). 2012;2013:1–7.
125. Azimi I, Roberts-Thomson SJ, Monteith GR. Calcium influx pathways in breast cancer: opportunities for pharmacological intervention. Br J Pharmacol. 2014;171:945–60.
126. Deliot N, Constantin B. Plasma membrane calcium channels in cancer: alterations and consequences for cell proliferation and migration. Biochim Biophys Acta. 2015;1848:2512–22.
127. Cross BM, Breitwieser GE, Reinhardt TA, Rao R. Cellular calcium dynamics in lactation and breast cancer: from physiology to pathology. Am J Physiol Cell Physiol. 2014;306:C515–26.
128. Bonavita E, Gentile S, Rubino M, Maina V, Papait R, Kunderfranco P, et al. PTX3 is an extrinsic oncosuppressor regulating complement-dependent inflammation in cancer. Cell. 2015;160:700–14.
129. Grcevic D, Sironi M, Valentino S, Deban L, Cvija H, Inforzato A, et al. The long pentraxin 3 plays a role in bone turnover and repair. Front Immunol. 2018;9:417.
130. Bonfiglio R, Scimeca M, Toschi N, Pistolese CA, Giannini E, Antonacci C, et al. Radiological, histological and chemical analysis of breast microcalcifications: diagnostic value and biological significance. J Mammary Gland Biol Neoplasia. 2018;23:89–99.
131. Smith JF, Shindel AW, Huang YC, Clavijo RI, Flechner L, Breyer BN, et al. Pentoxifylline treatment and penile calcifications in men with Peyronie's disease. Asian J Androl. 2011;13:322–5.
132. Bellahcene A, Castronovo V. Increased expression of osteonectin and osteopontin, two bone matrix proteins, in human breast cancer. Am J Pathol. 1995;146:95–100.
133. Talbot LJ, Mi Z, Bhattacharya SD, Kim V, Guo H, Kuo PC. Pharmacokinetic characterization of an RNA aptamer against osteopontin and demonstration of in vivo efficacy in reversing growth of human breast cancer cells. Surgery. 2011;150:224–30.
134. Shevde LA, Samant RS, Paik JC, Metge BJ, Chambers AF, Casey G, et al. Osteopontin knockdown suppresses tumorigenicity of human metastatic breast carcinoma, MDA-MB-435. Clin Exp Metastasis. 2006;23:123–33.
135. Dai J, Li B, Shi J, Peng L, Zhang D, Qian W, et al. A humanized anti-osteopontin antibody inhibits breast cancer growth and metastasis in vivo. Cancer Immunol Immunother. 2010;59:355–66.
136. Forlino A, Marini JC. Osteogenesis imperfecta. Lancet. 2016;387:1657–71.
137. Fidler AL, Boudko SP, Rokas A, Hudson BG. The triple helix of collagens - an ancient protein structure that enabled animal multicellularity and tissue evolution. J Cell Sci. 2018;131:jcs203950.
138. Marini JC, Cabral WA, Barnes AM, Chang W. Components of the collagen prolyl 3-hydroxylation complex are crucial for normal bone development. Cell Cycle. 2007;6:1675–81.
139. Marini JC, Forlino A, Cabral WA, Barnes AM, San Antonio JD, Milgrom S, et al. Consortium for osteogenesis imperfecta mutations in the helical domain of type I collagen: regions rich in lethal mutations align with collagen binding sites for integrins and proteoglycans. Hum Mutat. 2007;28:209–21.
140. Tauer JT, Robinson ME, Rauch F. Osteogenesis imperfecta: new perspectives from clinical and translational research. JBMR Plus. 2019;3:e10174.
141. Bronheim R, Khan S, Carter E, Sandhaus RA, Raggio C. Scoliosis and cardiopulmonary outcomes in osteogenesis imperfecta patients. Spine (Phila Pa 1976). 2019;44:1057–63.

142. Byers PH, Belmont J, Black J, De Backer J, Frank M, Jeunemaitre X, et al. Diagnosis, natural history, and management in vascular Ehlers-Danlos syndrome. Am J Med Genet C Semin Med Genet. 2017;175:40–7.
143. Kuivaniemi H, Tromp G, Prockop DJ. Mutations in collagen genes: causes of rare and some common diseases in humans. FASEB J. 1991;5:2052–60.
144. Lim J, Grafe I, Alexander S, Lee B. Genetic causes and mechanisms of osteogenesis imperfecta. Bone. 2017;102:40–9.
145. Chang SW, Shefelbine SJ, Buehler MJ. Structural and mechanical differences between collagen homo- and heterotrimers: relevance for the molecular origin of brittle bone disease. Biophys J. 2012;102:640–8.
146. Garibaldi N, Contento BM, Babini G, Morini J, Siciliani S, Biggiogera M, et al. Targeting cellular stress in vitro improves osteoblast homeostasis, matrix collagen content and mineralization in two murine models of osteogenesis imperfecta. Matrix Biol. 2021;98:1–20.
147. Sarathchandra P, Pope FM, Kayser MV, Ali SY. A light and electron microscopic study of osteogenesis imperfecta bone samples, with reference to collagen chemistry and clinical phenotype. J Pathol. 2000;192:385–95.
148. Nijhuis WH, Eastwood DM, Allgrove J, Hvid I, Weinans HH, Bank RA, et al. Current concepts in osteogenesis imperfecta: bone structure, biomechanics and medical management. J Child Orthop. 2019;13:1–11.
149. Vetter U, Weis MA, Morike M, Eanes ED, Eyre DR. Collagen crosslinks and mineral crystallinity in bone of patients with osteogenesis imperfecta. J Bone Miner Res. 1993;8:133–7.
150. Traub W, Arad T, Vetter U, Weiner S. Ultrastructural studies of bones from patients with osteogenesis imperfecta. Matrix Biol. 1994;14:337–45.
151. Landis WJ. Tomographic imaging of collagen-mineral interaction: implications for osteogenesis imperfecta. Connect Tissue Res. 1995;31:287–90.
152. Fratzl P, Paris O, Klaushofer K, Landis WJ. Bone mineralization in an osteogenesis imperfecta mouse model studied by small-angle x-ray scattering. J Clin Invest. 1996;97:396–402.
153. Vanleene M, Porter A, Guillot PV, Boyde A, Oyen M, Shefelbine S. Ultra-structural defects cause low bone matrix stiffness despite high mineralization in osteogenesis imperfecta mice. Bone. 2012;50:1317–23.
154. Cho TJ, Lee KE, Lee SK, Song SJ, Kim KJ, Jeon D, et al. A single recurrent mutation in the 5′-UTR of IFITM5 causes osteogenesis imperfecta type V. Am J Hum Genet. 2012;91:343–8.
155. Semler O, Garbes L, Keupp K, Swan D, Zimmermann K, Becker J, et al. A mutation in the 5′-UTR of IFITM5 creates an in-frame start codon and causes autosomal-dominant osteogenesis imperfecta type V with hyperplastic callus. Am J Hum Genet. 2012;91:349–57.
156. Moffatt P, Gaumond MH, Salois P, Sellin K, Bessette MC, Godin E, et al. Bril: a novel bone-specific modulator of mineralization. J Bone Miner Res. 2008;23:1497–508.
157. Hanagata N, Li X, Morita H, Takemura T, Li J, Minowa T. Characterization of the osteoblast-specific transmembrane protein IFITM5 and analysis of IFITM5-deficient mice. J Bone Miner Metab. 2011;29:279–90.
158. Shapiro JR, Lietman C, Grover M, Lu JT, Nagamani SC, Dawson BC, et al. Phenotypic variability of osteogenesis imperfecta type V caused by an IFITM5 mutation. J Bone Miner Res. 2013;28:1523–30.
159. Dobrocky I, Seidl G, Grill F. MRI and CT features of hyperplastic callus in osteogenesis imperfecta tarda. Eur Radiol. 1999;9:665–8.
160. Besio R, Chow CW, Tonelli F, Marini JC, Forlino A. Bone biology: insights from osteogenesis imperfecta and related rare fragility syndromes. FEBS J. 2019;286:3033–56.
161. Liu Y, Wang J, Ma D, Lv F, Xu X, Xia W, et al. Osteogenesis imperfecta type V: genetic and clinical findings in eleven Chinese patients. Clin Chim Acta. 2016;462:201–9.
162. Cabral WA, Chang W, Barnes AM, Weis M, Scott MA, Leikin S, et al. Prolyl 3-hydroxylase 1 deficiency causes a recessive metabolic bone disorder resembling lethal/severe osteogenesis imperfecta. Nat Genet. 2007;39:359–65.

163. Fratzl-Zelman N, Morello R, Lee B, Rauch F, Glorieux FH, Misof BM, et al. CRTAP deficiency leads to abnormally high bone matrix mineralization in a murine model and in children with osteogenesis imperfecta type VII. Bone. 2010;46:820–6.
164. Ito S, Nagata K. Roles of the endoplasmic reticulum-resident, collagen-specific molecular chaperone Hsp47 in vertebrate cells and human disease. J Biol Chem. 2019;294:2133–41.
165. Shaheen R, Alazami AM, Alshammari MJ, Faqeih E, Alhashmi N, Mousa N, et al. Study of autosomal recessive osteogenesis imperfecta in Arabia reveals a novel locus defined by TMEM38B mutation. J Med Genet. 2012;49:630–5.
166. Venturi E, Matyjaszkiewicz A, Pitt SJ, Tsaneva-Atanasova K, Nishi M, Yamazaki D, et al. TRIC-B channels display labile gating: evidence from the TRIC-A knockout mouse model. Pflugers Arch. 2013;465:1135–48.
167. Robinson C, Shore RC, Wood SR, Brookes SJ, Smith DA, Wright JT, et al. Subunit structures in hydroxyapatite crystal development in enamel: implications for amelogenesis imperfecta. Connect Tissue Res. 2003;44:65–71.
168. Smith CEL, Poulter JA, Antanaviciute A, Kirkham J, Brookes SJ, Inglehearn CF, et al. Amelogenesis imperfecta; genes, proteins, and pathways. Front Physiol. 2017;8:435.
169. Kantaputra PN. Dentinogenesis imperfecta-associated syndromes. Am J Med Genet. 2001;104:75–8.
170. Napierala D, Sun Y, Maciejewska I, Bertin TK, Dawson B, D'Souza R, et al. Transcriptional repression of the Dspp gene leads to dentinogenesis imperfecta phenotype in Col1a1-Trps1 transgenic mice. J Bone Miner Res. 2012;27:1735–45.
171. Rocha CT, Nelson-Filho P, Silva LA, Assed S, Queiroz AM. Variation of dentin dysplasia type I: report of atypical findings in the permanent dentition. Braz Dent J. 2011;22:74–8.
172. Chen D, Li X, Lu F, Wang Y, Xiong F, Li Q. Dentin dysplasia type I-A dental disease with genetic heterogeneity. Oral Dis. 2019;25:439–46.
173. Colombo M, Moita C, van Niel G, Kowal J, Vigneron J, Benaroch P, et al. Analysis of ESCRT functions in exosome biogenesis, composition and secretion highlights the heterogeneity of extracellular vesicles. J Cell Sci. 2013;126:5553–65.
174. Sanchez-Duffhues G, Garcia de Vinuesa A, van de Pol V, Geerts ME, de Vries MR, Janson SG, et al. Inflammation induces endothelial-to-mesenchymal transition and promotes vascular calcification through downregulation of BMPR2. J Pathol. 2019;247:333–46.
175. Li X, Yang HY, Giachelli CM. BMP-2 promotes phosphate uptake, phenotypic modulation, and calcification of human vascular smooth muscle cells. Atherosclerosis. 2008;199:271–7.
176. Abdelbaky A, Corsini E, Figueroa AL, Subramanian S, Fontanez S, Emami H, et al. Early aortic valve inflammation precedes calcification: a longitudinal FDG-PET/CT study. Atherosclerosis. 2015;238:165–72.
177. Tintut Y, Patel J, Territo M, Saini T, Parhami F, Demer LL. Monocyte/macrophage regulation of vascular calcification in vitro. Circulation. 2002;105:650–5.
178. Li R, Mittelstein D, Lee J, Fang K, Majumdar R, Tintut Y, et al. A dynamic model of calcific nodule destabilization in response to monocyte- and oxidized lipid-induced matrix metalloproteinases. Am J Physiol Cell Physiol. 2012;302:C658–65.
179. Liu X, Kang H, Shahnazari M, Kim H, Wang L, Larm O, et al. A novel mouse model of trauma induced heterotopic ossification. J Orthop Res. 2014;32:183–8.
180. Schmitt JM, Hwang K, Winn SR, Hollinger JO. Bone morphogenetic proteins: an update on basic biology and clinical relevance. J Orthop Res. 1999;17:269–78.
181. Jani P, Zhang H, Benson MD, Qin C. Noggin inhibition of mouse dentinogenesis. J Oral Biosci. 2020;62:72–9.
182. Suzuki HI. MicroRNA control of TGF-β signaling. Int J Mol Sci. 2018;19:1901.
183. Nikoletopoulou V, Markaki M, Palikaras K, Tavernarakis N. Crosstalk between apoptosis, necrosis and autophagy. Biochim Biophys Acta. 2013;1833:3448–59.
184. Scherz-Shouval R, Elazar Z. Regulation of autophagy by ROS: physiology and pathology. Trends Biochem Sci. 2011;36:30–8.

185. Li L, Tan J, Miao Y, Lei P, Zhang Q. ROS and autophagy: interactions and molecular regulatory mechanisms. Cell Mol Neurobiol. 2015;35:615–21.
186. Yousefi S, Perozzo R, Schmid I, Ziemiecki A, Schaffner T, Scapozza L, et al. Calpain-mediated cleavage of Atg5 switches autophagy to apoptosis. Nat Cell Biol. 2006;8:1124–32.
187. Nollet M, Santucci-Darmanin S, Breuil V, Al-Sahlanee R, Cros C, Topi M, et al. Autophagy in osteoblasts is involved in mineralization and bone homeostasis. Autophagy. 2014;10:1965–77.
188. Bohensky J, Shapiro IM, Leshinsky S, Terkhorn SP, Adams CS, Srinivas V. HIF-1 regulation of chondrocyte apoptosis: induction of the autophagic pathway. Autophagy. 2007;3:207–14.
189. Maeda S, Nobukuni T, Shimo-Onoda K, Hayashi K, Yone K, Komiya S, et al. Sortilin is upregulated during osteoblastic differentiation of mesenchymal stem cells and promotes extracellular matrix mineralization. J Cell Physiol. 2002;193:73–9.
190. Murphy CM, Duffy GP, Schindeler A, O'Brien FJ. Effect of collagen-glycosaminoglycan scaffold pore size on matrix mineralization and cellular behavior in different cell types. J Biomed Mater Res A. 2016;104:291–304.
191. Tintut Y, Alfonso Z, Saini T, Radcliff K, Watson K, Bostrom K, et al. Multilineage potential of cells from the artery wall. Circulation. 2003;108:2505–10.
192. de Souza LE, Malta TM, Kashima Haddad S, Covas DT. Mesenchymal stem cells and pericytes: to what extent are they related? Stem Cells Dev. 2016;25:1843–52.
193. Tapscott SJ, Davis RL, Thayer MJ, Cheng PF, Weintraub H, Lassar AB. MyoD1: a nuclear phosphoprotein requiring a Myc homology region to convert fibroblasts to myoblasts. Science. 1988;242:405–11.
194. Hu DP, Ferro F, Yang F, Taylor AJ, Chang W, Miclau T, et al. Cartilage to bone transformation during fracture healing is coordinated by the invading vasculature and induction of the core pluripotency genes. Development. 2017;144:221–34.
195. Merrell AJ, Stanger BZ. Adult cell plasticity in vivo: de-differentiation and transdifferentiation are back in style. Nat Rev Mol Cell Biol. 2016;17:413–25.
196. Samavarchi-Tehrani P, Golipour A, David L, Sung HK, Beyer TA, Datti A, et al. Functional genomics reveals a BMP-driven mesenchymal-to-epithelial transition in the initiation of somatic cell reprogramming. Cell Stem Cell. 2010;7:64–77.
197. Vining KH, Mooney DJ. Mechanical forces direct stem cell behaviour in development and regeneration. Nat Rev Mol Cell Biol. 2017;18:728–42.
198. Guo L, Zhou Y, Wang S, Wu Y. Epigenetic changes of mesenchymal stem cells in three-dimensional (3D) spheroids. J Cell Mol Med. 2014;18:2009–19.
199. Handy DE, Castro R, Loscalzo J. Epigenetic modifications: basic mechanisms and role in cardiovascular disease. Circulation. 2011;123:2145–56.
200. Quarles LD. Role of FGF23 in vitamin D and phosphate metabolism: implications in chronic kidney disease. Exp Cell Res. 2012;318:1040–8.
201. Walker MD, Silverberg SJ. Primary hyperparathyroidism. Nat Rev Endocrinol. 2018;14:115–25.
202. Ketteler M, Westenfeld R, Schlieper G, Brandenburg V. Pathogenesis of vascular calcification in dialysis patients. Clin Exp Nephrol. 2005;9:265–70.
203. Gade PS, Tulamo R, Lee KW, Mut F, Ollikainen E, Chuang CY, et al. Calcification in human intracranial aneurysms is highly prevalent and displays both atherosclerotic and nonatherosclerotic types. Arterioscler Thromb Vasc Biol. 2019;39:2157–67.
204. Li Z, Zhao Z, Cai Z, Sun Y, Li L, Yao F, et al. Runx2 (Runt-related transcription factor 2)-mediated microcalcification is a novel pathological characteristic and potential mediator of abdominal aortic aneurysm. Arterioscler Thromb Vasc Biol. 2020;40:1352–69.
205. Vesely I, Barber JE, Ratliff NB. Tissue damage and calcification may be independent mechanisms of bioprosthetic heart valve failure. J Heart Valve Dis. 2001;10:471–7.
206. Xue S, Zhao Q, Tai M, Li N, Liu Y. Correlation between breast ultrasound microcalcification and the prognosis of breast cancer. J Healthc Eng. 2021;2021:6835963.
207. Churchill W. Speech at the Lord Mayor's luncheon. London: The Mansion House; 1942.

Epilogue

The chapters of this book together have now presented a compendium of information which provides a mechanistic understanding of the mineralization of skeletal and dental tissues. The foundation for this work rests with the premise that the formation of mineral in the complex extracellular matrices of vertebrates results effectively from coordinated events orchestrated by groups of cells comprising these tissues. While many such events are known in great detail, not all their aspects have been described even in a rudimentary fashion. This Epilogue, then, identifies some of the critical topics and questions about mineralization which remain poorly understood, thereby framing directions for future research to elucidate a more comprehensive knowledge of the mechanisms of vertebrate mineralization.

The number of indefinite and unexplored topic areas in the mineralization field is extensive. The authors could select far too many of these to be included in this final book section. We have instead addressed a limited few issues with hope that forthcoming investigations of these concerns will be the most revealing to broaden the scientific basis of the mineralization process in a significant manner. Our comments relate to the following two major themes of this book, first, intracellular ion accumulation and release and, second, extracellular matrix control of mineral formation and growth.

Formation and Role of Cell Palisades

The major ions involved in mineral formation, calcium and phosphate, gain entry to cells through the action of various plasma membrane channels and pumps. To protect cell viability and mediate the activities of a host of signal transduction pathways, calcium ion access to cells and intracellular micromolar calcium ion levels are extraordinarily tightly controlled. In this regard, it appears that many vertebrate hard tissue cells form extensive polarized palisades which separate those cells from their extracellular milieu containing millimolar concentrations of calcium ions. These palisades are most evident in forming bone (osteoblasts), dentin (odontoblasts), and enamel (ameloblasts). However, the driving force(s) promoting such exquisite cell alignment and the precise means by which palisades of cells regulate intracellular entry and egress of calcium (and other ions) are poorly understood.

I. M. Shapiro, W. J. Landis, *Mechanisms of Mineralization of Vertebrate Skeletal and Dental Tissues*, https://doi.org/10.1007/978-3-031-34304-9

Thus, a central unknown relates to the putative role of cell palisades in mineral and extracellular matrix deposition. In this context, it would be important to determine, for example, how the palisades might guide and control the intracellular flow of calcium ions from the systemic circulation to the mineralization front, whether palisading influences paracellular calcium ion transit as is commonly found in other systems such as the kidney, if palisades confer a new function on the cells such as promoting exosome release, and whether the presence of a palisade impacts phosphate ion delivery to cells and extracellular matrices.

Phosphate Ion Tracking

The trafficking and fate of intracellular calcium ions has been well described and includes their loading principally in the endoplasmic reticulum and distribution to mitochondria, endosomes, and other cellular organelles. Calcium ion efflux from cells is mediated by exosomes and the plasma membrane sodium calcium exchanger. But what of phosphate ions, whose contribution to vertebrate mineralization is surely as important as that of calcium? Much of what is known of phosphate transport relates to the endocrinal system where communication between bone-derived FGF23 and kidney-derived α-klotho regulates phosphate homeostasis and where dysregulation of this interplay can induce aberrant mineralization. However, the origin of phosphate ions for mineral formation is not as well reported as that of calcium, and relatively little is understood regarding the pathways and intracellular transiting of $HPO_4^{2-}/H_2PO_4^-$ ions. Currently, the plasma membrane entry to cells and the intracellular routing of these ions leading to their presence in mitochondria and other cellular organelles are not fully defined, and only the XRT transporter, sensitive to inositol hexakisphosphate, is known to support phosphate ion egress from cells. Additionally, it is uncertain as to whether phosphate ions are clustered and ferried together with calcium ions along intracellular pathways in common vesicles or other organelles, destined for extracellular export. It would be valuable to establish more complete details concerning the influx, movement, distribution, and efflux of phosphate ions, alone or complexed with other ions, in the cells of vertebrate mineralized tissues. Another question relates to possible pathways by which these ions may be transferred to mitochondria, an organelle associated with the presence of calcium phosphate particles, without disturbing their energy-generating function. Moreover, mitochondria were reported to transfer droplets of "liquid-like" mineral precursors to autophagosomes or lysosomes during mitophagy, an observation suggesting mitochondrial interaction with the endosomal pathway [1], and perhaps the possibility that mitochondrial phosphate contributes to intravesicular calcium phosphate granule formation and ultimately extracellular matrix mineralization. Given the great number of mitochondria per cell, these organelles would accommodate the transfer of a considerable quantity of mineral particles from the cell to the extracellular matrix without a loss of energy status or the inability of the cell to protect its enzyme systems from high cytosolic calcium flux. From this perspective, it would be valuable to establish whether both the endoplasmic reticulum and the mitochondria mediate intracellular calcium phosphate transport through membrane-invested vesicles and other organelles and promote the secretion and liberation of this cargo to the extracellular milieu for mineralization.

Clarity to the Endomembrane System of Cells

Calcium-loaded intracellular vesicles originating at the endoplasmic reticulum and the autophagic pathway form the endomembrane system of cells which is functionally integrated with the plasma membrane. Integral to endomembrane function, proteins like sortilin bind and sort ligands for transport into lysosomes, endosomes, exosomes, and the plasma membrane, and these organelles traffic calcium and phosphate ions from the cell to the extracellular matrix. In this complex scheme, the role of sortilin is yet to be fully described and details are needed to define more completely the interactions between the various organelles and their contents of cargo, including calcium and phosphate ions, as keys to vesicle and ion transport and exocytosis.

Influence of the pH Status of Mitochondria and Intracellular Transport Organelles on Mineral Ion and Particle Formation and Dissolution

Based on the results of many studies recounted in this book, calcium and phosphate ions are trafficked intracellularly by organelles such as mitochondria, lysosomes, and endosomal and autophagic vacuoles. The internal environment of each of these organelles unquestionably varies, particularly the pH, and under this circumstance the milieu will affect both the chemical and physical nature of the organelle cargo. This consideration leads to several questions concerning the transit of cargo and possible pathways to vertebrate mineralization. More specifically, it would be interesting to know the form of the cargo in different organelles and stages of transport, whether the cargo consists of ions, prenucleation clusters, or an amorphous or poorly crystalline mineral phase serving as a potential intermediate or precursor to apatite. These several calcium phosphate species may undergo transitions or be stabilized under different pH conditions. Of course, pH is only one of several factors that can influence the cargo carried by intracellular organelles. Thus, there is a further need to understand how other parameters, including the energy landscape, ion crowding, and the possible presence of oxygen radicals, may modulate the chemistry and structure of nascent ions and mineral phases destined to reach extracellular matrices and drive apatite formation.

Extracellular Mineralization at Extended Distances from Cells

Liberated ion-loaded organelles or mineral particles at the plasma membrane encounter the extracellular milieu as an initial step in matrix mineralization. This secretory process occurs through blebbing, budding, or extravasation by cells of matrix vesicles/exosomes into the extracellular matrix. Additionally, the movement or transport of such organelles along the lacuno-canalicular network system provides an extracellular ion source. In each of these interactions between cells and their surrounding environment, however, it is not at all clear how the organelles and/or their cargo reach the farthest regions of the matrix to promote mineral deposition. In this regard, the recent reports of an extensive nanochannel system penetrating and traversing the extracellular matrices of bone and cartilage may support a means by which ions and small molecules gain access to extracellular regions remote from cells [2, 3]. This concept requires validation as does the possibility that such nanochannels provide space for cell processes which may accommodate intact vesicles carrying cargo responsible for apatite nucleation and growth.

The Precise Mechanism of Vertebrate Mineralization

Extensive numbers of studies in vivo of a variety of vertebrate mineralized tissues and investigations in vitro of mineralizing collagenous solutions have led to well-accepted understanding of certain aspects of extracellular mineral deposition. These points include the presence of mineral nodules of a size range consistent with that of matrix vesicles which find their way between type I collagen fibrils in bone and certain mineralizing avian tendons; the appearance of mineral nodules leading to mineralization of these fibrils first on their surfaces, then within the fibrils and ultimately through, around and over the fibrils; and the nucleation of mineral platelets principally in the gap zone of type I collagen. While these findings are of considerable significance in gaining insight into the mechanism of vertebrate mineralization, there are large gaps of knowledge which preclude a more comprehensive description of this process. For example, the exact physicochemical and structural relationship in vivo between calcium and phosphate ion-laden matrix vesicles and collagen leading to extracellular matrix mineralization is still quite unclear. Similarly, the structure of the type I and II collagen gap and overlap zones which presumably dictate apatite nucleation is yet to be defined; and the possible roles in matrix mineral deposition of non-collagenous proteins, namely, bone sialoprotein, osteopontin, and osteocalcin, which bind to collagen have not been specified. In a related context, the charged amino acid sites which characterize collagen fibril and fiber surfaces have not been investigated as they, too, may mediate extracellular mineralization. As an added uncertainty, although type X collagen is required for endochondral bone growth and development, the function of this protein has not been strictly determined. Finally, the mechanism by which enamel mineralizes remains elusive: Whether ion flow is transcellular or intercellular, the absence in this tissue of an extracellular matrix resembling the osteoid in bone or predentin in dentin primed for mineralization is unexplained. Moreover, it remains to be determined if the formation of apatite is attributable to calcium and phosphate ion interaction with arrays of nanospheres of amelogenin proteins, structures resembling amyloid, or some other molecule(s) with its own particular chemistry, arrangement, and organization.

Cellular Events Leading to Cortical Bone Mineralization

It may be recalled that, while bones of the appendicular and axial skeleton are formed by endochondral ossification, the postnatal skeleton is predominantly comprised of lamellar bone, much of which is organized into arrays of bone cylinders or osteons. In turn, these structures undergo considerable remodeling and mineralization in response to mechanical forces and several other environmental factors. One of the most glaring current omissions in skeletal biology is failure to understand the process by which such lamellar or osteonal bone, comprising both the embryonic and mature skeleton, is laid down, mineralized and re-mineralized. In the absence of definitive information, many scholars have presumed or conflated processes characteristic of matrix vesicle-mediated endochondral bone formation with lamellar or osteonal bone mineralization. To date, there is no evidence that matrix vesicles/exosomes are involved with this type of bone mineralization. However, some recent clues have been found which may relate to cellular events leading to cortical bone formation. As has been mentioned in other parts of this book, FIB-SEM analysis of human cortical bone and cartilage shows that very fine nanochannels infiltrate the extracellular matrices of these two tissues [2, 3]. Further studies of the nanochannel networks, considering, for instance, possible cell-mediated changes in local pH and cellular control and regulation of protein secretion, exosome liberation, and/or calcium or phosphate ion efflux into the extracellular milieu, may provide greater insight into the relationships between cells and cortical bone growth and development.

Regulation of Mineralization by Local and Systemic Factors

Local and systemic factors regulate vertebrate mineral deposition as documented in numerous reports describing the existence of several hierarchical levels of control. At the systemic level, the most important endocrine factors involved with mineralization are insulin-like growth factor, fetuin, growth and thyroid hormones, retinoids, fibroblast growth factor 23, ASARM, PTH, OCN, 1,25-dihydroxycholecalciferol, and estrogen. Within the BMU, local control is mediated by PTH, IHH, WNT/β-catenin, RANKL, OPG, semaphorins, SOST, and EPHB4-EPHRINB2. At the bone surface, where the calcium and phosphate activity product is very high, the charged anion, pyrophosphate, and calcium-binding proteins like MGP and the SIBLINGs serve to prevent aberrant mineralization. Not surprisingly, despite decades of work focused on the identification and potential function of these many proteins, peptides, growth factors, sterols, and charged anions in mineral regulation, details of the mechanisms by which each of these agents coordinates and integrates with one another to maintain skeletal mineral homeostasis during development and turnover are only fragmentary. While additional factors will likely be isolated which prevent undue or aberrant mineralization, it will be of major importance to understand the molecular feedforward, feedback, and autoregulatory mechanisms by which these agents interact and mediate mineralization events at cellular and extracellular levels. Further within the BMU, the roles of non-resident cells and organelles, in particular macrophages, lymphocytes, megakaryocytes and platelets, remain to be integrated with those of resident bone, bone marrow and other cells of vertebrate mineralized tissues.

As a summary comment, we two authors would submit that the morphological, biochemical, and genetic attributes of even the most poorly differentiated cell are of enormous complexity and as such difficult to comprehend. There is an obvious need for a global theory to bring together and explain many of the findings discussed in this book and thereby delineate a mechanism or mechanisms for the mineralization of vertebrate tissues. It would seem that there are two possible means to achieving this goal. The first involves Reductionism, a systematic approach which looks for associations between phenomena to generate a theory. An example of Reductionism would involve the centrality of matrix vesicles in mediating the initial formation of mineralizing nodules which then fuse together to form the mineral phase of a tissue. However, even if matrix vesicles were nodes which promoted mineral deposition, we now know that many other cellular and extracellular events contribute to this activity and that feedback and feedforward loops exist which incorporate many other intracellular events into regulating vesicle biogenesis and function. Thus, if we carry Reductionism to its conclusion, that matrix vesicles are nodes around which vertebrate tissues mineralize, then we have only a partial understanding of the mineralization process, based on a myriad of other studies.

Recognition of the fact that Reductionism fails to explain many complex phenomena has given rise to the concept of "emergence," in which a natural phenomenon—say mineralization—is governed by laws reflecting hierarchical order in a tissue. This topic discussed earlier with reference to Edith and the Lot family considered at each hierarchical level how emergent properties were exhibited which characterized a particular level and were not evident at lower order levels. The underlying philosophy of the Non-reductionist approach is to define emergent properties of a system so as to understand complex phenomena which in many cases are nonlinear and circuitous and receive input from many diverse sources. In the case of a theory of mineralization, Non-reductionism would take into account an encompassing number of factors, including the activities of ion storage organelles; protein-ion interactions; ion transport phenomena; energy and metabolic landscapes; liquid-liquid crystal formation and crystal phase transitions; restricted intracellular pathways; the impact of growth factors, hormones and cytokines; genetic and epigenetic processes, and more. With so many molecules, molecular platforms and pathways involved in the process and their numerous redundancies, it is not surprising that the quest for a holistic understanding of vertebrate mineralization is far from complete.

Over the past decade, emergent properties of a system combined with probability theory and entropy maximization have been repeatedly employed for the modeling and analysis of large-scale biological data sets. The type of insight derived from these mathematical models is remarkably diverse, including fundamental organizational principles in structured populations to specific gene-gene interaction networks, macromolecular structures, and signaling networks; analysis of DNA sequences; and the onset of collective behavior in large animal groups [4]. With the accumulation of new and existing data sets relevant to mineral deposition, there is no reason why similar approaches cannot be applied to understand holistically the mechanism of mineralization of skeletal tissues and to use this information to inform clinicians of the hierarchical components linked to the inception of the disease state. We suggest that such an avenue into biology may lead to new and needed perspectives into the phenomena of vertebrate mineralization, and it may possibly even provide an explanation of the events by which Edith, a lithe and vital woman with functioning bones and teeth, was precipitated into a block of mineralized tissue in the Dead Sea Valley over 4000 years ago.

References

1. Pei D, Sun J, Zhu C, Tian F, Jiao K, Anderson MR, et al. Contribution of mitophagy to cell-mediated mineralization: revisiting a 50-year-old conundrum. Adv Sci (Weinh). 2018;5:1800873.
2. Tang T, Landis W, Raguin E, Werner P, Bertinetti L, Dean M, et al. A 3D network of nanochannels for possible ion and molecule transit in mineralizing bone and cartilage. Adv Nanobiomed Res. 2022;2:2100162.
3. Tang T, Landis W, Blouin S, Bertinetti L, Hartmann MA, Berzlanovich A, et al. Subcanalicular nanochannel volume is inversely correlated with calcium content in human cortical bone. J Bone Miner Res. 2023;38:313–25.
4. De Martino A, De Martino D. An introduction to the maximum entropy approach and its application to inference problems in biology. Heliyon. 2018;4:e00596.

Printed by Printforce, the Netherlan